Shallow Lakes in a Changing World

Developments in Hydrobiology 196

Series editor
K. Martens

Shallow Lakes in a Changing World

Proceedings of the 5th International Symposium on Shallow Lakes, held at Dalfsen, The Netherlands, 5–9 June 2005

Edited by

Ramesh D. Gulati[1], Eddy Lammens[2], Niels De Pauw[3] & Ellen Van Donk[1]

[1]*NIOO/Centre of Limnology, Rijksstraatweg 6, 3631 AC Nieuwersluis, The Netherlands*
[2]*RIZA, PO Box 17, 8200 AA Lelystad, The Netherlands*
[3]*Ghent University, Faculty of Bioscience Engineering, Department of Applied Ecology and Environmental Biology, Laboratory of Environmental Toxicology and Aquatic Ecology, J. Plateaustraat 22, B-9000 Gent, Belgium*

Reprinted from Hydrobiologia, Volume 584 (2007)

Springer

Library of Congress Cataloging-in-Publication Data

A C.I.P. Catalogue record for this book is available from the Library of Congress.

ISBN-13 978-90-481-7626-7 e-ISBN-13 978-1-4020-6399-2

Published by Springer,
P.O. Box 17, 3300 AA Dordrecht, The Netherlands

Cite this publication as Hydrobiologia vol. 584 (2007).

Cover illustration: Lake Loosdrecht, the Netherlands. Photo: Lowie van Liere

Printed on acid-free paper

TABLE OF CONTENTS

LIMNOLOGY/PALEOLIMNOLOGY

FOOD WEBS

LAKE RESTORATION

CLIMATIC CHANGE

Hydrobiologia (2007) 584:1–2
DOI 10.1007/s10750-007-0576-y

SHALLOW LAKES

Preface

**Ramesh D. Gulati · Eddy Lammens ·
Niels De Pauw · Ellen Van Donk**

Shallow lakes, especially in Europe and North America, have been receiving increased attention since the late 1980s, not only from lake managers, fishery biologists and water-works authorities but also from limnologists and aquatic ecologists. Climatic change and unanticipated and sometimes drastic changes in water levels, both surface and underground, have further contributed to water-quality issues. Therefore, both the ecology and management of these ecosystems have been receiving greater attention during the last few decades. Overall concerns over deteriorating water quality, due to pollution and the increased pace of eutrophication of many lakes in northern European countries, have given a great impetus to both fundamental ecological as well as applied studies on many of these threatened lakes. There is also a noticeable increase in the number and frequency of international conferences and meetings dealing with the ecology of shallow lakes, leading to greater international contacts among researchers.

Published studies vividly demonstrate that the water-quality deterioration in many of these lakes, particularly since the 1950s, is primarily a sequel to the intensification of urban activities and agriculture in the lake catchment areas, and overexploitation of fish. The studies have given us deeper insights into the functioning of food-webs in shallow waters, nutrient dynamics in open water, and in the complexities of interactions in littoral and pelagic regions, and of the sediment-water interphase. We are now starting to implement the knowledge gained in order to rehabilitate some lakes, if not restore them.

Concomitantly with the above-mentioned developments, there is a spurt of publications on shallow lakes in the peer-reviewed, international journals, especially *Hydrobiologia* where traditionally many of the conference proceedings have been published. It is perhaps opportune to mention here that 1989 was an important year. In this year, the proceedings of the first such international conference *Biomanipulation As Tool For Water Management* was held at Amsterdam (8–11 August 1989). The proceedings of this scientific

Guest editors: R. D. Gulati, E. Lammens, N. De Pauw & E. Van Donk
Shallow lakes in a changing world

R. D. Gulati (✉) · E. Van Donk
Centre for Limnology, NIOO-KNAW, P.O. Box 1299, 3600 BG Maarssen, The Netherlands
e-mail: r.gulati@nioo.knaw.nl

E. Lammens
Department for Ecology, Institute for Inland Water Management and Wastewater Treatment (RIZA), Merlant 17, 8200 AA Lelystad, The Netherlands

N. De Pauw
RUG - Vakgroep Toegepaste Ecologie en, Milieubiologie, J. Plateaustraat 22, 9000 Gent, Belgium

meeting which were published in 2000 in *Hydrobiologia* (Vols. 200/201: 628 pp.), triggered great interest among freshwater ecologists and fishery biologists. Starting in the early nineties, four such triennial international conferences have been held in Europe, at Silkeborg (Denmark), Mikolajki (Poland), Berlin (Germany) and Balatonfüred (Hungary). The proceedings papers of these conferences deal with a variety of issues facing lakes, especially changes in water chemistry and nutrients, plankton biology relating to deterioration of water quality caused by increased eutrophication and pollution, perennial blooms of cyanobacteria and loss of biodiversity. Many of the threatened lakes have responded to restoration works by showing a notable decrease in turbidity, i.e., an improvement in underwater light climate, following a decrease in phytoplankton abundance, and the subsequent increase in macrophytes. Even though the improvement in water-quality is generally only transitory, i.e., a shift from a turbid-water state dominated by phytoplankton to a clear-water state generally dominated by macrophytes, it has given us grounds for further theoretical speculations on the existence of so-called *alternative stable states* in these lakes undergoing restoration.

Following up the traditions of the preceding four symposia, the '5th International Symposium on Shallow Lakes' was held at Dalfsen, in the eastern part of The Netherlands, from 5 to 9 June 2005, at the Conference Centre 'De Bron' situated on the bank of the River Vecht. About 250 participants from 33 countries attended this meeting, which had the theme *Shallow Lakes in a Changing World.* The theme is obviously a reflection on the ongoing climatic change, the increased occurrence of bio-invasions, and the decrease in biodiversity of lake biota in general. The conference schedule included plenary lectures (8 invited keynote speakers), which dealt with harmful biota, biodiversity, global change, ecological restoration, evolutionary ecology, and water fowl. The oral presentations (111) were divided into parallel sessions. There were 80 poster presentations, and five thematic workshops in the evenings. The subject matter of both the oral sessions and poster presentations related to general limnology, food-webs, biodiversity, climate change, the EU Water Framework Directive, nutrients and evolution, fishes and birds. The evening workshops dealt with five topics: lake biomanipulation, resistance of lakes to changes, Lakes Peipsi and IJsselmeer, the EU Water Framework Directive and Mediterranean shallow lakes. There was a full day of mid-conference excursions to various sites of interest, including protected water reserves (Ramsar Sites).

The participants were given until 31 October 2005 to prepare and submit manuscripts based on their presentations at the conference. In total 63 manuscripts were submitted. Each manuscript was sent to two or three anonymous reviewers and in several cases there was a second review round. This proceedings volume contains 41 peer-reviewed manuscripts that were finally accepted by Dr. Koen Martens, the Editor-in-Chief of *Hydrobiologia.*

The guest editors are thankful to the manuscript authors for their cooperation in keeping to the deadlines for the review process. We are highly indebted to the reviewers, some of whom reviewed more than one manuscript and also helped in improving the language. Finally we thank the members of the local organising committee (Paul Boers, conference chairman; Lowie van Liere, conference secretary; Maarten Ouboter; Tabe Tietema and Bas van der Wal) for their help in many ways. The financial help from the Ministry of Environment and Public Health, Ministry of Transport and Public Works, and Foundation of Applied Water Research (STOWA) is gratefully acknowledged. Petra Angelone and Anette Bisseling-Visser helped in the administrative work of the conference.

The *Shallow Lakes 2008* conference will be held in November 2008 in Urguay. It is being organized by the Faculty of Sciences, Universidad de la República at Punta del ste, Uruguay. For information see website: http://www.shallowlakes2008.org.

Hydrobiologia (2007) 584:3–12
DOI 10.1007/s10750-007-0584-y

SHALLOW LAKES

Phosphorus reference concentrations in European lakes

A. C. Cardoso · A. Solimini · G. Premazzi · L. Carvalho · A. Lyche · S. Rekolainen

Abstract As part of an assessment of the ecological quality of surface waters, recent European water legislation, the water framework directive (WFD), requires the setting of environmental objectives for particular chemicals. As part of this, many European countries are choosing to develop a quality classification scheme for total phosphorus (TP) concentration. The assessment of ecological quality and its component parts, such as TP, must be based on the degree of divergence of a water body from an appropriate baseline, or 'reference condition'. For this reason, it is important to determine TP reference conditions for all lake types, or alternatively, models for predicting reference TP concentrations on a site-specific basis. With this purpose in mind, we have assembled a large dataset of European lakes considered to be in reference condition; 567 lakes in total. Data has been collated on TP concentration, mean depth, alkalinity, humic type, altitude, area, and geographical region. Reference TP concentrations have been derived from this dataset using two empirical approaches. Firstly, lake-type specific TP reference conditions were derived simply from descriptive statistics (median and percentiles) of TP concentrations by lake type. Secondly, empirical models were developed for estimating site-specific reference TP concentrations from a set of potential predictor variables. TP concentrations were found to vary with lake type and by geographical region. TP increased with colour and alkalinity, and decreased with lake depth and altitude. There was no clear relationship between TP and lake area. Altitude, mean depth and alkalinity were selected as independent explanatory variables for prediction of site-specific reference TP concentrations, with different models developed for humic and non-humic lakes. A simpler site-specific predictive model based solely on the morphoedaphic index (MEI) (a combination of a site's alkalinity and mean depth) was also shown to be comparably effective.

Guest editors: R. D. Gulati, E. Lammens, N. De Pauw & E. Van Donk
Shallow lakes in a changing world

A. C. Cardoso (✉) · A. Solimini · G. Premazzi
European Commission – DG Joint Research Centre, Institute for Environment and Sustainability, TP 300, Via E. Fermi 1, 21020 Ispra, Italy
e-mail: ana-cristina.cardoso@jrc.it

L. Carvalho
Centre for Ecology and Hydrology, Edinburgh, UK

A. Lyche
Norwegian Institute for Water Research, Oslo, Norway

S. Rekolainen
Finnish Environment Institute, Helsinki, Finland

Keywords Reference lake · Total phosphorus reference concentration · Morphoedaphic index · WFD

Introduction

The estimation of reference conditions is crucial in any ecological assessment programme (e.g. Moss et al., 1996). These provide the baseline from which to determine lake change with time, and are necessary to evaluate a lake's current status or potential for change.

The most recent European water legislation, the water framework directive (WFD), prescribes the assessment of ecological quality of surface waters using an ecological quality ratio (EQR). The EQR is defined as the ratio between the reference value and the observed value for a given metric.

As part of the assessment of ecological quality, many European countries are choosing to develop a quality classification scheme for total phosphorus (TP) concentration as a relatively simple measure of water quality, in response to eutrophication pressures (e.g. Søndergaard et al., 2005). A large scale formal assessment of the comparability of national assessment schemes is also being carried out as part of the implementation of the WFD—a process known as intercalibration (IC). TP has been selected as a key parameter in this IC process in what concerns lakes because it has been considered as the main pressure to which the selected biological metrics are responsive in lakes. For this reason, it is essential to determine TP reference conditions for all WFD European lake types, or alternatively, models for predicting reference TP concentrations on a site-specific basis.

Reference conditions are expected to vary across Europe resulting from geographical differences in catchments (e.g. biogeography of species, geology and altitude), and lake factors (e.g. depth, area, water colour). To account for these differences, the WFD requires waterbodies to be differentiated into types and to derive type-specific reference conditions for the appropriate ecological quality elements.

A number of approaches can be used to establish reference conditions (see e.g. Hughes, 1995; Reynoldson et al., 1997, Nielsen et al., 2003) and these have been broadly summarised in published guidance for the WFD (Anonymous, 2003). This guidance suggests that the decision of which method/approach to take for the determination of reference conditions is dependent on the condition of the sites available for a certain lake type: (a) where undisturbed or nearly undisturbed conditions prevail, a validated spatial network is preferred; (b) if degraded conditions prevail then the preferred modelling approach would be based on establishing a good relationship between stress and ecological response; (c) expert judgement should be used as the last resort and be accompanied by an acceptable validation process.

The TP status of a lake is mainly determined by its external nutrient supply. Phosphorus export from a catchment in an undisturbed area will depend on the natural fertility of the local rocks or glacial deposits, and the soils derived from them. This geological difference in phosphorus content and availability between catchments will also be largely represented in the concentrations of the major ions leaching out of the catchment. It is for this reason that measures of total ionic concentration, such as alkalinity or conductivity, are often positively correlated with phosphorus concentrations in water bodies with relatively undisturbed catchments (Ryder et al., 1974). Because both alkalinity and conductivity are generally less impacted by anthropogenic activity in the catchment and by biological activity within the lake, either one may be useful as a potential predictor of reference TP concentrations (Ryder et al., 1974), with the exception being alkalinity values in catchments impacted by acidification.

The other factor that is widely acknowledged as affecting a lake's TP concentrations (and trophic status in general) is mean depth (Rawson, 1952; OECD, 1982). Ryder et al. (1974) provide a comprehensive review of why mean depth is a useful simple measure for a number of processes that potentially affect lake productivity. The reasons are complex, but one of the main ones appears to be due to the fact that as depth increases, lakes will lose a greater proportion of nutrients through the process of sedimentation, i.e. a negative correlation. The importance of both total ionic concentrations and mean depth as predictors of a lake's reference productivity, and specifically TP concentrations, has led to the development of a simple index that incorporates both aspects, the morphoedaphic index (MEI). The MEI is simply the ratio between the total dissolved solids (edaphic factor) in lake water (as measured by alkalinity or conductivity) to the lake mean depth (morphometric factor). Several early studies suggested that the MEI index was correlated with fish and phytoplankton production of lakes (e.g. Rawson, 1952; Ryder et al., 1974; Oglesby, 1977). Most recently, the MEI index has been extended to predict phosphorus concentrations resulting from natural background loading in undisturbed watersheds (Vighi & Chiaudani, 1985). The use of the MEI-TP model is simple and appears to be highly successful in predicting phosphorus concentrations from a limited set of undisturbed temperate lakes. The MEI-TP model of Vighi & Chiaudani (1985) is, however, based on only 53 lakes, of which only 12 were in Europe, and none were shallow (mean depth was 36.6 m (±26.4 m).

In this paper we derive type specific reference conditions for TP based on a large dataset of >500 European reference lakes gathered together within the EC FP6 Project REBECCA. Furthermore, we explore the relationships between TP concentrations and potential predictor variables (alkalinity, depth, altitude, area) or combinations of these predictor variables (e.g. MEI), in order to develop empirical models for predicting reference TP concentrations in European lakes on a site-specific basis. This analysis was conducted with reference to the lake typology developed for the purpose of the WFD intercalibration exercise (Van de Bund et al., 2004).

Materials and methods

Data set assemblage

We assembled lake data on TP concentration, mean depth, alkalinity, humic type, altitude, area and Geographical region. The latter was based on Geographical Intercalibration Groups (GIGs) created for the purposes of the WFD intercalibration exercise (see Van de Bund et al., 2004). There are five lake GIG regions (Northern, Atlantic, Alpine, Central-Baltic and Mediterranean). Lake data were gathered from REBECCA partners (see http://www.rbm-toolbox.net/) and from the Central-Baltic GIG secretariat. The MEI was calculated using alkalinity data, according to the following equation:

$$MEI_{alk} = \text{alkalinity (m equiv. } l^{-1}\text{)/mean depth (m)}$$

The final dataset was rather heterogeneous as some lakes were sampled quite frequently over several years whereas others were sampled only once or a few times per year. For our analysis, we only used those lakes sampled on at least 3 separate months over a single growing season (April–September) and averaged these growing season means for all years for a given lake.

The selection of reference sites was carried out by individual Member States using criteria laid down in WFD guidance (Anonymous, 2003). Some countries (e.g. Norway, Finland) used explicit pressure criteria (e.g. population in catchment, % intensive agriculture), others (e.g. UK and Ireland) have used palaeolimnology to identify undisturbed sites. In many countries, expert judgement was also used in site selection to some extent. Some countries selected sites that locally may be considered in very good condition biologically, but had relatively high mean TP concentrations (>100 μg l^{-1}) compared with other countries with lakes of a similar type. For the latter reason it was decided to select a threshold TP concentration of 35 μg l^{-1} above which sites chosen largely on expert judgement would be excluded from the analysis. This resulted in 41 sites being excluded out of a total of 608 (i.e. approx 7%) We acknowledge that a small number of 'naturally eutrophic' lakes, with background concentrations slightly higher than 35 μg l^{-1}, may have been excluded, but this would have a minor impact on our model compared to retaining dubious reference sites with much higher TP concentrations. Lakes with TP concentrations above 35 μg l^{-1} are classified according to the OECD (1982) as eutrophic and, therefore, it was considered an acceptable, albeit subjective, widely-held opinion of an impacted site.

The final dataset contained 567 lakes identified, by the data providers, as being only slightly impacted (i.e. reference lakes), of these 103 were shallow lakes. In the IC typology these are called "very shallow", having a depth ≤3 m which is considered as the boundary between non-stratified shallow lakes and temporarily or permanently stratified deep lakes (Søndergaard et al., 2003).

Statistical analysis

To derive type-specific reference TP concentrations we computed descriptive statistics for TP by each lake type. Median values for a type could potentially be considered as the type-specific reference value by stakeholders, with the 75th percentile potentially being a threshold for sites at high status. Note that for this analysis we only needed lakes categorized into type classes and not raw typology data (e.g. mean depth).

To derive models for estimating site-specific reference TP concentrations, we ran a General Linear Model (GLM) to estimate the best model to predict mean total phosphorous from several predictor variables either independently, or with some in combination (e.g. MEI_{alk}). Prior to the GLM, Pearson correlations were computed for each pair of variables to select the potential predictor variables. Lake area had a highly significant correlation with lake depth ($r^2 = 0.44$, $P < 0.001$, $n = 531$) and so was not considered further in the analysis in order to minimise the variance inflation factor. Altitude, alkalinity, mean depth, MEI_{alk}, and TP were all log transformed to help stabilize variance. Only lakes with raw typology data that could be allocated to a GIG type were included in the analysis. All the statistical analysis were performed with the statistics software SPSS (version 12, 2003).

Results

Type-specific TP reference conditions

We report TP reference conditions by GIG and lake type based on the dataset of all accepted reference lakes (Table 1). The mean TP concentrations for individual lakes ranged from less than 3 μg l^{-1} to over 30 μg l^{-1}. The low alkalinity types had an average TP concentration of 8.6 μg l^{-1} in the Northern GIG region to 15.5 μg l^{-1} in the Central-Baltic GIG region. The 4 peaty lake types had average TP concentrations from 9.4 μg l^{-1} to 13.0 μg l^{-1}, indicating little difference independently of GIG region and type. In the dataset there were only two deep lake types which were of two different geology types, L-AL3 (moderate to high alkalinity) and L-N2b (low alkalinity) with mean TP concentrations of, respectively, 5.3 μg l^{-1} and 6.4 μg l^{-1}. A total of 103 shallow lakes were included in the dataset, from 4 different GIG regions (Alpine, Atlantic, Central-Baltic and Northern). For the majority of these, however, we had insufficient typology information.

As described above and more clearly seen in Fig. 1, TP concentration does not only vary with lake type but also by GIG region. Reference lakes in the Central-Baltic GIG region had the highest TP concentrations, closely followed by the few Mediterranean lakes. In all other GIGs the median of the lake's mean TP was below 10 μg l^{-1}.

Humic type, depth and alkalinity (all categorized following IC type criteria) all had a certain effect on TP. TP was negatively correlated with altitude and depth (Pearson correlation, respectively, $r = -0.14$, $P < 0.001$, $n = 534$ and $r = -0.38$, $P < 0.001$, $n = 525$), positively correlated with alkalinity ($r = 0.17$, $P < 0.001$, $n = 492$) but not significantly correlated with lake area ($r = -0.06$, $P = 0.19$, $n = 563$).

Site-specific TP reference conditions

The results of the GLM model using independent predictors (altitude, alkalinity and mean depth), are reported in Tables 2 and 3. All the variables were significant predictors of the mean TP and the overall variance explained by the model was 51.4%.

MEI model

Figure 2 shows the relationship between annual mean TP and the MEI_{alk} using an appropriate scale for expected linearity (log). The graph shows some scatter, with reasonable linearity, but high expected variability resulting from

Table 1 Total phosphorus concentrations in European reference lakes

GIG region	GIG type	Mean	St dev	Min	Max	Median	25%	75%	*n*	IC-description
Atlantic	L-A1	9.1	9.7	3.0	20.4	4.0	3.5	12.2	3	Lowland, shallow, calcareous, small
	L-A2	7.7	4.7	1.0	13.5	8.0	2.5	11.5	9	Lowland, shallow, calcareous, large
	L-A3	9.7	4.3	5.7	20.0	9.0	6.4	12.0	10	Lowland, shallow, peat, small
	L-AX	8.3	4.5	3.0	12.0	9.0	3.8	12.0	4	Outside IC-types
Alpine	L-AL3	5.3	3.9	2.0	16.0	4.0	3.0	6.0	19	Lowland or mid-altitude, deep, moderate to high alkalinity. (alpine influence), large
	L-AL4	8.2	2.5	5.0	10.9	10.0	6.0	10.5	5	Mid-altitude, shallow, moderate to high alkalinity (alpine influence), large
	L-ALX	11.9	7.7	2.5	34.5	10.0	7.5	14.2	21	Outside IC-types
Central-Baltic	L-CB1	20.8	7.5	10.0	34.0	18.8	14.3	29.0	35	Lowland, shallow, stratified, calcareous
	L-CB2	17.7	12.1	0.0	33.3	17.8	7.4	30.6	12	Lowland, very shallow, calcareous
	L-CB3	15.5	7.2	3.9	29.5	15.8	9.1	21.0	16	Lowland, shallow, siliceous, vegetation dominated by Lobelia
Mediterranean	L-MX	16.6	1.6	15.0	18.1	16.6	15.8	17.4	3	Outside IC-types
Nordic	L-N1	10.3	3.8	4.6	19.0	9.1	7.5	13.5	17	Lowland, shallow, moderate alkalinity, clear large.
	L-N2a	7.5	3.0	3.0	16.5	6.7	5.5	8.8	51	Lowland, shallow, low alkalinity, clear large.
	L-N2b	6.4	3.1	3.1	20.0	5.6	4.5	7.5	48	Lowland, deep, low alkalinity, clear, large.
	L-N3	13.0	6.2	3.9	31.0	11.3	9.0	16.3	47	Lowland, shallow, low alkalinity, humic, large
	L-N5	6.6	2.6	2.5	13.2	6.3	5.0	7.3	25	Mid-altitude, shallow, low alkalinity, clear, large
	L-N6	9.4	3.0	4.8	16.0	9.1	6.7	11.5	17	Mid-altitude, shallow, low alkalinity, humic, large
	L-N8	13.7	6.8	5.4	29.1	12.7	9.1	16.7	9	Lowland, shallow, moderate alkalinity, humic large
	L-NX	11.2	7.4	2.4	34.0	8.8	5.9	15.5	183	Outside IC-types

Descriptive statistics by intecalibration (IC) geographical intercalibration group (GIG) type is reported

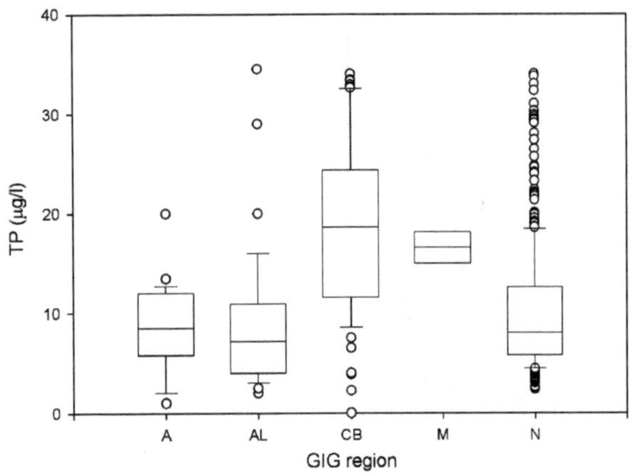

Fig. 1 Box-Wiskers plot of TP per GIG region. The middle line indicates the median TP value and the lower and higher wiskers indicate, respectively, the 25th and 75th percentiles. A = Atlantic, Al = Alpine, CB = Central-Baltic, M = Mediterranean, N = Northern

heterogeneity in the data. The results of the GLM model are presented in Table 4, revealing a significant effect of altitude, GIG region and humic type on the TP but not on the MEI_{alk}-TP relationship. Overall, the variance explained by the model was 51.3%. The resulting regression models, per GIG region and humic type are provided in Table 5.

Discussion

Regional and type-specific reference conditions

Our findings are in accordance with those already documented in the literature, that is humic (e.g. Nürnberg & Shaw, 1999), high alkalinity (Ryder et al., 1974) and shallow (Rawson, 1952) lakes have in general higher natural TP concentrations. The importance of altitude was more unexpected. Through its effect on temperature, altitude is widely recognised as affecting a lake's biology. In terms of lowland lakes having naturally higher TP concentrations, it could also be due to higher temperatures causing increased weathering of catchments in lowlands, but we cannot exclude

Table 2 General linear model results for TP variation in European reference lakes

Source	Sum of squares	df	Mean square	*F*	Sig.
Intercept	12.86	1	12.86	467.60	0.000
GIG region	1.62	3	0.54	19.65	0.000
Humic type	1.03	1	1.03	37.39	0.000
Altitude	0.40	1	0.40	14.68	0.000
Depth	0.34	1	0.34	12.31	0.001
Alkalinity	1.30	1	1.30	47.10	0.000
Error	6.30	229	0.03		
Total	205.99	237			

GIG region and humic type were entered as dummy variables

TP, altitude, alkalinity and depth were log transformed. df = degree of freedom, Sig. = level of significance

Table 3 Equations predicting TP reference concentration in European lakes using humic type, altitude, alkalinity and depth as independent predictors

Humic type	GIG region	Equation
Humic	N	Log(TP) = 1.58(0.06) – 0.08(0.02) Log(alt) – 0.13(.04) Log(depth) + 0.24 (0.03) Log(alk)
Non humic	N	Log(TP) = 1.35(0.01) – 0.08(0.02) Log(alt) – 0.13(.04) Log(depth) + 0.24 (0.03) Log(alk)
Humic	A	Log(TP) = 1.29(0.01) – 0.08(0.02) Log(alt) – 0.13(.04) Log(depth) + 0.24 (0.03) Log(alk)
Non humic	A	Log(TP) = 1.13(0.01) – 0.08(0.02) Log(alt) – 0.13(.04) Log(depth) + 0.24 (0.03) Log(alk)
Non humic	AL	Log(TP) = 1.03(0.01) – 0.08(0.02) Log(alt) – 0.13(.04) Log(depth) + 0.24 (0.03) Log(alk)
Humic	CB	Log(TP) = 1.65(0.01) – 0.08(0.02) Log(alt) – 0.13(.04) Log(depth) + 0.24 (0.03) Log(alk)
Non humic	CB	Log(TP) = 1.49(0.01) – 0.08(0.02) Log(alt) – 0.13(.04) Log(depth) + 0.24 (0.03) Log(alk)

Standard errors are in brackets. N = Northern; A = Atlantic; Al = Alpine; CB = Central-Baltic; alt = altitude; alk = alkalinity

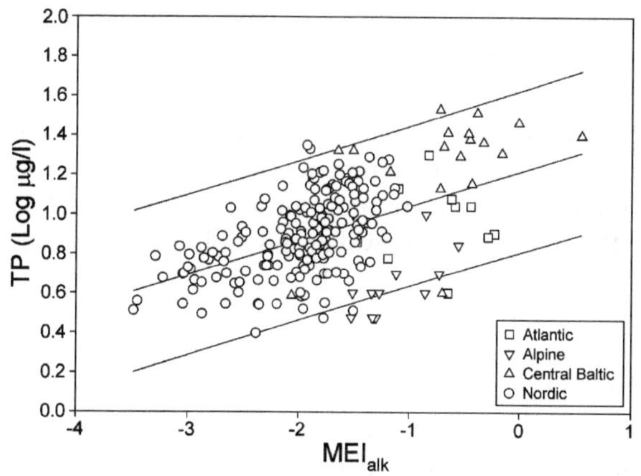

Fig. 2 Scatter plot of TP against MEI_{alk}. Lakes in different GIGs are identified

some possible correlated anthropogenic impact associated with greater disturbance in lowlands (i.e. reference lakes in lowlands being slightly more disturbed than those in uplands).

Exceptions are some lowland, shallow, high alkalinity lakes, located in catchments influenced by limestone, known as 'marl' lakes, and most common in Ireland and the UK, which are naturally rich in carbonate particles that precipitate phosphorus (Otsuki & Wetzel, 1972).

The results also highlighted that lakes in the Central-Baltic and Mediterranean GIGs had on average more TP than lakes in the Alpine, Atlantic and Northern GIGs. These differences do probably include a type effect (e.g. higher alkalinity, shallower and lowland lakes are relatively more frequent in the former regions) but there may also be a latitudinal effect related to temperature (and its effect on mineralization rates in catchment). Another explanation is that there was some bias in the identification of reference lakes in these regions. The lakes in the dataset were identified as being in reference condition either through the evaluation of the pressures in the lake's catchment (i.e. the area of the lake drainage basin occupied by agriculture and urban developments, against the area remaining in natural condition) or by expert judgement. Northern regions of Europe (i.e. Norway, Sweden and Finland) are, in general, considerably less

Table 4 General linear model results for TP variation in European reference lakes

Source	Sum of squares	df	Mean square	F	Sig.
Intercept	6.90	1	6.90	246.88	0.00
GIG region	0.54	3	0.18	6.43	0.00
Humic type	0.14	1	0.14	4.90	0.03
Altitude	0.44	1	0.44	15.82	0.00
MEI_{alk}	0.32	1	0.32	11.45	0.00
GIG region × MEI_{alk}	0.06	3	0.02	0.77	0.51
Humic type × MEI_{alk}	0.03	1	0.03	1.00	0.32
Error	6.31	226	0.03		
Total	205.99	237			

GIG region and humic type were entered as dummy variables

TP, MEI_{alk} and altitude were log transformed. df = degrees of freedom, Sig. = level of significance

Table 5 Equations predicting TP reference concentration in European lakes using humic type, altitude, MEI_{alk} as independent predictors

Humic type	GIG region	Equation
Humic	A, N	$\text{Log(TP)} = 1.62\ (0.12) - 0.09\ (0.02)\ \text{Log(alt)} + 0.24\ (0.06)\ \text{Log}(MEI_{alk})$
Non humic	A, N	$\text{Log(TP)} = 1.36\ (0.03) - 0.09\ (0.02)\ \text{Log(alt)} + 0.24\ (0.06)\ \text{Log}(MEI_{alk})$
Non humic	AL	$\text{Log(TP)} = 0.93\ (0.06) - 0.09\ (0.02)\ \text{Log(alt)} + 0.24\ (0.06)\ \text{Log}(MEI_{alk})$
Humic	CB	$\text{Log(TP)} = 1.81\ (0.02) - 0.09\ (0.02)\ \text{Log(alt)} + 0.24\ (0.06)\ \text{Log}(MEI_{alk})$
Non humic	CB	$\text{Log(TP)} = 1.55\ (0.04) - 0.09\ (0.02)\ \text{Log(alt)} + 0.24\ (0.06)\ \text{Log}(MEI_{alk})$

Standard errors are in brackets

N = Northern; A = Atlantic; Al = Alpine; CB = Central-Baltic

populated than Central European and Mediterranean regions and, therefore, lakes in Northern regions are, in general, less exposed to anthropogenic pressure and have as consequence less TP in their waters (EEA, 2003, Nixon et al., 2003). In many Northern European countries there are, therefore, plenty of undisturbed reference lakes that can be selected based on strict pressure criteria. In Central and Mediterranean Europe this is less true, there are very few relatively undisturbed catchments and expert judgement of a country's most undisturbed sites may have been used more widely than explicit pressure criteria. It may, therefore, be that there is some bias towards the inclusion of slightly impacted lowland, shallow lakes with a TP above a true reference TP concentration. The effect of this bias would have a differential effect on lakes of different geologic nature. Dillon & Kirchner (1975) determined the export ratio of TP from 34 watersheds in Ontario and found significantly higher ratios from agricultural lands (forest associated with pasture areas) in comparison to forested areas both in igneous and sediment rock watersheds. The differences between watersheds of different geology, but similar land use were also highly significant.

Site-specific models

The simple regression model of the annual mean TP against MEI_{alk} established by Vighi & Chiaudani (1985) for lakes with negligible anthropogenic phosphorus loads has been applied successfully for lake management purposes (e.g. lakes in North America, US Environment Protection Agency, 2000; lakes Vortsjarv and Peipsi, Noges, 2006). However, its applicability in Europe, in the context of setting reference TP concentrations for the WFD, required further testing and calibration to a wider variety of lakes. Our analysis has highlighted that a lake's altitude, geographical (GIG) region and humic type all have a significant relationship with TP, and, therefore, several different regression models for predicting reference TP concentrations are required across Europe to take account of the significant effects of these predictor variables. Our analysis also highlighted that more complex multiple regression models, using independent predictors rather than the MEI, were of very similar predictive strength to the MEI_{alk} models with the overall explained variance being 51.4% for the complex model using independent predictors and 51.3% for the MEI_{alk} model. We cannot clearly recommend one model over the other since they are more or less equivalent. The MEI models are, however, possibly more appealing to water managers because of their simpler parameterisation.

In the original model developed by Vighi & Chiaudani (1985) MEI_{alk} explained 83% of the variance in TP. It may be that our findings are affected by some bias in the dataset, which contains predominately Northern lakes, and possibly bias in the identification of reference lakes. The main basic assumption of the MEI model is that annual mean TP concentration relates to the MEI index only in non impacted lakes (i.e. where phosphorus loading approximately equals the natural background loading). This means that the variance explained by the MEI index would be greatly reduced with the false identification of reference lakes. Although, the data used in the analysis was restricted to sites with mean TP concentrations below 35 $\mu g\ l^{-1}$, there may still have been impacted nutrient poor lakes included in the dataset which have erroneously been identified as in reference condition. Also, because we have assembled a much larger dataset of varied provenance, gathered under monitoring programs using different sampling and analytical methods with differing accuracies, it must be expected that the variability associated with many of the model parameters (e.g. TP, alkalinity, colour) would increase. Nevertheless, the variability explained by the model is clearly acceptable and compares very favourably with variability explained, for example, by diatom-inferred TP (47%, Bradshaw & Anderson, 2001) or Chironomid-inferred TP (29%, Brooks et al., 2001) using weighted averaging regression and calibration TP models (Ter Braak & Juggins, 1993).

Typically the error associated with prediction of TP concentrations in reference conditions is high in the absence of reference sites, and there may be a big discrepancy in the results given from different methods. As an example, reference conditions for TP in the Swedish lake Mälaren have been estimated to be between 20 and 40 μg l^{-1} using historical data, 6 μg l^{-1} through a regional model and between 40 and 60 μg l^{-1} using paleolimnological data (R. Johnson, oral communication). Hence, it is always important to cross-validate results from different methods to ensure greater confidence in estimates of reference TP.

Conclusions

We have estimated lake type-specific reference TP concentrations for a set of European lake types. Models have also been presented for determining site-specific reference TP concentrations. Simpler models using the MEI_{alk} performed as effectively as models using independent predictors. The predictive strength of the models was relatively high, compared with other approaches for setting reference conditions (e.g. palaeolimnology). The dataset ideally needs further inclusion of lakes from central and southern Europe and, in particular, a more thorough checking of the criteria used for site selection of the reference lakes. Further validation of these European models is recommended using an independent test set of lakes in reference condition.

Acknowledgements This work has been carried out within the framework of the EU FP6 Research Project REBECCA 'Relationships Between Ecological and Chemical Status of Surface Waters'. We would like to thank our colleagues in the Project, in particular for those that supplied the data and those responsible for the database. Namely we would like to thank Jannicke Moe for her invaluable work building up the REBECCA Project lake database. We would also like to thank GIG representatives and data providers for their cooperation and comments on several aspects of the work.

References

Anonymous, 2003. River and Lakes – Typology, Reference Conditions and Classification Systems. Guidance No 10. CIS Working Group 2.3 REFCOND, European Communities, Luxembourg, 87 pp.

Bradshaw, E. G. & J. Anderson, 2001. Validation of a diatom-phosphorus calibration set for Sweden. Freshwater Biology 46: 1035–1048.

Brooks, S. J., H. Bennion & H. J. B. Birks, 2001. Tracing lake trophic history with a chironomid-total phosphorus inference model. Freshwater Biology 46: 513–533.

Dillon, P. J. & W. B. Kirchner, 1975. The effects of geology and land use on the export of phosphorus from watersheds. Water Research 9(2): 135–148.

EEA, 2003. Europe's environment: the third assessment. Environment assessment report No 10, EEA, Luxembourg, 343 pp.

Hughes, R. M., 1995. Defining acceptable biological status by comparing with reference conditions. In Davis, W. S. & T. P. Simpson (eds), Biological assessment criteria. Lewis Publishers, Boca Raton, Florida, 31–47.

Moss, B., P. Johnes & G. Phillips, 1996. The monitoring of ecological quality and the classification of standing water in temperate regions: A review and proposal based on a worked scheme for British waters. Biological Review 71: 301–339.

Nielsen, K., B. Sømod C. Ellegaard & D. Krause-Jensen, 2003. Assessing reference conditions according to the European Water Framework Directive using modelling and analysis of historical data: an example from Randers Fjord, Denmark. Ambio 32: 287–294.

Nixon, S., Z. Trent, C. Marcuello & C. Lallana, 2003. Europe's water: an indicator-based assessment. Topic report N.1/ 2003, EEA, Copenhagen, 97 pp.

Nõges, P. & T. Nõges, 2006. Indicators and criteria to assess ecological status of the large shallow temperate polymictic lakes Peipsi (Estonia/Russia) and Võrtsjäv (Estonia). Boreal Environment Research 11(1): 67–80.

Nürnberg, G. K. & M. Shaw, 1999. Productivity of clear and humic lakes: nutrients, phytoplankton, bacteria. Hydrobiologia 382: 97–112.

OECD, 1982. Eutrophication of Waters: Monitoring, Assessment and Control. OECD, Paris, 154 pp.

Oglesby, R. T., 1977. Phytoplankton summer standing crop and annual productivity as functions of phosphorus loading and various physical factors. Journal of Fishery Research Board Canada 34: 2255–2270.

Otsuki, A. & R. G. Wetzel, 1972. Coprecipitation of phosphate with carbonates in a marl lake. Limnology and Oceanography 17: 763–767.

Reynoldson, T. B., R. H. Norris, V. H. Resh, K. E. Day & D. M. Rosenberg, 1997. The reference condition: a comparation of multimetric and multivariate approaches to assess water-quality impairment using benthic invertebrates. Journal of North American Benthological Society 16: 833–852.

Rawson, D. S., 1952. Morphometry as a dominant factor in the productivity of large lakes. Verhangen International Verein Limnology 12: 164–175.

Ryder, R. A., S. R. Kerr, K. H. Loftus & H. A. Regier, 1974. The Morphoedaphic Index, a fish yield estimator – review and evaluation. Journal of the Fisheries Research Board of Canada 31: 663–688.

Søndegaard, M., J. P. Jensen & E. Jeppesen, 2003. Role of the sediment and internal loading of phosphorus in shallow lakes. Hydrobiologia, 506–509: 135–145.

Søndergaard, M., E. Jeppesen, J. P. Jensen & S. L. Amsinck, 2005. Water Framework Directive: ecological classification of Danish lakes. Journal of Applied Ecology, 42: 616–629.

Ter Braak, C. J. F. & S. Juggins, 1993. Weighted averaging partial least squares regression (WA-PLS): an improved method for reconstructing environmental variables from spaecies assemblages. Hydrobiologia 269–270: 485–502.

US Environment Protection Agency, 2000. Nutrient Criteria Technical Guidance Manual. Lakes and reservoirs. EPA, Office of Water and Technology, Washington DC.

Van de Bund, W., A. C.Cardoso, A. -S. Heiskanen & P. Nõges, 2004. Overview of Common Intercalibration Types. Electronic document. Publicly available at: http://wfd-reporting.jrc.cec.eu.int/Docs/typesmanual.

Vighi, M. & G. Chiauani, 1985. A simple method to estimate lake phosphorus concentrations resulting from natural, background, loadings. Water Research 19: 987–991.

Hydrobiologia (2007) 584:13–25
DOI 10.1007/s10750-007-0582-0

SHALLOW LAKES

Benthic nutrient fluxes in a eutrophic, polymictic lake

D. F. Burger · D. P. Hamilton · C. A. Pilditch · M. M. Gibbs

Abstract Sediment release rates of soluble reactive phosphorus (SRP) and ammonium (NH_4) were determined seasonally at three sites (water depth 7, 14 and 20 m) in Lake Rotorua using in situ benthic chamber incubations. Rates of release of SRP ranged from 2.2 to 85.6 mg P m^{-2} d^{-1} and were largely independent of dissolved oxygen (DO) concentration. Two phases of NH_4 release were observed in the chamber incubations; high initial rates of up to 2,200 mg N m^{-2} d^{-1} in the first 12 h of deployment followed by lower rates of up to 270 mg N m^{-2} d^{-1} in the remaining 36 h of deployment. Releases of SRP and NH_4 were highest in summer and at the deepest of the three sites. High organic matter supply rates to the sediments may be important for sustaining high rates of sediment nutrient release. A nutrient budget of Lake Rotorua indicates that internal nutrient sources derived from benthic fluxes are more important than external nutrient sources to the lake.

Keywords Nitrogen · Phosphorus · Sediment · Benthic chambers · Internal load · External load · Lake Rotorua

Guest editors: R. D. Gulati, E. Lammens, N. De Pauw & E. Van Donk
Shallow lakes in a changing world

D. F. Burger (✉) · D. P. Hamilton · C. A. Pilditch
Centre for Ecology and Biodiversity Research, University of Waikato, Private Bag 3105, Hamilton, New Zealand
e-mail: david.burger@wldelft.nl

M. M. Gibbs
National Institute for Water and Atmospheric Research, Private Bag 11115, Hamilton, New Zealand

Introduction

Sediment fluxes of nitrogen (N) and phosphorus (P) to the overlying water column may support a significant fraction of the total nutrient requirements for primary productivity in lotic systems (Marsden, 1989; Søndergaard et al., 1999). Prolonged external loading may ultimately produce elevated levels of nutrients in bottom sediments that, under certain environmental conditions such as anoxia, are remobilised and returned to the water column (Boström et al., 1988). Thus sustained high concentrations of water column nutrients have been observed in many lakes despite large reductions in external loads, particularly from point source inputs (Marsden, 1989; Søndergaard et al., 2003). Studies of lake nutrient cycling, particularly in the Northern Hemisphere, emphasise P release because of its impact on phytoplankton production (e.g., Kleeberg & Kozerski, 1997; Søndergaard et al., 2003). However, N can also limit or co-limit phytoplankton growth (White et al., 1986; Philips et al., 1997; Hameed et al., 1999), and may be as common as P limitation (Elser et al., 1990). Quantification of sediment N and P fluxes and their contribution to the total nutrient load is therefore important in eutrophic lakes and is a prerequisite to targeted nutrient management and lake restoration programmes.

Sediment nutrient release rates are mediated by interactions of many physical, chemical and biological processes (Boström et al., 1988). Factors influencing P include desorption and dissolution of P bound to iron, manganese and other inorganic complexes under reducing conditions (Mortimer, 1941, 1942), molecular diffusion from sediment porewaters to the water column along steep concentration gradients (Boström et al., 1988; Søndergaard, 1989), and mineralisation of organic material by bacteria (Gächter et al., 1988; Marsden, 1989). Sediment resuspension may also mediate P release by displacing of sediment porewaters (Søndergaard et al., 1992). Nitrogen release from bottom sediments occurs predominantly as ammonium (NH_4), as a result of particulate organic matter decomposition by bacterial mineralisation (Forsberg, 1989). Autolysis or hydrolysis of organic material leads to the production of dissolved organic N, which may be further mineralised to NH_4 via bacterial deamination (Hargreaves, 1998). Ammonium may be further oxidised to nitrate if the bottom sediments are well oxidised (Hargreaves, 1998; Beutel, 2001).

Direct measurements of sediment nutrient release rates are generally obtained from sediment core incubations conducted in the laboratory (e.g., Boström and Pettersson, 1982; Nürnberg, 1987; Jensen et al., 1992; Krivtsov et al., 2001). Incubations typically incorporate only a small sediment area (<0.01 m^{-2}), but allow for tight environmental controls and provide opportunities to manipulate cores to address specific questions. Removal of cores from the lake may also alter the physical, chemical and biological characteristics of the sediments through sediment re-oxygenation, porewater displacement, loss of the benthic boundary layer, and changes in temperature and light regimes. The use of benthic chambers allows nutrient fluxes to be determined in situ with minimal disturbance to the sediments and overlying water (see review by Tengberg et al., 1995), while incorporating a large sediment surface area as well as the natural light and temperature conditions of the lake.

The primary objective of this study was to quantify seasonal and spatial variations in sediment release of NH_4 and soluble reactive phosphorus (SRP) in eutrophic, polymictic Lake Rotorua, using in situ benthic chamber deployments. A further objective was to use this information to provide comparisons between the magnitude of internal and external nutrient loads to the lake, so that an understanding could be developed of the effects and response time related to catchment management actions.

Study site

Lake Rotorua is a large (79.8 km^2), shallow (mean depth 10.5 m), polymictic lake in central North Island, New Zealand (Fig. 1). It is eutrophic (Rutherford et al., 1996) and annual mean water column concentrations of total phosphorus (TP) and total nitrogen (TN) are 0.055 mg L^{-1} and 0.814 mg L^{-1}, respectively (Burger et al., 2005). The Lake Rotorua catchment area of 425 km^2 is dominated by agriculture (48%) and plantation forestry (23%). There are nine major inflows to the lake (mean discharge 0.22–2.75 $m^3 s^{-1}$) as well as 17 minor inflows, including urban drains and geothermal springs (mean discharge <0.06 $m^3 s^{-1}$). Until 1991 Lake Rotorua received discharges of wastewater from Rotorua city (population 60,000), which contributed annual nutrient loads of 35 tonnes TP yr^{-1} and 150 tonnes TN yr^{-1} (White et al., 1992). While removal of wastewater discharges was expected to improve lake water quality (Rutherford et al., 1996), nutrient concentrations have remained high and there are frequent summer blooms of cyanobacteria (Burger, 2006).

Methods

Sediment nutrient fluxes

Benthic chambers were deployed at three sites (Fig. 1) on four occasions between February 2003

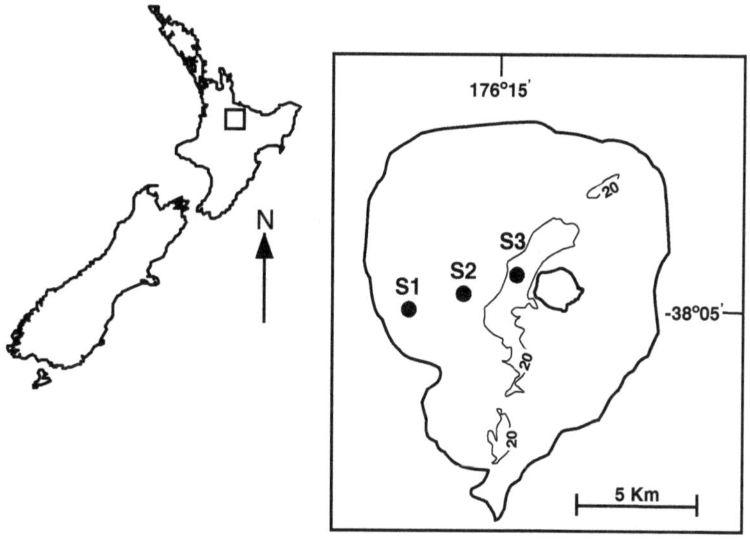

Fig. 1 Map of Lake Rotorua showing chamber deployment sites: Site 1 (depth 7 m), Site 2 (depth 14 m) and Site 3 (depth 20 m). The 20 m depth contour is also shown

and January 2004 (Table 1). Sites were chosen to reflect the natural variability in mixing regime, from permanently well mixed and oxygenated in shallow regions (site 1: depth 7 m), to periodically stratified and anoxic in deeper regions (site 2: depth 14 m and site 3: depth 20 m). At sites 1 and 2 both light chambers (circular, 6 mm acrylic plastic, encompassing sediment area 0.111 m^2 and water volume 17.76 L) and dark chambers (circular, 8 mm PVC, encompassing sediment area 0.116 m^2 and water volume 18.53 L) were deployed in duplicate. At site 3 (depth 20 m), where light at the sediment surface was <1% of photosynthetically available radiation (PAR) at the water surface (Table 1), only light chambers were deployed. Two chambers were deployed at site 3 in all periods except February 2004, when four chambers were used.

On each sampling date chambers were deployed between 08:30 and 14:00 h, depending on weather conditions. Deployment was assisted by SCUBA divers to minimise sediment disturbance during insertion of the chambers into the sediments. Chambers were pushed into the sediments to a depth of 80 mm, corresponding to the position of a flange on the outside of each

Table 1 Mean concentrations of ammonium (NH_4), soluble reactive phosphorus (SRP), nitrate (NO_3) and dissolved oxygen (DO) in bottom waters at sites 1, 2 and 3 at the start of each incubation

Site	Date	NH_4 (mg L^{-1})	SRP (mg L^{-1})	NO_3 (mg L^{-1})	DO (mg L^{-1})	T (°C)	PAR (μmol quanta m^{-2} s^{-1})
1	10 Feb. 2003	0.097	0.017	0.015	7.5	21.7	4.31
	13 Aug. 2003	0.037	0.011	0.015	10.2	10.2	9.19
	10 Nov. 2003	0.023	0.020	0.011	9.0	16.5	6.53
	27 Jan. 2004	0.056	0.024	0.007	8.5	21.8	0.45
2	10 Feb. 2003	0.187	0.027	0.015	4.9	20.1	0.09
	13 Aug. 2003	0.049	0.010	0.008	10.5	9.7	0.44
	10 Nov. 2003	0.037	0.023	0.004	8.9	18.2	0.12
	27 Jan. 2004	0.413	0.029	0.004	6.7	20.4	0.11
3	10 Feb. 2003	0.374	0.055	0.010	3.4	19.5	0.10
	13 Aug. 2003	0.040	0.015	0.005	9.1	9.6	0.09
	4 Nov. 2003	0.058	0.015	0.005	8.9	16.8	0.08
	27 Jan. 2004	0.263	0.081	0.015	3.6	20.3	0.12

Temperature (T) and photosyntheticaly available radiation (PAR) are mean values in bottom waters derived from daily CTD casts during each chamber deployment

chamber. After chamber insertion, taps (Ø = 9 mm) in the chamber lids were left open for 20 min before the start of each experiment. Water in the chambers was mixed throughout the experiment with a submersible 6 V DC pump (LVM Ltd), which circulated water for 5 s each minute. Laboratory trials showed that this flow regime maintained well-mixed conditions inside the chambers without inducing sediment resuspension. Bottom water samples were also collected during all deployments, and incubated in duplicate dark and light 1 L PET® bottles alongside the chambers, to measure water column nutrient regeneration rates in the absence of bottom sediments.

Chambers were sampled four times daily, depending on incubation start time and weather conditions; usually at 07:00, 11:00, 15:00 and 19:00 h, over 2 days. This deployment period was sufficient to create anoxic conditions in the chambers at sites 2 and 3 during summer incubations, thereby simulating release rates normally observed during natural stratification events when bottom waters become anoxic (Burger et al., 2005). Water was transported from the chambers to the surface under reduced pressure through clear plastic tubing (Ø = 4 mm) and collected in an in-line trap after flushing each line. A small external opening (Ø = 4 mm) in the chamber lid allowed replenishment of water sampled from the chamber with water from the depth of deployment. A sample volume of up to 60 ml was collected from each chamber on each sampling occasion, of which 25 ml was immediately separated for analysis of dissolved oxygen (DO) concentration (YSI Instruments, Model 50, probe electrode model 5739). The remaining sample was filtered through GF/C 25 mm diameter syringe filters and placed on ice before return to the laboratory, where filtrate was frozen before analysis for NH_4, NO_3 and SRP on a Lachat Instruments flow injection analyser (FIA, Zellweger Analytics, 2000). On three sampling occasions (August and November 2003, and January 2004), in situ DO sensors (Van Essen Instruments) were also used to record changes in DO concentration at 15 min intervals in a light and a dark chamber (sites 1 and 2) or in all chambers (site 3).

Sediment NH_4 and SRP fluxes were calculated from the slope of linear regressions of chamber nutrient concentrations with time (Gibbs et al., 2002). After correcting for the effect of dilution associated with sample removal, rates of change of nutrient concentrations in the chamber were divided by the sediment surface area in the chambers, to give an aerial release rate. Changes in nutrient concentrations in control bottles incubated alongside the chambers during each deployment were <0.020 and <0.006 mg L d^{-1} for NH_4 and SRP, respectively, and were subtracted from the final sediment release rate. Fluxes of NO_3 not calculated as concentrations in the chambers were less than 5% of those of NH_4 and were often below analytical detection limits (0.001 mg L^{-1}). Where in situ DO sensors were deployed inside the chambers, sediment oxygen demand (SOD) was calculated from the slope of linear regressions of chamber DO concentrations over the time period 0.25–2.25 h. For remaining deployments, SOD was calculated from the rate of change of DO between the first and second measurements (i.e., c. 4 h). Differences in SRP and NH_4 release rates and SOD between sites, period, and sites × period were examined using a two-way analysis of variance (ANOVA) with post-hoc analysis using the Tukey HSD test with a confidence interval of 95%. Prior to ANOVA analyses, data were examined for normality and homogeneity of variance by visual inspection of residuals and no transformations were necessary.

Chamber water displacement

Rates of water dilution in each chamber were assessed on one sampling occasion (February 2003) by injecting the chambers with a bromine tracer (LiBr at 5 μg L^{-1}) at the start of each deployment, and measuring changes in Br concentration at 0.5 and 48 h. Concentrations of Br were analysed by mass spectroscopy and varied little (<0.003 μg Br L^{-1}) from the dilution rate calculated using the total volume of water displaced from each chamber during sampling. Groundwater flows were also assessed using observations by divers of water bladders placed on the chamber outlet for

20 min at 48 h during the same deployment. The bladders did not inflate, indicating that groundwater influxes were likely to be low.

Water column measurements

Vertical profiles of conductivity-temperature-depth (CTD) profiles (Seabird Electronics) with an additional CTD-mounted sensor for DO concentration and PAR were taken at the start of each deployment and during subsequent collection of nutrient samples from the chambers. Concentrations of NH_4, NO_3 and SRP were determined on bottom water samples collected with a Schindler-Patalas trap at the start of each deployment. Further samples for nutrient analysis were collected daily at each site during the chamber incubations.

Results

Water column variables

Water column profiles of temperature and DO concentration, collected from each site at the start of the four chamber incubations, are shown in Fig. 2. The water column was thermally stratified at sites 2 and 3 in February 2003 and again in January 2004. The mean thermocline depth ($d\rho/dz$ = minimum) on these two occasions was between 7 and 8 m, with a temperature difference of 2°C between surface and bottom waters (Fig. 2). Stratification was accompanied by reduced DO concentrations in the hypolimnion during both periods, particularly at site 3 (<3.6 mg L^{-1}, Fig. 2, Table 1). During August (winter) and November (early summer) 2003, the water column was isothermal at all sites, and DO concentrations were nearly homogeneous and close to saturation. Mean bottom water temperatures across all sites during each chamber deployment were highest in January 2004 (20.8°C) and lowest in August 2003 (13.2°C, Table 1).

Bottom water concentrations of NH_4, SRP and NO_3, corresponding to initiation of each chamber deployment, varied between seasons and sites (Table 1). Mean concentrations of SRP over all seasons were highest at site 3 (0.042 mg L^{-1}) followed by site 2 and site 1 (0.022 and 0.018 mg L^{-1}, respectively). Nutrient concentrations in bottom waters were lowest in August

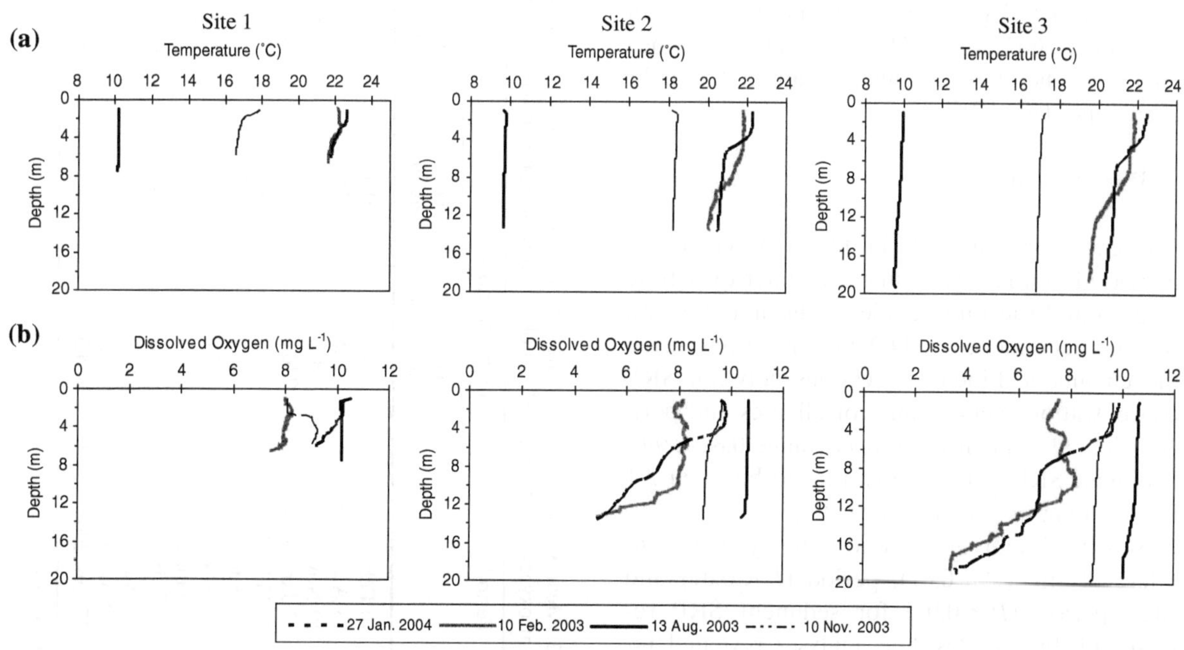

Fig. 2 Water column profiles of (**a**) temperature and (**b**) dissolved oxygen concentration at sites 1, 2 and 3 on 10 February, 13 August and 10 November 2003, and 27 January 2004

2003 (winter) and highest in February 2003, when DO concentrations were low. Mean concentrations of NH_4 were also highest at site 3 (0.184 mg L^{-1}), followed by site 2 (0.171 mg L^{-1}) and site 1 (0.053 mg L^{-1}). Mean concentrations of NO_3 over the four periods were highest at site 1 (0.012 mg L^{-1}) and lowest at site 2 (0.008 mg L^{-1}). Over all periods, mean values of PAR at the depth of the chamber incubations at each site were highest at site 1 (mean 5.12 μmol quanta m^{-2} s^{-1}) followed by site 2 (0.19 μmol quanta m^{-2} s^{-1}) and site 3 (0.10 μmol quanta m^{-2} s^{-1}, Table 1).

Sediment oxygen demand

Sediment oxygen demand ranged from 0.3 g m^{-2} d^{-1} (site 2, August 2003) to 4.0 g m^{-2} d^{-1} (site 3, November 2003, Table 2). Mean SOD calculated over all periods was highest at site 3 (mean 1.9 g m^{-2} d^{-1}) followed by site 2 (mean 1.5 g m^{-2} d^{-1}) and site 1 (mean 0.9 g m^{-2} d^{-1}, Table 2). Values of SOD at sites 2 and 3 were low in February 2003 and 2004, partly due to reduced initial DO concentrations associated with stratification (Fig. 2). At sites 1 and 2, differences in mean SOD between light and dark chambers were low (<0.3 g m^{-2} d^{-1}), except at site 1 in November 2003, when SOD was nearly five times higher in the dark chambers than in the light chambers (Table 2).

SRP benthic fluxes

Concentrations of SRP in the light chambers increased in a relatively linear manner over time (Fig. 3) and the rate of release did not show an obvious dependence on DO concentrations inside the chambers. Linear regressions between SRP concentrations versus time for all sites and periods yielded a mean R^2 of 0.93 (range 0.82–0.99). Sediment SRP release rates ranged between 2.1 and 85.6 mg m^{-2} d^{-1} (Table 2, Fig. 4).

Statistical analyses showed a highly significant effect of site ($P < 0.01$), period ($P < 0.01$) and site × period ($P < 0.01$) for sediment SRP releases (Table 3). Post-hoc analyses revealed release rates at site 3 (mean 44.3 mg m^{-2} d^{-1}) were significantly higher than release rates at site 2

Table 2 Mean sediment release rates for ammonium from 0 to 12 h (NH_4 1) and 12–48 h (NH_4 2), and soluble reactive phosphorus (SRP, 0–48 h), as well as sediment oxygen uptake rate (SOD) obtained with light and dark benthic chambers on four occasions between February 2003 and January 2004

Site	Period	*n*	Light chambers				Dark chambers			
			NH_4 1 (mg m^{-2} d^{-1})	NH_4 2 (mg m^{-2} d^{-1})	SRP (mg m^{-2} d^{-1})	SOD (g m^{-2} d^{-1})	NH_4 1 (mg m^{-2} d^{-1})	NH_4 2 (mg m^{-2} d^{-1})	SRP (mg m^{-2} d^{-1})	SOD (g m^{-2} d^{-1})
1	10 Feb. 2003	2	414.6	70.3	10.2	0.75	415.5	109.2	6.5	0.80
	13 Aug. 2003	2	43.8	48.8	2.1	0.49	57.6	70.5	2.3	0.70
	10 Nov. 2003	2	190.9	81.8	5.8	1.09	278.6	84.3	5.4	5.14
	27 Jan. 2004	2	230.0	168.8	12.9	1.16	266.5	283.4	13.8	1.36
2	10 Feb. 2003	2	578.2	80.1	9.0	1.73	1320.8	188.2	29.4	1.63
	13 Aug. 2003	2	60.7	32.6	3.8	0.33	237.1	105.4	6.3	0.49
	10 Nov. 2003	2	95.6	59.7	11.5	0.67	563.7	67.4	4.2	1.42
	27 Jan. 2004	2	351.4	75.1	12.7	3.23	375.2	34.9	10.6	3.83
3	10 Feb. 2003	2	1957.8	223.9	85.6	1.75				
	13 Aug. 2003	2	93.5	52.8	5.6	1.26				
	4 Nov. 2003	2	2212.8	94.5	55.3	3.95				
	27 Jan. 2004	4	484.5	172.8	30.7	0.67				

dark chambers were not used at site 3

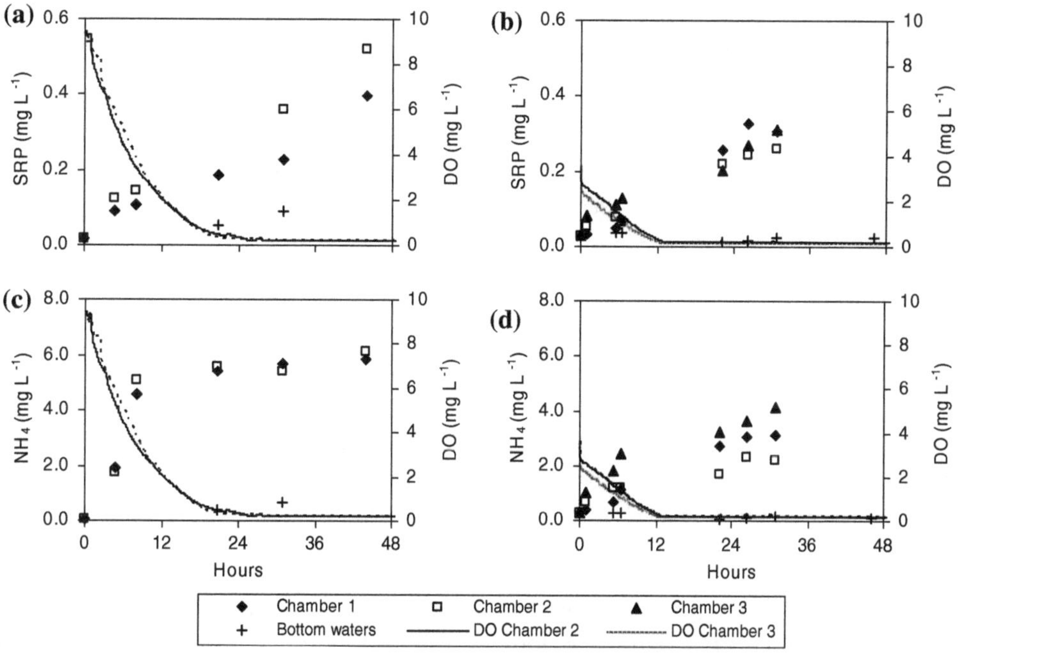

Fig. 3 Changes in concentrations of phosphorus (SRP) and ammonium (NH_4), and dissolved oxygen (DO) in replicate benthic chambers and bottom waters at site 3 in November 2003 (**a**, **c**) and January 2004 (**b**, **d**)

(mean 9.3 mg m^{-2} d^{-1}) and site 1 (mean 7.7 mg m^{-2} d^{-1}) for all periods except August 2003 (Tables 2 and 3, Fig. 4). Release rates at site 3 were also significantly different between periods ($P < 0.01$), and were highest in February 2003 and lowest in August 2003. The observed rate in August 2003 was nearly 10 times lower than in all other periods (Table 2). At sites 1 and 2, SRP release rates were highest in January 2004 and lowest in August 2003 but were not significantly different between periods ($P > 0.05$).

Differences in SRP release rates between light and dark chambers were <3.7 mg m^{-2} d^{-1} at site 1 (Table 2, Fig. 5), which was not statistically significant (Table 4). In February 2003, SRP release rates at site 2 were three times higher in dark chambers (29.4 mg m^{-2} d^{-1}) than in light chambers (9 mg m^{-2} d^{-1}, Fig. 5).

NH_4 benthic fluxes

NH_4 fluxes were characterised by two rates of release (Fig. 3). The first phase from 0 to 12 h of deployment had very high NH_4 release, with maximum values in the light chambers of 415, 578

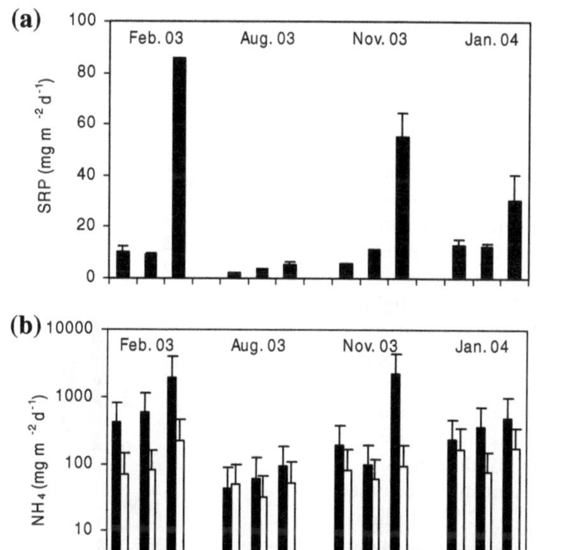

Fig. 4 Fluxes of (**a**) soluble reactive phosphorus (SRP) and (**b**) ammonium (NH_4), for sites 1 (S1), 2 (S2) and 3 (S3) from four periods of light chamber deployment. Error bars represent standard errors (n = 2–4). For NH_4, dark bars represent release initial release rates (0–12 h) and light bars represent secondary release rates (12–48 h)

Table 3 Two way analyses of variance between site, period and site × period for fluxes of soluble reactive phosphorus (SRP), ammonium in the phase 0–12 h (NH_4 1) and subsequent phase 12–48 h (NH_4 2)

Fluxes	Factor	*df*	MS	*F*	*P*
SRP	Site	2	1,554	58.07	<0.01
	Period	3	656	24.51	<0.01
	Site × Period	6	336	12.58	<0.01
	Error	13	27		
NH_4 1	Site	2	27,97,711	201.37	<0.01
	Period	3	887,278	63.86	<0.01
	Site × Period	6	740,314	53.29	<0.01
	Error	13	13,893		
NH_4 2	Site	2	11,650	5.09	0.023
	Period	3	11,625	5.08	0.015
	Site × Period	6	3,554	1.55	0.238
	Error	13	2,291		

df is degrees of freedom, MS is mean squares, *F* is the test statistic and *P* is probability

and 2,213 mg m^{-2} d^{-1} at sites 1, 2 and 3, respectively (Fig. 3). In the second phase NH_4 release was substantially lower, with a maximum value across all sites of 224 mg m^{-2} d^{-1} (Fig. 3).

For initial rates of NH_4 release, there was a significant effect of site ($P < 0.01$), period ($P < 0.01$) and site × period ($P < 0.01$) (Table 3). Initial NH_4 fluxes at site 3 (mean 1,187 mg m^{-2} d^{-1}) were significantly higher ($P < 0.05$) than at site 2 (mean 272 mg m^{-2} d^{-1}) and site 1 (mean 220 mg m^{-2} d^{-1}) (Tables 2 and 3, Fig. 4). Release rates between sites 1 and 2 were not significantly different. At all sites, release rates were significantly lower in August 2003 than in all other periods. Release rates at sites 1 and 2 were highest in February 2003 and at site 3 in November 2003. The secondary release rate of NH_4 was also highest at site 3 in all four deployments (mean 136 mg m^{-2} d^{-1}), followed by site 1 (mean 92 mg m^{-2} d^{-1}) and site 2 (mean 62 mg m^{-2} d^{-1}, Table 2, Fig. 5). Factorial analysis of variance ($P < 0.05$) showed an effect of site and of period, but not site × period (Table 3). Values of NH_4 release at site 3 were significantly different from site 2 over all periods. At site 1, there were no significant differences between light and dark

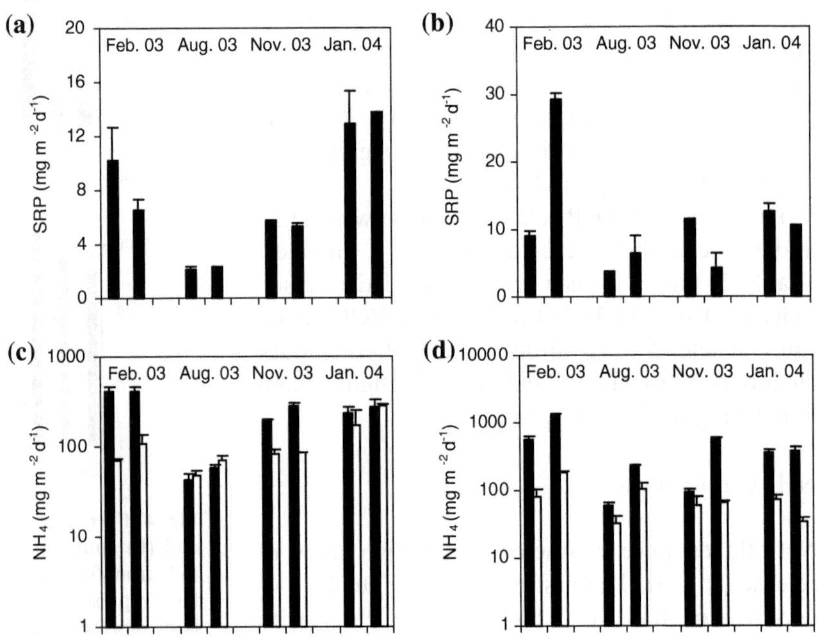

Fig. 5 Fluxes of soluble reactive phosphorus (SRP) at (**a**) site 1 and (**b**) site 2, and fluxes of ammonium at (**c**) site 1 and (**d**) site 2 for light (L) and dark (D) benthic chambers for four periods of chamber deployment. Error bars represent standard errors (n = 2–4). For NH_4, dark bars represent initial release rates (0–12 h) and light bars represent secondary release rates (12–48 h)

Table 4 Two way analyses of variance between period, chamber type (light or dark) and period × chamber for fluxes of soluble reactive phosphorus (SRP) and ammonium in the phase 0–12 h (NH_4 1) and 12–48 h (NH_4 2) at site 1 and site 2

Fluxes	Factor	Site 1				Site 2			
		df	MS	*F*	*P*	*df*	MS	*F*	*P*
SRP	Period	3	93	1.88	0.211	3	152	18.58	<0.01
	Chamber	1	7	0.14	0.722	1	44	5.36	0.05
	Period × Chamber	3	11	0.22	0.883	3	142	17.39	<0.01
	Error	8	49			8	8		
NH_4 1	Period	3	88,717	30.81	<0.01	3	259,983	56.58	<0.01
	Chamber	1	4,824	1.68	0.232	1	810,886	176.48	<0.01
	Period × Chamber	3	1,464	0.51	0.687	3	240,907	52.43	<0.01
	Error	8	2,880			8	4,595		
NH_4 2	Period	3	22,754	12.55	<0.01	3	71,205	9.93	<0.01
	Chamber	1	7,899	4.36	0.070	1	72,473	10.11	0.01
	Period × Chamber	3	2,410	1.33	0.331	3	60,865	8.49	<0.01
	Error	8	1,813			8	7,170		

df is degrees of freedom, MS is mean squares, *F* is the test statistic and *P* is probability

chambers for both phases of NH_4 release. At site 2, initial rates of NH_4 release were significantly higher ($P < 0.05$) in dark chambers than in light chambers.

Discussion

Sediment nutrient fluxes

Release rates of SRP of up to 86 mg m^{-2} d^{-1} estimated from this study are considerably higher than those of 7–50 mg m^{-2} d^{-1} observed in other eutrophic lakes (Nürnberg, 1988; Marsden, 1989), although rates between 100 and 200 mg P m^{-2} d^{-1} have been obtained from sediment cores incubated in the laboratory at high summer temperatures (Søndergaard, 1989; Jensen & Andersen, 1992). Previous summer estimates of P release in Lake Rotorua, based on changes in hypolimnion concentration during a stratification event, were within the range observed here (20–40 mg SRP m^{-2} d^{-1}, White et al., 1978).

Sediment release rates of SRP remained relatively invariant through time in all chambers, irrespective of initial DO concentration or changes in DO. While SRP release rates are generally reported to be much higher during water column anoxia than under well-oxygenated conditions (Andersen & Ring, 1999; Nowlin et al., 2005), there is evidence in sediment core incubations for high release rates under aerobic conditions (Søndergaard 1989; Jensen & Andersen, 1992; Krivtsov et al., 2001). It has been suggested that SRP release under aerobic conditions is associated with high rates of organic material decomposition at the sediment-water interface (Marsden, 1989; Kleeberg & Kozerski, 1997). Rapid decomposition utilises oxygen and nitrate and may create localised reducing conditions leading to desorption of P from metal cation complexes (Kleeberg & Kozerski, 1997; Krivtsov et al., 2001).

Gächter et al. (1988) suggest that SRP release rates may also be partly controlled by changes in sediment microbial physiology, including uptake, storage and release of P, as well as production and decomposition of bacterial biomass. Rates of P sedimentation are high in Lake Rotorua (0.5–2.7 g TP m^{-2} d^{-1}, Burger, 2006) and are therefore likely to be important in sustaining the high release rates observed in this study, particularly at the deeper sites. The large seasonal differences in release rates observed may be explained in part by changes in temperature, which controls rates of biological activity, as well as oxygen consumption rates and redox potential (Boström & Pattersson, 1982; Søndergaard, 1989; Søndergaard et al., 2003). Seasonal variability in sedimentation rates observed in the lake (Burger, 2006) may also be important.

A two-phase release rate was observed in the chamber incubations for NH_4. A sustained high initial rate of release in the first 12 h of chamber deployment was followed by a lower secondary release rate over the remaining 12 to 48 h. Increases in DIN concentration in sediment core incubations are typically found to be linear through time (Fukuhara & Sakamoto, 1988), although often only one measurement is collected following the first 24 h of incubation, which would not have captured the trends observed here. It is possible that our NH_4 fluxes may have been artificially enhanced by isolation of the sediments or disturbance of the porewaters during deployment, though there was no evidence of this phenomenon in association with SRP release. Further, the absence of NO_3 in the chambers suggests that nitrification did not occur or that if there was nitrification, the rate of denitrification was sufficiently rapid to rapidly remove NO_3. The release rate of NH_4 may have been associated with a decrease in bacterial metabolism and regeneration of NH_4 as the water column becomes progressively deoxygenated during each deployment. For example, before deployment of the chambers, high ammonium regeneration rates may be coupled to rapid nitrification and denitrification at the sediment–water interface. As DO decreases inside the chambers immediately after deployment, nitrification and denitrification may be shut down, leading to rapid build up of ammonium. Further, the secondary release rate observed in our chamber deployments may therefore represent the natural organic matter degradation rate normally observed under low DO. Without further experimental work involving artificial control of DO inside the chambers during incubations, it is difficult to ascertain the exact mechanisms leading to the high variability of ammonium release observed over time within chambers.

Release rates of NH_4 of up to 2,212 mg m^{-2} d^{-1} estimated in this study are also much higher than those observed elsewhere. In a literature review of NH_4 release rates in freshwater systems, Hargreaves (1998) lists a highest value of 185 mg m^{-2} d^{-1} while Fukuhara & Sakamoto (1988) found dissolved inorganic nitrogen (DIN) release rates of between 0.7 and 240 mg m^{-2} d^{-1}. Previous sediment release estimates for Lake Rotorua of 250–530 mg NH_4 m^{-2} d^{-1} (White et al., 1978), based on changes in hypolimnion concentration during a stratification event, were within the range observed here. High rates of NH_4 release have been found to coincide with high organic matter sedimentation rates (Fukuhara & Sakamoto, 1988) and in Lake Rotorua, sedimentation rates of up to 228 mg TN m^{-2} d^{-1} have been observed, with rates increasing significantly with increasing water column depth (Burger, 2006). The lack of a significant difference in sediment SRP and NH_4 release rates between light and dark benthic chambers at site 1 suggests that primary productivity at the sediment water interface may not have an important influence in mediating nutrient fluxes in shallow regions, perhaps as a result of high rates of advective transport and sediment disturbance not allowing an opportunity for significant accumulation of periphyton biomass. By contrast, at intermediate depths where there are high rates of deposition of chlorophyll (Burger, 2006), there may still be sufficient light for production and respiration by benthic algae to influence nutrient uptake rates between light and dark chambers (e.g. Dodds, 2003). For example, despite very low light levels at site 2 (< 1% of surface irradiance), rates of SRP and NH_4 release, and SOD were all lower in light than in dark chambers.

Verification of rates

A 19-day stratification event commencing 1 February 2003 coincided with chamber deployments commencing 10 February 2003. Sediment release rates of SRP during the stratification event, calculated from increases in SRP concentrations below the thermocline, and after accounting for sedimentation, inflows and diffusion across the thermocline, were estimated to be 13.8 mg m^{-2} d^{-1} (Burger et al., 2005). For this study, a mean SRP release rate beneath the thermocline of 17.4 mg m^{-2} d^{-1} was calculated for the same period using an estimate of SRP release rate at each 1 m depth interval beneath the thermocline, derived from linear interpolations between light chamber measurements at the three sites. The similarity between SRP release rates

Table 5 Source and description of nutrient load components used in Fig. 6

Component	Reference	Period	Method
External nutrient load	Burger, 2006	Jan. 01–Dec. 03	Daily loads associated with 9 major, 17 minor and ungauged inflows, and rainfall
Sedimentation rates	Burger, 2006	Feb. 03–Jan. 04	Sediment trap deployments coinciding with chamber deployments in this study
Outflows	Beyá et al., 2005	Jan. 02–Dec. 02	Monthly nutrient loads in single outflow

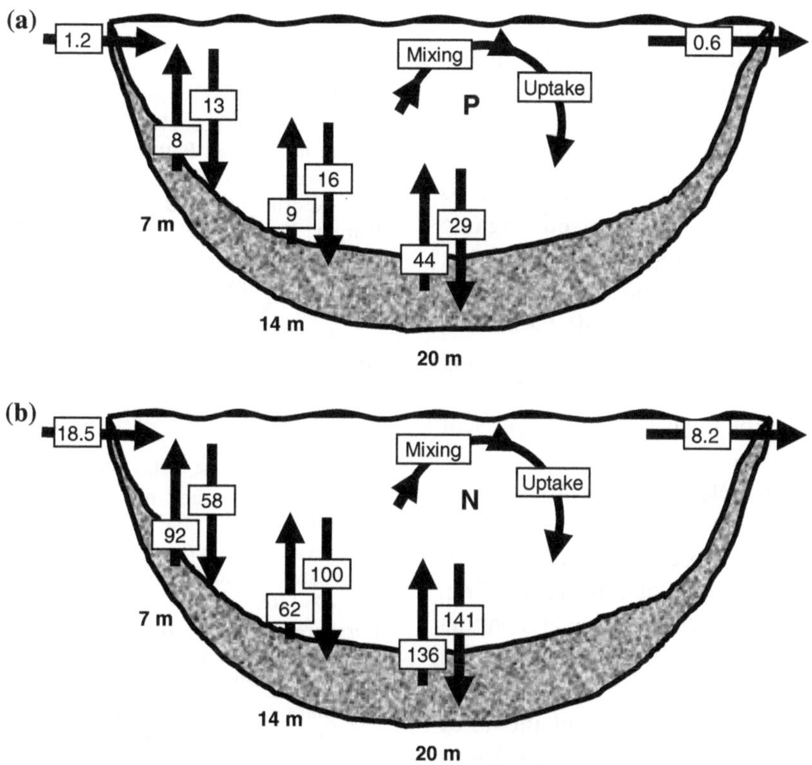

Fig. 6 Cycling of (**a**) phosphorus and (**b**) nitrogen in Lake Rotorua. All units are expressed as aerial rates (mg m^{-2} d^{-1}). Inflow, outflow and sedimentation rates represent total concentrations (TP or TN) and sediment release rates represent soluble reactive phosphorus (SRP) or ammonium (NH_4, secondary release rate). Sedimentation and sediment release rates are expressed as a seasonal mean calculated over the four sampling periods. Inflow and outflow concentrations are derived from Burger (2006) and Beyá et al. (2005), respectively, and sedimentation rates are derived from Burger (2006). See Table 5 for sources of inflow, outflow and sedimentation rates

derived from our chamber measurements and those estimated using changes in hypolimnion concentration suggests that chamber measurements are indeed representative at the lake scale.

Hypolimnion NH_4 concentrations during the same stratification event in February 2003 increased from 0.046 to 0.291 mg L^{-1}, representing a mean release rate of 63.9 mg m^{-2} d^{-1} for bottom sediments below the thermocline. This rate is substantially lower than that estimated beneath the thermocline from the chamber deployments for the same period (131.2 mg m^{-2} d^{-1}), which was derived from the second rate of NH_4 release (12–48 h) observed in the light chambers and calculated as for SRP. However, estimates from the chambers do not incorporate loss of NH_4 due to nitrification, as may occur higher in the water column, or uptake by benthic organisms.

Internal verses external loads

Fluxes of nutrients in Lake Rotorua, based on data presented in Table 5, indicate that bottom sediments are the dominant source of N and P (Fig. 6). Mean SRP and NH_4 sediment release rates at all sites, calculated from the current study, were at least three times greater than the mean daily external load determined by Burger (2006, Table 5). Mean release rates of SRP at site 3 exceeded mean TP sedimentation rates at the same site (Burger, 2006, Table 5), indicating the importance of the deeper sediments as a source of SRP to the water column. The fluxes did not include estimates of porewater SRP and NH_4 remobilisation in the shallow regions of the lake due to resuspension, or releases of dissolved organic phosphorus and dissolved organic nitrogen, that would have resulted in elevation of TP and TN releases above those measured for SRP and NH_4 Rates of water column denitrification, which may have important implications for the loss of N from the lake, were also not measured in the present study.

The high nutrient release rates observed in Lake Rotorua reflect the large pool of nutrients accumulated in the lake's sediments as a result of several decades of high rates of external loading, particularly in association with wastewater inputs to the lake. High nutrient release rates can be expected to continue in this lake, given the high organic matter sedimentation rates and anoxia of bottom sediments, and only a severe reduction in the external load may effectively 'break' the depositional cycle.

Acknowledgements We thank Eloise Ryan, Dudley Bell and Alex Ring (University of Waikato), Brian Coxhead (SeaQuest Ltd.), Andrew Lang (Rotorua District Council), Gerald Inskeep (Eastern Region Fish and Game) and Craig Putt (Environment Bay of Plenty) for assistance with field work. Eastern Region Fish and Game provided on-site laboratory facilities and accommodation. Drs Julie Hall and Kit Rutherford (National Institute for Water and Atmospheric Research) provided valuable discussion. This project was assisted financially by Environment Bay of Plenty and by a University of Waikato Hillary Jolly Memorial Ph.D. Scholarship to DFB.

References

Andersen, F. & P. Ring, 1999. Comparison of phosphorus release from littoral and profundal sediments in a shallow, eutrophic lake. Hydrobiologia 408/409: 175–183.

Beutel, M. W., 2001. Oxygen consumption and ammonia accumulation in the hypolimnion of Walker Lake, Nevada. Hydrobiologia 466: 107–117.

Beyá, J., D. P. Hamilton & D. F. Burger, 2005. Analysis of catchment hydrology and nutrient loads for Lakes Rotorua and Rotoiti. Report to the Rotorua Lakes Water Quality Society. Centre for Biodiversity of Ecology Research, University of Waikato, Hamilton.

Boström, B., J. M. Andersen, S. Fleischer & M. Jansson, 1988. Exchange of phosphorus across the sediment-water interface. Hydrobiologia 170: 229–244.

Boström, B. & K. Pettersson, 1982. Different patterns of phosphorus release from lake sediments in laboratory experiments. Hydrobiologia 92: 415–429.

Burger, D. F., D. P. Hamilton, C. P. Pilditch & M. M. Gibbs, 2005. Sediment phosphorus release during stratification in polymictic Lake Rotorua, New Zealand. Verhandlung Internationale Vereingung de Limnologie 29: 811–814.

Burger, D. F., 2006. Benthic-pelagic coupling of nutrients in Lake Rotorua. Ph.D. Thesis, Department of Biological Sciences, University of Waikato, Hamilton, 100 pp.

Elser, J. J., E. R. Marzolf & C. R. Goldman, 1990. Phosphorous and nitrogen limitation of phytoplankton growth in the freshwaters of North America: A review and critique of experimental enrichments. Canadian Journal of Fisheries and Aquatic Sciences 47: 1468–1477.

Forsberg, C., 1989. Importance of sediments in understanding nutrient cyclings in lakes. Hydrobiologia 176/177: 263–277.

Fukuhara, H. & M. Sakamoto, 1988. Ecological significance of bioturbation of zoobenthos community in nitrogen release from bottom sediments in a shallow eutrophic lake. Archiv für Hydrobiologie 113: 425–445.

Gächter, R., J. S. Meyer & A. Mares, 1988. Contribution of bacteria to release and fixation of phosphorus in lake sediments. Limnology and Oceanography 33: 1542–1558.

Gibbs, M., A. Ross & M. Downes, 2002. Nutrient cycling and fluxes in Beatrix Bay, Pelorus Sound, New Zealand. New Zealand Journal of Marine and Freshwater Research 36: 675–697.

Hameed, H. A, S. Kilinc, S. McGowan & B. Moss, 1999. Physiological tests and bioassays: aids or superfluities to the diagnosis of phytoplankton nutrient limitation? A comparative study in the Broads and the Meres of England. European Journal of Phycology 34: 253–269.

Hargreaves, J. A., 1998. Nitrogen biogeochemistry of aquaculture ponds. Aquaculture 166: 181–212.

Jensen, H. S. & F. Andersen, 1992. Imporatnce of temperature, nitrate and pH for phosphate release from aerobic sedimnets of four shallow, eutophic lakes. Limnology and Oceanography 37: 577–589.
Jensen, H. S., P. Kristensen, E. Jeppesen & A. Skytthe, 1992. Iron:phosphorus ratio in surface sediment as an indicator of phosphate release from aerobic sediments in shallow lakes. Hydrobiologia 235/236: 731–743.
Kleeberg, A. & H. P. Kozerski, 1997. Phosphorus release in Lake Grosser Muggelsee and its implications for lake restoration. Hydrobiologia 342/343: 9–26.
Krivtsov, V., D. C. Sigee, E. G. Bellinger & G. Porteous, 2001. Determination of P release from Rostherne Mere sediment cores. Acta Hydrochimica et Hydrobiologica 29: 111–117.
Marsden, M. W., 1989. Lake restoration by reducing external phosphorus loading: the influence of sediment phosphorus release. Freshwater Biology 21: 139–162.
Mortimer, C. H., 1941. The exchange of dissolved substances between mud and water in lakes, I. Journal of Ecology 29: 280–239.
Mortimer, C. H., 1942. The exchange of dissolved substances between mud and water, II. Journal of Ecology 30: 147–201.
Nowlin, W. H., J. L. Evarts & M. J. Vanni, 2005. Release rates and potential fates of nitrogen and phosphorus from sediments in a eutrophic reservoir. Freshwater Biology 50: 301–322.
Nürnberg, G. K., 1987. A comparison of internal phosphorus loads in lakes with anoxic hypolimnia: Laboratory incubation verses in situ hypolimnetic phosphorus accumulation. Limnology and Oceanography 32: 1160–1164.
Nürnberg, G. K., 1988. Prediction of phosphorus release rates from total and reductant soluble phosphorus in anoxic lake sediments. Canadian Journal of Fisheries and Aquatic Science 45: 453–462.
Philips, E. J., M. Cichra, K. E. Havens, C. Hanlon, S. Badylak, B. Rueter, M. Randall & P. Hansen, 1997. Relationships between phytoplankton dynamics and the availability of light and nutrients in a shallow subtropical lake. Journal of Plankton Research 19: 319–342.
Rutherford, J. C., S. M. Dumnov & A. H. Hoss, 1996. Predictions of phosphorus in Lake Rotorua following load reductions. New Zealand Journal of Marine and Freshwater Research 30: 383–396.
Søndergaard, M., 1989. Phosphorus release from a hypertrophic lake sediment: experiments with intact sediment cores in a continuous flow system. Archiv für Hydrobiologie 116: 45–59.
Søndergaard, M., J. P. Jensen & E. Jeppesen, 1999. Internal phosphorus loading in shallow Danish Lakes. Hydrobiologia 408/409: 145–152.
Søndergaard, M., J. P. Jensen & E. Jeppesen, 2003. Role of sediment and internal loading of phosphorus in shallow lakes. Hydrobiologia 506–509: 135–145.
Søndergaard, M., P. Kristensen & E. Jeppesen, 1992. Phosphorus release from resuspended sediment in the shallow and wind-exposed Lake Arresø, Denmark. Hydrobiologia 228: 91–99.
Tengberg, A. et al., 1995. Benthic chamber and profiling landers in oceanography–A review of design, technical solutions and functioning. Progress in Oceanography 35: 253–294.
White, E., B. J. Don, M. T. Downes, L. J. Kemp, A. L. MacKenzie & G. W. Payne, 1978. Sediments of Lake Rotorua as sources and sinks for plant nutrients. New Zealand Journal of Marine and Freshwater Research 12: 121–130.
White, E., G. W. Payne & S. Pickmere, 1992. The restoration of Lake Rotorua – comment on progress. Taupo Research Laboratory, DSIR Marine and Freshwater, Taupo. File Report No. 138, 28 pp.
White, E., G. W. Payne, S. Pickmere & P. Woods, 1986. Nutrient demand and availability related to growth among natural assemblages of phytoplankton. New Zealand Journal of Marine and Freshwater Research 20: 199–208.

Hydrobiologia (2007) 584:27–36
DOI 10.1007/s10750-007-0583-z

SHALLOW LAKES

Modeling nitrogen cycling in a coastal fresh water sediment

Richard W. Canavan · Anniet M. Laverman · Caroline P. Slomp

Abstract Increased nitrogen (N) loading to coastal marine and freshwater systems is occurring worldwide as a result of human activities. Diagenetic processes in sediments can change the N availability in these systems, by supporting removal through denitrification and burial of organic N (N_{org}) or by enhancing N recycling. In this study, we use a reactive transport model (RTM) to examine N transformations in a coastal fresh water sediment and quantify N removal rates. We also assess the response of the sediment N cycle to environmental changes that may result from increased salinity which is planned to occur at the site as a result of an estuarine restoration project. Field results show that much of the N_{org} deposited on the sediment is currently remineralized to ammonium. A rapid removal of nitrate is observed in the sediment pore water, with the resulting nitrate reduction rate estimated to be 130 μmol N cm^{-2} yr^{-1}. A model sensitivity study was conducted altering the distribution of nitrate reduction between dissimilatory nitrate reduction to ammonium (DNRA) and denitrification. These results show a 40% decline in sediment N removal as NO_3^- reduction shifts from denitrification to DNRA. This decreased N removal leads to a shift in sediment-water exchange flux of dissolved inorganic nitrogen (DIN) from near zero with denitrification to 133 μmol N cm^{-2} yr^{-1} if DNRA is the dominant pathway. The response to salinization includes a short-term release of adsorbed ammonium. Additional changes expected to result from the estuarine restoration include: lower NO_3^- concentrations and greater SO_4^{2-} concentrations in the bottom water, decreased nitrification rates, and increased sediment mixing. The effect of these changes on net DIN flux and N removal vary based on the distribution of DNRA versus denitrification, illustrating the need for a better understanding of factors controlling this competition.

Keywords denitrification · dissimilatory nitrate reduction to ammonium (DNRA) · ammonium adsorption · diagenetic model · sediment-water exchange · salinity

Guest editors: R. D. Gulati, E. Lammens, N. De Pauw & E. Van Donk
Shallow lakes in a changing world

R. W. Canavan (✉) · A. M. Laverman · C. P. Slomp
Department of Earth Sciences – Geochemistry, Faculty of Geosciences, Utrecht University, 80.021, 3508 TA Utrecht, The Netherlands
e-mail: R.Canavan@geo.uu.nl

Present Address:
A. M. Laverman
UMR 7619 Sisyphe, Université Pierre et Marie Curie, 4 Place Jussieu, 75252 Paris, France

Introduction

Nitrogen (N) inputs to freshwater and near shore marine systems have increased strongly due to human activities, leading to eutrophication (Cloern, 2001; de Jonge et al., 2002). A significant fraction of this excess N may be removed in the sediment through denitrification or burial of organic N (N_{org}). This nitrogen removal has been shown to correlate to N input loads and discharge rates of water (Windolf et al., 1996; Saunders & Kalff, 2001). Aquatic systems with high discharge rates have lower removal rates as sediment-water interactions are limited.

Removal through denitrification may also be limited by competition with an alternative sediment pathway, dissimilatory nitrate reduction to ammonium (DNRA). DNRA has been found to be a significant pathway of nitrate reduction in a variety of sediments and saturated soils (Gilbert et al., 1997; Ogilvie et al., 1997; Revsbech et al., 2005). It is thought that an abundance of organic carbon relative to NO_3^- favors DNRA over denitrification, but little is known about the competition between the organisms responsible for the two pathways (Megonigal et al., 2003). The difference between these two pathways is significant for water quality as denitrification produces a form of gaseous N that is generally unavailable to primary producers, while the NH_4^+ produced through DNRA is readily bioavailable (Fig. 1). In addition to diffusion from the overlying water, nitrification of NH_4^+ in the oxic surface sediment can also be an important source of NO_3^- for reduction (Seitzinger, 1988).

Ammonium can be retained in sediment due to adsorption at cation exchange sites, which are present on the surface of organic matter and clay minerals (Berner, 1980). Salinization of fresh water sediments can cause the desorption of NH_4^+ from cation exchange sites as a result of competition with Na^+ ions (Seitzinger et al., 1991). Lower nitrification rates (Gardner et al., 1991; Rysgaard et al., 1999), increased NO_2^- production, and increased rates of organic matter mineralization (Nyvang, 2003) have also been associated with increased salinity. Salinization of fresh waters, and the associated changes in N cycling, may occur in the coastal zone due to

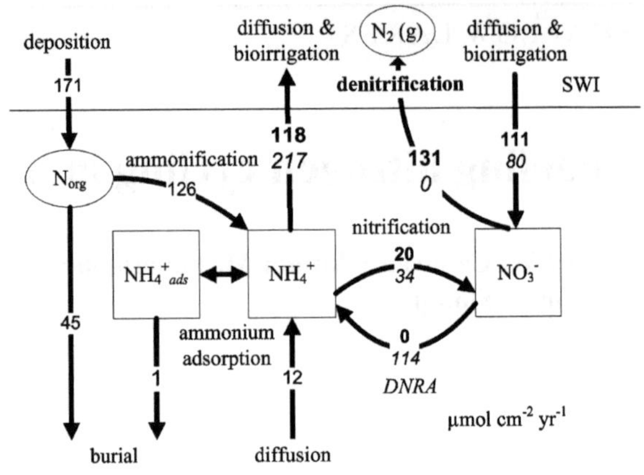

Fig. 1 A schematic diagram of the sedimentary N cycle with both reaction and transport processes labeled. Model derived values of rates and fluxes for the upper 30 cm of sediment are presented on the arrows in μmol N cm^{-2} yr^{-1}. Results of two steady-state simulations are shown, in one simulation all nitrate reduction occurs via denitrification (bold), in the other via DNRA (italics). The influx of organic matter and mineralization rates were kept constant

rising sea level, dam construction, and ground water withdrawal.

Our study site is currently a freshwater lake that will become brackish from 2008 onwards as a result of an ecological restoration project. In this study we adapt a previously developed reactive transport model (Canavan et al., 2006), to examine the response of the sediment N-cycle to changes that may result from salinization of this fresh water lake. Possible changes in site conditions are represented by changing parameters in steady-state and transient simulations. To allow for a more detailed model representation of the sediment N processes the existing model was adapted to include DNRA, in addition to denitrification, as experimental results suggest a possible role for DNRA at the site (Laverman et al., 2006). In this model sensitivity study, we specifically focus on the consequences for net sediment dissolved inorganic nitrogen (DIN) efflux rates.

Methods

Study Site

Haringvliet Lake (The Netherlands) is a eutrophic freshwater lake in the southwestern Netherlands. The lake was created as a result of damming of the mouth of a tidal estuary in 1970. Prior to dam construction, the Haringvliet was an outlet of the Meuse-Rhine River system to the North Sea. The closure of the Haringvliet caused physical and chemical changes in the water body, including the disappearance of the salinity gradient and the accumulation of river derived suspended matter (Smit et al., 1997). The lake is relatively shallow; the water depth at the sample site was 7.5 m. The lake retains some fluvial characteristics, such as high flow rates. The residence time of water in Haringvliet Lake is typically on the order of several days (Smit et al., 1997), however periods of increased residence time occur when river levels are low and discharge through the dam is stopped. Thermal stratification and bottom water anoxia are not observed. A partial restoration of estuarine conditions in Haringvliet Lake is proposed to increase the diversity and availability of estuarine habitat in the area (Anonymous, 1998). Restoration would be achieved by changing the opening and closing of gates in the dam that separates the lake from the North Sea. Our sample location was located near the dam in an area that would be affected by the restoration (51.50.080 N, 04.04.328 E). The sediment at the site was fine-grained and highly porous. A previous study at this site indicates organic matter mineralization rates on the order of 764 μmol cm^{-2} yr^{-1} where oxic degradation (55%), nitrate reduction (21%) and sulfate reduction (17%) are the important degradation pathways (Canavan et al., 2006).

Sample collection and analysis

Field sampling was carried out in November 2001, September 2002, and April 2003. These sampling times are referred to in the text as fall, late-summer, and spring, respectively. Sediment was collected using a cylindrical box corer (31 cm i.d.) deployed from RV Navicula. Sub-cores were taken with polycarbonate tubes (10 cm i.d.) and immediately sectioned in a N_2 purged glove box on board the ship. Sediment was centrifuged at 2500 g for 10 to 30 minutes in polyethylene tubes to collect pore water. After centrifugation, tubes were transferred to a N_2 purged glove box, and the pore water was filtered through 0.2 or 0.45 μm pore size filters. Pore water was then sub-divided and preserved for the different analyses. Sub-samples for NO_3^- and NH_4^+ were stored frozen until analysis at the laboratory and were determined colorimetrically on a nutrient auto-analyzer (Bran and Luebbe). Dissolved oxygen microprofiles were determined on-board the sampling ship with a Clark-type oxygen sensor as described in Canavan et al. (2006). Sediment C and N contents were determined on freeze dried sediments using a Carlo Erba CN analyzer and a LECO CS analyzer.

Model development

RTM calculations of 1-D sediment profiles were carried out with the Biogeochemical Reaction Network Simulator (BRNS; Aguilera et al., 2005; Jourabchi et al., 2005). The development and calibration of the model, which includes a reaction network of 24 chemical species and 32 reactions, is presented in detail in Canavan et al. (2006). The current study focuses solely on the sediment N-cycle, therefore the model description and results are limited to N transformation processes. The RTM includes 6 N-species: NH_4^+, NO_3^-, adsorbed NH_4^+ (NH_{4ads}^+), and three pools of N_{org} (Table 1), which are linked to the organic carbon pools through C:N ratios. The different organic matter pools are abbreviated as OM1, a highly reactive pool, OM2, a less reactive pool, and OM3, a refractory (non-reactive) pool (Westrich & Berner, 1984). Upper boundary conditions for the N-species are given in Table 1. All lower boundaries are set to no-flux conditions,

Table 1 Model nitrogen species and upper boundary conditions

solute	μM	solid	(μmol cm^{-2} yr^{-1})	CN ratio
NO_3^-	154	N_{org}-OM1	113	5.6
NH_4^+	14.5	N_{org}-OM2	32	10
		N_{org}-OM3	27	20
		NH_{4ads}^+	0	–

NO_3^- and NH_4^+ upper boundary concentrations are mean values of overlying water from the three sampling events. Input fluxes of N_{org} were calculated from the input fluxes for the three organic matter pools and CN ratios that were defined for those pools where the lower boundary of the simulation was 100 cm.

The decomposition of organic matter, which is modeled as a first order process defined by a rate constant kOM (yr^{-1}), results in the mineralization of N_{org} (Table 2). Organic matter mineralization occurs with a series of terminal electron acceptors: O_2, NO_3^-, Mn-oxides, Fe-oxides, SO_4^{2-}, and organic carbon. The distribution of the total organic matter decomposition rate over the different terminal electron acceptor pathways follows the approach of Van Cappellen & Wang, (1996). The model reaction network, which initially included only denitrification, was adapted to include both nitrate reduction pathways. The DNRA reactions were added to the reaction network (Table 2) and a term *f*DNRA was included to distribute the total nitrate reduction between the two pathways. This *f*DNRA term is set empirically and equals the fraction of total nitrate reduction occurring via DNRA. The fraction of total nitrate reduction occurring by denitrification is defined as 1-*f*DNRA. Because this study uses the boundary conditions calibrated in Canavan et al. (2006) the initial value of *f*DNRA is set to zero. Sensitivity of sediment processes to changes in the distribution of nitrate reduction pathways is examined by varying the *f*DNRA term. With the exception of Kelly-Gerreyn et al. (2001), existing sediment diagenetic models ignore DNRA and assume all nitrate reduction to occur via denitrification.

The oxidation of NH_4^+ with O_2 (nitrification) is described with a bimolecular rate law (Table 2; Van Cappellen & Wang, 1996). Adsorption of NH_4^+ to cation exchange sites is represented by a linear equilibrium isotherm, with a constant adsorption coefficient, K_N (Berner, 1980; Van Cappellen & Wang, 1996). K_N is a dimensionless adsorption coefficient:

$$K_N = \left(\frac{1-\phi}{\phi}\right)\rho K^* \quad (1)$$

where ϕ is porosity (vol. % porewater), ρ is sediment density (g cm^{-3}), and K^* (cm^3 g dw^{-1}) is the ratio of NH_{4ads}^+ (μmol g dw^{-1}) to NH_4^+ (mM). The concentration of NH_{4ads}^+ is estimated to be 3 μmol g dw^{-1} based on the release of NH_4^+ from sediment flow through reactors exposed to increased salinity (Laverman, unpublished results). Using this estimate and the corresponding pore water NH_4^+ concentration of 0.077 mM, gives a value of 39 (cm^3 g dw^{-1}) for K^* and of 10 for K_N. Both values fall within the range reported for fresh water sediments by Seitzinger et al. (1991).

The model includes transport of solutes by molecular diffusion, bioirrigation, bioturbation, and advection (burial). Transport of solids occurs by bioturbation and advection. Molecular

Table 2 Nitrogen containing reactions in the model

Reactions	Kinetic or equilibrium formulation
Ammonification[a,b]	
$[(CH_2O)_y(\boldsymbol{NH_3})_z]$ + yTEAox → z$\boldsymbol{NH_4^+}$ + + yCO_2 + yH_2O + $yTEA_{red}$	k_{OM} [OM] f_{TEA}
Denitrification	
$[(CH_2O)_y(\boldsymbol{NH_3})_z] + 0.8y\boldsymbol{NO_3^-} \rightarrow z\boldsymbol{NH_4^+} + 0.4y\boldsymbol{N_2} + (0.2y - z)CO_2$ $+(0.8y + z)HCO_3^- + (0.6y - z)H_2O$	k_{OM} [OM] f_{NO3} (1– *f*DRNA)
DNRA	
$[(CH_2O)_y(\boldsymbol{NH_3})_z] + 0.5y\boldsymbol{NO_3^-} + zCO_2 + (0.5y + z)H_2O \rightarrow (0.5y+ z)\boldsymbol{NH_4^+} + (y + z)HCO_3^-$	k_{OM} [OM] f_{NO3} *f*DRNA
Nitrification	
$\boldsymbol{NH_4^+} + 2O_2 + 2HCO_3^- \rightarrow \boldsymbol{NO_3^-} + 2CO_2 + 3H_2O$	k_{NH4OX} [NH_4^+] [O_2]
Ammonium adsorption[c]	
$\boldsymbol{NH_{4\,ads}^+} \leftrightarrow \boldsymbol{NH_4^+}$ (K_N $\varphi/\rho(1 - \varphi)$)	K_N [NH_4^+]

[a] Organic matter is represented by the formula $[(CH_2O)_y(NH_3)_z]$ where y/z is the C/N ratio. [b]A generalized mineralization reaction is depicted for ammonification where TEA_{ox} and TEA_{red} represent the oxidized and reduced terminal electron acceptor. The term f_{TEA} defines the fraction of the total carbon mineralization by each TEA pathway (Van Cappellen and Wang, 1996), for example f_{NO3} represents the nitrate reduction pathways. [c]K_N is the dimensionless adsorption coefficient as defined in eq. 1

diffusion and the associated tortuosity and temperature effects are included as described in Van Cappellen & Wang (1996). Bioirrigation is represented as a non-local exchange with the surface water of which the intensity is controlled by the coefficient, α (Boudreau, 1984); and bioturbation is parameterized using an additional diffusion term, D_b (Berner, 1980). The advective velocity of solids and solutes is determined from the sediment accumulation rate, ω (cm yr^{-1}) and porosity (φ) as described by Berner (1980). Table 3 includes a list of parameter values used in the model including the depth distributions of α, D_b, and φ.

Results

The oxygen penetration depth was between 0.3 to 0.6 cm in the sediment (Fig 2). Nitrate concentrations in the overlying water ranged from 120–180 μM, while NO_3^- in pore water from the 0–0.5 cm layer was always below 5 μM. The relatively coarse scale of the pore water sampling (0.5 cm sections in the upper 2 cm) and the rapid reduction of nitrate in the samples, limits the resolution of measured pore water NO_3^-. Typically the NO_3^- penetration should extend below that of O_2 (Stief et al., 2003). The resolution of the O_2 profiles is improved by the use of the microelectrode. Pore water NH_4^+ concentrations increased with depth, as a result of ammonification. Seasonal differences in NH_4^+ were observed in the upper 17 cm with highest concentrations in spring and lowest in fall, where high NH_4^+ corresponded with shallower O_2 penetration. The NH_4^+ profiles exhibit evidence of bioirrigation in the upper 17 cm with somewhat homogeneous concentrations, while below this depth upward diffusion is evident. Sediment N_{org} concentrations decline with depth due to organic matter decomposition.

Discussion

The model derived profiles of O_2, NO_3^-, NH_4^+, and N_{org} capture the vertical trends observed in the measured values (Fig. 2). The modeled NH_4^+ concentrations are greater when *f*DNRA is 1.

Table 3 Initial parameter values and distributions

Parameter	Value	Unit	Description
k_{OM1}	25	yr^{-1}	degradation rate constant OM1
k_{OM2}	0.01	yr^{-1}	degradation rate constant OM2
k_{OM3}	0	yr^{-1}	degradation rate constant OM3
K_N	10	–	NH_4^+ adsorption coefficient (eq. 1)
*f*DNRA	0	–	fraction of nitrate reduction occurring as DNRA
k_{NH4OX}	20	$\mu M^{-1} yr^{-1}$	nitrification rate constant
ω	1.0	cm yr^{-1}	sediment accumulation rate
ρ	2.1	g cm^{-3}	sediment density
α_0	10	yr^{-1}	bioirrigation coefficient at surface
D_{b0}	5	cm^2 yr^{-1}	bioturbation coefficient at surface
λ	2.5	cm^{-1}	D_b depth attenuation coefficient
φ_o	0.89	cm^3 cm^{-3}	porosity at surface
φ_∞	0.79	cm^3 cm^{-3}	porosity at depth
τ	0.2	cm^{-1}	φ depth attenuation coefficient
x	0–30	cm	sediment depth

Depth distributions	Description
$\alpha = \begin{cases} \alpha_0\left(1 - e^{(x-17)}\right)(x \leq 17\,cm) \\ 0\ (x > 17\,cm) \end{cases}$	Distribution of bioirrigation coefficient α
$D_b = D_{b0}\, e^{-\left(\frac{x}{\lambda}\right)}$	Distribution of bioturbation coefficient D_b
$\varphi(x) = \varphi_\infty + (\varphi_o - \varphi_\infty)e^{-(\tau\ x)}$	porosity distribution

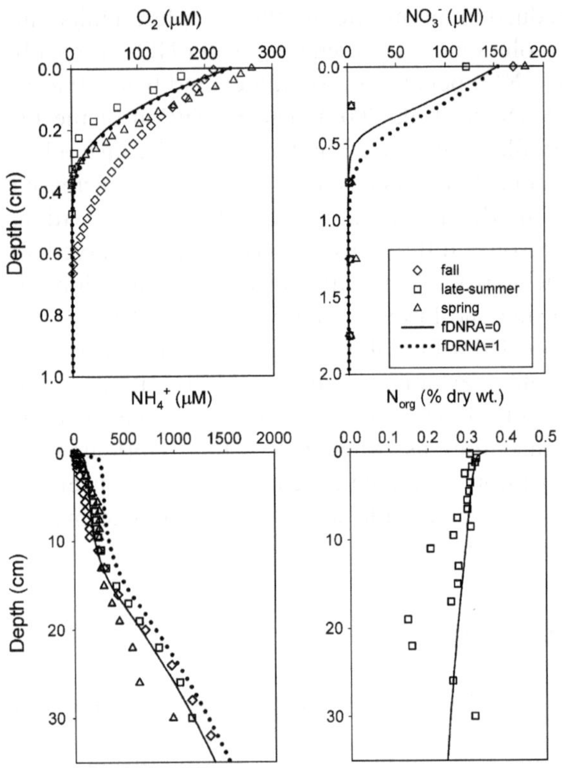

Fig. 2 Sediment pore water O_2, NO_3^-, and NH_4^+ concentrations and sediment organic nitrogen (N_{org}) concentrations from fall (◇),late-summer (□), and spring (Δ). Steady-state model derived profiles for porewater species are plotted for *f*DNRA = 0 (solid line) and for *f*DNRA = 1 (dotted line). The modeled N_{org} profile is not sensitive to changes in *f*DNRA which is depicted with a solid line

The model was calibrated for the condition *f*DRNA = 0 (Canavan et al, 2006). It is not possible to estimate a possible value of *f*DNRA directly from the NH_4^+ profiles given the uncertainty in the input fluxes of organic matter. Model-derived rates of transport and reaction in the sediment are presented in the schematic representation of N-cycling (Fig. 1) including results of two examples where all nitrate reduction occurs either by denitrification (in bold) or DNRA (in italics). The decomposition of organic matter releases 73% of the deposited N_{org} as NH_4^+ in the upper 30 cm of sediment. Much of the remaining N_{org} is buried in the sediment, however, decomposition does continue deeper in the sediment resulting in the upward diffusion of NH_4^+ in the pore water. The major part of the NH_4^+ generated by ammonification is transported to the overlying water, while approximately 15–26% is oxidized to NO_3^- (nitrification). Limitation of nitrification in the sediment may be due to the shallow O_2 penetration depth (Fig. 2), and competition from other processes such as oxic organic matter decomposition, and the oxidation of sulfides and methane (Canavan et al., 2006). Low rates of nitrification have been observed in coastal sediment with significant DNRA activity (Gilbert et al., 1997, shellfish farm site; An & Gardner, 2002; Magalhães et al., 2005, muddy sediment site). Conditions favoring DNRA over denitrification, such as highly reduced sediments, and low NO_3^- concentrations relative to organic substrate availability, may also contribute to low nitrification rates (Megonigal et al., 2003).

Role of Denitrification versus DNRA

Experimental results from flow through reactor (FTR) experiments show that potential rates of denitrification were approximately 50% less than those of NO_3^- reduction at our site (Laverman et al., 2006). Conditions in the FTR experiments favor denitrification over DNRA when compared to the in-situ conditions as NO_3^- concentrations in the FTRs were greater (Sørensen, 1987). Initial experimental evidence from the site (Laverman et al. 2006; unpublished results of N_2O microprofiling) suggests that a significant portion of the nitrate reduction in the sediment occurs by processes other than denitrification. To examine how the competition between DNRA and denitrification affects the N-cycle at our site we conducted several steady state simulations while varying *f*DNRA from 0 (all denitrification) to 1 (all DNRA).

The removal of N entering the sediment (NO_3^- influx and N_{org} deposition) ranges from 63% to 18%, with maximum removal occurring at *f*DNRA = 0 (Fig. 3a). The decreased N removal with increasing prevalence of DNRA leads to a shift from a situation where sediment is essentially neutral with respect to DIN efflux to one there the sediment is a net source with an efflux rate of 133 μmol cm^{-2} yr^{-1}. The NH_4^+ efflux ranged from 115 to 213 μmol cm^{-2} yr^{-1}. Such high NH_4^+ efflux rates have been found in other

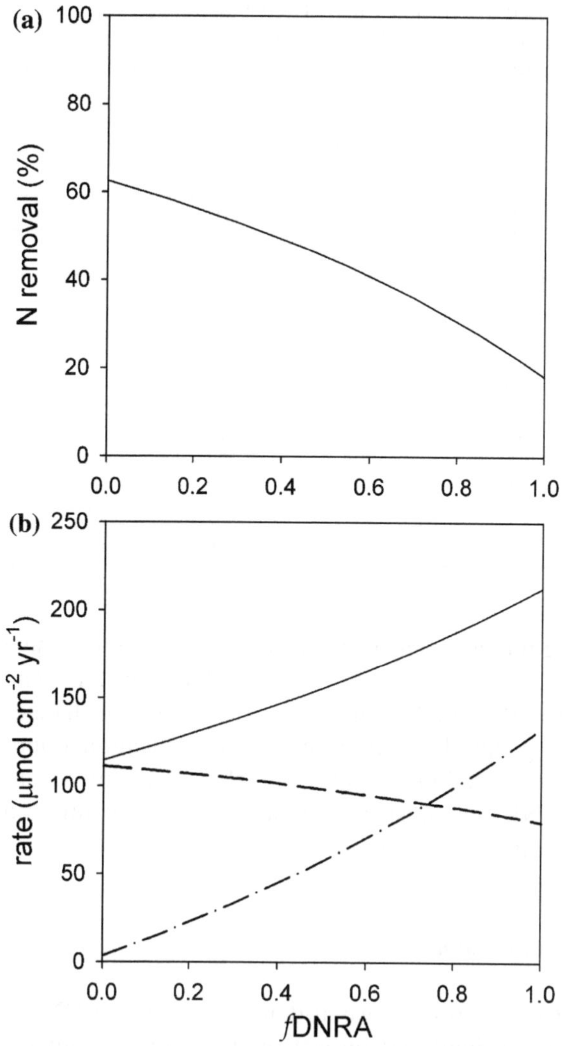

Fig. 3 The N removal as a percentage of the total N influx to the sediment for steady-state simulations where *f*DNRA was varied from 0–1 (**a**). N removal is calculated from denitrification and burial (at 30 cm) and sediment input is N_{org} deposition and NO_3^- influx. The rates of sediment ammonium efflux (solid line), nitrate influx (dashed line), and net DIN efflux (dash dot line) versus *f*DNRA are presented in plot (**b**)

freshwater sediments (65–307 µmol cm^{-2} yr^{-1}; Gardner et al., 2001). However, in sediments with a greater rate of coupled nitrification-denitrification, the NH_4^+ efflux is typically lower as more N is released as N_2 (Seitzinger, 1988). Nitrate influx decreased with increasing values of *f*DNRA (Fig. 3b), due to an accompanying increase in nitrification rate. In our simulations, much of the NO_3^- reduced in the sediment is transported from the overlying water, while only 15–30% of reduced NO_3^- is produced by nitrification (Fig. 1).

Ammonium adsorption response to salinization

The release of NH_{4ads}^+ in response to increasing salinity was examined with a transient model simulation. Results from the steady-state simulation using the parameter values listed in Table 3 were used as the initial conditions in a simulation where the value of K_N was reduced from 10 to 1.3 based on an estimate for marine sediments (Mackin & Aller, 1984). The response of the sediment is very rapid. Over the first day, the NH_4^+ efflux was approximately 2.5 times greater than was observed for the initial conditions, primarily driven by the diffusive efflux (Fig. 4). Ammonium released to the pore water at depth is transported to the overlying water via bioirrigation, which responds more slowly. Ammonium efflux declines to rates that are 15% greater than found for the initial conditions after a 6-month simulation period (not shown).

Estuarine Restoration

Restoration of the estuary will result in changes to the bottom water chemistry and sediment pore water. In addition to greater Na^+ and Cl^- concentrations, the coastal water adjacent to the

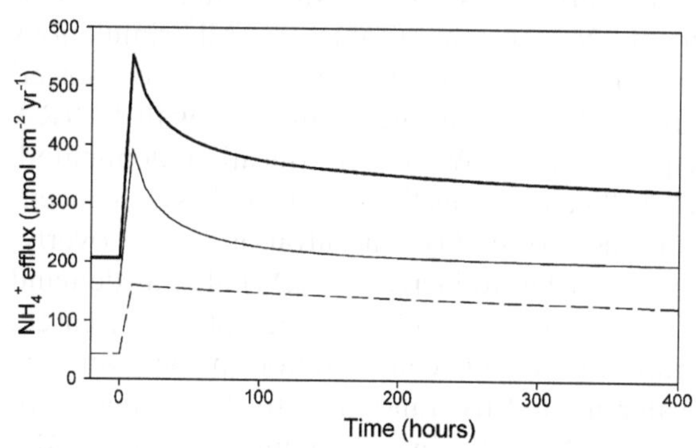

Fig. 4 Ammonium efflux from freshwater sediment subject to salinization. Results determined in a transient simulation where K_N was initially 10 and reduced to 1.3 at the start of the simulation (time = 0). Efflux by diffusion is the thin solid line, efflux via bioirrigation is the dashed line and the sum of the two processes is the thick solid line

Table 4 Sediment DIN flux (NH_4^+ efflux – NO_3^- influx) and N removal (denitrification + N_{org} burial) from steady-state simulations made to approximate possible changes resulting from estuarine restoration. Simulations were made with the initial conditions of only denitrification (*f*DNRA = 0) and where DNRA accounted for 95% of nitrate reduction (*f*DNRA = 0.95). Additional simulations where the input flux of labile organic matter (OM1) was decreased are also presented

Parameter[a]	Initial value	New value	*f*DNRA = 0		*f*DNRA = 0.95	
			DIN flux	N removal	DIN flux	N removal
			(μmol cm^{-2} yr^{-1})			
Initial conditions	–	3	177	124	55	
NO_3^- (μM)	154	90	39	141	126	53
SO_4^{2-} (mM)	0.64	10	3	177	124	55
O_2 (μM)	238	120	–42	222	121	57
K_N	10	1.3	7	177	129	55
D_{b0} (cm^2 yr^{-1})	5	25	–2	178	110	55
k_{NH4OX} (μM^{-1}yr^{-1})	20	10	7	173	124	54
kOM1&2 (yr^{-1})	25, 0.01	50, 0.02	14	164	130	46
All of the above changes			7	165	127	47
input flux OM1 (μmol cm^{-2} yr^{-1})	630	420	6	177	90	52

[a] see table 3 for a description of parameters

dam contains less NO_3^- and more SO_4^{2-}. Salinity stratification in the water column is expected and this could further affect the chemistry of the bottom water, for example, by resulting in lower oxygen concentrations. Additionally, changes in the microbial population and benthic fauna are expected, which will affect rates of reaction and transport in the sediment. To examine the sediment response to these possible changes, steady-state simulations were run where parameters were changed to represent the new conditions. Many of the salinity-related changes result in small changes to the net DIN flux (positive values are efflux to overlying water) and sediment N removal (Table 4). All simulations conducted with *f*DNRA equal to zero showed greater N removal rates than those with *f*DNRA equal to 0.95. When denitrification dominated, DIN flux was sensitive to O_2 and NO_3^- concentrations. Lower O_2 concentration in the overlying water led to increased NO_3^- influx as demand for other terminal electron acceptors increased, while lower NO_3^- concentration in the overlying water limited its transport into the sediment. In the DRNA dominated simulations, net DIN efflux was lowered by decreasing the input flux of highly reactive organic matter (OM1) and by increasing the amount of bioturbative mixing (D_{b0}). Both of these processes reduce the rate of ammonification in the upper most part of the sediment, which in turn, lowers the diffusive efflux of NH_4^+ to the overlying water. The response of N removal to environmental changes is generally greater in the denitrification dominated sediment, since in the sediment where DNRA is most important, removal is controlled mainly by burial.

Conclusions

Competition between denitrification and DNRA has a significant impact on N removal in aquatic sediments as only denitrification removes N from the system. Salinization results in a rapid and transient release of adsorbed NH_4^+. When DNRA is a key pathway at our site, changes in the chemistry of the overlying water and in reaction and transport parameters expected upon salinization do not significantly change the N removal rates. The ratio of DNRA to denitrification and the rates of N_{org} input and ammonification are the most sensitive sediment processes for N removal in this study. Our results show that DNRA needs to be included in diagenetic models of the sediment N-cycle. Additionally, a better understanding of the controls on the interaction between DNRA and denitrification is needed before such models can be used as predictive tools.

Acknowledgements We gratefully acknowledge the crew of the RV Navicula and members of the Utrecht University geochemistry research group for their help in the field. We thank Debby Los and Dineke van de Meent for analysis, and Parisa Jourabchi for the use of the reaction transport model. Philippe Van Cappellen, Gerard van den Berg, and Gertjan Zwolsman are acknowledged for their contributions to this project. The Netherlands Institute for Inland Water Management and Waste Water Treatment (RIZA) supported the fieldwork and RWC (contract no. RI-3494). AML was supported by a VENI fellowship of the Netherlands Organization for Scientific Research (NWO). CPS was supported by a fellowship of the Royal Netherlands Academy of Arts and Sciences (KNAW).

References

Aguilera, D. R., P. Jourabchi, C. Spiteri & P. Regnier, 2005. A knowledge-based reactive transport approach for the simulation of biogeochemical dynamics in earth systems. Geochemistry, Geophysics, Geosystems 6: Q07012.

An, S. M. & W. S. Gardner, 2002. Dissimilatory nitrate reduction to ammonium (DNRA) as a nitrogen link, versus denitrification as a sink in a shallow estuary (Laguna Madre/Baffin Bay, Texas). Marine Ecology-Progress Series 237: 41–50.

Anonymous, 1998. MER beheer Haringvlietsluizen report (Environmental Impact Report- Haringvliet dam management). Rijkswaterstaat directie Zuid-Holland, APV 98.186 (in Dutch).

Berner, R. A., 1980. Early diagenesis. A theoretical approach. Princeton University Press, Princeton NJ.

Boudreau, B. P., 1984. On the equivalence of nonlocal and radial-diffusion model for porewater irrigation. Journal of Marine Research 42: 731–735.

Canavan, R. W., C. P. Slomp, P. Jourabchi, P. Van Cappellen, A. M. Laverman & G. A. van den Berg, 2006. Organic matter mineralization in sediment of a coastal freshwater lake and response to salinization. Geochimica et Cosmochimica Acta 70: 2836–2855.

Cloern, J. E., 2001. Our evolving conceptual model of the coastal eutrophication problem. Marine Ecology-Progress Series 210: 223–253.

de Jonge, V. N., M. Elliott & E. Orive, 2002. Causes, historical development, effects and future challenges of a common environmental problem: eutrophication. Hydrobiologia 475: 1–19.

Gardner, W. S., S. P. Seitzinger & J. M. Malczyk, 1991. The effects of sea salts on the forms of nitrogen released from estuarine and freshwater sediments: Does ion pairing affect ammonium flux. Estuaries 14: 157–166.

Gardner, W. S., L. Y. Yang, J. B. Cotner, T. H. Johengen & P. J. Lavrentyev, 2001. Nitrogen dynamics in sandy freshwater sediments (Saginaw Bay, Lake Huron). Journal of Great Lakes Research 27: 84–97.

Gilbert, F., P. Souchu, M. Bianchi & P. Bonin, 1997. Influence of shellfish farming activities on nitrification, nitrate reduction to ammonium, and denitrification at the water-sediment interface of Thau lagoon, France. Marine Ecology Progress Series 151: 143–153.

Jourabchi, P., P. Van Cappellen & P. Regnier, 2005. Quantitative interpretation of pH distributions in aquatic sediments: A reaction-transport modeling approach. American Journal of Science 305: 919–956.

Kelly-Gerreyn, B. A., M. Trimmer & D. J. Hydes, 2001. A diagenetic model discriminating denitrification and dissimilatory nitrate reduction to ammonium in a temperate estuarine sediment. Marine Ecology-Progress Series 220: 33–46.

Laverman, A. M., P. Van Cappellen, D. van Rotterdam-Los, C. Pallud & J. Abell, 2006. Potential rates and pathways of microbial nitrate reduction in coastal sediments. FEMS Microbiology Ecology 58: 178–192.

Mackin, J. E. & R. C. Aller, 1984. Ammonia adsorption in marine sediments. Limnology and Oceanography 29: 250–257.

Magalhães, C. M., W. J. Wiebe, S. B. Joye & A. A. Bordalo, 2005. Inorganic nitrogen dynamics in intertidal rocky biofilms and sediments of the Douro River Estuary (Portugal). Estuaries 28: 592–607.

Megonigal, J. P., M. E. Hines & P. T. Visscher, 2003. Anaerobic metabolism: Linkages to trace gases and aerobic processes. In Schlesinger, W. H. (ed.), Biogeochemistry. 8,Treatise on Geochemistry. H. D. Hollands and K. K. Turekian (eds.). Elsevier-Pergamon, Oxford: 317–424.

Ogilvie, B., D. B. Nedwell, R. M. Harrison, A. Robinson & A. Sage, 1997. High nitrate, muddy estuaries as nitrogen sinks: The nitrogen budget of the River Colne estuary (United Kingdom). Marine Ecology-Progress Series 150: 217–228.

Revsbech, N. P., J. P. Jacobsen & L. P. Nielsen, 2005. Nitrogen transformations in microenvironments of river beds and riparian zones. Ecological Engineering 24: 447–455.

Rysgaard, S., P. Thastum, T. Dalsgaard, P. B. Christensen & N. P. Sloth, 1999. Effects of Salinity on NH_4^+ Adsorption Capacity, Nitrification, and Denitrification in Danish Estuarine Sediments. Estuaries 22(1): 21–30.

Saunders, D. L. & J. Kalff, 2001. Nitrogen retention in wetlands, lakes and rivers. Hydrobiologia 443: 205–212.

Seitzinger, S. P., 1988. Denitrification in freshwater and coastal marine ecosystems: Ecological and geochemical significance. Limnology and Oceanography 33: 702–724.

Seitzinger, S. P., W. S. Gardner & A. K. Spratt, 1991. The effect of salinity on ammonium sorption in aquatic sediments: Implications for benthic nutrient recycling. Estuaries 14: 167–174.

Smit, H., R. Smits, G. van der Velde & H. Coops, 1997. Ecosystem responses in the Rhine-Meuse delta during two decades after enclosure and steps toward estuary restoration. Estuaries 20: 504–520.

Sørensen, J., 1987. Nitrate reduction in marine sediment: pathways and interactions with iron and sulfur cycling. Geomicrobiology Journal 5: 401–421.

Stief, P., A. Schramm, D. Altmann & D. de Beer, 2003. Temporal variation of nitrification rates in experimental freshwater sediments enriched with ammonia or nitrite. FEMS Microbiology Ecology 46: 63–71.

Van Cappellen, P. & Y. F. Wang, 1996. Cycling of iron and manganese in surface sediments: A general theory for the coupled transport and reaction of carbon, oxygen, nitrogen, sulfur, iron, and manganese. American Journal of Science 296: 197–243.

Westrich, J. T. & R. A. Berner, 1984. The role of sedimentary organic matter in bacterial sulfate reduction: the G-model tested. Limnology and Oceanography 29: 236–249.

Windolf, J., E. Jeppesen, J. P. Jensen & P. Kristensen, 1996. Modelling of seasonal variation in nitrogen retention and in-lake concentration: A four-year mass balance study in 16 shallow Danish lakes. Biogeochemistry 33: 25–44.

Hydrobiologia (2007) 584:37–48
DOI 10.1007/s10750-007-0610-0

SHALLOW LAKES

Sediment phosphorus cycling in a large shallow lake: spatio-temporal variation in phosphorus pools and release

Bryan M. Spears · Laurence Carvalho · Rupert Perkins · Alex Kirika · David M. Paterson

Abstract Sediment and water column phosphorus fractions were recorded monthly for one year (April 2004–April 2005) in a shallow lake recovering from nutrient pollution (Loch Leven, Scotland). Equilibrium phosphate concentration (EPC0) and gross sediment phosphorus (P) release rates were estimated from laboratory experiments. Pore water and organic P pools were lowest during warm water periods whereas bottom water P was lowest during cold water periods. Reductant-soluble, organic, metal oxide-adsorbed, residual and sediment total phosphorus pools all varied significantly with overlying water depth. Short-term, high magnitude, redox initiated P release events occurred in late summer and winter as a result of anoxic sediment conditions. Lower magnitude long-term release conditions were maintained for most of the year, most likely as a result of organic P cycling and maintenance of high concentration gradients between the pore and bottom water P pools. Estimates of summer P uptake/release rates, across an intact sediment-water interface, suggested that maximum gross internal release was ~12 mg SRP m^{-2} lake surface area d^{-1} with EPC0 values ranging between 180 and 270 µg P L^{-1}. This study highlights the biological mediation of internal loading in shallow eutrophic lakes, and in particular, the role of sediment algae in decreasing, and sediment bacteria in enhancing, sediment P release.

Keywords Sediment · Phosphorus · Mobility · Bio-regulation · Seasonality

Guest editors: R. D. Gulati, E. Lammens, N. De Pauw & E. Van Donk
Shallow lakes in a changing world

B. M. Spears · L. Carvalho · A. Kirika
Centre for Ecology and Hydrology Edinburgh, Penicuik, Midlothian EH26 OQB, Scotland, UK

R. Perkins
School of Earth, Ocean and Planetary Sciences, Cardiff University, Main Building, Park Place, Cardiff CF10 3YE, Wales, UK

B. M. Spears (✉) · D. M. Paterson
Sediment Ecology Research Group, Gatty Marine Laboratory, University of St. Andrews, Fife KY16 8LB, Scotland, UK
e-mail: bs29@st-andrews.ac.uk

Introduction

Slow recovery from eutrophication, following a reduction in external phosphorus loading, is common in shallow lakes (Sas, 1989). Attention has been focussed on the physico-chemical processes that maintain poor water quality after the reduction of external loads. Early studies highlighted the role of lake sediment as a redox mediated phosphorus (P)-buffer, where P is released under conditions of anoxia and retained under oxic conditions (Mortimer, 1941, 1942). Additionally, adsorption/desorption mechanisms play an important role in P cycling within shallow lakes (House et al., 1998; Perkins et al., 2001). Despite the fact that these processes were recognised as being biologically-mediated, very little attention has been paid to the biological communities acting on sediment-P release. The total phosphorus (TP) pool in sediment is composed of several operationally-defined components, each with their own release pathways. Temperature is commonly regarded as an important driver of internal loading through elevated microbial remineralisation (depletion of organic sediment P), the resultant onset of anoxia (depletion of reductant-soluble sediment P), and raised diffusion rates (depletion of labile and pore water P).

Benthic communities (autotrophs, heterotrophs, invertebrates etc.) may, both directly and indirectly, regulate sediment P release (Paterson and Black, 1999). With sufficient light, benthic micro-algae proliferate and cap internal loading from sediment P pools sensitive to anoxia (Woodruff et al., 1999), inhibit resuspension through increased sediment stability and reduce P availability through direct uptake; Van Luijn et al., 1995). However, where light is limited, heterotrophs dominate biofilms and can enhance regeneration of previously unavailable organic and refractory P, release P from redox sensitive fractions, and absorb pore water P (Gächter & Meyer, 1993). Finally, sedimenting phytoplankton cells and plankton migrating back into the water column may further regulate organic P-partitioning between the sediment and the water column.

Sediment P uptake/release is a complex process, and affected by equilibrium conditions between the surface sediment and the overlying water. Traditional estimates of equilibrium phosphate concentrations (EPC0; SRP concentration in overlying water that results in no sediment P uptake/release) have often used sediment suspensions and, therefore, only represent equilibrium conditions of resuspended sediment. Recently, intact sediment cores, incubated under laboratory-controlled conditions, have been used to assess sediment P uptake/release (e.g. Søndergaard, 1989). The aims of this study were to (1) investigate the seasonality of concentration gradients between water column and sediment P pools in terms of internal loading mechanisms; (2) assess the effects of seasonal temperature on sediment P composition; (3) assess the effects of overlying water depth on the spatial distribution of sediment P composition; and (4) to use laboratory controlled intact sediment core experiments to estimate the current uptake/release conditions (including EPC0) of a shallow lake recovering from nutrient pollution (Loch Leven, Scotland).

Methods

Study site

Loch Leven (Fig. 1) is a shallow eutrophic lake (mean depth 3.9 m) situated in the southeast of Scotland (latitude 56°10′ N, longitude 3°30′ W) with a surface area of 13.3 km^2 the majority of which remains well mixed throughout the year, with intermittent stratification in two deep basins of ~25 m. Water chemistry has been recorded approximately every two weeks since 1968 providing a detailed history of recovery from eutrophication. External loading to Loch Leven was reduced from 4.05 mg TP m^{-2} d^{-1} in 1985 to 1.62 mg TP m^{-2} d^{-1} in 1995 (Bailey-Watts and Kirika, 1999). However, prolonged internal loading coupled with a low flushing rate (2–3 lake

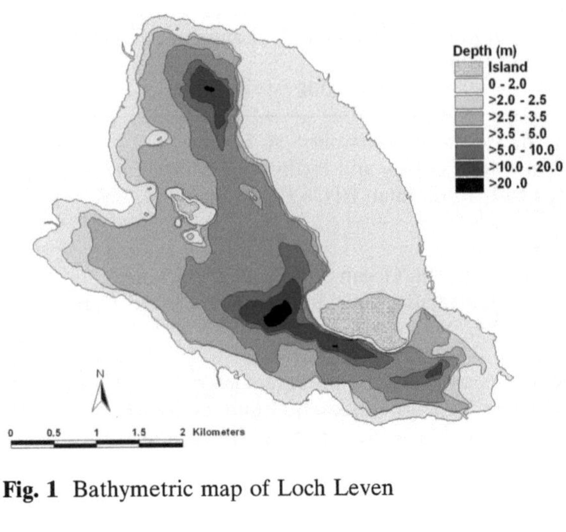

Fig. 1 Bathymetric map of Loch Leven

volumes per year) resulted in poor water quality being maintained, with an improvement in water column SRP concentration only being observed in recent years (Carvalho & Kirika, 2003). This improvement in water quality has not been mirrored in the sediment P composition (Farmer et al., 1994; Spears et al., 2006).

Field measurements, water chemistry, and sediment P fractionation

Monthly sampling was conducted between April 2004 and April 2005 from six sites along a depth transect (2, 2.5, 3.5, 5.5, 10, 22 m overlying water depth), to allow comparison of shallow and deep regions of the loch throughout the year. Detailed sampling locations are described in Spears et al. (2006). Surface waters were collected and stored in acid-washed Azlon bottles. Depth profiles of water temperature, dissolved oxygen % saturation (DO%), pH and conductivity were recorded on each visit (Hydrolab Water Quality Monitoring System, Hydrolab Corporation, Colorado, USA). On each date, single sediment cores were collected from each site using a Jenkin surface-sediment sampler (internal diameter 67 mm) and transported to the laboratory for processing. The cores had about 30 cm overlying water and 20 cm sediment.

In the laboratory, bottom water was sampled in the cores from 1 cm above the sediment surface. Both bottom and lake surface waters were filtered (Whatman GF/C) and stored at 4°C in acid washed Nalgene containers for chemical analysis. The filters were frozen and stored for analysis of photosynthetic pigments. The upper 3 cm of sediment from each core was removed, homogenised, then split, preceding both centrifugation and filtration (as above) for pore water chemical analysis, or storage for pigment and chemical analysis. Sediment for chlorophyll *a* analysis was filtered (GF/C) before dark extraction in 90% v/v cold acetone for 24 h at 0°C. Chlorophyll *a* was quantified spectrophotometrically, with correction for phaeopigment concentration by acidification (APHA, 1992). Inorganic nutrient analyses were carried out using the methods of Wetzel and Likens (2000). Sediment P fractionation was carried out using a sequential chemical extraction technique based on Psenner et al. (1988) and Farmer et al. (1994). Two sub-samples of homogenised slurry, from each site, were subject to the following extraction procedure on each sample date: (1) repeated extraction in 1 M NH_4Cl for 2 h and quantification of SRP [labile P]; (2) repeated extraction with 0.11 M $NaHCO_3$/0.11 M $Na_2S_2O_4$ for 1 h and quantification of SRP [reductant-soluble P] and total soluble phosphorus (TSP) [reductant-soluble soluble un-reactive phosphorus (SURP)]; (3) repeated (for 1 h) extraction in 1 M NaOH for 18 h followed by quantification of SRP (metal-oxide adsorbed P), and TSP; (4) repeated (for 1 h) extraction with 0.5 M HCl for 18 h and quantification of SRP [apatite-bound P]; (5) digestion with 30% (v/v) H_2SO_4 and 8% $K_2S_2O_4$ at 121°C for 30 min followed by SRP quantification [residual P]. Organic P was quantified by subtracting NaOH-SRP from NaOH-TSP. Sediment slurry and extraction medium were continually shaken in 50 ml centrifuge tubes in the dark at 20°C throughout each extraction. Supernatants were collected following centrifugation and filtration (GF/C filter) at the end of each extraction step.

Sediment release experiments

EPC0 experiments were conducted using intact sediment cores on the 6th August and the 8th September 2005. On each date, 12 sediment cores were collected from a site at 3.5 m overlying water depth, selected to represent an average water depth. Cores were held in a water bath at 17°C with an irradiance of 63.3 μmol m^{-2} s^{-1} for 1 h. The temperature of the overlying water on collection of the cores was 15.4°C on the 8th of August 2005 and 16.2°C on the 6th September 2005. Following acclimatisation, three sets of triplicate cores were spiked with KH_2PO_4 to achieve overlying water column phosphorus concentrations of 0.5, 5, and 10 mg P L^{-1}. A fourth set of triplicate cores had no additional phosphate added and represented ambient lake conditions. To maintain oxic conditions at the sediment-water interface, air was bubbled through the overlying water with care being taken to ensure minimal sediment disturbance.

Water samples (10 ml) were removed from the bottom water (0–1 cm above the sediment) at the beginning and end of a 72 h incubation. A pilot study indicated that, under these conditions, phosphorus equilibrium across the interface was achieved over this time period. Uptake/release values were calculated as the difference between initial and final overlying water SRP concentration. The range of concentrations used resulted in a linear plot of uptake/release versus overlying-water SRP concentration, enabling the EPC0 values to be estimated using linear regression. Estimates of daily sediment P release were taken from the three cores that had had no P additions, after 24 h of incubation and are expressed as mg SRP per m^{-2} lake surface area d^{-1}.

Statistical analysis

Seasonal changes in average lake sediment P composition were assessed by comparing the average monthly conditions of the six sites. The spatial (depth) and temporal (seasonal) variation in bottom water temperature and DO%, pore and bottom water SRP, labile P, reductant-soluble SRP, organic P, and sediment TP was assessed using two-way analysis of variance (ANOVA; total df = 59; $\alpha = 0.05$). Comparisons undertaken were: (1) between warm water (11th May–7th October 2004; $n = 5$) and cold water (23rd November 2004–14th April 2005; $n = 5$) conditions (seasonal variation df = 1), and (2) along a gradient of overlying water depth (spatial variation; df. = 5). Interaction effects between season and depth were also considered (df = 5).

Regression analysis was used to examine the spatial variation (along a transect of increasing water depth) of a number of P fractions (the absolute concentrations of labile P, reductant-soluble SRP, reductant-soluble SURP, metal-oxide adsorbed P, organic P, apatite-bound P, residual P, and total sediment P) in terms of annual, cold water season, and warm water season means. Stepwise regression was first used to identify the variables that explained significant additional independent variance (p to enter = 0.15) for each of the P fractions. Explanatory variables included in the stepwise regression were: pore, surface, and bottom water SRP; surface water TP; surface water, bottom water and sediment chlorophyll *a* concentration; bottom water (in situ measurements conducted at ~30 cm above sediment surface) DO%, temperature, pH and conductivity. Only significant relationships ($P < 0.05$) were included in the final regression models. All statistical analyses were carried out using Minitab statistical software version 14 (Minitab Ltd., Coventry, UK).

Results

Physical variables measured

Seasonal and spatial variation

The water column was generally well mixed throughout the year, with very weak temperature and oxygen gradients present in the warm water period. Weak stratification was only observed in May 2004, with a thermocline between 8 and 9 m water depth. The DO% saturation at 0.5 m above the deepest sediment sampled (22 m overlying water depth) was never below 62%. Average bottom water temperature (Table 1) was 14.0°C in the "warm water period" (May 2004–October 2004) and 4.9°C in the "cold water period" (November 2004–April 2005). DO% decreased with increasing water depth and was significantly lower in the warm water period (Table 1).

Water column phosphorus and chlorophyll *a* concentrations

Temporal variation

Bottom water SRP concentrations (Fig. 2a) exceeded surface water concentrations for the majority of the year and both showed peaks in both winter (November 2004–February 2005) and summer (July 2004–September 2004). Bottom water chlorophyll *a* concentrations were higher than surface chlorophyll *a* concentrations throughout the year (Fig. 2b) and similar seasonal variation was observed in both.

Seasonal and spatial variation

Bottom water SRP concentrations increased with overlying water depth under both warm and cold water conditions (Fig. 4 and Table 1). Pore water

Table 1 Effects of season (warm vs. cold water) and overlying water depth on bottom water dissolved oxygen % saturation, temperature and phosphorus concentrations in bottom water and sediments

Variable	Season		Depth		Interaction	
	P value	*F* value	*P* value	*F* value	*P* value	*F* value
Bottom water						
DO%	<0.01	21.74	<0.01	5.52	<0.01	4.61
Temperature	<0.01	243.01				
SRP	0.02	2.95				
Sediment						
Pore water SRP	<0.01	26.77				
Reductant-soluble P			<0.01	3.95		
Organic P	0.04	4.35	0.01	3.28		
Apatite bound P	<0.01	14.23	0.01	3.25	0.03	2.79
Residual P			0.05	2.42		
Sediment TP			<0.01	4.40		

SRP concentrations were higher in all sites under cold water conditions, showing little spatial variation throughout. Regression analysis (Table 3) indicated no significant relationships between pore water SRP and other variables under cold water conditions. Under warm water conditions, however, a significant positive relationship between pore water SRP and sediment chlorophyll *a* was observed. Bottom water SRP decreased with increasing DO% and sediment TP under cold water conditions but solely with increasing DO%, under warm water conditions.

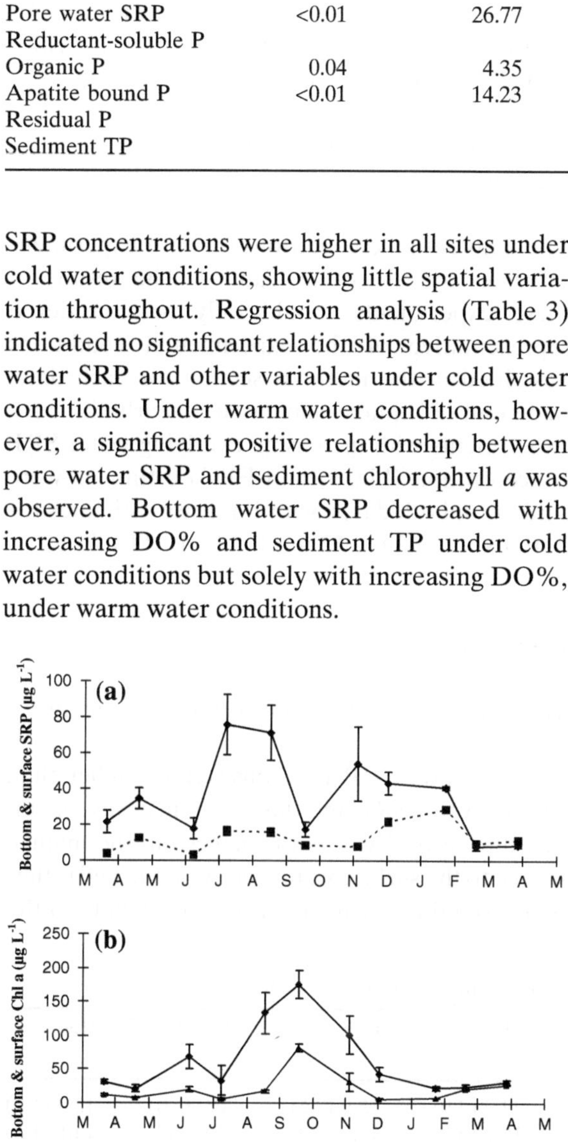

Fig. 2 Temporal variation in (**a**) bottom (full line) and surface (dashed line) water SRP concentration and (**b**) bottom (full line) and surface (dashed line) water chlorophyll *a* concentration. Each value represents the average value of the six sites sampled. Error bars represent the standard error of the mean

Sediment phosphorus fractions

Temporal variation

Monthly-average concentrations of labile, reductant-soluble, organic and total sediment P pools are shown in Fig. 3. Labile P peaked in May and remained low for the rest of the year. Reductant-soluble SRP decreased in April, August, and December but was above 0.80 mg g^{-1} dw in all other months. The low values in August and December corresponded with the lowest concentrations of labile P and the peaks in surface and bottom water SRP concentrations. Organic P varied little with the exception of a marked decrease in September corresponding with peaks in both labile and reductant-soluble P fractions. Total sediment P showed little temporal variation with the exception of a peak in mid-winter.

Seasonal and spatial variation

The spatial variation in each P pool is summarised in Table 2. The reductant-soluble SRP pool was the most abundant fraction throughout the lake. The labile P pool was the smallest fraction of sediment TP. On an annual basis, the spatial variation in sediment TP was significantly related to organic P (Table 3).

Reductant-soluble P varied significantly with water depth with the highest values observed at intermediate depths (Fig. 4 and Table 1). Organic P was consistently higher under cold water

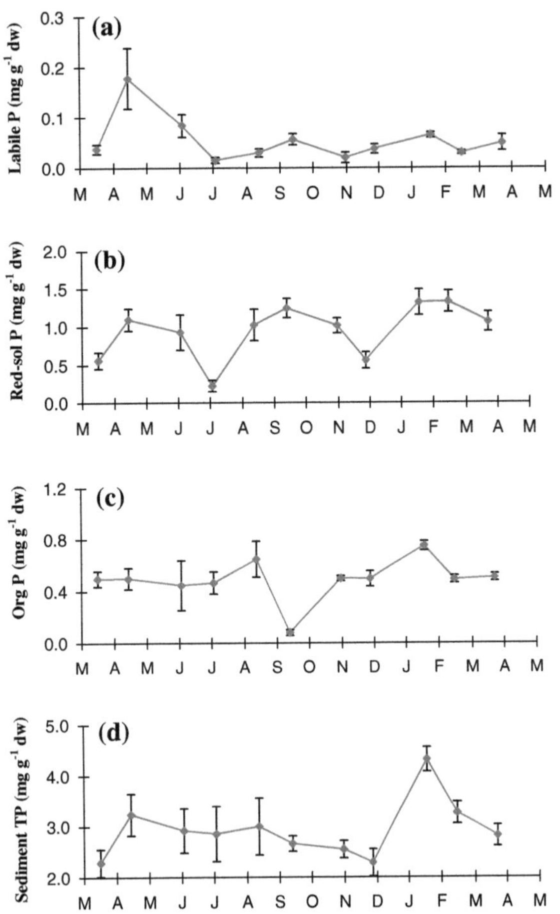

Fig. 3 Temporal variation in (**a**) labile P, (**b**) reductant-soluble P, (**c**) organic P, and (**d**) total sediment P. Each value represents the average value of the six sites sampled. Error bars represent the standard error of the mean

conditions and increased with overlying water depth. Apatite bound P varied significantly with season and depth. Residual P and Sediment TP both varied spatially but not seasonally.

Reductant-soluble P was positively related to organic P under cold water conditions but negatively related to a combination of apatite-bound and labile P fractions under warm water conditions (Table 3). Labile P was negatively correlated with bottom water SRP under cold water conditions but positively correlated with organic P under warm water conditions. Organic P was positively correlated with reductant-soluble P under cold water conditions and positively correlated with a combination of sediment TP and bottom water chlorophyll *a* under warm water conditions.

Equilibrium phosphorus concentration experiments

The ambient lake conditions on both experimental dates are shown in Table 4. The results of these experiments should only be considered representative of the experimental conditions described in the methods section for each date. Sediment P uptake increased linearly in both experiments with increasing overlying water SRP concentration (Fig. 5) and EPC0 estimates were similar on both dates, with P release estimates for both experiments ~12 mg P m^{-2} d^{-1} (Table 5).

Discussion

Temporal variation of concentration gradients

Bottom water SRP was consistently higher than surface water SRP and this difference was magnified during peaks of both, suggesting the main P source to the surface water was from the bed rather than from the catchment. The troughs in both the

Table 2 Composition of sediment P fractions expressed as % annual composition of total sediment P

Water column depth (m)	Labile	Red Sol	Red Sol SURP	Met ox	Org	Apatite	Res
2.0	2.3 (0.8)	23.2 (4.4)	2.9 (1.3)	22.9 (2.9)	15.9 (2.5)	20.9 (5.0)	12.0 (1.9)
2.5	2.1 (0.7)	28.2 (4.4)	5.5 (2.8)	18.1 (3.0)	16.0 (2.3)	15.9 (2.2)	14.1 (1.8)
3.5	2.4 (1.0)	30.6 (4.6)	4.3 (1.2)	16.0 (4.0)	19.0 (2.8)	13.0 (1.2)	14.7 (1.6)
5.5	1.5 (0.4)	39.5 (5.0)	3.3 (1.1)	16.2 (5.3)	16.8 (1.8)	9.5 (0.6)	13.2 (1.5)
10.0	1.1 (0.3)	33.4 (5.4)	4.7 (1.5)	16.2 (4.8)	16.5 (2.5)	14.4 (1.3)	13.6 (1.5)
22.0	1.4 (0.4)	34.2 (5.0)	5.1 (1.8)	17.7 (4.5)	15.3 (3.6)	12.2 (1.6)	14.0 (1.1)

Values in brackets represent the standard error of the mean (n = 11). Red Sol = reductant-soluble SRP; Red Sol SURP = reductant-soluble soluble un-reactive P; Met ox = Metal oxide adsorbed P; Org = organic P; Res = residual P

Table 3 Summary of the multiple regression analysis designed to explain differences in the mechanisms controlling pore water and bottom (bot) water SRP (μg L^{-1}), organic P (org), labile P, reductant-soluble P (Red), and sediment TP (sed TP) (all as mg g^{-1} dw)

Relations	*P*-value	Equation of line *y* = ...	R^2
Annual			
Sed TP	<0.01	0.98 + 4.11 org	0.96
Cool			
Pore SRP	N/A	N/A	N/A
Bot SRP	<0.01	1804–17.1 DO%–16.3 sed TP	0.97
Red	0.02	–0.59 + 3.0 org	0.76
Labile	0.04	0.06–0.0006 bot SRP	0.69
Warm			
Pore SRP	<0.01	19.7 + 1.04 sed chl a	0.87
Bot SRP	<0.01	228.1–2.03 DO%	0.96
Red	0.02	1.45–0.05 Apat–2.97 labile	0.92
Org	<0.01	–0.4 + 0.26 sed TP + 0.0008 bot Chl a	0.98
Labile	0.02	–0.001 + 0.17 org	0.79

Sed chl a = sediment chlorophyll *a* (μg g^{-1} dw), DO% = dissolved oxygen % saturation, bot chl a = bottom water chlorophyll *a* (μg L^{-1})

bottom and surface water SRP concentrations were associated with peaks in chlorophyll *a* concentrations in bottom and surface waters. However, using chlorophyll *a* to interpret processes occurring in the epilimnion and hypolimnion requires caution since bottom water chlorophyll *a* concentrations may represent both settling and upwardly migrating phytoplankton (Petterson et al., 1993) and resuspended benthic algae. Also, cyanobacteria capable of migrating between the sediment and the water column characteristically dominate the phytoplankton during mid-summer in Loch Leven (Carvalho and Kirika, 2005).

Both water column SRP peaks corresponded with the lowest values of reductant-soluble P and labile P indicating that the main source of P was redox-sensitive and sediment-derived. Additionally, the multiple regression analysis indicated a negative correlation between bottom water SRP and DO% saturation under both warm and cold conditions. This uptake/release pathway agrees with the classical P mobility model where SRP is released from the sediment under anoxic conditions (Mortimer, 1941, 1942; Boström et al., 1988). It, therefore, seems evident that the P fraction responsible for the large, short-term, P release events from the sediments of Loch Leven is the reductant-soluble P pool.

The effects of increased temperature may be important in causing the mid-summer peak where high P release rates have been observed (Kamp-Nielson, 1975b). However, other factors must be driving the winter peaks in water column SRP. Part of the winter peak in water column SRP may be derived from increased external loads, however, the synchronised occurrence of the winter reductant-soluble P decrease and water column SRP peaks suggests that a redox initiated release from sediments may also make an important contribution. Although the bottom water DO% levels are higher (on average) in winter than in summer, this finding agrees with studies of internal loading in deep stratifying lakes. Penn et al. (2000) observed two main release events, the largest of which occurred in mid-summer with a lesser magnitude event occurring in early winter. This seasonal pattern has also been reported by Kamp-Neilson (1975a). Anoxia may occur in winter as a result of an overall shift from an autotrophically dominated system in summer (high nutrients, high intensity/duration light, high temperatures) to a heterotrophically dominated system in winter (low nutrients, low intensity/duration light, low temperatures). Thus, under summer conditions, temperature will drive DO concentrations down through a mixture of decreased solubility and increased microbial demand, with photosynthetic production of DO acting to buffer anoxia. However in winter, low rates of photosynthetic production may be insufficient to maintain sediment DO levels, resulting in localised anoxia at the sediment surface.

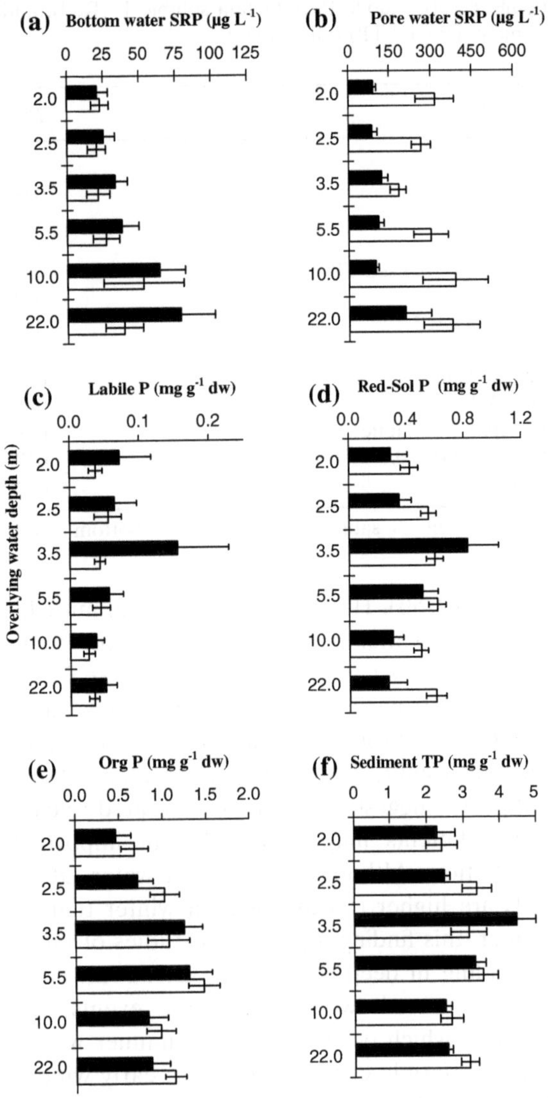

Fig. 4 Plot showing the spatial (depth gradient) versus seasonal (warm versus cool bottom water) variation in **(a)** bottom water SRP concentration, **(b)** pore water SRP concentration, **(c)** labile P, **(d)** reductant-soluble P, **(e)** organic P, and **(f)** total sediment P. Each bar represents either the cold (white bar) or warm (black bar) water period average value at a given depth. Error bars represent the standard error of the mean

Temperature regulated P cycling

Pore-bottom water concentration gradients

Pore water SRP concentrations generally exceeded bottom water SRP concentrations suggesting chemical gradients supportive of P release throughout the year. A comparison of the seasonal variations in pore and bottom water SRP concentrations revealed that concentration gradients were reduced during the warm water months where pore water SRP concentrations were lowest and bottom water SRP concentrations were highest. This suggests that the transfer of P from pore to bottom waters was greater under warm water conditions.

Under cold water conditions, pore water SRP concentrations were highest and bottom water SRP concentrations were lowest. This suggests less release from pore to bottom waters under cold water conditions. In this colder period, however, concentration gradients were greatest in the two deepest sites indicating that the key release mechanisms acting to facilitate the transfer of P from pore to bottom waters in winter were greatest in shallower waters and least in deeper waters. It is likely that the release from shallow water sediments is enhanced by sediment resuspension (Marsden, 1989).

Table 4 Ambient conditions of bottom water on the 8th of August and the 6th of September 2005

Date	DO%	Conductivity ($\mu S\ cm^{-2}$)	pH	Temperature (°C)	Secchi depth (m)
8/8/05	94.4	217	8.92	15.38	0.98
6/9/05	90.4	233	8.37	16.20	0.93

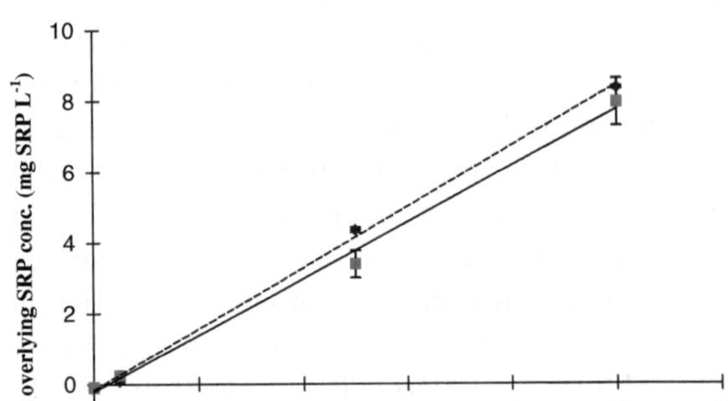

Fig. 5 Results of the intact sediment core EPC0 experiments. Square tabs represent the SRP uptake/release (Δ overlying SRP concentration) versus initial overlying water SRP concentration plots for the 8th August 2005 and triangular tabs represent the results for the 6th of September 2005

The positive relationship between pore water P and sediment chlorophyll *a* under warm water conditions indicates that, even though P release from pore to bottom waters is highest in the warm water period, release is less in areas of the lake bed where sediment autotrophic biomass is highest. There are two plausible explanations for this capping of pore water P by autotrophic biofilms: firstly, this community helps maintain an oxygenated boundary layer at the sediment-water interface, and secondly there is strong evidence that benthic algae increase sediment stability and, therefore, decrease resuspension-related release (e.g. Underwood and Paterson, 1993; Van Luijn et al., 1995; Perkins et al., 2004).

Sediment P fractions

Reductant-soluble P was the dominant form of phosphorus within the sediment. Regression analysis indicated that, under cold water conditions, reductant-soluble P varied positively with organic P. This relationship may result from increased microbial remineralisation under aerobic conditions. Under warm water conditions, reductant-soluble P varied negatively with apatite-bound P and labile P. Again this suggests P transfer between sediment pools. The driving mechanism may be pH. Labile P and pH were significantly and positively correlated under cold water conditions. This agrees with Penn et al. (2000) where release rates of P from the labile P pool were highest at lower pH conditions. The release mechanism here is expected to be the pH-induced dissolution of $CaCO_3$ (Driscoll et al., 1993). Additionally, Søndergaard (1989) observed a positive relationship between labile P and pore water pH. If release from any pool occurs under oxic conditions, P is likely to be transferred into the reductant-soluble P pool.

The effects of temperature in regulating internal P release have been well documented (e.g. Søndergaard, 1989; Boström et al., 1988; Jensen and Anderson, 1992). It is generally agreed that, under warm conditions, the organic P pool is mineralised by increased microbial activity. Deoxygenation of the sediment water interface through increased biological oxygen demand, as demonstrated here, is well documented (Boström et al., 1988). Another temperature-driven organic P related process is the translocation of P through the recruitment of cyanobacterial tychoplankton (Karlsson-Elfgren et al., 2004). Upward translocation rates of TP from the sediment, mostly attributable to cyanobacterial recruitment, have been estimated at 2–40 mg P m^{-2} d^{-1} for a range of sites (Pettersson et al., 1993; Barbiero and Kann, 1994, Head et al., 1999). Further evidence of cyanobacterial recruitment is the sudden reduction in organic P corresponding with the peak in water column chlorophyll *a* concentration and SRP reduction in late summer. This explanation is supported by Pettersson et al. (1993) who estimated that the P contribution of Gloeotrichia echinulata in sediment (before migration) could be 45% of sediment TP.

Regression analyses of the spatial variation in organic P also highlight the importance of biogenic P cycling. Under warm water conditions organic P is significantly related to bottom water chlorophyll *a*. This suggests that algae may be the main component of the organic P pool under warm water conditions. The bottom chlorophyll *a* concentrations also indicate that translocation of organic P (through both settling and recruitment) is highest under warm water conditions.

Sediment P release estimates

The EPC0 experiments indicated that Loch Leven sediment was still capable of significant P release. The release rates measured may not, however, represent the average conditions throughout the year. The temperatures on the days of the experiments were the maximum temperatures recorded throughout the sediment P fractionation survey. However, under these experimental conditions, the sediment continually released phosphorus and so physico-chemical adsorption into the sediment was unlikely. Instead the main source of P to the sediment was likely to be particulate (e.g. planktonic detritus). This argument is supported by the high content of organic P in the sediment of this and many other eutrophic shallow lakes (e.g. Phillips et al., 1994).

The linearity of the uptake/release plots would suggest that Loch Leven sediments are currently far below their adsorptive saturation levels (Hwang et al., 1976). This finding agrees with previous work where estimates of saturation levels were <16% (Spears et al., 2006) indicating that Loch Leven has a very high P buffering capacity and is capable of storing much more P within the sediment. The estimated EPC0 values indicate that under experimental conditions, bottom water SRP concentration would have to exceed at least 180 μg L^{-1} for physico-chemical sediment uptake to occur. Surface water SRP concentrations have never exceeded this value in recorded history (Carvalho and Kirika, 2003). Estimates from traditional (i.e. shaken sediment) EPC0 measurements for Loch Leven sediment yielded EPC0 estimates of between 9 and 30 μg L^{-1} (Spears et al., 2006). These estimates may be considered representative of re-suspended sediment suggesting that, under conditions of re-suspension, P would be stripped from the water column at SRP concentrations in excess of 30 μg L^{-1}. The difference between EPC0 estimates taken from intact cores and continuously shaken slurry clearly highlights the importance of an undisturbed sediment-water interface in regulating internal P release.

Estimated release rates, though probably indicative of near maximum rates in warm conditions, varied little between the two experiments (~12 mg P m^{-2} d^{-1}). With an external P load of 1.62 mg P m^{-2} d^{-1} (in 1995), internal loading plays an important role in the maintenance of eutrophic conditions within the loch (Bailey-Watts and Kirika, 1999). These release rates are in agreement with other comparable studies, under warm, aerobic conditions, for example 10.3 mg P m^{-2} d^{-1} (hypereutrophic Onondaga Lake, N.Y. USA, Penn et al., 2000) and 278 mg P m^{-2} d^{-1} (South Walsham Broad, UK, Phillips et al., 1994).

Conclusions

This study highlighted several factors that drive the seasonality of P release. Firstly the sedimentation and mineralisation of organic P is important in replenishing the release-sensitive P fractions. Under aerobic conditions, P released as a result of microbial mineralisation is transferred to the reductant-soluble, labile and pore water P pools. This process maintains release conditions for much of the year with concentration gradients indicative of negative retention present for 10 months. In shallow waters, pore water P concentrations remain low, probably due to wind-induced sediment resuspension. Large short-term release events result from the onset of anoxia where the reductant-soluble P pool is greatly depleted in mid-summer and winter. We need to extend the measurements further to determine whether the gross release rates reported here are characteristic of annual average conditions. Evidence is provided here on the importance of heterotrophic and autotrophic microbial communities in regulating sediment P release. The specific roles of functional groups within both communities and the environmental conditions under which they operate require further attention.

These results provide valuable lake management information where a large proportion of the release sensitive sediment P is transferred from the sediment to the water column in mid-summer, either via redox-regulated processes or through organic P translocation. In systems where flow regulation is achievable, increasing the flushing

Table 5 Summarised results of the intact core sediment P release experiments.

Date	Equation of line	P	R^2	EPC0	P release
8-8-2005	Y = 0.798x–215.7	0.0024	0.995	270.4	12.91 (8%)
6-9-2005	Y = 0.862x–155.3	0.0009	0.998	180.2	12.60 (10%)

Table includes the linear relationship between the initial overlying water column concentration (x) and P uptake/release (y), the equilibrium phosphate concentration (EPC0 in μg L–1), and the gross sediment P release values (mg SRP m–2 d–1). Values in brackets represent the percent error of the estimate

rates to correspond with the above transfer of P from the sediment to the water column will help recovery from eutrophication.

Acknowledgements We gratefully acknowledge the logistical assistance provided by the staff at the Loch Leven Fishery. We also wish to acknowledge Bernie Dudley for graciously ensuring the quality of our arithmetic. BMS was funded by NERC (NER/S/A/2003/11324).

References

APHA, 1992. Standard methods for the examination of water and wastewater. In Greenberg, A. E., L. S. Clesceri & A. D. Eaton (eds), American Public Health Association, Washington, DC.

Bailey-Watts, A. E. & A. Kirika, 1999. Poor water quality in Loch Leven (Scotland) in 1995, in spite of reduced phosphorus loading since 1985: the influences of catchment management and inter-annual weather variation. Hydrobiologia 403: 135–151.

Barbiero, R. P. & J. Kann, 1994. The importance of benthic recruitment to the population development of Aphanizomenon flos-aquae and internal loading in a shallow lake. Journal of Plankton Research 16: 1581–1588.

Boström, B., J. M. Anderson, S. Fleischer & M Jansson, 1988. Exchange of phosphorus across the sediment-water interface. Hydrobiologia 170: 229–244.

Carvalho, L. & A. Kirika, 2003. Changes in shallow lake functioning: response to climate change and nutrient reduction. Hydrobiologia 506–509: 789–796.

Carvalho, L. & A. Kirika, 2005. Loch Leven 2003: physical, chemical, and algal aspects of water quality. Report to Scottish Natural Heritage. Centre for ecology and hydrology, Edinburgh.

Driscoll, C. T., S. D. Effler, M. T. Auer, S. M. Doer & M. R. Penn, 1993. Supply of phosphorus to the water column of a productive hard water lake: controlling mechanisms and management considerations. Hydrobiologia 253: 61–72.

Farmer, J. G., A. E. Bailey-Watts, A. Kirika & C. Scott, 1994. Phosphorus fractionation and mobility in Loch Leven sediments. Aquatic Conservation: Marine and Freshwater Ecosystems 4: 45–56.

Gächter, R. & J. S. Meyer, 1993. The role of microorganisms in mobilization and fixation of phosphorus in sediments. Hydrobiologia 253: 103–122.

Head, R. M., R. I. Jones & A. E. Bailey-Watts, 1999. Vertical movements by planktonic cyanobacterial and the translocation of phosphorus: implications for lake restoration. Aquatic Conservation: Marine and Freshwater Ecosystems 9: 111–120.

House, W. A., T. D. Jickells, A. C. Edwards, K. E. Praska & F. H. Denison, 1998. Reactions of phosphorus with sediments in fresh and marine waters. Soil Use Management 14: 139–146.

Hwang, C. P., T. H. Lackie & P. M. Huang, 1976. Adsorption of inorganic phosphorus by lake sediments. Journal of the Water Pollution Control Federation 48: 2754–2760.

Jenson, H. S. & F. Ø. Anderson, 1992. Importance of temperature, nitrate, and pH for phosphate release from aerobic sediments of four shallow, eutrophic lakes. Limnology and Oceanography 37: 577–589.

Kamp-Neilson, L., 1975a. Seasonal variation in sediment-water exchange of nutrient ions in Lake Esrom. Verhandlungen internationale der Vereinigung für theoretische und angewandte Limnologie 19: 1057–1065.

Kamp-Nielson, L., 1975b. A kinetic approach to the aerobic sediment-water exchange of phosphorus in Lake Esrom. Ecological Modelling 1: 153–160.

Karlson-Elfgren, I., K. Rengefors & S. Gustafsson, 2004. Factors regulating recruitment from the sediment to the water column in the bloom-forming cyanobacterium Gloeotrichia echinulata. Freshwater Biology 49: 265–273.

Marsden, M. W., 1989. Lake restoration by reducing external phosphorus loading- the influence of sediment phosphorus release. Freshwater Biology 21: 139–162.

Mortimer, C. H., 1941. The exchange of dissolved substances between mud and water in lakes 1. Journal of Ecology 29: 280–329.

Mortimer, C. H., 1942. The exchange of dissolved substances between mud and water in lakes 2. Journal of Ecology 30: 147–201.

Paterson, D. M. & K. S. Black, 1999. Water flow, sediment dynamics and benthic biology. Advances in Ecological Research 29: 155–193.

Penn, M. R., M. T. Auer, S. M. Doerr, C. T. Driscoll, C. M. Brookes & S. W. Effler, 2000. Seasonality in phosphorus release rates from the sediments of a hypereutrophic lake under a matrix of pH and redox conditions. Canadian Journal of Fisheries and Aquatic Sciences 57: 1033–1041.

Perkins, R. G. & G. J. C. Underwood, 2001. The potential for phosphorus release across the sediment-water interface in an eutrophic reservoir dosed with ferric sulphate. Water Research 35: 1399–1406.

Perkins, R. G., D. M. Paterson, H. Sun, J. Watson & M. A. Player, 2004. Extracellular polymeric substances: quantification and use in erosion experiments. Continental Shelf Research 24: 1623–1635.

Pettersson, K., E. Herlitz & V. Istvanovics, 1993. The role of Gloeotrichia echinulata in the transfer of phosphorus from sediment to water in Lake Erken. Hydrobiologia 253: 123–129.

Phillips, G., R. Jackson, C. Bennett & A. Chilvers, 1994, The importance of sediment phosphorus release in the restoration of very shallow lakes (The Norfolk Broads, England) and the implications for biomanipulation. Hydrobiologia 275/276: 445–456.

Psenner, R., B. Bostrom, M. Dinka, K. Petterson, R. Pucsko & M. Sager, 1988. Fractionation of phosphorus in suspended matter and sediment. Archive für

Hydrobiologie. Bieh Ergebnisse der Limnologie 30: 98–103.
Sas, H., 1989. Lake Restoration by Reduction of Nutrient Loading. Academic Verlag, Richarz.
Søndergaard, M., 1989. Phosphorus release from a hypertrophic lake sediment: Experiments with intact sediment cores in a continuous flow system. Archive fùr Hydrobiologie 116: 45–59.
Spears, B. M., L. Carvalho, R. Perkins, A. Kirika & D. M. Paterson, 2006. Spatial and historical variation in sediment phosphorus fractions and mobility in a large shallow lake. Water Research 40: 383–391.
Underwood, G. J. C. & D. M. Paterson, 1993. Seasonal-changes in diatom biomass, sediment stability and biogenic stabilization in the Severn estuary. Journal of the Marine Biological Association of the United Kingdom 73: 871–887.
Van Luijn, F., D. T. Van der Molen, W. J. Luttmer & P. C. M. Boers, 1995. Influence of benthic diatoms on the nutrient release from sediments of shallow lakes recovering from eutrophication. Water Science and Technology 32: 89–97.
Wetzel, R. G. & G. E. Likens, 2000. Limnological Analyses 3rd edn. Springer-Verlag, New York.
Woodruff, S. L., W. A. House, M. E. Callow & B. S. C. Leadbeater, 1999. The effects of biofilms on chemical processes in surficial sediments. Freshwater Biology 41: 73–89.

Hydrobiologia (2007) 584:49–58
DOI 10.1007/s10750-007-0589-6

SHALLOW LAKES

Nitrate dynamics in a reed belt of a shallow sand dune lake in Japan: Analysis of nitrate retention using stable nitrogen isotope ratios

Haruo Fukuhara · Fumiko Nemoto · Yoko Takeuchi · Nobushige Toda

Abstract A stable nitrogen isotope analysis was used to clarify the relative importance of denitrification and nitrate uptake by plants in the nitrate reduction in a reed belt of L. Kamisagata (N 37°49′, E 138°53′, alt. 4.5 m, depth 30–80 cm, area 0.025 km^2), one of about 20 sand dune lakes in Japan. A very high concentration of NO_3^--N with 19.0 ± 5.9 mg N l^{-1} in spring sources decreased during passage through the reed belt along two set transect lines about 120 m long in any season, whereas progressive enrichment in ^{15}N-NO_3^- in flowing water was detected. Loss rate of nitrate ranged from 38.4 to 73.1% with an average of 56.7 ± 11.6%. Enrichment factors calculated using a Rayleigh curve method ranged from –1.03 to –5.12‰. The contribution of denitrification to nitrate loss ranged from 6 to 28%, with a mean of 19.5% (±7.0), whereas that of plant uptake was from 72 to 94%, with a mean of 80.5% (±7.0), indicating the importance of vegetation in a sand dune riparian zone. A technique using the variation of natural abundance of ^{15}N may provide useful information on the nitrate dynamics in artificial or natural wetlands under a non-destructive condition.

Guest editors: R. D. Gulati, E. Lammens, N. De Pauw & E. Van Donk
Shallow lakes in a changing world

H. Fukuhara (✉) · F. Nemoto · Y. Takeuchi
Faculty of Education and Human Sciences, Niigata University, Ikarashi 2-8050, Niigata 950-2181, Japan
e-mail: fukuhara@ed.niigata-u.ac.jp

N. Toda
Faculty of Science, Shinshu University, Asahimachi 1-1, Matsumoto, Japan

Keywords ^{15}N · Nitrate · Enrichment factor · Wetland · Sand dune lake · *Phragmites australis*

Introduction

Nitrate contamination from agricultural and urban sources of various water bodies, wetlands, riparians, lakes, streams, rivers, fens and also groundwater poses serious problems in many regions throughout the world. Increasing nitrate loading has often triggered lake eutrophication with heavy algal bloom, and a high nitrate concentration in drinking water over a specific level is known to adversely affect human health. It has been demonstrated that natural or constructed wetlands vegetated by macrophytes, including in the broad sense the riparian zones, surface water/groundwater interface under the ecotone concept (Dahm et al., 1998), or nitrogen farming (Hey, 2002), have a very important role in the decrease of pollutants, especially nitrogen, using various-sized experimental tanks, semi-fields or fields (Jansson et al., 1994;

Hill, 1996; Dahm et al., 1998; Hill et al., 2000; Cronk & Fennessy, 2001; Saunders & Kalff, 2001; Fukuhara et al., 2003; Huett et al., 2005; Naiman et al., 2005). Sanders & Kalf (2001) compared the magnitude of nitrogen retention via nitrogen loading of 43 freshwater bodies, demonstrating the result of higher nitrogen retention in wetlands with an average 64% in total loading compared to lakes and rivers.

Essential processes for nitrate removal in wetlands are denitrification, uptake by plants, immobilization by bacteria and nitrate reduction to ammonium (Mariotti et al., 1988; Hill, 1996; Kendall, 1998; Dhondt, 2003). The contribution of each of these processes to nitrate removal may differ greatly in accordance with wetland types, but denitrification and uptake by plants are considered to be the principal and effective means of nitrate removal by many investigators (Xue et al., 1999; Lund et al., 2000; Saunders & Kalff, 2001; Lin et al., 2002; Dhondt et al., 2003; Huett et al., 2005). There is no agreement on the relative importance of denitrification and nitrate uptake by plants in nitrogen retention processes in riparian zone (Dhondt et al., 2003). Lin et al. (2002) demonstrated the importance of denitrification in nitrate removal of 89–96% of nitrogen loaded in constructed wetlands. Lund et al. (2000) reported a high rate of plant uptake of 77–90% in nitrogen removal in an artificial wetland. It is important to distinguish the relative contributions of denitrification and uptake by aquatic plants in nitrate removal in any wetlands, in order to control the optimizing utilization.

Natural abundances of the rare stable isotope nitrogen, ^{15}N, are now being used widely in research on N cycling in organisms and ecosystems (Mariotti et al., 1981; Khol & Shearer, 1995; Kendall, 1998; Robinson, 2001; Toda et al., 2002). ^{15}N analysis is one of the useful tools for investigating nitrate dynamics, as when estimating the cause of nitrate removal in many wetlands including riparian buffer zones (Smith et al., 1991; Lund et al., 2000; Kellman & Hillaire-Marcel, 2003; Clement et al., 2003; Dhondt et al., 2003; Chen & MacQuarrie, 2004).

The isotope technique is also an effective method from the viewpoint of non-destructive measurement, whether harvesting plants in experimental tanks or natural fields, or postulating black boxes often used in a mass budget calculation. Lund et al. (2000) estimated the contribution of both denitrification and plant assimilation in large constructed wetlands with rich macrophytes using the enrichment factor of ^{15}N in retained NO_3^-. Some improvements concerning a plant uptake fractionation factor were conducted by Dhondt et al. (2003), and they quantified nitrate retention processes in a riparian zone using ^{15}N in NO_3^-.

We used the method of Lund et al. (2000) to evaluate the respective contributions of denitrification and plant uptake in nitrate removal. In the present study, we describe the relative importance of plant uptake in nitrate removal in spring water flowing through a wide reed belt riparian zone along a sand dune lake in Japan, using a stable nitrogen isotope technique. Three hypotheses were tested: (1) Nitrate concentrations in spring water decrease during its flow through a reed belt; (2) $\delta^{15}N$-NO_3^- values in flowing water increase in terms of the nitrate reduction, and (3) the relationship between nitrate reduction and increase of $\delta^{15}N$-NO_3^- obey a Rayleigh curve with a significant enrichment factor of ^{15}N. The contribution of plant uptake in nitrogen removal was duly calculated under these hypotheses.

Study area and sampling points

The study area was described in detail in a previous paper (Fukuhara et al., 2003). L. Kamisagata (N 37°49′, E 138°53′, alt. 4.5 m, depth 30–80 cm, area 0.025 km^2), one of about 20 sand-dammed lakes in Japan, was used for the field survey. L. Kamisagata is fed by rainwater and spring water reaching the open lake through a rich macrophyte belt (maximum 150 m with 0.12 km^2), which is mainly constituted by *Phragmites australis* (Cav.) Trin. ex Steud. with the density of 67.5 (±20.0, n = 20, October 02) culms m^{-2} around the lake. Many spring sources are distributed along the periphery of the reed belt. A watermelon or a tobacco plant crop in spring and a Japanese radish plant crop in autumn are grown yearly, and manured with more fertilizers than other cultivation areas in Japan (Nonaka et al.,

1997). Consequently, groundwater containing a high nitrate concentration of 5–30 mg NO_3^--N l^{-1} permeate the form of springs around the lake, and flow through a reed belt as small streams to the open lake (Fukuhara et al., 2003).

Two transect lines (I and II) about 90–120 m in length, originating from spring sources up to the reed belt outlets, located on the south side of the lake, were established in spring (June), summer (August) and autumn (October) of 2002 and 2004. There is a distance of about 200 m between the two lines I and II. Water samples were taken from 6 and 5 sampling points along transect I and II, respectively. Although we took only one bottle of water sample at a given sampling point, we set several sampling points to analyse statistically the decrease of nitrate concentration. The mean water depths of sampling points were 11.3 (June) and 4.6 cm (August) for line I and 8.8 (June) and 4.0 cm (August) for line II in 2002. Water flows on the surface of bottom substrate largely composed from sand, on which a thin layer of dead leaves of reeds accumulated.

Methods

Water temperature (stick thermometer), pH (TOA HM5B), and dissolved oxygen (Winkler method) were determined at each point along the transect lines. Nitrate and ammonium ion concentrations were determined in the filtered water (GF/C) by an ion chromatography (Shimadzu HIC-6A). Nitrite in filtrate was analysed after Strickland and Parsons (1972). A part of the filtered water samples was provided for measurement of natural abundance of ^{15}N (Thermo Quest DELTA plus, precision level; 0.5‰) in NO_3^- after concentrating to a nitrogen content of 0.35 mg ml^{-1} (Toda et al., 2002). Though we tried to determine ^{15}N in water from all sampling points, we missed to measure water from four cases (line I: Spring, 14 June 02; line II: Spring, 14 June 02, Summer, 31 August 02, Fall, 17 October 04) .The respective contributions of plant uptake and denitrification to the decrease in nitrate in water flowing through the reed belt were obtained from the analysis of a Rayleigh curve following Lund et al. (2000), under the hypothesis that only denitrification causes enrichment of ^{15}N in the remaining nitrate in passing water. According to Mariotti et al. (1981), we can express a Rayleigh curve as follows;

$$\delta s = \delta o + \varepsilon \times \ln(f)$$

δ s; $\delta^{15}N$-NO_3^- value in the residual NO_3^- in water at each sampling point along line

δ o; $\delta^{15}N$-NO_3^- value in water at a spring source

ε; enrichment factor,

f; the fraction of the remaining nitrate in water at each sampling point

If we put $\Delta = \delta s - \delta o = \varepsilon \times \ln(f)$, we can calculate an enrichment factor as the value of the slope of a significant regression line. This factor is sometimes called an apparent enrichment factor in field study, since the reaction involves all related reactions including a major process.

We presumed the enrichment factor of ^{15}N by plant uptake as 0 ‰, since the effect of plant uptake for ^{15}N fractionation in nitrate might exert an extremely minor influence. It is known that vegetation fractionates ^{15}N minimally or not at all in nitrate uptake (Mariotti et al., 1980; Lund et al., 2000). Mariotti et al. (1980) reported an average value of –0.25 ± –0.10 ‰ for 38 plants. An enrichment factor of –22.5 ‰ for denitrification by Dhondt et al. (2003) was used in our analyses, because this factor was obtained from a carefully designed laboratory experiment using riparian soil slurries under a NO_3^--N concentration of 15–20 mg l^{-1} similar to our field. Loss rate of nitrate was calculated from the differences between nitrate concentration at spring mouths (in two cases, at next sampling point) and outlets.

Results

The concentrations of NH_4^+-N and NO_2^--N along the transect lines were only one-twentieth of the NO_3^--N concentrations at the outlets (Table 1). Changes in water temperature in each transect were small and within 3°C, whereas average water temperature ranged from 15.7°C to 20.8°C (Table 1). Dissolved oxygen (DO) concentrations at springs showed low values of 39.6–82.1% (3.54–7.62 mg O_2 l^{-1}, Table 1). Oxygen content

Table 1 Changes in nitrate concentration, trend analysis of nitrate decrease, $\delta^{15}N$ values and nitrate loss rate between springs and lake inlets, DO saturation decrease from maximum to inlet of lake, average water temperature (WT), pH, NH_4^+-N and NO_2^--N concentrations along transect lines. Trend of nitrate decrease hypothesized a first-order reaction evaluated statistically by a regression analysis

Line	Date	NO_3^--N ($mg\ l^{-1}$) spring-lake	Trend of nitrate decrease significant level, number	Loss rate (%)	$\delta^{15}N$ (‰) spring-lake	WT (±SD) (°C)	DO (%) max.-lake	pH (±SD) (unit)	NH_4^+-N (±SD) ($mg\ l^{-1}$)	NO_2^--N (±SD) ($mg\ l^{-1}$)
I	Spring, 14 June 02	13.49-8.31	$P < 0.05$, 5	38.4	4.8-7.2	19.0 ± 0.8	77.6-37.5	6.68 ± 0.04	0.089 ± 0.031	0.035 ± 0.010
II	Spring, 14 June 02	17.90-4.81	$P < 0.01$, 5	73.1	3.7-4.4	16.0 ± 0.8	69.9-32.2	6.70 ± 0.13	0.208 ± 0.096	0.030 ± 0.016
I	Summer, 31 August 02	29.76-10.20	$P < 0.025$, 5	65.7	3.9-6.4	21.1 ± 0.6	93.8-62.2	6.81 ± 0.07	0.018 ± 0.010	0.013 ± 0.005
II	Summer, 1 August 04	13.94-6.88	$P < 0.01$, 6	50.6	3.8-7.0	18.1 ± 0.6	65.2-30.3	6.59 ± 0.09	0.067 ± 0.026	0.016 ± 0.006
II	Summer, 31 August 02	20.11-10.14	$P < 0.01$, 5	49.6	3.9-6.8	19.2 ± 1.3	82.1-51.8	6.98 ± 0.10	0.021 ± 0.008	0.014 ± 0.002
II	Fall, 17 October 04	18.61-6.88	$P < 0.025$, 6	63.0	3.0-5.9	15.7 ± 0.7	72.4-48.3	6.81 ± 0.05	0.049 ± 0.068	0.026 ± 0.008

in water increased somewhat after about 20 m flow (in mostly at the following sampling station), and again decreased toward the outlet of streams, ranging from 2.73 to 5.31 mg $O_2\ l^{-1}$ (saturation rate: 30.3–62.2%; Table 1) at outlets. Average pH of water ranged from 6.59 to 6.98 with very small changes in each transect line.

The highest concentrations of NO_3^- were measured at the source of springs except for line I in spring and summer of 2002 (Fig. 1, Table 1). The next sampling point at line I showed the highest nitrate level, revealing another spring source with a high concentration of nitrate. Data on spring mouth at line I were excluded from the following analyses. Average NO_3^--N concentration at the spring sources was 19.0 ± 5.9 mg NO_3^--N l^{-1}. Concentrations of NO_3^- in spring water decreased gradually during its passage through the reed belts toward the lake in any season, demonstrating the consumption of nitrate (Fig. 1). The tendency of the nitrate decrease, when a first-order reaction was hypothesized, was statistically significant (Table 1). The loss rate of nitrate showed small difference in seasons, ranged from 38.4 to 73.1% with an average of 56.7 ± 11.6% (Table 1).

The values of $\delta^{15}N$ in NO_3^- were low in water at spring sources, ranging from 3.0 to 4.8‰ (Table 1). These values increased gradually at each line toward lake inlets, reflecting progressive enrichment of ^{15}N in retained nitrate in water passing through the *Phragmites* belt. The values of $\delta^{15}N$ in retained nitrate ranged from 4.4–7.2‰ at reed bed outlets.

The coincidence of the increase in the values of $\delta^{15}N$-NO_3^- with the nitrate reduction in the water flowing through the reed belt, was shown by the results of plotting the significant regression lines appearing in Fig. 2. This also indicates that the relationship between nitrate reduction and increase of $\delta^{15}N$-NO_3^- follows a Rayleigh curve with a significant enrichment factor of ^{15}N. The apparent enrichment factors in the present research were calculated for each transect line by the method of Mariotti et al. (1981), ranging from –1.03 to –5.12‰ (Table 2).

The higher contribution of plant uptake, ranging from 72 to 94%, with a mean of 80.5 ± 7.0%, estimated by the means of Lund et al. (2000), was measured using the apparent enrichment factors

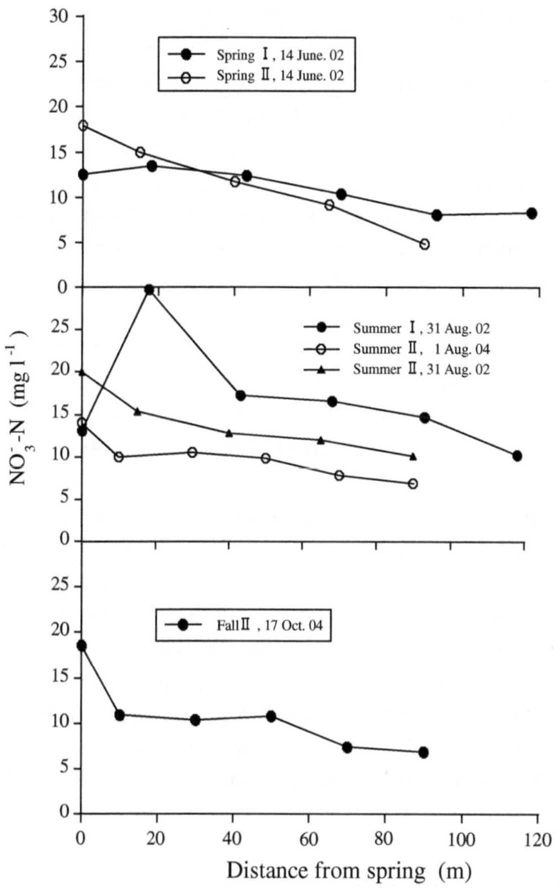

Fig. 1 Changes in nitrate concentrations along transect lines from a spring source to the inlets of a lake

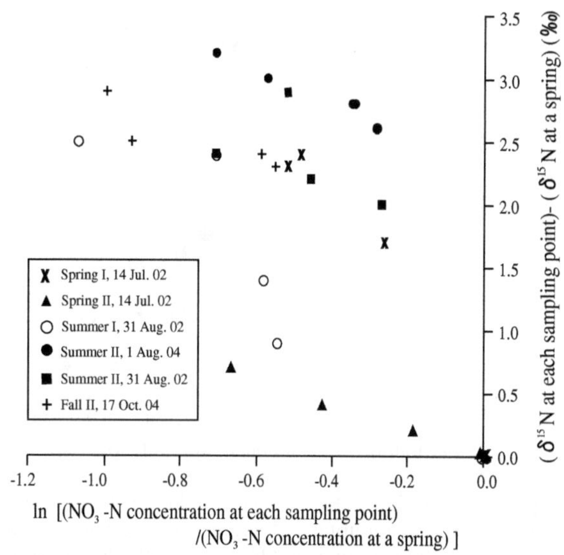

Fig. 2 Relationships between NO_3^- remaining and $\delta^{15}N$-NO_3^- in water flowing through a reed belt

given in Table 2, whereas the contribution of denitrification ranged from 6 to 28% with a mean of 19.5 ± 7.0% (Table 2). Plant uptake showed no significant difference between spring and summer ($P = 0.39$, *t*-test).

A statistically week ($P = 0.060$) but a positive correlation ($r = 0.792$, $n = 6$) was observed between the enrichment factor and the rate of contribution by plant uptake in nitrogen loss in our study (Fig. 3).

Discussion

Nitrate is a major ion in inorganic nitrogen in spring waters in our study area. Organic nitrogen in spring water was negligible (Fukuhara et al., unpublished). Therefore, almost all ^{15}N measured in our determination may be derived from NO_3^-. Nitrification might be considered to have little effect on the change in nitrate concentrations throughout the reed belts, since there was virtually no change in the NH_4^+-N and NO_2^--N concentrations.

Table 2 ^{15}N enrichment factors of NO_3^- and contribution of denitrification and plant uptake to nitrate loss. Significant level of regression and number of calculated data in each line were shown

Transect line	Date	Enrichment factor (‰)	Regression coefficient (R^2)	Significant level, number	Contribution (%)	
					Denitrification	Plant uptake
I	Spring, 14 June 02	–4.52	0.949	$P < 0.05$, 4	24	76
II	Spring, 14 June 02	–1.03	0.994	$P < 0.005$, 4	6	94
I	Summer, 31 August 02	–2.52	0.853	$P < 0.05$, 5	24	76
II	Summer, 1 August 04	–4.13	0.717	$P < 0.05$, 6	17	83
II	Summer, 31 August 02	–5.12	0.928	$P < 0.05$, 4	28	72
II	Fall, 17 October 04	–2.72	0.870	$P < 0.05$, 5	18	82

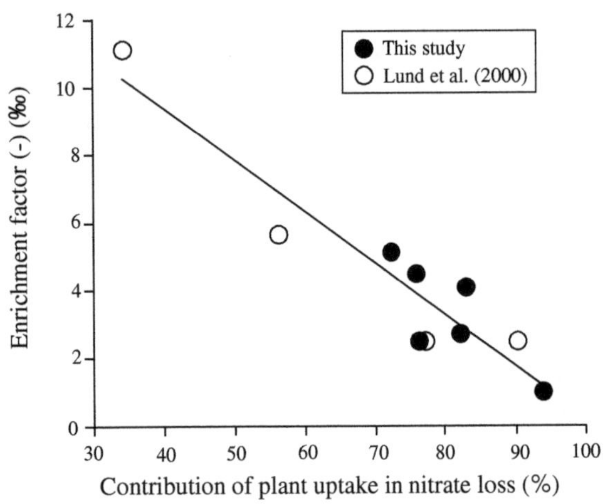

Fig. 3 Relationship between ^{15}N enrichment factors and contribution of plant uptake. A regression line was written using all data by our study and by Lund et al. (2000). Note in the figure, the enrichment factors express negative values, but the figure indicates positive values

Though differences in the average water temperature within transect lines were not large about 5°C, they might affect fractionation of ^{15}N (Mariotti et al., 1981). We took the view of Dhondt et al. (2003), who found no significant temperature effect on isotope fractionation during denitrification in their laboratory study.

Under-saturation of DO in flowing water in the reed belts was detected at all sampling points of transect lines, and DO decreased through the spring water flow. Organic matter from reeds, fallen leaves, dead roots and underground stems, might be supplied to heterotrophic bacteria, thereby causing the decrease of oxygen in water and sediment by their respiration. Although oxygen content in overlying water was low, it did not fall below 30% because the shallow water depth made for a ready supply of oxygen from air.

Three hypotheses presented here were substantiated by our measurements. A significant decrease of nitrate in spring water was observed through the reed belts of about 100 in width with a high rate of nitrate reduction averaging of 56.7 ± 11.6% (Table 1, Fig. 1). This nitrate loss may be mainly caused by denitrification and plant uptake, as pointed out by many authors studying wetlands including riparian zones (Xue et al., 1999; Saunders & Kalff, 2001; Lin et al., 2002;

Table 3 Apparent ^{15}N enrichment factor in nitrate retention at water systems dominated by denitrification and/or plant uptake of nitrate

Enrichment factor (‰)	Type of study site (condition)	References
–28 to –33	Laboratory	Robinson (2001)
–24.6 to –29.4	Laboratory (soil)	Mariotti et al. (1981)
–22.5	Laboratory (riparian soil)	Dhondt et al. (2003)
–14, –22	Laboratory (agricultural soil)	Blackmer and Bremmer (1977)
–3.6, –17.8	Laboratory (river sediment, flow, mixing)	Sebilo et al. (2003)
–4.78	Laboratory (paddy soil)	Park and Kumazawa (1998)
–27.6	Riparian groundwater (loam, stony fine sand, well and piezometer)	Menges et al. (1999)
–17.9	River side groundwater (sand and gravel, well)	Tsushima et al. (2002)
–15.9	Groundwater (well)	Bottcher et al. (1990)
–13.9	Groundwater (sand and gravel, well)	Smith et al (1991)
–13.62	River side groundwater (sand silt and grave, borehole)	Fukada et al. (2003)
–8.38	Riparian groundwater (silty clay loam, well and piezometer)	Clement et al. (2003)
–2 to –8	Riparian groundwater (piezometer)	Chen and Macquarrie (2004)
–5.7, –7.7	Groundwater (well, piezometer)	Spalding et al. (1993)
–4.0 to –5.2	Groundwater (well and piezometer)	Fustec et al. (1991)
–4.8 to –5.0	Groundwater (chalk, well)	Mariotti et al. (1988)
–11.2, –20.7	Lake (hyplimnion, model)	Lehmann et al. (2003)
–6.2 to –16.6	Riparian with grasses (surface flow)	Dhondt et al. (2003)
–6.1, –10	Spring, pond with algae	Recalculation in Lund et al. (2000)
–1.03 to –5.12	Reed belt (surface flow)	This study
–2.5	Constructed wetland with macrophytes (pond)	Lund et al. (2000)
–1.5	Reservoir (bottom water)	Sebilo et al. (2003)

Huett et al., 2005). ^{15}N in retained nitrate in water was enriched progressively passing through the *Phragmites* belt (hypothesis (2)). The relationship between nitrate reduction and increase of $\delta^{15}N$-NO_3^- obeyed statistically the Rayleigh curve as indicated in Fig. 2. We obtained apparent enrichment factors ranging from –1.03 to –5.12‰ (Table 2).

It is clear from the literatures that the apparent ^{15}N enrichment factors in NO_3^- have shown great variation (–1.5 to –33‰) in many water systems, groundwater, wetlands and various riparian buffer zones, in laboratory experiments using culture mediums or field soils, under flow or stagnant conditions (Table 3). Laboratory experiments (~–33‰) presented the highest upper limit value of enrichment factor, followed by groundwaters (~–27.6‰), wetlands and ponds (~–16.6‰). Lund et al. (2000) showed low values of –2.5‰ for a constructed wetland with macrophytes. Our data shows similar low values as seen in Table 3. Though great variations of ^{15}N enrichment factor have been reported, large enrichment factors tended to be obtained from groundwater research in fields, whereas wetlands inhabited by macrophytes or ponds with heavy algal bloom showed a much lower enrichment factor of ^{15}N in NO_3^- retention.

The lower apparent enrichment factors of ^{15}N in NO_3^- retention were caused by certain factors, including the following: relatively rapid removal of nitrate by denitrification (Mariotti et al., 1988; Kendall, 1998), relatively reduced removal of nitrate by denitrification due to low electron donor concentration (Hill et al., 2000; Bastviken et al., 2005), dispersion or diffusion effect on denitrification activity (Mariotti et al., 1988; Kawanishi et al., 1993), and other nitrogen sinks with the enrichment factors that are lower than the enrichment factor for denitrification such as plant uptake, microbial nitrate immobilization, and nitrate reduction to ammonium (Dhondt et al., 2003).

Little effect of microbial nitrate loss except denitrification on low enrichment factors was expected in our study, since many authors have suggested that significant loss of nitrate in wetlands is largely responsible for plant uptake and denitrification (Xue et al., 1999; Lund et al., 2000; Huett et al., 2005). The rapid removal of nitrate triggered by high denitrifying activity causes small enrichment factors, but this does not apply to our results that denitrification plays a small role, as mentioned below. The dispersion effect might largely influence the smaller enrichment factor values obtained in this study. Unfortunately, we failed to measure the stream flow speed, though water flowed slowly through the reed belt. Our data on enrichment factors may be an underestimate, given the slow flow. If the dispersion process works to the maximum, our data recalculated give maximum values of –2.06 to –10.24‰, albeit small, when corrected by the method of Kawanishi et al. (1993).

Since it is known that vegetation fractionates ^{15}N minimally or not at all in nitrate uptake (Mariotti et al., 1980; Lund et al., 2000), the effect of plant uptake for ^{15}N fractionation in nitrate

Table 4 Contribution of denitrification and plant uptake to nitrate nitrogen loss in various wetland systems calculated employing a Rayleigh curve method

Study area	Season	Contribution (%)		References
		Denitrification	Plant uptake	
Artificial wetland	Fall	23	77	Lund et al., (2000)
Artificial wetland	Winter	10	90	Lund et al., (2000)
Pond	Spring	66	34	Recalculation by Lund et al., (2000)
Pond	Spring	44	56	Recalculation by Lund et al., (2000)
River riparian zone	Spring	49	51	Dhondt et al., (2003)
River riparian zone	Summer	53	47	Dhondt et al., (2003)
River riparian zone	Fall	75	25	Dhondt et al., (2003)
Reed belt of lake	Spring	15	85	This study
Reed belt of lake	Summer	23	77	This study
Reed belt of lake	Fall	18	82	This study

may be small. If the contribution of plant uptake of nitrate is larger than that of denitrification for nitrate loss inside the reed belt, low apparent enrichment factors of ^{15}N in NO_3^- will be expected in remaining water. A statistically significant ($P < 0.001$) strong correlation ($r = 0.937$, $n = 10$) was observed between the enrichment factor and the plant uptake of nitrate, when data by Lund et al. (2000) was added to our data (Fig. 3).

In Table 4, data are summarized on the contribution of denitrification and plant uptake in nitrate loss at wetlands, calculated by applying the Rayleigh curve method used here. The denitrification contribution in the present study is similar to that of an artificial wetland with rich macrophytes, and less than that of ponds (Lund et al., 2000) and river riparian zone (Dhondt et al., 2003). The contribution rate of denitrification and plant uptake to nitrogen loss varies greatly, indicating that site-specific factors exert much influence on target wetlands. The low contribution of denitrification at the reed belt may possibly be characteristic of a sand dune lake riparian zone. Water flows through the reed belt on the surface of substrate essentially composed of sand. Decomposition of dead *Phragmites australis* accumulated on sand surface with a very thin layer may supply a low concentration of DOC for denitrifying bacteria, although we failed to determine the DOC concentration in flowing water. Denitrification activity is largely affected by the supplying electron donor (Hill et al., 2000; Bastviken et al., 2005). The slightly higher contribution rate of denitrification in summer may be due to the abundant supply of DOC by decomposition at a higher water temperature. An oxic condition above 30% oxygen contents could also suppress the denitrification reaction on the sediment surface of a very shallow water layer. As a result of low contribution of denitrification, plant uptake shows a relatively higher position in nitrate reduction in a sand dune lake.

Conclusion

The present findings serve to substantiate three hypotheses, (1) Nitrate in spring water decreases during its flow through a reed belt; (2) $\delta^{15}N\text{-}NO_3^-$ values in flowing water increase in terms of the nitrate reduction, and (3) the relationship between nitrate reduction and increase of $\delta^{15}N\text{-}NO_3^-$ obey a Rayleigh curve. Moreover, apparent enrichment factors of ^{15}N to estimate the contribution of denitrification and plant uptake in nitrogen removal were duly calculated. The larger enrichment factors were reported from groundwater researche, whereas the low enrichment factors measured in our study were obtained from the wetlands inhabitated by macrophytes. The low apparent enrichment factors might be caused by higher plant uptake of nitrate rather than loss by denitrification. It is again indicated that the reed belt has an important role to remove nitrate with a high loss rate through largely physiological uptake by the plants in a sand dune lake riparian zone. This technique using the variation in natural abundance of ^{15}N may provide useful information about the overall nitrate dynamics in a macrophyte area such as artificial or natural wetlands under a non-destructive condition.

Acknowledgements The authors are indebted to members of the Sand Dune Lakes Research Group, Ms. A. Saito and Ms. E. Niino, who provided water samples for stable isotope analyses. We also thank Mr. Akamatsu and Ms. M. Ohota (Shinshu University) for their assistance in the nitrogen isotope measurement, and also Dr. Yoshioka (Research Institute for Humanity and Nature) for his incisive and kind advice on the isotope methodology. This study was supported in part by a Grant-in-Aid for Scientific Research (C) No.17510193 from the Japan Society for the Promotion of Science.

References

Bastiviken, S. K., P. G. Eriksson, A. Premrov & K. Tonderski, 2005. Potential denitrification in wetland sediments with different plant species detritus. Ecological Engineering 25: 183–190.

Blackmer, A. M. & J. M. Bremner, 1977. Nitrogen isotope discrimination in denitrification of nitrate in soils. Soil Biology and Biochemistry 9: 73–77.

Bottcher, J., O. Strebel, S. Voerkelius & H. –L. Schmidt, 1990. Using isotope fractionation of nitrate-nitrogen and nitrate-oxygen for evaluation of microbial denitrification in a sandy aquifer. Journal of Hydrology 114: 413–424.

Chen, D. J. Z. & K. T. B. MacQuarrie, 2004.Numerical simulation of organic carbon, nitrate, and nitrogen isotope behaviour during denitrification in a riparian zone. Hydrology 293: 235–254.

Clement, J. -C., R. M. Holmes, B. J. Peterson & G. Pinay, 2003. Isotopic investigation of denitrification in a riparian ecosystem in Western France. Journal of Applied Ecology 40:1035–1048.

Cronk, J. K. & M. S. Fennessy, 2001. Wetland Plants: Biology and Ecology. Lewis Publishers, New York.

Dahm, C. N., N. B. Grimm, P. Marmonier, H. M. Valett & P. Vervier, 1998. Nutrient dynamics at the interface between surface waters and groundwaters. Freshwater Biology 40: 427–451.

Dhondt, K., P. Boeckx, O. V. Cleemput & G. Hofman, 2003. Quantifying nitrate retention processes in a riparian buffer zone using the natural abundance of ^{15}N in NO_3^-. Rapid Communication in Mass Spectrometry 17:2597–2604.

Fukada, T., K. M. Hiscock, P. F. Dennis & T. Grischek, 2003. A dual isotope approach to identify denitrification in groundwater at river-bank infiltration site. Water Research 37: 3070–3078.

Fukuhara, H., A. Kawakami & T. Shimogaito, 2003. Characteristics of nutrient dynamics in Lake Sagata (Niigata Prefecture) Japan, a shallow sand-dammed lake supplied by spring water with special reference to nitrates. Hydrobiologia 506: 93–99.

Fustec, E., A. Mariotti, X. Grillo & J. Sajus, 1991. Nitrate removal by denitrification in alluvial groundwater: Role of a former channel. Journal of Hydrology 123: 337–354.

Hey, D., 2002. Nitrogen farming: Harvesting a different crop. Restoration Ecology 10: 1–10.

Hill, A. R., 1996. Nitrate removal in stream riparian zones. Journal of Environmental Quality 25: 743–755.

Hill, A. R., K. J. Devito, S. Campagnolo & K. Sanmugadas, 2000. Subsurface denitrification in a forest riparian zone: Interactions between hydrology and supplies of nitrate and organic carbon. Biogeochemistry 51: 193–223.

Huett, D. O., S. G. Morris, G. Smith & N. Hunt, 2005. Nitrogen and phosphorus removal from plant runoff in vegetated and unvegetated subsurface flow wetlands. Water research 39: 3529–3272.

Jansson, M., R. Andersson, H. Berggren & L. Leonardson, 1994. Wetlands and lakes as nitrogen traps. Ambio 23: 320–325.

Kawanishi, T., Y. Hayashi, N. Kihou, T. Yoneyama & Y. Ozaki, 1993. Dispersion effects on the apparent isotope fractionation factor associated with denitrification in soil; evaluation by mathematical model. Soil Biology and Biochemistry 25: 349–354.

Kellman, L. M. & C. Hillaire -Marcel, 2003. Evaluation of nitrogen isotopes as indicators of nitrate contamination sources in an agricultural watershed. Agriculture Ecosystems & Environment 95: 87–102.

Kendall, C., 1998. Tracing nitrogen sources and cycling in catchment. In Kendal, C. & J. J. McDonnell (eds), Isotope tracers in catchment hydrology, Elsevier, Amsterdam, 519–576.

Khol, D. H. & G. B. Shearer, 1995. Using variations in natural ^{15}N abundance to investigate N cycle processes. In Wada, E., T. Yoneyama, M. Minagawa, T. Ando & B. D. Fry (eds), Stable isotopes in the biosphere. Kyoto University Press, Kyoto, Japan, 103–130.

Lehmann, M. F., P. Reichert, S. M. Bernasconi, A. Barbieri & J. A. McKenzie, 2003. Modelling nitrogen and oxygen isotope fractionation during denitrification in a lacustrine redox-transition zone. Geochemica et Cosmochemica Acta 67: 2529–2542.

Lin, Y. -F., S. -R. Jing, T. -W. Wang & D. -Y. Lee, 2002. Effects of macrophytes and external carbon sources on nitrate removal from groundwater in constructed wetlands. Environmental Pollution 119: 413–420.

Lund, L. J., A. J. Horne & A. E. Williams, 2000. Estimating denitrification in a large constructed wetland using stable nitrogen isotope ratios. Ecological Engineering 14: 67–76.

Mariotti, A., A. Landreau & B. Simon, 1988. ^{15}N isotope biogeochemistry and natural denitrification process in groundwater: Application to the chalk aquifer of northern France. Geochimica et Cosmohimica Acta 52: 1869–1878.

Mariotti, A., J. C. Germon, P. Hubert, P. Kaiser, R. Letolle, A. Tardieux & P. Tardieux, 1981. Experimental determination of nitrogen kinetic isotope fractionation: Some principles; illustration for the denitrification and nitrification processes. Plant and Soil 62: 413–430.

Mariotti, A., F. Mariotti, N. Amarger, G. Pizelle, J. -M. Ngambi, M.-L. Champigny & A. Moyse, 1980. Fractionnements isotopiques de l'azote lors des processus d'absorption des nitrates et de fixation de l'azote atmosphe'rique par les plantes. Physiologie Vegetale 18: 163–181.

Mengis, M., S. L. Schiff, M. Harris, M. C. English, R. Aravena, R. J. Elgood & A. MacLean, 1999. Multiple geochemical and isotopic approaches for assessing ground water NO_3^- elimination in a riparian zone. Ground water 37: 448–457.

Naiman, R. J., H. Decamps & M. E. McClain, 2005. Riparia: Ecology, conservation, and management of streamside communities. Elsevier, Tokyo, 430 pp.

Nonaka, M., R. Abe & H. Tanabe, 1997. Increasing groundwater nitrate nitrogen concentration in sand dune upland soil by fertilizer application. Sand Dune Research 44: 23–29 (In Japanese with English summary).

Park, K. L. & K. Kumazawa, 1998. Variation of natural abundance of nitrogen isotope due to denitrification. Journal of the Science of Soil and Manure, Japan 69: 293–295 (in Japanese).

Robinson, D., 2001. $\delta^{15}N$ as an integrator of the nitrogen cycle. Trends in Ecology & Evolution 16: 153–162.

Saunders, D. L. & J. Kalff, 2001. Nitrogen retention in wetlands, lake and rivers. Hydrobiologia 443: 205–212.

Sebilo, M., G. Billen, M. Grably & A. Mariotti, 2003. Isotopic composition of nitrate-nitrogen as a marker of riparian and benthic denitrification at the scale of the whole Seine River system. Biogeochemistry 63: 35–51.

Smith, R. L., B. L. Howes & J. H. Duff, 1991. Denitrification in nitrate-contaminated groundwater: Occurrence in steep vertical geochemical gradients. Geochimica Cosmohimica Acta 55: 1815–1825.

Spalding, R. F., M. E. Exner, G. E. Martin & D. D. Snow, 1993. Effects of sludge disposal on groundwater nitrate concentration. Journal of Hydrology 142: 213–226.

Strickland J. D. H. & T. R. Parsons, 1972. A practical handbook of sea water analysis. Bulletin Fisheries Research Board of Canada 167: 207–211.

Toda, H., U. Uemura T. Okino T. kawanishi & H. Kawashima, 2002. Use of nitrogen stable isotope ratio of periphyton for monitoring nitrogen sources in a river system. Water Science & Technology 46: 431–435.

Tsushima, K., S. Ueda & N. Ogura, 2002. Nitrate loss for denitrification during high frequency research in floodplain groundwater of the Tama River. Water Air and Soil Pollution 137: 167–178.

Xue, Y., D. A. Kovacic, M. B. David, L. E. Gentry, R. L. Mulvaney & C. W. Lindau, 1999. In situ measurements of denitrification in constructed wetlands. Journal Environmental Quality 28: 263–269.

Hydrobiologia (2007) 584:59–68
DOI 10.1007/s10750-007-0605-x

SHALLOW LAKES

Temporal trends of ion contents and nutrients in three Kenyan Rift Valley saline–alkaline lakes and their influence on phytoplankton biomass

S. O. Oduor · M. Schagerl

Abstract This study, carried out between November 2003 and February 2005, aimed to investigate the temporal trends of conductivity, ions, nutrient concentrations and phytoplankton biomass expressed as chlorophyll *a* in the Kenyan Rift Valley saline-alkaline lakes namely Nakuru, Bogoria and Elmentaita. The influence of environmental variables on phytoplankton biomass has always been of much interest in understanding phytoplankton dynamics. Being shallow and endorheic, these lakes' chemical, physical and biological properties were found to be strongly influenced by the hydrologic cycle within their catchment area. The lakes are characterised by high nutrient concentrations but with low Ntot : Ptot ratios. Significant differences between surface and near-bottom samples for water temperature, chlorophyll *a* and some nutrients were found in these lakes. A stepwise Discriminant Analysis with lakes as defined groups resulted in a significant model with SRP, nitrate–N, conductivity and light supply being of major importance. A significant correlation between specific conductivity and total alkalinity (Kendalls $\tau = 0.85$, $n = 132$) was calculated. While L. Bogoria showed the least temporal variation in conductivity – (65–73 mS cm^{-1}), larger variations were observed in L. Elmentaita (21–77 mS cm^{-1}). Na^+ and K^+ form the main cations with Cl^-, HCO_3^- and CO_3^{2-} being the major anions in all the three lakes. Flouride was detected in high quantities (mean values L. Bogoria 72 meq l^{-1}, L. Nakuru 17 meq l^{-1} and L. Elmentaita 71 meq l^{-1}). A PCA followed by multiple regression analysis with chlorophyll *a* as dependent variable showed that nitrate-N, conductivity, phosphorus and light supply were the key variables influencing algal biomass in these lakes.

Keywords Saline · Alkaline · Chlorophyll *a* · Trona · Rift Valley

Guest editors: R. D. Gulati, E. Lammens, N. De Pauw & E. Van Donk
Shallow lakes in a changing world

S. O. Oduor
Department of Biological Sciences, Egerton University, Njoro, Kenya

M. Schagerl (✉)
Department of Marine Biology, University of Vienna, Vienna, Austria
e-mail: michael.schagerl@univie.ac.at

Introduction

Saline-alkaline lakes tend to be very simple in their biodiversity and therefore offer ideal ecosystems for studies of biological aquatic dynamics and foodwebs (Vareschi, 1982; Talling, 2001). Inspite of this simplicity, their highly stochastic temporal dynamics in their environmental variables make them very challenging in trying to establish trends of these variables variations. The key driving force appears to be the hydrologic cycle which influences the chemical dynamics of the water column and therefore their biota (Sondergaard et al., 1990). The high surface area to volume ratio in the shallow lakes makes them to easily and rapidly respond to the physical changes in their environments. This in turn rapidly affect the biota, forcing it to respond to the changes through changes in composition and biomass (Vareschi, 1987). Such changes may at times be extreme resulting in drastic algal biomass crashes and big changes in community composition structure (Melack, 1976; Vareschi et al., 1981).

The features mentioned above fit the characteristics of the saline-alkaline lakes in the Eastern arm of the African Great Rift Valley in Kenya as well. Though simple in their biodiversity, these lakes offer very interesting relationship between the primary producers and the abiotic world in terms of nutrient supply and the influence that changes in physical and chemical parameters have on the lakes productivity. High primary productivity and large phytoplankton biomass with paucity in biodiversity in these lakes have been observed by various authors (Grant et al., 1990; Wood & Talling, 1988; Melack, 1981; Vareschi, 1982). Key reasons attributed to this phenomenon include: (i) unlimited availability of dissolved inorganic carbon in these carbonate-rich waters, (ii) elevated temperatures and (iii) massive growth of cyanobacteria dominated by *Arthrospira fusiformis* Voronichin especially in L. Bogoria and other species such as *Synechococcus* sp., *Anabaenopsis arnoldii, A. abijatae, Anabaena* sp. and the chlorophyte *Monoraphidium minutum* (Ballot et al., 2004, Vareschi, 1982; Schagerl & Oduor, 2007). Being in the tropics, these soda lakes enjoy an almost constant daily cycle of 12 h light and 12 h darkness. Though the inverse relationship between increase in salinity and biodiversity has been noted (Vareschi, 1982), individual species adaptations have allowed for building up large biomass of the well adapted organisms to develop due to the large resource availability with minimal competition. However, the availability of these resources, especially nutrients may sometimes be hampered by the extreme conditions operating in these lakes. Both phosphorus and nitrogen availability is known to be influenced by oxygen concentration and the alkalinity level through pH (Sondergaard et al., 1990).

This paper gives information of nutrient and ions concentrations of the Kenyan Rift Valley saline-alkaline lakes Nakuru, Bogoria and Elmentaita between November 2003 and February 2005. It characterises these lakes according to their environmental variables and relate the fluctuations in these variables to the changes in phytoplankton biomass. Statistical models were performed to establish the extent to which changes in chemical and physical properties of these lakes may influence phytoplankton biomass dynamics.

Study area and sites description

The three endorheic athallasic lakes Bogoria, Nakuru and Elmentaita were formed through tectonic and volcanic activities and lie on the Eastern arm of the great Rift Valley in Kenya (Fig. 1, Table 1). They are located in semi-arid regions with Lake Bogoria experiencing the harshest dry conditions among them. Annual rainfall in the lake areas is highly variable and low, mostly concentrated between April and July for the long rains and October and November for the short rains. The water budget is therefore strongly influenced by precipitation, evaporation and the highly seasonal small inflows. Air temperatures around the lakes are variable with Lake Bogoria having between 30°C and 34°C during day time while Lake Nakuru and Elmentaita may register a range of 25°C and 32°C. In the evening the temperature falls to below 15°C. Correspondingly high water temperatures are recorded during the day which may cause slight thermal stratification during midday. This stratification is rare in Lake Elmentaita due to its very shallow nature but prominent in the deeper Lake Bogoria. In the evening, consistent wind occurs in the lakes causing mixing (isothermal conditions in the water column from evening up to late morning hours). Humidity in these regions is normally very

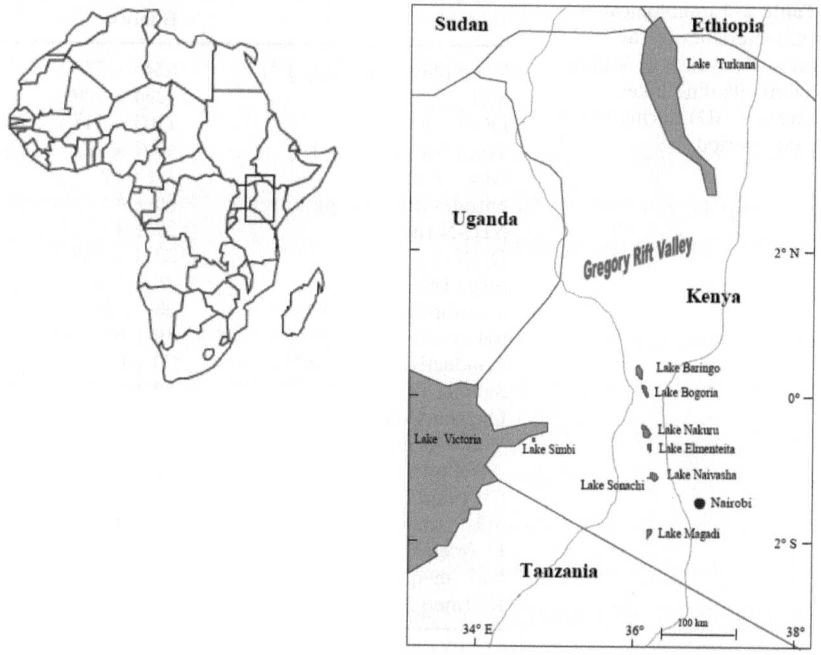

Fig. 1 Map showing the location of Kenya in Africa, the Kenyan Rift Valley lakes and the three studied saline-alkaline lakes Bogoria, Nakuru and Elmentaita (after Ballot, 2004)

Table 1 Geographical and morphometric features of the Lakes Bogoria, Nakuru and Elmenteita (after McCall, 1967; Vareschi, 1982; Melack, 1988; Mwaura & Moore, 1991, Harper et al., 2003)

	Lake Bogoria	Lake Nakuru	Lake Elmentaita
Geographical position	N 00°15′ E 36°05′	S 00°20′E 36°05′	S 00°27′E 36°15′
Altitude (m a. s. l)	990	1,759	1,782
Surface area (km^2)	34	40	20
Mean depth (cm)	580	110	46
Maximum depth (cm)	1,020	450	300
Catchment area (km^2)	1,500	1,800	500
Rainfall (mm)	500–1,000	800–1,000	600–1,200
Sampling site (North)	N00°18′39E036°4′41	S00°16′07E036°′40	S00°25′12E036°4′38
Sampling site (Center)	N00°15′43E036°5′32	S00°21′12E036°′50	S00°26′06E036°14′44
Sampling site (South)	N00°11′34E036°6′47	S00°23′18E036°05′37	S00°26′31E036°15′15

low thereby enhancing evaporation. The endorheic nature of these lakes, coupled with their shallowness and daily wind influence have combined forces making them experience high concentrations in most of the solute variables due to net accumulation and resuspension from the sediments (Table 2).

Lake Nakuru and Bogoria are important Ramsar sites in Kenya, being habitats for diverse water fowls, mainly the lesser flamingo *Phoeniconaias minor* Gregory and other hundreds of bird species (Owino et al., 2001). Both of them are protected wildlife parks and are important revenue sources through tourism to the government of Kenya. Lake Elmentaita offers an important alternative sanctuary for the flamingos and other water birds which migrate between all the three lakes. This lake is also an important breeding ground for the pelican *Pelecanus onocrotalus* that is found in large numbers between it and the neighbouring Lake Nakuru.

Lake Bogoria

Lake Bogoria is an endorheic narrow long lake that lies on the north-western part of Kenya. It is fed by few freshwater hot springs from the escarpment above it, some small impermanent tributaries on the west, including Emsoss river fed by hot springs, and the Wasagess, flowing from the north. The lake receives rain erraticaly which falls as a storm within a short time resulting into

Table 2 Limnological characteristics of the three Kenyan Rift Valley Saline-alkaline lakes (mean ± SD) during the study period

Variable	Bogoria	Nakuru	Elmentaita
Total phosphorus ($\mu g\ l^{-1}$)	6246 ± 72	2426 ± 85	1228 ± 79
SRP ($\mu g\ l^{-1}$)	4254 ± 129	589 ± 28	98 ± 10
DOP ($\mu g\ l^{-1}$)	1103 ± 117	1359 ± 77	881 ± 106
Total Nitrogen ($\mu g\ l^{-1}$)	4067 ± 225	11853 ± 605	12171 ± 653
Ntot : Ptot ratio	1 ± 0.02	5 ± 0.24	10 ± 0.77
Nitrate-nitrogen ($\mu g\ l^{-1}$)	66 ± 4	322 ± 19	72 ± 7
NH_4-N ($\mu g\ l^{-1}$)	37 ± 4	157 ± 22	347 ± 4
DON ($\mu g\ l^{-1}$)	2331 ± 145	11377 ± 589	8416 ± 469
Silica ($mg\ l^{-1}$)	28 ± 1	48 ± 5	51 ± 5
Chlorophyll *a* ($\mu g\ l^{-1}$)	388 ± 26	646 ± 34	267 ± 29
pH	10 ± 0	10 ± 0	10 ± 0
Conductivity ($mS\ cm^{-1}$)	70 ± 0	29 ± 1	39 ± 2
Salinity (‰)	49 ± 0	18 ± 0	26 ± 0
Oxygen Conc ($mg\ l^{-1}$)	13 ± 1.3	17 ± 0.9	8 ± 0.8
Water Temperature °C	27.9 ± 1.7	24.1 ± 1.6	23.1 ± 2.3
Alkalinity ($meq\ l^{-1}$)	1132 ± 14	327 ± 10	376 ± 37
Cl^- ($meq\ l^{-1}$)	176.0 ± 59.9	67.3 ± 15.0	215.0 ± 136.9
SO_4^{2-} ($meq\ l^{-1}$)	8.2 ± 0.3	5.0 ± 1.4	14.1 ± 9.5
F^- ($meq\ l^{-1}$)	72.3 ± 16.5	17.4 ± 5.3	70.5 ± 49.1
Na^+ ($meq\ l^{-1}$)	1348.7 ± 50.6	413.1 ± 98.4	718.8 ± 537.4
K^+ ($meq\ l^{-1}$)	19.2 ± 2.3	8.5 ± 2.6	15.9 ± 12.7

erosion of the nearby scarcely vegetated areas around the lake and washing lots of sediment into the lake. Being highly alkaline, minimal macrophytes growth is found along its shores with salt tolerant grasses such as *Sporobolus spicata* and sedges such as *Cyperus rubicundus* and *C. laevigatus* on the western shores near the hotsprings.

Lake Nakuru

Lake Nakuru is a shallow endorheic pan lying next to the city of Nakuru surrounded by the world famous Lake Nakuru National park. The lake is recharged mainly by rainfall and inflows from three seasonal surface streams (Njoro, Makalia and Nderit Rivers), which drain from the Mau Forest. The lake is also fed by the Baharini springs at the northern end and the city's sewage system. Direct precipitation on the lake now appears to contribute more to its recharge than the river inflows.

Lake Elmentaita

Lake Elmentaita lies in a shallow pan floored by coarser salt-impregnated sedimentary material. It is fed by three seasonal small streams, the Mbaruk to the north, Kariandus and Memeroni to the south. Some hot springs along the southern end supply minimal amount of water into the lake. The dry spells and its endorheic nature make it vulnerable to evaporation resulting into large fluctuations in its volume.

Materials and methods

Three off - shore sampling sites, at least 300 m from the nearest shoreline were selected from each of the lakes with monthly samplings done from November 2003 to February 2005. The geographical points for these sites were located using GPS (Fig. 1, Table 1; Model Magellan, Sportrak Thales Navigation). Transparency was estimated by means of a Secchi disk. Water temperature, oxygen saturation and concentration, specific conductivity and pH were measured at the sampling sites using a multiprobe (WTW Multi 340i Wissenschaftlich Technische Werkstätten Weilheim, Germany). Water samples for chemical analysis were collected using a 2 l Ruttner sampler from an integrated depth of 40 cm from the surface and near the bottom in L. Nakuru and Bogoria, while in L. Elmentaita one integral was enough. The samples were stored in plastic containers at 4°C while being transferred to the laboratory for further analyses at Egerton University Njoro (Kenya) where they were filtered (Ederol BM/C) within 10 h after collection. The samples for ionic composition determination were preserved with 4% formaline solution and stored at 4°C until analysed by ion chromatography (Metrohm IC Compact 761).

Standard procedures (APHA, 1995; Legler, 1988) were used in the determination of alkalinity, nutrient concentrations i.e. total phosphorus (Ptot), total nitrogen (Ntot), nitrate–N (NO_3–N), nitrite–N (NO_2–N), ammonium–N (NH_4–N) and soluble reactive silica (SiO_2–Si) with dilution of the samples and pH adjustments where necessary to minimise the effect of the high pH on the analyses. Suspended solids dry mass was estimated gravimetrically on pre-combusted and pre-weighed glass-fibre filters (Ederol BM/C filters) after drying at 95°C with the ash mass being determined after 2 h combustion at 450°C. Chlorophyll-*a* (Chl *a*) was extracted using 90% acetone after gentle vacuum filtration through Ederol BM/C (0.47 μm pore size) filters. The filters were stored at –20°C to aid in the bursting of the cells – then homogenised in a tissue grinder covered with the extraction solvent. Using spectrophotometer (Pharmacia Biotech Novaspec II) the absorbance of the clarified extract was measured at 663 nm and 750 nm and the biomass worked out according to Talling and Driver (1961).

Statistical analyses were carried out on the obtained data in order to determine the interrelationship between these variables and the extent to which they influence the phytoplankton biomass. The data collected were analysed for normality (Kolmogorov-Smirnov test) and where necessary ln-transformed before further statistical analyses. For bivariate correlations, Kendalls τ was calculated. Obtained surface and near-bottom data were tested for significant differences with the WILCOXON-Test (pairwise dependent variables). A stepwise Discriminant Analysis (DA) extracted functions containing variables with major influence on separating the lakes. To establish the variables having the highest influence on the phytoplankton biomass, Principal Component Analyses (PCA) with varimax rotation were performed (criterion for extraction: eigenvalue higher than one). In following Multiple Linear Regression Analyses (LR), Chl *a* was explained by the PC obtained in the PCA. SPSS 12.0 statistical package was used in all the statistical analyses.

Results

Environmental variables

Dissolved oxygen concentration was always high during the day, reaching supersaturation levels over 300% and a concentration of between 25 and 31 mg l^{-1} around midday in all the three lakes; mean surface values are given in Table 2. Depth profile for oxygen concentration for Lake Bogoria shows presence of anoxic layers beyond 5 m during most of the days. At the sediment-water interface anoxic conditions were recorded always in both L. Bogoria and Nakuru.

High values for pH, total alkalinity and conductivity were recorded in all the three lakes (Table 2) and a significant correlation between specific conductivity and total alkalinity (Kendalls $\tau = 0.85$, $n = 132$) was calculated. Fluctuations in these variables corresponded to changes in the hydrologic cycles and water levels in the lakes with low values for conductivity and alkalinity recorded during the rainy season and high values during the dry spells. A gradual increase in conductivity in all the three lakes was observed during the study period (Fig. 2). High total alkalinity was recorded with both carbonate and bicarbonate ions found in all three lakes. Bicarbonates concentration was significantly ($P < 0.01$) higher than the carbonates ions in all three lakes.

Total ion content was highly correlated with the specific conductivity ($r = 0.99$, $n = 29$). The ion composition in all the three lakes was dominated by the Na^+ cations and CO_3^{2-}, HCO_3 and Cl^- anions (Fig. 3, Table 2). K^+ ions, though present in all the three lakes, form a very small portion to the total cations concentration. Mg^{2+} and Ca^{2+} cations were found only in trace amounts. F^- was detected in high quantities, too (mean values L. Bogoria 72 meq l^{-1}, L. Nakuru 17 meq l^{-1} and L. Elmentaita 71 meq l^{-1}).

High concentration of the major nutrients, characteristic of eutrophic lakes was evident in all the three lakes (Table 2). The high concentration of total nitrogen in L. Nakuru and

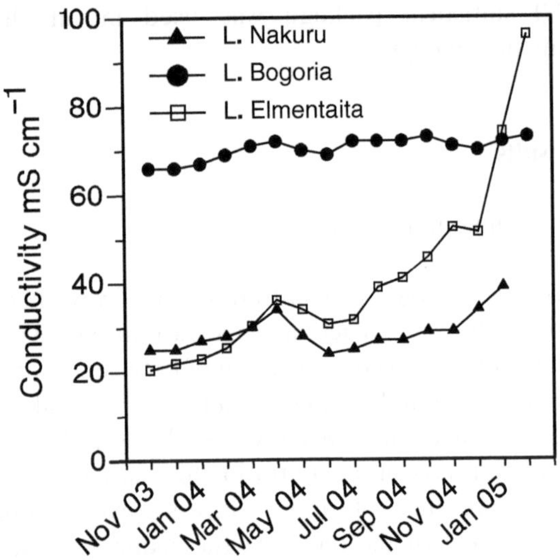

Fig. 2 Temporal trend of specific conductivity in the three Rift Valley saline-alkaline lakes thoughout the study period

L. Elmentaita was contributed by the dissolved organic nitrogen component with extremely low inorganic nitrogen (Table 2). Ammonium concentration was low in all the lakes with L. Bogoria giving an unexpectedly very low value inspite of its anoxic conditions below 5 m depth. High silica concentration in all the three lakes ranging between 28 mg l^{-1} and 51 mg l^{-1} was recorded. When analyzing surface and near-bottom samples SRP, total alkalinity, temperature and Chl *a* showed significant differences in L. Nakuru, while in L. Bogoria additionally Ptot, and nitrate–N were different (Table 3).

All the three lakes had high phytoplankton biomass expressed as Chl *a* with Lake Nakuru having a range of 470–718 μg l^{-1}, Lake Bogoria (74–653 μg l^{-1}) and Elmentaita with 58–602 μg l^{-1} (Fig. 4, bottom). However, a very low total nitrogen to total phosphorus ratio was observed especially for the Lakes Bogoria and Nakuru (Fig. 4, top). Phytoplankton biomass dropped from over 600 μg L^{-1}–74 μg l^{-1} in L. Bogoria between July and September 2004. The lake recovered by October 2004 to a biomass of over 200 μg l^{-1}. Though the contribution of *A. fusiformis* to the total biomass was overwhelming in all the three lakes, in L. Elmentaita diatoms and *Monoraphidium* sp. contributed significantly too.

PCA for Bogoria extracted three PC's which explained 77% of the total variation (Table 4). High loadings of nitrate–N and conductivity were found in PC1, PC2 consisted mainly of Ptot and SRP, while PC3 was highly determined by ammonium–N and light attenuation (negative Secchi depth). The LR resulted in a significant model that explained 36% of the variations of Chl *a* by the PC1 and PC2 (Table 4). In L. Nakuru, the PCA extracted three PC's that explained 73% of the total variation. The variables with the highest factor loadings were nitrate concentration and ammonium–N for PC1, conductivity for PC2 and SRP for PC3. Chl *a* significantly coincided by PC1. PCA for L. Elmentaita, extracted two PC's with an explanation of the total variance of 76% (Table 4). The following LR included both PC's. PC1 was mainly controlled by conductivity, ammonium–N and SRP, while PC2 consisted of light attenuation and nitrate–N.

Fig. 3 Ion field diagrams for the three saline alkaline Kenyan Rift Valley lakes

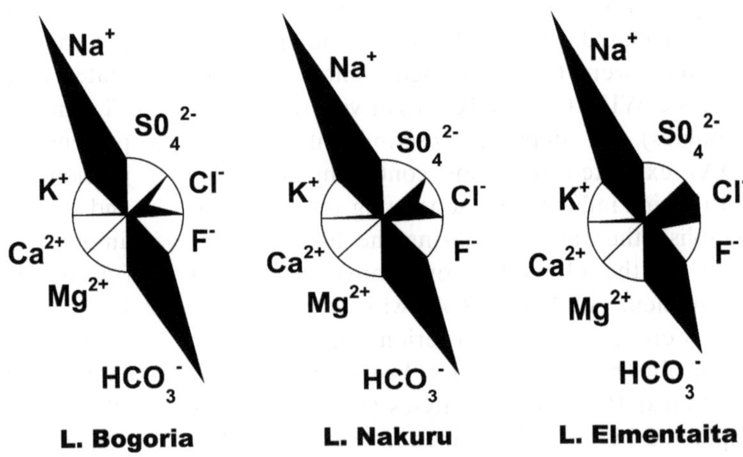

Table 3 Results of the surface-bottom comparison of the Lakes Bogoria and Nakuru (Wilcoxon-test; n = 14)

Variable	Nakuru	Bogoria
Ptot	0.508	**0.009**
SRP	**0.017**	**0.002**
Nitrate-N	0.114	**0.016**
Ammonium–N	0.169	0.162
Silica	0.878	0.084
Alkalinity	**0.041**	**0.016**
Chl *a*	**0.037**	**0.041**
Temperature	**0.005**	**0.001**
pH	0.386	0.176
Conductivity	0.332	0.572

Significant differences ($P < 0.05$) are indicated in bold

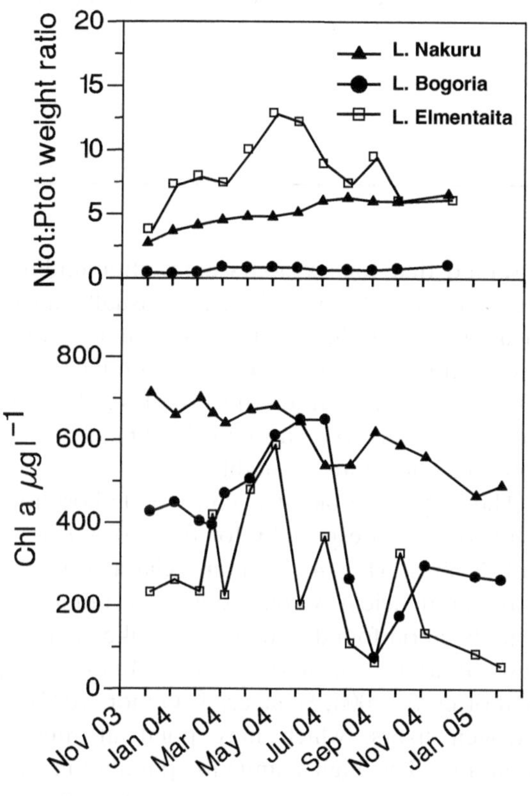

Fig. 4 Temporal trends for the Total nitrogen:Total phosphorus ratio on a weight basis (top) for Chlorophyll *a* concentration (bottom) in the three Kenyan Rift Valley saline-alkaline lakes between October 2003 and February 2005

The DA resulted in a significant model, in which the lakes were clearly separated (Table 5, Fig. 5). SRP, nitrate–N, conductivity and light supply were found to be major parameter for the grouping.

Discussion

Shallow lakes, especially those found in dry regions experience large fluctuations in their variables which are strongly linked to climatic changes and especially rainfall (Talling, 2001). The three investigated lakes show such fluctuations causing temporal variations in their chemistry and nutrients concentrations and ultimately in their biota. Being endorheic, the ranges of these fluctuations and the biological consequences may be very large (Talling & Lemoalle, 1998). These fluctuations are accompanied by changes in the concentrations or proportions of all major ions which comprise the anions Cl^-, $HCO_3^- + CO_3^{2-}$, F^- and SO_4^{2-}, and the cations which are mainly Na^+ and K^+. Figure 1 clearly points out such fluctuations. Bogoria showing the lowest surface/volume ratio (Table 1) was rather stable throughout the investigation period, whereas Nakuru exhibit bigger changes in its ion content followed by Elmentaita with the mostly pronounced fluctuations. L. Elmentaita is extremely shallow and shows a pan-like characteristic that makes it loose water rapidly through rapid heating of its sediments that further enhances the evaporation (the big increase in conductivity at the end of the study period was caused by desiccation).

The very low concentration of Ca^{2+} and Mg^{2+} in these lakes may be due to precipitation as carbonates, formation of aluminosilicate either through ion exchange or by incorporation into clays by the process of "reverse weathering" as reported by Von- Damm & Edmond (1984) and Rippey & Wood (1985). Concentrations of SO_4^{2-} in saline lakes are generally low and tend to be influenced by reduction to sulphide in anoxic hypolimnia, which may also escape as hydrogen sulphide to the atmosphere (Wood & Talling, 1988).

Very high nutrients concentration is characteristic of the saline-alkaline Rift Valley lakes. This may suggest that growth limitation of algae caused by nutrient defiency may not be important in these lakes. While this may be true for L. Bogoria and L. Nakuru which show very high amounts of SRP up to 4 mg l^{-1} (Table 2), L. Elmentaita sometimes tends to suffer

Table 4 Principal component analyses with six environmental variables. Factor loadings over 0.750 are bold, explained variances are in parentheses. Regressions of chl-a (ln-transformed) on the principal components contain only the significant ($P < 0.05$) coefficients

Bogoria	PC1	PC2	PC3	Multiple regression analysis
	(28.1 %)	(26.1 %)	(22.4 %)	ln Chl $a = -0.28 * PC1 + 0.26 * PC2 + 5.82$
lnNitrate	**0.856**	−0.005	0.023	$n = 45, r = 0.60, P \leq 0.000$
lnCond	**0.814**	0.369	−0.032	
lnPtot	0.042	**0.800**	0.337	
lnSRP	0.185	**0.881**	−0.158	
Lnsecchi	0.298	−0.115	**−0.791**	
lnNH$_4$-N	0.406	−0.019	**0.762**	
Nakuru	PC1	PC2	PC3	
	(27.7 %)	(24.5 %)	(20.6 %)	ln Chl $a = -0.11 * PC1 + 6.41$ $n = 42$,
lnNitrate	**−0.802**	0.509	0.05	$r = 0.36, P = 0.018$
lnNH$_4$-N	**0.877**	0.307	0.044	
lnCond	0.128	**0.908**	−0.042	
lnSRP	0.03	−0.099	**0.928**	
lnPtot	0.479	0.387	0.489	
Lnsecchi	0.056	−0.367	−0.363	
Elmentaita	PC1	PC2		
	(48.2 %)	(27.9 %)		ln Chl $a = -0.33 * PC1 + 0.36 * PC2 + 5.50$
Lnsecchi	0.151	**−0.86**		$n = 36, r = 0.68, P \leq 0.000$
lnSRP	**−0.911**	−0.23		
lnNitrate	0.233	**0.753**		
lnCond	**0.971**	0.069		
lnNH$_4$-N	**0.794**	−0.103		
lnPtot	0.639	0.545		

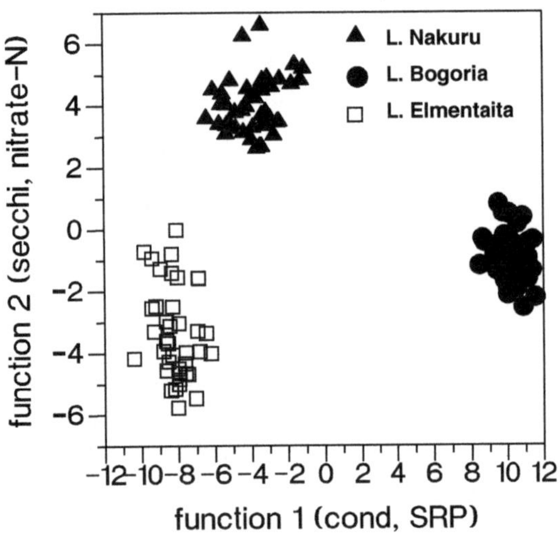

Fig. 5 Groupings of the three saline alkaline Kenyan Rift Valley lakes according to the discriminant functions

phosphorus deficiency. Comparatively low values recorded during our study support the observations from Peters & MacIntyre (1976) in which they observed in this lake very rapid turn-over rates in SRP. High phosphorus concentrations in L. Bogoria and Nakuru may be coming from the sediments, being anoxic in L. Bogoria, coupled with the high pH. For L. Nakuru, Vareschi (1982) observed that for six hours daily after midnight the water close to the sediments is usually anoxic. In addition, another source for enhanced phosphorus concentrations in L. Nakuru might be the sewage plant discharging into the lake. Unfortunately, no data on nutrients loading from this potential source are available.

The very low total nitrogen to total phosphorus ratios observed especially for the Lakes Bogoria and Nakuru (Fig. 4, top) are characteristic of nitrogen deficient water bodies. A number of cyanobacteria found in these lakes like *Anabaenopsis arnoldii, A. abijatae* and *Anabaena* sp. (Ballot et al., 2004, Schagerl & Oduor, 2007) are nitrogen fixers, which may point at nitrogen depletion. The high mammals' population in L. Nakuru may be another source of nitrogen in the lake through their droppings which are washed into the lake by rain storm. Furthermore, recycled ammonia from the zooplankton communities could also contribute towards alleviating low nitrogen concentration as observed by Ganf & Blazka (1974) in L. George, Uganda and James (1987) in lakes Taupo and Rotongaio in New Zealand. While ammonium concentration may be expected to be higher in Lake Bogoria due to its anoxic state below 5 m depth, lowest values of

Table 5 Stepwise Discriminant Analysis resulted in a model with eight environmental variables. Canonical discriminant function coefficients (unstandardized and z-standardized) of the discriminant functions. Canonical coefficients of correlation are indicated in brackets

Variable	function 1 (cancorr = 0.99)	z-standardized function 1	function 2 (cancorr = 0.95)	z-standardized function 2
lnPtot	1.91	0.37	2.47	0.48
lnSRP	1.98	1.00	0.35	0.18
lnNitrate	–0.50	–0.24	1.74	0.84
$\ln NH_4$–N	–0.70	–0.51	0.46	0.34
lndry mass	–0.99	–0.29	–0.72	–0.21
lnChl *a*	0.32	0.19	0.73	0.42
Lnsecchi	–0.02	–0.01	2.83	0.73
lnConductivity	7.85	1.29	–3.83	–0.63
(constant)	–48.48		–27.33	

about 40 μg l^{-1} were recorded in this lake with no significant differences between surface and near-bottom samples (Table 3). One probable reason is the high pH, which shifts the equilibrium from ammonium to volatile ammonia resulting into its rapid loss from the water (Jones and Grant, 1999). High concentration of organic nitrogen recorded in these lakes may be associated to the high algal biomass coupled with the large bacterioplankton community as reported by Kilham (1981) and Yasindi et al. (2002) in these lakes. Temporal fluctuations of Chl *a* observed in these lakes coincided with certain environmental variables (Table 4). The performances of the LR models are high especially for the lakes Bogoria and Elmentaita. However, it can be assumed that the inclusion of other factors like bacterial production, zooplankton and fish grazing (*Oreochromis grahami* was introduced to L. Nakuru in the 1970-ies) or sediment-water interactions will increase the explanatory value of such models.

It could be questioned, if each of the lakes situated closely together is a unique ecosystem, because until 6000 years B. C. a big freshwater lake ranging from the recent L. Naivsha upto L. Nakuru including L. Elmentaita covered the whole region (Burgis & Morris, 1987; Milbrink, 1974; Richardson & Richardson, 1972). Because of tectonic forces and climate changes, this big lake disappeared leaving only the remnants in terms of endorheic salt pans. As shown by the DA, these lakes developed their own unique characteristics. It clearly shows that each of the three studied lakes (the freshwater lake Naivasha was not considered here) could be separated and has its own characteristics (Table 5, Fig. 5). SRP, nitrate–N, conductivity and light supply are of major importance for their differentiation. A second DA considering major ions conspicuously presented a significant model with mainly Na^+ and HCO_3^- responsible for the lakes grouping (data not shown). Another interesting fact could be found in our study: although being shallow, the lakes Nakuru and Bogoria showed significant differences between surface and near-bottom samples for water temperature, Chl *a* and some nutrients (Table 3), which were notably pronounced in L. Bogoria with its deeper basin and lee exposure. A probable reason for the observed vertical differences could be seen in the *Arthrospira fusiformis* blooms. *Arthrospira* filaments contain gas vesicles which allow vertical movement to the light-supplied surface layer. Because of its enormous productivity, *Arthrospira* cells have to assimilate high amounts of nutrients therefore effecting the vertical differences.

Acknowledgement This study was funded by the Austrian government through the OEAD a wing of the Austrian Development Cooperation to whom we are most grateful. We acknowledge and appreciate the assistance given by Geoffrey Ongondo during the sampling exercise. We also thank Hubert Kraill of Limnology Department, Vienna University for the ion analyses and Christian Fesl for many valuable comments and his attendance for discussion.

References

APHA, 1995. Standard methods for the Examination of Water and Wastewater, 19th edn. American Public Health Association, Washington, DC, USA.

Ballot, A., 2004. Cyanobacteria in Kenyan Rift Valley lakes. A Ph.D Dissertaion submitted to the Freie Universitat–Berlin, Germany, 146 pp. http://www.diss.fu-berlin.de/2004/116/.

Ballot, A., L. Krienitz, K. Kotut, C. Wiegand, J. S. Metcalf, G. A. Codd & S. Pflugmacher, 2004. Cyanobacteria and cyanobacterial toxins in three alkaline Rift Valley lakes of Kenya-Lakes Bogoria, Nakuru and Elmenta-ita. Journal Plankton Research 26: 925–935.

Burgis, M. J. & P. Morris, 1987. The Natural History of Lakes. Cambridge University Press, Cambridge, 218.

Ganf, G. G. & P. Blazka, 1974. Oxygen uptake, ammonia and phosphate excretion by zooplankton of a shallow equatorial lake (Lake George, Uganda). Limnology and Oceanography 19: 313–325.

Grant, W. D., W. E. Mwatha & B. E. Jones, 1990. Alkaliphiles: ecology, diversity and applications. FEMS Microbiology Reviews 75: 255–270.

Harper, D. M., R. B. Childress, M. M. Harper, R. R. Boar, P. Hickey, S. C. Mills, N. Otieno, T. Drane, E. Vareschi, O. Nasirwa, W. E. Mwatha, J. P. E. C. Darlington & X. Escute-Gasulla, 2003. Aquatic biodiversity and saline lakes: Lake Bogoria National Reserve, Kenya. Hydrobiologia 500: 259–276.

James, M. R., 1987. Seasonal changes in metabolic rates of zooplankton and their importance to phytoplankton production in two New Zealand lakes of contrasting trophic state. New Zealand Journal of Marine and Freshwater Research 21: 573–583.

Jones, B. E. & W. D. Grant, 1999. Microbial diversity and ecology of the Soda Lakes of East Africa. In Bell, C. R., Brylinsky M. & J. P. Johnson-Green (eds), Microbial Biosystems: New Frontiers: Proceedings of the 8th International Symposium for Microbial Ecology (2000). Atlantic Canada Society for Microbial Ecology, Halifax, Canada, 681–687.

Kilham, P., 1981. Pelagic bacteria: extreme abundances in African saline lakes. Naturwissenschaften 68: 380–381.

Legler, C., 1988. Ausgewählte Methoden der Wasseruntersuchung. Band 1. Gustav Fischer Verlag Jena, 517 pp.

McCall, J. G. H., 1967. Geology of the Nakuru-Thomson's Fall - Lake Hannington Area. Republic of Kenya; Ministry of Natural Resources: Geological Survey of Kenya. Report Number 78, 122 pp.

Melack, J. M., 1976. Temporal variability of phytoplankton in tropical lakes. Oecologia 44: 1–7.

Melack, J. M., 1981. Photosynthetic activity of phytopalnkton in tropical African soda lakes. Hydrobiologia 81: 71–85.

Melack, J. M., 1988. Primary producer dynamics associated with evaporative concentration in a shallow, equatorial soda lake (Lake Elmenteita, Kenya). Hydrobiologia 158: 1–14.

Milbrink, G., 1974. On the limnology of two alkaline lakes (Nakuru and Naivasha) in the Rift Valley System in Kenya. Interantionale Revue der gesamten Hydrobiologie 62: 1–17.

Mwaura, F. & T. R. Moore, 1991. Forest and woodland depletion in the Lake Elementeita Basin, Kenya. Geoforum 22: 17–26.

Owino, A. O., J. O. Oyugi, O. O. Nasirwa & L. A. Bennun, 2001. Patterns of variation in waterbird numbers on four Rift Valley lakes in Kenya, 1991–1999. Hydrobiologia 458: 45–53.

Peters, H. R. & S. MacIntyre, 1976. Orthophosphate turnover in East African Lakes. Oecologia. 25: 313–319.

Richardson, J. L. & A. E. Richardson, 1972. History of an african rift valley lake and its climatic implications. Ecological monographs 42: 499–534.

Rippey, B. & R. B. Wood, 1985. Trends in major ion composition of five Bishoftu crater lakes. Ethiopian Journal of Science 8: 9–28.

Schagerl, M. & S. O. Oduor, 2007. Phytoplankton communities relationships to environmental variables in three Kenyan Rift Valley saline-alkaline lakes. Submitted to Limnology and Oceanography.

Sondergaard, M., E. Jeppesen, P. Kristensen & O. Sortkjaer, 1990. Interactions between sediment and water in a shallow and hypertrophic lake: a study on phytoplankton collapses in Lake Sobygard, Denmark. Hydrobiologia 191: 139–148.

Talling, J. F & D. Driver, 1961. Some problems in the estimation of chlorophyll-a in phytoplankton. In Proceedings of the conference on primary productivity measurement, marine and freshwater, held at University of Hawaii, August 21–September 6, 1961.

Talling, J. F. & J. Lemoalle, 1998. Ecological dynamics of tropical inland waters. Cambridge University Press, 441 pp.

Talling, J. F., 2001. Environmental controls on the functioning of shallow tropical lakes. Hydrobiologia 458: 1–8.

Vareschi, E., 1987. Saline lake ecosystems. In Schulze, E. D. & H. Zwölfer (eds). Potentials and Limitations of Ecosystem Analysis. Springer-Verlag Publishers, Berlin, 435 pp.

Vareschi, E., 1982. The ecology of Lake Nakuru (Kenya) III. Abiotic factors and primary production. Oecologia 55: 81–101 pp.

Vareschi, E., J. M. Melack & P. Kilham, 1981. Saline waters. In Symoens, J. J., M. J. Burgis & J. J. Gaudet (eds), The Ecology and Urilization of African Inland Waters. UNEP Reports and Proceeding Series No. 1, Nairobi, 93–102.

Von Damm, K. L. & J. M. Edmond, 1984. Reverse weathering in The closed basin lakes of the Ethiopian Rift. American Journal of Science 284: 835–862.

Wood, R. B. & J. F. Talling, 1988. Chemical and algal relationships in a salinity series of Ethiopian inland waters. Hydrobiologia 158: 29–67.

Yasindi, A. W., D. H. Lynn & W. D. Taylor, 2002. Ciliated protozoa in Lake Nakuru, a shallow alkaline-saline lake in Kenya: Seasonal variation, potential production and role in the food web. Archiv für Hydrobiologie 154: 311–325.

Hydrobiologia (2007) 584:69–76
DOI 10.1007/s10750-007-0578-9

SHALLOW LAKES

Elongation and mat formation of *Chara aspera* under different light and salinity conditions

Irmgard Blindow · Manuela Schütte

Abstract Former laboratory results indicate that shoot elongation at low light intensities of *Chara aspera* is absent already at 10 psu which is within the physiologically optimal salinity range for brackish water populations. To investigate if similar restrictions occur in the field, density and morphology of *C. aspera* were compared between three freshwater and three brackish water sites along its depth range.

The lower depth limit of *C. aspera* varied considerably among sites (30–600 cm) related to turbidity. Light availability at the lower depth limit corresponded to about 15% of surface irradiance in freshwater and brackish water with lower salinity (3.4 psu). Total length increased and fresh weight:length ratio decreased with depth at these sites indicating shoot elongation related to lower light availability. Due to shoot elongation, light availability was far higher at the upper parts of the shoot than at the bottom in the turbid sites. Light availability at the lower depth limit was higher (about 40%) at two sites with higher salinity (7–8 psu), where no shoot elongation was observed at the lower depth limit. Instead, the plants were stunted and often covered with filamentous algae or shaded by other rooted submerged macrophytes indicating competitive disadvantages of *C. aspera* at higher salinities.

As growth in high densities (mat formation) exposes the plants to severe self-shading, it is suggested that shoot elongation is a prerequisite to mat formation. Dense vegetation of *C. aspera* was found only in freshwater and brackish water with lower salinity. Single, richly branched plants occurred in clearwater sites with higher salinity. *C aspera* was not found in "double stress" environments with both high turbidity and high salinity: We asume that the species is a poor competitor under these conditions.

Our results indicate that morphological differences between freshwater and brackish water populations of *C. aspera* are at least partly explained by salinity rather than genetic differences.

Keywords Shoot elongation · Density · Alternate states · Freshwater · Brackish water

Guest editors: R. D. Gulati, E. Lammens, N. De Pauw & E. Van Donk
Shallow lakes in a changing world

I. Blindow (✉) · M. Schütte
Biological Station of Hiddensee, University of Greifswald, Biologenweg 15, 18 565 Kloster, Germany
e-mail: blindi@uni-greifswald.de

Introduction

Chara aspera Willd. is a cosmopolitan (Croy, 1982) which occurs over a wide salinity range (Hasslow, 1931; Olsen, 1944; Blindow, 2000).

Recent laboratory investigations revealed that salinity optima differ between freshwater and brackish water populations even after a longer adaptation period (Blindow et al., 2003). A certain level of genetic isolation between these groups can therefore be assumed which seems to be supported by the existence of morphological differences such as a more "stunted" growth of brackish water plants (Olsen, 1944) and formation of mats which is a common phenomenon in freshwater populations (Blindow, 1992a; van den Berg et al., 1998), but seems to lack at higher salinities (own observations).

In one aspect, however, freshwater and brackish water populations responded equally to different salinity conditions (Blindow et al., 2003): When incubated under low light intensities (7% of natural light), plants collected from both freshwater and brackish water elongated at lower salinities (0 and 5 psu), but shoot elongation was lacking or small under higher salinities (10 and 20 psu). Salinity stress is not a satisfying explanation as weight gain, photosynthesis rates and fertility were higher at 5 and 10 psu compared to 0 psu for plants collected from brackish water (Blindow et al., 2003). These observations confirm laboratory results by Winter & Kirst (1990) showing that salinity increase resulted in a failure of internode elongation in *Chara vulgaris* before cell division rates were affected.

In freshwater lakes, charophytes can dominate among submerged macrophytes and act as a "key factor" which has major influence on all trophic levels (Hargeby et al., 1994; van den Berg et al., 1998) due to mat formation and high biomass (Blindow, 1992b). Under clearwater conditions, these mats can reach far down, and charophytes can be found on much deeper water than angiosperms (Chambers & Kalff, 1985). In brackish water, the ecological role of charophytes is poorly understood. Though some studies (Martin & Torn, 2004; Munsterhjelm, 2005) show that these plants also can form substantial biomass in shallow brackish lagoons, inhibition of shoot elongation and / or mat formation could restrict these plants from higher salinities and reduce their competitive abilities especially under more turbid conditions. Based on the laboratory studies mentioned above (Winter & Kirst, 1990; Blindow et al., 2003), we hypothesize that low light availability and high salinity represent a double stress situation for *C. aspera*. To test for the relevance of these findings for the field situation, we investigated occurrence, density and morphology of *C. aspera* along depth gradients in different freshwater and brackish water habitats.

Material and methods

Site description

Chara aspera was collected from three freshwater and three brackish water sites. Lake Storacksen (60°56′ N, 15°12′ E), province of Dalarna (middle Sweden), is a calcium-rich, oligotrophic lake. Lake Krankesjön (55°42′ N, 13°29′ E), province of Scania, south Sweden, is a calcium-rich, meso- to eutrophic lake (see Blindow et al., 2002). Lake Börringesjön (55°29′ N, 13°20′ E), Scania, is highly eutrophic with algal blooms throughout the vegetation period. The brackish water sites were Höllviken bay (55°25′ N, 12°56′ E) and Edenryd bay (56°02′ N, 14°32′ E), both situated in Scania, and Redensee bay (54°22′ N, 12°35′ E) in the Darß-Zingster bodden chain, Germany.

Sampling and analysis

Apart from Lake Storacksen and Redensee bay which were only visited once the sites were sampled twice during the vegetation period of 2002. Light was measured with a photometer (LI-COR) at 20 and 40 cm depth. Out of three replicate measurements, the vertical light attenuation (K) was obtained out of the slope of the log-transformed light measurements and used to calculate light availability as percent surface light (% SI) at the sediment surface and at the shoot tips. The salinity of all brackish water sites was measured directly in the surface water with a conductometer (WTW, LF 235) with automatic temperature compensation.

On the first visit at each site, the depth range of *C. aspera* was investigated. A transect was laid from the upper range (= minimum depth of

C. aspera) down to the lower range (= maximum depth of *C. aspera*). Three plots were selected and marked along the transect: Upper range, lower range and middle range (between upper and lower range). During each sampling occasion, composition and percentage coverage of the submerged vegetation was investigated by means of frames (1 × 1m) which were put at random inside each plot (five replicates). Five individual plants of *C. aspera* were collected at random from each plot for determination of length and weight parameters. All plant samples were stored in a cooling box or refrigerator, respectively, and brought to the laboratory within a few days. The total length of individual plants was measured (distance from sediment surface to the top of the plants). Fresh weight was determined after drying the plants surficially (household tissue).

Statistics

Values for coverage (%), plant length and plant fresh weight were not normally distributed, and homoscedascidity (*F*-test) could not be achieved even after applying different transformations (log, root, arcsine). Therefore, non-parametric tests (Spearman rank, two-tailed) were chosen to calculate the relationships between these factors and environmental variables (salinity, depth, light), while two-factor univariate ANOVA and Scheffé post hoc were used to calculate the relationship between log-transformed weight: length ratios and salinity and depth range, respectively (all values tested for homoscedasticity, F-test, $P > 0.05$). Linear regression analysis was used to calculate the dependency of the weight: length ratio from depth. All tests were performed with SPSS (version 12.0.1).

Results

Water transparency varied considerably among sites (Table 1). While Lake Storacksen, Höllviken bay and Edenryd bay had clear water, Lake Börringesjön was highly turbid, and Lake Krankesjön as well as Redensee bay had intermediate values. Among brackish water sites, salinity was lower in Redensee bay compared to Höllviken and Edenryd bays (Table 1).

The maximum depth of *C. aspera* varied considerably among sites, ranging from 30–40 cm (Lake Börringesjön) to 6 m (L. Storacksen) (Fig. 1). Percentage coverage of *C. aspera* (all sites and sampling occasions combined) was negatively correlated with salinity ($R = -0{,}464$, $P < 0.01$). In the three freshwater lakes and Redensee bay, *C. aspera* formed dense mats covering between 60 and 100% of the lake bottom. In the brackish water sites Höllviken and Edenryd bays, the plants grew single, and their coverage did not exceed 20%. While *C. aspera* was the dominant submerged macrophyte along its depth range in all three freshwater lakes, other plants (*Zostera marina* and other angiosperms, filamentous algae) dominated in Höllviken and Edenryd bays especially in deeper water.

Total length of *C. aspera* was negatively correlated with both salinity and light availability ($P < 0.01$). Total length increased with depth on all three freshwater sites ($P < 0.001$, all sites combined) and in Redensee bay ($P < 0.01$), while no such increase was observed for plants from

Table 1 Light and salinity characteristics as well as sampling dates for both freshwater sites (F) and brackish water sites (B).

Site	Date	Salinity (psu)	Light attenuation (K)
L. Storacksen (F)	23 August		0.9
L. Krankesjön (F)	14 June		1.7
	2 August		1.7
L. Börringesjön (F)	12 June		9.8
	30 July		8.9
Höllviken bay (B)	11 June	8.5	0.6
	31 July		0.8
Edenryd bay (B)	13 June	7.6	0.7
	1 August		0.6
Redensee bay (B)	30 August	3.4	2.2

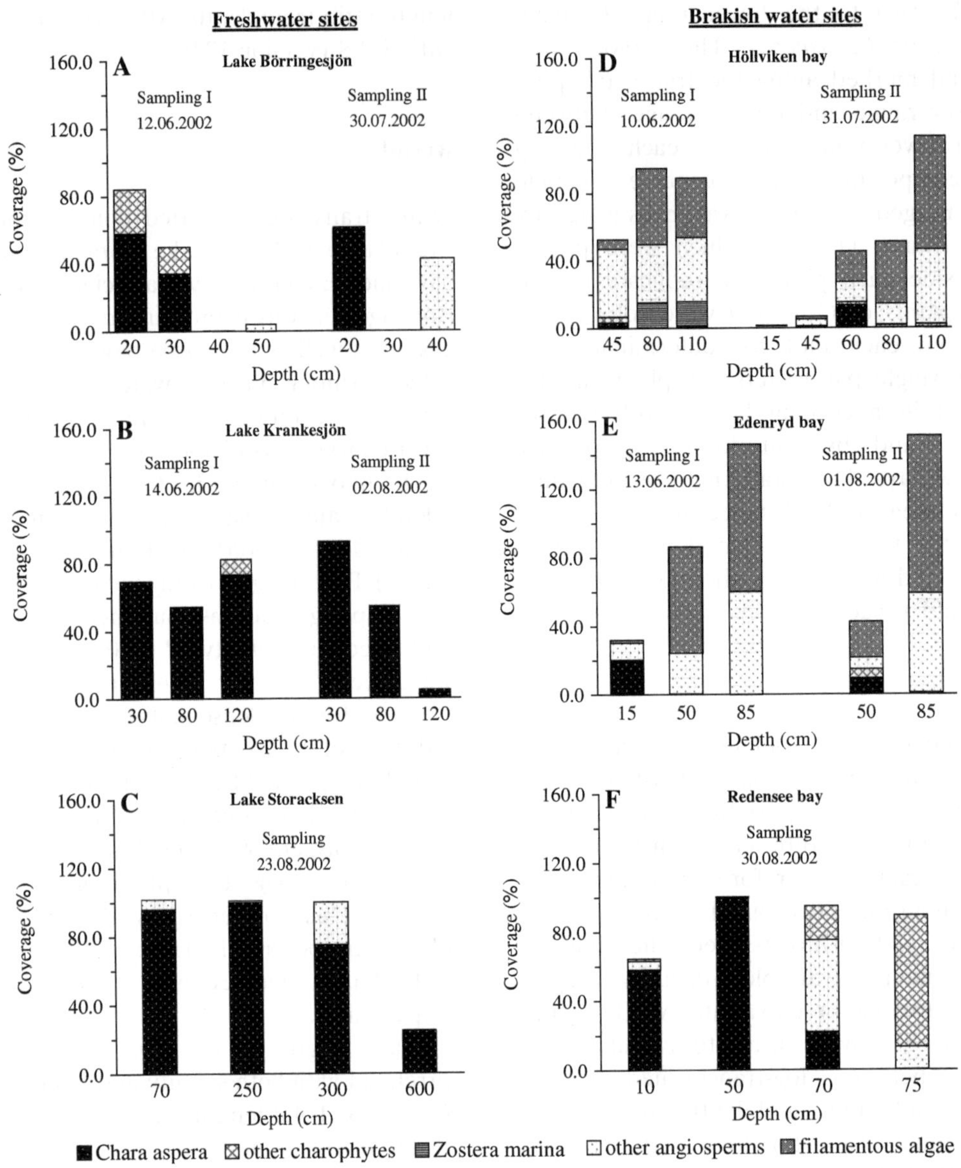

Fig. 1 Vegetation structure along the depth range of *C. aspera*. Left panel: freshwater sites, right panel: brackish water sites. Total coverage can reach more than 100% as coverage values for single species, which can have shaded each other, have been added. All values shown are mean values of five replicate 1 × 1 m vegetation plots

Höllviken and Edenryd bays ($P = 0.28$, both sites combined) (Fig. 2).

Individual plant fresh weight was positively correlated with depth, salinity and light availability ($P < 0.01$). While salinity had a significant effect on weight:length ratio ($P < 0.001$), the effect of depth was not significant ($P = 0.717$, Fig. 3). The interaction effect of salinity * depth on weight:length ratio was significant ($P = 0.005$). There was no significant difference in weight: length ratios between plants from Höllviken and Edenryd bays ($P = 0.208$), but plants collected from any other site had lower weight:length ratios ($P < 0.05$ for all comparisons).

While plants from Höllviken and Edenryd bays as well as plants from Lake Börringesjön had about the same length and weight:length ratio along their depth range, plants from Redensee

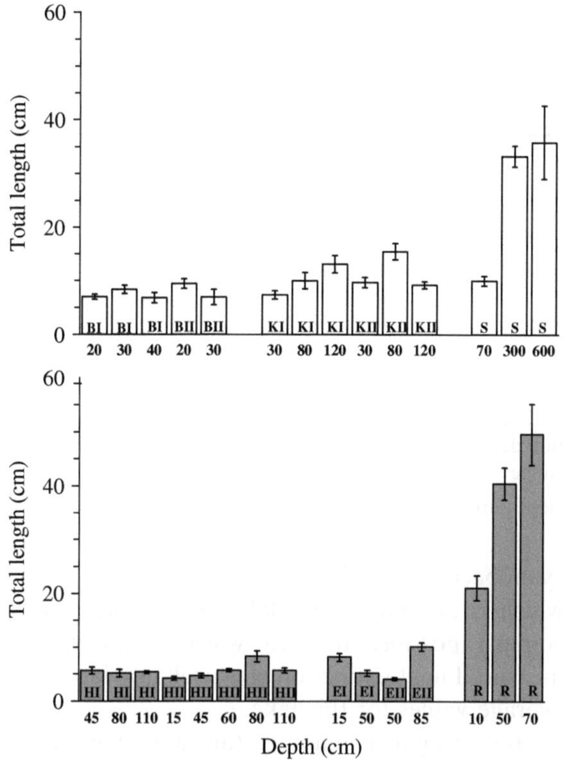

Fig. 2 Total length of *C. aspera* (mean values ± S.E., $n = 5$), determined at all sites, depth ranges and sampling occasions. Above: freshwater sites, below: brackish water sites. K = Lake Krankesjön, B = L. Börringesjön, S = L. Storacksen, E = Edenryd bay, H = Höllviken bay, R = Redensee bay. I and II = first and second sampling occasion, respectively

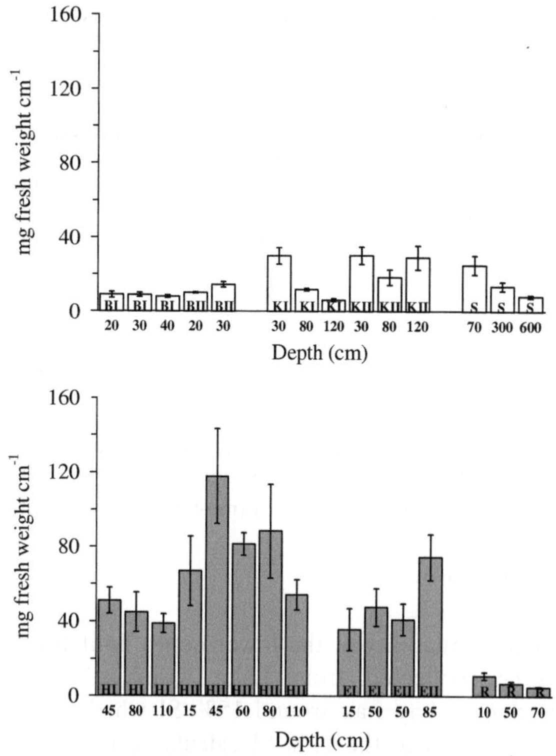

Fig. 3 Fresh weight:total length ratio of *C. aspera*, determined at all sites, depth ranges and sampling occasions. For further explanations, see Fig. 2

bay, Lake Storacksen and L. Krankesjön (first sampling occasion) became thinner and longer with increasing depth. Plants from Lake Krankesjön (second sampling occasion) became longer with depth, but not thinner (Fig. 2, 3).

The weight:length ratio decreased with decreasing light availability for plants collected from freshwater ($P < 0.01$), but did not change significantly ($P = 0.462$) for plants collected from brackish water (Fig. 4). Weight:length ratios were lower for plants collected from Redensee bay compared to plants collected from the two other brackish water sites.

Light availability at the sediment surface along the lower depth limit of *C. aspera* was lower than 20% SI in the freshwater sites and Redensee bay, but about 30–65% SI in Höllviken and Edenryd bays (Table 2). Especially in turbid sites (Redensee bay, Lake Börringesjön) the shoot tips received far more light than the lower plant parts. The lowest values both at upper and lower plant parts (<1% SI) were recorded for Lake Storacksen (Table 2).

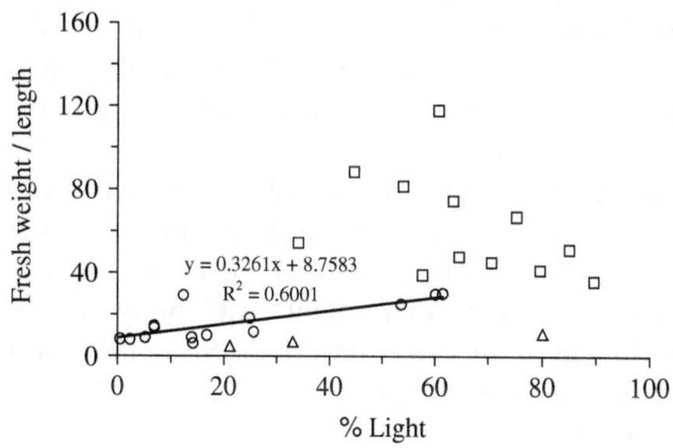

Fig. 4 Fresh weight:total length ratio plotted against the light availability (% of surface irradiance) for all sites, sampling occasions and depth ranges. Circles: freshwater sites, squares: brackish water sites except Redensee bay, triangles: Redensee bay. The regression line and equation indicate the dependency of weight:length ratio on light availability (freshwater sites only). All symbols represent means (n=5)

Table 2 Light availability (% of surface irradiance, calculated from measured light attenuation) for different sampling sites and sampling occasions

Site	Date	Max range (cm)	Light at max range (%)	Light at plant surface (%)
L. Krankesjön	14 June	115	14	18
	2 August	120	13	15
L.Börringesjön	12 June	38	2.4	4.5[a]
	30 July	30	5.2	12
L. Storacksen	23 August	600	0.5	0.7
Höllviken bay	11 June	95	58	59
	31 July	140	34	35
Edenryd bay	13 June	60	65	69
	1 August	80	64	67
Redensee bay	30 August	70	16	58

Light at max range = light availability at the bottom of the maximum depth range of *C. aspera*. Light at plant surface = light availability at the shoot tips of *C. aspera* at its maximum depth range.

[a] = the plants had disappeared (most probably died) from this depth after the first sampling occasion.

Discussion

Light availability at the lower depth limit for *C. aspera* collected from Lake Krankesjön and Redensee bay was around 15% SI which corresponds well to the 11% SI calculated for charophytes by Chambers & Kalff (1985). Much lower values were measured in turbid Lake Börringesjön. During June, plants were found down to 40 cm depth where light availability was about 2% SI. No *C. aspera* was found on this depth during the second sampling occasion (end of July), suggesting that the plants had died back due to low light availability. At 30 cm depth, light availability was only around 5% SI at the sediment surface, but around 12% at the shoot tips which obviously was sufficient to maintain the dense mat of *C. aspera* on this depth. The advantage which submerged macrophytes can gain from shoot elongation in turbid water (Chambers & Kalff, 1987) also becomes clear in Redensee Bay where light availability at the sediment surface was only 16% SI at the lower depth range of *C. aspera*, but about 60% SI at the shoot tips.

Light availability was less than 1% SI at the lower depth limit (6 m) in clearwater Lake Storacksen. *C. aspera* hibernates as green plant on this depth, and new shoot tips were short (just a few cm) during end of August indicating that this perennial growth form may extend the depth range. Additionally, there were high densities of cyanobacteria during the sampling occasion, which is unusual in this lake (K. Leander, pers. comm.) possibly due to warm weather, and measured light extinction was probably far above average values for this lake.

Higher light availability (around 35–65% SI) was observed at the maximum depth limit of *C. aspera* in Höllviken and Edenryd bays, which had the highest salinities of all sites studied. These values correspond well to the lower light limits (around 40–45% SI) given for other charophytes in brackish water (Domin et al., 2004). Higher energy demand due to the necessity of turgor regulation is just one explanation for this restriction of charophytes from deeper water. Absence of shoot elongation may be another important factor: Stunted growth not only reduces light availability at the shoot tips, but also allows for coverage by filamentous algae and shading by other rooted submerged macrophytes. In Höllviken bay, *C. aspera* was shaded by *Zostera marina* on its lower depth range and in deeper water replaced by this angiosperm, which is adapted to higher salinities and does not show growth inhibition at the relatively low salinities of the Baltic Sea (Kamermans et al., 1999; Moore et al., 2000). We were not able to find any *C. aspera* at sites with both high salinity and high turbidity (unpublished results) which indicates that the species can not compete with other primary producers under these conditions.

Shoot elongation was observed at all freshwater sites and Redensee bay, which had lower salinity (3.4 psu) than Höllviken and Edenryd bays, and was indicated by an increasing total length, and decreasing weight:length ratio with increasing depth. In Lake Krankesjön, weight: length ratio decreased with depth only during the first sampling occasion, but not during the second, suggesting that deep-water plants improve their light availability early in the season by favouring elongation and gain in weight when the shoot tips have reached higher light availabilities.

Absence of shoot elongation at higher salinities corresponds well to former laboratory results which show that shoot elongation occurs at 0 and 5 psu, but not at 10 and 20 psu (Blindow et al., 2003). Mat formation was observed on all sites where shoot elongation occurred, but not in Höllviken and Edenryd bays where *C. aspera* had stunted growth. As growth in high densities (mat formation) exposes the plants to severe self-shading, it is possible that shoot elongation is a prerequisite to mat formation. Such mat formation has also been observed in other brackish water lagoons with lower salinities like the Uppland archipelago in middle Sweden (Wallström & Persson, 1997). While laboratory results indicate genetic differentiation between freshwater and brackish water populations of *C. aspera* (Blindow et al., 2003), this study shows that environmental factors sufficiently can explain the morphological differences which have been described between these habitats (Olsen, 1944).

Though *C. aspera* specimens from brackish water have their highest photosynthesis rate and weight gain at 5–10 psu (Blindow et al., 2003), and perfect turgor regulation up to 8 psu (Winter & Kirst, 1992), this plant is not well adapted to higher salinities. We suggest that *C. aspera* and most other charophytes have competitive disadvantages at higher salinities and can not be a dominant part of the submerged vegetation like in freshwater and brackish water of lower salinities. This seems to contradict assumptions derived from analysis of turgor regulation (Winter & Kirst, 1991) that charophytes may originate from marine environments and first thereafter, colonized brackish water and freshwater.

Conclusion

Our results indicate that both turbidity and salinity act as stress factors on *C. aspera*. In clearwater lakes, the species can occur in high densities and reach deep water, where the ability to hibernate as a green plant together with shoot elongation may further extend the lower depth limit. In turbid lakes, the plants can still form dense mats, but are restricted to shallow water by poor light availability, although shoot elongation may allow a certain extension of the depth range. At salinities above 6 psu, *C. aspera* can be common under clearwater conditions, but does not form mats. Absence of shoot elongation exposes the plants to coverage by filamentous algae and shading by other submerged macrophytes and restricts it from deeper water.

Acknowledgements We thank Nils Möllmann for assistance in the field and laboratory and Peter Hansson for diving assistance in Lake Storacksen. The study was financed by the German Research Foundation (DF-G, project no. BL 559/3–1).

References

Blindow, I., 1992a. Decline of charophytes during eutrophication: comparison with angiosperms. Freshwater Biology 28: 9–14.

Blindow, I., 1992b. Long- and short-term dynamics of submerged macrophytes in two shallow eutrophic lakes. Freshwater Biology 28: 15–27.

Blindow, I., 2000. Distribution of charophytes along the Swedish coast in relation to salinity and eutrophication. International Reviews in Hydrobiology 85: 707–717.

Blindow, I., J. Dietrich, N. Möllmann & H. Schubert, 2003. Growth, photosynthesis and fertility of *Chara aspera* under different light and salinity conditions. Aquatic Botany 76: 213–234.

Blindow, I., A. Hargeby & G. Andersson, 2002. Seasonal changes of mechanisms maintaining clear water in a shallow lake with abundant *Chara* vegetation. Aquatic Botany 72: 315–334.

Chambers, P. A. & J. Kalff, 1985. Depth distribution and biomass of submersed aquatic macrophytes communities in relation to Secchi depth. Canadian Journal of Fisheries and Aquatic Sciences 42: 701–709.

Chambers, P. A. & J. Kalff, 1987. Light and nutrients in the control of aquatic plant community structure. I. In situ experiments. Journal of Ecology 75: 611–619.

Croy, C. D., 1982. *Chara aspera* (Charophyta). Breeding pattern in the northern hemisphere. Phycologia 21: 243–246.

Domin, A., H. Schubert, J. C. Krause & U. Schiewer, 2004. Modelling of pristine depth limits for macrophyte growth in the southern Baltic Sea. Hydrobiologia 514: 29–39.

Hargeby, A., G. Andersson, I. Blindow & S. Johansson, 1994. Trophic web structure in a shallow eutrophic lake during a dominance shift from phytoplankton to submerged macrophytes. Hydrobiologia 279/280: 83–90.

Hasslow, O. J., 1931. Sveriges characéer. Botaniska Notiser 1931: 63–136.

Kamermans, P., M. A. Hemminga & D. J. de Jong, 1999. Significance of salinity and silicon levels for growth of a formerly estuarine eelgrass (*Zostera marina*) population (Lake Grevelingen, the Netherlands). Marine Biology 133: 527–539.

Martin, G. & K. Torn, 2004. Classification and description of phytobenthic communities in the waters of the West-Estonian Archipelago Sea. Hydrobiologia 514: 151–162.

Moore, K. A., D. J. Wilcox & R. J. Orth, 2000. Analysis of the abundance of submersed aquatic vegetation communities in the Chesapeake Bay. Estuaries 23: 115–127.

Munsterhjelm, R., 2005. Natural succession and human-induced changes in the soft-bottom macrovegetation of shallow brackish bays on the southern coast of Finland. Walter and Andrée de Nottbeck Foundation. Scientific Reports No. 26. Helsinki.

Olsen, S., 1944. Danish Charophyta. Det Kongelige Danske Videnskabernes Selskab. Biologiske Skrifter, Bind III, Nr. 1. Copenhagen.

van den Berg, M., H. Coops, M. -L. Meijer, M. Scheffer & J. Simons, 1998. Clear water associated with a dense Chara vegetation in the shallow and turbid Lake Veluwemeer, The Netherlands. In Jeppesen, E., Ma. Søndergaard, Mo. Søndergaard & K. Christoffersen (eds), The Structuring Role of Submerged Macrophytes in Lakes. Springer, 339–352.

Wallström, K. & J. Persson, 1997. Grunda havsvikar i Uppsala Län. Västra Öresundsgrepen. Stencil Nr. 12 (Report). Upplandsstiftelsen, Uppsala, 47 pp (in Swedish).

Winter, U. & G. O. Kirst, 1990. Salinity response of a freshwater charophyte, *Chara vulgaris*. Plant Cell and Environment 13: 123–134.

Winter, U. & G. O. Kirst, 1991. Partial turgor pressure regulation in *Chara canescens* and its implications for a generalized hypothesis of salinity response in charophytes. Botanica Acta 104: 37–46.

Winter, U. & G.O. Kirst, 1992. Turgor pressure regulation in *Chara aspera* (Charophyta): the role of sucrose accumulation in fertile and sterile plants. Phycologia 31: 240–245.

Hydrobiologia (2007) 584:77–88
DOI 10.1007/s10750-007-0591-z

SHALLOW LAKES

Searching for allelopathic effects of submerged macrophytes on phytoplankton—state of the art and open questions

Elisabeth M. Gross · Sabine Hilt (nee Körner) · Paola Lombardo · Gabi Mulderij

Abstract Allelopathy, here defined as biochemical interactions between aquatic primary producers, has always been intriguing as a process explaining the dominance of certain plant or algal species over others. Negative chemical interference has been invoked as one of the steering mechanisms behind mutual dominance of either submerged macrophytes or phytoplankton in shallow eutrophic lakes. Yet, despite much effort, convincing evidence for allelopathic interactions in situ is still missing. Also, laboratory approaches often lack reality. Inspired by a series of talks at the Shallow Lakes 2005 meeting in Dalfsen, the Netherlands, we argue that there is circumstantial but strong evidence that allelopathic interference between submerged macrophytes and phytoplankton may indeed exist in aquatic ecosystems despite the problems associated with research in this field. We first discuss experimental approaches combining laboratory and field studies, based on examples presented at this meeting. We then discuss the impact of nutrient status of both producing and target organism and biotic factors such as herbivory or pathogens that might affect allelopathy. Further topics are the potential seasonality of effects and the species-specificity of certain allelochemicals. We conclude with some thoughts why a final

Guest editors: R. D. Gulati, E. Lammens, N. De Pauw & E. Van Donk
Shallow lakes in a changing world

E. M. Gross (✉)
Limnological Institute, Biology, University of Konstanz, P.O. Box 5560 M659, 78457 Konstanz, Germany
e-mail: Elisabeth.Gross@uni-konstanz.de

S. Hilt (nee Körner)
Leibniz-Institute of Freshwater Ecology and Inland Fisheries, Müggelseedamm 301, 12587 Berlin, Germany

P. Lombardo
Department of Genetics and Molecular Biology – Ecology Area, La Sapienza State University of Rome, via dei Sardi 70, 00185 Rome, Italy

G. Mulderij
Netherlands Institute of Ecology, Centre for Limnology, Rijksstraatweg 6, 3631 AC Nieuwersluis, The Netherlands

Present Address:
P. Lombardo
Department of Environmental Sciences, Ecology Section, Coppito Campus, University of L'Aquila, via Vetoio, Coppito, 67100 L'Aquila, Italy

Present Address:
G. Mulderij
Koeman en Bijkerk, ecological research and advise, Kerklaan 30, 9750 AA Haren, The Netherlands

proof for allelopathy in situ might remain difficult or even inaccessible in some cases, and why we nevertheless should not abandon this idea.

Keywords Allelochemical interaction · Algicidal activity · Chemical interference · Culture filtrate · Exudation

Introduction

Even before the Vienna botanist Hans Molisch coined the term allelopathy to describe biochemical interactions among plants and between plants and microorganisms (Molisch, 1937), some studies reported negative effects of one primary producer on another not related to resource competition (Harder, 1917; Akehurst, 1931), and multiple studies followed (e.g., Pratt et al., 1944; Keating, 1977). Although by its original definition, stimulatory and inhibitory allelopathic effects might occur, most studies report negative interactions. Some early prominent reports on detrimental chemical effects of aquatic plants on microalgae are the studies by Hasler & Jones (1949), Fitzgerald (1969), Kogan & Chinnova (1972). Further information is compiled in the reviews by Gopal & Goel (1993) and Gross (2003a).

We now have ample evidence that low phytoplankton and epiphyte densities in shallow eutrophic lakes might be the result of complex abiotic and biotic feedback mechanisms, and allelopathy is considered as one alternative mechanism (Scheffer et al., 1993). The importance of pelagic and benthic trophic interactions, as well as physical and chemical conditions such as resuspension or sedimentation, shading or increased carbonate (and phosphate) precipitation have all been investigated at a wide range of spatial and temporal scales, with laboratory, mesocosm and field experiments.

In contrast, proving evidence for allelopathy still remains difficult. According to Willis (1985), six prerequisites have to be met to unequivocally show the occurrence of allelopathy: (1) a pattern of inhibition of target plant(s) or alga(e), (2) allelopathical compound(s) produced by donor plants, (3) the release of these compounds by the producing plant, (4) their transport and/or accumulation in the environment, (5) the uptake by the target organism(s) and (6) the inhibition cannot be explained solely by other physical or biotic factors, especially herbivory and competition. Consequently, experimental approaches on aquatic allelopathy need to address all these aspects.

Studies investigating macrophyte-phytoplankton interactions in shallow, eutrophic lakes have suggested allelopathy as a (co-)factor in explaining plant influence, up to dominance, over phytoplankton at high (i.e., non-limiting) nutrient levels, at least when known allelopathic plant species and/or sensitive phytoplankton species are involved (Phillips et al., 1978; Körner & Nicklisch, 2002; van Donk & van de Bund, 2002). Other studies cast doubts on the natural strength, or even the existence of in situ allelopathic interference among primary producers (Forsberg et al., 1990). An understanding of the occurrence, strength, and mode(s) of action of allelopathy in natural situations may therefore result in a major contribution to the understanding and management of eutrophic, shallow lakes. Though allelopathic interactions may exist between all types of primary producers, we specifically focus on plant allelopathy on phytoplankton, as this aspect has the highest applicability to shallow lake ecology.

Experimental approaches, possible insights and inherent problems

The following section will present frequently used methods to address allelopathy between plants and algae in aquatic systems. While any one method alone cannot be used as proof of allelopathic effects in situ, a combination of those methods might present insight into the strength of allelopathic interactions, the mechanisms involved and explain the variability of responses.

Plant homogenates or extracts

Some authors use plant homogenates or extracts to test for allelopathy. Fresh or dried macrophytes might be homogenated with water, incubated for a certain period, filtered and then the filtrate is added to target organisms. Controls receive only water or nothing.

Water is assumed to be the most natural solvent since the plants live in water and have no well-developed cuticle, thus should leach active compounds into the water. Elakovich and colleagues (Elakovich & Wooten, 1989, 1995; Wooten & Elakovich, 1991) used fresh macrophytes homogenised with water and the filtrate was then used in bioassays against lettuce seedlings and duckweed. Besides the problems mentioned below when using water as solvent, the utilised target species did not allow ecologically important extrapolations. Problems when using water might arise from osmotic changes caused by the addition of fresh tissue homogenates, including e.g., sugars. If heterotrophic bacteria are present in the target organism culture, they might use these substrates and probably also the inhibitory compounds as co-substrates.

Gross (1999) argued that moderately lipophilic solvents such as methanol, ethanol and acetone or mixtures of those solvents with water are also appropriate to extract more lipophilic compounds that might be transferred by direct cell-cell contact or transported via micelles or aggregates. Using concentrated solvent extracts avoids some of the above mentioned problems. Extracts might be separated further by liquid–liquid or solid phase extraction to distinguish whether hydrophilic or lipophilic compounds are active. Further separation by chromatography (TLC - thin layer chromatography or HPLC - high performance liquid chromatography) guided by bioassays may ultimately lead to the identification of active compounds and their mode(s) of action (Gross et al., 1996; Leu et al., 2002).

Extract-based methods offer the chance to compare the allelopathic potential of different macrophyte species or seasonal differences in one species in a standardised way. The direct application of extracts also enables comparison of the response of nutrient limited and non-limited phytoplankton in order to test whether additional stress of target organisms results in a higher sensitivity against allelopathic substances (Reigosa et al., 1999). Additionally, the producing plants can be cultivated under different nutrient and light conditions to study the impact of resource availability on the production of allelopathic compounds (Gross, 2003b).

Bioassay-directed extractions of *Ceratophyllum demersum*, *Elodea* spp., *Myriophyllum* spp., *Najas marina* and *Stratiotes aloides* with different solvents and solvent mixtures (water, methanol, and acetone; pure or mixed with water) often revealed the highest activity against cyanobacteria when solvents with intermediate lipophilic properties were used, e.g., 50% (v/v) acetone or 70% (v/v) methanol in water (Gross et al., 1996, 2003; Erhard & Gross, 2006; Mulderij et al., 2007). Under these conditions both hydrophilic and moderately lipophilic compounds can be extracted from the plant tissue. This method will result in compounds that in general have a low molecular weight, are fairly water-soluble, and can bind to and/or pass through cell membranes. Further fractionation and separation of extracts by solid phase extraction (SPE) and HPLC can ultimately lead to the bioassay-guided isolation of the active compound(s). These compounds can then also be tracked in culture filtrate or extracts from culture filtrates.

Experiments conducted with extracts certainly give no convincing ecological proof for allelopathic interactions. The inherent difficulties to extract active compounds from a complex matrix of organic and inorganic compounds in the water, however, do not allow the easy identification of released active compounds. Since all exuded active compounds, or their precursors, should be present in much higher concentrations in the plant tissue, working with extracts and a bioassay-directed fractionation using realistic concentrations (i.e., comparable to plant biomass per volume in situ) might provide information about active compounds, and facilitate their structure elucidation. This in turn may facilitate the evaluation of the mode of action (for discussion see Gross, 1999, 2003a). Such studies should only be presented accompanied with experiments using culture filtrate (exudate) from the active macrophytes.

Coexistence experiments

Comparing growth of co-cultivated plants and/or algae to growth in single-species controls is a classical set-up to test for reciprocal interactions. Especially when varying densities of both species are applied, the combined results provide information on positive or negative interactions, facilitation or inhibition. Such experiments have been widely applied in plant ecology as "De Wit replacement series" (Gurevitch et al., 2002), and have also been used to study macrophyte-macrophyte interactions (Agami & Waisel, 1985). Batch cultures require defined experimental conditions of density and nutrient availability. In order to study the shading effect of coexisting species, either plastic structures resembling macrophytes placed into the growth vessels or nylon stockings mantled around the growth vessels can be used (Körner & Nicklisch, 2002; Erhard & Gross, 2006).

Coexistence experiments per se do not provide evidence—either direct or circumstantial—that allelopathy occurs. However, while coexistence experiments quantify the occurrence and strength of competitive-like interspecific interactions, parallel experiments with extracts (or filtrates) may shed light on the possible involvement of allelopathy.

Allelopathy has been invoked as a (co-)factor for changes in algal biomass or assemblage composition when target algae grown in the presence of macrophytes exhibited patterns different from similar cultures grown in the absence of macrophytes (Jasser, 1995; Nakai et al., 1999; Lombardo, Mjelde, Källqvist & Brettum, unpublished). Similar methods have been adopted for algal-algal (Sukenik et al., 2002) and plant-plant interactions (Elakovich & Wooten, 1995; Vance & Francko, 1997).

Though coexistence studies provide a direct evaluation of donor-target species interactions, allelopathy sensu strictu is difficult to isolate as a mechanism behind the observed patterns. Other factors such as presence of other organisms in non-axenic cultures, or competition for resources between coexisting species, may be involved, thus potentially confounding the results.

Only few studies describe the use of axenic macrophytes (e.g., Pringsheim & Pringsheim, 1962; Forsberg, 1965; Wetzel & McGregor, 1968; Ervin & Wetzel, 2000). Although axenic plants represent an artificial system, they can serve as controls that can then be compared to non-axenic plants. The use of axenic cultures for example enables the investigation of exudates (culture filtrates) that have not been subjected to microbial breakdown. Axenic cultures of *Myriophyllum spicatum* (surface-sterilised plants) have been established to investigate allelopathic interactions without interference of epiphytic bacteria, cyanobacteria and algae (Gross et al., 1996). Both light and nitrogen availability influenced the content and exudation of allelopathic polyphenols in axenic *M. spicatum* (Gross, 2003b).

Possible competition for light in coexistence experiments may be minimised by using sufficiently small aquaria that can be uniformly permeated by light or by using plastic plants in controls. Potential competition for nutrients may be a thornier issue, as both donor and target species may be actively absorbing, metabolising, and leaking nutrients during experiments. Knowledge about nutrient dynamics in coexistence experiments, coupled with direct measurements of potentially allelopathic compounds (e.g., extracts experiments), may greatly assist in the search for evidence of allelopathy. Though many studies report nutrient concentrations deliberately set at non-limiting levels to exclude competition for nutrients as a factor in donor-target species interactions (e.g., Nakai et al., 1999), coexistence investigations seldom address this issue directly.

Nutrient mediation, whether negatively affecting target species through competition or counteracting inhibition by allelopathy, depending on the physiological status donor and/or target species, cannot be excluded with certainty even at clearly non-limiting nutrient concentrations (Sukenik et al., 2002). Though plants may not be a nutrient (especially phosphorus) sink in the medium- or long-term (weeks to months: Lombardo & Cooke, 2003), net foliar uptake in the short term (days), at which allelopathy experiments are often performed, may be quite high (Pelton et al., 1998; Lombardo & Cooke, 2003). Though including nutrient determinations in coexistence experiments may not reveal competition for nutrients as a (co-)factor involved in investigations targeting allelopathy, it nonetheless offers some insight when gathering different pieces of evidence in multiple-approach, comprehensive allelopathy investigations.

Even more difficult is the determination of micronutrient availability. Micronutrients such as vitamins, required by some algae and cyanobacteria, including some species commonly used in bioassays (Wetzel, 2001), have never been addressed in allelopathy studies. The involvement of low-weight dissolved organic matter (DOM), continuously leaked by living plants in the surrounding medium (Søndergaard, 1981), also remains little explored. Multiple-approach allelopathy investigations that address at least some of these issues may provide useful insight in the mechanisms behind macrophyte-phytoplankton interactions.

Dialysis bag experiments

Dialysis bag experiments may be seen as a type of coexistence experiments in which organisms are physically separated by dialysis membranes that allow free movement of low molecular weight compounds. Advantages of this method are the possibility to test the influence of non-axenic plants on unialgal or axenic cultures without the risk of contamination with other algae and the applicability for in situ studies. Either individual species or whole lake samples are put in water into dialysis bags and placed into aquaria or a lake containing the other primary producer(s) under study.

Difficulties may arise from toxic effects of some commercially available dialysis membranes, a problem solved for example by using thoroughly rinsed synthetic sausage skins (Körner & Nicklisch, 2002). Knowledge about the potential size of the active compounds to select the appropriate pore size of dialysis membranes is also important. Even without a detailed chemical characterisation, we might expect most active compounds to be low molecular weight secondary metabolites (<1000 Da).

To control for shading effects, controls receive plastic plants or other shades such as perforated metal plates or nylon stockings. Alternatively, target specimens can be kept just below the water surface to minimise shading (Jasser, 1995). Further problems might arise from using plastic plants, which might not be an inert control since they can exude plasticisers or other potential toxic compounds.

In a controlled laboratory experiment using aquaria, the allelopathic effect of submerged macrophytes on different phytoplankton species kept in dialysis bags was investigated (Körner & Nicklisch, 2002). PAM fluorometry revealed that *M. spicatum* and *C. demersum* but not *Potamogeton pectinatus* interfered with PS II activity of four cyanobacteria, a diatom and a green alga in a species-specific manner. *M. spicatum* also inhibited the growth of certain target species. Recently, in situ dialysis bag experiments in a small lake with a dense stand of *M. verticillatum* revealed seasonal and species-specific allelopathic effects on different phytoplankton species (Hilt et al., 2006). Laboratory experiments, performed in parallel to this field study, investigating allelopathic interactions both under phosphorus limited and non-limiting conditions showed that phosphorus limitation might mask potential allelopathic effects for some of the target species. Two other studies investigating allelopathic interactions between submerged macrophytes and phytoplankton pinned down nutrient effects. *Synechococcus leopoliensis* growth was enhanced in the presence of *Elodea canadensis* in a small-scale coexistence study, and inhibitory effects of allelochemicals from *Stratiotes aloides* declined when target species were nutrient limited (potassium or phosphorus limitation; Mulderij et al., 2007).

Culture filtrates

Frequently, culture filtrates of macrophytes cultivated in aquaria or mesocosms, or even water from dense stands in the field, have been added to monospecific algal or cyanobacterial cultures (e.g., Kogan & Chinnova, 1972; Mulderij et al., 2003, 2005a, b) or to natural phytoplankton communities (Jasser, 1995; Mulderij et al., 2006). While addition to single cultures mainly investigated growth and performance of target species, changes in community composition due to differential sensitivity can be tested with lake phytoplankton samples. For controls, the same water or culture medium without macrophyte exudate, or water from the same lake, but outside the macrophyte stands can be used. In general, adjustment for nutrient uptake by plants is made by addition of at least the major nutrients such as phosphorus and nitrogen to the culture filtrate, which might lead to a slightly higher concentration in the treatments compared to the controls.

Exudate experiments often involve a one-time inoculum of the filtrate (e.g., Lombardo et al., unpublished), which contains high concentrations of the alleged allelochemicals under examination, to a culture of target species. Growth patterns of target species are compared with those in exudate-free controls, and if differences are observed, they are ascribed to allelochemicals. Though seemingly fail-proof, such approaches often fail to detect allelopathic effects, even if parallel coexistence experiments suggest at least a partial implication of such a mechanism (Lombardo et al., unpublished). Such shortcomings may be derived from flaws in the filtration procedure, or from loss of inhibitory effects due to adsorption to non-target surfaces or to volatility of some allelochemicals (Gross, 2003a).

At least some of these problems may be circumvented with multiple, sequential exudate additions during an experiment. This method should account also for potential microbial or photochemical degradation of active compounds. In fact, several authors (Nakai et al., 1999; Mulderij et al., 2003, 2005a, b, 2006) showed that only one initial addition of culture filtrate underestimated the allelopathic effect compared to coexistence assays or the (semi)-continuous addition of active culture filtrate.

As in other types of experiments, nutrient concentrations are rarely determined in exudate experiments. High nutrient concentrations (as often used in exudate experiments) may, however, counteract allelopathic inhibition of target species, thus potentially confounding the results. On the other hand, allelopathy experiments carried out at low nutrient levels may fail to show effects because target species may be nutrient limited in controls (Mulderij et al., 2007). As for other types of experiments, determinations of the availability of major nutrients (P, N), and possibly also micronutrients, is highly desirable in studies with exudates.

Extracting active compounds from the culture filtrate

To avoid difficulties in adjusting appropriate nutrient concentrations in treatments and controls, some authors extracted culture filtrate and control water with either organic solvents or solid phases. Previous studies showed that most active compounds are at least moderately lipophilic, thus can be extracted in e.g., diethyl ether or ethyl acetate, or can be trapped with reversed phase sorbens (e.g., C18) of SPE cartridges and filters. Such techniques have been successfully used to prove the allelopathic activity of culture filtrates from *C. demersum, Elodea nuttallii, M. spicatum* and *N. marina* (Gross et al., 1996, 2003; Erhard & Gross, 2006). C18-coated filters prevent the fast clogging of cartridges by solid or colloidal particles. These techniques do not require large volumes of organic solvents, which may themselves act inhibitory or contain toxic impurities, and can be adjusted to extracted water volumes from a few centilitres to a few litres. C18-bound active compounds can be eluted with solvent, usually methanol or ethanol. Enrichment of culture filtrate 5,000- to 20,000-fold is possible, allowing a controlled addition of exudate to defined media without differences in nutrient concentration. Using a step-wise elution of bound compounds with increasing methanol:water ratios in SPE or fractionation of exudates by HPLC ultimately leads to the isolation of active compounds and comparison to active compounds found in extracts.

Discussion

Proving allelopathic interaction in an ecologically meaningful way is not easy. None of the currently available methods per se addresses all of Willis' (1985) pre-requisites (i.e., inhibition, production, release, transport, uptake, exclusion of other factors; see introduction) to prove allelopathy. Some methods, such as extracts experiments, may provide strong evidence for some but not all of those requisites (e.g., allelopathy production vs. transport/uptake). Others (e.g., coexistence experiments) may address most of Willis' (1985) requirements, but none in a complete, conclusive way (Fig. 1). Also, exclusion of potential involvement of nutrients in macrophyte-phytoplankton interactions is typically more difficult to address in experiments with a higher degree of realism (Fig. 1). It is therefore important to keep in mind alternative explanations for the observed patterns and choose the right methods. The above mentioned approaches can be modified and integrated in a plethora of ways in order to reflect the complexity of the situation in situ.

Modifications of the main approaches used to study allelopathy should reflect natural physical, chemical, and biological factors as closely as possible. The following approaches could be used:

- Changes in nutrient status and light availability of both target and donor species.
- Testing a wide range of target species, including blends of allopatric and sympatric populations to account for possible co-evolutionary effects.
- Investigating the seasonality of allelopathic effects. This implies also that allelopathic interference might differ depending on the physiological status of both donor and target species.
- Testing the impact of other stressors, such as the presence of herbivores and/or pathogens, on allelopathic interactions.

Space-limitations restrict a thorough discussion of all of those aspects, but we would like to focus on some important issues in more detail.

One of the main problems of in situ studies on allelopathic interactions is the interference with nutrient competition. Several authors wondered whether separating resource competition from allelopathy is at all realistic (Inderjit & del Moral, 1997; Lürling et al., 2006; Lombardo et al., unpublished). Phosphorus and nitrogen limitation may affect the production and release of allelopathic compounds by submerged macrophytes (Gross, 2003b), as well as the response of the target organism (Reigosa et al., 1999).

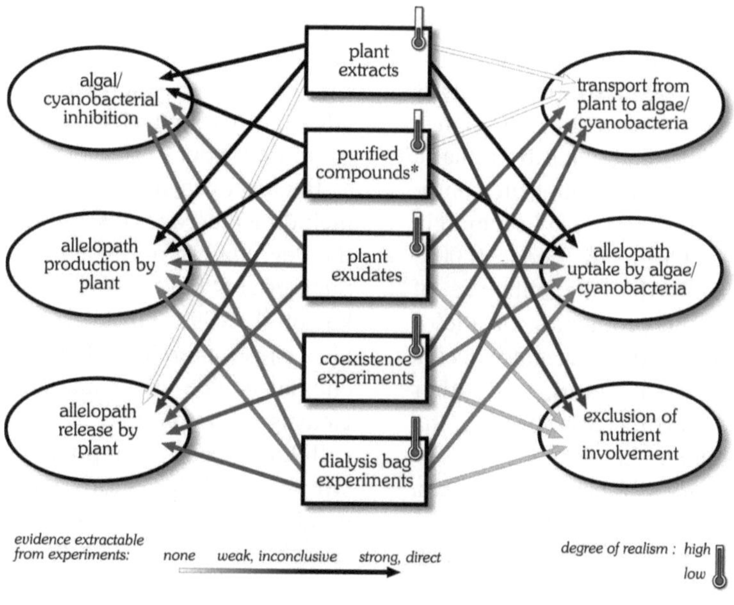

Fig. 1 Schematic representation of the evidence extractable from typical allelopathy experimental set-ups (boxes) addressing different aspects of plant-algal allelopathic interactions as requested by Willis (1985; ellipses). The sixth prerequisite, i.e., the exclusion of other factors potentially involved in plant-algae interactions, has been simplified towards nutrient involvement *sensu latu*, as exclusion of other factors (i.e., shading, herbivory) is easier to accomplish experimentally. The degree of realism, represented by a "thermometer", is illustrated for each experiment type. Details of the separate experimental approaches and their pros and cons can be found in the text. *purified compounds = purified active compounds from extracts or exudates

As macrophytes have the potential to reduce or increase nutrient levels in shallow lakes (van Donk & van de Bund, 2002), competition for nutrients or an excess nutrient supply may occur. Thus, field studies will never allow discrimination between allelopathy and resource competition. Knowledge of the nutrient status of both donor and target species seems crucial for allelopathy experiments, as e.g., phosphorus limitation enhanced inhibition of alkaline phosphatase in cyanobacteria mediated by polyphenols released from *M. spicatum* (Gross et al., 1996), or low nitrogen availability resulted in an increased tellimagrandin II content in this macrophyte (Gross, 2003b). Phosphorus-limited *Chlorella minutissima* was, however, not more sensitive to *Chara* exudates than non-limited cultures (Mulderij et al., 2003), but *Scenedesmus obliquus* grown under lower light intensities was more sensitive to allelopathic substances from *Stratiotes* than cultures grown at high light intensities (Mulderij et al., 2005a). Lombardo et al. (unpublished) have shown that, though the relative importance of competition for nutrients and allelopathy could not be quantified, algal inhibition by coexisting plants occurred only when sprigs grew actively (i.e., during net accrual of biomass and net nutrient sequestration). This is indirectly supported by observations of higher production of secondary metabolites by aquatic plants under increased light and nutrient conditions (Cronin & Lodge, 2003). The seemingly contradictory results of Cronin & Lodge (2003) vs. Gross (2003b) suggest that the influence of physiological status on the production of allelochemicals may be influenced by species-specific responses to stress, nutrient status, and other environmental factors, further complicating the assessment of allelopathic effects.

Additionally, herbivory on macrophytes can influence allelopathy. Allelopathic effects may be severely dampened when tissue containing high concentrations of allelochemicals is removed (Gross et al., 2001). Thus, herbivory might also cause changes in the dominance of certain macrophytes with allelopathic activity. On the other hand, snails that do not eat living macrophytes often enhance their growth even under high-nutrient conditions that would otherwise favour algae (Underwood, 1991; Brönmark & Vermaat, 1998; McCollum et al., 1998; Lombardo & Cooke, 2001). Thus, the assessment of herbivore influence may be very important for in situ allelopathy experiments.

The picture gets more complex as secondary metabolites with allelopathic activity may possess further biological activity, such as antiherbivore or antimicrobial effects. The hydrolysable polyphenols in *M. spicatum* comprise all those activities (Choi et al., 2002; Leu et al., 2002; Walenciak et al., 2002). The cyclic sulphur compounds considered responsible for allelopathic activity in *Chara* have insecticidal properties (Anthoni et al., 1982; Jacobsen & Pedersen, 1983). In plant-herbivore and plant-pathogen interactions, defensive compounds might be constitutively present in the plant or can be induced upon feeding damage. Further, nutrient status has often been considered to influence plant content of especially carbon-based secondary metabolites sensu the 'Carbon-Nutrient Balance Hypothesis' (Bryant et al., 1983; Koricheva, 2002; Gross, 2003b). If more than one factor may affect the content of defensive compounds, we have to expect variation in allelopathic activity. Studies investigating more than two groups of organisms, e.g., concurrent herbivory (plant-animal) and allelopathy (plant-plant) experiments, are needed to find answers to these questions.

Bacteria can not only be pathogens of plants, including macrophytes, but may also contribute to the degradation of released bioactive compounds. Such effects have frequently been observed (Nakai et al., 1999; Gross, 2003a). (Sun)light might also cause chemical changes of released allelochemicals due to oxidation, polymerisation or cleavage, comparable to effects on other extracellular organic compounds (Cole, 1999). We strongly call for more studies on changes of exuded bioactive compounds caused by bacteria or light.

Coming back to prerequisite no. 4 of Willis (1985), cited in the introduction, a major problem of most studies on aquatic allelopathy is the dilution of active compounds and no clear insight into concentrations of active compounds in situ. The problem may occur at any scale, but may be particularly acute in large-scale investigations.

Several authors have independently suggested a possible involvement of allelopathy behind observed phytoplankton patterns in whole-lake studies of vegetated, shallow lakes (Mjelde & Faafeng, 1997; Blindow et al., 2002; Lombardo, 2005). However, none could provide any evidence in favour of or against allelopathy, and the debate on the role of allelopathy at ecosystem level remains open.

Inhibition at a close range, i.e., in macrophyte-epiphyte interactions, seems much more probable, but has seldom been studied, also because of the problems associated with cultivating such organisms. Bioassays used to study such interactions would not necessarily rely on exudates and liquid cultures, but rather on 2-dimensional assays, such as agar-diffusion assays (e.g., Gross et al., 2003; Erhard & Gross, 2006). Trials testing the impact of *M. spicatum* and *M. verticillatum* on suspended cultures of common epiphytic species (*Stigeoclonium tenue*, *Gomphonema parvulum* and *Oscillatoria limosa*) did not show significant effects (Hilt, 2006). Some authors suggested that allelopathic interactions are more likely between organisms from different environments because of possible adaptations of the target species towards donors (Reigosa et al., 1999). Lower inhibition or even stimulation of epiphytes isolated from donor plants such as *Elodea nuttallii* support this view (Erhard & Gross, 2006).

Allelopathic inhibition between different phytoplankton species had been regarded impossible due to the vast distances between donor and target species (Lewis, 1986). Yet, active compounds might not be equally distributed in the water column, i.e., they might bind easily to surfaces because of their chemical characteristics (lipophilic compounds should bind to cell membranes easily). In terrestrial and aquatic studies, so-called micelles (Fischer & Quijano, 1985; Perez & Martin, 2001) have been discussed as vehicles for transport of allelopathical compounds. Again, more detailed studies are needed to address the question of release, distribution and physical range of action of allelochemicals in aquatic systems.

The explanatory power gained from the available methods (Fig. 1) depends on how much evidence is gained for each of Willis' (1985) requirements. The complex interactions strongly demand a combination of experimental approaches to maximise the experimental evidence and degree of realism.

Conclusion

Conclusive evidence for allelopathy in action in aquatic systems has eluded ecologists for years, and may continue to do so. Despite absence of a final proof in favour of (or against) allelopathy, the information gathered so far strongly suggests that allelopathy in aquatic systems indeed may be real, though it may be influenced by many factors, from environmental conditions to genetically based species-specific responses, possibly interacting synergistically and/or in non-linear ways. Our knowledge in this field has increased in recent years thanks to improvements in investigative techniques, also refined by the few answers and many questions gathered from earlier research. We believe that different pieces of evidence can be combined from multiple-approach allelopathy investigations, each targeting specific questions, eventually leading us to the answers that we seek. New developments in technology and methodology will further assist ecologists in their quest to assess the extent of allelopathy as a factor behind mutual exclusion of macrophytes and phytoplankton/epiphyton.

Acknowledgments We greatly appreciate helpful discussion with Marit Mjelde during the Shallow Lakes 2005 meeting. We very much acknowledge constructive comments by Miquel Lürling and an anonymous reviewer on a previous version of this manuscript. This project was in part supported by the German Science Foundation (DFG) in Project SFB454-A2 to EMG and by a grant by the Berliner Programm zur Förderung von Frauen in Forschung und Lehre to SH.

References

Agami, M. & Y. Waisel, 1985. Inter-relationship between *Najas marina* L. and three other species of aquatic macrophytes. Hydrobiologia 126: 169–173.

Akehurst, S. C., 1931. Observations on pond life. Journal of the Royal Microscopic Society, London 51: 236–265.

Anthoni, U., C. Christophersen, N. Jacobsen & A. Svendsen, 1982. Synthesis of 4-methylthio-1,2-dithiolane and 5-methylthio-1,2,3-trithiane. Two naturally occurring bioactive compounds. Tetrahedron 38: 2425–2427.

Blindow, I., A. Hargeby & G. Andersson, 2002. Seasonal changes of mechanisms maintaining clear water in a shallow lake with abundant *Chara* vegetation. Aquatic Botany 72: 315–334.

Brönmark, C. & J. E. Vermaat, 1998. Complex fish-snail-epiphyton interactions and their effects on submerged freshwater macrophytes. In Jeppesen, E., M. Søndergaard, M. Søndergaard & K. Christoffersen (eds), The Structuring Role of Submerged Macrophytes in Lakes, Springer, New York. Ecological Studies 131: 47–68.

Bryant, J. P., F. S. Chapin III & D. R. Klein, 1983. Carbon/nutrient balance of boreal plants in relation to vertebrate herbivory. Oikos 40: 357–368.

Choi, C., C. Bareiss, O. Walenciak & E. M. Gross, 2002. Impact of polyphenols on the growth of the aquatic herbivore *Acentria ephemerella* (Lepidoptera: Pyralidae). Journal of Chemical Ecology 28: 2223–2235.

Cole, J. J., 1999. Aquatic microbiology for ecosystem scientists: new and recycled paradigms in ecological microbiology. Ecosystems 2: 215–225.

Cronin, G. & D. M. Lodge, 2003. Effect of light and nutrient availability on the growth, allocation, carbon/nitrogen balance, phenolic chemistry, and resistance to herbivory of two freshwater macrophytes. Oecologia 137: 32–41.

Elakovich, S. D. & J. W. Wooten, 1989. Allelopathic potential of 16 aquatic and wetland plants. Journal of Aquatic Plant Management 27: 78–84.

Elakovich, S. D. & J. W. Wooten, 1995. Allelopathic, herbaceous, vascular hydrophytes. In Inderjit, K. M. M. Dakshini & F. A. Einhellig (eds), Allelopathy—organisms, processes, and applications, ACS Symposium series 582. American Chemical Society, Washington, DC, pp 58–73.

Erhard, D. & E. M. Gross, 2006. Allelopathic activity of *Elodea canadensis* and *E. nuttallii* against epiphytes and phytoplankton. Aquatic Botany 85: 203–211.

Ervin, G. N. & R. G. Wetzel, 2000. Allelochemical autotoxicity in the emergent wetland macrophyte *Juncus effusus* (Juncaceae). American Journal of Botany 87: 853–860.

Fischer, N. H. & L. Quijano, 1985. Allelopathic agents from common weeds. In A. C. Thompson (ed.), The Chemistry of Allelopathy, ACS Symposium Series. American Chemical Society, Washington, D.C. 268: 133–147.

Fitzgerald, G. P., 1969. Some factors in the competition or antagonism among bacteria, algae and aquatic weeds. Journal of Phycology 5: 351–359.

Forsberg, C., 1965. Nutritional studies of *Chara* in axenic cultures. Physiologia Plantarum 18: 275–290.

Forsberg, C., S. Kleiven & T. Willen, 1990. Absence of allelopathic effects of *Chara* on phytoplankton in situ. Aquatic Botany 38: 289–294.

Gopal, B. & U. Goel, 1993. Competition and allelopathy in aquatic plant communities. Botanical Review 59: 155–210.

Gross, E. M., 1999. Allelopathy in benthic and littoral areas: Case studies on allelochemicals from benthic cyanobacteria and submersed macrophytes. In Inderjit, K. M. M. Dakshini & C. L. Foy (eds), Principles and practices in plant ecology: allelochemical interactions. CRC Press, LLC, Boca Raton, 179–199.

Gross, E. M., 2003a. Allelopathy of aquatic autotrophs. Critical Reviews in Plant Sciences 22: 313–339.

Gross, E. M., 2003b. Differential response of tellimagrandin II and total bioactive hydrolysable tannins in an aquatic angiosperm to changes in light and nitrogen. Oikos 103: 497–504.

Gross, E. M., D. Erhard & E. Ivanyi, 2003. Allelopathic activity of *Ceratophyllum demersum* L. and *Najas marina* ssp. *intermedia*. Hydrobiologia 506: 583–589.

Gross, E. M., R. L. Johnson, N. G. Hairston Jr., 2001. Experimental evidence for changes in submersed macrophyte species composition caused by the herbivore *Acentria ephemerella* (Lepidoptera). Oecologia 127: 105–114.

Gross, E. M., H. Meyer & G. Schilling, 1996. Release and ecological impact of algicidal hydrolyzable polyphenols in *Myriophyllum spicatum*. Phytochemistry 41: 133–138.

Gurevitch, J., S. M. Scheiner & G. A. Fox, 2002. The Ecology of Plants. Sinauer Associates Inc., Sunderland, MA, USA. See p. 193 for De-Wit replacement series.

Harder, R., 1917. Ernährungsphysiologische Untersuchungen an Cyanophyceen, hauptsächlich dem endophytischen *Nostoc punctiforme*. Zeitschrift für Botanik IX: 145–242.

Hasler, A. D. & E. Jones, 1949. Demonstration of the antagonistic action of large aquatic plants on algae and rotifers. Ecology 30: 359–364.

Hilt, S., 2006. Allelopathic inhibition of epiphytes by submerged macrophytes. Aquatic Botany 85: 252–256.

Hilt, S., M. Ghobrial & E. M. Gross, 2006. In situ allelopathic potential of *Myriophyllum verticillatum* (Haloragaceae) against selected phytoplankton species. Journal of Phycology: 42:1189–1198.

Inderjit & R. del Moral, 1997. Is separating resource competition from allelopathy realistic? Botanical Review 63: 221–230.

Jacobsen, N. & L. E. K. Pedersen, 1983. Synthesis and insecticidal properties of derivatives of propane-1,3-dithiol (analogs of the insecticidal derivatives of dithiolane and trithiane from the alga *Chara globularis* (Thuillier). Pesticide Science 14: 90–97.

Jasser, I., 1995. The influence of macrophytes on a phytoplankton community in experimental conditions. Hydrobiologia 306: 21–32.

Keating, K. I., 1977. Allelopathic influence on blue-green bloom sequence in a eutrophic lake. Science 196: 885–887.

Kogan, S. I. & G. A. Chinnova, 1972. Relations between *Ceratophyllum demersum* (L.) and some blue-green algae. Hydrobiological Journal (Ghidrobiol. Zh.) 8: 14–19 (21–27).

Koricheva, J., 2002. The Carbon-Nutrient Balance Hypothesis is dead: long live the Carbon-Nutrient Balance Hypothesis? Oikos 98: 537–539.

Körner, S. & A. Nicklisch, 2002. Allelopathic growth inhibition of selected phytoplankton species by submerged macrophytes. Journal of Phycology 38: 862–871.

Leu, E., A. Krieger-Liszkay, C. Goussias, E. M. Gross, 2002. Polyphenolic allelochemicals from the aquatic angiosperm *Myriophyllum spicatum* L. inhibit photosystem II. Plant Physiology 130: 2011–2018.

Lewis, W. M. Jr., 1986. Evolutionary interpretation of allelochemical interactions in phytoplanktonic algae. American Naturalist 127: 184–194.

Lombardo, P. & G. D. Cooke, 2001. Effects of freshwater gastropods on epiphyton, macrophytes, and water transparency under meso- to eutrophic conditions. Chapter from PhD Dissertation. Kent State University, Kent, OH, USA.

Lombardo, P. & G. D. Cooke, 2003. *Ceratophyllum demersum*: phosphorus interactions in nutrient enriched aquaria. Hydrobiologia 497: 79–90.

Lombardo, P., 2005. Applicability of littoral food-web biomanipulation for lake management purposes: snails: macrophytes, and water transparency in northeast Ohio shallow lakes. Lake and Reservoir Management 21: 186–202.

Lürling, M., G. van Geest & M. Scheffer, 2006. Importance of nutrient competition and allelopathic effects in suppression of the green alga *Scenedesmus obliquus* by the macrophytes *Chara*, *Elodea* and *Myriophyllum*. Hydrobiologia 556: 209–220.

McCollum, E. W., L. B. Crowder & S. A. McCollum, 1998. Complex interactions of fish, snails, and littoral zone periphyton. Ecology 79: 1980–1994.

Mjelde, M. & B. A. Faafeng, 1997. *Ceratophyllum demersum* hampers phytoplankton development in some small Norwegian lakes over a wide range of phosphorus concentrations and geographical latitude. Freshwater Biology 37: 355–365.

Molisch, H., 1937. Der Einfluss einer Pflanze auf die andere—Allelopathie. Fischer, Jena.

Mulderij, G., B. Mau, E. van Donk & E. M. Gross, 2007. Allelopathic activity of *Stratiotes aloides* on phytoplankton—towards identification of allelopathic substances. Hydrobiologia 584: 89–100.

Mulderij, G., W. M. Mooij, A. J. P. Smolders & E. van Donk, 2005a. Inhibition of phytoplankton by allelopathic substances from *Stratiotes aloides*. Aquatic Botany 82: 284–296.

Mulderij, G., W. M. Mooij & E. van Donk, 2005b. Allelopathic growth inhibition and colony formation of the green alga *Scenedesmus obliquus* by the aquatic macrophyte *Stratiotes aloides*. Aquatic Ecology 39: 11–21.

Mulderij, G., A. J. P. Smolders & E. van Donk, 2006. The allelopathic effect of *Stratiotes aloides* on phytoplankton under natural conditions. Freshwater Biology 51: 554–562.

Mulderij, G., E. van Donk & J. G. M. Roelofs, 2003. Differential sensitivity of green algae to allelopathic substances from *Chara*. Hydrobiologia 491: 261–271.

Nakai, S., Y. Inoue, M. Hosomi & A. Murakami, 1999. Growth inhibition of blue–green algae by allelopathic effects of macrophytes. Water Science & Technology 39: 47–53.

Pelton, D. K., S. N. Levine & M. Braner, 1998. Measurements of phosphorus uptake by macrophytes and epiphytes from the LaPlatte River (VT) using ^{32}P in stream microcosms. Freshwater Biology 39: 285–299.

Perez, E. & D. F. Martin, 2001. Critical micelle concentrations of allelopathic substances produced by *Nannochloris oculata* which affect a red tide organism, *Gymnodinium breve*. Cytobios 106: 163–170.

Phillips, G. L., D. Eminson, & B. Moss, 1978. A mechanism to account for macrophyte decline in progressively eutrophicated freshwaters. Aquatic Botany 4: 103–126.

Pratt, R., R. C. Daniels, J. J. Eiler, J. B. Gunnison, W. D. Kumler, J. F. Oneto, L. A. Strait, H. A. Spoehr, G. J. Hardin, H. W. Milner, J. H. C. Smith & H. H. Strain, 1944. Chlorellin, an antibacterial substance from *Chlorella*. Science 99: 351–352.

Pringsheim, E. G. & O. Pringsheim, 1962. Axenic culture of *Utricularia*. American Journal of Botany 49: 898–901.

Reigosa, M. J., A. Sanchez-Moreiras & L. Gonzalez, 1999. Ecophysiological approach in allelopathy. Critical Reviews in Plant Sciences 18: 577–608.

Scheffer, M., S. H. Hosper, M. L. Meijer, B. Moss & E. Jeppesen, 1993. Alternative equilibria in shallow lakes. Trends in Ecology and Evolution 8: 275–279.

Søndergaard, M., 1981. Kinetics of extracellular release of ^{14}C-labelled organic carbon by submersed macrophytes. Oikos 36: 331–347.

Sukenik, A., R. Eshkol, A. Livne, O. Hadas, M. Rom, D. Tchernov, A. Vardi & A. Kaplan, 2002. Inhibition of growth and photosynthesis of the dinoflagellate *Peridinium gatunense* by *Microcystis* sp. (cyanobacteria): a novel allelopathic mechanism. Limnology and Oceanography 47: 1656–1663.

Underwood, G. J. C., 1991. Growth enhancement of the macrophyte *Ceratophyllum demersum* in the presence of the snail *Planorbis planorbis*: the effect of grazing and chemical conditioning. Freshwater Biology 26: 325–334.

van, Donk E. & W. J. van de Bund, 2002. Impact of submerged macrophytes including charophytes on phyto- and zooplankton communities: allelopathy versus other mechanisms. Aquatic Botany 72: 261–274.

Vance, H. D. & D. A. Francko, 1997. Allelopathic potential of *Nelumbo lutea* (Willd.) Pers. to alter growth of *Myriophyllum spicatum* L. and *Potamogeton pectinatus* L. Journal of Freshwater Ecology 12: 405–409.

Walenciak, O., W. Zwisler, E. M. Gross, 2002. Influence of *Myriophyllum spicatum*-derived tannins on gut microbiota of its herbivore *Acentria ephemerella*. Journal of Chemical Ecology 28: 2045–2056.

Wetzel, R. G. & D. L. McGregor, 1968. Axenic culture and nutritional studies of aquatic macrophytes. American Midland Naturalist 80: 52–64.

Wetzel R. G., 2001. Limnology—Lake and River Ecosystems, 3rd ed. Academic Press, San Diego, 1006 pp.

Willis, R. J., 1985. The historical bases of the concept of allelopathy. Journal of the History of Biology 18: 71–102.

Wooten, J. W. & S. D. Elakovich, 1991. Comparisons of potential allelopathy of seven freshwater species of spikerushes *(Eleocharis)*. Journal of Aquatic Plant Management 29: 12–15.

Hydrobiologia (2007) 584:89–100
DOI 10.1007/s10750-007-0602-0

SHALLOW LAKES

Allelopathic activity of *Stratiotes aloides* on phytoplankton—towards identification of allelopathic substances

Gabi Mulderij · Birgit Mau · Ellen van Donk · Elisabeth M. Gross

Abstract The allelopathic activity of the aquatic macrophyte, *Stratiotes aloides*, was determined with laboratory experiments. Active compounds exuded in the medium or present in plant tissue were extracted using standard procedures and solid phase extraction (SPE). The activity towards various cyanobacteria and chlorophytes was tested in two different bioassay systems using agar plates and liquid cultures of phytoplankton. Extracts and exudates of *S. aloides* affected phytoplankton growth. SPE-enriched exudates and enriched water from a natural *Stratiotes* stand caused inhibition of target species, however, also some controls were active. Phytoplankton species exhibited differential sensitivity to extracts of *S. aloides*. We observed inhibitory and stimulatory effects on phytoplankton. In general, more cyanobacteria than other phytoplankton species were inhibited, and the inhibition of cyanobacteria was stronger. In most cases, nutrient (P or K) limitation of *Synechococcus elongatus* and *Scenedesmus obliquus* decreased the sensitivity of these species towards allelochemicals from *Stratiotes aloides*, except for P-limited cultures of *Scenedesmus*. The allelopathically active compound(s) from *Stratiotes* are moderately lipophilic and most likely no phenolic compounds. Our results indicate that allelopathy (besides nutrient interference and shading) might also account for the low phytoplankton and filamentous algae densities in the vicinity of *Stratiotes* plants, at least during certain phases of the life-cycle of *Stratiotes*.

Keywords Allelopathy · Aquatic macrophyte · Exudate · Growth inhibition · Nutrient limitation

Guest editors: R. D. Gulati, E. Lammens, N. De Pauw & E. Van Donk
Shallow lakes in a changing world

G. Mulderij (✉) · E. van Donk
Department of Food Web Studies, NIOO-KNAW, Centre for Limnology, Nieuwersluis 3631 AC, The Netherlands
e-mail: g.mulderij@nioo.knaw.nl

E. van Donk
e-mail: e.vandonk@nioo.knaw.nl

G. Mulderij · E. van Donk
Department of Ecology, Aquatic Ecology and Environmental Biology Section, University of Nijmegen, PO Box 9010, Nijmegen 6500 GL, The Netherlands

B. Mau · E. M. Gross
Limnological Institute, University of Konstanz, PO Box M659, Konstanz 78457, Germany

B. Mau
e-mail: Mau.Birgit@web.de

E. M. Gross
e-mail: Elisabeth.Gross@uni-konstanz.de

Introduction

Aquatic macrophytes play a crucial role in stabilizing the clear water state in shallow mesotrophic and eutrophic lakes (Scheffer, 1998; Burks et al., in press). They can stimulate changes to clear water situations by a range of mechanisms (see review Van Donk & Van de Bund, 2002). One of those mechanisms is the excretion of inhibitory substances that reduce phytoplankton growth (= allelopathy, e.g., Gross, 2003). The release of these compounds may be an effective trait of submerged macrophytes to gain competitive advances against other photoautotrophs.

Only a few allelochemicals from submerged aquatic macrophytes are structurally elucidated (Wium-Andersen et al., 1982; Gross et al., 1996). Previous experiments (Mulderij et al., 2005a, b) provided evidence for allelopathic interference of *Stratiotes aloides* with phytoplankton species, but the nature of the active compound(s) was not investigated. *S. aloides* is a dioecious, perennial aquatic macrophyte, native to Europe and Siberia (Cook & Urmi-König, 1983). A part of its life cycle is submerged (Bloemendaal & Roelofs, 1988) and the plants overwinter as turions or entire (but rootless) plants on the bottom of the lake. The macrophyte becomes buoyant in spring as a result of increased photosynthesis and remains floating on the water surface until autumn (De Geus-Kruyt & Segal, 1973). Especially during its submerged phase, *Stratiotes* might be favoured by the ability to produce allelopathic compounds that inhibit phytoplankton growth.

Other mechanisms for the exclusion of phytoplankton in the proximity of water soldier have been proposed. The availability of potassium or sodium might be limiting phytoplankton growth in *S. aloides* stands and further, co-precipitation of phosphates with calcium might be an important mechanism (Brammer, 1979; Brammer & Wetzel, 1984). Other authors, however, indicated allelopathic activity of *Stratiotes*. Extracts of the macrophyte showed inhibitory effects even at the lowest concentration tested (2 g FM/l, Jasser, 1995). Usenko et al. (2002) showed that phenolic acids might be responsible for algicidal effects of *Stratiotes*. Both studies, however, did not prove that *Stratiotes* really excreted allelopathic compounds, an important issue to asses the ecological importance of allelopathic interactions. Recent laboratory experiments showed that exudates of *Stratiotes* inhibited the green alga *Scenedesmus obliquus* and induced colony formation in this alga (Mulderij et al., 2005a). Further, allelopathic effects of *Stratiotes* exudates on the cyanobacterium *Microcystis aeruginosa* and the eustigmatophyte *Nannochloropsis limnetica* were observed (Mulderij et al., 2005b).

We performed bioassay-directed extractions and fractionations to elucidate the chemical characteristics of allelopathic substances in extracts and exudates of *Stratiotes*, and were especially interested whether the active compounds are hydro- or lipophilic. We also tested if the sensitivity of target organisms to *Stratiotes* would increase when they were cultured under nutrient limiting conditions, as Reigosa et al. (1999) stated that target organisms under stress (e.g., nutrient limitation) become more susceptible to additional stressors (e.g., allelopathy) than equivalent target organisms not further stressed.

Materials and methods

Macrophyte culture

The *Stratiotes* plants originated from a ditch in Tienhoven (52°10′0″ N, 5°4′60″ E), the Netherlands, and were collected in July 2003. The plants were carefully rinsed with tap water and transferred into 500 l aquaria filled with 10 cm sediment originating from the same ditch and with copper-free tap water. The macrophytes were grown at a density comparable to moderate densities in the field (ca. 10 g FM/l), and water was renewed fortnightly. The aquaria were illuminated (30 μmol m^{-2} s^{-1}) 16 h per day and the water temperature was 18 ± 1°C. At the start of an experiment, plants were transferred to sediment-free aquaria with macrophyte medium (Mulderij et al., 2005a).

Target organisms

We used axenic monocultures of twelve phytoplankton species, consisting of cyanobacteria, green algae and an eustigmatophyte. *Anabaena* sp. PCC 7120 and *A. variabilis* P9 ATCC 29413 were used in agar diffusion assays (*ADAs*, see *Bioassays*), while the four *Microcystis aeruginosa* strains (NIVA-CYA 140 (toxic), NIVA-CYA 43, V131 (from E. Kardinaal), and PCC 7820, the latter two non-toxic), the cyanobacterium *Synechococcus elongatus* (SAG 89.79), the green algae *Chlorella minutissima* (CCAP 211/52), *Chlamydomonas reinhardtii* (NIVA-CHL 13), *Scenedesmus obliquus* (CCAP 276/3A), *Selenastrum capricornutum* (NIVA-CHL 1) and the eustigmatophyte *Nannochloropsis limnetica* (Krienitz 1998/3) were used in liquid culture assays (*LCAs*, see *Bioassays*). All cyanobacteria were cultured in cyanobacteria medium (Jüttner et al., 1983; with modifications as in Gross et al., 1991), while all green algae and *Nannochloropsis* were cultured in modified WC medium (Guillard & Lorentzen, 1972; modified by Lurling, 1999).

Phosphate- (P) or potassium- (K) limited cultures of *S. obliquus* and *S. elongatus* were established by centrifuging (6 min., 2,300 rpm) aliquots of non-limited cultures twice. After each centrifugation step the supernatant was removed and cells were resuspended in P- or K-limited medium. P-limited medium was created by substituting 50 μM K_2PO_4 by 1 μM K_2PO_4 and 40 μM KCl. K-limited medium was created by substituting 50 μM K_2HPO_4 with 1 μM K_2HPO_4 and 40 μM Na_2HPO_4 (Alahari & Apte, 2004). P- or K-limited batch cultures (300 ml) were grown for at least 9 days on an orbital shaker (110 rpm) at 22°C, 50 μmol PAR m^{-2} s^{-1} with a photoperiod of 16:8 (L:D).

Extraction

Stratiotes plants were carefully rinsed with tap water, shock-frozen with liquid nitrogen, and immediately lyophilized for 24 h. This material was homogenized and stored in the dark at room temperature until further use. Later, this plant material was extracted (1 ml solvent per 10 mg plant dry mass) for 2 h at room temperature under continuous stirring. As solvents we used water, methanol and acetone in different mixtures (50% or 70% [v/v] in water, and 100%). Extracts were filtered (Whatman GF/F) to remove plant particles, evaporated under vacuum to dryness and resuspended in 50% [v/v] aqueous methanol at a final concentration equivalent to 100 mg extracted dry mass (DM) per ml. Extracts were stored at –20°C. Solvent controls were performed using the same procedure without adding plant material.

Fractionation of crude extracts

Crude extracts were fractionated using solid phase extraction (SPE). An aliquot of crude extract (20 mg extracted DM) dissolved in 2 ml 50% [v/v] aqueous methanol, was diluted 1:25 with water and passed over a preconditioned SPE-C18 cartridge (Varian Bond Elute, 3cc, 2 g sorbens) and the eluate was collected. Then the cartridge was stepwise eluted with each two reservoir volumes (10 ml) of 40, 60, 70, 80, 90 and 100% [v/v] methanol (in water). Those seven fractions were collected separately, evaporated to dryness and redissolved in 50% [v/v] aqueous methanol for further use in bioassays.

Heat stability and polyvinylpolypyrrolidone (PVPP) test

Crude extracts of *Stratiotes* were heated for 5 min at 95°C. Immediately thereafter the extracts were cooled down in ice, evaporated to dryness, and resuspended in 50% [v/v] aqueous methanol. To test for the presence of phenolic compounds, we conducted a polyvinylpolypyrrolidone (PVPP, Sigma P6755) test modified after Loomis & Battaile (1966). A suspension of PVPP in water (2 g PVPP/20 ml water) was added to an aliquot of the crude extract (v/v 1:1) and stored overnight at 4°C. Thereafter, the suspension was centrifuged (5 min., 4,000 g) and the supernatant used in the bioassays.

Exudation

Possible allelopathic compounds, exuded into the water by *Stratiotes*, were investigated with three approaches. Laboratory approaches (set-ups 1 & 2): Incubating single or multiple plants in microcosms. Field approach (set-up 3): Collecting water from a natural *Stratiotes* stand in Lake Naardermeer (nature reserve). In set-up 1 (single plant culture), we filled 1.5-l glass vials (Ø 10 cm, 20 cm height) with 1 l macrophyte medium (Mulderij et al., 2005a) and one *Stratiotes* plant (25–30 g FM) from the laboratory culture. Control vials only received medium. All vials were kept at 16°C and approx. 60 μmol PAR m^{-2} s^{-1} (L:D 14:10). After 1, 2 and 5 days, the culture water of three vials (one control and two with *Stratiotes*) was used for analyses. In set-up 2 (microcosms), two 50 l-aquaria were filled with 30 l macrophyte medium. One aquarium served as control and was filled with medium, while the other aquarium was filled with medium and eight *Stratiotes* plants (ca. 200 g FM) originating from Giethoorn (52°45′0″ N 6°5′0″ E, The Netherlands), collected the week before the start of the experiment. After 1, 2 and 3 days, 1l water of each aquarium was collected for further analysis. For set-up 3, water samples (1 l) were collected in the vicinity of *Stratiotes* plants. As a control, two sites without *Stratiotes* were sampled: one in the same ditch (Ditch 1) just outside the *Stratiotes* stand and the other in an other ditch (Ditch 2) in the nature reserve. In ditch 2, *Stratiotes* plants were absent during the whole experiment. Samples were collected between 7 April and 14 June 2004.

All samples (set-ups 1, 2 & 3) were filtered over Whatman GF/F and then over 0.2 μm celluloseacetate membrane filters (Schleicher & Schuell). Potentially exuded allelopathic compounds were trapped on preconditioned C18–filters (Ø 47 mm, 3M Empore™, Phenomenex) and eluted with 100% methanol (set-up 1) or with 50, 90, and 100% methanol (set-ups 2 & 3). All fractions were evaporated to dryness and resuspended in 50% [v/v] aqueous methanol.

Bioassays

We conducted two types of bioassays under sterile conditions: Agar diffusion assays, (*ADAs*, Flores & Wolk, 1986; Gross et al., 1991) and liquid culture assays, (*LCAs*, Schrader et al., 1997; slightly modified by Erhard & Gross, submitted).

Agar diffusion assay (*ADA*): Extracts or SPE fractions (0.5 to 3 mg extracted plant DM) were spotted in multiple intervals onto 1% agar plates so that the diameter never exceeded 5 mm. Subsequently, a second layer containing cyanobacteria medium, target cells and heated 4% agar (final agar strength: 1%) was added. Target organisms were inoculated at an optical density (OD) of 40 mAU at 530 nm. The agar plates were incubated at 28°C and 80 μmol PAR m^{-2} s^{-1} (L:D 16:8) for one week. Exudates were spotted in concentrations equivalent to 40, 120, and 240 ml culture water (set-up 1, equal to 1.2, 3.6 and 7.2 g FM) or 600 ml (set-up 2 & 3, equal to 3.96 g FM for set-up 2). Extracts, exudates and SPE fractions with algicidal activity caused clearing areas in the algal lawn. All ADAs were carried out at least in duplicate.

Liquid culture assay (*LCA*): Each of the four rows on a 24-well plate was assigned six different concentrations of A) extract + target cells, B) extract, C) solvent control + target cells, or D) solvent control. Extracts were pipetted into the wells and the solvent was evaporated. Then, per well, 2 ml culture with target cells ($OD_{530\ nm}$: 40 mAU) in the respective medium were added. Plates were incubated for one week at either 21°C and 60 PAR m^{-2} s^{-1} L:D 16:8 (NIOO) or at 28°C and 60 PAR m^{-2} s^{-1} under continuous illumination (University of Konstanz), depending on the phytoplankton species. All LCAs were conducted in triplicate and the growth of target cells, as percentage relative to the control, was calculated for each extract concentration as follows:

$$\text{Growth [as \% of control]} = \frac{Row\ A - Row\ B}{Row\ C - Row\ D} \times 100\%$$

The first LCA was conducted with four different concentrations of *Stratiotes* extract (0, 0.75, 1.5 and 3.75 mg DM/ml; *LCA 1*) and all target species. In the second LCA (*LCA 2*) lower extract concentrations (0, 0.5, 1.0 and

1.5 mg DM/ml) and only some target species were used. In this way we determined which extract concentration and which test organisms should be used in the following assays. The two most sensitive species of cyanobacteria and green algae were chosen as target organisms for the next LCAs.

We conducted two short term LCAs with 3 mg DM/ml *Stratiotes* extract, to test when allelopathic effects of *Stratiotes* extract become apparent and how long they persist without extra addition of fresh extract. The first short term LCA with *S. elongatus* and *S. obliquus* was sampled after 0, 2, 4, 6, 8, 24, and 48 h incubation (*LCA 3*). The second short term LCA with only *S. elongatus* was sampled after 24, 48, 72 and 216 h incubation (*LCA 4*).

The next LCA (*LCA 5*) was conducted with extract concentrations ranging from 0 to 5.5 mg DM/ml at increments of 0.5 mg DM/ml). This enabled the determination of threshold levels of inhibition/stimulation of the growth of *S. elongatus* and *S. obliquus* in the presence of *Stratiotes* extract. *LCAs 3-5* were conducted with nutrient (K or P) limited and non-limited target cells.

Two additional LCAs were conducted with SPE fractions of the crude extract (*LCA SPE*) and PVPP treated extracts (*LCA PVPP*) added in four concentrations (0, 1, 3 and 5 mg DM/ml).

Statistical analysis

The influence of *Stratiotes* extracts on the growth of target cells in LCAs ($OD_{530\,nm}$ values) was assessed by means of one-way ANOVA ($\alpha = 0.05$, Fowler et al., 1998). Prior to the ANOVA, all data were tested for normality (Kolmogorov–Smirnov test, $\alpha = 0.05$) and homoscedasticity (Levene's test, $\alpha = 0.05$). All data followed a normal distribution, but sometimes heteroscedasticity was observed. Therefore, log-transformations were carried out. If these transformations did not remove heteroscedasticity, an α-value of 0.01 was used (Fowler-Walker & Connell, 2002). When the ANOVA revealed significant effects, a Tukey multiple comparisons test ($\alpha = 0.05$) was applied to group homogeneous means.

Results

Allelopathic activity of *Stratiotes* extracts

ADAs with *Anabaena* sp. and *A. variabilis* showed that *Anabaena* sp. was not strongly inhibited by any of the extracts. *A. variabilis* was much more sensitive and strongly inhibited by 50 and 70% methanol or acetone extracts. Increasing extract concentrations caused increased clearing zones. The strongest inhibition of *A. variabilis* was observed when *Stratiotes* was extracted with 50% [v/v] aqueous acetone. This solvent was subsequently used for further extractions.

M. aeruginosa V131 was the most sensitive cyanobacterium, exhibiting a significant ($P < 0.001$) growth inhibition in *LCA 1* of 32% already at 1 mg DM/ml (Fig. 1a). *S. elongatus* was inhibited (56%) by extract concentrations above 1.5 mg DM/ml (Fig. 1b). Other cyanobacteria showed no significant response (*LCA 1*) or even a significant growth stimulation (*LCA 2*), depending on the experiment (Table 1). While all tested green algae and the eustigmatophyte *N. limnetica* exhibited no significant inhibition or stimulation in *LCA 1*, we observed an inhibition above 1 mg DM/ml (Fig. 1c) with *N. limnetica* and a stimulation of all chlorophytes at 0.5 or 1 mg DM/ml in *LCA 2* (Table 1).

Short term *LCA 3* showed a significant ($P < 0.001$) growth stimulation of *S. elongatus* (130%) and *S. obliquus* (282%) after 48 h incubation in *Stratiotes* extract, while short term *LCA 4* exhibited a significant ($P = 0.002$) inhibition on *S. elongatus* only after 216 h (45 ± 2%).

Fractionation of crude extracts

The ADA with *A. variabilis* and SPE fractions of crude extract showed strong inhibitory effects of the 80 and 90% methanol SPE fractions (Table 2). An additional LCA with crude extract

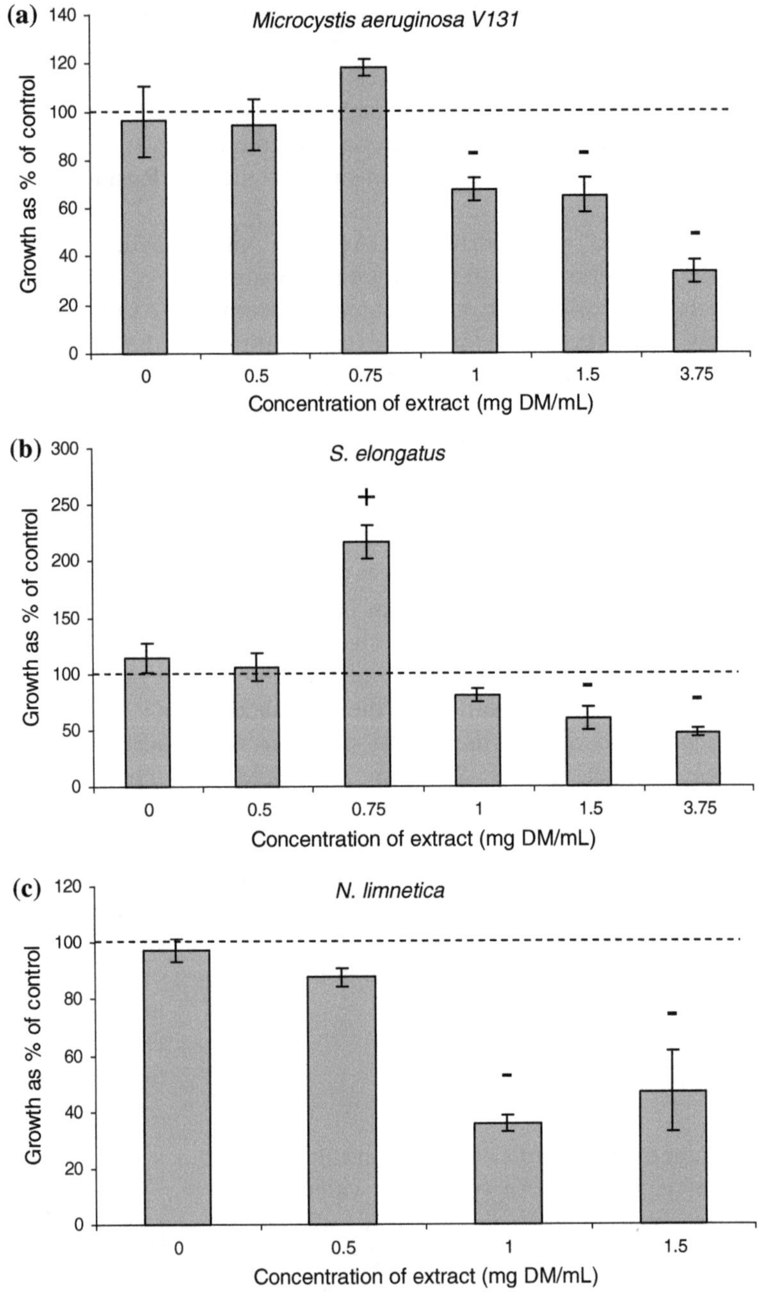

Fig. 1 Growth relative to the control of *Microcystis aeruginosa* (**a**), *Synechococcus elongatus* (**b**), and *Nannochloropsis limnetica* (**c**) at different concentrations of *Stratiotes* extract in the first two liquid assays. Bars indicate average values ($n = 3$), error bars are 1 SE. + and – indicate significant inhibitory and stimulatory effects on the growth of the phytoplankton species

and its SPE fractions (*LCA SPE*) showed significant ($P < 0.001$) effects on the growth of *S. elongatus* (Fig. 2b). Crude extract and the 90 and 100% methanol SPE fractions showed significant inhibitory effects at concentrations of 3 and 5 mg DM/ml, and the 100% fraction also already at 1 mg DM/ml, between 36 and 94% (Fig. 2b). The 50% SPE fraction, on the contrary, showed significant stimulatory effects at concentrations of 1 and 3 mg DM/ml (Fig. 2b).

Heat stability and polyvinylpolypyrrolidone (PVPP) test

ADAs showed that *Stratiotes* extract remains allelopathically active after heating. The clearing

Table 1 Results statistical analyses of liquid culture assays 1 and 2 (*LCA 1* & *LCA 2*). Effects of extract addition were either stimulatory (+), inhibitory (–) or not clear (0) with threshold values indicated in parenthesis

	LCA 1		LCA 2	
	0–3.75 mg DM/ml		0–1.5 mg DM/ml	
	P-value	Effect	*P*-value	Effect
Green algae				
C. minutissima	0.603	0	<0.001	+ (0.5)
C. reinhardthii	0.402	0	<0.001	+ (0.5)
S. obliquus	0.057	0	<0.001	+ (0.5)
S. capricornutum	0.635	0	<0.001	+ (1.0)
Eustigmatophyte				
N. limnetica	0.222	–	<0.001	– (1.0)
Cyanobacteria				
M. aeruginosa CYA 43	0.044*	–	<0.001	+ (0.5)
M. aeruginosa CYA 140	0.317	0	<0.001	+ (0.5)
M. aeruginosa V131	0.056	–	<0.001	– (1.0)
M. aeruginosa PCC 7820	0.230	–	<0.001	+ (0.5)
S. elongatus	0.033	–(1.5)	0.100	– (1.5)

The one-way ANOVA of the growth as percentage of control, was carried out on a series of 24 values per phytoplankton species (3 replicates, 2 treatments: control and Stratiotes, and four extract concentrations: 0, 0.5, 1.0 and 1.5 mg DM/ml or 0, 0.75, 1.5 and 3.75 mg DM/ml)

* Criterion for significance was 0.01, see *Statistical analysis*

zone of untreated or heat-treated extracts did not differ. Further, three independent ADAs showed that PVPP treatment of *Stratiotes* crude extract did not or only slightly remove inhibitory effects. PVPP treatment neither removed the inhibitory activity of the extract in an LCA (*LCA PVPP*) with *S. elongatus* (Fig. 2a versus b).

Phosphorus and potassium limitation of S. elongatus and S. obliquus

LCA 5 with *S. elongatus* showed significant ($P < 0.001$) stimulatory effects of P-limited cultures above extract concentrations of 2.5 mg DM/ml (Fig. 3a), while non-limited cultures were significantly ($P < 0.001$) inhibited (45%) by concentrations above 4.0 mg DM/ml (Fig. 3a). P-limited and non-limited *S. obliquus* cultures, were significantly ($P < 0.001$) stimulated (up to 1600%) by *Stratiotes* extract at concentrations above 0.5 mg DM/ml (Fig. 3b), with the pronounced effects for non-limited cultures (Fig. 3b).

The short term *LCA 4* showed significant stimulatory effects of *Stratiotes* extract on P-limited cultures of *S. elongatus* and *S. obliquus*. The stimulatory effects on P-limited cultures were stronger (130 vs. 414% for *S. elongatus*; 282 vs. 844% for *S. obliquus*) and occurred earlier (after 24 instead of 48 h) than on non-limited cultures.

K-limited cultures of *S. elongatus* were less strongly inhibited than non-limited ones. Whereas the non-limited *S. elongatus* cultures were significantly ($P < 0.001$) inhibited by crude extract and the 90 and 100% SPE fraction (Fig. 2b), no significant inhibitory effects were observed with K-limited cultures (Fig. 2c). The

Table 2 Algicidal activity of SPE-fractions of *Stratiotes* extract (50% acetone) on *A. variabilis*

Fraction	Clearing area diameter (cm)	
	2 mg DM	3 mg DM
Flow through	–	–
40% Methanol	–	–
60% Methanol	(+)	(+)
70% Methanol	–	–
80% Methanol	(0.7)	(0.8)
90% Methanol	0.8	1.1
100% Methanol	(+)	(+)

Aliquots were applied in two concentrations in the agar diffusion assay

( ) inhibition less strong. Algal cells not completely absent in the clearing areas, only lower densities observed

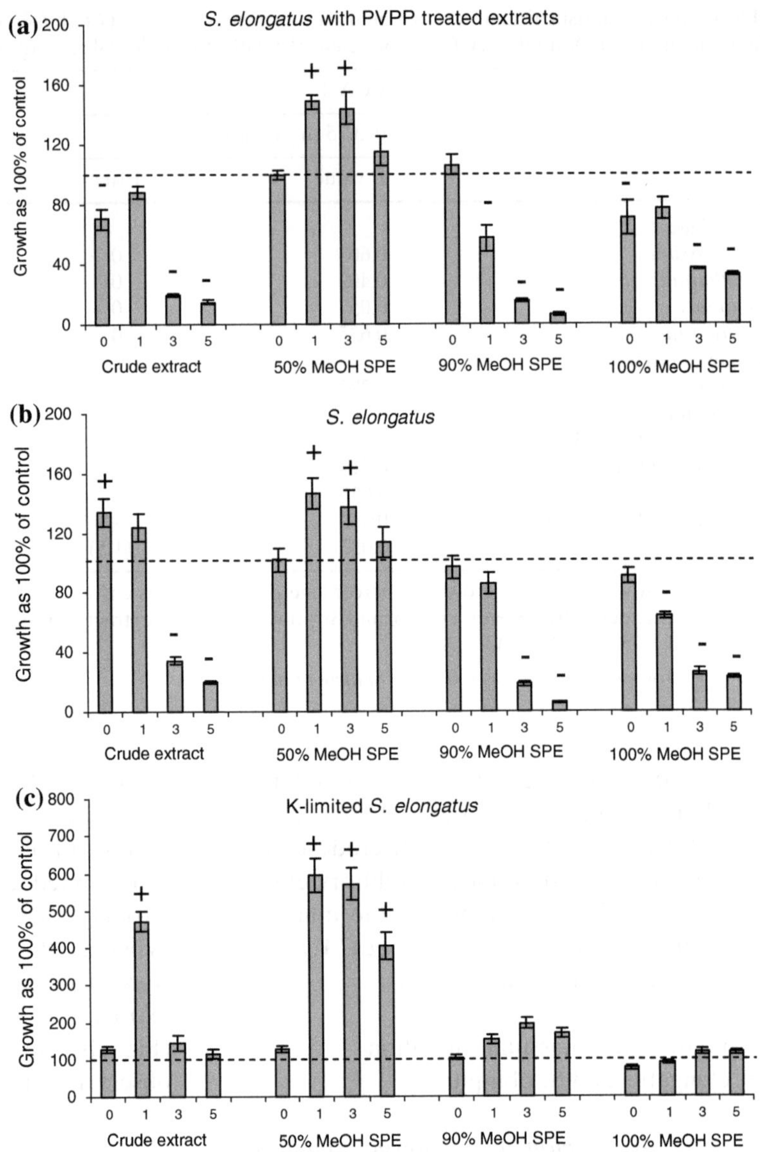

Fig. 2 Growth relative to the control in non-limited (**a, b**) and K-limited cultures (**c**) of *Synechococcus elongatus* when exposed to different concentrations (0, 1, 3 and 5 mg DM/ml) of *Stratiotes* extract (crude extract) or SPE fractions (50-, 90-, and 100% Methanol) either treated (**a**) or not treated (**b, c**) with polyvinylpolypyrrolidone (PVPP). Bars indicate average values (n = 3), error bars are 1 SE. + and – indicate significant inhibitory and stimulatory effects on the growth of the phytoplankton species

50% SPE fraction exhibited a significant growth stimulation on both non-limited and K-limited cultures of *S. elongatus*.

Exudation

The ADA with *A. variabilis* showed clearing zones (>6 mm) for SPE-enriched medium equivalent to 6–7 g FM plants (240 ml medium) after 5 days incubation (set-up 1). An other ADA (set-up 2) showed inhibitory effects of SPE-enriched medium equivalent to 4–5 g FM plants (600 ml medium) after one day incubation (Table 3), but some controls also showed slight clearing areas.

ADAs with exudates from Lake Naardermeer (set-up 3) also showed a strong inhibition of *A. variabilis* (Table 4), with the strongest inhibitory effect found at 20 April (when *Stratiotes* plants were still submerged). Controls, however, also showed inhibitory effects, but generally less strong than the exudates.

Discussion

In general, our results showed both stimulatory and inhibitory effects of *Stratiotes* extract and exudates on phytoplankton growth, with more

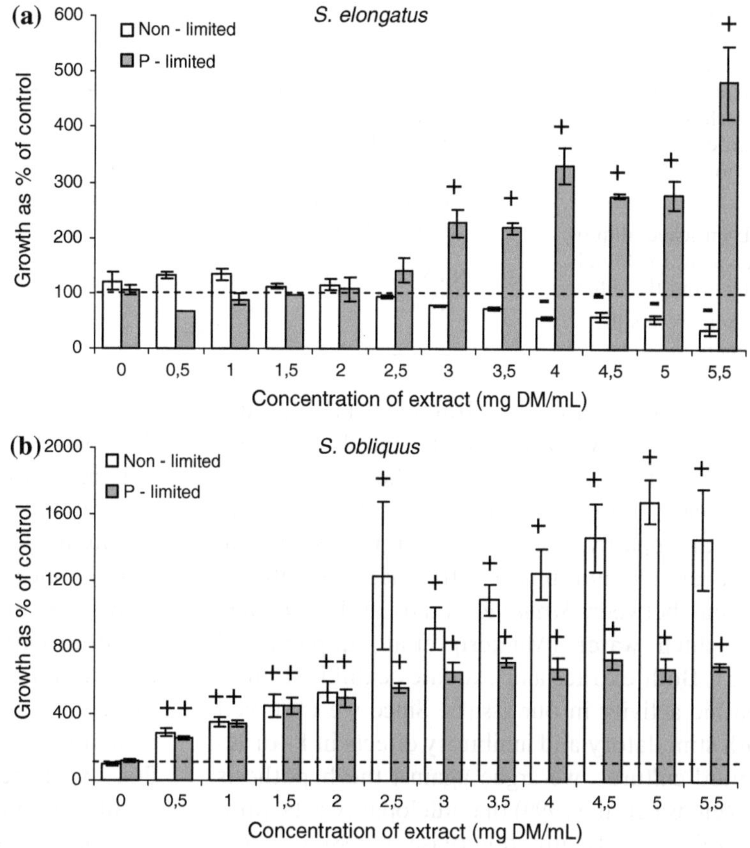

Fig. 3 Growth relative to the control in non-limited (white bars) and P-limited cultures (grey bars) of *Synechococcus elongatus* (**a**) and *Scenedesmus obliquus* (**b**) over a range of *Stratiotes* extract concentrations. Bars indicate average values (n = 3), error bars are 1 SE. + and – indicate significant inhibitory and stimulatory effects on the growth of the phytoplankton species

cyanobacteria than green algae being sensitive to allelopathic substances. Jasser (1994) and Körner & Nicklisch (2002) showed similar results, but Mulderij et al. (2005b) observed no significant differences in the sensitivity of cyanobacteria and green algae to allelopathic exudates from *Stratiotes*.

Table 3 Results agar diffusion assays with *Stratiotes* exudates from microcosms (set-up 2)

Incubation time [day]	Methanol fraction (%)	Clearing area diameter [cm]	
		Exudate	Control
1	50	(+)/(+)	0.6
	90	0.8/–	0.7
	100	–/+	–
2	50	–/+	–
	90	–/–	1
	100	–/–	–
3	50	(+)/–	(+)
	90	1.1/–	–
	100	–/–	–

SPE-enriched aliquot equivalent to 600 ml water (= 3.96 g FM) from the microcosms

( ), Inhibition less strong. Algal cells not completely absent in the clearing areas, only lower densities observed; –, no inhibition; +, slight inhibition

Similar to Fitzgerald (1969), P-limitation of *S. elongatus* did not increase the sensitivity of this cyanobacterium to *Stratiotes* extracts, while P-limited *S. obliquus* cultures showed opposite effects, corresponding with the hypothesis that nutrient-stressed organisms are more susceptible to allelopathy (Reigosa et al. 1999). K-limitation of *S. elongatus* did not intensify its sensitivity to *Stratiotes* extract. The stimulatory effects on K-limited target cells may have occurred because the extract acted as K-source. The extract might have been a P-source, because we observed both stimulatory and inhibitory effects of P-limited target cells upon extract addition. Organic phosphorus compounds might bind to C18 and may have been present in the 50% fraction, causing a stimulation of phytoplankton growth, while inhibitory compounds present in the 90–100% fractions resulted in adverse effects on

Table 4 Results agar diffusion assay with *A. variabilis* and *Stratiotes* exudates from three sites in ditches near Lake Naardermeer (C18 fractions)

SPE-enriched aliquot equivalent to 600 ml water from the ditch

[a] plants were still submerged

Site description	Methanol fraction (%)	Clearing area diameter (cm)				
		7 April[a]	20 April[a]	7 May	21 May	14 June
Inside *Stratiotes* (Ditch 1)	50		>3.0	–	0.5	(1)
	90		>2.5	1.0	(0.7)	0.9
	100		0.8	–	1.5	–
Outside *Stratiotes* (Control, Ditch 1)	50		1.5	–	1.0	(0.5)
	90		2.2	(+)	0.8	(0.5)
	100		1	(+)	–	1
No *Stratiotes* (Control, Ditch 2)	50		2.4	(+)	–	–
	90	>2.0	2.1	(+)	1.4	0.9
	100		0.8	–	0.5	–

phytoplankton growth. Brammer (1979) and Brammer & Wetzel (1984) stated that the absence of phytoplankton in the presence of *Stratiotes* is caused by nutrient limitation rather than allelopathy. Recent measurements in situ, however, did not reveal differences in nutrient content between *Stratiotes* stands and *Stratiotes*-free ditch water (Mulderij et al., unpublished data). Both extract and exudates exhibited allelopathic activity in our assays. Since we observed both stimulatory and inhibitory effects in P- or K-limited cultures, we argue against the hypothesis of Reigosa et al. (1999) that allelopathy acts more severely on already otherwise stressed target cells.

ADAs with *Stratiotes* exudates from laboratory cultures showed inhibitory effects after an incubation period of at least 3 days. Furthermore, exudates from a natural *Stratiotes* stand in Lake Naardermeer showed inhibitory effects, but some controls exhibited also a slight activity, which was, however, in general less strong than the exudates. Inhibitory effects were strongest when *Stratiotes* was still submerged, indicating that plants might start excreting allelopathic compounds at an early life-stage. This would be an effective strategy, because early in spring the competition with other photoautrophs is probably strongest. There are more indications for higher allelopathic activity in younger *Stratiotes* plants than in older ones (Mulderij et al., 2005a).

Compared to other studies (e.g., Gross et al., 1996; Gross et al., 2003) and based on the biomass of *Stratiotes* needed in both extract and exudate to achieve an inhibition, the allelopathic effects of *Stratiotes* shown in our study, are relatively weak. Mulderij et al. (2005a, b), on the contrary, observed inhibitory effects of *Stratiotes* exudates on phytoplankton growth ranging between 8 and 51%. The allelopathic activity of plant extracts does not reflect the potential allelopathic activity of the same substances once they are excreted. The allelopathic activity of macrophytes further depends on the chemical nature of allelopathic substance(s) and on the rate at which they are produced and excreted. As a consequence, experiments with exudates are ecologically more relevant. Our first experiment with *Stratiotes* exudates showed significant inhibitory effects of SPE-enriched *Stratiotes* medium on the cyanobacterium *A. variabilis*, indicating that *Stratiotes* may excrete the allelopathic substances. However, exudation experiments overall did not yet show a clear effect and differed from previous experiments using macrophyte filtrates (Mulderij et al., 2005a, b). This might be due to an incomplete binding on the C18 filters used, or because the active compounds are otherwise lost during the SPE process. The inhibitory effects found in field exudates of macrophyte-free control stations might be caused by humic compounds, which are also known to have allelopathic properties (Serrano, 1992; Mulderij et al., unpublished data).

Knowing the chemical nature of allelopathically active compounds, facilitates studies of plant content, exudation and mode of action. Bioassays with PVPP treated extracts showed that the active substance(s) in *Stratiotes* are most likely not of phenolic nature. Smolders et al. (2000) showed that the phenolic content of *Stratiotes* leaves is relatively low compared to other macrophytes. Our results contradict those of Usenko et al.

(2002) who proposed that the algicidal activity in *Stratiotes* is based on phenolic acids. Our further characterization of the allelopathic compounds in *Stratiotes* indicated that the allelopathic substance is heat-stable and moderately lipophilic.

Conclusion

Our assays showed both stimulatory and inhibitory effects of *Stratiotes* extract and exudates on phytoplankton, where cyanobacteria were more sensitive than other phytoplankton species. This differential sensitivity of phytoplankton species may affect the biomass and composition of phytoplankton populations under natural conditions. Nutrient limitation of phytoplankton cells did not increase their sensitivity to allelopathic extracts of *Stratiotes*. The substance(s) responsible for allelopathic inhibition of phytoplankton are moderately lipophilic and most likely not phenolic compounds.

Acknowledgements Parts of this study were supported by the Schure-Beijerink-Popping foundation (SBP 2001-29). We are grateful to Claudia Feldbaum for her assistance in the laboratory.

References

Alahari, A. & S. K. Apte, 2004. A novel potassium deficiency-induced stimulon in *Anabaena torulosa*. Journal of Biosciences 29: 153–161.

Anthoni, U., C. Christophersen, J. Madsen, S. Wium-Andersen & N. Jacobsen, 1980. Biologically active sulphur compounds from the green alga *Chara globularis*. Phytochemistry 19: 1228–1229.

Bloemendaal, F. H. J. L. & J. G. M. Roelofs, 1988. Waterplanten en Waterkwaliteit [in Dutch]. Stichting Uitgeverij KNNV, Utrecht, The Netherlands.

Brammer, E. S., 1979. Exclusion of phytoplankton in the proximity of dominant water soldier (*Stratiotes aloides*). Freshwater Biology 9: 233–249.

Brammer, E. S. & R. G. Wetzel, 1984. Uptake and release of K^+, Na^+ and Ca^{2+} by the water soldier, *Stratiotes aloides* L. Aquatic Botany 19: 119–130.

Burks, R. L., G. Mulderij, E. M. Gross, J. I. Jones, L. Jacobsen, E. Jeppesen & E. Van Donk (in press). Center stage: The crucial role of macrophytes in regulating trophic interactions in shallow lake wetlands. In Bobbink, R., B. Beltman, J. T. A Verhoeven & D. F. Whigham (eds), Wetlands: functioning, biodiversity, conservation and restoration. Ecological Studies.

Cook, C. D. K. & K. Urmi-König, 1983. A revision of the genus *Stratiotes* (Hydrocharitaceae). Aquatic Botany 16: 213–249.

De Geus-Kruyt, M. & S. Segal, 1973. Notes on the productivity of *Stratiotes aloides* in two lakes in the Netherlands. Polish Archive for Hydrobiology 20: 195–205.

Fitzgerald, G. P., 1969. Some factors in the competition or antagonism among bacteria, algae and aquatic weeds. Journal of Phycology 5: 351–359.

Flores, E. & C. P. Wolk, 1986. Production by filamentous, nitrogen-fixing cyanobacteria of a bacteriocin and of other antibiotics that kill related strains. Archives of Microbiology 145: 215–219.

Fowler, J., L. Cohen & P. Jarvis, 1998. Practical Statistics for Field Biology. John Wiley and Sons Ltd. Chichester, UK.

Fowler-Walker, M. J. & S. D. Connell, 2002. Opposing states of subtidal habitat across temperate Australia: consistency and predictability in kelp canopy-benthic associations. Marine Ecology Progress Series 240: 49–56.

Godmaire, H. & D. Planas, 1986. Influence of *Myriophyllum spicatum* L. on the species composition, biomass and primary productivity of phytoplankton. Aquatic Botany 23: 299–308.

Gross, E. M., 2003. Allelopathy of aquatic autotrophs. Critical Reviews in Plant Sciences 22: 313–339.

Gross, E. M., D. Erhard & E. Ivanyi, 2003. Allelopathic activity of *Ceratophyllum demersum* L. and *Najas marina* ssp. *intermedia* (Wolfgang) Casper. Hydrobiologia 506/509: 583–589.

Gross, E. M., H. Meyer & G. Schilling, 1996. Release and ecological impact of algicidal hydrolysable polyphenols in *Myriophyllum spicatum*. Phytochemistry 41: 133–138.

Gross, E. & R. Sütfeld, 1994. Polyphenols with algicidal activity in the submerged macrophyte *Myriophyllum spicatum* L. Acta Horticultura 381: 710–716.

Gross, E. M., C. P. Wolk & F. Jüttner, 1991. Fischerellin, a new allelochemical from the freshwater cyanobacterium *Fischerella muscicola*. Journal of Phycology 27: 686–692.

Guillard, R. R. L. & C. J. Lorenzen, 1972. Yellow-green algae with chlorophyllide *c*. Journal of Phycology 8: 10–14.

Jasser, I., 1995. The influence of macrophytes on a phytoplankton community in experimental conditions. Hydrobiologia 306: 21–32.

Jasser, I, 1994. Influence of *Ceratophyllum demersum* on phytoplankton community in experimental conditions. Verhandlungen der Internationalen Vereinigung für Theoretische und Angewandte Limnologie 25: 2291–2295.

Jüttner, F., J. Leonhardt & S. Mohren, 1983. Environmental factors affecting the formation of mesityloxide, dimethylallylic alcohol and other volatile compounds excreted by *Anabaena cylindrica*. Journal of General Microbiology 129: 407–412.

Kornatowski, J., 1938/84. Morphological forms of the water soldier. Acta Hydrobiologica 25/26: 145–156.

Körner, S. & A. Nicklisch, 2002. Allelopathic growth inhibition of selected phytoplankton species by submerged macrophytes. Journal of Phycology 38: 862–871.

Loomis, W. D. & Battaile J., 1966. Plant phenolic compounds and the isolation of plant enzymes. Phytochemistry 5: 423–438.

Lürling, M., 1999. The smell of water. Grazer-induced Colony Formation in Scenedesmus. PhD-thesis, University of Wageningen, The Netherlands.

Molisch, H., 1937. Der Einfluss einer Pflanze auf die andere - Allelopathie, Fisher, Jena.

Mulderij, G., W. M. Mooij & E. van Donk, 2005a. Allelopathic growth inhibition and colony formation of the green alga *Scenedesmus obliquus* by the aquatic macrophytes *Stratiotes aloides*. Aquatic Ecology 39: 11–21.

Mulderij, G., W. M. Mooij, A. J. P. Smolders & E. van Donk, 2005b. Inhibition of phytoplankton by allelopathic substances from *Stratiotes aloides*. Aquatic Botany 82: 284–296.

Mulderij, G., E. van Donk & J. G. M. Roelofs, 2003. Differential sensitivity of green algae to allelopathic sub-stances from *Chara*. Hydrobiologia 491: 261–271.

Planas, D., F. Sarhan, L. Dube, H. Godmaire & C. Cadieux, 1981. Ecological significance of phenolic compounds of *Myriophyllum spicatum*. Verhandlungen der Internationalen Vereinigung für Theoretische und Angewandte Limnologie 21: 1492–1496.

Reigosa, M. J., A. Sánchez-Moreiras, L. González, 1999. Ecophysiological approach in allelopathy. Critical Reviews in Plant Sciences 18: 577–608.

Scheffer, M, 1998. Ecology of shallow lakes. Chapman & Hall, London. 357 pp.

Schrader, K. K., M. Q. DeRegt, C. S. Tucker & S. O. Duke, 1997. A rapid bioassay for selective algicides. Weed Technology 11: 767–774.

Serrano, L., 1992. Leaching from vegetation of soluble polyphenolic compounds and their abundance in temporary ponds in the Donana National Park (SW Spain). Hydrobiologia 229: 43–50.

Smolders, A. J. P., L. H. T. Vergeer, G. van der Velde & J. G. M. Roelofs, 2000. Phenolic contents of submerged, emergent, and floating leaves of aquatic and semi-aquatic macrophyte species: why do they differ? Oikos 91: 307–310.

Suikkanen, S., G. O. Fistarol & E. Granéli, 2005. Effects of cyanobacterial allelochemicals on a natural plankton community. Marine Ecology Progress Series 287: 1–9.

Usenko, O. M., A. I. Sakevich & V. D. Palamarchuk, 2002. Influence of phenol acids of hydrophytes on the development of plankton algae. International Journal on Algae 4: 20–28.

van Donk, E. & W. J. van de Bund, 2002. Impact of submerged macrophytes including charophytes on phyto- and zooplankton communities: allelopathy vs other mechanisms. Aquatic Botany 72: 261–274.

Willis, R. J., 1985. The historical basis of the concept of allelopathy. Journal of the History of Biology 18: 71–102.

Wium-Andersen, S., U. Anthoni, C. Christophersen & G. Houen, 1982. Allelopathic effects on phytoplankton by substances isolated from aquatic macrophytes (Charales). Oikos 39: 187–190.

Wium-Andersen, S., 1987. Allelopathy among aquatic plants. Archiv für Hydrobiologie, Beiheft Ergebnisse der Limnologie 27: 167–172.

Hydrobiologia (2007) 584:101–109
DOI 10.1007/s10750-007-0595-8

SHALLOW LAKES

The role of aquatic macrophytes in microhabitatual transformation of physical-chemical features of small water bodies

Tomasz Joniak · Natalia Kuczyńska-Kippen · Barbara Nagengast

Abstract The paper presents the results of an examination of the phycical-chemical parameters of water together with an analysis of the chlorophyll *a* concentration of 12 small water bodies situated within urban and suburban areas of the city of Poznań (mid-west Poland)—typical mid-forest, strongly anthropogenically modified in the urban landscape, strongly antropogenically modified in an agricultural area and clay-pits. There were zones of open water (Unvegetated Zone) as well as zones of rush and aquatic vegetation (Vegetated Zone) distinguished in the examined ponds. The influence of the rush vegetation, nymphaeids and elodeids on the abiotic parameters of an aquatic environment was examined. Water samples were taken during the summer of 2004 from 12 stations within the open water and 24 within macrophytes. The plant matter was randomly collected in triplicate from the central part of the vegetated stand. The influence of macrophytes on the abiotic features of water was estimated using the parameter of the plant length (cm l^{-1}) and the plant biomass (g l^{-1}). In the studied ponds 12 aquatic macrophyte communities were distinguished. A salient feature of submerged macrophytes was a great density of plant stems along with considerebly low biomass, however, the rush vegetation (*Phragmitetum communis*, *Typhetum latifoliae*) when compared to nymphaeids (*Polygonetum natantis*, *Potametum natantis*) and elodeids (*Potametum lucentis*) was characterised by lower stem densities and higher biomass. The water bodies were alkaline and of pronounced hardness. In most of them high trophy conditions were found with especially high concentrations of phosphorus (96 μg l^{-1} on average). There was significant differentiation in the water chemistry (mainly in respect to mineral compounds) between the Vegetated and Unvegetated Zones as well as between particular aquatic macrophyte communities.

Keywords Shallow lakes · Macrophyte habitats · Water chemistry · Nutrients · Water trophy

Guest editors: R. D. Gulati, E. Lammens, N. De Pauw & E. Van Donk
Shallow lakes in a changing world

T. Joniak (✉) · N. Kuczyńska-Kippen · B. Nagengast
Department of Water Protection, Adam Mickiewicz University, Umultowska 89, 61-614 Poznan, Poland
e-mail: tjoniak@wp.pl

Introduction

The vegetation of water and swamp habitats creates a complex area in the ecotone zone between water and land, where the transport of allochtonic matter takes place. The nearshore or meadow vegetation may be a barrier restricting the distribution of anthropogenic contamination (e.g., Szpakowska, 1999). In relation to aquatic macrophytes, opinions considering the role of macrophytes in modyfying the physical-chemical features of their environment have as yet not been defined. On one hand the dynamics of water in shallow reservoirs is often determined by irregular processes e.g., floods, wind mixing etc., which have a great influence on water parameters (Joniak et al., 2000). On the other hand overgrowing macrophytes may intake nutrients from water or restore them from sediments into the water (Granéli & Solander, 1988). Nowadays, particular attention is paid to the relative role of aquatic vegetation, which influences both physical, chemical and biological parameters.

The role of macrophytes in nutrient circulation depends on the biotic and abiotic conditions of the environment and also on the taxonomical structure, biomass and life cycles of plants (Lampert & Sommer, 2001). In the case of the stage of plant infestation, macrophytes may have a positive effect on the ecosystem, increasing its biodiversity, e.g., through creating a refuge for small fish or invertebrate animals or providing inhabiting organisms with cooler and better oxygenated, (compared to lower parts of the reservoir), waters. However, the abundant development of water vegetation decreases the water surface, lowers the temperature and cumulates biogenic substances, thereby restricting their availability to phytoplankton (Kajak, 2001).

Biophilic elements and organic substances as well as their abiotic and biotic relations originate in the complexes of rush and water vegetation of small water bodies undergoing the anthropogenic pressure (Wetzel, 1995; Gołdyn, 2000). The aim of the study was to determine the differentiation of the physical and chemical parameters of habitats within macrophytes (Vegetated Zone) and the open water (Unvegetated Zone) zones of small water bodies located in the urban and suburban area of a large city aglomeration (Poznań, western Poland). Various ecological types of aquatic vegetation were included: helophytes, nymphaeids and elodeids. A hypothesis was proposed that ponds as water bodies of small area and shallow depth cannot be treated as a homogenic unit, since particular habitats created by macrophytes may differentiate the physical–chemical features of an environment.

Materials and methods

The examination of macrophytes and the physical-chemical features of water were carried out (ones for each pond) in June/July 2005 on 12 small water bodies of an area of between 0.5 and 5 ha. Due to the diurnal dynamics of oxygen, temperature and pH the research was undertaken in the mornings (between 8–10 a.m.). The analysis included stands located in the open water zone as well as among different ecological types of rush and water vegetation. According to the character of the catchment area the following types of water bodies were distinguished: clay-pits (No. 1, 2, 3), strongly antropogenically modified in the urban landscape (No. 4, 5, 6), strongly antropogenically modified in an agricultural area (No. 7, 8, 9) and typical mid-forest (No. 10, 11, 12). Most of the ponds were closed units in the hydrological sense (possessing neither inflows nor outflows).

The water samples for physical–chemical analysis were taken seperately from each VZ (24 station) and from UZ (12 station). Water temperature, dissolved oxygen, oxygen saturation, pH, conductivity (Sension 156 Hach) and Secchi depth were measured directly at the sampling sites. Water samples for chemical analysis were collected in polyethylene flasks washed in 10% HCl and rinsed several times in distilled water. Later the water samples were analysed in the laboratory to determine total phosphorus (TP), total reactive phosphorus (TRP), nitrate (NO_3), nitrite (NO_2), ammonium (NH_4) and total hardness ($CaCO_3$). These analyses were carried out following standard methods as reported in Hermanowicz et al. (1999). Dissolved inorganic nitrogen (DIN) concentration was calculated by summing nitrate, nitrite and ammonium concentration. Water colour was measured on filtrate (0.45 μm) at 436 nm and dissolved organic substances (DOM) at 254 nm using a Cadas 200 UV–VIS (Dr Lange) spectrophotometer with a 1 cm (A_{436}) and 5 cm (A_{254}) glass cell against deionised water. The chlorophyll *a* concentration was determined with a spectrophotometer (at 663 and 750 nm) following extraction after 24 h in 4°C acetone (Wetzel & Likens, 2000). Trophic conditions of water were assessed according to the classification suggested by Nürnberg (1996).

The plant material was collected in triplicate from randomly chosen central parts of each macrophyte stand. Macrophyte stems were cut out from an area of 0.25 × 0.25 m and 0.2 m depth. In the case of helophytes only the underwater parts of plants were sampled. Later the length of stems was measured in order to determine the plant densities and the stems were dried and weighed to obtain the biomass. The length of particular macrophyte stems and their biomass adequate to 1 l of water were estimated afterwards.

In order to determine the influence of macrophyte habitats on the chemical and physical parameters of water the Canonical Correspondence Analysis (CCA) was applied. CANOCO 4.5 software was used to perform the analyses (Ter Braak & Šmilauer, 1998).

Statistic calculations were made using STATISTICA 6.0 software.

Results

Macrophyte habitats

In the investigated water bodies rush and water vegetation (elodeids and nymphaeids) zones were established. The influence of aquatic vegetation on the abiotic parameters of water were measured by applying two parameters—the plant length (cm l^{-1}) and its biomass (g l^{-1}). A great density followed by considerably low biomass were observed in the case of submerged macrophyte communities—*Zannichellietum palustris* (233 ± 66.7 cm l^{-1}, 0.4 ± 0.1 g l^{-1}), *Ceratophylletum demersi* (99.4 ± 38.4 cm l^{-1}, 0.4 ± 0.2 g l^{-1}) and *Myriophylletum spicati* (74.9 ± 7.3 cm l^{-1}, 1 ± 0.3 g l^{-1}). The rush vegetation (*Phragmitetum communis*, *Typhetum latifoliae*) compared to nymphaeids (*Polygonetum natantis*, *Potametum natantis*) and elodeids (*Potametum lucentis*) was characterised by significantly lower densities (18.8 ± 12.8 and 78.9 ± 61.3 cm l^{-1}, respectively). In the case of biomass this ratio was opposite (1.11 ± 0.5 and 0.6 ± 0.4 g l^{-1}).

Within the submerged macrophytes the longest stems were found for plants among *Z. palustris* and *Ceratophylletum demersi* in pond No. 8. In the group of plants of floating leaves (*Polygonetum natantis*, *Potametum natantis*) the plant stems were of low and similar lengths, while the highest biomass was recorded in the *Polygonetum natantis* from both elodeids and nymphaeids. With respect to the rush vegetation much longer stems and therefore greater densities were obtained in the stands of *Phragmitetum communis*, compared to *Typhetum latifoliae*. Also the biomass of *Phragmites australis* and *Schoenoplectus lacustris* were higher than that of *Typha latifolia*.

Water chemistry

In the UZ of examined water bodies considerable differences in the oxygen conditions of the surface waters were recorded—from the lowest in pond No. 7 (3.9 mg O_2 l^{-1}) to the highest in No. 10 (16.0 mg O_2 l^{-1}). The oxygen saturation in only four ponds was lower than 100% (min. 43% relative saturation), and in the remaining water bodies it exceeded this value (max. 170%). Comparing the saturation of water in VZ with UZ no statistically significant differences were found. However, according to the rule of statistical conclusions, nonascertainment of statistical significance does not confirm its lack.

Most of studied reservoirs had an alkaline reaction, with the exception of mid-forest pond No. 12 and field pond No. 8, which had a neutral and slightly acid reaction (Table 1). The reaction of water in the VZ and UZ was similar and differences between phytocoenosis, even though they were often considerable, were not statistically significant ($P > 0.05$). In the examined water bodies a high differentiation concerning electrical conductivity was recorded—from low values in the mid-forest ponds to higher in the ponds located within the borders of the city (Table 2). The strong linear correlation found between conductivity and water hardness ($r = 0.86$, $P < 0.0004$, $n = 12$) suggests a distinct influence of bivalent cations on the degree of water mineralisation. Analysing water hardness, three

Table 1 Characteristics of macrophyte habitats

Pond No.	Habitat	L cm l^{-1}	V cm^{-3}	B g l^{-1}	Cond µS cm^{-1}	pH	Oxygen mg O_2 l^{-1}	DOM abs cm^{-1}	TP µg P l^{-1}	TRP µg P l^{-1}	NH_4 µg l^{-1}	Hardness mg $CaCO_3$ l^{-1}	Chl a µg l^{-1}
1	M.s	75	5.6	1	566	9.9	10.9	0.061	27	3	0	132	0.8
	Ph.a	19.3	3.5	1.4	572	9.5	8.3	0.061	34	21	26	133	12.8
2	C.d	57.1	3.3	0.4	2627	9.2	13	0.058	124	104	120	811	1.9
3	Ph.a	15.3	6.1	1.4	1318	8.4	7.44	0.013	74	65	739	556	6.2
4	Ph.a	32.2	9.8	1.5	219	9.5	9.73	0.168	17	2	1058	16	81.3
	P.a	27.8	4.6	1.6	116	9.6	8.75	0.148	13	8	1027	14	57.7
5	T.l	6.9	18.9	1.2	1090	8.3	7.51	0.308	196	129	887	560	44.9
6	Ph.a	14.8	4.8	1.3	1002	9.9	14.6	0.371	258	176	1914	281	396.7
7	Ph.a	27.1	9.7	1.9	785	7.7	2.53	0.089	169	140	91	341	3.4
	Z.p	233	2.5	0.38	784	7.9	4.84	0.034	134	126	113	385	1.9
	P.p	84.4	5.3	0.6	769	8.4	9.46	0.024	154	134	113	343	3.8
8	P.n	18.7	7.8	0.4	424	6.7	10.5	0.981	56	12	1836	118	19.2
	C.d	97.2	8	0.5	393	7.4	8.6	0.926	75	15	1447	–	33.6
	T.l	1.9	10.2	0.3	417	6.7	8.2	1.03	63	9	1564	112	4.5
9	P.c	27.3	7.1	0.4	1293	9.8	10.65	0.477	203	137	708	136	7
	T.l	4.9	24.6	0.9	1300	9.2	9.05	0.478	265	184	319	97	32.7
	C.d	84	7	0.5	1300	9.2	10.32	0.479	236	168	692	101	35.7
10	P.l	34.8	9.8	0.5	401	8.9	13.4	0.824	37	8	918	190	1.1
	Ph.a	15.2	6.9	0.9	459	7.7	13.6	0.594	44	5	817	202	5.6
	B.c	89.2	7.7	0.7	455	8.5	13	0.5	43	8	817	246	2.6
	S.l	37.5	34.8	0.9	444	8.6	13.2	0.526	40	32	1190	204	1.7
11	Ph.a	32	104	1.2	471	7.9	10.2	0.463	64	5	1042	37	0.4
	P.l	18	6	0.5	401	8.6	13.4	0.426	46	9	1253	26	1.3
12	C.d	99.5	10.2	0.4	463	7.2	7	0.659	103	60	739	91	20.5

(L, macrophyte length; V, volume; B, biomass) and physico–chemical parameters of water collected from particular stations of vegetated zones (S.l, Schoenoplectus lacustris; Ph.a, Phragmites australis; T.l, Typha latifolia; P.n, Potamogeton natans; P.a, Polygonum amphibium f. natans; Z.p, Zannichellia palustris; M.s, Myriophyllum spicatum; B.c, Batrachium circinatum; P.c, Potamogeton crispus; P.l, Potamogeton lucens; P.p, Potamogeton pectinatus; C.d, Ceratophyllum demersum)

groups of reservoirs were distinguished—mid-forest ponds of soft (85.5–170.8 mg $CaCO_3$ l^{-1}), field ponds of moderately hard (170.9–341.6 mg $CaCO_3$ l^{-1}) and urban of hard water (>341.7 mg $CaCO_3$ l^{-1}). The only exception was mid-city pond No. 4, whose waters contained only a sparce quantity amounts of cations. Among UZ areas this parameter was generally higher compared to VZ, where it was characterised by great variation both in the case of ecological group of aquatic plants (max. values in the rushes), and in particular phytocoenosis (max. in *Ceratophylletum demersi*).

High concentrations of phosphorus were detected in the ponds which reflected the high level of trophy of their waters. Hypertrophy (>100 µg P l^{-1}) was found in the reservoirs No. 2, 5, 6, 7, 9, eutrophy (31–100 µg P l^{-1})—No. 3, 8, 10, 11, 12, while meso-eutrophy (10–30 µg P l^{-1}) only in two – No. 1 and 4. The middle percentage TRP in a pool of TP among VZ water bodies lowered along with trophy and accounted for 71 ± 13% in hypereutrophic waters, 40 ± 34% in eutrophic and 34 ± 27% in mesoeutrophic.

The obtained results revealed a significant distinction of chemical features of the waters of reservoirs with the rush zone, where the TRP concentrations were usually lower in the TP. Plant communities arranged in the increasing gradient of participation of TRP in the total phosphorus created the following sequence: *Typha latifolia*/*Phragmites australis* > *Myriophyllum spicatum* > *Batrachium circinatum* > *Ceratophyllum demersum* > *Potamogeton lucens* > *Polygonum amphibium* > *Potamogeton crispus* > *Schoenoplectus palustris* > *Potamogeton pectinatus* > *Zannichella palustris*.

The second biophilic element—nitrogen—was mainly represented by ammonium form. The concentrations of its mineral forms in UZ belonged to high—80% DIN on average (Table 2). Ammonium appeared in the highest

Table 2 Characteristics of physical and chemical parameters of water in unvegetated zones of typically mid-forest (I), strongly anthropogenically modified in the urban landscape (II), agricultural area (III) and clay-pits (IV) water bodies

	Cond ($\mu S\ cm^{-1}$)	pH	Oxygen ($mg\ O_2\ l^{-1}$)	DOM ($abs\ cm^{-1}$)	Colour ($mg\ Pt\ l^{-1}$)	TP ($\mu g\ P\ l^{-1}$)	TRP ($\mu g\ l^{-1}$)	NH_4 ($\mu g\ l^{-1}$)	NO_3 ($\mu g\ l^{-1}$)	Hardness ($mg\ CaCO_3\ l^{-1}$)	Chl *a* ($\mu g\ l^{-1}$)
I	463 ± 13	8 ± 0.7	11 ± 5.4	0.558 ± 0.322	36 ± 8	46 ± 14	10 ± 4	1161 ± 316	859 ± 84	117 ± 91	3 ± 4
II	769 ± 480	9.2 ± 0.9	11.8 ± 2.9	0.223 ± 0.11	52 ± 20	153 ± 131	87 ± 70	1242 ± 491	1288 ± 921	278 ± 271	162 ± 175
III	834 ± 441	7.8 ± 1.3	7.7 ± 3.9	0.483 ± 0.41	33 ± 16	119 ± 62	88 ± 66	680 ± 660	1284 ± 947	169 ± 145	11 ± 8
IV	1503 ± 1034	9.1 ± 0.6	10.8 ± 2.7	0.087 ± 0.07	10 ± 5	67 ± 47	51 ± 42	203 ± 244	775 ± 443	504 ± 346	4 ± 3

concentrations reaching a maximum in pond No. 12—1525 μg $NH_4\ l^{-1}$ (90% DIN). A lower concentration of this form was found in clay-pit 1 and 2, where it constituted 18% and 31% DIN respectively. The concentrations of easily bioavailable nitrates were usually lower than the concentrations of ammonium—from 407 μg $NO_3\ l^{-1}$ (No. 3) to 2346 μg $NO_3\ l^{-1}$ (No. 6, 9), with the exception of clay-pits (No. 1, 2), where the distribution of values was opposite. Nitrites appeared in trace concentrations (from 0 to 36 μg $NO_2\ l^{-1}$).

The waters of the examined ponds were mostly rich in ammonium ions. In the rush communities, in comparison with nymphaeids and elodeids, the concentrations of ammonium were slightly higher—878 ± 581 μg $NH_4\ l^{-1}$ and 752 ± 563 μg $NH_4\ l^{-1}$ respectively. The maximum was found in the stand of *Phragmitetum communis* of field reservoir No. 6 (1914 μg $NH_4\ l^{-1}$), which contributed to the quality distinctiveness of the waters of this water body (Fig. 1). A complete lack of ammonium was noticed in the *Myriophylletum spicati* stand of clay-pit No. 1. In the case of nitrogen concentrations, which were lower than ammonium ions, no constant tendencies concerning different macrophytes were recorded.

The content of dissolved organic matter (DOM) expressed as the water absorbance value was differentiated in the studied ponds. Among the UZ of small water bodies the concentrations of DOM varied from very low in clay-pit No. 1

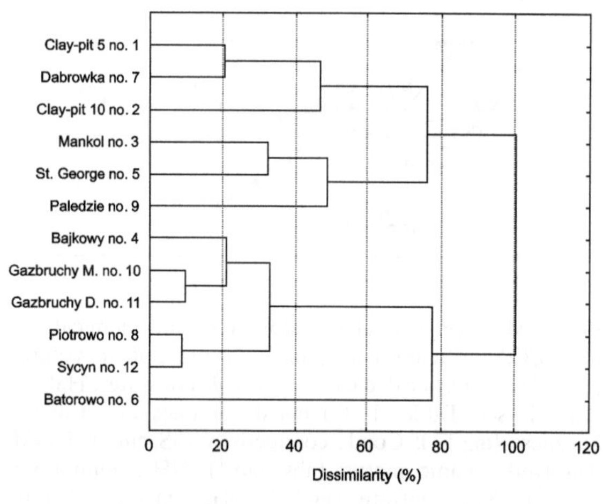

Fig. 1 Tree diagram for physical–chemical variables. Average linkage method with City distance (Manhattan)

(0.032) to very high in ponds No. 12 (0.929) and 8 (0.905). The reservoirs situated within the city borders were characterised by a lower DOM content when compared with the mid-forest reservoirs (Table 2). An interesting feature of some of the water bodies situated within the urban area were high values of absorbance indicating the occurrence of organic carbon bounded with high-molecular organic substances such as humic substances.

In the vegetated areas compared to UZ higher concentrations of DOM, especially in the rush zone with *Typha latifolia* and submerged macrophytes with *Ceratophyllum demersum*, were observed (Fig. 2). Moreover, the water colour of VZ was usually higher than in UZ. Record values were recorded in the pastoral pond No. 6 within the *Phragmitetum* community—176 mg l^{-1} Pt, which confirmed the distinctivness of this water body (Fig. 1).

Analysis of chlorophyll *a* concentrations confirmed a zonal differentiation of chemical parameters of waters connected with the presence of aquatic macrophytes. In UZ the mean concentrations of this pigment accounted for 45.5 μg l^{-1}, while in VZ for 34 μg l^{-1}. The maximum concentration was found in hypereutrophic pond No. 6 in the water of UZ (363.6 μg l^{-1}), as well as among VZ with *Phragmites australis* (396.7 μg l^{-1}). Analysing the spatial differentiation of chlorophyll *a* concentrations between various macrophyte communities it was noted that they were highest in the *Phragmites australis*, which created a habitat of strong light conditions due to the specific structure of the macrophyte shoots (Table 1).

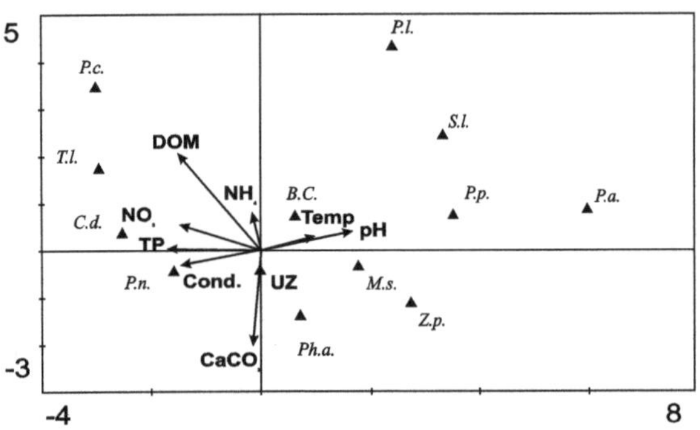

Fig. 2 Macrophytes habitats along physical–chemical factors: CCA ordination diagram with macrophyte volume (cm l^{-1}) and physical–chemical variables of water. Habitat symbols see Table. 1. Chemical variables are: $CaCO_3$, hardness (mg l^{-1}); Cond., conductivity (μS cm^{-1}); DOM, dissolved organic matter (abs. cm^{-1}); NH_4, ammonium (μg l^{-1}); NO3, nitrate (μg l^{-1}); pH, reaction; Temp., temperature (°C); TP, total phosphorus (μg P l^{-1}); UZ, unvegetated zone of water

The analysis of CCA did not reveal explicit and clear solutions for the size and direction of abiotic parameter changeability within a habitat as a result of the effect of aquatic macrophytes. However, the quality distinctiveness of waters within areas lacking in macrophytes (UZ) compared to VZ was found, which could confirm the hypothesis according to which small water bodies should be treated as non-homogenic units in relation to abiotic features of the water environment. On analysing the physical and chemical characteristics of vegetated areas using CCA-ordination, it was predicted that three groups of parameters had the strongest relationship with the type of macrophyte habitat. In the first group were temperature and pH which affected five communities representing submerged macrophytes as well as floating-leaved plants. The second group comprised parameters relating to the content of a mineral compound of nitrogen, phosphorus and DOM, which underwent differentiation in the habitats of plant communities—being characterised by high nutritional demands—belonging to the submerged type (*Ceratophyllum demersum, Potamogeton crispus*) and rushes (*Typha latifolia*). Conductivity and hardness occurred in the last group and they had the strongest relationship with the nymphaeid *Potamogeton natans.* Communities such as *Phragmites australis, Zannichellia palustris* and *Myriophyllum spicatum* revealed no relationship with abiotic parameters.

Discussion

Small water bodies located within large urban agglomerations undergo strong anthropogenic pressure—recreational and exploatative, which influences the floristic and community structure of aquatic macrophytes. In such conditions the accelaration of eutrophy may favour the dominance of plant species with high nutritional demands (Wetzel & Hough, 1973). At the same time the small and shallow water bodies favour the creation of a mosaic structure of water vegetation, which in turn may contribute to the differentiation of the physical–chemical features of a habitat. The character of surface waters depends to a large extent on the allochtonic parameters connected with the character and size of the catchment area. The morphology of the surrounding area (high trees, hills, banks of soil) strongly influence the mixing of water by wind (Joniak et al., 2000). In small water reservoirs also an important role, if not the most important, is played by autochtonic parameters, among which besides physical (sedimentation) and chemical (biodegradation of organic matter) the abundance and structure of aquatic macrophytes should be encountered (Asaeda et al., 2000). A high level of eutrophy, characteristic for small reservoirs, does not eliminate the fluctuations of environmental factors, however, it can change their causes and character (Wetzel, 1995). The increase in the speed of eutrophy processes is not a rule in the case of small water bodies, however, it is often accelerated by the inflow of the biogenic substances from the catchment area and by covering the pond bottom by the organic sediments (Hongve, 1999). On the other hand, the chemical content of soils, the landform features of the catchment area as well as the presence of submerged and emergent aquatic vegetation may contribute to the retainment and immobilization of great amounts of biogenic compounds (Joniak, 2005).

In small, shallow water bodies a mosaic structure of water vegetation was created, enhancing the differentiation of the chemical conditions of waters within even a small area. The character of surface waters depends considerably on the physical and chemical parameters of soils in the catchment area. The rush vegetation together with submerged and floating leaved plants, which are a frequent component of vegetation cover of most stagnant freshwater bodies, usually create the first transitional zone of land-water vegetation that overgrows not only the shallow parts of littoral but also the lake-shore (Podbielkowski & Tomaszewicz, 1996). In small water bodies the littoral area may cover the whole pond bottom. They comprise therefore an ecotone zone between land and water that undergoes the effect of allo- and autochtonic processes. The examination of physical and chemical features revealed their spatial differentiation in relation to UZ and VZ, as well as to particular macrophyte communities. This may be due to the fact that the aquatic plants influence the concentrations of biogenic elements in water by increasing or decreasing their contents and therefore they may affect other physical or chemical parameters of water. The quality distinctiveness of the water from among the rush vegetation, especially in the community of *Typha latifolia*, compared with other types of macrophytes (elodeids and nymphaeids) was probably due to the stronger modification of the habitat by helophytes that were characterised by the greatest biomass as well as due to the higher pressure of plants on the water environment in the range of intensity of biogen intake for the biomass production and of physical condition stabilization (Gołdyn, 2000).

A feature of aquatic macrophytes is their impact on the water chemism through releasing high—mollecular organic compounds (Wetzel & Hough, 1973) as well as multicomponent products of metabolism. The analysis of DOM confirmed the influence of aquatic macrophytes on the content of organic substances in water (Wetzel, 1995). However, it should also be added that in waters of high trophic conditions the presence of substantial concentrations of DOM may also result from the primary production of phytoplankton (Joniak et al., 2003). The mid-forest and pastoral ponds are exposed to the additional inflow of great amounts of organic carbon from the humus compounds, which contribute the main part of DOM in the underground waters of field and forest soils (Szpakowska, 1999). After reaching the surface waters they may significantly change the abiotic conditions, e.g., decrease the light availability and also turn phosphorus, nitrogen and metal cations into chelates compounds (Jones, 1992). Such reactions would also affect the physiological processes of vascular plants (Steinberg et al., 2003).

The source of high concentrations of ammonia nitrogen is mainly their inflow with rain and underground waters from the catchment area as well as some chemical processes e.g., dissimilative reduction of nitrate or the changing, together with the dissolved oxygen concentration, direction of nitrification. The factor that accelarated the content of ammonium could have been an abundance of organic matter in the sediments as an effect of ammonification (Joniak, 2005). The spatial differentiation of nitrate concentrations suggest the great significance of their assimilation not only by aquatic macrophytes but also by epiphytic organisms, which cover the underwater parts of aquatic plants and other substrata (Gołdyn, 2000).

To sum up it should be said that the aquatic vegetation of the examined small water bodies had a moddifying effect on their chemism and especially on their mineral element content. The spatial differentiation of the physical–chemical parameters of waters in relation to open water zones and those infested by macrophytes (UZ versus VZ), as well as differentiation between particular plant communities confirm the hypothesis of the internal differentiation and heterogeneity of a water body. However, as the amount of the ponds taken into analysis was not fully comprehensive it is difficult to present explicit settlements in relation to the aquatic macrophyte impact on the concentration of dissolved organic matter. However, it should be underlined that aquatic vegetation, irrespective of the water body size and the way of its use, significantly contributes to an increase of biocoenotic variation (Kuczyńska-Kippen et al., 2003; Kuczyńska-Kippen & Nagengast, 2006). Small water bodies create specific microecosystems, where seasonal changes in the thermical and chemical conditions reflect the dynamic balance between organisms and the environment in relation to numerous influential external factors.

Acknowledgements Research work has been financed by Polish State Committee for Scientific Research in 2005–2008 as research project 2 P06S 008 29.

References

Asaeda, T., V. K. Trung & J. Manatunge, 2000. Modelling the effects of macrophytes growth and decomposition on the nutrient budget in shallow lakes. Aquatic Botany 68: 217–237.

Gołdyn, R., 2000. Changes in Biological and Physico–Chemical Parameters of River Water Quality as a Result of Its Damming in Preliminary Lowland Reservoirs. Adam Mickiewicz University Press, Poznań.

Granéli, W. & D. Solander, 1988. Influence of aquatic macrophytes on phosphorus cycling in lakes. Hydrobiologia 170: 245–266.

Hermanowicz, W., J. Dojlido, W. Dożańska, B. Koziorowski & J. Zerbe, 1999. Fizyczno-chemiczne badanie wody i ścieków [The Physical–Chemical Analyses of Water and Wastewater]. Arkady Press, Warszawa.

Hongve, D., 1999. Production of dissolved organic carbon in forested catchments. Journal of Hydrology 224: 91–99.

Jones, R. I., 1992. The influence of humic substances on lacustrine planctonic food chains. Hydrobiologia 229: 73–91.

Joniak, T., A. Kamińska & R. Gołdyn, 2000. Influence of the Wind on Vertical Changes of Water Properties in a Shallow Reservoir. In Gurgul, H. (ed.), Physicochemical Problems of Natural Waters Ecology, Vol. II. Scient. Publ. Univ, Szczecin: 67–76.

Joniak, T., R. Gołdyn & A. Kozak, 2003. The primary production of phytoplankton in the Maltański Reservoir in Poland. Hydrobiologia 506–509: 311–316.

Joniak, T., 2005. Struktura i funkcjonowanie ekosystemów jezior humusowych w Drawieńskim Parku Narodowym [The biological structure and functioning of humic lake ecosystems in Drawieński National Park], Ph.D. thesis. Adam Mickiewicz University of Poznań.

Kajak, Z., 2001. Hydrobiologia-Limnologia. Ekosystemy wód śródlądowych [Hydrobiology–Limnology. Ecosystems of Inland Water]. PWN-Polish Scientific Publisher, Warszawa.

Kuczyńska-Kippen, N., B. Messyasz & B. Nagengast, 2003. Plankton communities within Chara fragilis of the mid-forest pond. Biologia, Bratislava 58/4: 555–561.

Kuczyńska-Kippen, N. & B. Nagengast, 2006. The influence of the spatial structure of hydromacrophytes and differentiating habitat on the structure of the rotifer and cladoceran communities. Hydrobiologia 559: 203–212.

Lampert, W. & U. Sommer, 2001. Ekologia wód śródlądowych [Ecology of freshwater]. PWN-Polish Scientific Publisher, Warszawa.

Nürnberg, G. K., 1996. Trophic state of clear and coloured, soft- and hardwater lakes with special consideration of nutrients, anoxia, phytoplankton and fish. Lake Reservation Management 12: 432–447.

Podbielkowski, Z. & H. Tomaszewicz, 1996. Zarys hydrobotaniki [An Outline of Hydrobotanics]. PWN-Polish Scientific Publisher, Warszawa.

Steinberg, Ch. E. W., P. S. Pflugmacher, T. Meinelt, R. Klocking & C. Wiegand, 2003. Pure humic substances have the potential to act as xenobiotic chemicals—a review. Fresenius Enir. Bull. 12: 391–401.

Szpakowska, B., 1999. Występowanie i rola substancji organicznych rozpuszczonych w wodach powierzchniowych i gruntowych krajobrazu rolniczego [Occurrence and Role of Organic Compounds Dissolved in

Surface and Ground Waters of Agricultural Landscape]. Nicolai Copernici University Press, Toruń.
Ter Braak, C. J. F. & P. Šmilauer, 1998. CANOCO Reference Manual and User's Guide to CANOCO for Windows. Microcomputer Power, Ithaca.
Wetzel, R. G., 1995. Death, detritus, and energy flow in aquatic ecosystems. Freshwater Biology 33: 83–89.
Wetzel, R. G. & R. A. Hough, 1973. Productivity and role of aquatic macrophytes in lakes. An assessment. Polish Archives of Hydrobiology 20:9–19.
Wetzel, R. G. & G. E. Likens, 2000. Limnological Analyses. Springer-Verlag.

Hydrobiologia (2007) 584:111–119
DOI 10.1007/s10750-007-0612-y

SHALLOW LAKES

Assessment in two shallow lakes of a hydroacoustic system for surveying aquatic macrophytes

I. J. Winfield · C. Onoufriou · M. J. O'Connell · M. Godlewska · R. M. Ward · A. F. Brown · M. L. Yallop

Abstract A technique for the rapid surveying of submersed aquatic vegetation by post-processing of data collected using a high frequency (420 kHz) digital echo sounder (BioSonics DT4000) has recently been developed and successfully tested in an estuarine environment by Sabol et al. [Sabol, B. M., R. E. Melton, R. Chamberlain, P. Doering & K. Haunert, 2002. Evaluation of a digital echo sounder system for detection of submersed aquatic vegetation. Estuaries 25: 133–141.], where it was used to map the cover and height of freshwater tape grass (*Vallisneria americana*) and seagrasses (*Thalassia testudinum*, *Haladule wrightii* and *Syringodium filiforme*). This technique, which is also spatially referenced by input from a global positioning system, has many potential applications in macrophyte studies in shallow lakes, although it has not yet been extensively tested in such habitats using systems of lower sound frequency. This paper reports such a test in two shallow (maximum depth c. 5.9 m) lakes of the Cotswold Water Park, U.K., using a 200 kHz digital echo sounder (BioSonics DT6000 and DT-X upgrade) and post-processing analysis using the now commercially available software EcoSAV, which incorporates the algorithms of Sabol et al. (2002). Hydroacoustic assessment of the coverage by macrophytes, mainly Nuttall's pondweed (*Elodea nuttallii*) and charophytes (*Chara* spp.), showed high agreement with those recorded during a simultaneous visual survey by underwater video recording ($r^2 = 0.8478$, $n = 74$, $P < 0.001$). Assessment of macrophyte height was also apparently consistent between the two systems, although the video system could not produce quantitative data and so statistical assessment of the agreement was not possible. Repeated hydroacoustic surveys over the course of the winter of 2003–2004 were conducted in one lake and illustrate the application of this new macrophyte survey technique.

Guest editors: R. D. Gulati, E. Lammens, N. De Pauw & E. Van Donk
Shallow lakes in a changing world

I. J. Winfield (✉)
Centre for Ecology & Hydrology, Lancaster Environment Centre, Library Avenue, Bailrigg, Lancaster LA1 4AP, UK
e-mail: ijw@ceh.ac.uk

C. Onoufriou · M. L. Yallop
School of Biological Sciences, University of Bristol, Woodland Road, Bristol BS8 1UG, UK

M. J. O'Connell · R. M. Ward
Wildfowl & Wetlands Trust, Slimbridge, Gloucestershire GL2 7BT, UK

M. Godlewska
International Centre for Ecology, ul. Tylna 3, Lodz 90 364, Poland

A. F. Brown
English Nature, Northminster House, Peterborough PE1 1UA, UK

Such applications include the rapid measurement of mean Percentage Volume Inhabited (PVI), which fell from 12.15% (95% confidence limits, ±0.55%) to 7.10% (±0.40%) over the course of the winter.

Introduction

Established manual techniques for characterising and monitoring aquatic vegetation are labour-intensive and generate observations of very limited spatial extent. Alternative optical techniques, such as aerial photography, provide large synoptic assessments of spatial patterns but are highly dependent on uncontrollable environmental factors such as water clarity, water surface roughness and cloud cover (see review by Madsen (1993)). In contrast, hydroacoustic techniques are largely free of these limitations and are now widely used in the assessment and study of other components of lake ecosystems, providing rapid, extensive and spatially-referenced data on fish, zooplankton and bottom sediments (e.g. Godlewska et al., 2002; Hoffman et al., 2002; Godlewska et al., 2004). The use of this approach for the assessment of macrophyte populations in lakes and rivers is now receiving increasing attention.

Hydroacoustic methods developed for macrophyte surveys include the use of horizontally-aimed side scanning sonar systems for delineating macrophyte beds (Bozzano et al., 1998; Moreno et al., 1998) and vertically-aimed echo sounders for quantifying vegetation height and density (Sabol & Burczyński, 1998). Although a number of researchers have reported success in detecting and qualitatively characterising macrophytes using hydroacoustics for over two decades (Maceina & Shireman, 1980; Duarte, 1987; Thomas et al., 1990; Fortin et al., 1993; Tegowski et al., 2003), fully quantitative assessment has been hampered by hardware and software limitations.

Following technological advances including the advent of highly portable Global Positioning Systems (GPS), the development of fully quantitative macrophyte assessment has recently made major advances as described by Sabol et al. (2002) in an estuarine environment, where one such system was used to map the cover and height of freshwater tape grass (*Vallisneria americana*) and seagrasses (*Thalassia testudinum*, *Haladule wrightii* and *Syringodium filiforme*). The software component of this system is now commercially available as EcoSAV (BioSonics Inc, Seattle, U.S.A., www.biosonicsinc.com) and is further described by Hoffman et al. (2002).

However, this recent development and its subsequent application have almost exclusively used only relatively high sound frequencies of 420 kHz (Hoffman et al., 2002; Sabol et al., 2002) or 430 kHz (Valley et al., 2005). In contrast, most hydroacoustic systems currently used in fresh waters, which are deployed primarily in fish studies, operate at considerably lower frequencies of c. 70 to 200 kHz (e.g. Jurvelius, 1991; Elliott et al., 1996; George & Winfield, 2000; Wanzenböck et al., 2003; Schmidt et al., 2005). Only two EcoSAV studies have used such frequencies. Firstly, Schneider et al. (2001) employed a sound frequency of 208 kHz in an apparently successful survey of the seagrasses *Zostera marina* and *Z. noltii* in the estuary of the River Ason, Spain, although no rigorous assessment was made of the efficacy of the system at this sound frequency. Secondly, Hoffman et al. (2002) used a frequency of 70 kHz, together with one of 420 kHz, in an EcoSAV study of unspecified milfoil species and elodeids in Lake Washington, U.S.A. When results were compared from the two sound frequencies, Hoffman et al. (op. cit.) concluded that the higher frequency system performed significantly better than that of 70 kHz, with the latter resulting in a horizontal difference in the placement of macrophyte boundaries of over 100 m. Given the prevalence in lake studies of hydroacoustic systems operating at c. 200 kHz or less, further assessment of the performance of EcoSAV at such relatively lower frequencies is highly desirable. Changes in sound frequency are potentially technically significant for EcoSAV for a number of reasons, including influences on reflectivity, vertical resolution and penetration into bottom sediments.

The objectives of the present study were to test the efficacy of EcoSAV when used with a 200 kHz hydroacoustic system in two shallow lakes of the Cotswold Water Park, U.K., by comparing hydroacoustic and underwater video assessments, and to use it to survey the distribution and abundance of macrophytes in a shallow lake over the course of a winter as an illustration of the application of this technique.

Materials and methods

Study site

The Cotswold Water Park in south-west England, U.K., covers over 100 km^2 and includes over 130 shallow lakes of varying age and size created by gravel extraction. Many of the lakes support extensive growths of macrophytes, two of which were selected for study in the present investigation: Lake 31 (surface area 10.3 ha, maximum depth c. 4.5 m, latitude 51°, 39.870′ N, longitude 1°, 57.640′ W) and Lake 32 (surface area 20.0 ha, maximum depth c. 5.9 m, latitude 51°, 39.520′ N, longitude 1°, 57.587′ W).

Hydroacoustic system

The hydroacoustic system was based on a BioSonics DT6000 (upgraded to a DT-X in November 2004) echo sounder with a 200 kHz split-beam vertical transducer of circular beam angle 6.5° operating under the controlling software Visual Acquisition Version 4.0.2 (upgraded to Version 5.0.4 in November 2004) (BioSonics Inc, Seattle, U.S.A., www.biosonicsinc.com). Throughout the comparisons and surveys, data threshold was set at –130 dB, pulse rate at 5 pings s^{-1}, pulse duration at 0.1 ms, and data were recorded from a range of 0 m from the transducer. Positional data were inputted from a Magellan SporTrak Color (EU basemap) GPS (www.magellangps.com) with accuracy to less than 7 m, which was upgraded in November 2004 to a JRC Model DGPS212 GPS (www.jrc.co.jp) with accuracy to less than 5 m and a fix update interval of 1 s. In addition to the real-time production of an echogram through a colour display on a laptop computer, data were also recorded to hard disc. The system was deployed from a rigid punt powered by an outboard engine and moving at a speed of c. 4 km h^{-1}. The transducer was positioned approximately 0.5 m below the surface of the water. Prior to the comparisons and surveys, the hydroacoustic system had been calibrated using a tungsten carbide sphere of target strength (TS) –39.5 dB at a sound velocity of 1470 m s^{-1}.

Data were subsequently processed using EcoSAV Version 1.0 (BioSonics Inc, Seattle, U.S.A., www.biosonicsinc.com), using default values for all parameters with the exception of Bottom Thickness Limit (see BioSonics (2004)) which was increased from 12 to 30 on the basis of a pilot study carried out at Derwent Water, Cumbria, U.K. (Godlewska et al., 2004). The EcoSAV algorithm outputs lake bottom depth, macrophyte cover (expressed as a percentage), macrophyte height and location (latitude and longitude) summarised by 10-ping segments within each data file. Macrophyte cover values were subsequently averaged over arbitrary 1 min intervals (corresponding to c. 67 m segments along transects) to facilitate their comparison with video data. For surveys, lake bottom depth (LBD) (corrected for transducer depth), macrophyte cover (MC) (expressed as a percentage) and macrophyte height (MH) were subsequently used to calculate Percentage Volume Inhabited (PVI) for each 10-ping sequence according to the equation

$$PVI = ((MH*(MC/100))/LBD)*100$$

which simplifies to

$$PVI = (MH/LBD)*MC$$

Comparison of hydroacoustic and underwater video assessments

When water clarity made such work viable, an underwater video camera system (Simrad OE 1372 Miniature High Definition Colour Underwater Camera recording to a Sony Video Walkman GV-S50E) orientated vertically downwards was attached to the transducer of the hydroacoustic system and simultaneous recordings made during transects as described below. In the laboratory, the video recordings were digitised using the hardware and software system WinTV-USB (Hauppauge Computer, Inc., www.hauppauge.com) before being reviewed and scored for macrophyte presence or absence within the part of the image insonified by the hydroacoustic system. Such scores were made at 2 s intervals for the duration of each transect, before being summed into 1 min segments (corresponding to c. 67 m segments along transects) and macrophyte cover calculated as a percentage for each segment. This procedure produced totals of 50 such segments from a total of 16 transects from Lake 31 between c. 10.15 and 12.00 h on 1 October 2004 (8 transects) and 12 January 2005 (8 transects), and 24 such segments from 10 transects from Lake 32 between c. 14.45 and 15.30 h on 12 January 2005. Grapnel and Ekman Grab samples were also taken at each lake to allow identification of the dominant macrophyte species.

Statistical assessments of the degree of agreement between the measure of macrophyte cover produced by the hydroacoustic and underwater video systems in the two lakes were made by linear regression. The video system could not produce quantitative data for macrophyte height and so statistical assessment of agreement was not possible for this parameter, although a qualitative comparison was made by comparing appropriate EcoSAV output, echograms and video recordings from different areas of the two lakes.

Surveys

Although full surveys were carried out for both lakes, only those of Lake 31 are considered here. Prior to undertaking surveys, transects were planned such that they followed a discrete systematic parallel design as far as field conditions allowed. Fourteen transects were followed, running west-east across the lake (Fig. 1). Navigation was facilitated by a Magellan SporTrak Pro (EU basemap) GPS (www.magellangps.com) with accuracy to less than 7 m, preloaded with appropriate way points. Immediately before each survey, inshore surface water temperature was taken to an accuracy of 0.1°C and entered into the hydroacoustic system. Total transect length for each survey was 3,313 m, giving a ratio of coverage (length of surveys: square root of research area) of 10.3:1.

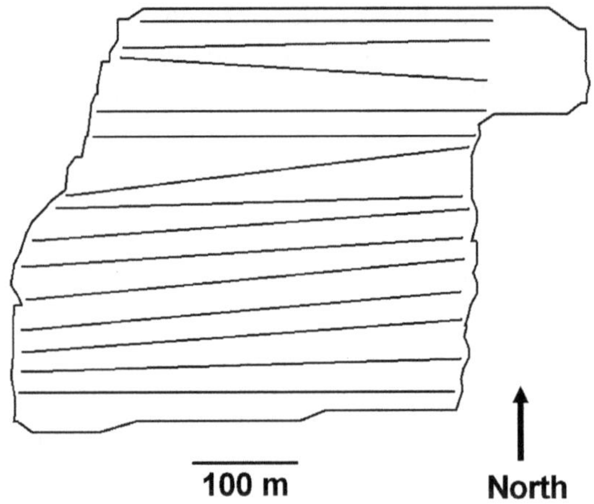

Fig. 1 Outline map of Lake 31 of the Cotswold Water Park, U.K., showing the locations of 14 hydroacoustic transects (straight lines) undertaken during surveys on 29 October 2003, 20 January 2004 and 30 April 2004. At the time of the surveys, access to the small north-east extension to the lake was denied by a boom

Such surveys, each of which took c. 50 min to complete, were performed between 09.00 and 14.00 h on 29 October 2003, 20 January 2004 and 30 April 2004.

Spatial patterns in selected EcoSAV output variables were examined using the software package Surfer (Version 8.05, Golden Software, Inc., Colorado, U.S.A., www.goldensoftware.com) to perform point kriging with a linear variogram model (slope = 1, anisotropy ratio = 1 and angle = 0).

Results

Comparison of hydroacoustic and underwater video assessments

Fig. 2 shows that macrophyte coverages in both lakes as assessed by underwater video recording and by EcoSAV were similar and fell around the line of equality. Specific regressions for Lake 31 (EcoSAV = 0.9417(Video) + 4.8248; $r^2 = 0.8749$, $n = 50$, $P < 0.001$) and Lake 32 (EcoSAV = 1.2099(Video)–19.8152; $r^2 = 0.7932$, $n = 24$, $P < 0.001$) were analysed by a variance ratio test following Mead & Curnow (1983). This revealed that these relationships were not significantly different ($F = 2.613$, df = 22.48, $0.05 > P > 0.10$),

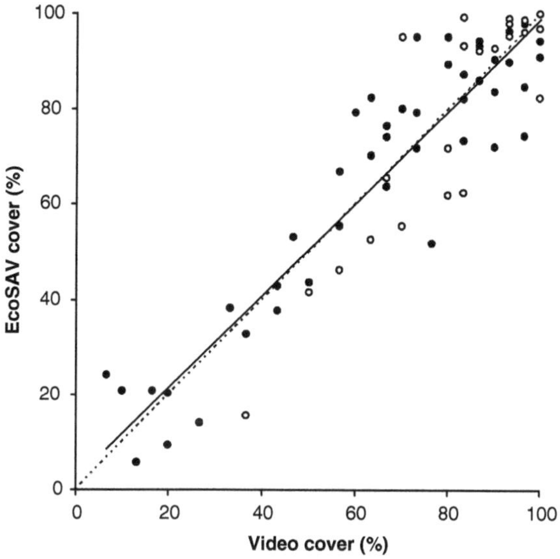

Fig. 2 The relationship between macrophyte cover assessed by underwater video recording and by EcoSAV in Lake 31 (derived from 50 1-minute segments of 16 transects surveyed on 1 October 2004 and 12 January 2005, closed symbols) and Lake 32 (derived from 24 1-minute segments of 10 transects surveyed on 12 January 2005, open symbols). The line of equality is indicated by a broken line. Statistics for the overall regression (continuous line) are $r^2 = 0.8478$, $n = 74$, $P < 0.001$. Other regression statistics are given in the text

and so the relationship is most appropriately described by an overall regression (EcoSAV = 0.9695(Video) + 1.9577; $r^2 = 0.8478$, $n = 74$, $P < 0.001$). For this overall regression, the intercept was not significantly different from 0 (t-test; $t = 0.5242$, $P > 0.10$) and the slope was not significantly different from 1 (lower and upper 95% confidence limits of 0.8730 and 1.0660, respectively).

Qualitative assessment of the efficacy of EcoSAV at estimating macrophyte height also indicated a good agreement, as far as this parameter could be assessed by examination of simultaneous echograms and video recordings. In some areas of the lakes, echograms showed occasional tall strands of macrophytes protruding from lower-growing masses, while corresponding macrophyte heights produced by EcoSAV increased notably. In the same areas, video recording showed tall strands of macrophytes, primarily *Elodea nuttallii* or *Lagarosiphon major*, emerging from lower beds. Even though height measurements could not be made, the relatively greater height of these protruding individual strands was evident as the video camera passed over them.

Macrophyte species recorded by grapnel and Ekman Grab samples in Lake 31 during the comparison of techniques included *Chara curta*, *C. virgata*, *C. globularis*, *Nitella flexilis*, *Tolypella intricata*, *Elodea nuttallii*, *Myriophyllum spicatum*, *Ranunculus circinatus*, *Potamogeton trichoides* and *P. pusillus*. In Lake 32 they included *Chara curta*, *C. contraria*, *C. globularis*, *Elodea nuttallii*, *Lagarosiphon major*, *Myriophyllum spicatum* and *Ranunculus circinatus*.

Surveys

Figure 3 illustrates trends in macrophyte cover, macrophyte height and PVI over the winter of 2003–2004 in Lake 31. Macrophyte cover fell from a mean of 60.26% (95% confidence limits, ± 2.16%) in October at the beginning of the winter to a low of 45.64% (±2.40%) in January, before increasing slightly to 49.18% (±2.18%) at the end of the winter in April. Mean macrophyte height fell successively from 0.54 m (±0.01 m), through 0.48 m (±0.01 m), to 0.46 m (±0.01 m) over the same time period. Mean PVI, which is influenced by changes in both of the above parameters, fell markedly between October and January from 12.15% (±0.55%) to 7.36% (±0.49%), with a further slight decrease to 7.10% (±0.40%) in April.

Finally, Fig. 4 shows examples of further analyses that can be performed rapidly with EcoSAV output in the form of contour maps of lake bottom depth and macrophyte cover for Lake 31 on 29 October 2003.

Macrophyte species recorded by grapnel and Ekman Grab samples during the surveys of Lake 31 included *Chara curta*, *C. virgata*, *C. globularis*, *Nitella flexilis*, *Tolypella intricata*, *Elodea nuttallii*, *Myriophyllum spicatum*, *Ranunculus circinatus*, *Potamogeton trichoides* and *P. pusillus*.

Discussion

The present assessment in two shallow lakes of a hydroacoustic system using a sound frequency of 200 kHz against simultaneously collected visual

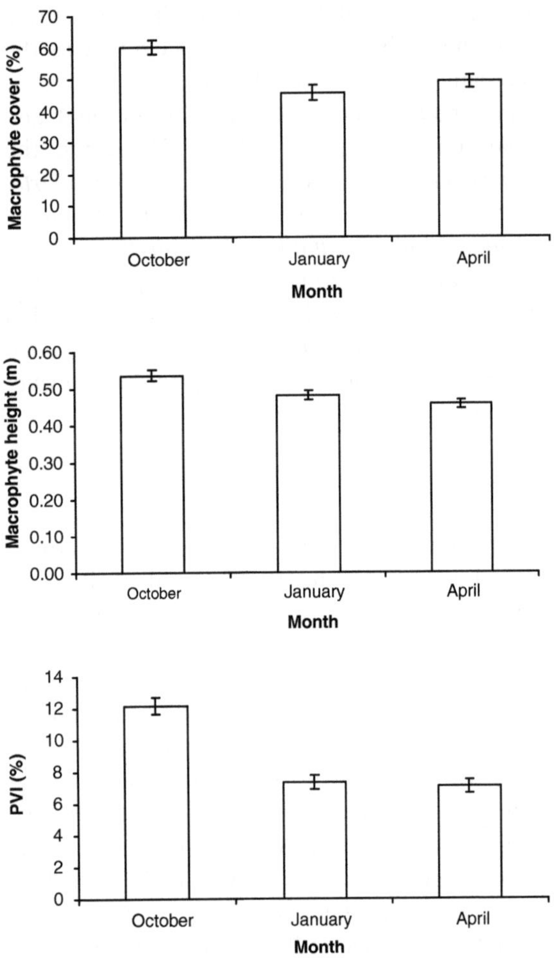

Fig. 3 Changes in macrophyte cover, height and PVI (Percentage Volume Inhabited) in Lake 31 over the course of the winter of 2003–2004 as measured by or calculated from EcoSAV analysis. Data are presented as means with 95% confidence limits

data revealed a high degree of agreement in terms of macrophyte cover. Furthermore, the degree of agreement observed was comparable with that recorded by Sabol et al. (2002) in a sandy estuary using a 420 kHz system and a more sophisticated underwater video system. In fact, the agreement found here was considerably better than that found by Sabol et al. (op. cit.), although this was probably because the present analysis averaged data into 1 min segments rather than used single data points as did the former study.

It is also encouraging that the present agreement between hydroacoustic and visual surveys was statistically indistinguishable between the two study lakes, suggesting that the described relationship may be generically applicable to any lake with a predominantly hard bottom and macrophyte species with growth forms similar to those observed here. Future studies should seek to examine additional hard-bottomed lakes, as well as expanding to lakes with softer substrates. However, the latter conditions can be expected to present not only a considerable challenge because of the greater hydroacoustic similarity between a soft bottom and macrophytes, but they are also likely to make visual assessment more difficult because of a reduced visual contrast between macrophytes and organic sediments. Soft-bottomed lakes or areas of lakes are also frequently associated with higher levels of suspended sediments in the water column itself, the levels of which although relatively low in the present study lakes were still sufficient to compromise video recordings on some transects and thus preclude their data from the present analysis.

The present assessment was incapable of a quantitative examination of the efficacy of EcoSAV and a sound frequency of 200 kHz at estimating macrophyte height due to the technical limitations of the visual recording system. However, in their estuarine study Sabol et al. (2002) found a good agreement between EcoSAV estimates of this parameter using 420 kHz sound and direct measurements made by a diving team. Given that detecting the top of a macrophyte is technically much easier than determining the true bottom depth below a fully developed macrophyte canopy (see Sabol et al. (2002)), it is likely that the present system is also capable of accurate measurement of macrophyte height although this should be established in future assessments. Certainly, the present transects showed qualitative agreement between macrophyte heights estimated by EcoSAV and those apparent from the video recordings. The technical limitations of the video recording system, particularly its resolution, also prevented a quantitative assessment of the ability of the hydroacoustic system to distinguish between macrophyte species. However, it was clear that macrophytes of substantially different growth forms such as low-growing *Chara* species and tall-growing *Elodea*, *Myriophyllum* and *Lagarosiphon* species could be readily distinguished.

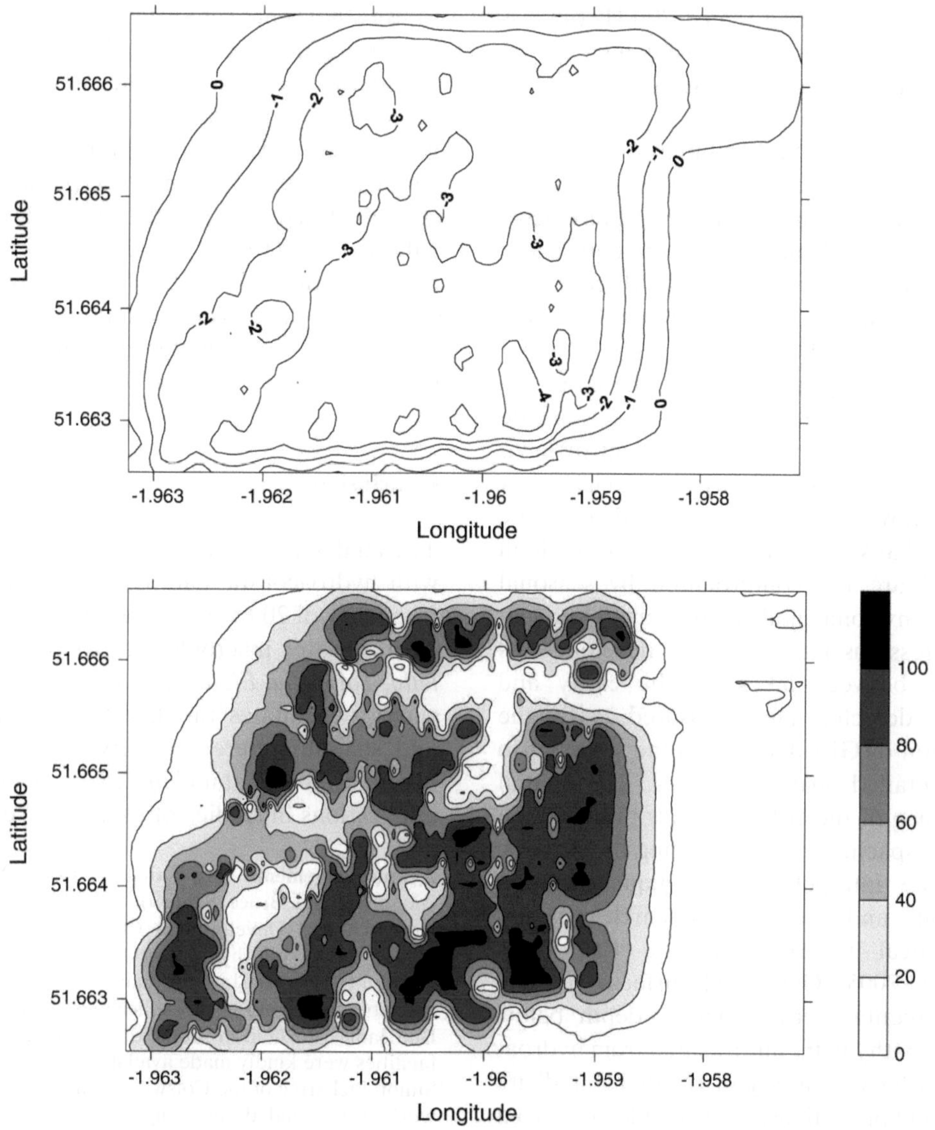

Fig. 4 Examples of the use of EcoSAV output to produce contour maps of lake bottom depth (in metres, upper) and macrophyte cover (as percentage, lower) for Lake 31 on 29 October 2003

This study demonstrates that EcoSAV works effectively with hydroacoustic data collected using a sound frequency of 200 kHz in shallow lakes for a range of previously untested freshwater macrophytes including *Chara*, *Nitella*, *Tolypella*, *Elodea* and *Lagarosiphon* spp. Furthermore, this success was achieved with very little alteration to the default analysis parameters, i.e. just an increase from 12 to 30 in the Bottom Thickness Limit (see BioSonics (2004)). A requirement for the latter is to be expected from the underlying physics of sound in water, specifically its differential penetration into bottom sediments as a function of frequency. Nevertheless, with a lowering of sound frequency there is an inherent consequent loss of resolution and Hoffman et al. (2002) found that an EcoSAV system using 70 kHz performed significantly poorer than one using 420 kHz, although it should be noted that pulse duration also differed and was longer for the former system. Given this observation and the present results, it can be concluded that sound frequencies suitable for use in both macrophyte and fish studies lie in the upper range of those used for the latter alone in fresh waters, i.e. greater than c. 70 kHz but less than c. 200 kHz. Given the relatively high capital cost of hydroacoustic systems, such dual use has great practical benefit.

The successful ground-truthing of EcoSAV with a sound frequency of 200 kHz enables this system to be used with confidence in full macrophyte surveys. The demonstration data presented from Lake 31 illustrate some of the direct (macrophyte cover, macrophyte height) and indirect (Percentage Volume Inhabited) types of data that can be generated by such surveys. The latter are both relatively fast, e.g. a c. 10 ha lake can be surveyed in c. 50 min, and independent of environmental conditions such as water clarity. A particularly powerful feature of the hydroacoustic approach is that surveys can be repeated with high fidelity and are not compromised by seasonal changes in environmental parameters. Thus seasonal progressions in macrophyte developments within and between lakes can be easily and objectively detected and measured, with the incorporation of GPS data within the system also allowing detailed spatial analyses. The spatial error variance of the latter depends primarily on the transect spacing employed during the surveys, although the interpolation technique, transect point density and interpolation search area are also of critical importance (Guan et al., 1999; Valley et al., 2005). Generated 'surfaces' of macrophyte percentage cover, height, depth below surface and bathymetry interpolated from hydroacoustic data have a variety of ecological and site management applications. For example, data from the work described in this paper are being used as part of an integrated study to model waterbird disturbance risk as a function of a variety of environmental variables, in particular the availability of food resources in the form of aquatic macrophytes (O'Connell et al., in press).

As concluded by Schneider et al. (2001), Hoffman et al. (2002), Sabol et al. (2002) and Valley et al. (2005) working in other aquatic systems, recent technological developments in the field of hydroacoustics now make this a viable technique for the survey of aquatic macrophytes in shallow lakes. This does not mean that biological sampling is redundant, anymore than the more mature application of hydroacoustic techniques to fish studies has removed the need for sampling by netting in order to secure biological specimens for species identification and other examination. However, hydroacoustic techniques facilitate more quantitative and spatially-referenced studies of macrophyte abundance, paralleling recent studies of lake fish populations such as those by Elliott et al. (1996), George & Winfield (2000) and Schmidt et al. (2005). As such, hydroacoustic techniques have many potential applications in studies of macrophytes within shallow lake systems.

Conclusion

The analysis software EcoSAV works effectively with hydroacoustic data collected using a sound frequency of 200 kHz in shallow lakes for a range of freshwater macrophytes including *Chara*, *Nitella*, *Tolypella*, *Elodea* and *Lagarosiphon* spp. As such, it facilitates rapid, extensive, quantitative and spatially-referenced surveys of macrophyte distribution and abundance, with many potential applications in studies of shallow lake systems.

Acknowledgements We thank Bob McClure of BioSonics and Bruce Sabol of the U.S. Army Engineer Research and Development Center for their help and advice during our hydroacoustic analyses. We are also grateful to Bob McClure for making the EcoSAV software available to us, and to our colleagues Janice Fletcher and Ben James for help in the field. Boat and other logistical facilities were kindly made available by Gareth Harris and Simon Pickering of the Cotswold Water Park Society. This work was funded by English Nature, the Natural Environment Research Council, the University of Bristol, and the Wildfowl & Wetlands Trust.

References

BioSonics, 2004. EcoSAV User Guide. BioSonics, Inc., 4027 Leary Way NW, Seattle, WA 98107, USA.

Bozzano, R., L. Castellano & A. Siccardi, 1998. Characterization of submerged aquatic vegetation by a sector-scanning sonar. In Alippi, A. & G. B. Canelli (eds), Proceedings of the 4th European Conference on Underwater Acoustics. Italian National Research Council, Rome, Italy, pp 159–164.

Duarte C. M., 1987. Use of echosounder tracing to estimate aboveground biomass of submersed plants in lakes. Canadian Journal of Fisheries and Aquatic Science 44: 732–735.

Elliott, J. M., J. M. Fletcher, J. A. Elliott, P. R. Cubby & E. Baroudy, 1996. Changes in the population density of pelagic salmonids in relation to changes in lake enrichment in Windermere (northwest England). Ecology of Freshwater Fish 5: 153–162.

Fortin, G., L. Saint-Cyr & M. LeClerc, 1993. Distribution of submersed macrophytes by echo-sounder tracings in Lake Saint-Pierre, Quebec. Journal of Aquatic Plant Management 31: 232–240.

George, D. G. & I. J. Winfield, 2000. Factors influencing the spatial distribution of zooplankton and fish in Loch Ness, UK. Freshwater Biology 44: 557–570.

Godlewska, M., N. Górska, Z. Klusek, J. Szczucka, A. Świerzowski & J. Tęgowski, 2002. Hydroacoustics as a tool for monitoring the quality of aquatic ecosysytems. Ecology and Ecotechnologies, Proceedings of the Review Conference on the Scientific Cooperation between Austria and Poland, February 24–28, 2002, Scientific Centre of the Polish Academy of Sciences, Vienna, Austria, 2: 287–295.

Godlewska, M., A. Świerzowski & I. J. Winfield, 2004. Hydroacoustics as a tool for studies of fish and their habitat. Ecohydrology & Hydrobiology 4: 417–427.

Guan, W., R. H. Chamberlain, B. M. Sabol & P. H. Doering, 1999. Mapping submersed aquatic vegetation with GIS: evaluation of different interpolation methods. Marine Geodesy 22: 69–91.

Hoffman, J. C., J. Burczynski, B. Sabol & M. Heilman, 2002. Digital acoustic system for ecosystem monitoring and mapping: assessment of fish, plankton, submersed aquatic vegetation, and bottom substrata classification. White Paper. BioSonics, Inc., 4027 Leary Way NW, Seattle, WA 98107, USA. Available at: http://www.biosonicsinc.com/doc_library/docs/lake_wa_assessment.pdf.

Jurvelius J., 1991. Distribution and density of pelagic fish stocks, especially vendace (*Coregonus albula* (L.)), monitored by hydroacoustics in shallow and deep southern boreal lakes. Finnish Fisheries Research 12: 45–63.

Maceina, M. J. & J. V. Shireman, 1980. The use of a recording fathometer for determination of distribution and biomass of hydrilla. Journal of Aquatic Plant Management 18: 34–39.

Madsen J. D., 1993. Biomass techniques for monitoring and assessing control of aquatic vegetation. Lake and Reservoir Management 7: 141–154.

Mead, R. & R. N. Curnow, 1983. Statistical methods in agriculture and experimental biology. Chapman & Hall, London.

Moreno, A., P. Siljestrom & J. Rey, 1998. Benthic pharenogam species recognition in side scan sonar images: Importance of the sensor direction. In Alippi A. & G. B. Canelli (eds), Proceedings of the 4th European Conference on Underwater Acoustics. Italian National Research Council, Rome, Italy 173–178.

O'Connell, M. J., R. M. Ward, C. Onoufriou, I. J. Winfield, G. Harris, R. Jones, M. L. Yallop & A. F. Brown, in press. Integrating multi-scale data to model the trade off between food resource quality and disturbance risk: preliminary findings & potential uses. Ibis.

Sabol, B. & J. Burczyński, 1998. Digital echo sounder system for characterising vegetation in shallow-water environments. In Alippi, A. & G. B. Canelli (eds), Proceedings of the 4th European Conference on Underwater Acoustics. Italian National Research Council, Rome, Italy: 165–171.

Sabol, B. M., R. E. Melton, R. Chamberlain, P. Doering & K. Haunert, 2002. Evaluation of a digital echo sounder system for detection of submersed aquatic vegetation. Estuaries 25: 133–141.

Schmidt, M. B., H. Gassner & E. I. Meyer, 2005. Distribution and biomass of an underfished vendace, *Coregonus albula*, population in a mesotrophic German reservoir. Fisheries Management and Ecology 12: 169–175.

Schneider, P., J. Burczynski, A. Monteoliva & A. Valle, 2001. Results from submerged aquatic plant assessment using digital echosounder technique. International Council for the Exploration of the Sea, Joint Session of FTFB and FAST Working Groups, 23–27.

Tęgowski, J., N. Górska & Z. Klusek, 2003. Statistical analysis of acoustic echoes from underwater meadows in the eutrophic PuckBay (southern Baltic Sea). Aquatic Living Resources 16: 215–221.

Thomas, G. L., S. I. Thiesfield, S. A. Bonar, R. N. Crittenden & G. B. Pauley, 1990. Distribution of submergent plant bed biovolume using acoustic range information. Canadian Journal of Fisheries and Aquatic Sciences 47: 805–812.

Valley, R. D., M. T. Drake & C. S. Anderson, 2005. Evaluation of alternative interpolation techniques for the mapping of remotely-sensed submersed vegetation abundance. Aquatic Botany 81: 13–25.

Vis, C., C. Hudon & R. Carignan, 2003. An evaluation of approaches used to determine the distribution and biomass of emergent and submerged aquatic macrophytes over large spatial scales. Aquatic Botany 77: 187–201.

Wanzenböck, J., T. Mehner, M. Schulz, H. Gassner & I. J. Winfield, 2003. Quality assurance of hydroacoustic surveys: the repeatability of fish-abundance and biomass estimates in lakes within and between hydroacoustic systems. ICES Journal of Marine Science 60: 486–492.

Hydrobiologia (2007) 584:121–132
DOI 10.1007/s10750-007-0597-6

SHALLOW LAKES

Combining limnological and palaeolimnological approaches in assessing degradation of Lake Pskov

Mihkel Kangur · Külli Kangur · Reet Laugaste · Jaan-Mati Punning · Tõnu Möls

Abstract Limnological monitoring data and palaeorecords from large shallow Lake Pskov were used to study the long-term dynamic pattern of the lake ecosystem and to identify the start of its degradation. Lake Pskov is the southern part of Lake Peipsi *s.l.*, the largest transboundary lake in Europe. The limnological monitoring data collected in the years 1956–2005 show that the water quality of L. Pskov has deteriorated and caused adverse changes in the whole ecosystem (*e.g.* excessive growth of algae, increased cyanobacterial blooms, silting of the lake bottom, fish kills). Doubled total phosphorus (P_{tot}), dissolved inorganic P (PO4-P) and chlorophyll *a* (Chl-*a*) contents, increased total alkalinity (HCO_3^-) and pH, as well as decline in water transparency and oxygenation conditions, indicate a clear increase in the trophic level of the lake. However, the limnological studies do not show when the degradation started. To understand long-term dynamics of the lake ecosystem, a 52 cm sediment core taken from one monitoring station of L. Pskov was studied and dated by the ^{210}Pb method. Palaeodata show that substantial changes in the L. Pskov ecosystem started already in the 1930s when mesotrophic conditions in the lake turned increasingly eutrophic. Since that time, the content of P, nitrogen (N), carbon (C) and relative abundance (RA%) of planktonic diatoms in the sediment have increased significantly ($P < 0.005$–0.030). Comparison of water variables with the sediment variables in five to eight time points, coinciding in palaeorecords and monitoring data, reveals several significant correlations. Strong and highly significant Spearman correlations ($r > 0.9$ or $r < -0.9$, $P < 0.001$) were observed between the contents of P, N, C and sulphur (S) in the sediment and the mean water level (WL) and temperature (WT) registered one to five years earlier in the lake. The C content of the sediment showed a negative relationship with WL and a positive correlation with WT. A significant positive relationship was recorded also between WT, and N and P content in the sediment. It is hypothesized that a higher WT and lower WL result in an increased organic mater accumulation in the sediment in the coming years.

Guest editors: R. D. Gulati, E. Lammens, N. De Pauw & E. Van Donk
Shallow lakes in a changing world

Electronic supplementary material The online version of this article (doi:10.1007/s10750-007-0597-6) contains supplementary material, which is available to authorized users.

M. Kangur (✉) · J.-M. Punning
Institute of Ecology at Tallinn University,
Uus-Sadama 5, Tallinn 10120, Estonia
e-mail: mihkel.kangur@tlu.ee

K. Kangur · R. Laugaste · T. Möls
Centre for Limnology, Institute of Agricultural and Environmental Sciences, Estonian University of Life Sciences, 61101 Rannu, Estonia

Keywords Large shallow lake · Eutrophication · Time-series data · Palaeorecords · Diatoms

Introduction

To understand better the long-term development of a lake and its dependence on natural biogeophysical conditions and human impact, it is necessary to have observations for different temporal scales. Contemporary monitoring data are most correct but very limited in time extent; such data could be described as snapshots. For monitoring the long-term development of the lake ecosystem the palaeorecords obtained by studying lake sediments can be used. The biological, geochemical and lithological information stored in accumulative deposits allows us to follow the changes in the processes affecting sediment deposition over certain (longer) time periods (Chambers, 1993; Leavitt, 1993). Sediment records extend time-scales and integrate subannual variability but they suffer from taphonomic biases and occasionally from uncertain chronology. Therefore palaeolimnological data and limnological time-series data are highly complementary (Battarbee et al., 2005). The integration of limnological and palaeoecological approaches is of prime importance for effective ecosystem management (Smol, 1992). Furthermore, the information concerning the natural situation before disturbance is necessary for assessment of the present ecological state of the lake according to the Water Framework Directive (Directive 2000/60/EC, 2000).

The present work is the first attempt to compare long-term limnological monitoring data with palaeolimnological records obtained by the study of the sediment core from Lake Pskov (also known as L. Pihkva) in order to follow the long-term changes in the lake and to determine when the ecosystem began to change. The lake is one of the most thoroughly studied water bodies in Estonia and a rather extensive set of data has been gathered during the last 50 years. The limnological time-series data indicate that the L. Pskov ecosystem is under threat of eutrophication, which has led to an undesirable growth of algae, massive blooms of cyanobacteria accompanied by fish kills, low water transparency and silting of the lake bottom (Laugaste et al., 2001; Kangur et al., 2003; 2005). Previous assessments have shown that most of the nutrients are carried into L. Pskov via the Velikaya River (Stålnacke et al., 2002). Unfortunately, at present long-term data on the external phosphorus and nitrogen load from the Russian part of the catchment area are not available. As L. Pskov is a shallow lake, its ecological state is also strongly influenced by periodical fluctuations of water level and water temperature. However, current monitoring programmes cover a too short time-span, therefore the start of the lake ecosystem degradation is ecologically undocumented and information on the pre-perturbation habitat (reference condition) is not available. No palaeoecological studies have been carried out on L. Pskov till now.

We combined limnological time-series data and palaeolimnological data from L. Pskov to provide a long-term record of lake history. The purposes of our investigation were (1) to get an overview of the dynamics of the L. Pskov ecosystem, (2) to identify the start and trends of the ecological degradation of the lake, and (3) to determine the relationship between water monitoring variables and sediment variables.

Study site

Lake Pskov is a shallow lowland lake (Table 1). It is the southern and most eutrophicated part of the lake Peipsi *s.l.*, situated on the border of Estonia and Russia (Fig. 1). The lake catchment is a gently undulating glaciolacustrine or till-covered plain. Forests and semi-natural areas dominate in the lake drainage basin, next to agricultural areas (~14%). The lake basin is mostly sparsely

Table 1 Morphometric data on L. Pskov at the mean water level (after Jaani, 2001a)

Altitude, m a.s.l.	30
Surface area, km^2	708
Greatest depth, m	5.3
Mean depth, m	3.8
Water volume, km^3	2.68
Retention time, years	~2

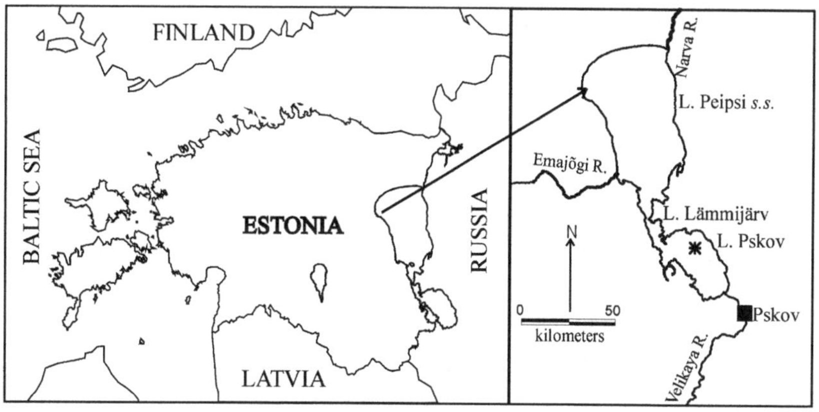

Fig. 1 Location of Lake Pskov. *Location of monitoring station no. 51 and the sediment sampling site

populated (about 11 inhabitants/km^2). The city of Pskov (Russia), with approximately 200,000 inhabitants, is located in the immediate vicinity of the lake and is the largest point source of nutrients in the drainage area (Stålnacke et al., 2002). The main source of inflow to the lake is the Velikaya River (catchment area 25,200 km^2) (Jaani, 2001a). Due to the large surface area and relative shallowness of the lake, temperature stratification is unstable and can be disturbed already by a moderate wind or undulation. Therefore, the water of L. Pskov is usually rich in oxygen during the open water period. The lake is, as a rule, covered with ice from December till April, and during that period the near-bottom water frequently suffers from oxygen deficiency. Natural water level fluctuations have shown an overall range of 3.04 m over the last 80 years, with an average annual range of 1.15 m (Jaani, 2001b). The mean values and the variability of key physical and chemical characteristics of water in L. Pskov over the current monitoring period are given in Electronic supplementary material.

Methods

Sampling and analysis of water

In the current paper the data of yearly monitoring and sediment core analysis from station number 51 (co-ordinates 58°04′13″ N and 27°53′44″ E) are used (Fig. 1). The monitoring data for L. Pskov cover the period from 1956 (hydrochemical) and 1962 (phytoplankton) to 2005. Since 1992 mainly the Estonian part has been studied but several joint Estonian–Russian expeditions to the whole lake have also been carried out. Depending on the year, the number of monitored sampling sites varied between 4 and 25 in L. Pskov (Kangur et al., 2002). Seasonal (or monthly) samples for routine hydrochemical analysis (Starast et al., 2001) were taken from surface water (from a depth of 0.1–1.0 m) and from the near-bottom layer of water (0.5 m from bottom), both with a Ruttner sampler.

Integrated water samples for plankton analysis were taken from the surface and with 1-metre intervals from the water column. Samples were preserved in Lugol's solution in the 1990s and in formaldechyde in earlier years. Preserved materials were identified and counted using the Utermöhl (1958) technique in the 1990s and on striped microscope slide (0.01 ml of sample) in earlier years. The methods of collecting and treating hydrochemical and phytoplankton samples are described in detail in Laugaste et al. (2001) and Starast et al. (2001).

Data on water level and surface water temperature, collected by the Estonian Institute of Hydrology and Meteorology, were available for the period from the 1920s to 2004.

Sampling and analysis of sediments

The sediment core (52 cm long) was taken from the northern part of L. Pskov at monitoring station 51 (Fig. 1). The sampling was performed with a modified Livingstone-Vallentyne piston corer in summer 2003. Sampling was continuous, with intervals of 1 cm. The content of dry matter in sediment was determined by drying the samples at 105°C to constant weight. The contents of C, N and S were measured in the Institute of Chemistry at Tallinn University of Technology with a Perkin Elmer Elemental Analyser, type PE 2400/2. The results were expressed as percentage of dry weight at 105°C. On the basis of these data C/N ratios were calculated (Punning et al., 2003). The phosphorus content (P_{AL}) of the sediment samples was determined by the ammonium lactate method in the Laboratory of Plant Biochemistry at the Estonian University of Life Sciences with a Flow Injection Analyser type FiaStar5000.

The radiometric dating of the core was carried out at the Centre for Environmental Monitoring and Technology, Ukrainian Hydrometeorological Research Institute by direct gamma assay using an EG&G Ortec HPGe GWL series well-type coaxial low background germanium detector. The data were calculated by using the CRS model (Appleby et al., 1986).

Diatom slides were prepared by applying sulphuric acid treatment (Vilbaste et al., 2000) after the calcareous compounds had been removed from the samples with chloric acid. Hyrax (refractive index 1.65) was used as the mounting medium. On each slide 400–500 valves were identified and counted under a bright field microscope MBL 2000 using a 100X oil immersion objective with the total magnification of 1600–2000. The absolute number of the counted taxa was converted to relative abundance (RA,%). Diatom taxonomy follows the handbooks of Krammer and Lange-Bertalot (1986, 1988, 1991).

Data treatment

Water chemical and plankton variables were $\log_2$-transformed to improve their statistical distribution. Prior to this, to enable calculation of logarithms, zero values were replaced with appropriate small positive numbers. The whole data array was filtered by inspecting studentized residuals calculated from an appropriate statistical model. Arithmetic means of logarithmic variables were transformed back to natural scale and presented as geometric means.

The main statistical analysis was carried out with the general linear model technique provided by the SAS System, Release 8.2 (SAS Institute Inc., 1999), especially by the SAS GLM procedure. A large regression model (Möls, 2005, p. 125) was used for estimating chemical water variables from the raw monitoring data series. With this model, the water variables were predicted for the studied sampling site and times. The predicted values and related 95% confidence limits were calculated with tailored parametric functions or by using the OUTPUT option of the GLM procedure. The estimates were used thereafter as if they were really measured values of water variables. This modelling approach enabled us to use all monitoring data in the Peipsi dataset for making water variable estimates more precise for a particular time and site (Möls, 2005).

We studied the synchrony between the monitoring records and sediment data with the Spearman correlation analysis, presenting monitoring data by various water level and temperature averages, and by the water chemical variables. The group of averages consisted of variables $WL_{0,0}$, $WL_{0,1}$,..., $WL_{5,8}$ which represent the mean water levels, and $WT_{0,0}$, $WT_{0,1}$,...,$WT_{5,8}$ which represent the mean water temperatures. Indexes denote here the time periods used for averaging: the first index shows the year of averaging (0,current year; 1,previous year; etc.), the second index stands for the averaging period within the year (values 0 to 7 mean averaging over days 1–100, 100–130, 130–165, 165–200, 200–240, 240–280, 280–320 and 320–365, respectively, composite values like 012 mean averaging over the union of time intervals shown in the index).

We compared the WT and WL variables as well as the water chemical variables with the sediment variables. Of correlations with WL and WT variables only those having $P < 0.001$ are presented as significant, for the other monitoring variables the cut-off *P*-value was 0.005. This choice of significance levels follows the Bonferroni method, when it is used in two separate groups of multiple tests, first for testing correlations with WL and WT variables and, second, for testing correlations with other water monitoring variables.

Results

Time series of water data

The long-term monitoring data for L. Pskov indicate essential changes in water quality (Table 2), which can be related to eutrophication. Thus, the pH trend towards alkalinity can be observed and the mean HCO_3^- is clearly increasing (Table 2). The PO_4-P and Si contents have doubled during the last four decades, while the total organic matter content (COD_{Cr}) and dissolved inorganic N (NH_4-N + NO_3-N + NO_2-N = DIN) content showed only slight changes. The phosphorus and Chl-*a* contents have increased approximately twofold, demonstrating an increase in the trophic level of L. Pskov during the last two decades. The N_{tot} content, however, changed only slightly. The two main phytoplankton groups in L. Pskov are diatoms (BAC) and cyanobacteria (CY). Both the phytoplankton biomass (FBM) and especially the biomass of CY increased substantially in the 1990s, while the biomass of BAC showed no significant changes. In the last 50 years, the water transparency by the Secchi disc has decreased considerably.

Water chemistry and sediment data at monitoring station 51

The data for monitoring station 51 in L. Pskov are used in further analysis and for combining the limnological and palaeoecological datasets. The sediment core was collected at the same station.

The upper 25 cm of the core was dated using the ^{210}Pb method with the reference levels of ^{137}Cs and ^{210}Am, which record a maximum input of atmospherically derived nuclides in 1963 and 1986 (nuclear bomb testing and the Chernobyl accident, respectively). The age–depth curve given in Fig. 2a

Table 2 Long-term changes in key physical, chemical and plankton variables of surface water in L. Pskov in the open water period (days 100–310 within the year)

Parameter, Unit	Years	Geometric mean	95% confidence limits	
HCO_3^-, meq l^{-1}	1966–1970	1.7	1.6	1.9
	2001–2005	2.4	2.3	2.6
pH	1956–1960	7.9	7.7	8.0
	2001–2005	8.4	8.2	8.6
P_{tot}, mg P m^{-3}	1985–1990	62	58	66
	2001–2005	143	122	168
PO_4–P, mg P m^{-3}	1966–1970	14.1	5.5	36,1
	2001–2005	31.4	19.4	51.0
N_{tot}, mg N m^{-3}	1985–1990	982	925	1043
	2001–2005	1006	835	1212
DIN, mg N m^{-3}	1976–1980	178	134	236
	2001–2005	108	78	150
Si, mg l^{-1}	1966–1970	1.0	0.7	1.4
	2001–2005	2.0	1.6	2.5
COD_{Cr}, mg O l^{-1}	1966–1970	36.4	27.4	48.3
	2001–2005	32.3	22.1	47.2
Chl *a*, mg m^{-3}	1985–1990	20.5	18.3	22.9
	2001–2005	53.5	42.4	67.5
Secchi depth, m	1956–1960	1.2	1.1	1.4
	2001–2005	0.7	0.5	0.8
FBM, gWW m^{-3}	1961–1965	12.2	10.6	14.2
	2001–2005	20.8	15.4	28.1
CY, gWW m^{-3}	1961–1965	4.4	3.7	5.3
	2001–2005	8.6	3.9	19.0
BAC, gWW m^{-3}	1961–1965	5.6	4.7	6.8
	2001–2005	5.1	3.3	7.9

enables us to describe changes in the palaeoecological records in the calendar years and combine them with the monitoring data from station 51. As seen from Fig. 2a, the accumulation rate has increased gradually from a depth of 25 cm (accumulated AD 1890) and reached the values up to 0.07 g cm^{-2} y^{-1} in the surface layers. The water content in deeper sediment layers has permanent values (ca 84%) up to a depth of 18 cm, increasing then up to 98% in the surface layers (Fig. 2b).

Monitoring data

Long-term monitoring data show a two-phase change in the concentration of SO_4^{2-} and Cl^- (Fig. 3e, f) since the 1950s. During the first phase, the content of sulphates increased about 3–4 times till the late 1970s, whereas the content of chlorides increased about two times till the end of the 1980s. In the second phase, the content of these ions dropped back to the level of the 1960s by the end of the 1990s. However, SO_4^{2-} content increased in the early 2000s.

The mean P_{tot} and PO4–P contents in water decreased in the late 1980s, but have been increasing since the beginning of the 1990s (Fig. 3h, i). Since 1985 the content of Chl-*a* (Fig. 3j) has been increasing in parallel with P_{tot}. The dissolved oxygen content was quite high in the 1970s but decreased significantly at the beginning of the 2000s. The greatest decrease in oxygen content was recorded in the near-bottom water during winter (Fig. 3k).

Palaeolimnological records

In Fig. 3 the palaeolimnological records are presented in calendar years according to the age-depth curve. Variations in the C content of sediment dry mass (DM) are complicated (Fig. 3a). A general tendency in carbon content variation shows its decrease in the layers that accumulated in the 1880s–1930s, an essential short-term increase in the 1950s and slow increase and stabilization in the layers that accumulated from 1970 up to the present. The N content is less variable (Fig. 3b). It decreases slightly from the bottom up to the layers that accumulated the 1920s, increases abruptly at the depth corresponding to the 1930s and is almost constant up to the surface layers. The C/N ratio was almost constant (9.6–9.8) in the interval from the deepest layers up to the layers that accumulated nearly in the 1930s (Fig. 3c). Then the C/N ratio started to decrease steadily and reached its minimum value (8.4) in the layers of the late 1950s. Upwards, the C/N ratio increases slightly, reaching the value of 9.2 in the surface layers. The S concentration is rather constant (0.25 mg kg^{-1}) in the lower part of the core, but fluctuates notably from the level of 18 cm upwards, reaching maximum values in the layers corresponding to the 1940s and the beginning of the 1990s (Fig. 3d). The content of P_{AL} is also rather constant from the bottom layer up to a depth of 18 cm, with one sharp peak of 76 mg/kg in 1950 (Fig. 3g).

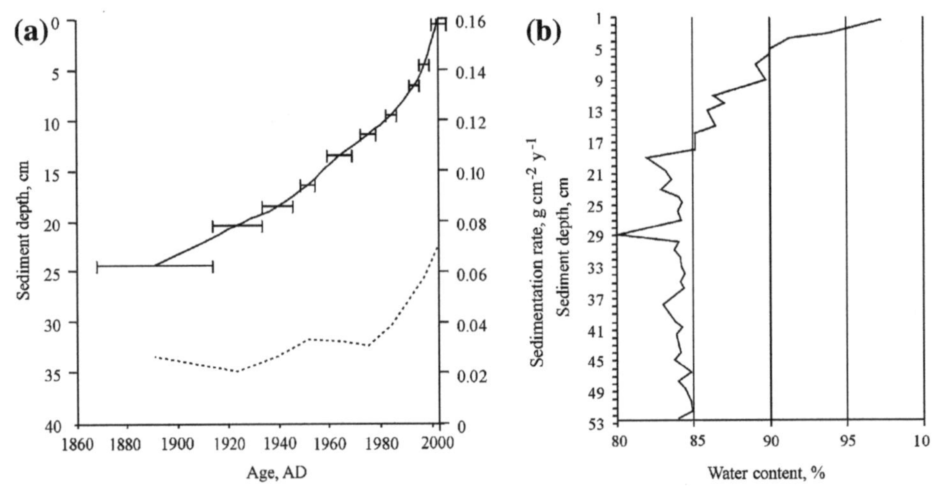

Fig. 2 (**a**) Dates (solid line) and sediment accumulation rates (dashed line) in the L. Pskov sediment core calculated using CRS model. (**b**) water content of the L. Pskov sediment core

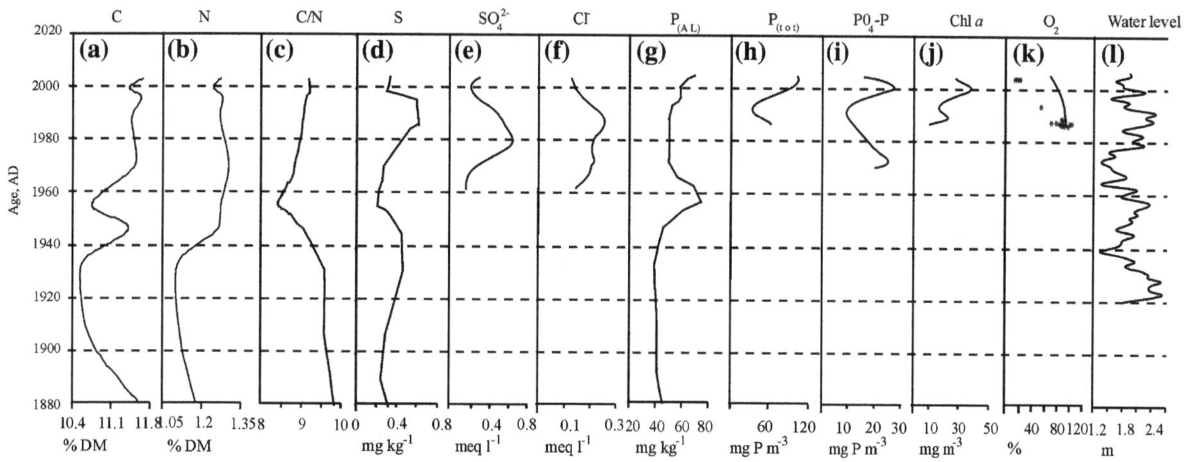

Fig. 3 Palaeorecords from the L. Pskov sediment core (**a**, **b**, **c**, **d** and **g**) and monitoring data (**e**, **f**, **h**, **i**, **j**, **k** and **l**) from station no. 51. k – O_2 content of water; line – O_2 content of surface water in the open water period (days 100–310 within the year), *O_2 content of near bottom water in winter

Comparison of sediment variables and water monitoring variables

The longest monitoring dataset available describes the water level changes in the studied lake system since 1921, while water temperature data are available since 1924. Long-term variability of water level in the lake has a cyclic nature with the alternation of high- and low-water periods (Fig. 3l). A great decrease in water level was recorded in the 1930s, with the lowest annual lake level in 1940. A low-water period was also documented at the beginning of the 1970s, while in the 1980s the water level was relatively high. Comparison of palaeoecological data with long-term water level dynamics indicates that important shifts in the lake ecosystem in the 1930s can be related to a sharp decline in water level. Comparison of water variables with the sediment variables in five to eight time points (1932, 1946, 1955, 1971, 1987, 1996, 1999 and 2003), coinciding in palaeorecords and monitoring data, reveals several significant correlations. The Spearman correlation analysis revealed a significant negative relationship between the mean water level (WL) in spring (days 130–165 within the year) and C content of sediment of the current year, and a positive correlation between mean water temperature (WT) (days 130–165 within the year) and C content of sediment with a time lag of one year (lag 1) (Table 3). A significant positive relationship appeared between WT (days 130–165) and N content (lag 1) as well as between WT (days 1–165) and P content of sediment (lag 5). A

Table 3 Statistically significant Spearman correlations *r* between sediment variables and water monitoring variables

Sediment variable	Water variable	n	r	*P*-value
C	$WT_{1,012}$	8	0.95	0.0003
	$WL_{0,2}$	8	–0.93	0.0009
N	$WT_{1,2}$	8	0.93	0.0007
P	$WT_{5,012}$	8	0.98	<0.0001
Planktonic diatoms RA %	N_{tot}	4	–1.0	<0.0001
Planktonic diatoms RA %	SO_4^-	6	–0.94	0.005

WT and WL with indexes denote mean water temperature and water level of current or previous time periods: the first index shows lag in years, the second index stands for the averaging period within the year. Of correlations with WL and WT variables only those having $P < 0.001$ are presented, for other monitoring variables the cut-off *P*-value is 0.005

negative correlation was recorded between RA% of planktonic diatoms in the sediment and N_{tot} and SO_4^{2-} content of water.

Diatom assemblages in water and sediment

The dominant diatom species in the water column in different time periods and appropriate sediment layers are given in Table 4. A comparison of phytoplankton monitoring data with diatom assemblages was possible in the upper 14–15 cm of the sediment core.

Altogether 138 species were found in the sediment core. The genus *Aulacoseira* (*A. ambigua, A. islandica, A. granulata*) is dominant in the whole core and the genus *Fragilaria* (*F. construens* and varieties, *F. brevistriata*) among epiphytic/benthic species. *Gyrosigma* and *Navicula* dominate among benthic algae. Relative importance of planktonic algae is more than 60% at a depth of 53–54 cm, it is descending smoothly up to a depth of 20 cm and increases abruptly in the upper layers (Fig. 4). Relative frequency of littoral algae (epiphytic and benthic species) is above 50% at a depth of 49–20 cm, having peaks at depths 41–42 and 22–28 cm.

On the basis of the composition and relative abundance of diatoms, the sediment core can be divided as follows:

Zone I: 54–50 cm oligo-mesotrophic zone, prevalence of *Aulacoseira islandica*, and 49–44 cm transient zone;

Zone II: 44–18 cm meso-eutrophic, prevalence of *Fragilaria* species, abundance of *Aulacoseira islandica* is decreasing;

Zone III: 18–0 cm eutrophic, prevalence of *Aulacoseira ambigua*, silting of the lake bottom.

Discussion

Eutrophication is well known as the main reason for degradation of almost all lakes in the lowlands (Gulati & van Donk, 2002; Moss et al., 2005). The natural lake evolution is regarded as "ontogeny", whereas "eutrophication" is a result of external nutrient loading from human activity (Battarbee et al., 2005). Both these processes may induce increase in bioproductivity and changes in the biological structure of lakes. The natural and human induced processes are hard to distinguish and can amplify each other (Padisák & Koncsos, 2002).

Eutrophication or nutrient enrichment from human activity is clearly a key environmental problem for L. Pskov (Kangur et al., 2002). Yearly data on water chemistry since the 1950s and on biota since the 1960s indicate a common trend of water environment deterioration during the whole monitoring period. The P_{tot}, and Chl-*a* content of water as well as biomass of CY has doubled during the last two decades. As a result, the water transparency has decreased significantly since the 1950s. Eutrophication of the lake is

Table 4 Algae in water and in the approximately corresponding sediment layer

	Lake water	Sediment core
1960s	Dominants: *Aulacoseira ambigua* (Grun.) Simonsen, *Asterionella formosa* Hassall, *Stephanodiscus binderanus* (Kütz.) Krieger Subdominants: *A. granulata* (Ehr.) Sim. *Tabellaria fenestrate*, *Fragilaria crotonensis* Kitton Notable: gen. *Synedra*, *Surirella*	Dominants: *A. ambigua*, *A. granulata* var. *angustissima* (O. Müller) Sim. Subdominants: *A. granulata Stephanodiscus* sp., Fragilaria construens (Ehr.) Grunow and varieties Notable: *Gyrosigma attenuatum* (Kütz.) Rabenh.
1980s	Dominants: *Aulacoseira valida* (Grun.) Krammer, *S. binderanus*, *F. crotonensis* Subdominants: *A. granulata*	Dominants: *A. ambigua*, *A. granulata* var. *angustissima* Subdominants: *A. valida*
1990s–2000s	Dominants: *Aulacoseira ambigua*, *A. granulata* Subdominants: *Stephanodiscus* spp. *Cyclostephanos dubius* (Fricke) Round	Dominants: *A. ambigua* Subdominants: *A. granulata*, *A. islandica* (O. Müller) Sim., *Cyclostephanos dubius Fragilaria brevistriata* Grunow

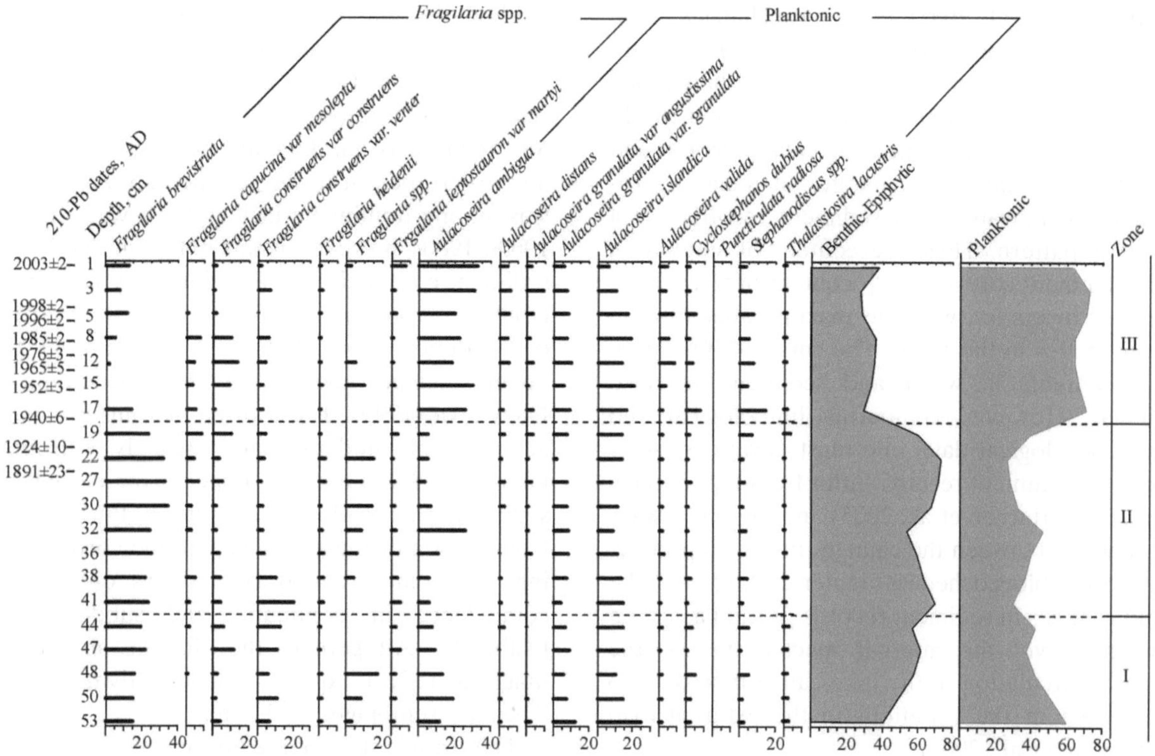

Fig. 4 Relative abundance of diatom species in the sediment core of L. Pskov

indicated also by an increase in total alkalinity and pH as well as worsening oxygen conditions during winter. The structure of the phytoplankton community changed considerably and phytoplankton biomass (particularly cyanobacteria) displayed an increasing trend (Kangur et al., 2003). Strong blooms of cyanobacteria induced fish kills and siltation of the lake bottom (Kangur et al., 2005).

It is hard to explain the causes of deterioration of the ecological state of the lake. As L. Pskov is a shallow lake with long water residence time, the water level changes might be one factor favouring the deterioration of its ecological state. Water level in the lake was low in the early 1970s and late 1930s, while 1940 was the driest year of the data series (Jaani, 2001b). Changes in many water and phytoplankton parameters are correlated with natural water level fluctuations (Kangur et al., 2003; Milius et al., 2005). The results of Laugaste et al. (2001) show the coherence of high phytoplankton biomass with periods of low water level. PO4–P concentration in L. Pskov increased considerably during the low water period in the 1970s and decreased at the beginning of the high water period in the 1980s (Fig. 3). Similar trends have been described by Rippey et al. (1997) in White Lough, North Ireland, where the eutrophication accelerated between 1973 and 1979, which they believed was due to reduced hydraulic flushing of the lake during a dry period and accelerated internal nutrient loading from sediments. Accelerated eutrophication periods of L. Pskov appeared in dry years with low water level when water residence time in the lake was longer. Our results indicate strong effect of water temperature and water level on sediment variables with a time lag of 0–5 years (Table 3).

Even if the common trend of deterioration of the ecological state of L. Pskov is observable during the whole monitoring period (Kangur et al., 2003) it is not clear when these processes started. Palaeolimnological records enable extension of the timescale for the lake outside the instrumentally documented range and follow general trajectories in the lake development

(Anderson et al., 2005). Long-term changes in the dynamics of the matter cycle in L. Pskov are reflected in palaeorecords of the sediment sequence collected from monitoring station 51. In the overlapping sequences of palaeorecords and monitoring data the dynamics of some hydrochemical indicators and sediment records has a similar pattern. Thus, the content of sulphur in the sediment core and SO_4^{2-} content of water have similar dynamics, with the maxima appearing in both curves in the 1980s. The curves of phosphorus contents in water and sediment are also similar. However, comparing limnological and palaeoecological data, one must always consider that the sediment records suffer from taphonomic biases (Battarbee et al., 2005) and there is always a time lag between the changes in the bioproductivity and biogeochemical mater cycling in a lake and changes in sediment records. Within a certain trophic level the internal mechanisms of the matter circulation in the lake can compensate to some extent the variations in the external load (Punning et al., 2003).

Significant changes in palaeorecords from L. Pskov have occurred in sediment layers that accumulated in the 1930s (depth 18–19 cm). All indicators measured from sediment layers beyond a depth of 18 cm are rather stable. The sediment accumulation rate and content of sulphur are increasing above this depth. The C/N ratio is minimal at a depth of 15 cm, in layers that accumulated in the 1950s, indicating an increase in the importance of planktonic material in the sediment. At the same depth also the maximum values of phosphorus content were measured. The C/N ratio starts to increase in the 1960s and the content of phosphorus begins to decrease in the early 1970s. At the end of the 1980s the phosphorus content in the sediment starts to increase again. At a depth of 18 cm also the water content of the sediment increases or the content of dry matter decreases, which is probably another indicator of eutrophication from this time period. However, it can implicate active interstitial water exchange and the palaeorecords may therefore reflect the dynamics of sediment diagenesis.

The changes in the trophic status of the lake are most reliably reflected in relative abundance diagrams of diatoms, which is a widely used method in palaeolimnology (e.g. Hall & Smol, 1999; Battarbee et al., 2005). It is obvious that some diatoms with thin and fragile shells, such as *Asterionella formosa, Fragilaria crotonensis* and *Stephanodiscus binderanus*, decompose quickly and do not persist in the sediment. *S. binderanus* was among dominants in L. Pihkva water in the 1960s–1980s, however, it was not identified in our sediment core, neither was it found by Davydova (1999). *F. crotonensis* and *A. formosa* were rare in our sediment core, but quite common in the upper sediment layer in the 1970s according to Davydova (1999). These three diatoms are characteristic of moderate and strongly eutrophic lakes. In 1962–1964, the upper layer of bottom sediment was studied (Laugaste & Pork, 1996). Comparison of these results with the corresponding layers in sediment core shows quite good coincidence in dominants and subdominants (Table 4). Altogether, the diatom abundance diagram reflects quite well the dynamics of diatom assemblages in the lake.

The trophic state of the lake can be the leading factor in the formation of diatom relative abundance profiles (Hall & Smol, 1999; Battarbee et al., 2001). *Aulacoseira islandica* is the most persistent dominant alga in the sediment core, being widespread in mesotrophic and oligotrophic rather deep lakes and attaining maximum development in mesotrophic water bodies. Its development can be stimulated by eutrophication, but only at first stages, and is reduced in highly eutrophic lakes (Trifonova & Genkal, 2001). Maximum relative abundance of *A. islandica* in deeper layers (53–54 cm) of the sediment core indicates the lowest trophy in the studied timescale; decrease in the upper layers (from 17–18 cm corresponding to the 1930s), on the contrary, reflects intensive eutrophication. *Aulacoseira granulata* prefers moderate trophy and is withdrawing in highly eutrophic lakes. Its relative importance is fluctuating in sediment core. *A. granulata* var. *angustissima* indicates more elevated trophy; its abundance increases abruptly in upper layers beginning from 17–18 cm. *Aulacoseira ambigua* is widely distributed in meso- and eutrophic lakes, attaining peak values in shallow eutrophic lakes. Its relative abundance in the sediment core rises sharply from the level of

17–18 cm upwards. In Estonian lakes, *A. ambigua* indicates a higher trophic state than *A. granulata* (Laugaste et al., 2001). Among planktonic subdominants, the genera *Cyclotella* and *Puncticulata* have inclination to oligo-mesotrophy, *Stephanodiscus* indicates eutrophy (some species show even hypertrophy) and *Cyclostephanos* strong eutrophy. The 17–18 cm level (and upwards) marks an abrupt increase in *Stephanodiscus* (*S. rotula* (Kütz.) Hendey, *S. neoastraea* Håkansson & Hickel and *S. hantzschii* (Grunow).

Epiphytic algae in the sediment core have no clear trophic preference. *Fragilaria brevistriata* is known as an alga growing on plants as well as on sandy and less on muddy bottom (Snoeijs, 1989). Its high abundance in the middle of the sediment core (22–31 cm) may indicate, besides relative scantiness of plankton, that the lake bottom was sandy, and its abrupt decrease above 17–18 cm may reflect the silting of the lake bottom. Another reason for the decline in *Fragilaria* may be a reduction in light transparency as a result of algal shading in the water column.

The relationship of planktonic and littoral diatom species in the sediment core from L. Pskov varies quite widely. According to Battarbee (1978), the changes in the plankton/periphyton ratio can be interpreted in two main ways: either caused by a change in the morphology of the lake basin as the lake fills in, and/or by a change in productivity. Dominance of oligo-mesotrophic planktonic species in the deepest layers (54–50 cm) and of eutrophic ones above 18 cm can show the corresponding trophic state of L. Pskov.

Thus, the most profound changes in the diatoms community structure occurred at a sediment depth of 18 cm, indicating a large shift from oligo-mesotrophic to eutrophic conditions in the lake in the 1930s. The diatom relative abundance profile confirms the results of palaeoecological analysis discussed above. The results of our study show that the ecological degradation of L. Pskov started in the 1930s and can be related to sharp decrease in water level. Our results confirm the assumption that natural processes have significant impact on the ecological state of shallow lakes (Punning et al., 2003; Battarbee et al., 2005; Milius et al., 2005). The effect of warm weather is especially strong when it synchronises with low water level.

Conclusions

Key hydrochemical and -biological parameters demonstrate eutrophication of L. Pskov and destability of its ecosystem during last decades. As the current monitoring programme on L. Pskov covers a too short time-span, the start of the degradation of the lake ecosystem is not documented. The first attempt to combine sediment records with monitoring data enabled us to extend the time scale and to demonstrate large shifts in the trophic status of the lake since the 1930s. Our results provide a better understanding of the development in this large shallow lake, which is fundamental for predicting its future response to human impact and climate related processes. However, interactions of natural processes (e.g. changes in water level and temperature) and changes in external nutrient loading are not fully understood and the distinction of these effects is a challenge for future research. Nevertheless, our study illustrates the importance of water level and temperature fluctuations in the assessment of the ecological state of a large shallow lake.

Acknowledgements The authors are indebted to the Estonian Science Foundation (grants 6855, 6820 and 6008) for financial support. This research was supported by the Estonian target financed projects SF 0362483s03 and SF SF0282120s02. We are grateful to the reviewers for their constructive comments.

References

Anderson, N. J., E. Jeppesen & M. Søndergaard, 2005. Ecological effects of reduced nutrient loading (oligotrophication) on lakes: an introduction. Freshwater Biology 50: 1589–1593.

Appleby, P. G., P. J. Nolan, D. W. Gifford, M. J. Goodfrey, F. Oldfield, N. J. Anderson & R. W. Battarbee, 1986. 210-Pb dating by low-background gamma counting. Hydrobiologia 143: 21–27.

Battarbee, R. W., 1978. Relative composition, concentration and calculated influx of diatoms from a sediment core from Lough Erne, Northern Ireland. Polish Archive Hydrobiology 25: 9–16.

Battarbee, R. W., V. J. Jones, R. J. Flower, N. G. Cameron & H. Bennion, 2001. Diatoms. In Smol, J. P., H. J. B. Birks & W. M. Last (eds), Tracking Environmental Change Using Lake Sediments. Vol. 3 – Terrestrial, Algal and Siliceous Indicators. Kluwer Academic Publishers, Dordrecht, 155–202.

Battarbee, R. W., N. J. Anderson, E. Jeppesen & P. R. Leavitt, 2005. Combining palaeolimnological and limnological approaches in assessing lake ecosystem response to nutrient reduction. Freshwater Biology 50: 1772–1780.

Chambers, F. M. (ed.), 1993. Climate Change and Human Impact on the Land. Chapman & Hall, London, Glasgow, New York, Tokyo, Melbourne, Madras.

Davydova, N., 1999. Flora in the surficial layer. Diatoms. In Miidel, A. & A. Raukas (eds), Lake Peipsi. Geology. Sulemees Publishers, Tartu, 80–86.

Directive 2000/60/EC of the European Parliament and of the Council of 23 October 2000 establishing a framework for Community Action in the field of water policy. Official Journal of the European Communities L327/1–72 (22.12.2000).

Gulati, R.D. & E. van Donk, 2002. Lakes in the Netherlands, their origin, eutrophication and restoration: state-of-the-art review. Hydrobiologia 478: 73–106.

Hall, R. I. & J. P. Smol, 1999. Diatoms as indicators of lake eutrofication. In Stoermer, E. F. & J. P. Smol (eds), The Diatoms: Applications for the Envisronmental and Earth Sciences. Cambridge University Press, Cambridge, 128–168.

Jaani, A., 2001a. The location, size and general characterization of Lake Peipsi and its catchment area. In Nõges, T. (ed.), Lake Peipsi. Meteorology, Hydrology, Hydrochemistry. Sulemees Publishers, Tartu, 10–17.

Jaani, A., 2001b. Water regime. In Nõges, T. (ed.), Lake Peipsi. Meteorology, Hydrology, Hydrochemistry. Sulemees Publishers, Tartu, 65–72.

Kangur, K., A. Milius, T. Möls, R. Laugaste & J. Haberman, 2002. Lake Peipsi: Changes in nutrient elements and plankton communities in the last decade. Journal of Aquatic Ecosystem Health and Management 5: 363–377.

Kangur, K., T. Möls, A. Milius & R. Laugaste, 2003. Phytoplankton response to changed nutrient level in Lake Peipsi (Estonia) in 1992-2001. Hydrobiologia 506–509: 265–272.

Kangur, K., A. Kangur, P. Kangur & R. Laugaste, 2005. Fish kill in Lake Peipsi in summer 2002 as a synergistic effect of cyanobacterial bloom, high temperature and low water level. Proceedings of the Estonian Academy of Sciences. Biology Ecology 54: 67–80.

Krammer, K. & H. Lange-Bertalot, 1986: 1988: 1991. Bacillariophyceae. Teil 1–4. Süsswasserflora von Mitteleuropa 2(1), 846 pp.; 2(2), 596 pp.; 2(3), 576 pp.; 2(4), 437 pp.

Laugaste, R. & M. Pork, 1996. Diatoms of Lake Peipsi-Pihkva: a floristic and ecological review. Hydrobiologia 338: 63–76.

Laugaste, R., T. Nõges, P. Nõges, V. V. Jastremskij, A. Milius & I. Ott, 2001. Algae. In Pihu, E. & J. Haberman (eds), Lake Peipsi. Flora and Fauna. Sulemees Publishers, Tartu, 31–49.

Leavitt, P. R., 1993. A review of factors that regulate carotenoid and chlorophyll deposition and fossil pigment abundance. Journal of Paleolimnology 9: 109–127.

Milius, A., R. Laugaste, T. Möls, M. Haldna & K. Kangur, 2005. Weather conditions and water level as factors determining phytoplankton biomass and nutrient content in Lake Peipsi. Proceedings of the Estonian Academy of Sciences. Biology Ecology 54: 5–17.

Moss, B., T. Barker, D. Stephen, A. E. Williams, D. Balayla, M. Beklioglu & L. Carvalho, 2005. Consequences of reduced nutrient loading on a lake system in a lowland catchment: deviations from the norm? Freshwater Biology 50: 1687–1705.

Möls, T., 2005. Linear statistical methods for Estonian freshwater waterbodies. Estonian Naturalists' Society, Tartu, 256 p. (in Estonian with English summary).

Padisák, J. & I. Koncsos, 2002. Trend and noise: long-term changes of phytoplankton in the Keszthely Basin of Lake Balaton, Hungary. Verhandlungen Internationale Vereinigung Limnology 28: 194–203.

Punning, J. M., M. Kangur, T. Koff & G. Possnert, 2003. Holocene lake-level changes and their reflection in the paleolimnological records of two lakes in northern Estonia. Journal of Paleolimnology 29: 167–178.

Rippey, B. N., N. M. Anderson & R. H. Foy, 1997. Accuracy of diatom-inferred total phosphorus concentrations and the accelerated eutrophication of a lake due to reduced flushing and increased internal loading. Canadian Journal of Fisheries and Aquatic Sciences 54: 2637–2646.

SAS Institute Inc. 1999. SAS OnlineDoc, Version 8. Cary, NC, SAS Institute Inc.

Smol, J. P., 1992. Paleolimnology: an important tool for effective ecosystem management. Journal of Aquatic Ecosystem Health 1: 49–58.

Snoeijs, P., 1989. A check-list of the benthic diatoms at Forsmark (northern Baltic Sea). I. Epilithic and epiphytic taxa. Annales Botanicy Fennicy 26: 427–387.

Starast, H., A. Milius, T. Möls & A. Lindpere, 2001. Hydrochemistry of Lake Peipsi. In Nõges, T. (ed.), Lake Peipsi. Meteorology, Hydrology, Hydrochemistry. Sulemees Publishers, Tartu, 97–131.

Stålnacke, P., Ü. Sults, A. Vasiliev, B. Skakalsky, A. Botina, G. Roll, K. Pachel, & T. Maltsman, 2002. An assessment of riverine loads of nutrients to Lake Peipsi, 1995–1998. Archiv für Hydrobiology 141: 437–457.

Trifonova, I. & S. Genkal, 2001. Species of the genus *Aulacoseira* Thwaites in lakes and rivers of northwestern Russia. Distribution and ecology. In Economou-Amilli, A. (ed.), Proceedings 16th International Diatom Symposium. University of Athens, Greece, 315–323.

Utermöhl, H., 1958. Zur Vervollkommung der quantitatieven Phytoplankton-Methodik. Mitteilungen der Internationale Vereinigung Limnologie 9: 1–38.

Vilbaste, S., K. Sundbäck, C. Nilsson, & J. Truu, 2000. Distribution of benthic diatoms in the littoral zone of the Gulf of Riga, the Baltic Sea. European Journal of Phycology 35: 373–384.

Hydrobiologia (2007) 584:133–143
DOI 10.1007/s10750-007-8615-2

SHALLOW LAKES

Sediment diatom assemblages and composition of pore-water dissolved organic matter reflect recent eutrophication history of Lake Peipsi (Estonia/Russia)

Atko Heinsalu · Tiiu Alliksaar · Aina Leeben · Tiina Nõges

Abstract A paleolimnological approach was used for the assessment of the recent eutrophication history and identification of possible reference conditions in the large, shallow, eutrophic Lake Peipsi. Lake Peipsi is the fourth largest lake by area, and the largest transboundary lake in Europe, being shared between Estonia and Russia. Lake Peipsi has been anthropogenically impacted over a longer time-scale than that covered by instrumental limnological monitoring. The ^{210}Pb record and down-core distribution of fly-ash particles in the 40-cm core from the middle part of the lake suggest 130 years of sediment accumulation. Diatom assemblages indicate alkaline mesotrophic conditions and a well-illuminated water column, sediment pore-water fluorescence index values suggest low autochthonous productivity and a stable aquatic ecosystem similar to natural reference conditions during the second half of 19th and early 20th century. Near-synchronous stratigraphic changes including the expansion of the eutrophic planktonic diatom *Stephanodiscus parvus*, the appearance of new species associated with eutrophic lakes and the decrease in the relative abundance of littoral diatoms, together with changes in the fluorescence properties of sediment pore-water dissolved organic matter, imply increased nutrient availability, enlarged phytoplankton crops, reduced water-column transparency and the onset of human-induced disturbances in the lake since the mid-20th century. The most conspicuous expansion of eutrophic planktonic diatoms and maximum concentration of siliceous microfossils occur simultaneously with changes in the fluorescence indexes of pore-water dissolved organic matter, indicating a pronounced increase in the contribution of autochthonous organic matter to the lake sediment. This implies that nutrient loading and anthropogenic impact was at a maximum during the 1970s and 1980s. Sedimentary diatom flora may reflect a reduction of phosphorus loading since the 1990s. However, the absolute abundance of planktonic diatoms and sediment pore-water fluorescence index values vary greatly implying that the lake ecosystem is still rather unstable.

Guest editors: R. D. Gulati, E. Lammens, N. De Pauw & E. Van Donk
Shallow lakes in a changing world

A. Heinsalu (✉) · T. Alliksaar
Institute of Geology, Tallinn University of Technology, Ehitajate tee 5, 19086 Tallinn, Estonia
e-mail: heinsalu@gi.ee

A. Leeben
Marine Systems Institute, Tallinn University of Technology, Akadeemia tee 21, 12618 Tallinn, Estonia

T. Nõges
Centre for Limnology, Institute of Agricultural and Environmental Sciences, Estonian Agricultural University, 61101 Rannu, Tartu County, Estonia

Keywords Paleolimnology · Sediment · Diatoms · Pore-water dissolved organic matter · Lake Peipsi · Estonia

Introduction

Lake Peipsi, with a surface area of 3,555 km^2 is the fourth largest lake and the largest transboundary lake in Europe, being shared between Estonia and Russia. The lake consists of three sub-basins: Lake Peipsi *sensu stricto* (s.s.) in the north, Lake Lämmijärv in the middle and Lake Pihkva in the south (Fig. 1). Lake Peipsi s.s. is a shallow (mean depth 7.1 m) unstratified eutrophic lake with a unique ecosystem and constitutes an important economic resource for the people living in its vicinity. Therefore the water quality of the lake is a matter of great public interest. Some early studies refer to blue–green algae blooms and fish-kills in late 19th century (Pihu & Haberman, 2001), but the generally accepted opinion is that marked eutrophication of the lake occurred after World War II due to intensive use of mineral fertilisers, industrial development and population growth in urbanised areas. Nutrient loading caused by various human activities increased most drastically during the 1980s (Nõges et al., 2007). A sharp decrease in nitrogen loading resulting from the collapse of intensive agriculture during the Soviet period took place in the early 1990s, while the phosphorus load did not decline significantly. This decrease in N:P ratio created favourable conditions for bloom-forming nitrogen-fixing cyanobacteria and caused severe environmental problems in the lake (Nõges et al., 2004).

Adequate information about the ecological status and historical development of Lake Peipsi is essential for the protection of its ecosystem and the implementation of the EU's Water Framework Directive (Directive, 2000). Lake Peipsi is one of the most studied lakes in Estonia, and different chemical and biological water quality characteristics have been monitored regularly for four decades (Nõges, 2001; Pihu & Haberman, 2001). However, as earlier studies applied different laboratory techniques and methods and a diverse range of measured variables, the exact timing of the onset and temporal development of the eutrophication in Lake Peipsi is still a matter for debate. For example total phosphorus (TP) measurements

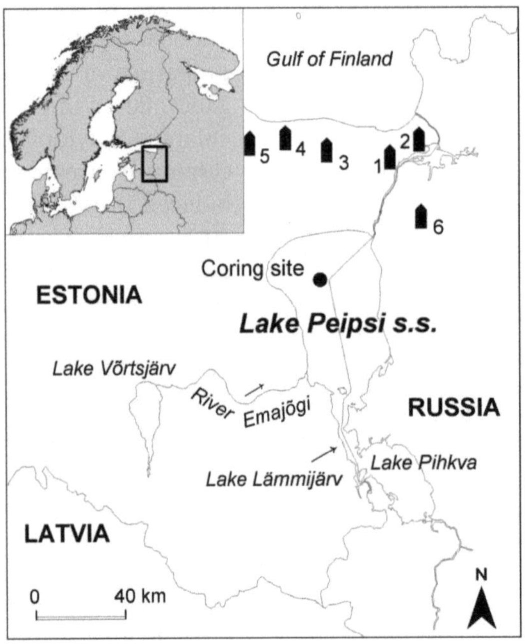

Fig. 1 Location of Lake Peipsi, the coring site (black dot) and nearby oil-shale-fired power plants (PP): (1) Estonian PP, (2) Baltic PP, (3) Ahtme PP, (4) Kohtla-Järve PP, (5) Kiviõli PP, (6) Slantsy PP

started since 1985, showing that mean TP in Lake Peipsi s.s. is 40 mg m^{-3} (Nõges & Nõges, 2006), and maximum TP values (higher than 120 mg m^{-3}) occurred in the late 1980s and early 1990s (Nõges et al., 2007). However, the available limnological data cover too short time-span to decide what the natural baseline conditions were and how the lake ecosystem behaved before significant human impact occurred.

In recent years several paleolimnological investigations on siliceous microfossils, mainly diatoms preserved in lake sediment, have provided reconstructions of ecological changes and the recent environmental histories of large lakes in northern Europe, e.g. Lake Ladoga (Kukkonen & Simola, 1999), Lake Saimaa (Sandman et al., 2000), Lake Päijänne (Meriläinen et al., 2001) and Lake Mälaren (Bradshaw & Anderson, 2001). These studies have indicated considerable changes in the diatom assemblages of all the lakes reflecting various levels of human induced perturbation.

Fluorescence spectroscopy is a simple, sensitive and fast tool, which provides information on the amount and chemical composition of dissolved organic matter (DOM) in the aquatic environment. The fluorescence method has been applied to seawater (e.g. Coble et al., 1990) as well as to pore-water of marine sediments (e.g. Sierra et al., 2001). Recently fluorescence analysis of pore-water DOM (pwDOM) was performed in lake sediments. The results indicated that the fluorescence approach can also be used for tracking the eutrophication history of a lake (Wolfe et al., 2002; Leeben et al., 2005).

The objectives of the present investigation are to provide insights concerning the recent history of human impact and trophic status of Lake Peipsi on the basis of sub-fossil diatom assemblages; to determine the possibilities of using fluorescence spectra of pwDOM for the identification of the properties of accumulated organic material; and to determine the relationship between these characteristics and lake productivity.

Methods

A 70-cm long sediment core was taken from the central part of Lake Peipsi s.s. (58°47.232′ N; 27°19.338′ E) at a water depth of 9.2 m (Fig. 1) in February 2002. Sampling was performed from the ice using a freeze corer (Wright, 1980). The recovered in situ frozen sediment core was sliced into continuous 1-cm thick sub-samples.

The chronology of the sediment core is based on ^{210}Pb radiometric dating. The sub-samples were analysed for ^{210}Pb and ^{226}Ra by direct gamma assay performed in the Radiometric Laboratory of the Ukrainian Hydrometeorological Research Institute (UHRI) using a low background germanium detector (Appleby et al., 1986). The results of the ^{210}Pb dates were compared to the sediment distributions of the artificial radionuclide ^{137}Cs (also analysed in UHRI) and spheroidal fly-ash particles (SFAP), products of fossil fuel burning. For SFAP analysis, sediment samples were subjected to sequential treatment with 30% H_2O_2, 3M HCl and 0.3 M NaOH in order to remove organic matter, carbonates and biogenic silica, respectively (Rose, 1990). A known number of commercially available *Lycopodium* spores were then added to the residue in order to calculate particle concentration. Particles were counted at 250 × magnification under a light microscope.

Sediment lithostratigraphic properties were analysed using standard procedures. The water content was measured by drying the wet sediment to constant weight at 105°C. The organic matter and carbonate content was determined by loss-on-ignition at 550°C for 4 h and at 950°C for 2 h, respectively (Heiri et al., 2001). Sediment wet and dry bulk densities were calculated according to Håkansson & Jansson (1983).

For diatom analysis the weighted sediment samples were treated with 30% H_2O_2 (Battarbee et al., 2001). Slides were mounted using Naphrax® and analysed for microfossils using a Zeiss Axiolab microscope (oil immersion, phase contrast, 1000 × magnification, numerical aperture – 1.30). A minimum of 450 diatom valves was counted for each sediment sample. Diatom concentrations were estimated by the addition of a known quantity of external *Lycopodium* marker spores to the cleaned sediment slurry (Kaland & Stabell, 1981). Diatom taxonomy was based primarily on Krammer & Lange-Bertalot (1986–1991).

For fluorescence analysis of pwDOM the frozen sediment samples were thawed slowly at

4°C. The water, emerging from the solid fraction of sediment was collected, centrifuged and filtered through a membrane filter with pore size of 0.22 μm (type Millex® GP, Millipore). Fluorescence spectra of pwDOM were analysed with a spectrofluorimeter (Fluo-Imager M51, LDI Ltd., Estonia) using a 1.4-ml semimicrocuvette. Fluorescence signal was recorded as 25 fluorescence emission spectra subsequently measured at 5 nm excitation intervals in the 240–360 nm range. The initial wavelength of emission scans (200 nm) was shifted by 10 nm at each subsequent excitation wavelength. As only relative changes of pwDOM in the sediment profile were considered, no spectral correction was performed. To track qualitative changes in pwDOM, two fluorescence indicators were chosen: (i) a fluorescence index (FI), which reflects the origin of DOM (McKnight et al., 2001), in our study we used the ratio of emission intensities at 460 nm and 510 nm at the excitation of 360 nm; (ii) the ratio of the fluorescence intensities H1:P, which in water studies has been shown to be related to biological activity (Parlanti et al., 2002).

Results

Lithostratigraphy

The freeze-coring device allowed to recover a 70-cm sediment core of loose, high-water-content gyttja without notable disturbance. A marked darkening in the uppermost part of the sequence is the only apparent lithostratigraphic change in the sediment column. Black spots and lenses (probably sulphide colouring) extend from the surface sediment down-core to 30 cm, whereas the most intensive darkening occurs at a depth of 10–21 cm. The base of the core consists of brownish–green homogeneous gyttja.

Sediments in Lake Peipsi are characterised by low dry bulk density gradually increasing from 0.02 to 0.09 g cm^{-3} down-core (Fig. 2). The sediment matrix consists mainly of silt grains, diatom silica and organic matter, while the proportion of $CaCO_3$ is low (2–3%) and stable (Fig. 2). There is a slight increase in organic matter from 23 cm core-depth to the sediment surface.

Sediment chronology

Radioisotope analyses show that ^{210}Pb concentration decreases almost monotonically down the sediment profile, suggesting that sediment perturbation has been negligible, and reach values close to ^{226}Ra activity (i.e. supported ^{210}Pb concentration) in the sample at 41.5 cm (Fig. 2). The only observed irregularity in unsupported ^{210}Pb activity is at 22.5 cm coinciding with changes in sediment dry bulk density (Fig. 2). The sedimentation pattern allows the assumption that supported ^{226}Ra and ^{210}Pb are in equilibrium

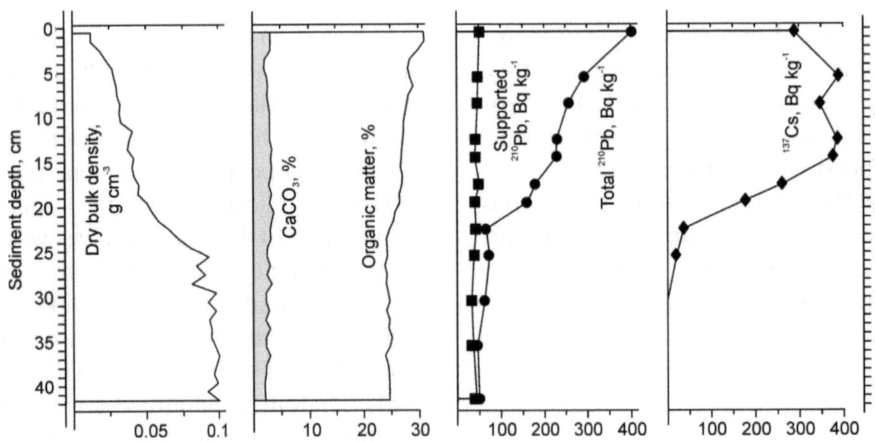

Fig. 2 Vertical distribution of dry bulk density (g cm^{-3}), $CaCO_3$ and organic matter (% per dry weight) as well as supported and total ^{210}Pb activities and ^{137}Cs activity (Bq kg^{-1}) in the studied sediment core from Lake Peipsi

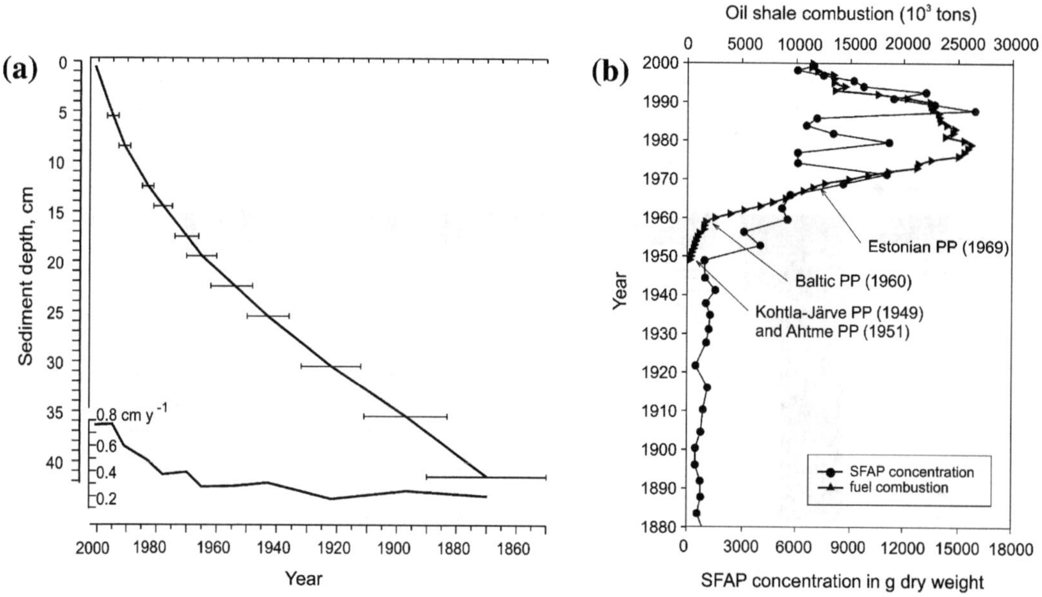

Fig. 3 (**a**) The age-depth profile from the radiometric chronology for the Lake Peipsi sediment core together with error margins of the ^{210}Pb dates and accumulation rate (cm y^{-1}) of the sediments; (**b**) The temporal distribution of the spheroidal fly-ash particle (SFAP) concentration in the studied core together with the oil-shale combustion history and the installation of individual power plants (PP) in the region (for location see Fig. 1)

at ca. 45 cm, corresponding to 150 years of accumulation. The constant rate of supply approach (Appleby & Oldfield, 1978) was used to calculate ^{210}Pb sediment ages for the uppermost 40 cm of the core (Fig. 3). Data show that sedimentation pattern has been fairly uniform over the last century with mass accumulation rate varying between 0.016–0.027 g cm^{-2} y^{-1}. There are no statistically significant changes in ^{226}Ra concentration (33–51 Bq kg^{-1}) over the studied sediment sequence revealing that the sediments accumulated over the last 100–130 years in Lake Peipsi have the same origin.

The concentration of ^{137}Cs is between 350–400 Bq kg^{-1} at 5–15 cm and below this depth gradually declines to zero (Fig. 2). The profile, which should reflect atomic bomb testing during 1953–1963 and the Chernobyl reactor accident in 1986, is quite broad, which may be due to post-depositional mobility of radiocesium. The broad ^{137}Cs peak and absence of corroborating reliable indicators (e.g. ^{134}Cs or ^{241}Am are below detection limit) do not allow to distinguish possible fall-out events and, therefore, use in the correction of the ^{210}Pb chronology.

However, the accuracy of ^{210}Pb dates for the Lake Peipsi sediment core is validated by independent chronological evidence from the down-core distribution of SFAP abundance linked to regional oil-shale combustion. A temporal sediment record of these particles, emitted from high-temperature combustion of fossil fuels, is a useful dating tool for the post-industrial era (Renberg & Wik, 1985). Moreover, as the region north of Lake Peipsi is the largest commercially exploited oil shale deposit in the world, and the main energy production area for Estonia, the well-documented history of oil-shale industry and fly-ash emissions enables the use of these particles as a reliable chronological tool (Alliksaar, 2000). An increase in SFAP concentration at 23 cm corresponds to a ^{210}Pb date in the early 1950s (Fig. 3) and correlates well with the installation of power plants in northeastern Estonia, at Kohtla-Järve in 1949 and at Ahtme in 1951. In sediments deposited since the 1960s a continuous and marked increase in the abundance of SFAP is observed, coinciding with the establishment of the large Baltic and Estonian Power Plants at Narva in 1960 and 1969, respectively. Since the beginning

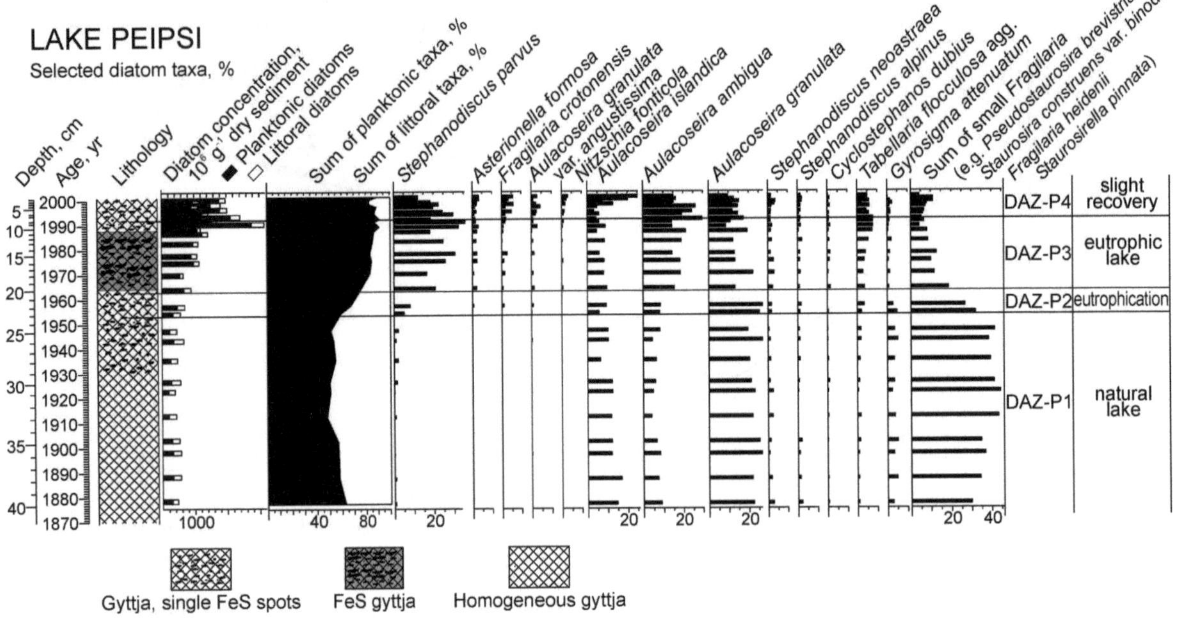

Fig. 4 Diatom stratigraphy of Lake Peipsi indicating the diatom concentration × 10^6 diatoms g^{-1} dry sediment, and relative frequencies of the most abundant diatom taxa are shown. DAZ – diatom assemblage zone

of 1990s SFAP concentration in sediments have decreased rapidly corresponding to the considerable reduction in energy production and decline in emission loads due to improved particle arresting techniques.

In summary, the good agreement of the ^{210}Pb chronology and the distribution of SFAP in the sediments (Fig. 3) suggests that the age-depth model for the core is reliable.

Diatom stratigraphy

Diatom analysis is carried out over the 0–40 cm sediment interval and is divided into four diatom assemblage zones (DAZ; Fig. 4). Diatoms are not perfectly preserved in the sediment and although there were no signs of dissolution, diatom breakage possibly related to grazing was a common feature.

Planktonic diatoms, typical of large lakes, *Aulacoseira islandica* (O.Müll.) Simonsen, *A. granulata* (Ehrenb.) Simonsen, and *A. ambigua* (Grunow) Simonsen predominate in DAZ P-1 (core-depth 23–40 cm), accumulated prior to mid-1950s and planktonic taxa make up 50–63%. *Stephanodiscus neoastraea* Håk. & Hickel, *S. alpinus* Hust. and *Puncticulata bodanica* (Grunow) Håk. are also present but less abundant. Besides these planktonic species, littoral diatoms are common (up to 50%), and mostly represented by small-size benthic taxa of the former genus *Fragilaria*, such as *Pseudostaurosira brevistriata* (Grunow in VanHeurck) D.M.Williams & Round, and *Fragilaria heidenii* Østrup, while a large, epipelic diatom *Gyrosigma attenuatum* (Kütz.) Rabenh. is also present. Total diatom concentrations range between 390–670 × 10^6 g^{-1} dry sediment.

In DAZ P-2 (20–23 cm), representing the mid-1950s to early 1960s, planktonic *Aulacoseira* species still predominate while *Stephanodiscus parvus* Stoermer & Håk., a planktonic diatom of eutrophic lakes starts to increase and eutrophic planktonic species like *Fragilaria crotonensis* Kitton and *Aulacoseira granulata* var. *angustissima* (O.Müll.) Simonsen appear. The share of littoral diatoms decreases and an increase in absolute abundance of planktonic diatoms takes place.

In the sediment accumulated in the 1970s and 1980s (DAZ P-3; 8–20 cm), *Stephanodiscus parvus* dominates and reaches maximum values at

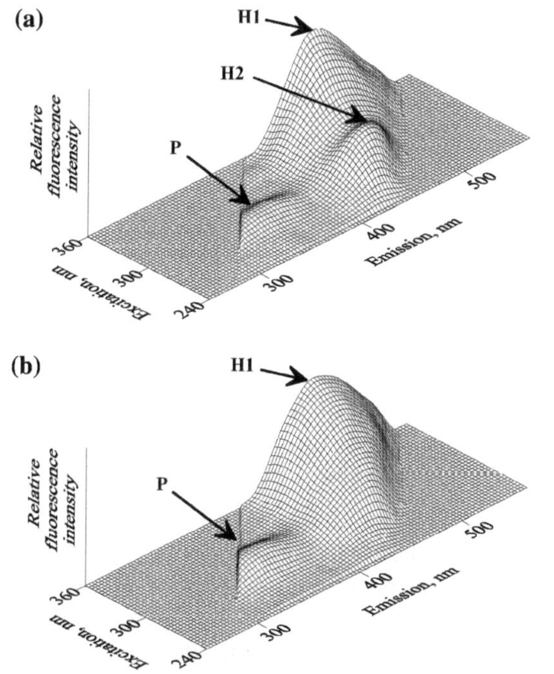

Fig. 5 Characteristic 3-D fluorescence spectra of dissolved organic matter (**a**) in the upper (0–25 cm) and (**b**) lower (26–50 cm) parts of the studied core. Spectra are scaled to 1 at the fluorescence maximum. H1 – humic-like fluorescence in near visible and H2 – in ultra violet part of spectrum, P – protein-like fluorescence

the top of the zone. The frequency of littoral diatoms decreases continuously towards the upper boundary of the zone while the absolute abundance of diatoms and especially the concentration of planktonic diatoms increases from 900 to 3100 × 10^6 g^{-1} dry sediment.

Towards the top of the zone DAZ P-4 (0–8 cm) deposited between the 1990s and early 2000s, the importance of *Stephanodiscus parvus* starts to decrease and *Aulacoseira islandica* increases. Littoral diatoms are present in low numbers. The absolute abundance of planktonic diatoms decreases and varies greatly.

Fluorescence of sediment pore water

Two distinctive types of 3-D fluorescence spectra of pwDOM from the Lake Peipsi sediment core are recorded (Fig. 5). Differences appear in the spectral shape of fluorescence attributed to humic-like substances (Coble, 1996). In pore-water samples below a core-depth of 26 cm, which represents sediment accumulated from the 1870s to the early 1940s, the distinctive ultra violet (UV) excitation maximum of humic-like fluorescence (H2) is missing. In the upper 25-cm of the sediment core, accumulated since the mid-1940s humic-like components have an excitation maximum in the near visible (H1) as well as in the UV (H2) part of spectrum. In all spectra a clear fluorescence maximum reported for protein-like substances (P) is noticeable (Traganza, 1969; Coble et al., 1990) and the position of the peak (excitation 289–297 nm, emission 346–360 nm) resembles tryptophane-like fluorescence (Yamashita & Tanoue, 2003).

The age-depth fluorescence profiles of humic substances show quite stable values within the 5–40 cm sediment layer, accumulated between 1870 and 1990, and rapid changes occur over the following 10 years (Fig. 6a). Protein fluorescence decreases slightly up to the middle part of the core, accumulated in the mid-1960s, and thereafter a higher variability is recorded (Fig. 6b).

Values of the FI and H1:P ratio are lower and quite stable in the lower part of the core up to the 1940s (Fig. 6c, d), while after this period the values start to increase. The increasing trend of the ratio H1:P is observed up to the top of the core, while the values of the FI decrease somewhat and vary greatly in the uppermost 10-cm layer formed since the 1990s.

Discussion

Diatom evidence for the history of human impact in Lake Peipsi

The diatom flora in the Lake Peipsi sediments, accumulated from ca. 1870s to the mid-1950s did not show striking changes either in species composition or diatom production. Low diatom abundances suggest that production was low, and compared with the modern situation, particularly low in planktonic communities. Relatively high proportions of small-size benthic fragilarioid diatoms suggest a well-illuminated water column, the light either penetrated down to the bottom, or the zone where periphytic diatoms might have

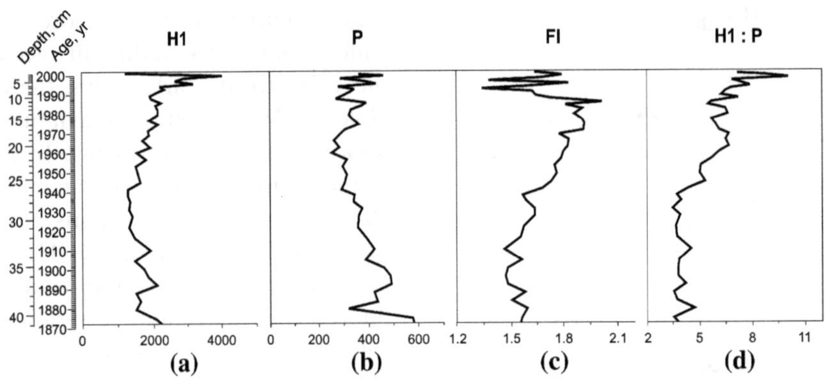

Fig. 6 Age-depth profile of (**a**) fluorescence intensity of humic substances (in near visible part of spectrum—H1); (**b**) proteins (P); (**c**) fluorescence index (FI); and (**d**) H1:P ratio in sediment pore-water of Lake Peipsi

grown was much closer to the coring site. Sediment records from smaller sub-basins of Lake Peipsi, i.e. Lake Lämmijärv and Lake Pihkva, indicate that planktonic diatoms like *Aulacoseira islandica*, *A. granulata* and *A. ambigua* together with littoral small-size fragilarioid taxa existed throughout post-glacial history (Davydova, 1999). A similar diatom flora was found in Holocene deposits of other large lakes in the region e.g. in Lake Ladoga and Lake Onega (Davydova, 1985). Therefore it is possible that it represents an assemblage typical for natural large lakes. Diatom evidence shows that the ecosystem of Lake Peipsi was stable and possibly close to natural baseline conditions for at least 70 years before the 1950s and that the influence of human activity in the catchment was still negligible. Instrumental observations of the water level of the lake, which started in 1885 show fluctuations of drier and wetter periods alternating most clearly within 21 to 33-year cycles (Jaani, 2001). However, it seems that the small, up to 1 m amplitude of the mean annual water level change had little influence on the diatom composition.

A substantial modification of the diatom flora in Lake Peipsi started in the mid-1950s and 1960s. The relative abundance of a small eutrophic diatom *Stephanodiscus parvus* increased although *Aulacoseira* species were still dominant. The appearance of new indicators of eutrophic conditions like *Fragilaria crotonensis* also indicates that nutrient availability and anthropogenic eutrophication increased. It is likely that this change reflects the onset of human-induced disturbance of the ecosystem after a long period of stability. Moreover, the abundance of periphytic diatoms was reduced, possibly due to the shading effects of larger plankton crops. An increase in the concentration of planktonic diatoms supports enhanced planktonic production.

Further changes in the diatom flora took place during the 1970s and 1980s. High relative abundance of planktonic diatoms suggests increased phytoplankton productivity and low water transparency and, hence, reflects progressive nutrient enrichment. The planktonic component of the assemblage also indicates progressive eutrophication, as *Stephanodiscus parvus*, an indicator of eutrophic to hypereutrophic conditions, became a significant part of the diatom plankton in the lake. Owing to its small size and thin frustule, *S. parvus* flourishes at high phosphorus availability and low Si:P ratios because of the better competitive ability for Si (Interlandi et al., 1999). The timing of the change in the diatom assemblage is in good agreement with hydrochemical observations, which indicate a peak in nutrient loading in the 1980s (Nõges et al., 2004).

A slight change in the diatom assemblage took place in the 1990s. Both the relative and absolute abundance of *Stephanodiscus parvus* decreased substantially during the 1990s, which may reflect a reduction in phosphorus loading. However, the absolute abundance of planktonic diatoms varied greatly, implying that the ecosystem was still rather unstable. Instrumental measurements revealed a decrease of phosphorus loading in the early 1990s and a recovery at the end of the decade while the in-lake P concentrations did not decrease substantially in the 1990s because of the remaining high pollution load from Russian catchment and in some extent also because of the P release from bottom sediments (Nõges et al., 2007).

Since the late-1990s the ecosystem of Lake Peipsi s.s. has destabilised and summer cyanobacteria blooms accompanied with fish-kills have become the most serious problem (Nõges et al., 2007). The great abundance and competitive success of nitrogen-fixing blue–greens is associated with a decreased nitrogen concentration and N:P ratio at stable phosphorus concentration (Nõges et al., 2004). Blue–greens have probably limited the abundance of small planktonic diatoms by competition for nutrients and shading (Bradbury et al., 2004). This assumption is supported by a corresponding increase in *Aulacoseira islandica* in the uppermost part of the sediment core as this species can grow in limited light and survive substantial periods without illumination (Stoermer et al., 1981).

Spectrofluorescence evidence of eutrophication in Lake Peipsi

The DOM fluorescence is usually characterised by two very distinct components and this is also the case for the sediment pore-water of Lake Peipsi (Fig. 5). One component with an emission wavelength maximum at 300–350 nm and optimal excitation in a range 250–280 nm is related to protein-like material and the other component with an emission maximum in a range of 400–480 nm has essentially been assigned to humic substances (Coble, 1996). From the total intensity of the latter maximum a significant quantity (60–90%) at the wavelengths 440–460 nm is reported to account for the fluorescence of fulvic acids (e.g. Green et al., 1992). Apparent changes in the 3-D fluorescence spectra of pwDOM at 26 cm depth (Fig. 5) suggest that the type of accumulated organic matter changed after the 1940s. There can be two reasons for this: (i) the composition of humic substances in the lake changed; (ii) a shift in the biogeochemical transformation pathway of organic material in the water column and/or the sediments has taken place. Although the nature of the components, which are responsible for the correspondence in the patterns of humic-like fluorescence, is still unclear (e.g. Chen et al., 2003) the temporal differences in the accumulation of humic substances of organic matter in Lake Peipsi are evident.

Fluorescence of pwDOM in Lake Peipsi show patterns (Fig. 6a, b), which may be accounted for by humic substances, mainly fulvic acids and protein-like substances that fluoresce at the wavelengths of tryptophane. Intensity of H1 demonstrates high and increasing values in the top of the core that could be related to the trophic state of the lake. Humic components from aquatic sources are generally more sensitive to bacterial degradation than those from terrestrial material, such as lignin. According to McKnight et al. (2001) spectral shape, i.e. FI, can indicate the origin of humic substances. The FI has a higher value (around 1.9) in the case of autochthonous material and a lower value (around 1.4) for components with allochthonous origin. The observed increase in FI values of Lake Peipsi (Fig. 6c) suggests that since the end of the 1940s the sources of humic substances have changed. The organic matter accumulated prior the 1940s has a more aromatic character compared to the sediments formed afterwards. While a higher aromaticity is typical for organic matter from terrestrial sources, compounds with an aliphatic structure dominate in organic matter produced by aquatic algae (McKnight et al., 2001). The obtained FI values suggest a progressive increase in the supply of autochthonous organic matter since the mid-20th century and maximum aquatic primary production during the 1970s and 1980s. This further supports the results from the optical properties of pwDOM indicating eutrophication of Lake Peipsi. During the last decade the FI of the lake pwDOM decreases and shows instability, probably reflecting the reduction of the autochthonous organic matter. The observed increase in FI values is synchronous with the increase of the H1:P ratio in the sediments of Lake Peipsi (Fig. 6d), although the latter value increases continuously to the top of the core. This ongoing increase in the 1990s indicates that algal production is still the dominant source of organic matter in the ecosystem.

The detailed investigation of sediment pore-water fluorescence properties in Lake Peipsi showed that aquatic production in the lake increased during the second half of the last century, resulting in greater autochthonous organic matter supply to the sediments and a change in accumulated humic substances. Moreover, the observed changes in pwDOM fluorescence properties are near-synchronous with shifts in the composition and absolute abundance of diatom flora indicating that pwDOM fluorescence spectroscopy can give useful information for sediment-based reconstructions of historical changes in lake paleoproductivity.

Conclusions

Paleolimnological analyses of lake sediment core have provided important insights into understanding of trophic history of Lake Peipsi especially outside the instrumentally documented range and evaluating natural background conditions. The study is mainly based on diatom analysis and fluorescence properties of pore-water dissolved organic matter. A chronology of the 40-cm core from the middle part of the lake was established on the basis of ^{210}Pb dating and cross-checked by SFAP particle counting and suggested 130 years of sediment accumulation. During the second half of 19th and early 20th century a very stable diatom assemblage indicates transparent water column and mesotrophic conditions and sediment pore-water fluorescence properties suggest low autochthonous productivity and a stable aquatic ecosystem similar to natural reference conditions. The onset of anthropogenic impact on the ecosystem of Lake Peipsi becomes evident in mid-1950s including the expansion of the eutrophic planktonic diatoms, the decrease in the relative abundance of littoral diatoms, together with changes in the fluorescence properties of pwDOM that implied increased nutrient availability, enlarged phytoplankton crops and reduced water-column illumination. Highly eutrophic conditions developed in the lake in the 1970s and 1980s. In the 1990s a slight recovery of the ecosystem may have taken place, however, the absolute abundance of planktonic diatoms and pwDOM properties vary greatly implying that the lake ecosystem is still rather unstable.

Acknowledgements This study was supported by the target financed research programmes of Estonian Ministry of Education (SF0332710s06 and 0822522s03), by Estonian Science Foundation (5923 and 5582), and by European Commission project MANTRA-East (EVK1-CT-2000-00076). We are grateful to Neil Rose for language check. We appreciate Laur Karu and Jaan Klõšeiko for their great help. We thank an anonymous referee and Holger Cremer for their comments and suggestions on the manuscript.

References

Alliksaar, T., 2000. Spatial and temporal variability of the distribution of spherical fly-ash particles in sediments in Estonia. Tallinn Pedagogical University, Dissertations on Natural Sciences 4: 1–44.

Appleby, P. G. & F. Oldfield, 1978. The calculation of ^{210}Pb dates assuming a constant rate of supply of unsupported ^{210}Pb to the sediment. Catena 5: 1–8.

Appleby, P. G., P. J. Nolan, D. W. Gifford, M. J. Godfrey, F. Oldfield, N. J. Anderson & R. W. Battarbee, 1986. ^{210}Pb dating by low background gamma counting. Hydrobiologia 141: 21–27.

Battarbee, R., V. J. Jones, R. J. Flower, N. G. Cameron, H. Bennion, L. Carvalho & S. Juggins, 2001. Diatoms. In Smol, J. P., H. J. B. Birks & W. M. Last (eds), Tracking Environmental Change Using Lake Sediments. Vol. 3: Terrestrial, Algal, and Siliceous Indicators. Kluwer Academic Publishers, Dordrecht, 155–202.

Bradbury, J. P., S. M. Colman & R. L. Reynolds, 2004. The history of recent limnological changes and human impact on Upper Klamath Lake, Oregon. Journal of Paleolimnology 31: 151–165.

Bradshaw, E. G. & N. J. Anderson, 2001. Validation of a diatom-phosphorus calibration set for Sweden. Freshwater Biology 46: 1035–1048.

Chen, J., E. J. LeBoeuf, S. Dai & B. Gu, 2003. Fluorescence spectroscopic studies of natural organic matter fractions. Chemosphere 50: 639–647.

Coble, P. G., 1996. Characterization of marine and terrestrial DOM in seawater using excitation-emission matrix spectroscopy. Marine Chemistry 51: 325–346.

Coble, P. G., S. A. Green, N. V. Blough & R. B. Gagosian, 1990. Characterization of dissolved organic matter in the Black Sea by fluorescence spectroscopy. Nature 348: 432–434.

Davydova, N. N., 1985. Diatoms as Indicators of Holocene Lake Environments. Nauka, Leningrad. (in Russian with English summary).

Davydova, N., 1999. Diatoms. In Miidel, A. & A. Raukas (eds), Lake Peipsi. Geology. Sulemees Publishers, Tallinn, 80–86.

Directive 2000/60/EC, 2000. Directive 2000/60/EC of the European Parliament and of the Council of 23 October 2000 establishing a framework for Community action in the field of water policy. Official Journal L 327: 1–72.

Green, S. A., F. M. M. Morel & N. V. Bluogh, 1992. Investigation of the electrostatic properties of humic substances by fluorescence quenching. Environmental Science & Technology 26: 294–302.

Håkansson, L. & M. Jansson, 1983. Principles of Lake Sedimentology. Springer-Verlag, Berlin.

Heiri, O., A. F. Lotter & G. Lemcke, 2001. Loss on ignition as a method for estimating organic and carbonate content in sediments: reproducibility and comparability of results. Journal of Paleolimnology 25: 101–110.

Interlandi, S. J., S. S. Kilham & E. C. Theriot, 1999. Responses of phytoplankton to varied resource availability in large lakes of the Greater Yellowstone Ecosystem. Limnology and Oceanography 44: 668–682.

Jaani ,A., 2001. Hydrological regime and water balance. In Nõges, T. (ed.), Lake Peipsi. Hydrology, Meteorology, Hydrochemistry. Sulemees Publishers, Tallinn: 38–53.

Kaland, P. E. & B. Stabell, 1981. Methods for absolute diatom frequency analysis and combined diatom and pollen analysis in sediments. Nordic Journal of Botany 1: 697–700.

Krammer, K. & H. Lange-Bertalot, 1986–1991. Bacillariophyceae. In Ettl, H., G. Gärtner, J. Gerloff, H. Heynig & D. Mollenhauer (eds), Süßwasserflora von Mitteleuropa, Vol. 2 (1–4). Gustav Fischer Verlag, Stuttgart.

Kukkonen, M. & H. Simola, 1999. Stratigraphy of diatoms in two ^{210}Pb-dated deep-water cores from Lake Ladoga: evidence of environmental change in the largest lake of Europe. In Mayama, S., M. Idei & I. Koizumi (eds), Proceedings of the 14th International Diatom Symposium, Toyko, Japan, September 2–8, 1996. Koeltz Scientific Books, Koenigstein, 427–435.

Leeben, A., A. Heinsalu, T. Alliksaar & L. Saarse, 2005. A high-resolution spectroscopic study of pore-water dissolved organic matter in annually laminated lake sediments: a new tool for reconstructing eutrophication history. Verhandlungen der Internationalen Vereinigung für Theoretische und Angewandte Limnologie 29: 457–460.

McKnight, D. M., E. W. Boyer, P. K. Westerhoff, P. T. Doran, T. Kulbe & D. T. Andersen, 2001. Spectrofluorometric characterization of dissolved organic matter for indication of precursor organic material and aromaticity. Limnology and Oceanography 46: 38–48.

Meriläinen, J. J., J. Hynynen, A. Palomäki, H. Veijola, K. Mäntykoski, K. Granberg & K. Lehtinen, 2001. Pulp and paper mill pollution and subsequent ecosystem recovery of a large boreal lake in Finland: a palaeolimnological analysis. Journal of Paleolimnology 26: 11–35.

Nõges, T. (ed.) 2001. Lake Peipsi. Hydrology, Meteorology, Hydrochemistry. Sulemees Publishers, Tallinn.

Nõges, T., I. Tõnno, R. Laugaste, E. Loigu & B. Skakalski, 2004. The impact of changes in nutrient loading on cyanobacterial dominance in Lake Peipsi (Estonia/Russia). Archiv für Hydrobiologie 160: 261–279.

Nõges, T., A. Järvet, A. Kisand, R. Laugaste, E. Loigu & P. Nõges, 2007. Reaction of large and shallow lakes Peipsi and Vortsjärv to the changes of nutrient loading. Hydrobiologia 584: 253–264.

Nõges, P. & T. Nõges, 2006. Indicators and criteria to assess ecological status of the large shallow temperate polymictic lakes Peipsi (Estonia/Russia) and Võrtsjärv (Estonia). Boreal Environment Research 11: 67–80.

Parlanti, E., B. Morin & L. Vacher, 2002. Combined 3D-spectrofluorometry, high performance liquid chromatography and capillary electrophoresis for the characterization of dissolved organic matter in natural waters. Organic Geochemistry 33: 221–236.

Pihu, E. & J. Haberman (eds), 2001. Lake Peipsi. Flora and Fauna. Sulemees Publishers, Tallinn.

Renberg, I. & M. Wik, 1985. Soot particle counting in recent lake sediments: an indirect dating method. Ecological Bulletins 37: 53–57.

Rose, N. L., 1990. A method for the selective removal of inorganic ash particles from lake sediments. Journal of Paleolimnology 4: 61–67.

Sandman, O., J. Meriläinen, H. Simola, J. Hynynen, J. Lahtinen, V. Marttila & P. Reinikainen, 2000. Short-core paleolimnological investigation of Lake Pihlajavesi in the Saimaa Lake complex, eastern Finland: assessment of habitat quality of an endemic and endangered seal population. Journal of Paleolimnology 24: 317–329.

Sierra, M. M. D., O. F. X. Donard, D. H. Etcheber, E. J. Soriano-Sierra & M. Ewald, 2001. Fluorescence and DOC contents of pore waters from coastal and deep-sea sediments in the Gulf of Biscay. Organic Geochemistry 32: 1319–1328.

Stoermer, E. F., R. G. Kreis Jr. & L. Sicko-Goad, 1981. A systematic, quantitative, and ecological comparison of *Melosira islandica* O. Müll. with *M. granulata* (Ehr.) Ralfs from the Laurentian Great Lakes. Journal of Great Lakes Research 7: 345–356.

Traganza, E. D, 1969. Fluorescence excitation and emission spectra of dissolved organic matter in sea water. Bulletin of Marine Science 19: 897–904.

Wolfe, A. P., S. S. Kaushal, I. R. Fulton & D. M. McKnight, 2002. Spectrofluorescence of sediment humic substances and historical changes of lacustrine organic matter provenance in response to atmospheric nutrient enrichment. Environmental Science & Technology 36: 3217–3223.

Wright, H. E. Jr., 1980. Cores of soft lake sediments. Boreas 9: 107–114.

Yamashita, Y. & E. Tanoue, 2003. Chemical characterization of protein-like fluorophores in DOM relation to aromatic amino acids. Marine Chemistry 82: 255–271.

Hydrobiologia (2007) 584:145–155
DOI 10.1007/s10750-007-0613-x

Factors influencing taxonomic composition and abundance of macrozoobenthos in extralittoral zone of shallow eutrophic lakes

Janusz Żbikowski · Jarosław Kobak

Abstract Zoobenthos is an essential part of shallow lake ecosystems, exerting a considerable impact upon their functioning. We studied 13 eutrophic, shallow, polymictic lakes from Northern Poland to find out which environmental factors influence taxonomic composition, abundance and biodiversity of their zoobenthos. The Canonical Correspondence Analysis allowed to distinguish three lake types: (1) macrophyte-dominated lakes, with high plant cover and well illuminated bottom, inhabited by abundant, diverse benthic taxa; (2) deeper phytoplankton-dominated lakes, with shaded bottom, high sediment oxygen demand (SOD) and rather sparse zobenthos community, dominated by *Chironomus* and *Chaoborus* larvae; (3) shallower phytoplankton-dominated lakes, with intermediate amount of light at the bottom and lower SOD values and comparatively diverse zoobenthos, but with lower number of taxa than in the first group. Apart from plant presence, distinguishing between macrophyte-dominated lakes and the other types, the most important variable in the CCA was amount of light reaching the bottom. Probably the impact of light on the bottom fauna was indirect: light stimulated development of macrophytes or phytobenthos (depending on its intensity) and thus improved food and oxygen conditions. Zoobenthos was also affected by oxygen conditions (mainly SOD), presumably by short-time oxygen depletions occurring in the deep phytoplankton-dominated lakes and preventing survival of some benthic taxa.

Keywords Macrozoobenthos · Shallow lakes · Open water zone · Bottom sediments · CCA · Light

Guest editors: R. D. Gulati, E. Lammens, N. De Pauw & E. Van Donk
Shallow lakes in a changing world

J. Żbikowski (✉)
Department of Hydrobiology, Institute of Ecology and Environmental Protection, Nicolaus Copernicus University, Torun, Poland
e-mail: jzbikow@biol.uni.torun.pl

J. Kobak
Department of Invertebrate Zoology, Institute of General and Molecular Biology, Nicolaus Copernicus University, Torun, Poland
e-mail: jkob@biol.uni.torun.pl

Introduction

Shallow lakes differ from deep ones with many respects. The main differences follow from the lack of long-term water stratification, which increases the impact of water-sediment interface processes upon a lake ecosystem (Scheffer, 1998). Shallow lakes are usually dominated either by macrophytes or phytoplankton (Scheffer, 1990; Scheffer et al., 1993), although Bayley & Prather (2003) distinguish two further types: lakes with high densities of both macrophytes and phytoplankton and lakes with low densities of both these groups.

The lack of permanent water stratification and shallow depth lead to development of a specific zone, characterised by muddy sediments with high water content, similar to those from the profundal of deep lakes and devoid of macrophytes, and littoral-like water parameters, e.g. light, temperature and oxygen conditions (Wolnomiejski & Papis, 1974; Wiśniewski, 1980a). This zone is inhabited by a specific benthic community (Wiśniewski, 1980b).

Events occurring in the top layer of sediments considerably affect functioning of a shallow lake. Zoobenthos is an essential factor influencing these events, especially in shallow lakes (Lindegaard, 1994). Its activity changes the physical-chemical parameters of sediments (Meadows & Tait, 1989; Jones & Jago, 1992) and extends the contact surface between water and sediments (Svensson & Leonardson, 1996). It also increases water dynamics in the near-bottom zone (Wood, 1975), thus influencing oxygen level (Svensson & Leonardson, 1996), which in turn determines processes occurring within the sediments and substance exchange across the water-sediment interface (Bostrom et al., 1988; Marsden, 1989; Sweerts, 1990). Moreover, these factors affect biodiversity, quantity and metabolic activity of microorganisms living in sediments (Krantzberg, 1985; van de Bund et al., 1994). These phenomena, indirectly influencing the element cycling in a lake, are much more important than nutrient excretion by zoobenthos (Fukuhara & Yasuda, 1985; Svensson, 1997; Svensson, 1998). Intensity and direction of substance exchange between water and sediments depends mainly on zoobenthos abundance, but its taxonomic composition is also important (Covich et al., 1999).

Therefore, studies on factors determining the quality and quantity of zoobenthos in the extra-littoral zone of shallow lakes are reasonable in view of its largest surface area. It has been shown that zoobenthos is affected by sediment quality (Reynoldson et al., 1995; Chapman et al., 1997), contaminants (Clements & Kiffney, 1993; Phipps et al., 1995), biotic factors (Kohler, 1992; MacKay, 1992), oxygen conditions (Dinsmore & Prepas, 1997) or macrophytes (Engel, 1988; Hargeby et al., 1994; Weatherhead & James, 2001). The impact of food is also important (Marsh & Tenore, 1990; Goedkoop & Johnson, 1996; Vos et al., 2000; Vos et al., 2004). However, most of this research does not concern shallow lakes, which, as we have shown above, differ considerably from deeper water bodies. Thus, we think that further studies on this topic, focusing on the shallow lakes, are still necessary.

The aim of our research was to find out which factors influence taxonomic composition, abundance and biodiversity of zoobenthos inhabiting the muddy bottom of shallow lakes. To accomplish this, we carried out a multivariate analysis of zoobenthos from 13 shallow lakes and several environmental parameters. Other studies have shown that zoobenthos in macrophyte-dominated lakes is more abundant and diverse than in phytoplankton-dominated ones, due to higher substratum heterogeneity (Hargeby et al., 1994; Diehl & Kornijów, 1999). We assumed that environmental conditions in the phytoplankton-dominated lakes (e.g. light, oxygen, temperature) would change with depth, which may affect zoobenthic community. Thus, we hypothesised that taxonomic composition and abundance of zoobenthos would also differ in the shallow phytoplankton-dominated lakes of various depths.

Materials and methods

Study area

We studied 13 eutrophic, shallow, polymictic lakes from Northern Poland (Fig. 1, Table 1). Most of them are small water bodies, with an area of a few dozen ha, except the Jeziorak Lake, which is the longest Polish lake. We examined its oldest, shallowed bay part.

Five of the studied lakes are dominated by macrophytes (Zielone, Ząbrowo, Karaś, Piotrkowskie, Głuchowskie). They differ from one another in depth, bottom coverage by plants and their taxonomic composition. The other lakes are dominated by phytoplankton. Due to the low water transparency, macrophytes (mainly

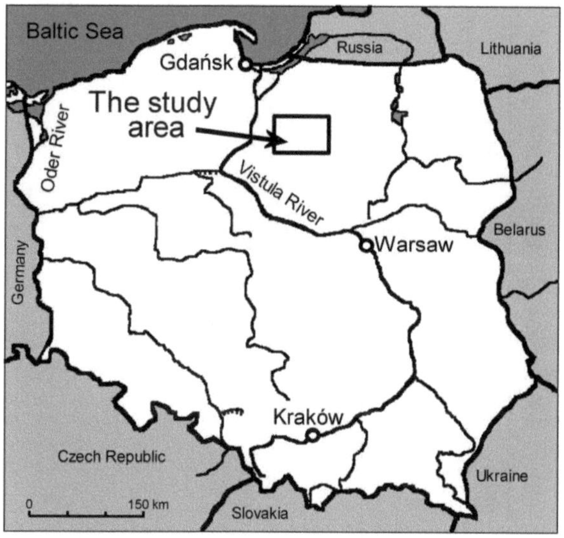

Fig. 1 Location of the studied lakes in Poland

helophytes) are present only near the shore. These lakes can be divided into shallower, 0.9–1.2 m in depth (Czerwica, Gardzień, Stęgwica, Jeziorak), and deeper, 2.2–3.8 m in depth (Tynwałd, Iławskie, Kolmowo, Silm). The Stęgwica Lake is a polyhumic, brown-water lake.

Collecting samples

We took samples in the 1998–2004 period. We sampled each lake for 2 or 3 years, except the Ząbrowo Lake, studied only from October 2000 to July 2001. Further sampling was impossible due to the total overgrowth of its bottom by macrophytes. In the first year of the study, we selected 3–6 stations in each lake, depending on its size. In the following years, we continued sampling at 1–2 stations, at which the zoobenthos community was the most typical for the lake. Each year we collected 8–10 series of samples in monthly intervals. In winter, sampling was less frequent, as the ice cover was sometimes too thin for walking and too thick for using a boat. In the macrophyte-dominated lakes, we used a core sampler (catching area: 40 cm^2) to take 7–10 replicate samples (depending on the zoobenthos abundance). We sampled sites devoid of plants because our main aim was to examine the bottom fauna and not the epiphytic one. In the phytoplankton-dominated lakes, we took 2–4 replicate samples (depending on the zoobenthos abundance) using an Ekman-Birge grab (catching area: 225 cm^2). We rinsed the samples using a 0.5 mm sieve and preserved in 4% formaldehyde. Diptera (except the family Ceratopogonidae and subfamily Orthocladiinae), Oligochaeta (except the family Naididae), Mollusca and Crustacea were idenfified to species or genus level when practical. Other taxa were not identified beyond order or class.

Measured parameters

Simultaneously with collecting the benthic samples, we monitored several physical and chemical parameters of water and sediments (Table 1).

To assess the light conditions we used a luxometer Slandi LX204. We also calculated the vertical attenuation coefficient of light under water (E) (Scheffer, 1998: 22, Eq. 2). Furthermore, to estimate the amount of light reaching the bottom, we multiplied the E value by the lake depth (D). This index (ED) is related to the shade level at the lake bottom (the higher ED value, the less light reaches the bottom) (Scheffer, 1998). Moreover, we calculated euphotic depth (Scheffer, 1998: 25, Eq. 7). This is the depth beyond which the light level falls below 1% of the surface irradiation and is considered too low to maintain a positive net photosynthesis of algae.

We measured temperature, conductivity and oxygen concentration of the near-bottom water layer with the core sampler and a MultiLine P4 (WTW) Universal Pocket Sized Meter.

We collected the bottom sediments (0–5 cm top layer) with the core sampler and measured their water content (by oven-drying sediments to a constant weight at 105°C) and organic mater content (dried sediments were heated at 550°C for 2 h). We also estimated sediment oxygen demand (SOD) by adding 300 ml of tap water oxygenated up to 100% to a dish containing 20 cm^3 of fresh sediments. The diameter of an oxygen sensor matched tightly the outlet of the dish, which prevented the oxygen exchange with the outside. Our preliminary trials showed that the oxygen concentration did not change ca. 15 min after the end of the oxygenation process, so we assumed that oxygen losses in the pure tap

Table 1 Morphometry of the studied lakes and mean values of the environmental variables

		macrophyte-dominated					deeper					shallower			
							phytoplankton-dominated					phytoplankton-dominated			
Parameter	Abbrev.	Zb	P	Z	Kr	G	S	C	Jtot	J	Gr	K	T	I	Si
Surface (ha)		12.7	54.4	20.2	423.3	24.1	39.7	37.8	3230		85.5	43.4	29.9	154.5	58.8
Max. depth (m)		1.6	2.5	2.4	2.8	4.5	1.2	0.9	11.0	2.0	1.1	5.7	2.2	2.8	3.8
Mean depth (m)		1.1	1.7	1.3	0.6	2.2	1.0	0.5	5.5	1.3	0.9	3.1	1.8	1.1	2.0
Euphotic depth (m)		2.2	3.2	3.6	3.1	2.6	1.6	1.3		1.7	1.2	1.6	1.2	1.6	2.5
Max. length (m)		960	1450	1375	4100	740	1880	1030	27400		4175	1050	1090	3700	1625
Max. width (m)		215	850	240	2250	475	290	685	1200		350	550	615	750	800
ED[a]	ED	2.7	2.1	2.5	2.6	7.2	3.2	3.1		3.7	4.1	9.4	8.0	7.4	7.1
Oxygen (mg l^{-1})	ox.	9.8	9.0	7.7	8.7	7.0	8.0	14.6		11.0	11.4	7.2	8.0	8.6	7.7
Temperature (°C)	temp.	11.1	14.5	14.7	12.4	12.2	12.5	16.2		13.0	13.2	11.5	13.3	12.4	12.1
Conductivity (–)	cond.	435	258	102	459	544	201	205		303	183	434	282	352	417
Water content (%)	WC	95.9	96.4	97.5	94.5	93.4	95.5	97.9		96.3	98.2	95.1	96.2	96.1	97.6
Organic matter (%)	OM%	49.8	63.1	78.0	42.8	32.7	52.0	57.2		53.4	68.8	35.2	49.6	37.7	57.0
Organic matter (mg)[b]	OMmg	201	219	186	204	225	254	119		171	87	185	176	141	143
Sediment oxygen demand	SOD	5.0	3.1	2.8	3.1	5.8	4.3	2.8		4.6	2.2	7.5	11.1	5.2	2.7
Amount of plants (1–10)	plants	7.5	6.2	6.0	5.7	3.7	2.1	1.0		2.2	1.0	1.0	1.0	1.0	1.0

[a] ED – a product of the vertical attenuation coefficient of light under water and the lake depth

[b] organic matter (mg) – Organic matter content in mg of dry weight per 10 cm^3 of fresh sediments

Lakes: Zb – Ząbrowo, P – Piotrkowskie, Z – Zielone, Kr – Karaś, G – Głuchowskie, S – Stęgwica, C – Czerwica, Jtot – total Jeziorak (not analysed), J – the bay part of the Jeziorak Lake, G – Gardzień, K – Kolmowo, T – Tynwałd, I – Iławskie, S – Silm. Abbrev. – an abbreviation standing for a given parameter in Fig. 2 and 3

water would be negligible. The oxygen uptake was measured during 1 h, at 20°C. The sediments were kept in permanent resuspension by means of a magnetic stirrer, to sustain the water flow around the membrane of the oxygen sensor.

We assessed the amount of macrophytes at the sampling stations using an arbitrary scale from 1–10 (1: total lack of plants, 10: presence of plants in the whole water column).

Data analysis

For each lake we calculated the mean zoobenthos density, biomass and biodiversity. As the numbers of sampled invertebrates differed strongly among various lakes, we applied the rarefaction technique (Hurlbert, 1971) to assess a theoretical number of taxa that would be found in a given number of collected individuals. We also calculated a traditional measure of biodiversity, Shannon-Wiener index. To find differences in these parameters among various lake types, we used one-way ANOVA of the log-transformed data (to reduce heteroscedasticity), followed by Tukey test. In the case of taxon richness, we compared values assessed for 450 individuals (the lowest number of individuals collected from a single lake, namely: Iławskie Lake). We removed the Głuchowskie Lake from these analyses, as it clearly differed from the other macrophyte-dominated lakes.

To find the relationships between benthic taxonomic composition and environmental conditions, we performed Canonical Correspondence Analysis (MVSP, Kovach Computing Services, 2004). The environmental variables were standardized and the densities of taxa were transformed using the $\log(x + 1)$ transformation to dampen the effects of a few most abundant taxa. To reduce noise, we removed rare taxa (with frequencies lower than 0.5%) from the data set. Lake depth and vertical attenuation coefficient of light under water (E) were removed from the analysis to avoid multicolinearity of data.

To find out which factors determine distribution of taxa within the phytoplankton-dominated lakes, we run CCA on the reduced data set, including only the samples from these lakes. We excluded the amount of plants from this analysis.

Results

The CCA run on the entire data set explained 12.4% of its variance (Fig. 2). The first CCA axis allowed to distinguish the three types of lakes: (1) macrophyte-dominated lakes, with high plant cover and comparatively well illuminated bottom (ED <3, Table 1), inhabited by macrophyte-associated bottom fauna (e.g. *Corynocera ambigua* (Zetterstedt), *Polypedilum nubeculosum* (Meigen), Ephemeroptera, Trichoptera, Mollusca); (2) deeper phytoplankton-dominated lakes, with shaded bottom (ED >7) and high sediment oxygen demand (SOD); (3) shallower phytoplankton-dominated lakes, with lower ED (3–4) and SOD values. The Głuchowskie Lake, though dominated by macrophytes, was grouped with the deeper phytoplankton-dominated lakes. The ED value in this lake was 7.2, much higher than in the other macrophyte-dominated lakes. The second CCA axis was correlated with conductivity and percentage of organic matter in sediments. These factors differentiated the deeper phytoplankton-dominated lakes from the shallower ones, with the latter having lower water conductivity and higher organic matter content in the sediments.

The two first axes of the CCA run on the samples from the phytoplankton-dominated lakes explained ca. 10% of variance (Fig. 3). Apart from ED and SOD, also conductivity, percentage organic matter content and oxygen concentration were correlated with the first axis. It allowed to separate taxa into those inhabiting deeper (e.g. *Chironomus* sp. or *Chaoborus flavicans* (Meigen)) and shallower (other chironomids, e.g. *Einfeldia* gr. *dissidens* (Walker), *Procladius* spp., *Tanytarsus* sp.) lakes. The latter preferred comparatively well-illuminated and oxygenated sites with low water conductivity and sediments having low SOD and high organic matter content. Along the second CCA axis, zoobenthos of the shallower lakes was divided into taxa inhabiting sites with high organic matter content in sediments (mainly in the Stęgwica Lake) and low conductivity and those preferring the opposite conditions.

The zoobenthos densities in various lake types differed from one another, but the significance of this difference was rather low (ANOVA:

$F_{2,\ 9} = 5.26$, $P = 0.031$), probably due to the high within-group variance. The density in the macrophyte-dominated lakes (ca. 6000 ± 4400 individuals per m^{-2} on average ±SD) was significantly higher (Tukey test) than in the deeper phytoplankton-dominated ones (ca. 760 ± 541). The density in the shallower lakes dominated by phytoplankton was intermediate (1074 ± 1160) and did not differ significantly from the other groups, though its difference from the density in the macrophyte-dominated lakes was only marginally insignificant. The differences in mean biomass (average for all lakes: 6.2 ± 6.2 g of wet weight m^{-2}) were not significant (ANOVA: $F_{2,\ 9} = 3.07$, $P = 0.096$).

The rarefaction curves (Fig. 4) showed the high taxon richness and biodiversity of the lakes dominated by macrophytes, except the Głuchowskie Lake. The biodiversity in the deeper lakes dominated by phytoplankton was the lowest. The curves for the shallower phytoplankton-dominated lakes were intermediate, with two of them being more similar to those for the macrophyte-dominated lakes and the two other closer to the curves for the deeper phytoplankton-dominated lakes. The rarified taxon numbers in all the lake types differed from one another (ANOVA: $F_{2,\ 9} = 18.18$, $P = 0.001$), with the highest values found in the macrophyte-dominated lakes (21 ± 1.8), intermediate in the the shallower phytoplankton-domi-

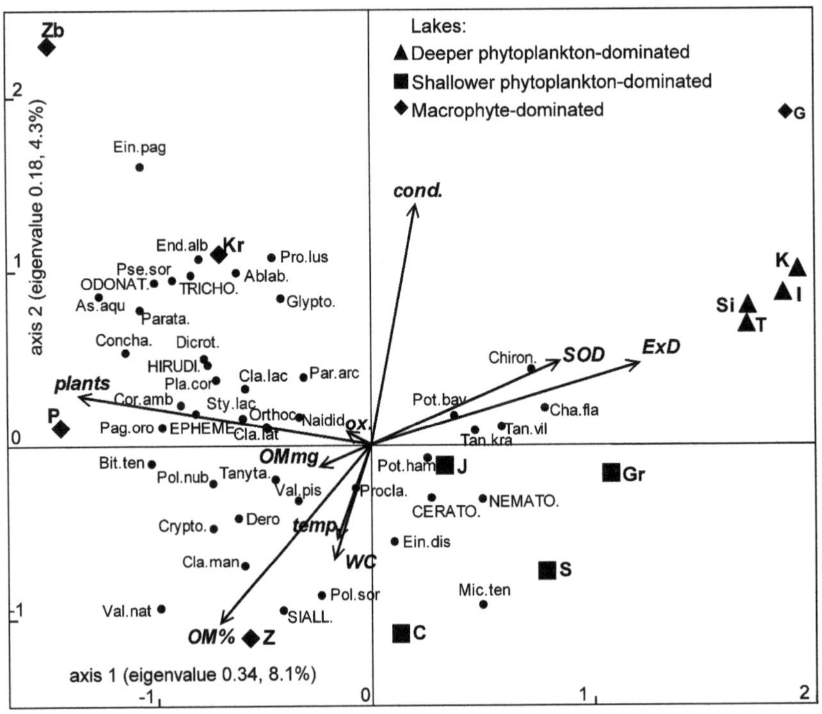

Fig. 2 The position of lakes and taxa constrained by the environmental variables (arrows) in the CCA ordination of the total data set. Site scores were averaged for each lake to avoid overcrowding the diagram. See Table 1 for the abbreviations of lake names and environmental variables. Taxa: Ablab.-*Ablabesmyia* sp.; As.aqu-*Asellus aquaticus*; Bit.ten-*Bithynia tentaculata*; CERATO.-Ceratopogonidae; Cha.fla-*Chaoborus flavicans*; Chiron.-*Chironomus* sp.; Cla.lac-*Cladopelma lacophila*; Cla.lat-*Cladopelma lateralis*; Cla.man-*Cladotanytarsus mancus*; Concha.-*Conchapelopia* sp.; Cor.amb-*Corynocera ambigua*; Crypto.-*Cryptochironomus* sp.; Dero-*Dero* sp.; Dicrot.-*Dicrotendipes* sp.; Ein.dis-*Einfeldia dissidens*; Ein.pag-*Einfeldia pagana*; End.alb-*Endochironomus albipennis*; EPHEME.-Ephemeroptera; Glypto.-*Glyptotendipes* sp.; HIRUDI.-Hirudinea; Mic.ten-*Microchironomus tener*; Naidid.-Naididae; NEMATO.-Nematoda; ODONAT.-Odonata; Orthoc.-Orthocladiinae; Pag.oro-*Pagastiella orophila*; Par.arc-*Parachironomus arcuatus*; Parata.-*Paratanytarsus* sp.; Pla.cor-*Planorbarius corneus*; Pol.nub-*Polypedilum nubeculosum*; Pol.sor-*Polypedilum sordens*; Pot.bav-*Potamothrix bavaricus*; Pot.ham-*Potamothrix hammoniensis*; Pro.lus-*Propsilocerus lusatiensis*; Procla.-*Procladius* spp.; Pse.sor-*Psectrocladius sordidellus*; SIALL.-*Sialis* sp.; Sty.lac-*Stylaria lacustris*; Tan.kra-*Tanypus kraatzi; Tan.vil-Tanypus vilipennis; Tanyta.-Tanytarsus sp.; TRICHO.-Trichoptera; Val.nat-Valvata naticina; Val.pis-Valvata piscinalis*

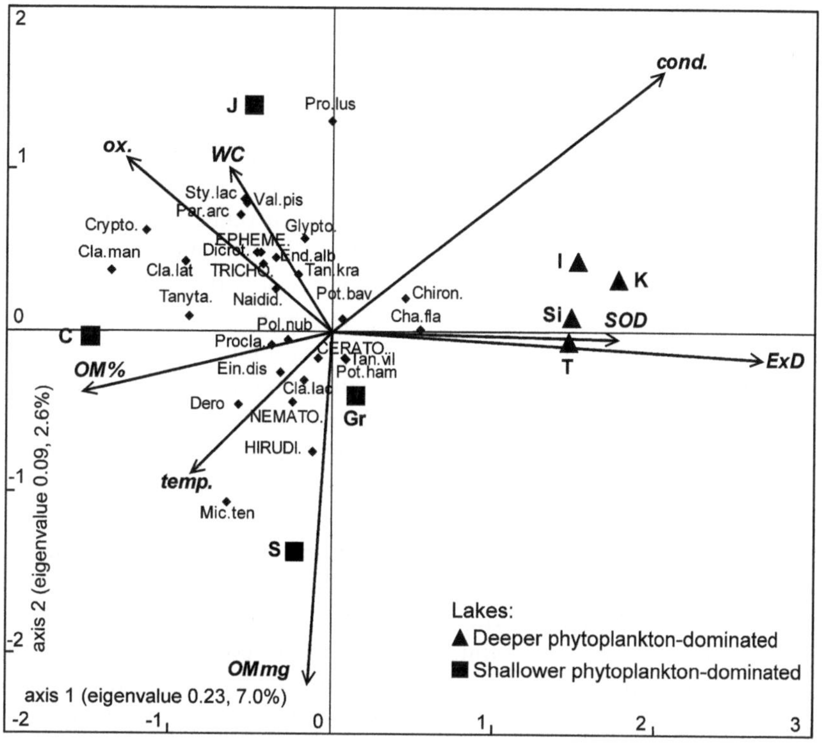

Fig. 3 The position of lakes and taxa constrained by the environmental variables (arrows) in the CCA ordination of the phytoplankton-dominated lakes only. See Table 1 and Fig. 2 for the abbreviation meaning

nated lakes (14 ± 5.4) and the lowest in the deeper phytoplankton-dominated lakes (8 ± 2.6). The Shannon-Wiener index values differentiated (ANOVA: $F_{2,\ 9} = 14.48$, $P = 0.002$) the deeper phytoplankton-dominated lakes (1.36 ± 0.21) from the other lake types (2.72 ± 0.71).

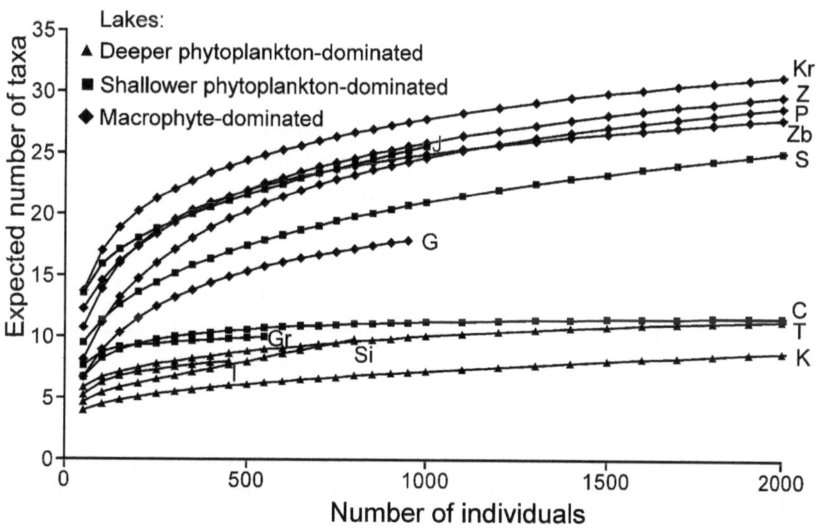

Fig. 4 The rarefaction curves for the studied lakes. See Table 1 for the abbreviations of lake names. The curve steepness is a function of the community taxon evenness, while its height indicates its taxon richness

Discussion

Our study confirmed the strong, positive effect of macrophytes upon abundance, taxonomic composition and taxon richness of zoobenthos. The macrophyte-dominated lakes were the most diverse group in our study (Fig. 2), probably due to domination of different plant taxa in various lakes. Macrophytes enhance environmental heterogeneity, provide protection from predators and improve food conditions for zoobenthos (Hargeby et al., 1994; Diehl & Kornijów, 1997; Van Den Berg et al., 1997; Gong et al., 2000). However, biodiversity (Shannon-Wiener index) of the shallower phytoplankton-dominated lakes did not differ from that measured in the macrophyte-dominated lakes, despite the lack of plants in the former. Only the biodiversity in the deeper lakes differed from those in the other lake types. Thus, there are some other factors controlling zoobenthos composition in the studied lakes.

The multivariate analysis revealed that the amount of light reaching the bottom was an important factor influencing zoobenthos. It was highly correlated with the CCA axis, which discriminated between the macrophyte-dominated and phytoplankton-dominated lakes, as well as between the two types of the phytoplankton-dominated ones (Fig. 2 and 3). ED values in the deeper phytoplankton-dominated lakes were above 7 and their bottoms were located below the euphotic depth, which made the development of phytobenthos impossible. The zoobenthos of these lakes consisted of few taxa, mainly *Chaoborus* and *Chironomus* larvae. A similar benthic community was found in the Głuchowskie Lake (Fig. 2), although this is a macrophyte-dominated lake. However, it is deeper than the other lakes of this type (ED value above 7). It suggests that light was a stronger factor determining zoobenthos distribution than macrophyte presence. The ED values in the shallower phytoplankton-dominated lakes were between 3 and 4. Their bottoms were located slightly above the euphotic depth, which enabled photosyntesis at the lake bottom but was insufficient for the plant growth. The zoobenthos composition in these lakes was different and their biodiversity was higher. ED in the macrophyte-dominated lakes was below 3 and their depth was much shallower than the euphotic depth, allowing for development of macrophytes and a rich, diverse zoobenthos community. One should note that the Czerwica lake is devoid of macrophytes despite its low ED value, indicating that not only light conditions determine macrophyte occurrence. Our study showed that phytobenthos development, which leads to a relatively higher zoobenthos diversity, was possible at ED value lower than 7. According to Scheffer (1998), the maximum shade level (ED) tolerated by phytoplankton in northern temperate, shallow lakes was 16 (in this case ED is a product of the vertical light attenuation coefficient and the depth of the mixed water layer, which in a shallow lake equals the lake depth). This value is much higher than that obtained in our study for phytobenthos, because in a shallow lake planktonic algae are mixed in the entire water column and can reach better light conditions in the upper water layer.

Taxonomic composition and abundance of bottom fauna depend mainly on food and oxygen conditions. One can assume that this statement applies also to shallow lakes. The caloric value of muddy sediments is low compared with that of living phytoplankton (Jonasson & Lindegaard, 1979), due to oxidation and bacterial decomposition (Ahlgren et al., 1997). Superficial sediments contain less fatty acids, which are a crucial component of an animal diet, than sedimenting matter (Goedkoop et al., 2000). That is why appearance of any additional food sources is beneficial for benthic invertebrates. Light is a key limiting factor for photosyntesis and production of benthic algae (Lassen et al., 1997). Thus, if the amount of light reaching the bottom allows for phytobenthos growth, food conditions are considerably improved (Gullberg et al., 1997). Moreover, benthic algae excrete extracellularly polysaccharides and glycoproteins, which may be used as food by zoobenthos (Madsen et al., 1993; Underwood & Paterson, 1993). During sampling, we noticed that the sediments from the shallower phytoplankton-dominated lakes, in contrast to the deeper ones, were greenish, suggesting the occurrence of benthic algae. This may indicate that it was the food quality, which determined comparatively high biodiversity in these lakes. However, more thorough examination of phytobenthos is needed to confirm this hypothesis.

Oxygen conditions are another factor influencing zoobenthos quality and quantity. In our study, the impact of oxygen concentration was visible only after removing the lakes dominated by macrophytes from the data set (Fig. 3). On the other hand, sediment oxygen demand was highly correlated with the first CCA axis, explaining most of the observed variance (Fig. 2 and 3). The near-bottom layer is often less oxygenated than the rest of the water column, due to mineralization occurring at the sediment surface (Sweerts, 1990). In our study, oxygen saturation of the near-bottom water layer (2–3 cm) was good (above 80%) but in summer, when the wind speed was low, we observed short oxygen depletions (to ca. 1 mg O_2 L^{-1}) in the deeper phytoplankton-dominated lakes. Furthermore, the SOD in the deeper phytoplankton-dominated lakes was the highest, suggesting the possibility of a fast decrease of the oxygen concentration, e.g. during a sediment resuspension. Such short-time events are difficult to detect in measurements of oxygen concentration, but may still strongly influence the survival of benthic animals (Heinis & Davids, 1993; Armitage et al., 1995), especially if they cannot migrate to better-oxygenated places, e.g. macrophytes (Kornijów & Moss, 2002). Probably this is why SOD was one of the most important variables explaining the zoobenthos composition in our study (Fig. 2 and 3). On the contrary, oxygen saturation of the shallower lakes was always above 80%. It could be a result of phytobenthos presence, sustaining good oxygen conditions at the sediment-water interface (Kelderman, 1984). Thus, it is possible that oxygen conditions, partly determined by illumination, directly influenced the bottom fauna composition and abundance in our study.

A phenomenon that often occurs in a shallow lake is sediment resuspension. Usually its effect on zoobenthos is negative (Raspopov et al., 1988; Gross et al., 1992), but sometimes periodical resuspension may enhance benthic biodiversity by inhibiting domination of a few most competitive species (Shin, 1989). By excreting organic compounds, benthic algae consolidate superficial layer of sediments and decrease the probability of resuspension (Madsen et al., 1993; Underwood & Paterson, 1993; Wiśniewski, 1995). This is another indirect mechanism of the impact of light on conditions experienced by zoobenthos.

The strong contribution of organic matter content to the second axis of the CCA run on the samples from the phytoplankton-dominated lakes (Fig. 3) probably resulted from the high concentration of humic substances in the Stęgwica Lake, which differentiated it from the other studied lakes.

Conclusions

According to our hypothesis, the zoobenthos composition and number of taxa in the shallow phytoplankton-dominated lakes depended on their depth. Light conditions were the most important factor influencing the bottom fauna in these lakes. The impact of light was probably indirect, resulting from stimulation of phytobenthos development and its photosynthesis at the lake bottom and improving food and oxygen conditions. Illumination is particularly important in shallow lakes, in which amount of light accessing the bottom strongly depends on their depth and water transparency. Another important factor influencing the zoobenthos of shallow lakes was oxygen, especially its short-time depletions that might be lethal to some taxa.

References

Ahlgren, G., W. Goedkoop, H. Markensten, L. Sonesten & M. Boberg, 1997. Seasonal variation in food quality for pelagic and benthic invertebrates in Lake Erken – the role of fatty acids. Freshwater Biology 38: 555–570.

Armitage, P. D., P. S. Cranston & L. C. V. Pinder (eds), 1995. The Chironomidae: Biology and ecology of non-biting midges. Chapman & Hall.

Bayley, S .E. & C. M. Prather, 2003. Do wetland lakes exhibit alternative stable states? Submersed aquatic vegetation and chlorophyll in western boreal lakes. Limnology & Oceanography 48: 2335–2345.

Bostrom, B., J. M. Andersen, S. Fleischer & M. Jansson 1988. Exchange of phosphorus across the sediment-water interface. Hydrobiologia 170: 229–244.

Chapman, P. M., B. Anderson, S. Carr, V. Engle, R. Green, J. Hameedi, M. Harmon, P. Haverland, J. Hyland, C. Ingersoll, E. Long, J. Rodgers Jr,

M. Salazar, P. K. Sibley & H. Windom, 1997. General guidelines for using the sediment quality triad. Marine Pollution Bulletin 34: 368–372.

Clements, W. H. & P. M. Kiffney, 1993. Assessing contaminant effects at higher levels of biological organization. Environmental Toxicology and Chemistry 13: 357–359.

Covich, A. P., M. A. Palmer & T. A. Crowl, 1999. The role of benthic invertebrate species in freshwater ecosystems. Bioscience 49: 119–127.

Dinsmore, W. P. & E. E. Prepas, 1997. Impact of hypolimnetic oxygenation on profundal macroinvertebrates in a eutrophic lake in central Alberta .1. Changes in macroinvertebrate abundance and diversity. Canadian Journal of Fisheries and Aquatic Sciences 54: 2157–2169.

Diehl, S. & R. Kornijów, 1997. Influence of submerged macrophytes on trophic interactions among fish and macroinvertebrates. In Jeppesen E., M. Sondergaard, M. Sondergaard & K. Christoffersen (eds), The structuring role of submerged macrophytes in lakes. Springer, 24–46.

Engel S., 1988. The role and interactions of submersed macrophytes in a shallow Wisconsin Lake USA. Journal of Freshwater Ecology 4: 329–342.

Fukuhara, H. & K. Yasuda, 1985. Phosphorus excretion by some zoobenthos in a eutrophic freshwater lake and its temperature dependency. The Japanese Journal of Limnology 46: 287–296.

Goedkoop, W. & R. K. Johnson, 1996. Pelagic-benthic coupling: profundal benthic community response to spring diatom deposition in mesotrophic Lake Erken. Limnology And Oceanography 41: 636–647.

Goedkoop, W., L. Sonesten, G. Ahlgren & M. Boberg, 2000. Fatty acids in profundal benthic invertebrates and their major food resources in Lake Erken, Sweden: Seasonal variation and trophic indications. Canadian Journal of Fisheries and Aquatic Sciences 57: 2267–2279.

Gong, Z., P. Xie & S. Wang, 2000. Macrozoobenthos in 2 shallow, mesotrophic Chinese lakes with contrasting sources of primary production. Journal of the North American Benthological Society 19: 709–724.

Gross, T. F., F. E. Werner & J. E. Eckman 1992. Numerical modeling of larval settlement in turbulent bottom boundary layers. Journal of Marine Research 50: 611–642.

Gullberg, K. R., W. Goedkoop & R. K. Johnson, 1997. The fate of diatom carbon within a freshwater benthic community – a microcosm study. Limnology & Oceanography 42: 452–460.

Hargeby, A., G. Andersson, I. Blindow & S. Johansson, 1994. Trophic web structure in a shallow eutrophic lake during a dominance shift from phytoplankton to submerged macrophytes. Hydrobiologia 280: 83–90.

Heinis, F. & C. Davids, 1993. Factors governing the spatial and temporal distribution of chironomid larvae in the Maarsseveen lakes with special emphasis on the role of oxygen conditions. Netherlands Journal of Aquatic Ecology 27: 21–34.

Hurlbert, S. M., 1971. The non-concept of species diversity: a critique and alternative parameters. Ecology 52: 577–586.

Jonasson, P. M. & C. Lindegaard, 1979. Zoobenthos and its contribution to the metabolism of shallow lakes. Archive Hydrobiology Beihefte Ergebnisse Limnology 13: 162–180.

Jones, S. E. & C. F. Jago, 1992. In situ assessment of modification of sediment properties by burrowing invertebrates. Marine Biology 115: 133–142.

Kelderman, P., 1984. Sediment-water exchange in Lake Grevelingen under different environmental conditions. Netherlands Journal of Sea Research 18: 286–311.

Kohler, S. L., 1992. Competition and the structure of a benthic stream community. Ecological Monographs 62: 165–188.

Kornijów, R. & B. Moss, 2002. On the survival of benthic macroinvertebrates during night hypoxia in a shallow lake. Verhandlungen Internationale Vereinigung für theoretische und angewandte Limnologie 28: 1–3.

Krantzberg, G., 1985. The influence of bioturbation on physical and biological parameters in aquatic environments; A review. Environmental Pollution 39: 99–122.

Lassen, C., N. P. Revsbech & O. Pedersen, 1997. Macrophyte development and resuspension regulate the photosyntesis and production of benthic microalgae. Hydrobiologia 350: 1–11.

Lindegaard, C., 1994. The role of zoobenthos in energy flow in two shallow lakes. Hydrobiologia 275/276: 313–322.

MacKay, R. J., 1992. Colonization by lotic macroinvertebrates: a review of processes and patterns. Canadian Journal of Fisheries and Aquatic Sciences 49: 617–628.

Madsen, K. N., P. Nilson & K. Sundback 1993. The influence of benthic microalgae on the stability of a subtidal sediment. Journal of Experimental Marine Biology and Ecology 170: 159–177.

Marsden, M. W., 1989. Lake restoration by reducing external phosphorus loading the influence of sediment phosphorus release. Freshwater Biology 21: 139–162.

Marsh, A. G. & K. R. Tenore, 1990. The role of nutrition in regulating the population dynamics of opportunistic, surface deposit feeders in a mesohaline community. Limnology & Oceanography 35: 710–724.

Meadows, P. S. & J. Tait, 1989. Modification of sediment permeability and shear strength by two burrowing invertebrates. Marine Biology (Berlin) 101: 75–82.

Phipps, G. L., V. R. Mattson & G. T. Ankley, 1995. Relative sensivity of three freshwater benthic macroinvertebrates to ten contaminants. Archives of Environmental Contamination and Toxicology 28: 281–286.

Raspopov, I. M., T. D. Slepukhina, F. F. Vorontzov & O.N. Dotzenko, 1988. Wave effects on the bottom biocenosis in the Onega Lake bays (USSR). Archiv fur Hydrobiologie 112: 115–124.

Reynoldson, T. B., R. C. Bailey, K. E. Day & R. H. Norris, 1995. Biological guidelines for freshwater sediment based on BEnthic Assessment of SedimenT (BEAST) using a multivariate approach for predicting biological state. Australian Journal of Ecology 20: 198–219.

Risnoveanu, G., C. Postolache & A. Vadineanu, 2004. Ecological significance of nitrogen cycling by tubificid communities in shallow eutrophic lakes of the Danube Delta. Hydrobiologia 524: 193–202.

Scheffer, M., 1990. Multiplicity of stable states in freshwater systems. Hydrobiologia 200/201: 475–486.

Scheffer, M., S. H. Hosper, M. L. Meijer & B. Moss 1993. Alternative equilibria in shallow lakes. Trends in Ecology and Evolution: 275–279.

Scheffer, M., 1998. Ecology of shallow lakes. Chapman & Hall.

Shin, P. K. S., 1989. Natural disturbance of benthic infauna in the offshore waters of Hong Kong. Asian Marine Biology 6: 193–207.

Svensson, J. M. & L. Leonardson, 1996. Effects of bioturbation by tube-dwelling chironomids larvae on oxygen uptake and denitrification in eutrophic lake sediments. Freshwater Biology 35: 289–300.

Svensson, J. M., 1997. Influence of *Chironomus plumosus* larvae on ammonium flux and denitrification (measured by the acetylene blockage - and the isotope pairing-technique) in eutrophic lake sediment. Hydrobiologia 346: 157–168.

Svensson, J. M., 1998. Emission of N_2O, nitrification and denitrification in a eutrophic lake sediment bioturbated by *Chironomus plumosus*. Aquatic Microbial Ecology 14: 289–299.

Sweerts, J-P R. A., 1990. Oxygen consumption processes, mineralization and nitrogen cycling at the sediment-water interface of north temperate lakes. Ph.D. Thesis.

Underwood, G. J. C. & D. M. Paterson, 1993. Recovery of intertidal benthic diatoms after biocide treatment and associated sediment dynamics. Journal of the Marine Biological Association of the UK 73: 25–45.

van de Bund, W. J., W. Goedkoop & R. K. Johnson, 1994. Effects of deposit-feeder activity on bacterial production and abundance in profundal lake sediment. Journal of the North American Benthological Society 13: 532–539.

Van Den Berg, M. S., H. Coops, R. Noordhuis, J. Van Schie & J. Simons 1997. Macroinvertebrate communities in relation to submerged vegetation in two Chara-dominated lakes. Hydrobiologia 342/343: 143–150.

Vos, J. H., M. A. G. Ooijevaar, J. F. Postma & W. Admiraal, 2000. Interaction between food availability and food quality during growth of early instar chironomid larvae. Journal of the North American Benthological Society 19: 158–168.

Vos, J. H., E. T. H. M. Peeters, R. Gylstra, M. H. S. Kraak & W. Admiraal, 2004. Nutritional value of sediments for macroinvertebrate communities in shallow eutrophic waters. Archiv fuer Hydrobiologie 161: 469–487.

Weatherhead, M. A. & M. R. James, 2001. Distribution of macroinvertebrates in relation to physical and biological variables in the littoral zone of nine New Zealand lakes. Hydrobiologia 462: 115–129.

Wiśniewski, R., 1980a. Pseudolittoral of Gopło lake. Part I. Characteristics of the environment. Limnological Papers 12: 61–81.

Wiśniewski, R., 1980b. Pseudolittoral of Gopło lake. Part II. Biological characteristics. Limnological Papers 12: 83–116.

Wiśniewski, R., 1995. Role of sediment resuspension in aquatic ecosystems functioning (in Polish). Wydawnictwo Uniwersytetu Mikołaja Kopernika, Toruń.

Wolnomiejski, N. & E. Papis, 1974. Pseudolittoral of the Jeziorak Lake – a separate bottom zone of a strongly eutrophied water body. Polskie Archiwum Hydrobiologii 21: 343–353.

Wood, L. W., 1975. Role of oligochaetes in the circulation of water and solutes across the mud-water interface. Verhandlungen Internationale Vereinigung für theoretische und angewandte Limnologie 19: 1530–1533.

Hydrobiologia (2007) 584:157–166
DOI 10.1007/s10750-007-0592-y

SHALLOW LAKES

The role of cladocerans reflecting the trophic status of two large and shallow Estonian lakes

Juta Haberman · Reet Laugaste · Tiina Nõges

Abstract The role of pelagic cladoceran communities is discussed on the basis of a comparative study conducted in two Estonian lakes, the moderately eutrophic Lake Peipsi (N_{tot} 700, P_{tot} 40 μg l^{-1} as average of ice-free period of 1997–2003) and in a strongly eutrophic Lake Võrtsjärv (N_{tot} 1600, P_{tot} 54 μg l^{-1}). The cladoceran community was found to reflect the differences in the trophic state of these lakes. In L. Peipsi, characteristic species of oligo-mesotrophic and eutrophic waters co-dominated (making up 20% or more of total zooplankton abundance or biomass), whereas in L. Võrtsjärv only species of eutrophic waters occurred. In L. Peipsi, the dominant cladocerans were *Bosmina berolinensis* and *Daphnia galeata*, while *Chydorus sphaericus* was the most abundant cladoceran in L. Võrtsjärv. The cladocerans of L. Peipsi (mean individual wet weight 25 μg) were significantly (threefold) larger than those of L. Võrtsjärv (8 μg). The mean wet biomass of cladocerans was higher and total cladoceran abundance was lower in L. Peipsi compared to L. Võrtsjärv (biomass varied from 0.133 to 1.570 g m^{-3}; mean value 0.800 g m^{-3} in L. Peipsi and from 0.201 to 0.706 g m^{-3}, mean 0.400 g m^{-3} in L. Võrtsjärv; the corresponding data for abundances were: 8,000–43,000 ind m^{-3}, mean 30,000 ind m^{-3} for L. Peipsi, 50,000–100,000, mean 52,000 ind m^{-3} for L. Võrtsjärv). Based upon differences in body size, cladocerans were more effective transporters of energy in L. Peipsi than in L. Võrtsjärv. Cladocerans proved to be informative indicators of the trophic status and of the efficiency of the food web in studied lakes.

Keywords Cladoceran community · Cladoceran mean weight · Lake trophic status

Guest Editors: R. D. Gulati, E. Lammens, N. De Pauw & E. Van Donk
Shallow lakes in a changing world

J. Haberman (✉) · R. Laugaste · T. Nõges
Centre for Limnology, Estonian University of Life Sciences, 61101 Rannu, Tartumaa, Estonia
e-mail: juta@zbi.ee

R. Laugaste
e-mail: reet@zbi.ee

T. Nõges
e-mail: tnoges@zbi.ee

Introduction

Zooplankton holds surrogate information about the situation of the main components of the ecosystem, from phytoplankton as the main food source for zooplankton up to fishes as the one of main consumers of zooplankton. Among zooplankton, cladocerans are considered one of the 'key players' in lake food webs (Kamjunke et al., 1999). It is quite common that in the course of eutrophication, large cladocerans are replaced by smaller ones (Jeppesen et al., 2000; Smakulska & Górniak, 2004). When the trophy of a water body rises, phytoplankton will be dominated by large filamentous algae, predominately cyanobacteria, which are not edible for cladocerans. As a result, large cladocerans, which feed mostly on small algae, start to disappear from the water body due to the low quality of food available to them, and the cladoceran group will be dominated by smaller species feeding mainly on bacteria and detritus. This process has already taken place in the strongly eutrophic L. Võrtsjärv but has not yet been begun in the moderately eutrophic L. Peipsi (Haberman, 1998, 2001).

Cladocerans are affected by a number of different factors, most universal and influential of them being water temperature (Hart, 2004; Vandekerkhove et al., 2005), quantity and quality of food (Gulati & DeMott, 1997; Alekseev & Lampert, 2004; von Elert, 2004), and predation by fishes (Gliwicz et al., 2004; Matthes, 2004). Herzig (1994) has noted that abiotic environmental factors (water temperature, wind, waves) influence zooplankton especially in spring and autumn, while the effect of biotic factors (food, predators) is usually more pronounced in summer. The efficiency of the zooplankter's feeding depends largely on water temperature and is highest in summer (July–August). The correlation between the food ration of herbivorous zooplankton and water temperature was fairly good ($r = 0.5$, $P < 0.0001$) in L. Peipsi (Haberman, 2001).

According to Knisely & Geller (1986), cladocerans prefer to eat small chrysophytes and cryptophytes, but in the absence of favourable food objects, daphniids are able to feed also on small cyanobacteria (Gulati et al., 2000), and even to split their large colonies (Kryuchkova, 1989). Both the quantity and quality of food affect the survival and reproduction of zooplankton (Gulati & DeMott, 1997; Dodson & Frey, 2001). Quantitative food limitation of zooplankton can be expected to be essential in water bodies of low trophy with low primary production. Under strongly eutrophic conditions, where phytoplankton is dominated by large forms of cyanobacteria that are not edible for zooplankton (Gliwicz & Lampert, 1990; DeMott et al., 2001), poor food quality, e.g. deficit of nutrients, particularly nitrogen and phosphorus (Ahlgren et al., 2000; Spaak & Keller, 2004; Lürling & van Donk, 1997), but also polyunsaturated fatty acids, in algae leads to reduced population growth in several cladoceran species (Boersma & Stelzer, 2000; von Elert, 2004). Phosphorus limitation influences the cell walls of green algae, making them thicker, which probably reduces their digestibility (Ahlgren et al., 2000). Dodson & Frey (2001) have found that most cladocerans prefer algae in the size range from 5 to 35 μm. The larger is the grazer, the more efficient is its grazing (Sartonov, 1995; Meijer et al., 1999).

The aim of the present study was to evaluate the role of cladocerans in reflecting the status of the ecosystem of two large and shallow Estonian lakes.

Study site

Lake Peipsi *s.l.* (57°52′–59°01′ N, 26°57′–28°21′ E) is located in Eastern Estonia, on the border of Estonia and Russia, and consists of three parts: the largest and deepest northern part L. Peipsi *s.s.* (2,661 km^2, mean depth 8.3 m), the middle strait-like part L. Lämmijärv, and the southern part L. Pihkva. The present paper treats only L. Peipsi *s.s.* (hereafter L. Peipsi). L. Võrtsjärv (58°05′–58°25′ N, 25°54′–26°09′ E) is situated in Central Estonia (Fig. 1). L. Peipsi *s.s.* is a moderately eutrophic lake, while L. Võrtsjärv is a strongly eutrophic lake (Table 1). A description of the phyto- and zooplankton communities of the lakes is given in Haberman (1998, 2001), Nõges & Laugaste (1998), and Laugaste et al. (2001).

Materials and methods

Zoo- and phytoplankton samples were collected monthly from the pelagial of L. Peipsi and L. Võrtsjärv during the vegetation period (May–October) of 1997–2003 in parallel with hydrochemical data (N_{tot}, P_{tot}) and water temperature. The methods of collecting and treating samples,

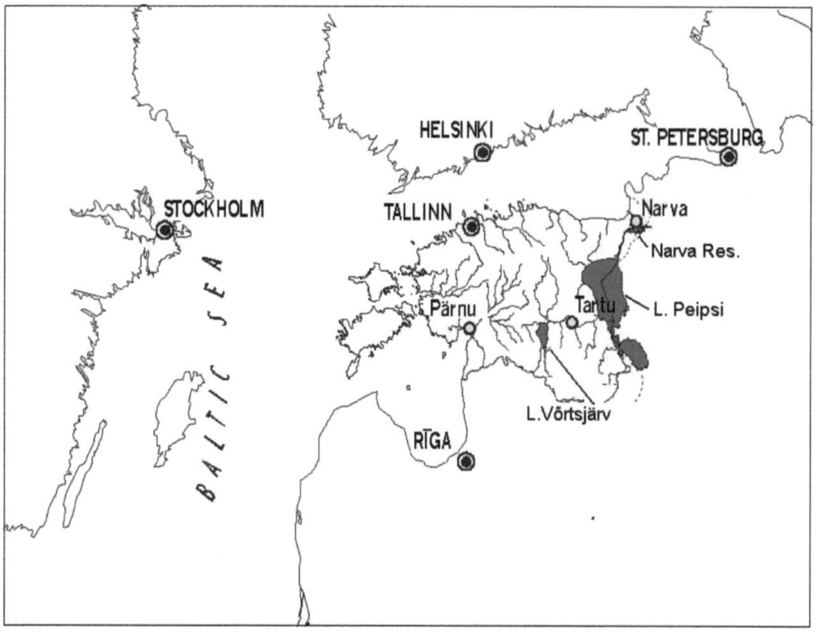

Fig. 1 Map showing the location of the lakes in Estonia and the location of Estonia in Europe

Table 1 Main morphometric features and trophic status of L. Peipsi *s.s* and L. Võrtsjärv

	L. Peipsi *s.s.*	L. Võrtsjärv
Area, km^2	2,611 (3,555 for L. Peipsi *s.l.*)	270
Catchment area, km^2	47,800	3,374
Mean depth, m	8.3	2.8
Maximum depth, m	12.9	6.0
Transparency, m	2.2	1.1
TP, $\mu g\ P\ l^{-1}$	35	54
TN, $\mu g\ N\ l^{-1}$	678	1,600
Chl a, $\mu g\ l^{-1}$	15	22

and biomass estimation are described in detail in Laugaste et al. (2001) and Haberman (2001). For the estimation of the quantitative parameters of cladocerans in both study lakes, the average values of a seven-year period (1997–2003) were used. Cladoceran taxa, which made up 20% or more of total zooplankton abundance and biomass, were considered the dominant zooplankters. Algae with a diameter less than 40 μm (mainly chlorophytes, cryptophytes and chrysophytes) were regarded as edible for zooplankton. The program STATISTICA for Windows was used for the calculation of descriptive statistics and for nonparametric (Spearman) correlation and factor analysis.

Results

A total of 57 cladocerans species have been identified from the moderately eutrophic Lake Peipsi. The characteristic species of both oligo-mesotrophic and eutrophic waters coexisted in the lake. Among the 47 cladoceran species found in the strongly eutrophic L. Võrtsjärv only the species indicating eutrophic conditions were present. Oligo-mesotrophic species (*Bosmina berolinensis*, *B. obtusirostris*, *Bythotrephes longimanus*, *Diaphanosoma brachyurum*) have disappeared from the L. Võrtsjärv plankton (Haberman, 1998). The dominant cladoceran species are presented in Table 2. Cladocerans were present throughout the vegetation period (May–October). In winter, only single individuals of *Chydorus sphaericus* and *Bosmina longirostris* were found (Haberman, 1998, 2001). Cladoceran abundance began to rise in May when water temperatures reach about 10°C. Cladoceran abundance maximized in summer (L. Peipsi—July, L. Võrtsjärv—June) and declined gradually thereafter

Table 2 Dominating cladoceran species in biomass of whole zooplankton in L. Peipsi and L. Võrtsjärv

Month	L. Peipsi	L. Võrtsjärv
May	*Bosmina berolinensis* Imhof	*Bosmina longirostris* (Müller)
June	*B. berolinensis*, Daphnia galeata Sars, *Daphnia cristata* Sars, *Daphnia cucullata* Sars	*Chydorus sphaericus* Müller
July	*D. galeata*, *D. cucullata*, *D. cristata*, *Diaphanosoma*, *brachyurum* (Liéven), *Limnosida frontosa* Sars, *Leptodora kindti* (Focke), *Bythotrephes longimanus* Leydig	*C. sphaericus*, *D. cucullata*
August	*D. cucullata*	*C. sphaericus*, *D. cucullata*
September	*B. berolinensis*, *Bosmina c. coregoni* Baird, *Bosmina gibbera* Schoedler, *D. galeata*	*C. sphaericus*
October	*B. berolinensis*, *B. c. coregoni*, *B. gibbera*, *D. galeata*	*C. sphaericus*, *B. longirostris*

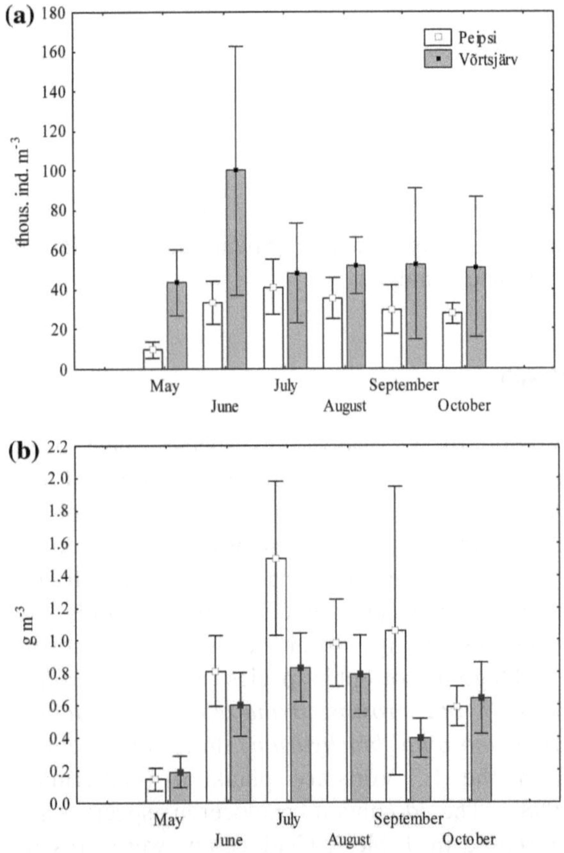

Fig. 2 Variability of cladoceran abundance and biomass during vegetation period (May–October of 1997–2003). (**a**) abundance, (**b**) biomass. On Figs. 2–4, means (small squares) with error plots, and bars of 95% confidence interval are presented

with decreasing water temperature until the total disappearance of the species (in October–November at water temperature about 5–6°C). The abundance of cladocerans in L. Võrtsjärv was up to double, whereas their biomass less than half, that found in L. Peipsi (Fig. 2). Cladocerans made up less than 20% of total zooplankton abundance in the studied lakes but, being relatively large-bodied, their share of the total zooplankton biomass was quite high (Table 3). The highest cladoceran biomass for the studied period in L. Peipsi was 4.64 g m^{-3} (in July 1999 at water temperature 22°C), and in L. Võrtsjärv 2.83 g m^{-3} (June 1999, 22°C). Commonly, July is characterized by the highest temperature in both lakes. The biomass of all cladocerans was positively correlated with water temperature (L. Peipsi: $r = 0.74$, $P < 0.0001$; L. Võrtsjärv: $r = 0.32$, $P = 0.04$). In both lakes, higher water temperature stimulated the growth of *Daphnia* species (in L. Peipsi: $r = 0.77$, $P < 0.0001$; L. Võrtsjärv: $r = 0.65$, $P < 0.0001$), while in L. Võrtsjärv lower temperature stimulated the growth of *Bosmina* species ($r = -0.6$, $P < 0.0001$).

Only one cladoceran, *Daphnia* (*D. cucullata*), and two *Bosmina* (*B. c. coregoni*, *B. longirostris*) species were found in L. Võrtsjärv, while the respective figures for L. Peipsi were three (*D. galeata*, *D. cucullata*, *D. cristata*) and ten (the most frequent *B. berolinensis*, *B. c. coregoni*, *B. gibbera*). In L. Peipsi, the abundance of the *Daphnia* was about twice and biomass up to four times higher than in L. Võrtsjärv. Also, the share of this genus in the cladoceran group as well as in whole zooplankton, calculated as a percentage, was higher for L. Peipsi. Similarly, the proportion of *Bosmina* in the zooplankton of L. Peipsi was

Table 3 Characteristic data (means of the vegetation period for 1997–2003 ± standard deviation SD) of cladocerans in L. Peipsi and L. Võrtsjärv (A—abundance, thousand ind. m^{-3}; B—biomass, g m^{-3}; Clad—Cladocera, ZPL—zooplankton)

	Võrtsjärv		Peipsi	
	Average	SD	Average	SD
$A_{Cladocera}$	51.8	52.6	30.7	33.6
$B_{Cladocera}$	0.3	0.3	0. 8	1.3
$A_{Daphnia}$	4.8	7.6	9.8	16.5
$B_{Daphnia}$	0.1	0.3	0.4	0.7
$A_{Bosmina}$	11.8	14.7	15.1	17.7
$B_{Bosmina}$	0.1	0.6	0.2	0.3
$A_{Chydorus}$	35.0	37.7	9.1	15.8
$B_{Chydorus}$	0.15	0.21	0.06	0.18
Clad % in ZPL A	19.8	16.4	13.4	16.1
Clad % in ZPL B	45.6	29.0	33.7	22.2
Daphnia % in ZPL A	1.2	1.5	2.26	5.5
Daphnia % in ZPL B	11.2	19.2	25.6	41.1
Daphnia % in Clad A	10.6	20.3	30.6	41.5
Daphnia % in Clad B	20.0	33.9	41.2	56.7
Bosmina % in ZPL A	1.4	1.1	2.5	5.3
Bosmina % in ZPL B	9.4	14.5	14.6	33.3
Bosmina % in Clad A	12.3	15.5	33.4	49.2
Bosmina % in Clad B	24.5	40.8	39.5	57.2
Chydorus % in ZPL A	8.8	6.8	2.2	6.2
Chydorus % in ZPL B	21.8	24.5	3.8	8.6
Chydorus % in Clad A	70.2	84.4	25.2	46.9
Chydorus % in Clad B	50.5	68.9	7.9	13.8

greater than in L. Võrtsjärv, although the between-lake difference was not as large as for *Daphnia.* In L. Võrtsjärv, both the abundance and biomass of *Ch. sphaericus* were more than three times higher than in L. Peipsi (Table 3). Also, the share of *Ch. sphaericus* in total zooplankton as well as in the cladoceran group was larger for L. Võrtsjärv compared to L. Peipsi. Cladoceran body size was significantly larger in L. Peipsi (mean wet weight 25 μg) than in L. Võrtsjärv (mean 8 μg). In both lakes, cladoceran size was the smallest in May, when there were few large thermophilic cladocerans, and the largest in July (Fig. 3). The peak in mean cladoceran weight for L. Peipsi was caused by the co-occurrence of several large-bodied thermophilic cladocerans (among them *B. longimanus*), while in L. Võrtsjärv it was due to the presence of the small sized *Daphnia cucullata*, the single representative of *Daphnia.* In L. Peipsi and in L. Võrtsjärv, mean *Daphnia* individual weight was 47 μg and 30 μg, and mean *Bosmina* weight was 21 and 8 μg, respectively. The dominant *Daphnia* species in L. Peipsi, *D. galeata* (30–80 μg), *D. cucullata* (40–50 μg) and *D. cristata* (30–40 μg), were larger than *D. cucullata* (25–35 μg) in L. Võrtsjärv. The mean individual weight of *Bosmina* in L. Peipsi was primarily established by the large

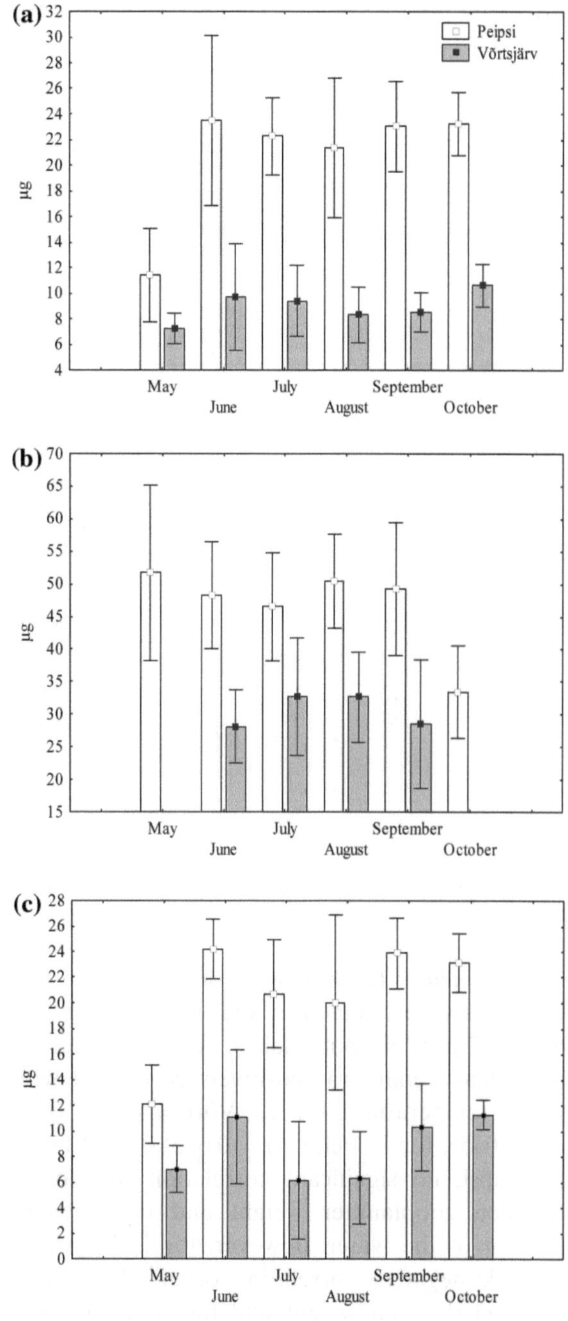

Fig. 3 (**a**) Mean weight of cladocerans, (**b**) Mean weight of *Daphnia*, (**c**) Mean weight of *Bosmina*

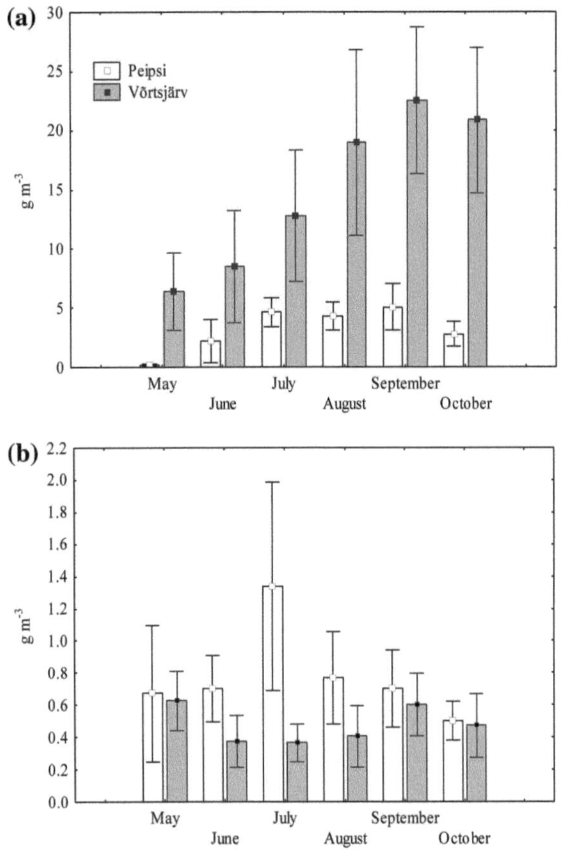

Fig. 4 (**a**) Biomass of cyanobacteria, (**b**) Biomass of small algae

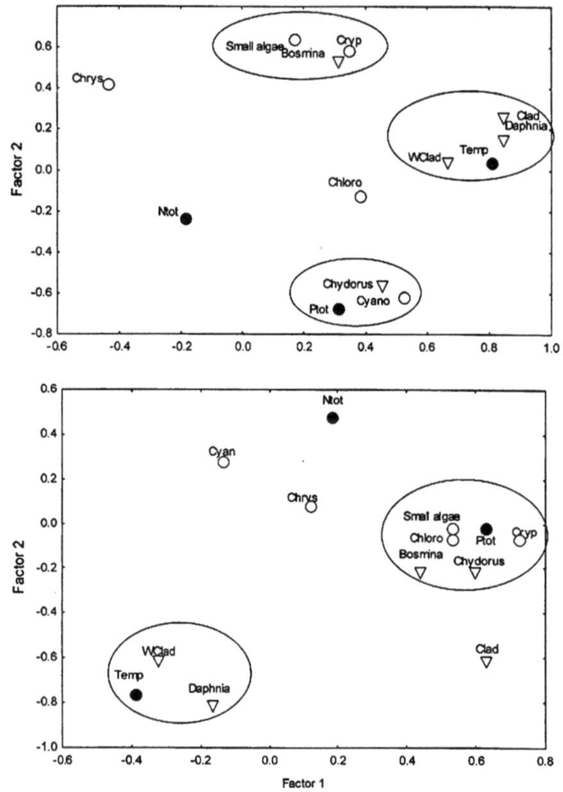

Fig. 5 Graph of the results of factor analysis reflecting the grouping of cladocerans with environmental parameters and phytoplankton indices: (**a**) L. Peipsi, (**b**) L. Võrtsjärv. Abbreviations: Temp—water temperature; Cyan—cyanobacteria; Chloro—chlorophytes, Cryp—cryptophytes; Chrys—chrysophytes; Wclad—mean weight of cladoceran. Clad (cladocerans), Daphnia, Bosmina and Chydorus with their biomass are presented

B. berolinensis (54 μg), while in L. Võrtsjärv it was largely determined by the small *B. longirostris* (5–7 μg). In both lakes, mean cladoceran individual weight was positively correlated with water temperature: for L. Võrtsjärv ($r = 0.32$, $P = 0.045$), for L. Peipsi ($r = 0.6$, $P < 0.0001$). In L. Peipsi, no significant correlation was found between zooplankter weight and the nutrient content of the water; however, for L. Võrtsjärv a weak negative correlation occurred between mean cladoceran weight and the total nitrogen content of the water ($r = -0.40$, $P = 0.01$). L. Võrtsjärv was characterized by far higher cyanobacterial biomass and lower biomass of small algae compared with L. Peipsi (Fig. 4). Correlation analysis revealed positive correlations between cladoceran biomass and small algae ($r = 0.045$) and between the abundance of cladocerans and the biomass of chlorophytes and cryptophytes ($r = 0.4$, $P < 0.01$). Factor analysis showed the grouping of *Daphnia* with water temperature and mean weight of cladocerans in both lakes. *Chydorus sphaericus* was related with total P in both lakes, with cyanobacteria in the L. Peipsi (evidently via temperature) and with food objects in L. Võrtsjärv (Fig. 5).

Discussion

Along with the characteristic species of oligo-mesotrophic waters (*Bosmina berolinensis*, *Daphnia galeata*, *Bythotrephes longimanus*, *Limnosida frontosa*), the dominants of L. Peipsi included also species preferring eutrophic waters such as *Daphnia cucullata* and *Bosmina c. coregoni* (Hakkari, 1972; Gulati, 1983; Pejler, 1983; Haberman, 1998;

Ostojić, 2000; Stich et al., 2005) (Table 2). Such coexistence is possible owing to the large size of the lake (with a different trophic state in different parts) and its transition stage from moderately eutrophic (mesotrophic until the early 1960s) to eutrophic. In the moderately eutrophic L. Peipsi, the most frequently dominant species were *B. berolinensis* and *D. galeata*, while in the strongly eutrophic L. Võrtsjärv, the indicator of eutrophy *Ch. sphaericus* dominated. The abundance of cladocerans in L. Võrtsjärv was up to double, while their biomass was less than half that in L. Peipsi. The causes of this discrepancy were the differences in the dominant species as well as the consequent differences in weights (Table 2, Fig. 3). The higher the trophic level of a water body, the smaller is the weight of cladoceran individuals (Jeppesen et al., 2000). In both lakes zooplankton abundance was dominated by rotifers; cladocerans accounted for less than 20% of total zooplankton abundance; however, as they are relatively large-bodied, their share in total zooplankton biomass was large. The smaller proportion of cladocerans in L. Peipsi compared with L. Võrtsjärv can be explained by the large amount of big copepods (which were lacking in L. Võrtsjärv) in total zooplankton biomass. The abundance of the indicator species of eutrophy, *Ch. sphaericus*, was also substantially higher in L. Võrtsjärv than in L. Peipsi (Table 3). Cladoceran individual weight is largely determined by the trophic state of a water body as well as by fish predation. Weight decreases with increasing trophy as small zooplankters begin to dominate (Jeppesen et al., 2000; Smakulska & Górniak, 2004). Larger cladocerans are more effective grazers of phytoplankton. Consequently, more energy, fixed in plankton photosynthesis, reaches the higher trophic levels. The higher individual weight of cladocerans in the less eutrophic L. Peipsi (Fig. 3) may indicate more effective food chains in this lake than in the strongly eutrophic L. Võrtsjärv.

The co-clustering of the biomass of one of the smallest cladocerans, *Chydorus sphaericus*, with the concentration of total phosphorus in the studied lakes (Fig. 5) agrees well with the results of a study of Danish lakes with varying trophy, which showed that a rise in the trophic level (increase in TP content from <0.05 to 0.4 mg P l^{-1}) was accompanied by a decrease in mean zooplankter weight, from 5.1 to 1.5 μg. This was caused by a decline in the abundance of large-sized zooplankters (*Daphnia* spp., calanoids). The share of *Daphnia* spp. in the biomass of the cladoceran group decreased from 63–70 to 30% (Jeppesen et al., 2000). Smakulska & Górniak (2004) reported that during eight years of progressive eutrophication in the Siemianówka Reservoir (Poland) the biomass and body size of *D. cucullata* clearly decreased. Cladocerans are significantly larger in L. Peipsi (25 μg) than in L. Võrtsjärv (8 μg). As a result of eutrophy, some larger cladocerans (*B. longimanus*, *B. berolinensis*, *B. obtusirostris*) have disappeared from L. Võrtsjärv. The small-bodied *Ch. sphaericus* and *B. longirostris*, which most frequently dominate the L. Võrtsjärv cladoceran population, are indicator species of eutrophy (Pejler, 1983). *D. cucullata*, which is also considered a species of eutrophic waters (Smakulska & Górniak, 2004), seems still to indicate lower trophic status compared with the two former species. *D. cucullata* has been reported as the dominant species from several mesotrophic Polish lakes (Karabin, 1985) and is a dominating cladoceran also in the moderately eutrophic L. Peipsi. In L. Peipsi, *D. cucullata* individuals (40–50 μg) are about twice larger than in L. Võrtsjärv (25–30 μg). *Daphnia* species are more vulnerable to mechanical interference by cyanobacteria than smaller cladocerans (Gliwicz & Lampert, 1990). Smakulska & Górniak (2004) have suggested that such vulnerability is largely dependent on their size and the carapace gape which decreases simultaneously with body size. The lower impact of large cyanobacteria on the feeding and growth of *D. cucullata* could be explained by its smaller body size. Also, it has been hypothesized that larger cladoceran species are more sensitive to the toxins of cyanobacteria compared with smaller species (Kirk & Gilbert, 1992). Experimental studies revealed that in L. Võrtsjärv, most herbivorous cladocerans consume bacteria; the most effective bacterivores are *Ch. sphaericus*, *B. longirostris* and *B. c. coregoni* (Nõges et al., 2004). Also *D. cucullata* and *Daphnia* spp. are not very fastidious about food and are able to consume food objects of various size (including bacteria), which serves as an advantage for them as filtrators (Jeppesen et al., 1999; Kamjunke et al., 1999). The cladoceran population in L. Peipsi was mostly dominated by larger cladocerans such as *B. berolinensis* (average weight 54 μg) and *Daphnia galeata* (up to 80 μg) (Table 2). *D. galeata* tends to be a species of oligo-mesotrophic lakes as it does not thrive in eutrophic waters (Korovchinsky, 2000). Its abundance in L. Peipsi gives evidence of the relatively good status of this water body (Table 1). The cladoceran data from L. Peipsi and in L. Võrtsjärv (Table 2, 3; Fig 3) suggest that cladocerans consume more algae in L. Peipsi than in L. Võrtsjärv. The food ration of cladocerans is determined directly by mean cladoceran weight; feeding efficiency is proportional to body size (Carpenter & Kitchell, 1992; Sartonov, 1995). Correlation analysis has revealed a positive relationship between mean cladoceran weight and grazing activity for both studied lakes (L. Peipsi: $r = 0.6$, L. Võrtsjärv: $r = 0.4$, $P < 0.0001$; Haberman & Nõges, 2003). The food ration of cladocerans is substantially higher in L. Peipsi than in L. Võrtsjärv, also it constitutes a much higher proportion in the total food ration of filter-feeding zooplankton as well as of primary production in L. Peipsi than in L. Võrtsjärv (Haberman & Nõges, 2003). About 10% of the algae in L. Võrtsjärv and 20% in L. Peipsi fall into the size fraction suitable as food for cladocerans. Elser & Goldman (1990), Carney & Elser (1990), and Jeppesen et al. (1999) also claim that the grazing pressure on phytoplankton is strong in mesotrophic lakes dominated by large-sized *Daphnia*, and weak in eutrophic waters dominated by small-sized zooplankters. Large-bodied cladocerans, dominating in L. Peipsi, feed mainly on living algae, and an effective grazing food chain (algae → zooplankton → fish) prevails in the ecosystem: on average 6% of the energy of algae reaches zooplankton (Nõges et al., 2001). The small-bodied cladocerans of L. Võrtsjärv are not able to consume large filamentous cyanobacteria dominating in the lake; they feed mainly on bacteria and detritus and dominating is an ineffective detrital food chain: only 2% of the energy of algae reaches zooplankton (Nõges et al., 2004). Cladocerans are effective energy transformers in L. Peipsi, but ineffective in this regard in L. Võrtsjärv. The fish production of L. Peipsi accounts for up to 0.4% of primary production, while the corresponding value for L. Võrtsjärv is only 0.2% (Nõges et al., 2001, 2004). This difference is also revealed in fish catches which are 25–34 kg ha^{-1} in L. Peipsi and 15–20 kg ha^{-1} in L. Võrtsjärv (Pihu 1998; Pihu & Kangur, 2001).

Besides recognizing the key role of the trophic status of the lake in shaping of the cladoceran community's composition and size structure, one also has to consider the important role of water temperature (Fig. 5). In both L. Peipsi and L. Võrtsjärv cladoceran body size was smallest in May when the share of large thermophilic cladocerans was negligible or lacking, and largest in July when several large-bodied thermophilic cladocerans co-occurred. Also one has to take into account the effect of water temperature when comparing cladoceran communities of different climatic regions.

In conclusion, cladocerans are informative indicators of the trophic status and of the efficiency of the food web in Estonian large shallow lakes. The mean cladoceran weight, as well as the relative importance of cladocerans in zooplankton biomass can be used as a marker characteristic in the qualification of the ecosystem of the water body.

Acknowledgements This study was supported by the Estonian target financed project SF 0362483s03. Use was made of the data of the Estonian State monitoring programme. We are indebted to Mrs. Ester Jaigma for linguistic revision of the manuscript. The contributions from the anonymous referees are highly appreciated.

References

Ahlgren, G., P. Hyenstrand, T. Vrede, E. Karlsson & S. Zetterberg, 2000. Nutritional quality of *Scenedesmus quadricauda* (Chlorophyceae) grown in different nitrogen regimes and tested on *Daphnia*. Verhandlungen Internat. Verein. Limnology 27: 1234–1238.

Alekseev, V. & W. Lampert, 2004. Maternal effects of photoperiod and food level on life history characteristics of the cladoceran *Daphnia pulicaria* Forbes. Hydrobiologia 526: 225–230.

Boersma, M. & C.-P. Stelzer, 2000. Response of a zooplankton community to the addition of

unsaturated fatty acids: an enclosure study. Freshwater Biology 45: 179–188.
Carney, J. J. & J. J. Elser, 1990. The strength of zooplankton-phytoplankton coupling in relation to trophic state. In Tilzer, M. M. (ed.), Ecology of Large Lakes. New York, 616–631.
Carpenter, S. R. & J. F. Kitchell, 1992. Trophic cascade and biomanipulation: interface of research and management - a reply to the comment by DeMelo et al. Limnology and Oceanography 37: 208–213.
DeMott, W. R., R. D. Gulati & E. van Donk, 2001. *Daphnia* food limitation in three hupereutrophic Dutch lakes: evidence for exclusion of large-bodied species by interfering filaments of cyanobacteria. Limnology and Oceanography 46: 2054–2060.
Dodson, S. I. & D. G. Frey, 2001. Cladocera and other Branchiopoda. In Thorp, J. H. & A. P. Covich (eds), Ecology and classification of North American Freshwater Invertebrates. Academic Press, London, 850–914.
Elser, J. J. & C. R. Goldman, 1990. Zooplankton effects on phytoplankton in lakes of contrasting trophic status. Limnology and Oceanography 36: 64–90.
Gliwicz, Z. M., A. Jawiński & R. M. Pawlowicz, 2004. Cladoceran densities, day-to-day variability in food selection by smelt, and the birth-rate-compensation hypothesis. Hydrobiologia 526: 171–186.
Gliwicz, Z. M. & W. Lampert, 1990. Food thresholds in *Daphnia* species in the absence and presence of blue-green filaments. Ecology 21: 691–702.
Gulati, R. D., 1983. Zooplankton and its grazing as indicators of trophic status in Dutch lakes. Environmental Monitoring and Assessment 3: 343–354.
Gulati, R. D. & W. R. DeMott, 1997. The role of food quality for zooplankton: remarks on the state-of-the-art, perspectives and priorities. Freshwater Biology 38: 753–768.
Gulati, R. D., M. Bronkhorst & E. van Donk, 2000. Feeding in *Daphnia galeata* on *Oscillatoria limnetica* and on detritus derived from it. J. Plankton Research 23: 705–718.
Haberman, J., 1998. Zooplankton of Lake Võrtsjärv. Limnologica 28: 49–65.
Haberman J., 2001. Zooplankton. In Pihu, E. & J. Haberman (eds), Lake Peipsi. Flora and Fauna, Sulemees Publishers, Tartu, 50–68.
Haberman, J. & T. Nõges, 2003. Cladocerans reflecting trophic state of the large Estonian lakes Peipus and Võrtsjärv. In Simola, H., A. Y. Terzhevik, M. Viljanen & I. J. Holopainen (eds), Proceedings of the Fourth International Lake Ladoga Symposium 2002. University of Joensuu, Publications of Karelian Institute, Joensuu, 143–147.
Hakkari, L., 1972. Zooplankton species as indicators of environment. Aqua Fennica 1972: 46–54.
Hart, R. C., 2004. Cladoceran periodicity patterns in relation to selected environmental factors in two cascading warm-water reservoirs over a decade. Hydrobiologia 526: 99–117.
Herzig, A., 1994. Predator-prey relationships within the pelagic community of Neusiedler See. Hydrobiologia 275(276): 81–96.
Jeppesen, E., J. P. Jensen, M. Søndergaard & T. Lauridsen, 1999. Trophic dynamics in turbid and clearwater lakes with special emphasis on the role of zooplankton for water clarity. Hydrobiologia 408(409): 217–231.
Jeppesen, E., J. P. Jensen, M. Søndergaard, T. Lauridsen & F. Landkildehus, 2000. Trophic structure, species richness and biodiversity in Danish lakes: changes along phosphorus gradient. Freshwater Biology 45: 201–218.
Kamjunke, N., A. Benndorf, C. Wilbert, M. Opitz, J. Kranich, M. Bollenbach & J. Benndort, 1999. Bacteria ingestion by *Daphnia galeata* in a biomanipulated reservoir: a mechanism stabilizing biomanipulation. Hydrobiologia 403: 109–121.
Karabin, A., 1985. Pelagic zooplankton (Rotatoria, Crustacea) variation in the process of lake eutrophication. I. Structural and quantitative features. Ecologia Polska 33: 567–616.
Kirk, K. L. & J. J. Gilbert, 1992. Variation in herbivore response to chemical defenses: zooplankton foraging on toxic cyanobacteria. Ecology 73: 2208–2217.
Knisely, K. & W. Geller, 1986. Selective feeding of four zooplankton species on natural lake phytoplankton. Oecologia 69: 86–94.
Korovchinsky, N. M., 2000. Species richness of pelagic Cladocera of large lakes in the eastern hemisphere. Hydrobiologia 434: 41–54.
Kryuchkova, N. M., 1989. Trophic relations between zoo- and phytoplankton. Nauka, Moskva, 124 (in Russian).
Laugaste, R., T. Nõges, P. Nõges, V. V. Yastremskij, A. Milius & I. Ott, 2001. Algae. In Pihu, E & J. Haberman (eds), Lake Peipsi. Flora and Fauna. Sulemees Publishers, Tartu, 31–49.
Lürling, M. & E. van Donk, 1997. Life history consequences for *Daphnia pulex* feeding on nutrient-limited phytoplankton. Freshwater Biology 38: 693–709.
Matthes M., 2004. Low genotypic diversity in a *Daphnia pulex* population in a biomanipulated lake: the lack of vertical and seasonal variability. Hydrobiologia 526: 33–42.
Meijer, M.-L., I. DeBoois, M. Scheffer, R. Portielje & H. Hosper, 1999. Biomanipulation in shallow lakes in the Netherlands: an evaluation of 18 case studies. Hydrobiologia 408(409): 13–30.
Nõges, P. & R. Laugaste, 1998. Seasonal and long-term changes in phytoplankton of Lake Võrtsjärv. Limnologica 28: 21–28.
Nõges, T., J. Haberman, V. Kisand, R. Laugaste & P. Zingel, 2001. Trophic relations and food web structure of plankton community in Lake Peipsi. In Pihu, E. & J. Haberman (eds), Lake Peipsi. Flora and Fauna. Sulemees Publishers, Tartu, 74–81.
Nõges, T., H. Agasild, J. Haberman, A. Kangur, P. Kangur, H. Timm, P. Zingel & P. Nõges, 2004. Food webs in Lake Võrtsjärv. In Haberman, J., A. Raukas & E. Pihu (eds), Lake Võrtsjärv. Estonian Encyclopaedia Publishers, Tallinn, 335–345.
Ostojić A. N., 2000. Effect of eutrophication on changes in the composition of zooplankton in the Grošnica

Reservoir (Serbia, Yugoslavia). Hydrobiologia 436: 171–178.

Pejler, B., 1983. Zooplanktic indicators of trophy and their food. Hydrobiologia 101: 111–114.

Pihu, E., 1998. Fishes and fisheries management in Lake Võrtsjärv. Limnologica 28(1): 91–94.

Pihu, E. & A. Kangur, 2001. Fishes and fisheries management. In Pihu, E. & J. Haberman (eds), Lake Peipsi. Flora and Fauna. Sulemees Publishers, Tartu, 100–111.

Sartonov, A., 1995. Effects of *Microcystis aeruginosa* on interference competition between *Daphnia pulex* and *Keratella cochlearis*. Hydrobiologia 307: 1–3.

Smakulska, J. & A. Górniak, 2004. Morphological variation in *Daphnia cucullata* Sars with progressive eutrophication of a polymictic lowland reservoir. Hydrobiologia 526: 119–127.

Spaak, P. & B. Keller, 2004. No evidence for adaptive micro-evolution to a decrease in phosphorus-loading of a *Daphnia* population inhabiting a pre-alpine lake. Hydrobiologia 526: 15–21.

Stich, H. B., Pfeiffer M. & G. Maier, 2005. Zooplankton communities in a large prealpine lake, Lake Constance: comparison between the Upper and the Lower Lake. Journal of Limnology 64: 129–138.

Vandekerkhove, J., S. Declerck, L. Brendonck, J. H. Conde-Porcuna, E. Jeppesen & L. DeMeester, 2005. Hatching of cladoceran resting eggs: temperature and photoperiod. Freshwater Biology 50: 96–104.

von Elert, E., 2004. Food quality constraints in *Daphnia*: interspecific differences in the response to the absence of a long chain polyunsaturated fatty acid in the food source. Hydrobiologia 526: 187–196.

Hydrobiologia (2007) 584:167–177
DOI 10.1007/s10750-007-0575-z

SHALLOW LAKES

Contribution of different zooplankton groups in grazing on phytoplankton in shallow eutrophic Lake Võrtsjärv (Estonia)

H. Agasild · P. Zingel · I. Tõnno · J. Haberman · T. Nõges

Abstract The grazing impact of different sized zooplankton on 'edible' and total phytoplankton biomass and primary production was measured in L. Võrtsjärv during a seasonal study in 1998 and 2000. The organisms of 48–100 μm size class, composed of ciliates and rotifers, contributed significantly to the total grazing of zooplankton community throughout the study period (average 68%). The average daily filtering and grazing rate of the whole zooplankton community (micro- and macro-zooplankton) remained low, corresponding to a filtration of 44% of the water volume, 4% of the total phytoplankton biomass and 29% of primary production. However, a strong grazing pressure on small-sized phytoplankton (<30 μm) was estimated in most of the study period (average 44% d^{-1}). Among size classes of 'edible' phytoplankton, the size range 5–15 μm was the most important algal food for the dominant zooplankton grazers (herbivorous ciliates, *Polyarthra* spp., *Chydorus sphaericus* and *Daphnia cucullata*) in L. Võrtsjärv.

Keywords Micro-zooplankton · Macro-zooplankton · Filtration rate · Grazing rate · Edible phytoplankton · Primary production

Guest editors: R. D. Gulati, E. Lammens, N. De Pauw & E. Van Donk
Shallow lakes in a changing world

H. Agasild (✉) · P. Zingel · I. Tõnno · J. Haberman · T. Nõges
Estonian University of Life Sciences, Institute of Agricultural and Environmental Sciences, Centre for Limnology, Rannu, Tartu County 61101, Estonia
e-mail: helen.agasild@emu.ee

Introduction

Cladocerans are typically considered as predominant phytoplankton grazers in lakes, mainly due to their high ingestion rates and abilities to filter particles of wide size spectrum and to consume a great range of food types (Lampert & Sommer, 1993). In zooplankton grazing estimations, mostly cladocerans and copepods are taken into account while the impact of rotifers and protozoa is usually neglected. In shallow eutrophic lakes, however, the zooplankton is often dominated by small-sized forms like the protozoa and the rotifers (Mayer et al., 1997; Zingel, 1999), while the large-sized cladocerans (i.e. *Daphnia* spp.) and the copepods are suppressed by the size selective fish predation (Jeppesen et al., 1996) and unfavourable feeding conditions like filamentous cyanobacteria and high seston concentration (Porter & McDonough, 1984; Tóth, 1992). In such lakes small-sized zooplankton like ciliates and rotifers may compose a considerable part of total zooplankton biomass (Zingel, 1999) and may even dominate in phytoplankton grazing (Quiblier-Lloberas et al., 1996), altering the frequency of distribution and amount of food available to larger zooplankton due to highly selective feeding (Starkweather, 1980).

The aim of the present study was to evaluate the zooplankton grazing impact on 'edible' and total phytoplankton biomass and primary production during a seasonal cycle, and to find out the relative importance of small (ciliates and rotifers) and large (mainly cladocerans and copepods) zooplankton in grazing on phytoplankton in L. Võrtsjärv. Nevertheless, the algal and zooplankton communities occurring in L. Võrtsjärv are well studied (Haberman, 1998; Nõges et al., 1998a), including grazing studies (Nõges, 1998; Agasild & Nõges, 2005), there is still debate as to whether small-bodied or large-bodied zooplankton have a greater impact on phytoplankton grazing.

Study site

Lake Võrtsjärv is a large (270 km^2) and shallow (mean depth 2.8 m, maximum depth 6 m) eutrophic lake in Central Estonia (58°05′—58°25′ N, and 25°55′–26°10′ E). The average total phosphorus concentration is 54 μg l^{-1} and total nitrogen concentration is 1.6 mg l^{-1} (Haberman et al., 1998). The shallowness of the lake and the wave-induced resuspension of bottom sediments contribute to the formation of high seston concentrations and high turbidity during summer.

Methods

Zooplankton grazing experiments, zooplankton and phytoplankton biomass estimations, and primary production measurements were conducted in 1998 and 2000 from April to November at bi-weekly intervals. One stationary sampling station was sampled for the integrated lake water obtained by mixing up the water collected by a 2-l Ruttner sampler from the entire water column at 1 m intervals from the surface to the bottom (maximum depth 3–4 m depending on the water level). Subsamples for phytoplankton, ciliates, primary production and chlorophyll *a* concentration analyses, and for zooplankton grazing experiments were taken from this water. Metazooplankton samples were collected with a quantitative Juday net of 85 μm from the same monitoring station.

Phytoplankton and ciliate samples were preserved with acidified Lugol's solution (0.5% final concentration), and studied using Utermöhl (1958) technique for the species composition and biomass. For determining the 'edible' phytoplankton biomass (B_{pe}), the algal samples were counted by splitting the presumably edible size fraction for zooplankton (<30 μm) into three size classes (SC5, SC15 and SC30) by the maximum linear dimension: 2–5 μm, 5–15 μm and 15–30 μm, respectively.

Chlorophyll *a* was measured spectrophotometrically in 96% ethanol extracts. Primary production (PP) of phytoplankton was estimated in situ using the $^{14}CO_2$ assimilation technique introduced first by Steeman-Nielsen (1952) on 13 times in 1998 and seven times in 2000. Depth integrated lake water was poured into 24 ml glass scintillation vials, 50 μl of sterile $NaH^{14}CO_3$ (VKI, Denmark) solution (1.7 μCi per vial) was added to achieve final activity 0.07 μCi ml^{-1}. The vials were incubated for 2 h at six depths: 0 m, 0.25 * Secchi depth (S), 0.5 * S, 1 * S, 2 * S and 3 * S in the lake. Then 6 ml of water from each sample was poured into a clean glass scintillation vial and acidified (pH < 2) by adding 150 μl of 0.5 N HCl. Inorganic ^{14}C was assumed to be removed during 24 h (Niemi et al., 1983). Next, 5 ml subsamples were poured into new plastic vials. The radioactivity was assessed by LSC RackBeta 1211 (Wallac, Finland) using external standardization for DPM calculations. Scintillation cocktail OptiPhase HiSafe 3 (Perkin Elmer) was applied. PP was calculated according to the standard formula (Guidelines, 1984). Non-photosynthetic carbon fixation was measured in dark vials and subtracted from light assimilation. The trapeze integration over depth was applied for calculating PP (mg C m^{-2} h^{-1}). Daily values (PP_{day}; mg C m^{-2} day^{-1}) were calculated according Nõges & Nõges (1998) as $PP_{day} = PP/(0.230–890 * 10^{-5} * DL)$, where DL is the length of the light day in hours.

Metazooplankton samples were fixed with acidified Lugol's solution (0.5% final concentration), counted under a binocular microscope in a Bogorov chamber and enumerated at ×32–56 magnification. For biomass calculations, average body lengths of at least 20 individuals of each taxon were measured. The individual weights of rotifers were estimated from average lengths according to Ruttner-Kolisko (1977). The lengths of crustaceans were converted to wet weights according to Studenikina & Cherepakhina (1969) for nauplii, and to Balushkina & Winberg (1979) for other groups. In order to estimate the biomass of herbivorous and non-herbivorous zooplankton, adult cyclopoid copepods (Hansen & Santer, 1995), cladoceran *Leptodora kindtii* (Focke) (Herzig, 1995), and rotifer *Asplanchna priodonta* Gosse (Thouvenot et al., 1999) were considered as predominantly non-herbivorous organisms among metazooplankton. Ciliates were divided into herbivores and non-herbivores using data, which was gained by the simultaneous grazing experiments using fluorescently labelled microspheres performed in L. Võrtsjärv in 2000.

To assess zooplankton community grazing, 50 ml of *Scenedesmus brasiliensis* Bohlin monoculture (cell size 2.5–4 μm) was labelled with $NaH^{14}CO_3$ during 36 h at 120 W m^{-2}. Then the cells were centrifuged at 3000 rpm, washed with filtered lake water (Whatman GF/C) and suspended in the same solution. The final quantity of the obtained algal suspension was 30 ml. Ten milliliters of this suspension was added into 3 l of integral lake water and incubated in the lake for 7 min. In experimental vessels the concentration of labelled *S. brasiliensis* monoculture did not exceed 5–10% of the total suspended solid concentration of the lake water that varied between 5.5 mg DW $l^{-1}$and 39.5 mg DW l^{-1}.

After incubation, water was filtered through the plankton net to collect the zooplankton. To obtain two size fractions of zooplankton: microzooplankton (MicroZ, 48–100 μm) and macrozooplankton (MacroZ, >100 μm), the incubated water was first filtered through the mesh size of 100 μm and the filtrate through the mesh size of 48 μm. The filtered zooplankton was anesthetized with carbonated water and stored on ice. To measure the radioactivity of food, 30 ml of filtrate was retained on the GF/C filters. In the laboratory, zooplankton was washed from the net to the GF/C filters. The radioactivity of the filters with zooplankton and food particles was measured by LSC RackBeta 1211 (Wallac, Finland) using scintillation coctail OptiPhase HiSafe 3 (Perkin Elmer).

Zooplankton community (MicroZ + MacroZ) filtration rate (F; ml l^{-1} h^{-1}) was calculated according to the formulae developed by Lampert & Taylor (1985):

$$F = \frac{R(\text{animals})}{[R(\text{water}) - R(\text{filtrate})]/v} * \frac{60}{t * V},$$

where F = community filtration rate (ml l^{-1} h^{-1}), t = feeding time (min), V = volume of water in experiment (l), R = measured radioactivity (becquerels), and v = volume of the water samples (ml).

One measurement series consisted of three replicates. Grazing rate (G; mg l^{-1} h^{-1}) was obtained by multiplying F with the biomass of the 'edible' phytoplankton (B_{pe}, 2.0–30.0 μm). Relative grazing rate of total phytoplankton biomass (RGR_b; % d^{-1}) was found as $RGR_b = 24 * G/B_p$, where B_p is total phytoplankton biomass (mgWW l^{-1}), and relative grazing rate of 'edible' phytoplankton biomass (RGR_{pe}; % d^{-1}) as $RGR_{pe} = 24 * G/B_{pe}$. Relative grazing rate of primary production (RGR_{pp}; % d^{-1}) was found as $RGR_{pp} = 24 * G/PP_{day}$ assuming 10% carbon content of phytoplankton wet biomass. Zooplankton grazing experiments were performed on 15 occasions in 1998 and on 14 times in 2000. The MicroZ grazing was not measured in the first part of the 1998 season until July.

Statistical analyses

The program STATISTICA FOR WINDOWS version 5.0 was used for statistical analyses. Spearman's correlation coefficients were used to determine the relations of the grazers and phytoplankton variables.

Results

Phytoplankton and zooplankton communities

During the study period, water temperature varied from 2.6°C to 20°C in 1998 and from 2.5°C to 19.3°C in 2000 (Fig. 1). The highest chlorophyll *a* concentrations were measured in September in both studied years (Fig. 1).

The average total phytoplankton biomass (B_p) was similar in both studied years (Fig. 2a), forming 21.8 gWW m^{-3} and 21.6 gWW m^{-3} in 1998 and 2000, respectively. Cyanobacteria *Limnothrix redekei* (Van Goor) Meffert and *L. planktonica* (Wolosz.) dominated in 1998 and *Planktolyngbya limnetica* (Lemm.) Kom.-Legn., *L. planktonica* (Wolosz.) Meffert., and *Aphanizomenon skujae* Kom.-Legn. et Cronb. dominated in 2000. Diatoms were the second most abundant group, mostly *Melosira* sp. in 1998 and *Aulacoseira* sp. in 2000.

In 1998 the 'edible' phytoplankton biomass (B_{pe}) formed on the average 14% of B_p but its share was considerably higher (up to 50%) in April and May due to low B_p at the same time (Fig. 2a). During the study period in 2000 the B_{pe} formed an average two times less (7%) of B_p compared to 1998. In 1998 the B_{pe} stayed relatively low until mid-August, in 2000 the period of minimum B_{pe} occurred from late August until the end of September (Fig. 2a).

In 1998 among the size classes of B_{pe}, the SC5 covered mainly green algae (*Scenedesmus* sp.) and crysophytes (*Dinobryon* sp.), SC15 comprised of mainly crysophytes (*Synura* sp.), cryptophytes (*Rhodomonas* sp.), diatoms (*Stephanodiscus* sp., *Fragillaria* sp.) and green algae (*Tetraedron* sp., *Pediastrum* sp.), and SC30 involved mainly diatoms (*Stephanodiscus* sp.), cyanobacteria and cryptophytes. In 2000 SC5 covered mainly small unidentified phytoplankton, SC15—mainly green algae (*Scenedesmus* sp.), crysophytes (*Dinobryon* sp.) and some cryptophytes, and SC30—mainly diatoms (*Stephanodiscus* sp., *Fragillaria* sp.), dinophytes (*Peridinium* sp.) and green algae (*Scenedesmus* sp.).

In 1998 the highest rates of primary production (PP) were measured in June and August. Five PP measurements available in 2000 yielded a maximum rate in May (Fig. 2b).

In 1998 and 2000 the mean zooplankton biomass was 2.7 gWW m^{-3} and ciliates contributed on average 58% of total zooplankton biomass (Fig. 3a). High biomasses of metazooplankton and ciliates occurred in spring from April until the end of June, and in July–August.

The ciliates dominated in MicroZ fraction (Fig. 3b) and cladocerans in MacroZ fraction (Fig. 3c). Herbivores contributed on average 66% and 68% to the MicroZ and MacroZ biomasses, respectively. In seasonal dynamics short seasonal delay between herbivorous micro- and macrozooplankton peak abundances was observed on several occasions. In spring of 1998 higher MicroZ biomass occurred than in spring of 2000, while in spring of 2000 MacroZ biomass was higher than in the spring of 1998 (Fig. 3b, c). During the spring period, MicroZ main grazers

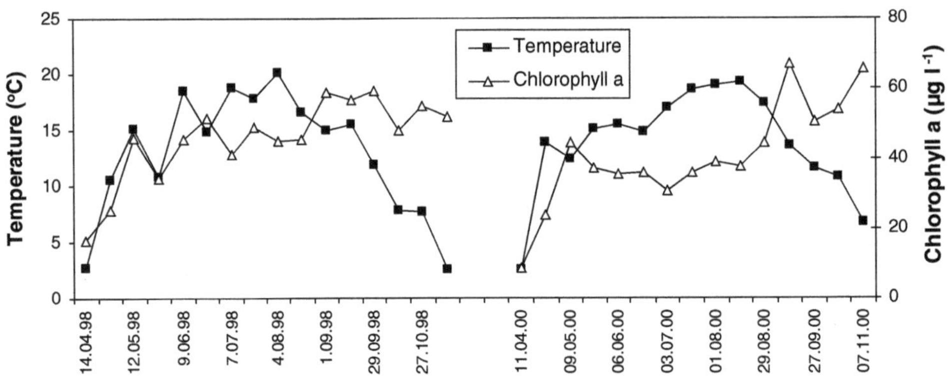

Fig. 1 Changes in temperature and chlorophyll *a* in L. Võrtsjärv in 1998 and 2000

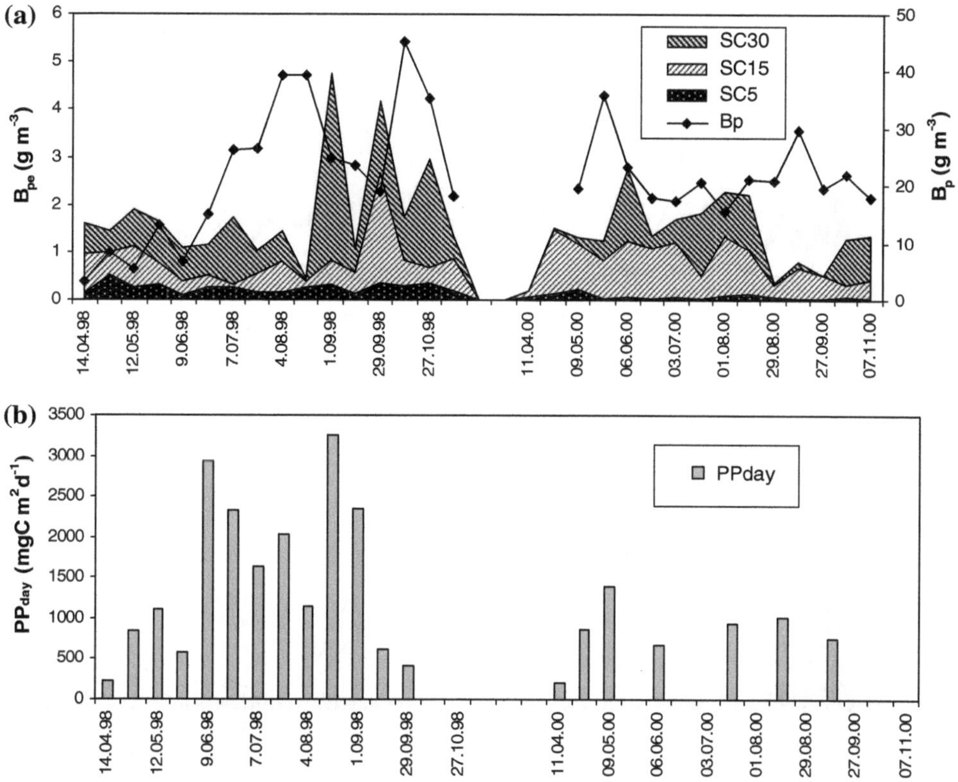

Fig. 2 (**a**) Temporal changes in biomasses of total phytoplankton (B_p) and 'edible' phytoplankton (B_{pe}, 2.0–30.0 μm) and its contributing size classes (SC5, SC15, SC30); (**b**) phytoplankton primary production rate (PP_{day}), in L. Võrtsjärv in 1998 and 2000

were *Polyarthra dolichoptera* Idelson and *Keratella cochlearis* (Gosse) among rotifers, and *Codonella cratera* (Leidy), *Rimostrombidium* sp., *Pelagostrombidium* sp. and *Tintinnidium fluviatile* (Stein) among ciliates. In the spring of 1998 small-sized picovores (mainly scuticociliates) were also abundant. Among MacroZ, cladocerans *Bosmina longirostris* (O.F. Müller) and *Chydorus sphaericus* (O.F. Müller) dominated in spring period, and rotifers also contributed significantly in May 1998 (Fig. 3c). In summer the peaks of MicroZ and MacroZ biomasses were lower than in spring, and the species composition was also different (Fig. 3b, c). In this period the community of ciliates was dominated by small-sized bacterivorous species like *Cyclidium* sp., *Uronema* sp. and *Halteria* sp., and *Anuraeopsis fissa* (Gosse), *Keratella c. tecta* (Gosse) among the rotifers. In August the larger filter-feeder cladoceran *Daphnia cucullata* Sars also peaked.

Grazing influence on 'edible' and total phytoplankton biomass and primary production

Zooplankton filtering rates (F) in both studied years were quite similar forming on average 41% and 47% of the water volume per day in 1998 and 2000, respectively (Fig. 4a).

The highest measured total zooplankton F in August 1998 and in September 2000 coincided with the highest relative grazing rates of 'edible' phytoplankton biomass (RGR_{pe}) (104% d^{-1} and 86% d^{-1}, respectively) (Fig. 4a, b), and with the decreased B_{pe} at the same time (Fig. 2a). According to F, on average 44% of B_{pe} was grazed daily by zooplankton in 1998 and 2000. Among the size classes of B_{pe}, a significant inverse relationship between SC15 and the main grazers: herbivorous ciliates ($r = -0.6$, $P < 0.05$), *Polyarthra* spp. ($r = -0.5$, $P < 0.05$), *C. sphaericus*

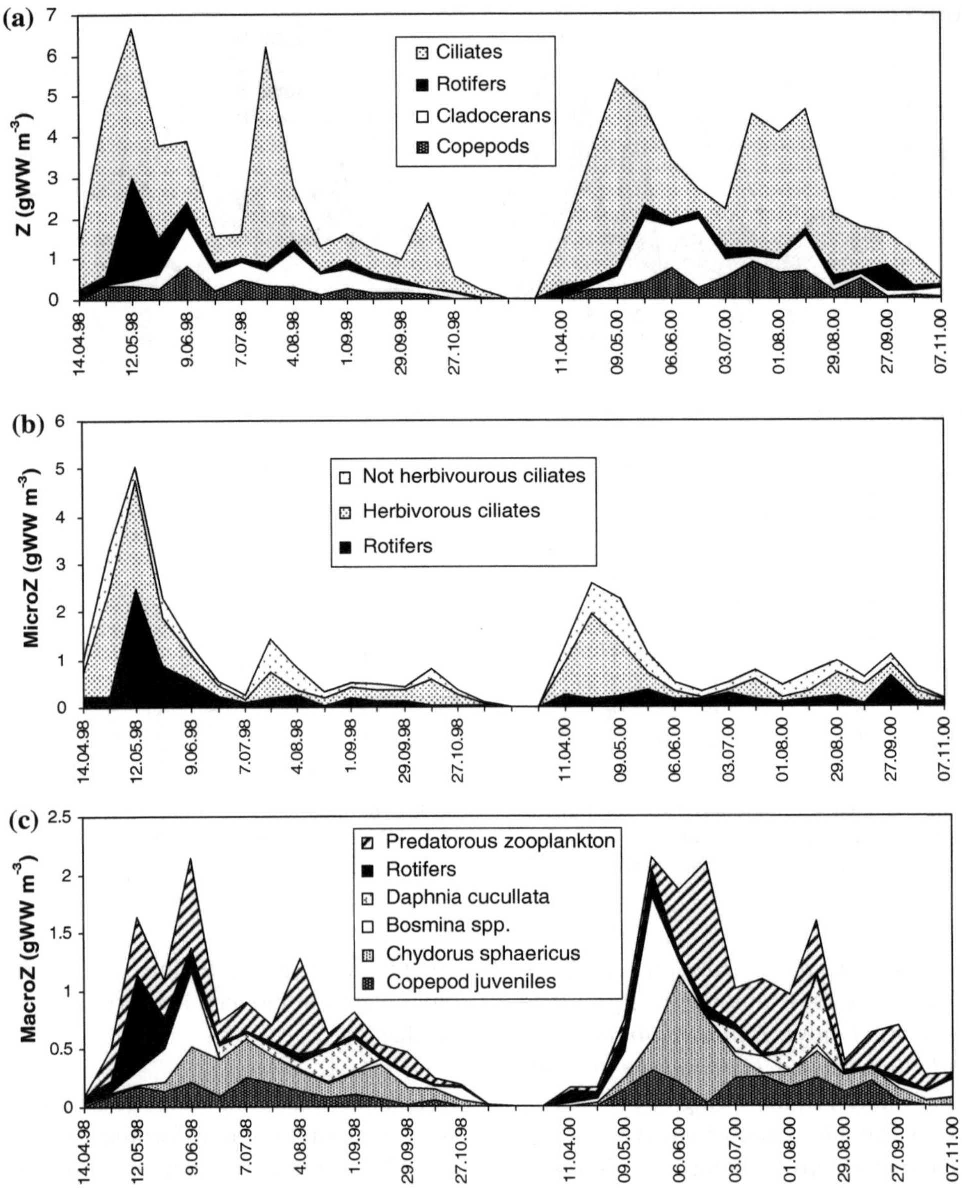

Fig. 3 Zooplankton biomasses in L. Võrtsjärv in 1998 and 2000. (**a**) Total zooplankton (Z) and the contributing groups (ciliates, rotifers, cladocerans, copepods); (**b**) micro-zooplankton (MicroZ, 48–100 μm) and the contributing groups (not herbivorous ciliates, herbivorous ciliates and rotifers); (**c**) macro-zooplankton (MacroZ, >100 μm) and the contributing taxa

($r = -0.6$, $P < 0.05$) and *D. cucullata* ($r = -0.5$, $P < 0.05$) was found.

In 1998 relative grazing rates of the total phytoplankton biomass (RGR_b) formed 0.85–16.4% d^{-1} (average 4.7% d^{-1}). In 2000 somewhat lower RGR_b values occurred (0.4–6.6% d^{-1}, average 3.3% d^{-1}). During the study period the highest RGR_b occurred on 14 April 1998 when 16% of B_p was consumed by the MacroZ alone (Fig. 4b). The highest RGR_b by both MicroZ and MacroZ formed 11% d^{-1} at the end of September 1998 and 6.6% d^{-1} in the middle of August in 2000.

In both studied years the relative grazing rate of phytoplankton primary production (RGR_{pp})

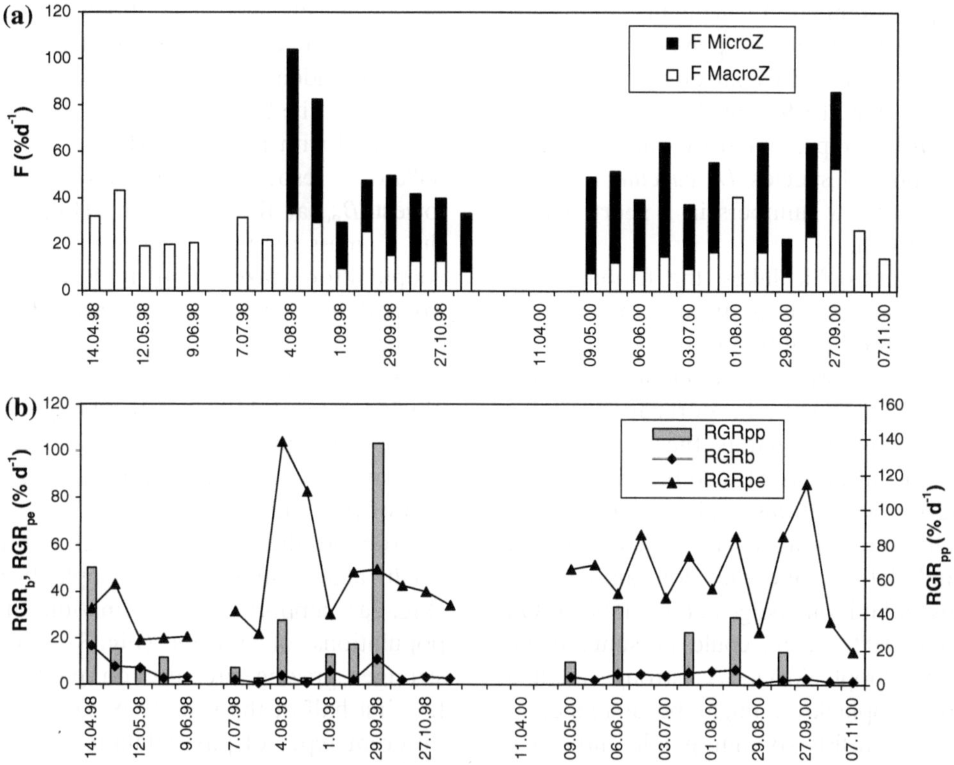

Fig. 4 (**a**) Filtration rates of micro-zooplankton (*F* MicroZ) and macro-zooplankton (*F* MacroZ); (**b**) relative zooplankton grazing rates of total phytoplankton biomass (RGR_b) and 'edible' phytoplankton biomass (RGR_{pe}), and relative grazing rate of primary production (RGR_{pp}) in L. Võrtsjärv in 1998 and 2000. Note that the MicroZ grazing was not measured in the first part of the 1998 season

formed on an average 29% d^{-1}. In 1998 the highest RGR_{pp} was measured at the end of September (137% d^{-1}) and 2000 yielded a maximum rate in June (39% d^{-1}) (Fig. 4b).

During the measurement period of both MicroZ and MacroZ feeding (August–November in 1998 and May–November in 2000), the grazer fraction 48–100 μm had major importance in zooplankton grazing. That group of organisms, including ciliates and rotifers, accounted for between 46% and 85% (average 68%) of the filtering and grazing rates by the total zooplankton (Fig. 4a).

Discussion

Although the phytoplankton and zooplankton communities occurring in L. Võrtsjärv are well studied (Haberman, 1998; Nõges et al., 1998a; Zingel, 1999), the relative importance of small-bodied and large-bodied zooplankton in grazing on phytoplankton has been under speculation. This study revealed MicroZ as the primary consumer on phytoplankton in L. Võrtsjärv. This fraction, formed by the ciliates and rotifers, dominated in zooplankton filtering and grazing activity over a major part of the seasonal cycle, while larger zooplankton (>100 μm) composed mainly of crustaceans formed on average one-third of the zooplankton feeding contribution (Fig. 4a).

Considering the zooplankton composition in L. Võrtsjärv, the greater impact of small-sized grazers in phytoplankton consumption was expected, as more than 60% of zooplankton biomass is formed by ciliates and rotifers (Zingel, 1999). Among the crustaceans, mainly cyclopoid copepods (mostly *Mesocyclops* spp.) and small-bodied filter-feeding cladocerans like *C. sphaericus* and

Bosmina spp. dominated. In many eutrophic lakes, particularly, *Daphnia* spp. have been shown to be the key species in controlling the phytoplankton (Sterner, 1989). In L. Võrtsjärv the genus *Daphnia* is represented by only one relatively small-bodied species *D. cucculata*, which occurs in rather low numbers in a short period (Haberman, 1998).

In shallow eutrophic lakes, the formation of zooplankton composition is strongly forced from two sides—by food quality and by size selective feeding of fish (Porter & McDonough, 1984; Jeppesen et al., 1996). In L. Võrtsjärv the zooplankton community inhabits the environment with high concentrations of inedible filamentous cyanobacteria and diatoms, while the food base for zooplankton contains large quantities of bacteria and detritus and only low amounts of 'edible' phytoplankton (Nõges et al., 1998b). On average, only 10% of B_p could presumably be ingested by zooplankton. A strong feeding pressure on zooplankton might be assumed, as the lake is rich in fish—bream, perch, and roach being the most abundant species (Nõges et al., 1998b).

As a result of algal and grazer composition, their grazing impact on total phytoplankton is quite weak in L. Võrtsjärv. On average only 4% of the B_p and 29% phytoplankton PP was consumed daily by zooplankton (larger than 48 μm) during the study period in 1998 and 2000. Although in some periods zooplankton filtered the whole water volume (104%) during 1 day and grazing exceeded the primary production rate (137%), the consumption was never enough to influence considerably the B_p. The maximum RGR_b (16% d^{-1}) occurred in April 1998 at very low level of B_p (Figs. 2a and 4b).

Quite similar results were obtained also in earlier studies in L. Võrtsjärv, where the consumption of only cladocerans was investigated. In 1984 and 1985 average cladoceran RGR_b and RGR_{pp} formed 2.5% d^{-1} and 28.5% d^{-1}, respectively, and the maximum values formed 12% d^{-1} and 133% d^{-1}, respectively (Nõges, 1998).

Assuming that zooplankton was feeding only on the 'edible' algae, however, heavy grazing on small-sized phytoplankton (<30 μm) was estimated in most of the study period (average 44% d^{-1}) (Fig. 4b). The direct phytoplankton–zooplankton relationship was observed with the clearest evidence in August 1998 and September 2000 when the highest zooplankton filtering rates coincided with maximum RGR_{pe} (104% d^{-1} and 86% d^{-1}, respectively) (Fig. 4), and with the lowest B_{pe} at the same time (Fig. 2a). In 1998 the biomass of small algae was kept low due to zooplankton grazing until mid August (Fig. 2a). MicroZ grazing was not measured during that period, but considering the high biomass of herbivorous ciliates and their theoretic daily grazing rate of 150% of the body mass (Arndt et al., 1990), most of the B_{pe} was assumed to be consumed by the MicroZ and macroZ in the first half of the study period.

Probably the low B_p and high contribution of small algae suitable for zooplankton grazing (Fig. 2a) supported the formation of abundant populations of grazers (Fig. 3) and therefore stronger grazing pressure on phytoplankton in the first half of the year 1998. During this period different types of phytoplankton grazers peaked in the zooplankton community. The spring population of ciliates (mainly *Rimostrombidium* sp., *Pelagostrombidium* sp. and *C. cratera*) and rotifers (mainly *P. dolichoptera, Keratella* spp.) formed the first peak of grazers in the middle of May, followed by the high biomass of cladocerans (*B. longirostris* and *C. sphaerius*) and copepods in early June (Fig. 3b, c).

Among the size classes of B_{pe}, the algal size class SC15 biomass from spring to August was twice lower in 1998 than in 2000 (Fig. 2a). Supported by the statistical analysis, the 5–15 μm sized phytoplankton is an important algal food source for the main zooplankton grazers as herbivorous ciliates, *Polyarthra* spp., and cladocerans *C. sphaericus* and *D. cucullata* in L. Võrtsjärv. Therefore, a strong food competition between ciliates and metazooplankton might be assumed over that size of algae. In 1998 the SC15 was largely composed of flagellates like *Rhodomonas*, which are more easily ingested by zooplankton than non-flagellated algae. The flagellates are supposed to be well grazable, because these cells may adhere by their flagella to the feeding appendages during the collection process, and because they don't often have sturdy cell walls (Knisely & Geller, 1986). In feeding studies with different algal types or with natural phytoplankton mixture, flagellates (like *Cryptomonas* and *Chlamydomonas*) are often found to be preferred algal food for zooplankton (Bogdan & Gilbert, 1982; Knisely & Geller, 1986). Larger sized (>5 μm) flagellated algae are an especially important component in the diet of *Polyarthra* (Bogdan & Gilbert, 1982). The high quality food base probably supported the development of abundant populations of small-sized grazers (herbivorous ciliates, *Polyarthra* spp.) responsible for the low SC15 of B_{pe} in 1998.

In 2000 the SC15 was composed primarily of less 'edible' algae (mainly *Scenedesmus* sp. and *Dinobryon* sp.), which may also be the reason for lower MicroZ biomass that year. Another cause may be the abundant population of crustaceans in 2000 (Fig. 3c). An exploitative food competition between ciliates and metazooplankton (Sanders et al., 1989; Weisse & Frahm, 2002) may cause the niche separation in seasonal dominance of grazer types. Metazooplankton are also known to prey intensively on ciliates and can affect their numbers (Adrian & Schneider-Olt, 1999; Jürgens et al., 1999).

Considering zooplankton community studies, there always remains a question about the method of measurement as the zooplankton community consists of variety of organisms, which are diverse in both feeding mechanisms and in modes of food selection (Pourriot, 1977; DeMott, 1986). In the present investigation commonly used radioactive method was applied to determine the zooplankton filtration rate. The tracer cells (2.5–4 μm diameter) were chosen to be of ingestible size for both smaller and larger zooplankton organisms in the studied lake. Though, the given sized tracer particles might not be equally preferred by all species and might be assumed to be more ingestible for MicroZ (Hansen et al., 1994). However, in crustacean community in L. Võrtsjärv, the dominating organisms are small cladocerans like *C. sphaericus* and naupliar stages of cyclopoid copepods, which are known as efficient grazers on small-sized food objects (Hwang & Heath, 1999; Hansen & Santer, 1995; Agasild & Nõges, 2005). Therefore we assume, that the tracer particle size used in the experiments did not seriously alter the outcome of MicroZ and MacroZ grazing results and general findings of this study.

In case of large lakes like L. Võrtsjärv, the horizontal heterogenity may influence the plankton communities between different parts. The grazing pattern studied here may be assumed to be valid in the northern and central plankton dominated part of the lake where the study was performed. Due to its shallowness and large opened area the water mass is well exposed to the wind action keeping it in movement and stirred. In the narrow and macrophyte covered southern part, however, the different species composition of plankton community (Nõges et al., 2004) may also involve different phytoplankton–zooplankton interactions.

In conclusion, the small-sized zooplankton such as ciliates and rotifers are the dominant phytoplankton consumers in L. Võrtsjärv. Due to specific algal (mainly filamentous forms) and grazer composition, the zooplankton community has generally low grazing impact on the total phytoplankton biomass in this lake. However, a strong feeding pressure is evident on the small-sized phytoplankton. Grazing impact is most significant on nano-plankton fraction, among which the size range of 5–15 μm seems to be the most important algal food for the main phytoplankton grazers (ciliates, *Polyarthra* spp., *C. sphaericus* and *D. cucullata*) in L. Võrtsjärv. Although a lot of attention has been paid on zooplankton grazing of phytoplankton community, there are few studies comparing the grazing impact of zooplankton of different size fractions or even taking into account protozooplankton feeding. The main results obtained in this study showing the major importance of MicroZ in phytoplankton grazing might be extrapolated to other eutrophic lakes, where small-sized organisms like protozoa and rotifers dominate in zooplankton assemblage.

Acknowledgements Funding for this research was provided by Estonian Science Foundation grants 4080 and 5738, by the European Union projects ECOFRAME (contract EVK1-CT-1999-39) and CLIME (EVK1-CT-2002-00121), and by Estonian Ministry of Education (0362480s03). The authors are grateful to the anonymous referees for their valuable advises and comments for improving the manuscript.

References

Adrian, R. & B. Schneider-Olt, 1999. Top-down effects of crustacean zooplankton in a mesotrophic lake. Journal of Plankton Research 21: 2175–2190.

Agasild, H. & T. Nõges, 2005. Cladoceran and rotifer grazing on bacteria and phytoplankton in two shallow eutrophic lakes: in situ measurement with fluorescent microspheres. Journal of Plankton Research 27/11: 1155–1174.

Arndt, H., J. Günter & N. Wasmund, 1990. Dynamics of pelagic ciliates in eutrophic estuarine waters: importance of functional groups among ciliates and responses to bacterial and phytoplankton production. Archiv für Hydrobiologie–Beiheft Ergebnisse der Limnologie 34: 239–245.

Balushkina, E. V. & G. G. Winberg, 1979. Relation between body mass and length in planktonic animals. In Winberg, G. G. (ed.), Obshchiye osnovy izucheniya vodnykh ekosistemi. Nauka, Leningrad, 169–172 (in Russian).

Bogdan, K. G. & J. J. Gilbert, 1982. Seasonal patterns of feeding by natural populations of Keratella, Polyarthra, and Bosmina: clearance rates, selectivities, and contributions to community grazing. Limnology and Oceanography 27: 918–934.

DeMott, R. W., 1986. The role of in food selection by freshwater zooplankton. Oecologia (Berlin) 69: 334–340.

Guidelines for the Baltic monitoring programme for the third stage, 1984. The Baltic Marine Biologists. Publ. 1. 2nd edn.

Haberman, J., 1998. Zooplankton of Lake Võrtsjärv. Limnologica 28: 49–65.

Haberman, J., P. Nõges, T. Nõges, E. Pihu, K. Kangur, A. Kangur & V. Kisand, 1998. Characterization of Lake Võrtsjärv. Limnologica 28: 3–11.

Hansen, A. M. & B. Santer, 1995. The influence of food resources on the development, survival and reproduction of the two cyclopoid copepods: *Cyclops vicinus* and *Mesocyclops leucarti*. Journal of Plankton Research 17: 631–646.

Hansen, B., P. K. Bjornsen & P. J. Hansen, 1994. The size ratio between planktonic predators and their prey. Limnology and Oceanography 39: 395–403.

Herzig, A., 1995. *Leptodora kindtii*: efficient predator and preferred prey item in Neusiedler See, Austria. Hydrobiologia 307: 273–282.

Hwang, S.-J. & R. Heath, 1999. Zooplankton bacterivory at coastal and offshore sites of Lake Erie. Journal of Plankton Research 21: 699–719.

Jeppesen, E., M. Søndergaard, J. P. Jensen, E. Mortensen & O. Sortkjær, 1996. Fish-induced changes in zooplankton grazing on phytoplankton and bacterioplankton: a long-term study in shallow hypertrophic Lake Søbygaard. Journal of Plankton Research 18: 1605–1625.

Jürgens, K., O. Skibbe & E. Jeppesen, 1999. Impact of metazooplankton on the composition and population dynamics of planktonic ciliates in a shallow, hypertrophic lake. Aquatic Microbial Ecology 17: 61–75.

Knisely, K. & W. Geller, 1986. Selective feeding of four zooplankton species on natural lake phytoplankton. Oecologia 69: 86–94.

Lampert, W. & U. Sommer, 1993. Limnoökologie. Thieme Verlag, Stuttgart-New York.

Lampert, W. & B. E. Taylor, 1985. Zooplankton grazing in a eutrophic lake: implications of diel vertical migration. Ecology 66: 68–82.

Mayer, J., M. T. Dokulil, M. Salbrechter, M. Berger, T. Posch, G. Pfister, A. K. T. Kirschner, B. Velimirov, A. Steitz & T. Ulbricht, 1997. Seasonal successions and trophic relations between phytoplankton, zooplankton, ciliate and bacteria in a hypertrophic shallow lake in Vienna, Austria. Hydrobiologia 342/343: 165–174.

Niemi, M., J. Kuparinen, A. Uusi-Rauva & K. Korhonen, 1983. Preparation of algal samples for liquid scintillation counting. Hydrobiologia 106: 149–159.

Nõges, T., 1998. Cladoceran grazing in Lake Võrtsjärv. Limnologica 28: 67–74.

Nõges, T. & P. Nõges, 1998. Primary production of Lake Võrtsjärv. Limnologica 28: 29–40.

Nõges, T., V. Kisand, P. Nõges, A. Põllumäe, L. Tuvikene & P. Zingel, 1998a. Plankton seasonal dynamics and its controlling factors in shallow polymictic eutrophic Lake Võrtsjärv, Estonia. International Review of Hydrobiology 83: 279–296.

Nõges, T., P. Nõges, J. Haberman, V. Kisand, K. Kangur, A. Kangur & A. Järvalt, 1998b. Food web structure in shallow eutrophic Lake Võrtsjärv (Estonia). Limnologica 28: 115–128.

Nõges, P., R. Laugaste & T. Nõges, 2004. Phyoplankton. In Haberman, J., E. Pihu & A. Raukas (eds), Estonian Encyclopaedia Publishers, Lake Võrtsjärv, 217–231.

Porter, K. G. & R. McDonough, 1984. The energetic cost of response to blue-green algal filaments by cladocerans. Limnology and Oceanography 29: 365–369.

Pourriot, R., 1977. Food and feeding habits of rotifera. Archiv für Hydrobiologie–Beiheft Ergebnisse der Limnologie 8: 243–260.

Quiblier-Lloberas, C., G. Bourdier, C. Amblard & D. Pepin, 1996. Impact of grazing on phytoplankton in Lake Pavin (France): contribution of different zooplankton groups. Journal of Plankton Research 18: 305–322.

Ruttner-Kolisko, A., 1977. Suggestion for biomass calculation of planktonic rotifers. Archiv für Hydrobiologie 8: 71–76.

Sanders, R. W., K. G. Porter, S. J. Bennett & A. E. DeBiase, 1989. Seasonal patterns of bacterivory by flagellates, ciliates, rotifers, and cladocerans in a freshwater planktonic community. Limnology and Oceanography 34: 673–687.

Starkweather, P. L., 1980. Aspects of feeding behaviour and trophic ecology of suspension feeding. Hydrobiologia 73: 63–72.

Steeman-Nielsen, E., 1952. The use of radioactive carbon (^{14}C) for measuring primary production in the sea. Journal du Conseil permanent international pour l'exploration del la mer 18: 117–140.

Sterner, R. W., 1989. The role of grazers in phytoplankton succession. In Sommer, U. (ed.), Plankton ecology: Succession in plankton communities. Springer-Verlag, Berlin, 107–170.

Studenikina, E. I. & M. M. Cherepakhina, 1969. Srednii ves osnovykh form zooplanktona Azavskogo morya [mean weight of basic zooplankton forms of the Azov Sea]. Gidrobiologicheskij Zhurnal 5: 89–91 (in Russian).

Thouvenot, A., D. Deboras, M. Richardot & J. Devaux, 1999. Impact of natural metazooplankton assemblage on planktonic microbial communities in a newly flooded reservoir. Journal of Plankton Research 21: 179–1999.

Tóth, L.G., 1992. Limiting effect of abioseston on food ingestion, postembronic development time and fecundity of daphnids in Lake Balaton (Hungary). Journal of Plankton Research 14: 435–446.

Utermöhl, H., 1958. Zur Vervollkommung der quantitativen Phytoplanktonmethodik. Mitteilungen Internationale Vereiningung fuer Theoretische und Angewandte Limnologie 9: 1–38.

Weisse, T. & A. Frahm, 2002. Direct and indirect impact of two common rotifer species (*Keratella* spp.) on two abundant ciliate species (*Urotricha furcata, Balanion planctonicum*). Freshwater Biology 47: 53–64.

Zingel, P., 1999. Pelagic ciliated protozoa in shallow eutrophic lake: community structure and seasonal dynamics. Archiv für Hydrobiologie 146: 495–511.

Hydrobiologia (2007) 584:179–189
DOI 10.1007/s10750-007-0599-4

SHALLOW LAKES

Horizontal dynamics of zooplankton in subtropical Lake Blanca (Uruguay) hosting multiple zooplankton predators and aquatic plant refuges

Carlos Iglesias · Guillermo Goyenola · Nestor Mazzeo · Mariana Meerhoff · Elena Rodó · Erik Jeppesen

Abstract In the subtropics, the effects of macrophytes on trophic interactions are more complex than in temperate lakes. Fish, particularly the smallest species and individuals, aggregate in high numbers in the vegetation, and a strong predation pressure on zooplankton by shrimps and invertebrates, such as *Chaoborus*, can occur in these systems. We studied seasonal and diel changes in zooplankton and their potential predators (both fish and invertebrates) and physical and chemical characteristics among open water and vegetated habitats (emergent and submerged plants (SP)), in the subtropical Lake Blanca (34°54′ S; 54°50′ W), a shallow system with an extensive and complex littoral area and high abundance of vertebrate and invertebrate predators on zooplankton. We found differential horizontal distribution of some zooplankton species under the scenario of high abundance of small omnivorous-planktivorous fish and *Chaoborus*, especially in the seasons with intermediate catch per unit effort of fish. We found indications of a diel horizontal migration (DHM) opposite than described for temperate systems, as the two main cladocerans *Bosmina longirostris* and *Diaphanosoma birgei* were found in higher densities in the submerged plant beds during night, in spring and autumn respectively. Although we need experiments to prove DHM, *Chaoborus* seemed to be the main trigger of the apparent DHM, followed by small omnivorous fish. During summer no spatial differences were found likely because of high densities of fish in all habitats. In absence of piscivorous fish, the distribution of fish *Jenynsia multidentata* seemed to be conditioned by food availability and by predation risk of waterfowl. The refuge capacity of aquatic plants for zooplankton in subtropical systems seems weak and with consequent weak or no cascading effects on water transparency, as under very high fish and invertebrate densities (summer) the refuge for zooplankton was lost.

Keywords Spatial use · Diel horizontal migration · Refuge effect · Invertebrate predation · Subtropical lakes

Guest editors: R. D. Gulati, E. Lammens, N. De Pauw & E. Van Donk
Shallow lakes in a changing world

C. Iglesias (✉) · G. Goyenola · N. Mazzeo · M. Meerhoff · E. Rodó
Departmento de Ecología, Facultad de Ciencias, Universidad de la República, Iguá 4225, C.P. 11400, Montevideo, Uruguay
e-mail: caif@fcien.edu.uy

M. Meerhoff · E. Jeppesen
National Environmental Research Institute
University of Aarhus, Vejlsøvej 25, P. O. Box 314, 8600 Silkeborg, Denmark

M. Meerhoff · E. Jeppesen
Department of Plant Biology, University of Aarhus, Ole Worms Alle, 8000 Aarhus C, Denmark

Introduction

The spatial distribution of animals is regarded as a compromise between several costs and benefits, especially the avoidance of predation and the search for optimal feeding and environmental conditions. In freshwater systems, aquatic plants play a relevant role in affecting the spatial distribution of several organisms and shaping predator-prey interactions (Jeppesen et al., 1997a). Submerged macrophytes can often provide daytime refuge for *Daphnia* and other cladocerans from fish predation in shallow temperate lakes (diel horizontal migration (DHM), Timms & Moss, 1984; Burks et al., 2002), although this pattern has not been observed in oligotrophic and mesotrophic lakes (Lauridsen et al., 1999). However, many juvenile fish often use the littoral zone as daytime refuge against predatory fish, which compromises the refuge capacity of macrophytes for zooplankton (Carpenter & Lodge, 1986). Also, invertebrate predators in the pelagic can affect zooplankton populations and DHM, but evidence is scarce yet (Burks et al., 2001; R. Burks et al., unpublished). It has been suggested that the effects of co-existing fish and pelagic invertebrate predators can be additive and increase the likelihood of large zooplankton seeking refuge in the littoral zone (Burks et al., 2002). The net effect of multiple predators, however, seems to depend on the direction of the response of the prey to each predator alone, and their relative risk for that prey. The response of the prey to combined predators would thus be similar to its response to the most risky predator in the pair (Relyea, 2003; Wojtal et al., 2003). Wicklum (1999) suggested that vertebrate and invertebrate predation act alternately because invertebrate predators are themselves subject to fish predation. As a consequence, the impact of invertebrate predation increases as fish predation declines (Mumm, 1997).

In the tropics and subtropics, the effects of macrophytes on trophic interactions are more complex than in temperate lakes, not least because all plant life forms (emergent, submerged, floating-leaved and large free-floating species) can be very abundant (Talling & Lemoalle, 1998), and the predation pressure on zooplankton seems stronger (Jeppesen et al., 2007). Fish are usually present in very high numbers (Scasso et al., 2001), and particularly the smallest species and individuals aggregate in the vegetation (Meschiatti et al., 2000, 2003). Besides, the zooplanktivorous phantom midge *Chaoborus* seems to be more abundant in tropical lakes (Lewis, 1996). This might reflect that tropical lakes become anoxic in the bottom more quickly than temperate lakes, providing *Chaoborus* with a fish predation refuge for longer periods (Lewis, 1996), which, in turn, permits a stronger predation pressure on zooplankton (Arcifa et al., 1992). Also shrimps can represent important predators on zooplankton in warm regions (Collins, 1999; Collins & Paggi, 1998), probably due to the dominance of small fish species with consequent lower risk of predation on shrimps, as seen for *Neomysis integer* Leach in north temperate brackish lakes with fish communities dominated by small sticklebacks (*Gasterosteus* spp.) (Søndergaard et al., 1997).

Our objectives were to describe the seasonal and diel variation in the horizontal distribution of zooplankton under the scenario of combined vertebrate and invertebrate predators and relate it to variations in predator abundance and the changes promoted by the aquatic plants, in a subtropical shallow lake with a very well-developed littoral zone hosting different plant life-forms and high abundance of small omnivorous fish and invertebrate predators (Mazzeo et al., 2003). Based on knowledge from eutrophic temperate lakes we hypothesised that predation-sensitive zooplankton would seek refuge during the day in those vegetated habitats with less potential predators, and move to open water at night to decrease the predation risk. However, an alternative scenario is that the plants are not feasible refuges for zooplankton due to a potential high density of small fish in the plant beds.

Study area

Lake Blanca (34°54′ S; 54°50′ W) is a shallow (total area = 40.5 ha, Z_{max} = 3.6 m; Mazzeo et al., 2003) lake used for water supply in the main tourist area of Uruguay. It is a warm polymictic (temperature range: 11.3–26.3°C) and eutrophic lake (in-lake annual mean TP, TN (Valderrama, 1981) and chlorophyll a (Nusch, 1980) concentrations: 93.6, 1010, and 31.9 μg l^{-1}, respectively). A drastic reduction of the lake volume and a consequent loss of piscivorous fish occurred during 1998–1999. Since then, turbid and clear water states have alternated in the system, and the fish community has been dominated by small (13 mm < SL < 86 mm) *Jenynsia multidentata* Jenyns (Anablepidae) (Mazzeo et al., 2003). This is an omnivorous-planktivorous fish (Escalante, 1983; Goyenola et al., unpublished data), with a mean whole lake density = 13 ind m^{-2} (density in littoral areas ≥ 100 ind m^{-2}, Mazzeo et al., 2003). The littoral zone of Lake Blanca is rather complex and dynamic, from the shore to the open water area there is an emergent plant zone (*Schoenoplectus californicus* L. and floating islands dominated by *Typha latifolia* L.), followed by submerged plants (SP) (*Egeria densa* Planch. and *Ceratophyllum demersum* Planch.), and finally by another zone of emergent plants (EP) (only *S. californicus*) (Fig. 1).

Materials and methods

The study was carried out seasonally from southern hemisphere winter 2003 to autumn 2004. The percent volume infested of the SP (PVI, according to Canfield et al., 1984) and the distribution of the EP were estimated from more than 100 sites homogeneously distributed over seven transects that were set perpendicular to the maximum length of the lake and covered all the system. Each season, we mapped the vegetation according to these observations and recorded the positions using a GPS. From the map, we randomly selected five sampling points per strata: EP, submerged plants (SP: PVI ≥ 25%), and the pelagic or open water area (OW: PVI = 0%). Water samples for physical, chemical and plankton analyses were collected with a vertical tube along the whole water column, at midday and midnight. Dissolved oxygen (DO), conductivity, pH, and transparency were registered in situ, using Horiba sensors and a Secchi minidisk. The euphotic zone depth (Z_{eu}, 1% of incident light) was determined from the photosynthetic active

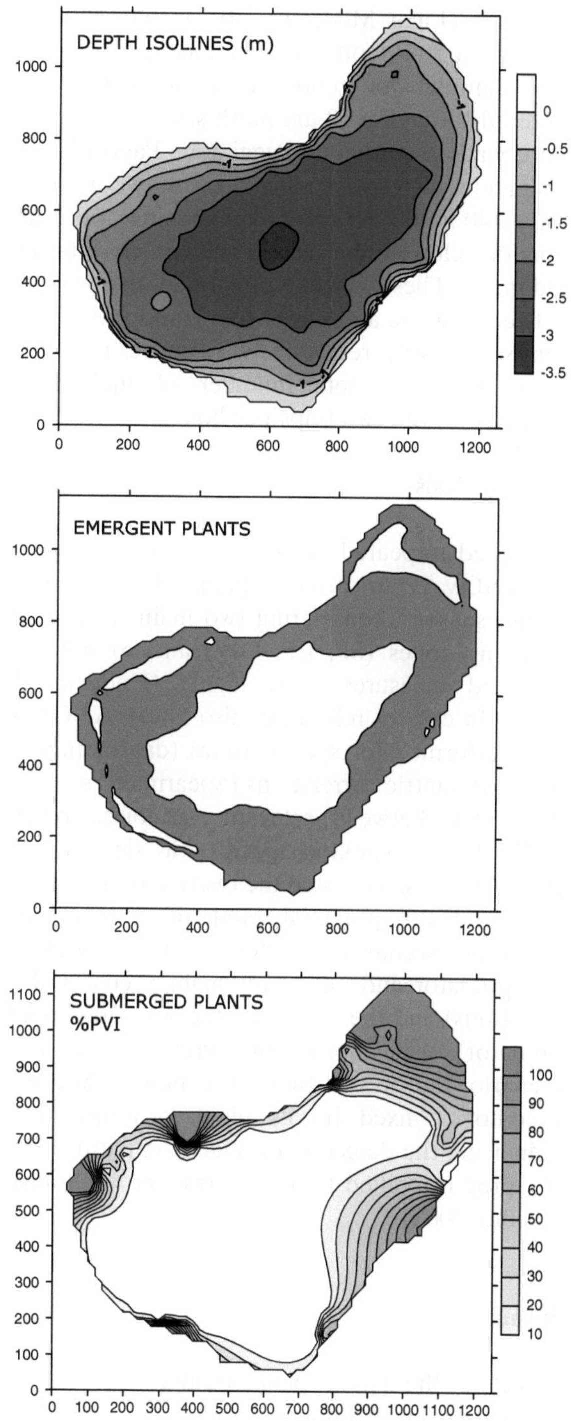

Fig. 1 Bathymetric map of Lake Blanca (upper panel), and spatial distribution of aquatic plants: area covered by emergent *Schoenoplectus californicus* (intermediate panel), and % PVI of submerged plants Egeria densa and *Ceratophyllum demersum* (lower panel). Distances are marked on the *x* and *y* axes (in metres). Data correspond to winter 2003

radiation (Licor Model Li-250). In each habitat we took a 20-l depth-integrated sample with the tube sampler for zooplankton and *Chaoborus* (filtered through a 50-μm mesh size net). Counts were made according to Paggi & de Paggi (1974). Fish and shrimps were sampled with two minnow traps (40 × 40 × 60 cm, 5-cm opening) per station, one close to the surface and one close to the sediment. These catches integrated the periods between sunrise and sunset (day), and sunset and sunrise (night), respectively (capture per unit effort (CPUE) = total number of individuals collected in the two traps per hour).

Data analysis

We used repeated measures 2-way ANOVA (rmANOVA) to detect significant differences within seasons, considering two main factors: a) sampling zones (SP, EP, OW) and b) time as repeated measures (day, night) (Underwood, 1997). In case of relevance, also 1-way ANOVA was performed for specific times (day or night). Non-parametric correlations (Spearman, rs) were performed between relevant variables. When DHM by zooplankton species was detected in the ANOVAs, we tested the overlap between the prey and the potential predator distributions under two scenarios: the "direct" (i.e. day prey-day predator and night prey-night predator distributions) and the "reverse" (i.e. day prey-night predator), using Spearman correlations. In this scenario, the prey is allowed to move while the predator is fixed (in its night location), thus enhancing the detection of migration patterns by the prey (modified from Bezerra-Neto & Pinto-Coelho, 2002).

Results

Effects of the aquatic plants on the physico-chemical variables

The mean area (annual averages) covered by the EP was 55% and 46% for SP, including some areas where both EP and SP were present (Fig. 1). We did not observe any clear seasonal pattern in coverage of the two plant types (authors, unpublished results). The SP sites had lower DO concentrations than OW and EP in spring (9.35, 10.41 and 10.12 mg l^{-1}, respectively), summer (5.28, 8.12 and 8.06 mg l^{-1}) and autumn (8.37, 11.82 and 10.51 mg l^{-1}). DO was always higher during day (Table 1). In SP, the sub-saturation dropped to 44% in summer during night. Intermediate pH values occurred in OW during day, while the same habitat had the highest pH during night (Table 1). The euphotic zone comprised the entire water column. Water transparency was most of the year higher within the submerged plant sites, especially in summer and autumn (Table 1). In summer, the water transparency was 46% higher in SP (35 cm Secchi minidisk) than in OW (21 cm Secchi minidisk), while in autumn the transparency was 34 and 14 cm (Secchi minidisk) in SP and OW respectively.

Structure of main communities

The zooplankton community was numerically dominated by small-sized specimens (mostly nauplii and rotifers) all year round, except in winter when medium-sized filtrators dominated (small cladocerans and calanoid copepods) (Fig. 2). The small *Bosmina longirostris* Müller and *Diaphanosoma birgei* Korinec were the most abundant cladocerans, but the calanoid copepod *Notodiaptomus incompositus* Brian was also common. The maximum total abundance occurred in autumn (2252 ind. l^{-1}) and the minimum in spring (856 ind l^{-1}) (whole system averages).

The phantom midge *Chaoborus* sp. occurred all year round, but was found only during night. Its abundance varied considerably through seasons, with minimum in winter (100 ind. m^{-3}) and maximum during spring and summer (*ca.* 1000 ind. m^{-3}) (whole system averages, Fig. 2). In terms of numbers, the nekton community was dominated by the fish *J. multidentata* in autumn and summer, when we found the highest CPUE (Fig. 2). The also small omnivorous fish *Cnesterodon decemmaculatus* Jenyns (Poecilidae) occurred in low abundances (<1.5% of total fish density and biomass). The shrimp *Palaemonetes argentinus* Nobili dominated in terms of abundance in winter and spring (Fig. 2).

Table 1 Summary of 2-way rmANOVA statistics

		Sites		Time	Site × Time
		Day	Night		
T	win	SP = OW = EP $2.34_{(2,12)}$ ns	SP = OW = EP $0.35_{(2,12)}$ ns	$30.2_{(1,12)}$ ***	$1.85_{(2,12)}$ ns
	spr	SP = OW = EP $0.90_{(2,12)}$ ns	SP = OW = EP $1.16_{(2,12)}$ ns	$1.16_{(2,12)}$ ns	$0.05_{(2,12)}$ ns
	sum	SP = EP > OW $9.72_{(2,12)}$ ***	SP = OW = EP $2.47_{(2,12)}$ ns	$2.47_{(2,12)}$ ns	$1.62_{(2,12)}$ ns
	aut	SP = OW > EP $3.08_{(2,12)}$ p = 0.08^{+}	SP = OW > EP $3.37_{(2,12)}$ p = 0.07^{+}	$11.7_{(1,12)}$ **	$3.46_{(2,12)}$ ns
DO	win	OW = SP = EP $0.60_{(2,12)}$ ns	OW = EP = SP $1.38_{(2,12)}$ ns	$34.7_{(1,12)}$ ***	$0.90_{(2,12)}$ ns
	spr	OW = SP = EP $0.15_{(2,12)}$ ns	OW = EP > SP $4.95_{(1,12)}$ **	$66.3_{(1,12)}$ ***	$1.82_{(2,12)}$ ns
	sum	OW = EP > SP $6.87_{(2,12)}$ **	OW = EP > SP $45.6_{(2,12)}$ ***	$41.8_{(1,12)}$ ***	$9.6_{(2,11)}$ **
	aut	OW = EP > SP $25.9_{(2,12)}$ ***	OW > EP = SP $5.14_{(2,12)}$ **	$101.9_{(1,12)}$ **	$3.9_{(2,11)}$ p = 0.05^{+}
PH	win	SP = OW = EP $0.2_{(2,12)}$ ns	EP = OW = SP $0.9_{(2,12)}$ ns	$93.6_{(1,12)}$ ***	$0.8_{(2,12)}$ ns
	spr	SP = OW = EP $1.0_{(2,12)}$ ns	OW = EP = SP $1.9_{(2,12)}$ ns	$5.3_{(1,11)}$ *	$9.2_{(2,11)}$ **
	sum	EP = OW = SP $0.5_{(2,12)}$ ns	OW = EP > SP $5.3_{(2,12)}$ *	$0.6_{(2,12)}$ ns	$2.6_{(2,12)}$ ns
	aut	EP = SP = OW $1.0_{(2,12)}$ ns	OW = EP > SP $9.8_{(2,12)}$ **	$33.7_{(1,12)}$ ***	$7.7_{(2,12)}$ **
Msd	win	EP = SP = OW $1.8_{(2,12)}$ ns			
	spr	OW = EP = SP $0.3_{(2,12)}$ ns			
	sum	SP > EP = OW $10.25_{(2,12)}$ **			
	aut	SP > EP = OW $3.1_{(2,12)}$ $P = 0.08^{+}$			

Effects of sites, time (day, night) and their interaction, on temperature (T), dissolved oxygen (DO), and pH, along seasons. The differences among sites for water transparency (mini Secchi Disk) correspond to midday (1-way ANOVA). Sites: OW = open water, SP = submerged plants, EP = emergent plants, ordered according to *post hoc* tests. The F values and the respective degrees of freedom are indicated in all cases.

Significance level: $P > 0.1$ ns, $P < 0.05$*, $P < 0.01$**, $P < 0.001$***. We indicate the P-value in the marginally significant cases ($0.05 < P < 0.1^{+}$)

The interseasonal variation in the total zooplankton abundance was negatively correlated with the CPUE of *P. argentinus* and *J. multidentata* (rs = –0.49, $P < 0.001$; rs = –0.29, $P < 0.05$, respectively), while there was no significant correlation with the abundance of the invertebrate predator *Chaoborus*.

Use of the space of zooplankton and potential predators

The diel use of the different habitats by most zooplankton species varied strongly among seasons. *Bosmina longirostris* appeared mainly in OW in winter (2-way ANOVA $F_{(2,12)} = 9.9$, $P < 0.05$), and in OW and EP in autumn (2-way ANOVA $F_{(2,12)} = 31.1$, $P < 0.001$), during both day and night. We found diel differences only in spring, when *Bosmina*, from a homogeneous distribution during day, became associated with SP and disappeared from OW at night (1-way ANOVA $F_{(2,12)} = 4.62$, $P < 0.05$) (Fig. 3A$_2$). *Diaphanosoma birgei* was collected mostly in OW in winter (2-way ANOVA $F_{(2,12)} = 136.2$, $P < 0.001$), while it was more equally distributed among habitats in spring and summer. In autumn, *Diaphanosoma* had the lowest density within the SP during day (1-way ANOVA $F_{(2,12)} = 5.21$, $P < 0.05$), but increased strongly its density in that habitat at night (from 2 to 67 ind. l^{-1}, Fig. 3B$_4$). *Chydorus sphaericus* Müller was associated with SP in winter (2-way ANOVA $F_{(2,12)} = 29.4$, $P < 0.001$) and spring ($F_{(2,12)} = 37.6$, $P < 0.001$). We found higher abundances of *Moina micrura* Kurz in OW in winter (2-way ANOVA $F_{(2,12)} = 4.9$, $P < 0.05$), and a homogenous spatial distribution during the rest of the year. Cyclopoids had higher abundances within the SP stands during all seasons (2-way ANOVA $F_{(2,117)} = 12.9$, $P < 0.001$), while the calanoid *N. incompositus* was homogeneously distributed among the three habitats. Nauplii showed different density patterns with seasons, being higher within vegetation (both plant types) in winter and summer (2-way ANOVA $F_{(2,12)} = 5.9$, $P < 0.01$; $F_{(2,12)} = 6.7$, $P < 0.01$, respectively).

Chaoborus was found only during night. It was more abundant in OW in winter and in spring (1-way ANOVA $F_{(2,12)} = 16.4$, $P < 0.01$; $F_{(2,12)} =$

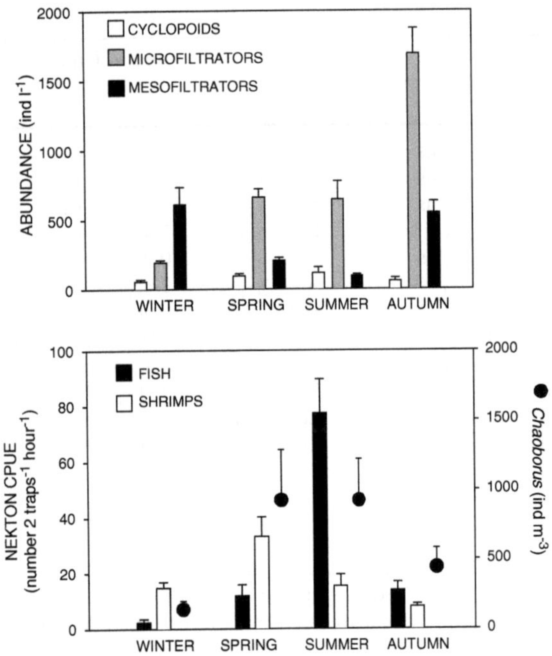

Fig. 2 Temporal variation of: Main zooplankton functional groups (according to diet and size): microfiltrators = rotifers + nauplii, mesofiltrators = calanoid copepods + cladocerans (*upper panel*); Potential vertebrates and invertebrate predators on zooplankton (nekton = fish *J. multidentata* + shrimps *P. argentinus*) (*lower panel*). Data represent whole system averages (error bars = 1 SE)

6.45, $P < 0.01$, respectively) (Fig. 3$C_{1,2}$). In summer and autumn it was mainly associated with EP and OW (1-way ANOVA $F_{(2,12)} = 4.3$, $P < 0.05$; $F_{(2,12)} = 3.5$, $P < 0.05$, respectively) (Fig. 3$C_{3,4}$). *Jenynsia multidentata* tended to occur mainly within the emergent vegetation. In winter, CPUE was higher within SP (though not significantly), while in spring it was collected primarily in EP sites (2-way ANOVA $F_{(2,12)} = 12.7$, $P < 0.001$), especially at night (Fig. 3D_2). In summer, the CPUE was higher at night (2-way ANOVA $F_{(1,12)} = 9.2$, $P < 0.05$), and the sequence was OW ≥ EP > SP during day and EP > SP > OW during night (significant interaction Sites × Time $F_{(2,12)} = 5.2$, $P < 0.05$) (Fig. 3D_3). The spatial distribution in autumn did not show a significant pattern (Fig. 3D_4).

The third potential predator, the shrimp *P. argentinus*, was collected in winter mostly from OW and during night (Fig. 3E_1), though not significantly. In spring and summer, the highest CPUE occurred during night (2-way ANOVA $F_{(1,12)} = 14.93$, $P < 0.01$ and $F_{(1,12)} = 9.6$, $P < 0.01$ respectively). For the summer, the sequence of the densities were SP ≥ OW ≥ EP during day and OW > SP > EP at night (significant interaction Sites × Time $F_{(2,12)} = 4.3$, $P < 0.05$). In autumn, no statistical differences were found in the habitat preference of shrimps. Only in summer there was an inverse spatial correlation between the CPUE of adult *J. multidentata* and *P. argentinus* (rs = –0.56, $P < 0.01$, during night) (Fig. 3D_3, E_3).

DHM analysis

We focus here on *B. longirostris* and *D. birgei*, as they were the only species showing significant diel variations between habitats in their densities (in spring and autumn, respectively). In spring the density and distribution of *B. longirostris* during day were positively, though not significantly, correlated with the distribution of *Chaoborus* during night. Contrary, the "direct" (day/day, night/night) correlation of both species during night was negative and significant (rs = –0.59, $P < 0.05$) (Fig. 3B_2, C_2). In the case of *J. multidentata*, both species were positively but not significantly correlated in the "direct" comparison, while their "reverse" correlation (day prey/night predator) was negative and also not significant (Fig. 3B_2, D_2). Besides, the distribution of *B. longirostris* was negatively correlated with DO concentrations in night (rs = –0.72, $P < 0.01$). In autumn, the day distribution of *D. birgei* was significant and positively correlated with the distribution of *Chaoborus* during night (rs = 0.81, $P < 0.001$), while the "direct" correlation of both species in the night was negative but not significant (rs = –0.33, n.s.). The "direct" correlations between *Diaphanosoma* and *J. multidentata* were positive and significant at night (rs = 0.79, $P < 0.001$), while the "reverse" correlation was negative but not significant. We did not detect any statistically significant relationship between the distribution of *D. birgei* and the main physico-chemical variables (DO, pH, T). In these intra-seasonal analyses, no significant correlation was found

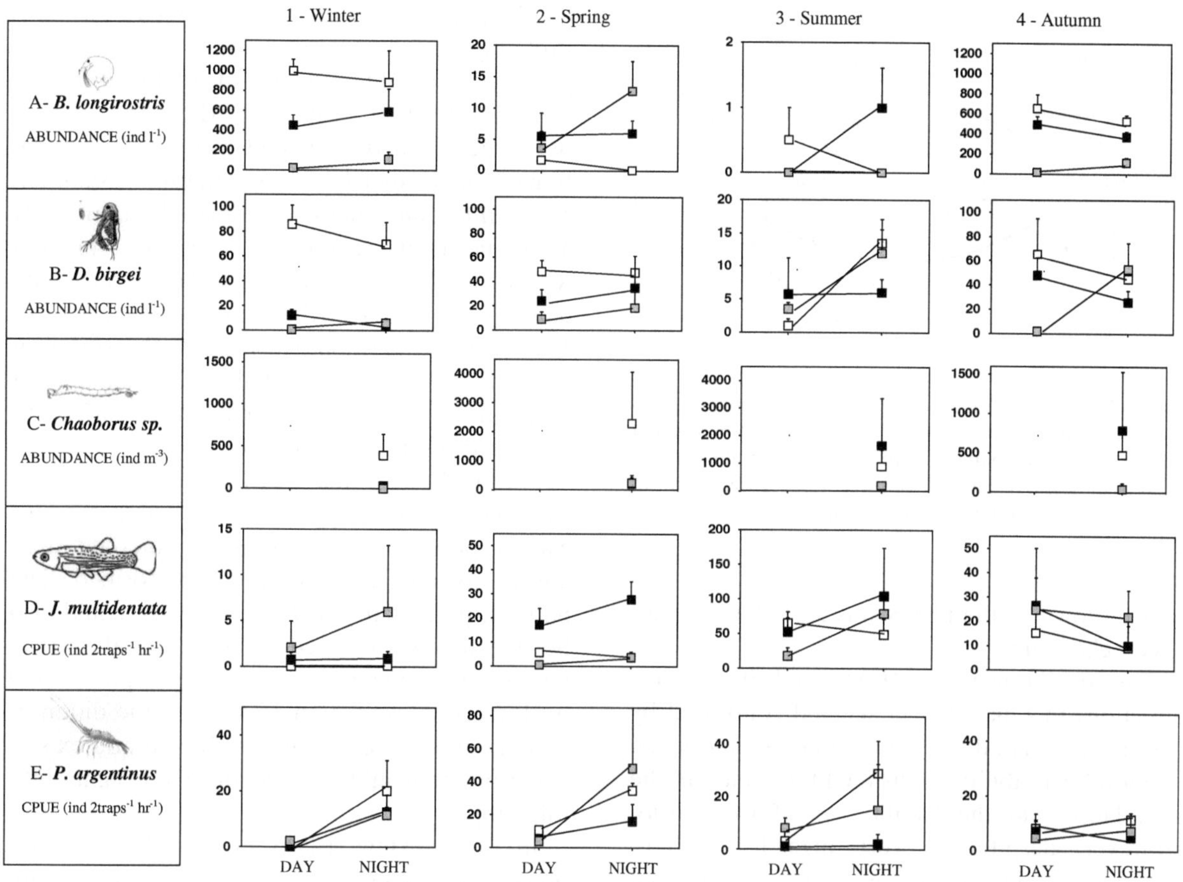

Fig. 3 Diel habitat choice in all seasons of *B. longirostris* (**A**), *D. birgei* (**B**), and main potential predators: *Chaoborus* (**C**), *J. multidentata* (**D**) and *P. argentinus* (**E**). From left to right, panels correspond to winter (1), spring (2), summer (3), and autumn (4). Symbols: □ = open water, ■ = emergent plants, ▪ = submerged plants. The cases of DHM are bold-framed. Note the different scales in the figures (error bars = 1 SE)

between the distribution of these cladocerans and the abundance (CPUE) of shrimps.

Discussion

Predators are thought to be the main factor controlling the spatial distribution, abundance and body size of zooplankton in shallow lakes (e.g. Scheffer, 1998; Burks et al., 2002). In Lake Blanca, the small size of the dominant cladocerans and the dominance by copepods and rotifers likely reflect the extremely high abundance of the potential predators, particularly the omnivorous-planktivorous fish *Jenynsia multidentata*. The most important finding of our study was the lack of fixed patterns in the horizontal distribution of the main zooplankton groups among seasons, particularly of cladocerans. Cyclopoid copepods were associated mainly with SP (day and night) as seen in other studies (Jeppesen et al., 1997b), while calanoids showed a homogeneous distribution.

However, in contrast to most studies from eutrophic shallow lakes in the temperate zone (Burks et al., 2002), the most abundant cladocerans *B. longirostris* and *D. birgei* preferred the vegetated areas (especially the SP) during night and not as expected during day (in seasons when distribution differed significantly). The significantly higher water transparency in sites with plants indicates that zooplankton would probably find a lower quantity of food resources in these habitats than in open water, which suggests that

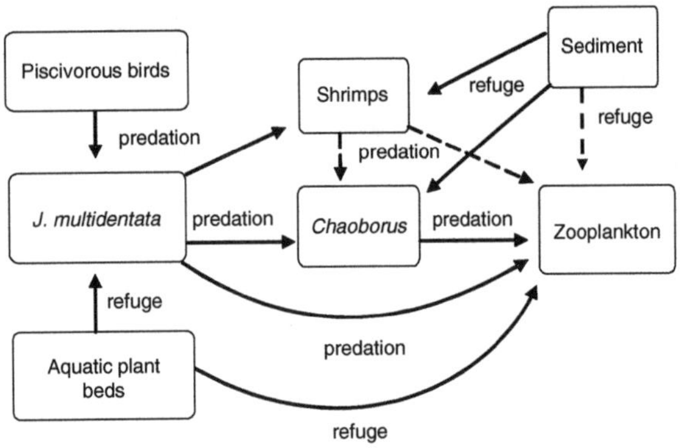

Fig. 4 Theoretical model of the main interactions (predation and refuge) among biotic and abiotic components that can influence the spatial distribution of zooplankton and nekton in Lake Blanca. Complete lines correspond to findings of this study while dotted lines refer to literature data

moving towards the plants was a predator-avoidance behaviour.

The occurrence of DHM (and its reverse direction) in Lake Blanca seemed controlled by two factors related to the potential predators: the temporal variability in fish CPUE, and in the abundance and the habitat use of *Chaoborus*. However, despite the very high fish CPUE, *Chaoborus*, occurring in high densities compared with those of temperate lakes (Liljendahl-Nurminen et al., 2002), seemed to play a more relevant role than fish. During winter, with the lowest densities of fish and *Chaoborus*, *B. longirostris* and *D. birgei* had the highest abundances and we did not find evidence of DHM as these cladocerans were mostly found in open waters. The reverse DHM occurred during the seasons with intermediate CPUE of fish (spring and autumn). From the "direct" and "reverse" correlation analyses, we can argue that if *Bosmina* did not migrate, from a relatively homogeneous diurnal distribution towards SP at night (in spring), it would encounter with *Chaoborus* in open waters, while with fish in the EP.

In the case of *D. birgei* the DHM pattern seemed more clear than in the case of *B. longirostris*, as day and night densities were balanced in the former. By migrating to SP at night (in autumn), *D. birgei* decreased the encounter rate with *Chaoborus* (which by then mostly appeared among EP), even though the overlapping with *J. multidentata* would increase. Probably, the costs of migrating to the submerged plant beds at night are high, but still lower than of migrating to the pelagic zone where the absence of physical structures would increase the predation efficiency by *Chaoborus* and fish. In summer, *J. multidentata*'s density increased enormously after the main reproductive season. Together with a remarkable decrease in cladoceran abundance, the spatial distribution of *B. longirostris* or *D. birgei* became homogeneous, with no indications of DHM. Clearly, no habitat seemed to offer a significant refuge for zooplankters under this enhanced-predation scenario.

Although we think DHM is the most likely explanation for the observed distributions, we cannot prove it as the sampling method used allows no mass balances to be performed. The pattern, particularly of *B. longirostris*, could also be due to, or enhanced by, differential vertical migration (stronger in the submerged plant sites) or differential predation pressure in the different habitats or their combination. In situ and ex situ experiments would provide further insights into this subject.

In their review, Burks et al. (2002) did not report examples of DHM in the scenario of both pelagic fish and active-at-night invertebrate predators (e.g. *Chaoborus*). Jeppesen et al. (1997a) suggested that DHM induced by invertebrates would probably be more important in lakes with low fish densities, and that the direction of this migration would depend on the habitat choice of the predacious invertebrates (whether pelagic or plant-associated). Likewise Lauridsen et al. (1997) suggested that the reverse DHM by *Daphnia* in the fishless Central Long Lake (USA) could be induced by pelagic nocturnal invertebrates, as *Chaoborus*. R. Burks et al. (unpublished), on the contrary, found direct DHM triggered by *Chaoborus* in a Danish lake with very low abundance of fish, as earlier observed by Kvam & Kleiven (1995) in a Norwegian lake. The advantage of DHM as an anti-predation mechanism depends on both the predators and the refuge capacity of the aquatic plants. Our results can be interpreted as a variation of the additive interaction between vertebrate and invertebrate predators, proposed by Burks et al. (2002). We did not find evidence that the occurrence of invertebrates can lead to higher densities of large-bodied zooplankton, by serving as alternative food for fish, as found by Sagrario & Balseiro (2003) in mesocosm experiments with some of the same predatory species that appeared in Lake Blanca. The physico-chemical differences among sites played a minor role in shaping these spatial patterns. Although we found a negative relationship between the spatial distribution of *B. longirostris* and the oxygen concentration in spring, the differences of DO among sites were low.

Also the predators showed temporal variability in habitat choice, on a diel and seasonal scale. Several authors reported that *Chaoborus* displays vertical migration to avoid fish predation, spending the day in the sediments and migrating upwards at night (Davidowicz et al., 1990; Voss & Mumm, 1999). The same behaviour was evidenced in Lake Blanca as *Chaoborus* was only caught during night, a likely less risky time to occupy the water column when visual predatory fish are present.

The habitat selection by *J. multidentata*, as top predator, probably depended on the search for food and shelter from the piscivorous waterfowl. The emergent plant zone is likely a safer place than the SP, as bird attacks can be avoided (e.g. from yellow-billed tern *Sterna superciliaris* Vieillot and olivaceous cormorant *Phalacrocorax olivaceus* Browning present in the lake). The preference of *J. multidentata* for the EP also by night, when predation risk by birds is absent, suggests that fish species also find a comparatively good availability of resources in this habitat. The differences in physical and chemical factors among sites seemed to play a minor role in the horizontal distribution of fish, especially when considering that both fish species, *J. multidentata* and *C. decemmaculatus*, have a broad tolerance to different and sometimes extreme environmental conditions (Ortubay et al., 2002). The habitat selection of fish and shrimps seemed to be mutually dependent, as we observed a clear spatial segregation between fish and shrimps. Contrary to earlier findings (Collins, 1999), there is no strong evidence that the shrimps were important predators on zooplankton in this lake (G. Goyenola et al., unpublished).

Final remarks

Studies on the behavioural response of preys to combinations of predators are still scarce (Relyea 2003). Based on experimental studies, Van de Meutter et al. (2005) proposed that the effects of particular predators on the spatial distribution of a prey (*Daphnia* in that case) can vary strongly in strength and direction, depending on the presence and activity of more than one predator. These authors also suggested that interactions and facilitation among predators can be relevant, and can change with time and environmental characteristics, as we found in Lake Blanca. According to our field evidence, the horizontal distribution of zooplankton was apparently conditioned by complex interactions with the potential predators and the architecture and spatial distribution of the aquatic plants, and indirectly probably also by the activity of piscivorous birds. We propose a conceptual model of the main interactions (predation versus refuge) in this lake, to be tested in experiments or under other field conditions (Fig. 4). We suggest that aquatic plants in subtropical shallow lakes offer refuge for zooplankton only at intermediate fish and invertebrate predator densities, and that refuge-mediated positive effects on water transparency therefore might be weak or rare.

Acknowledgements This study was partly funded by PEDECIBA (Uruguay), and supported by Aguas de la Costa S.A. EJ was supported by Eurolimpacs (GOCE-CT-2003-505540), the Danish Natural Science Research Council (CONWOY project) and CLEAR, and MM by the Danish Research Agency. We are very grateful to C. Boné for the construction of sampling equipment, and to G. Beyhaut, C. Bruzzone, C. Crisci, T. dos Santos, C. Fosalba, V. Pérez, M. Puppo, and the technical staff of Aguas de la Costa for field assistance, and to the family Vilar del Valle for lending their summer house during the sampling periods. We also wish to thank the generous remarks by three reviewers that helped us improve this manuscript.

References

Arcifa, M. S, E. A. Gomes & A. J. Meschiatti, 1992. Composition and fluctuations of the zooplankton of a tropical Brazilian reservoir. Archiv für Hydrobiologie 123: 479–495.

Bezerra-Neto, J. F. & R. M. Pinto-Coelho, 2002. A influencia da larva de *Chaoborus brasiliensis* (Theobald, 1901) (Diptera, Chaoboridae) na distribuçao vertical da comunidade zooplanctonica da lagoa do Nado, Belo Horizonte, Estado de Minas Gerais. Maringá 24: 337–344.

Burks, R. L., E. Jeppesen & D. M. Lodge, 2001. Pelagic prey and benthic predators: impact of odonate predation on *Daphnia* among complex structure. Journal of the North American Benthological Society 20: 683–696.

Burks, R. L., D. M. Lodge, E. Jeppesen & T. L. Lauridsen, 2002. Diel horizontal migration of zooplankton: costs and benefits of inhabiting the littoral. Freshwater Biology 47: 343–365.

Canfield, D. E. Jr., J. V. Shireman, D. E. Colle, W. T. Haller, C. E. Watkins & M. J. Maceina, 1984. Predictions of chlorophyll a concentrations in Florida lakes: importance of aquatic macrophytes. Canadian Journal of Fisheries and Aquatic Science 41: 497–501.

Carpenter, S. R. & D. M. Lodge, 1986. Effects of submerged macrophytes on ecosystem processes. Aquatic Botany 26: 341–370.

Collins, P., 1999. Feeding of *Palaemonetes argentinus* (Decapoda: Palaemonidae) from an oxbow lake of the Paraná River, Argentina. Journal of Crustacean Biology 19: 485–492.

Collins, P. & J. C. Paggi, 1998. Feeding ecology of *Macrobrachium borelli* (Nobili) (Decapoda: Palaemonidae) in the food valley of the River Paraná, Argentina. Hydrobiologia 363: 21–20.

Davidowicz, P., J. Paijanowska & K. Ciechomski, 1990. Vertical migration of *Chaoborus* larvae is induced by the presence of fish. Limnology and Oceanography 35: 1631–1637.

Escalante, A., 1983. Contribución al conocimiento de las relaciones tróficas de peces de agua dulce del área platense. III. Otras especies. Limnobios 2: 453–463.

Jeppesen, E., T. L. Lauridsen, T. Kairesalo & M. Perrow, 1997b. Impact of Submerged Macrophytes on Fish-zooplankton Interactions in Lakes. In Jeppesen, E., Ma. Søndergaard, Mo. Søndergaard & K. Christoffersen (eds), The Structuring Role of Submerged Macrophytes in Lakes. Ecological Studies, Springer Verlag, New York, 94–113.

Jeppesen, E., Ma. Søndergaard, Mo. Søndergaard & K. Christoffersen, 1997a. The Structuring Role of Submerged Macrophytes in Lakes. Ecological Studies Vol. 131. Springer-Verlag, New York.

Jeppesen, E., M. Meerhoff, B. A. Jakobsen, R. S. Hansen, M. Søndergaard, J. P. Jensen, T. L. Lauridsen, N. Mazzeo & C. Branco, 2007. Restoration of shallow lakes by nutrient control and biomanipulation—the successful strategy depends on lake size and climate. Hydrobiologia 581: 269–285.

Kvam, O. V. & O. T. Kleiven, 1995. Diel horizontal migration and swarm formation in *Daphnia* in response to *Chaoborus*. Hydrobiologia 307: 177–184.

Lauridsen, T. L., E. Jeppesen, M. Mitchell, S. F. Lodge, D. M. & R. L. Burks, 1999. Diel variation in horizontal distribution of *Daphnia* and *Ceriodaphnia* in oligotrophic and mesotrophic lakes with contrasting fish densities. Hydrobiologia 409: 241–250.

Lauridsen, T. L., E. Jeppesen, M. Søndergaard & D. M. Lodge, 1997. Horizontal Migration of Zooplankton: Predator-mediated Use of Macrophyte Habitat. In Jeppesen, E., M. Søndergaard, M. Søndergaard & K. Christoffersen (eds), The Structuring Role of Submerged Macrophytes in Lakes. Ecological Studies, Springer Verlag, New York, 233–239.

Lewis, W. M. Jr., 1996. Tropical Lakes: How Latitude Makes a Different. In Schiemer, F. & K. T. Boland (eds), Perspectives in Tropical Limnology. SPB Academic Publishing bv, Amsterdam, The Netherlands, 43–64.

Liljendahl-Nurminen, A., J. Horppila, P. Eloranta, T. Marinen & L. Uusitalo, 2002. The seasonal dynamics and distribution of *Chaoborus flavicans* larvae in adjacent lake basins of different morphometry and degree of eutrophication. Freshwater Biology 47: 1283–1295.

Mazzeo, N., L. Rodríguez-Gallego, C. Kruk, M. Meerhoff, J. Gorga, G. Lacerot., F. Quintans, M. Loureiro, D. Larrea & F. García-Rodríguez, 2003. Effects of *Egeria densa* Planch beds in a shallow lake without piscivorous fish. Hydrobiologia 506–509: 591–602.

Meerhoff, M., N. Mazzeo, B. Moss & L. Rodríguez-Gallego, 2003. The structuring role of free-floating versus submerged plants in a subtropical shallow lake. Aquatic Ecology 37: 377–391.

Meschiatti, A. J., M. S. Arcifa & N. Fenerich-Verani, 2000. Fish communities associated with macrophytes in Brazilian floodplain lakes. Environmental Biology of Fishes. 58: 133–143.

Mumm, H., 1997. Effects of competitors and *Chaoborus* predation on the cladocerans of a eutrophic lake: An enclosure study. Hydrobiologia 360: 253–264.

Ortubay, S., M. Lozada & V. Chusca, 2002. Aggressive behaviour between *Gymnocharacinus bergi* (Pisces, Characidae) and other neotropical fishes from a thermal stream in Patagonia. Environmental Biology of Fishes 63: 341–346.

Paggi, J. & S. De Paggi, 1974. Primeros estudios sobre el zooplancton de las aguas lóticas del Paraná medio. Physis 33: 94–114.

Relyea, R. A., 2003. How prey respond to combined predators: a review and an empirical test. Ecology 84: 1827–1839.

Sagrario, M. D. G. & E. Balseiro, 2003. Indirect enhancement of large zooplankton by comsumption of predacious macroinvertebrates by littoral fish. Archiv für Hydrobiologie 158: 551–574.

Scasso, F., N. Mazzeo, J. Gorga, C. Kruk, G. Lacerot, J. Clemente, D. Fabian & S. Bonilla, 2001. Limnological changes in a sub-tropical shallow hypertrophic lake during its restoration: two years of a whole-lake experiment. Aquatic Conservation: Marine and Freshwater Ecosystems 11: 31–44.

Scheffer, M., 1998. Ecology of Shallow Lakes. Population and Community Biology. Chapman & Hall, London.

Søndergaard, M., E. Jeppesen & H. F Aaser, 1997, *Neomysis integer* in a shallow hypertrophic brackish lake: distribution and predation by three-spined stickleback (*Gasterosteus aculeatus*). Hydrobiologia 428: 151–159.

Talling, J. F. & J. Lemoalle, 1998. Ecological Dynamics of Tropical Inland Waters. Cambridge University Press, Cambridge.

Timms, R. M. & B. Moss, 1984. Prevention of growth of potentially dense phytoplankton populations by zooplankton grazing, in the presence of zooplanktivorous fish, in a shallow wetland ecosystem. Limnology and Oceanography 29: 472–486.

Underwood, A. J., 1997. Experiments in Ecology. Cambridge University Press, Cambridge.

Valderrama, J., 1981. The simultaneous analysis of total N and P in natural waters. Marine Chemistry 10: 1009–1022.

Van de Meutter, F., R. Stoks & L. De Meester, 2005. Spatial avoidance of littoral and pelagic invertebrate predators by *Daphnia*. Oecologia 142: 489–499.

Voss, S. & H. Mumm, 1999. Where to stay by day and night: size-specific and seasonal differences in horizontal and vertical distribution of *Chaoborus flavicans* larvae. Freshwater Biology, 42: 201–213.

Wicklum, D., 1999. Variation in horizontal zooplankton abundance in mountain lakes: shore avoidance or fish predation? Journal of Plankton Research 21: 1957–1975.

Wotjal, A., P. Frankiewicz, K. Izydorczyk & M. Zalewski, 2003. Horizontal migration of zooplankton in a littoral zone of the lowland Sulejow Reservoir (Central Poland). Hydrobiologia 506–509: 339–346.

Hydrobiologia (2007) 584:191–204
DOI 10.1007/s10750-007-0580-2

SHALLOW LAKES

Functional response of *Anodonta anatina* feeding on a green alga and four strains of cyanobacteria, differing in shape, size and toxicity

Babette M. Bontes · Antonie M. Verschoor · L. Miguel Dionisio Pires · Ellen van Donk · Bas W. Ibelings

Abstract We studied the functional response of the freshwater unionid bivalve *Anodonta anatina*, feeding on five phytoplankton strains differing in food quality: the small green alga *Scenedesmus obliquus*, a toxic and a non-toxic strain of the filamentous cyanobacterium *Planktothrix agardhii* and a toxic and a non-toxic strain of the coccoid cyanobacterium *Microcystis aeruginosa*. On *S. obliquus*, *A. anatina* had a type II functional response with a maximum mass-specific ingestion rate (IR_{max}) of 5.24 mg C g DW^{-1} h^{-1} and a maximum mass-specific clearance rate (CR_{max}) of 492 (±38) ml g DW^{-1} h^{-1}, the highest values for all the phytoplankton strains that were investigated. On toxic and non-toxic *P. agardhii* filaments, *A. anatina* also had a type II functional response, but IR_{max} and CR_{max} were considerably lower (IR_{max} 1.90 and 1.56 mg C g DW^{-1} h^{-1}; CR_{max} 387 (±97) and 429 (±71) ml g DW^{-1} h^{-1}, respectively) than on *S. obliquus*. Toxicity of *P. agardhii* had no effect on the filtration rate of the mussels. On the non-toxic *M. aeruginosa* (small coccoid cells), we also observed a type II functional response, although a type I functional response fitted almost as good to these data. For the colonial and toxic *M. aeruginosa*, a type I functional response fitted best to the data: IR increased linearly with food concentration and CR remained constant. CR_{max} and IR_{max} values for the (colonial) toxic *M. aeruginosa* (383 (±40) ml g DW^{-1} h^{-1}; 3.7 mg C g DW^{-1} h^{-1}) demonstrated that *A. anatina* filtered and ingested this cyanobacterium as good as the other cyanobacterial strains. However, on the non-toxic *M. aeruginosa* we observed the lowest CR_{max} of all phytoplankters (246 (±23) ml g DW^{-1} h^{-1}, whereas IR_{max} was similar to that on toxic *M. aeruginosa*. The high maximum ingestion rates on *S. obliquus* and *M. aeruginosa* indicate a short handling time of these phytoplankton species. The high clearance rates on *S. obliquus*, toxic *M. aeruginosa* and *P. agardhii* reflect a high effort of the mussels to filter these particles out of the water

Guest editors: R. D. Gulati, E. Lammens, N. De Pauw & E. Van Donk
Shallow lakes in a changing world

B. M. Bontes · A. M. Verschoor · L. M. Dionisio Pires · E. van Donk · B. W. Ibelings
Department of Food Web Studies, Centre for Limnology, Netherlands Institute of Ecology (NIOO-KNAW), P.O. Box 1299, 3600 BG Maarssen, The Netherlands

Present Address:
B. M. Bontes (✉)
Royal Netherlands Institute of Sea Research (NIOZ), P.O. Box 59, 1790 AB Den Burg, The Netherlands
e-mail: bontes@nioz.nl

L. M. Dionisio Pires
Great Lakes Environmental Research Laboratory, NOAA, 2205 Commonwealth Blvd, Ann Arbor, MI 48105-2945, USA

column at low concentrations. The low clearance rates on non-toxic *M. aeruginosa* may be explained by the small size and coccoid form of this cyanobacterium, which may have impaired *A. anatina* to efficiently capture the cells. Although *A. anatina* had relatively high maximum clearance rates on non-toxic and toxic *P. agardhii*, this cyanobacterium does not seem to be a good food source, because of the observed high rates of pseudofaeces production and hence low ingestion rates.

Keywords Bivalves · Cyanobacteria · Functional response · Grazing · Pseudofaeces · *Unionidae*

Introduction

Shallow lakes in The Netherlands are inhabited by large freshwater bivalves of the family of Unionidae, such as *Anodonta anatina* L. (pond mussel). We were interested to see if *A. anatina* may contribute—like the zebra mussel *Dreissena polymorpha* Pallas (Reeders & Bij de Vaate, 1990)—to the improvement of water quality by reducing algal biomass and increasing water transparency. Enhanced grazing on phytoplankton may help shift shallow lake ecosystems with high and intermediate nutrient loading from a turbid, phytoplankton dominated, to a clear, water state dominated by macrophytes (Scheffer, 1998). One reason to predict a substantial impact of Unionid bivalve grazing on ecosystem processes is that, in marine and estuarine systems, similarly sized bivalves have been shown to have large ecosystem impacts on the benthic-pelagic coupling of energy cycles (Prins et al., 1998). A second reason to focus on Unionid species for the control of the effects of eutrophication is the potential risk of zebra mussel spreading in ecosystems. Although zebra mussels—with their high filtration rates on phytoplankton (Fahnenstiel et al., 1995)—have been promoted as a tool in biomanipulation of lakes in the Netherlands (Reeders & Bij de Vaate, 1990), they have also been shown to promote blooms of (toxic) cyanobacteria in North America (Vanderploeg et al., 2001). Moreover, we believe that for Dutch shallow lakes, with soft substrates, Unionids are better adapted to these habitats than zebra mussels that require hard substrate for settlement.

We studied the feeding of *A. anatina* on cultures of the green alga *S. obliquus* Turp. Kütz., toxic and non-toxic strains of the filamentous cyanobacterium *P. agardhii* Gomont and colonial *M. aeruginosa* Kütz., each at five different concentrations to determine the functional response. The functional response is the relation between feeding rate and food density and as such may provide indications of underlying mechanisms, such as an increase in handling time related to the size of food particles. When knowing the functional response of *A. anatina* for a certain type of food, we know more about which mechanism may be limiting to the food uptake and hence we can predict how the grazer may respond to this food type in the field. In bivalves particles are first filtered from the water column. In the mantle cavity (where gills and palps are situated), ingestible particles are directed straight to the mouth, whereas unwanted particles are sorted on the gills and palps, embedded in mucus and then expelled as pseudofaeces (PF) (Dionisio Pires, 2005).

Because the functional response of *Anodonta* has never been analysed before, and different types of functional responses have been described in bivalve literature, we compared three basic types of functional responses (Holling, 1959; Hassell, 1978). The simplest type, type I (rectilinear) functional response, shows a linear increase of the ingestion rate (IR) with the food concentration, up to a concentration where the maximum amount of food that can be ingested by an organism is reached. This concentration is called the incipient limiting level (ILL: Rigler, 1961). Above ILL, the IR remains constant and maximal irrespective of food concentration. This implies that above ILL, further increases in cell concentration result in higher PF production rates (Sprung & Rose, 1988). The clearance rate (CR), a measure for the efficiency with which food is filtered from the water column, decreases accordingly (Fig. 1). Often, such a discontinuous transition at ILL is not observed, unless the food is filtered from the environment with negligible handling of the food. For many organisms, handling time of food is a limiting factor for the

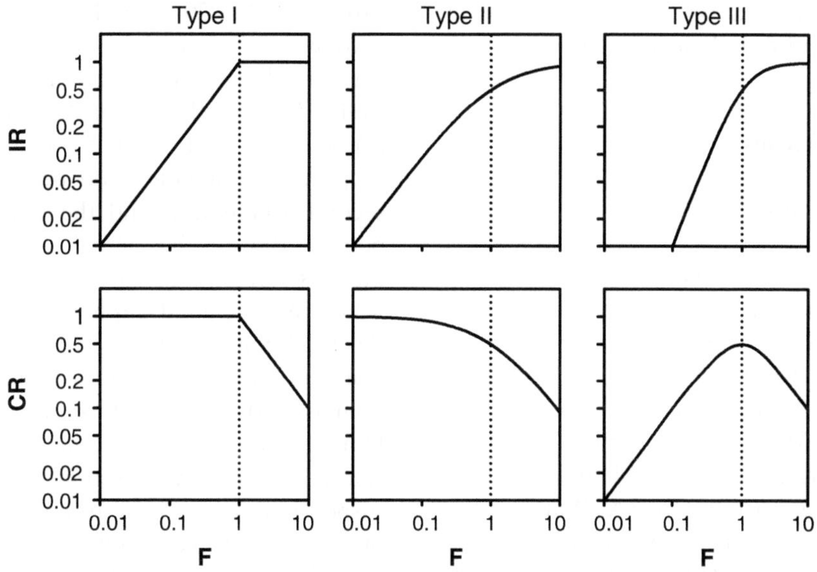

Fig. 1 Graphic representation of the three basic types of functional responses for ingestion rates (IR) and clearance rates (CR), respectively, as a function of food concentration (F) (*solid lines*). The dotted vertical lines indicate for Type I functional response the incipient limiting level (ILL), whereas they indicate the half-saturation concentration (*K*) for Type II and Type III functional responses. For comparability, all functional responses are scaled to a maximum IR or CR of 1, and to and ILL or *K* of 1, on axes with arbitrary units. Therefore, for type I, IR is maximum at ILL, and (for types II and III) IR is at half its maximum value (0.5) at *K*. Note that—for comparability with experimental data—graphs are plotted on logarithmic scales, which makes it difficult to immediately recognize the sigmoid shape of the Type III functional response

amount of food that can be ingested. In a type II (hyperbolic, curvilinear) functional response, this is visible in CR, which decreases progressively towards zero with increasing food concentrations. Consequently, IR decelerates at higher food concentrations and goes asymptotically towards the maximum ingestion rate. With a type III functional response, IR accelerates more than linearly at low food densities, which for example may be the case with reward-dependent feeding behaviour. The initial CR will be 0, and maximal at intermediate food densities (Fig. 1). Because IR will still saturate (and thus decelerate) at higher food concentrations, as in a type II functional response, this will result in a sigmoidal (S-shaped) functional response (Fig. 1). With a type III functional response, CR will be maximal at the half-saturation food concentration *K* (Fig. 1).

To investigate the functional response of *A. anatina*, we chose the different food items based on differences in food quality (green algae vs. cyanobacteria), shape (filamentous, colonial or single cells), and toxicity (toxic or non-toxic strains). Our null hypothesis (1) is that *A. anatina* has a negligible handling time of the food, meaning that it has a type I functional response. Therefore, (2) production of PF starts only above ILL and increases linearly with food concentration. We expect that *A. anatina* is able to discriminate between green algae and (often less nutritious) cyanobacteria so that: (3) CR on *S. obliquus* is higher than on all the cyanobacteria. Furthermore because retention efficiency is species specific and depends on particle size (Ward & Shumway, 2004); we expect that (4) small species like *M. aeruginosa* are more easily retained than large filaments of *P. agardhii*, which may congest the siphon. Finally (5) with respect to toxicity (i.e. microcystin-LR), CR and IR on microcystin producing cyanobacterial strains will be lower than on non-toxic strains. Our study adds new data on the functional response of the rarely studied freshwater bivalve *A. anatina* feeding on different food sources. These data are relevant for studies of behaviour and populations, as well as for the possible application of these bivalves in biomanipulation of turbid shallow lakes.

Materials and methods

Grazing experiments

Anodonta anatina mussels were collected in spring 2005 in the Babbelaar (Lake Lauwersmeer, The Netherlands). Mussels were kept in aquaria filled with filtered lake water (0.45 μm, Lake Maarsseveen) on a layer (10 cm) of river sand (Aqua-colisa, Gemert-Bakel). Mussels were kept at 17–20°C, under a light: dark regime of 16:8 h and fed every day with *S. obliquus* CCAP276/3A. Water in the aquaria was replaced once a week. Five phytoplankton strains were used in the experiments: four cyanobacteria (*P. agardhii* strains CYA116 and CYA126 and *M. aeruginosa* strains V131 and V40) from exponentially growing batch cultures, and one green alga (*S. obliquus* CCAP276/3a), which was cultured in a chemostat. For size characteristics, origin and toxicity of the strains we refer to Table 1. Before the experiments, the *M. aeruginosa* strains were filtered over a 70 μm filter, to remove extremely large colonies.

Grazing experiments were performed in the laboratory, with *A. anatina* feeding on the five phytoplankton strains as single food sources. For each strain, the mussels were exposed to a range of five food concentrations (0.5; 1; 2; 5; 10 mg C L^{-1}), with 5 replicates per strain per concentration. Food concentrations were based on regressions between ash-free dry weight of the strains (4 h at 550°C) and optical density at 750 nm (Helios-δ, Unicam, UK). Per grazing vessel, one mussel, with an average shell length between 8 and10 cm and a shell width between 5 and 6 cm, was cleaned with a brush under running tap water to remove phytoplankton adhered to the shell and incubated in 2 l of the appropriate food concentration (suspended in 0.45 μm filtered lake water). The grazing vessels were 2 l batch systems and contained no sediment, as a pilot study showed that sediment absence did not affect the grazing behaviour of *A. anatina*.

Acclimation with the appropriate food concentrations started 24 h prior to the actual grazing experiment (19–20°C; 10–15 μmol photons m^{-2} s^{-1}). Two hours before the grazing experiment food was added again to ensure that the mussels were well fed before the actual start of the experiment (Dionisio Pires et al., 2005). The 2 l were enough to prevent complete food depletion during the second acclimation and the grazing experiment period. Before the actual grazing period, the mussels were rinsed carefully under running tap water. We included duplicate vessels for each of the food concentrations without mussels as controls to correct for changes in the phytoplankton biomass other than those related to grazing. During the experiment the vessels were constantly mixed by aeration to keep food in suspension, without disturbing the mussels. Samples (15 ml) were taken from all vessels (both mussels and controls) at the start ($T_{0\ min}$), before the introduction of the mussels, and at the end ($T_{60\ min}$) of the experiment. After the grazing period the mussels were rinsed under running tap water and transferred to a vessel with 0.45 μm filtered lake water and samples were taken from these vessels after 30 min ($T_{90\ min}$) to assess PF production. Samples were stored in the dark, until being measured at the end of the experiment. Measurements (chlorophyll concentrations (μg l^{-1})) were done using a PhytoPAM (Walz, Germany). Carbon concentrations of the phytoplankton were measured on a UniQuant Universal Carbon and Nitrogen Analyzer (Calanus, Finland). After the grazing period, the mussels were stored at –20°C. At a later stage, the shells were removed, the soft tissue was freeze-dried, and weighed.

Determination of clearance rates, PF production rates and ingestion rates

Mass-specific clearance rates (CR, ml mg DW^{-1} h^{-1}) were determined from chlorophyll fluorescence measurements and calculated as follows (Coughlan, 1969):

$$CR = \frac{V}{nt} * \left\{ \ln\frac{A_0}{A_t} - \ln\frac{A'_0}{A'_t} \right\},$$

in which V is the volume in the grazing vessel (2,000 ml), n the dry weight of a single *A. anatina*

Table 1 Characteristics of the phytoplankton strains used in the experiments

Strain code	Shape	Origin	Size and volume	Mean (±SE)	Range	Toxin
Scenedesmus obliquus						
CCAP276/3A	Fusiform, mainly single cells	Collection of algae, culture and protozoa Windermere, UK	Length (μm)	9.8 (1.6)	9.1–18.1	N.T.
			Width (μm)	4.3 (2.2)	3.0–17.1	
			Biovol (μm^3)	162.45 (20.2)	42.1–1,816.2	
Planktothrix agardhii						
CYA116	Multicellular filaments	Norwegian Institute of Water Research, Oslo, Norway	Length (μm)	426.0 (18.9)	38.3–889.5	N.T.
			Diameter (μm)	3.7 (0.1)	3.1–4.5	
			Biovol (μm^3)	4,683.2 (207.4)	286.8–14,138.9	
CYA126	Multicellular filaments		Length (μm)	830.8 (80.1)	100.3–2,345.2	Microcystin-LR 2,581.7 μg g DW^{-1} (Dionisio Pires, 2005)
			Diameter (μm)	3.8 (0.2)	3.1–4.5	
			Biovol (μm^3)	9,438.0 (909.5)	754.8–37,295.6	
Microcystis aeruginosa						
V131	Coccoid single- and double cells	Culture collection University of Amsterdam; isolated from Lake Volkerak, The Netherlands.	Diameter (μm)	6.5 (0.3)	4.1–10.2	N.T.
			Biovol (μm^3)	164.4 (19.0)	33.5–523.6	
V40	Coccoid single cells & multicellular colonies		Diameter (μm)	15.1 (10.0)	4.5–70.2	Microcystin-LR 76.6 μg g DW^{-1} (Dionisio Pires, 2005)
			Biovol (μm^3)	463.3 (19.8)	129.8–2,021.9	

The toxin column indicates the main type of toxin that has been found, and it's cellular concentration. N.T. = no toxins described for this strain in the literature

mussel (mg), t the duration of the experiment (in h), A_0 the algae concentration (μg chlorophyll l^{-1}) in the vessels with mussels at $t = 0$, A_t the algal concentration in the vessels with mussels at time t, A_0' and A_t' the concentration of algae in the control vessels at $t = 0$ at time t, respectively. Note that each replicate contained one animal. The average algal carbon concentrations ($\bar{A}$ mg C l^{-1}) were determined by multiplying the geometric mean of initial and final algal concentrations with the slope (s) of previously made regressions for each phytoplankton species between carbon and chlorophyll concentrations:

$$\bar{A} = s \cdot \sqrt{A_t \cdot A_0}$$

Slopes of the regressions between chlorophyll and carbon were all forced through 0, and all had r^2 values higher than 0.99. The average algal carbon concentrations and clearance rates were used to calculate gross mass-specific ingestion rates (Lürling & Verschoor, 2003) (IR_g, mg C mg DW^{-1} h^{-1}):

$$IR_g = \bar{A} \cdot CR$$

To calculate net ingestion rates, we first calculated mass-specific PF production rates (PPR, mg C mg DW^{-1} h^{-1}) from the algal concentrations in the PF production treatments (A_P):

$$PPR = \frac{s \cdot V \cdot A_P}{\Delta t \cdot n}$$

so that net mass-specific ingestion rates (IR, mg C mg DW^{-1} h^{-1}) could be calculated by subtracting PPR from gross ingestion rates:

$$IR = IR_g - PPR$$

For ease of comparison among different treatments, individuals, and food concentrations, we also calculated PF production as a fraction of gross ingestion (F_P):

$$F_P = \frac{PPR}{IR_g}$$

Because for one strain (*S. obliquus* CCAP276/3A) PF data were available for only three concentrations (0.5, 1, and 5 mg C l^{-1}), we pooled the data for only these concentrations within each phytoplankton strain. Due to non-normality and heteroscedacity of these data, even after arcsin transformation, we applied nonparametric tests for comparisons among these strains. Overall differences of these fractions (F_P) between the different strains were compared using the Kruskall–Wallis test, followed by pairwise (two-tailed) Mann–Whitney U-tests for comparisons between strains.

For *S. obliquus* CCAP276/3A, the original PF data were considered to be too limited in terms of number of data points and range of food concentrations to make reliable model fits of the functional response models. Therefore, the average of these F_P values was used to estimate the net ingestion rates from gross ingestion rates.

Fitting and intercomparison of functional response models

Functional response models were fitted by iterative nonlinear regression of functional response models on mass-specific ingestion and clearance rates. Because variance increased with food concentration, we minimized the squared residuals between $\log_{10}$-transformed observations and model predictions. The functional response model that best fitted to a particular data set (i.e.: leaving the smallest squared residuals) was used for comparisons between phytoplankton strains.

The type I (rectilinear) functional response fitting functions used were:

$$\text{if } \bar{A} < ILL \text{ then } IR = IR_{max}\frac{\bar{A}}{ILL},$$
$$\text{and if } \bar{A} \geq ILL \text{ then } IR = IR_{max},$$

with ILL being the incipient limiting level: the food concentration where the ingestion rate becomes constant maximum, and IR_{max} being the maximum ingestion rate. Clearance rates were fitted similarly, and for reasons of comparability, maximum clearance rates (CR_{max}) were calculated as derived parameters:

$$\text{if } \bar{A} < ILL \text{ then } CR = \frac{IR_{max}}{ILL} = CR_{max},$$
$$\text{and if } \bar{A} \geq ILL \text{ then } CR = \frac{IR_{max}}{\bar{A}}$$

Type II (curvilinear) fitting functions used were

$$\mathrm{IR} = \mathrm{IR}_{\max}\frac{\bar{A}}{\bar{A}+K}, \quad \mathrm{CR} = \frac{\mathrm{IR}_{\max}}{\bar{A}+K}$$
$$\text{so } \mathrm{CR}_{\max} = \frac{\mathrm{IR}_{\max}}{K},$$

with K being the half-saturation food concentration at which half of $\mathrm{IR}_{\max}$ is reached, which is a measure for the initial slope and curviness of the functional response. With low K, the functional response starts steeply and saturates rapidly with increasing concentration, whereas at high K, the functional response rises and saturates slowly, and at very high K the functional response is practically linear.

Type III (sigmoid) fitting functions used were

$$\mathrm{IR} = \mathrm{IR}_{\max}\frac{\bar{A}^2}{\bar{A}^2+K^2}, \quad \mathrm{CR} = \mathrm{IR}_{\max}\frac{\bar{A}}{\bar{A}^2+K^2},$$
$$\text{so } \mathrm{CR}_{\max} = \mathrm{IR}_{\max}\frac{\mathrm{IR}_{\max}}{2K},$$

with K being the half-saturation food concentration, as well as the inflection point of the sigmoid curve, and the food concentration at which $\mathrm{CR}_{\max}$ is reached.

For the comparisons of $\mathrm{CR}_{\max}$ and $\mathrm{IR}_{\max}$ between phytoplankton strains, we performed pairwise comparisons between each possible strain pair. Our null hypothesis (H_0) for comparisons was that the separate functional response models were not significantly better predictors then when the model was fitted to the pooled observations for both strains. This hypothesis was compared against the alternative hypothesis (H_1) that the separate models were better predictors, using the significance of the maximum likelihood ratio statistic G (also known as G^2, Bishop et al., 1975),

$$G = 2N \times \ln\left(\frac{\mathrm{MS}_{\mathrm{among}}}{\mathrm{MS}_{\mathrm{within}}}\right)$$

with N being the total number of observations (i.e. samples), $\mathrm{MS}_{\mathrm{among}}$ the total variance of the residuals between the $\log_{10}$-transformed predicted and observed values of the alternative model (H_1), and $\mathrm{MS}_{\mathrm{within}}$ the variance of the residuals of the null model (H_0). G has a χ^2 distribution with degrees of freedom numbers corresponding to the difference between the numbers of parameters of the two models being compared.

Results

Clearance rates

CR decreased with increasing food concentrations, except for toxic *M. aeruginosa* (V40) where the CR was essentially constant (Fig. 2). The residuals, indicating the goodness of fit of the different functional response models (Table 2, lower indicates better fit), showed that clearance rates on *S. obliquus* (CCAP276/3A), both *P. agardhii* strains (non-toxic CYA116 and toxic CYA126) and non-toxic *M. aeruginosa* (V131) yielded a type II response. For toxic *M. aeruginosa* both type I and II fitted equally well (similar residuals) (Table 2). $\mathrm{CR}_{\max}$ decreased in the order *S. obliquus* > toxic *P. agardhii* > non-toxic *P. agardhii* > toxic *M. aeruginosa* > non-toxic *M. aeruginosa* (Fig. 2, Table 3). K is extremely large (4.7×10^{12}) for toxic *M. aeruginosa* (Table 3), meaning no decrease in CR at higher concentrations, and indicating that a linear (type I) functional response applies here. However, for reasons of comparability between strains, we used the type II functional response model with this extreme value for K, which yields a practically linear function. K is high for non-toxic *M. aeruginosa* and *S. obliquus*, and low for non-toxic and toxic *P. agardhii* (Fig. 2, Table 3). The functional responses based on clearance rates were all significantly different from each other, except for between toxic and non-toxic *P. agardhii* (Table 4).

Ingestion rates

On *S. obliquus* and both *P. agardhii* strains (non-toxic CYA116 and toxic CYA126), *A. anatina* had a type II response (Fig. 3 and Table 2), but for both *M. aeruginosa* strains (non-toxic V131 and toxic V40) a type I response was a better fit (smaller residuals for I than II; Table 2). For the type II functional response, K values as well as

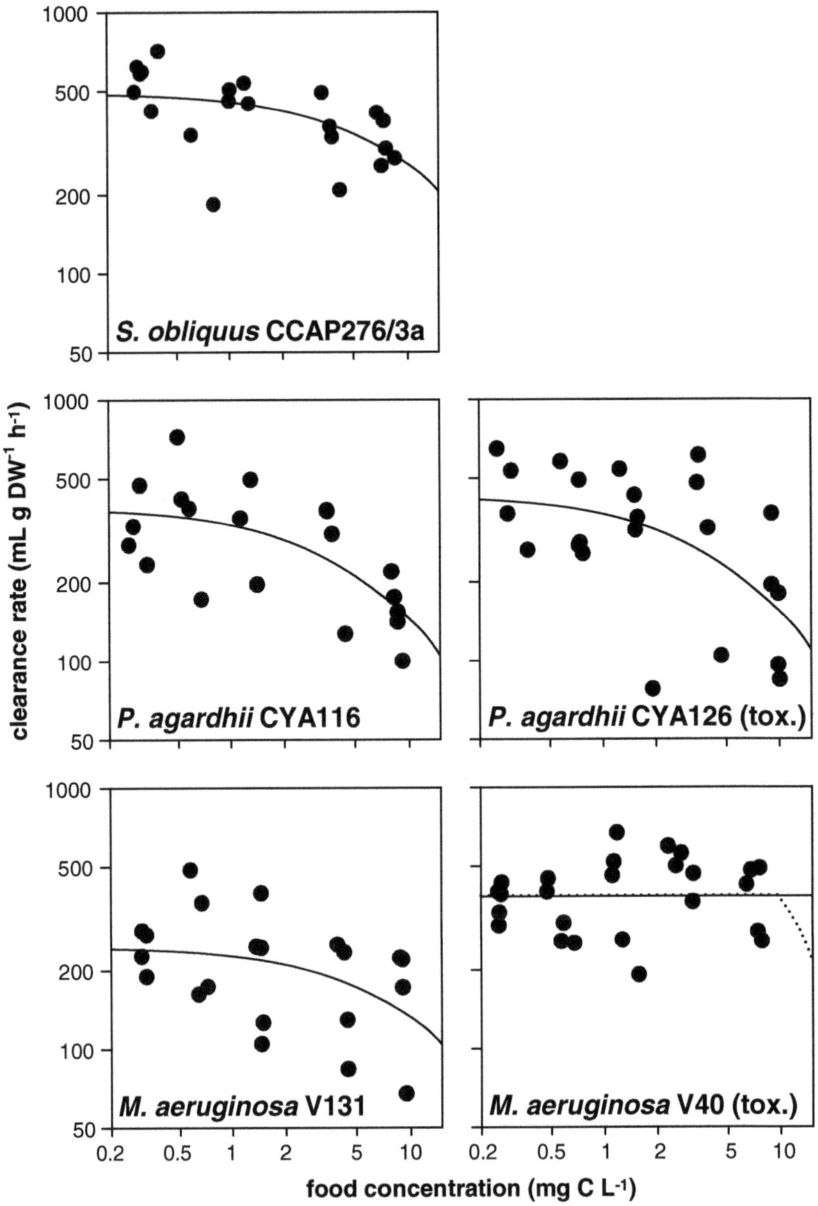

Fig. 2 Mass-specific clearance rates (CR) of *Anodonta anatina* grazing on different concentrations of five different phytoplankton strains. Food concentrations are geometric means of initial and final concentrations. Dots represent measured clearance rates, solid lines represent Type II functional response model regressions, dotted lines represent Type I functional response model regressions

IR_{max} decreased in the order toxic *M. aeruginosa* > non-toxic *M. aeruginosa* > *S. obliquus* > non-toxic *P. agardhii* > toxic *P. agardhii* (Table 3), meaning that handling times were shortest for both *M. aeruginosa* strains (non-toxic V131 and toxic V40). For both *M. aeruginosa* strains, *K* and IR_{max} were smaller with type I than with type II functional responses, yet they were still larger than for non-toxic and toxic *P. agardhii*. Strains that differed significantly in their functional response based on CR also differed significantly in their functional responses based on IR (Table 4), and if a type I functional response would be fitted on both *M. aeruginosa* strain data, these responses would still be significantly different (Table 4).

Pseudofaeces production

For all phytoplankton strains, PF production increased more than linearly with increasing food concentrations, and we did not observe absence

Table 2 Residual sum of squares for the fits of the three different functional response models to data of clearance rates (CR) and ingestion rates (IR) of *Anodonta anatina* on five different phytoplankton strains

Strains	N	Type I (rectilinear)	Type II (curvilinear)	Type III (sigmoidal)
CR				
S. obliquus CCAP276/3A	21	0.4179	**0.3514**	1.264
P. agardhii CYA 116	20	0.9712	**0.5331**	1.902
P. agardhii CYA 126 (toxic)	23	1.7411	**1.1939**	1.753
M. aeruginosa V131	21	0.9196	**0.7656**	1.326
M. aeruginosa V40 (toxic) V40	25	**0.4538**	**0.4538**	1.113
IR				
S. obliquus CCAP276/3a[a]	12	2.0248	**0.3365**	0.5960
S. obliquus CCAP276/3A[b]	21	0.4179	**0.3514**	1.2644
P. agardhii CYA116	20	1.2203	**0.7187**	0.7685
P. agardhii CYA 126 (toxic)	23	3.2579	**2.5232**	2.8503
M. aeruginosa V131	21	1.5167	**1.502**	2.2005
M. aeruginosa V40 (toxic) V40	25	**0.4803**	0.4804	1.1123

Also given are the number of data points used per regression (N). Bold numbers indicate best fits

[a] Original data CCAP276/3A

[b] Estimated data CCAP276/3A (see Materials and Methods)

Table 3 Parameter estimates of nonlinear regression of type II (curvilinear) and type I (rectilinear) functional response models of clearance and ingestion rates of *Anodonta anatina* on five different phytoplankton strains

		Clearance rates		Ingestion rates	
Type II	*N*	CR_{max}	*K*	IR_{max}	*K*
S. obliquus CCAP276/3A	21	492.70	11.44	5.24	11.44
P. agardhii CYA116	20	387.42	6.04	1.90	5.59
P. agardhii CYA126 (toxic)	23	429.27	5.60	1.56	4.51
M. aeruginosa V131	21	246.58	11.60	7.48	48.24
M. aeruginosa V40 (toxic)	25	382.92	4.77×10^{12}	4.23×10^{3}	1.14×10^{4}
Type I	*N*	CR_{max}	ILL	IR_{max}	ILL
M. aeruginosa V131	21	199.24	17.89	3.55	24.32
M. aeruginosa V40 (toxic)	25	382.92	9.74	3.73	10.01

Given are *N*, the number of data points used per regression; CR_{max}, the estimated maximum mass-specific clearance rates (ml g DW^{-1} h^{-1}); *K*, (in case of type II) the half-saturation food concentration (mg C l^{-1}), or (in case of type I) the incipient limiting food concentration (ILL, also mg C l^{-1}); and IR_{max}, the estimated maximum ingestion rates (mg C g DW^{-1} h^{-1})

of PF production at low food levels (i.e. below the incipient limiting level). The Kruskall–Wallis test revealed significant differences in PPR by *A. anatina*, when grazing on the five different strains ($p < 0.0001$, $H = 30.41$, d.f. = 4, $N = 64$). Pairwise comparisons (Mann–Whitney *U*-test) showed that *A. anatina* had significantly lower PPR for toxic *M. aeruginosa*, compared to the other cyanobacterial strains ($p < 0.05$), but PPR was not significantly different from *S. obliquus* ($p > 0.05$, Fig. 4). Significantly higher PPR were found for non-toxic *M. aeruginosa* than for *S. obliquus* ($p < 0.05$).

Discussion

The type I (rectilinear) functional response is often thought to be applicable for 'true' filter feeders such as bivalves, which is illustrated by the frequent use of the term 'incipient limiting level' in the literature. Unfortunately, such terminology is often not supported by a full functional response analysis. In our experiments, we found that *A. anatina* had a type II (curvilinear, hyperbolic) functional response on 4 out of 5 investigated phytoplankton strains, indicating that the handling time of the food plays an

Table 4 Pairwise comparisons of the best-fitting functional response models of clearance and ingestion rates of *A. anatina* on five different phytoplankton strains

Strain code	CYA116 *P. agardhii*		CYA126 *P. agardhii* (toxic)		V131 *M. aeruginosa*		V40 *M. aeruginosa* (toxic)	
Clearance rates (type II)								
S. obliquus CCAP276/3A	24.399	**[<0.001]**	11.699	**[0.003]**	51.372	**[<0.001]**	9.648	**[0.008]**
P. agardhii CYA116			0.685	[0.710]	12.172	**[0.002]**	32.424	**[<0.001]**
P. agardhii CYA126 (toxic)					14.318	**[<0.001]**	19.414	**[<0.001]**
M. aeruginosa V131							52.588	**[<0.001]**
Ingestion rates (type II)								
S. obliquus CCAP276/3A[a]	8.913	**[0.012]**	4.683	[0.096]	32.333	**[<0.001]**	5.429	[0.066]
S. obliquus CCAP276/3A[b]	28.575	**[<0.001]**	16.501	**[<0.001]**	55.830	**[<0.001]**	7.946	**[0.019]**
P. agardhii CYA116			0.244	[0.885]	18.538	**[<0.001]**	38.294	**[<0.001]**
P. agardhii CYA126 (toxic)					11.792	**[0.003]**	23.298	**[<0.001]**
M. aeruginosa V131							60.600	**[<0.001]**
Ingestion rates (type I)								
M. aeruginosa V131							56.863	**[<0.001]**

Shown are the values of the maximum likelihood ratio statistic (G) and corresponding p-values [in square brackets]. Model fits were considered to be significantly different when $p < 0.05$, shown in bold, indicating that the separate models were better predictors than when fitting a single model through the pooled observations of these pairs

[a] Original data CCAP276/3A

[b] Estimated data CCAP276/3A (see Materials and methods section)

important role in the ingestion process. Therefore, we reject hypothesis 1 (clearance and ingestion rates are according to a type I functional response) for all strains except for the toxic *M. aeruginosa* (V40).

The striking exception to the observed functional responses is the CR on the toxic *M. aeruginosa*, both on the basis of the functional response type (type I) and our parameter estimates. These suggest that toxic *M. aeruginosa* should be very small but in fact this strain consists of single cells and intermediately sized colonies (Table 1). Although these colonies are easily retained on the gills, maybe even better than the smaller sized single cells, ingestion of such particles would still involve some handling time. Visual observations of the colonies of toxic *M. aeruginosa* showed that these are not solid and packed in a thick mucus layer, as is often observed in large *M. aeruginosa* colonies but loose aggregates that disintegrate easily. It is not difficult to imagine that such colonies could disintegrate due to shear forces in the feeding process. Thus, the actual size of the particles once ingested could be much smaller, as small as single cells, leading to very short handling times. The combination of easy capture on the gills (of intermediately sized aggregates) and easy ingestion (of small cells) could thus explain why we found a type I functional response of *A. anatina* for this strain.

We did not observe absence of PF production at low food levels (i.e. below the incipient limiting level), so that hypothesis 2 is rejected as well. Since PF production increased more than linearly with food concentration, this again indicates that a type II functional response is more appropriate for the feeding behaviour of *A. anatina*. The differences in fractions of PF production relative to gross ingestion (Fig. 4) suggest that *A. anatina* is able to discriminate between food sources by varying the degree of excretion (Sprung & Rose, 1988). The high fractions of PF production for non-toxic *M. aeruginosa* (V131) (30% of gross ingestion) and both *P. agardhii* strains (non-toxic CYA116 and toxic CYA126) (15–20% of gross ingestion) indicate that these strains were the least preferred. Both PF and feeding data show that *S. obliquus* (CCAP276/3A) is the preferred food type, which is in concordance with our hypothesis 3.

Hypothesis 4, coccal/colonial *M. aeruginosa* is preferred over filamentous *P. agardhii*, is less easy to accept or reject. Maximum clearance rates were higher on both non-toxic and toxic

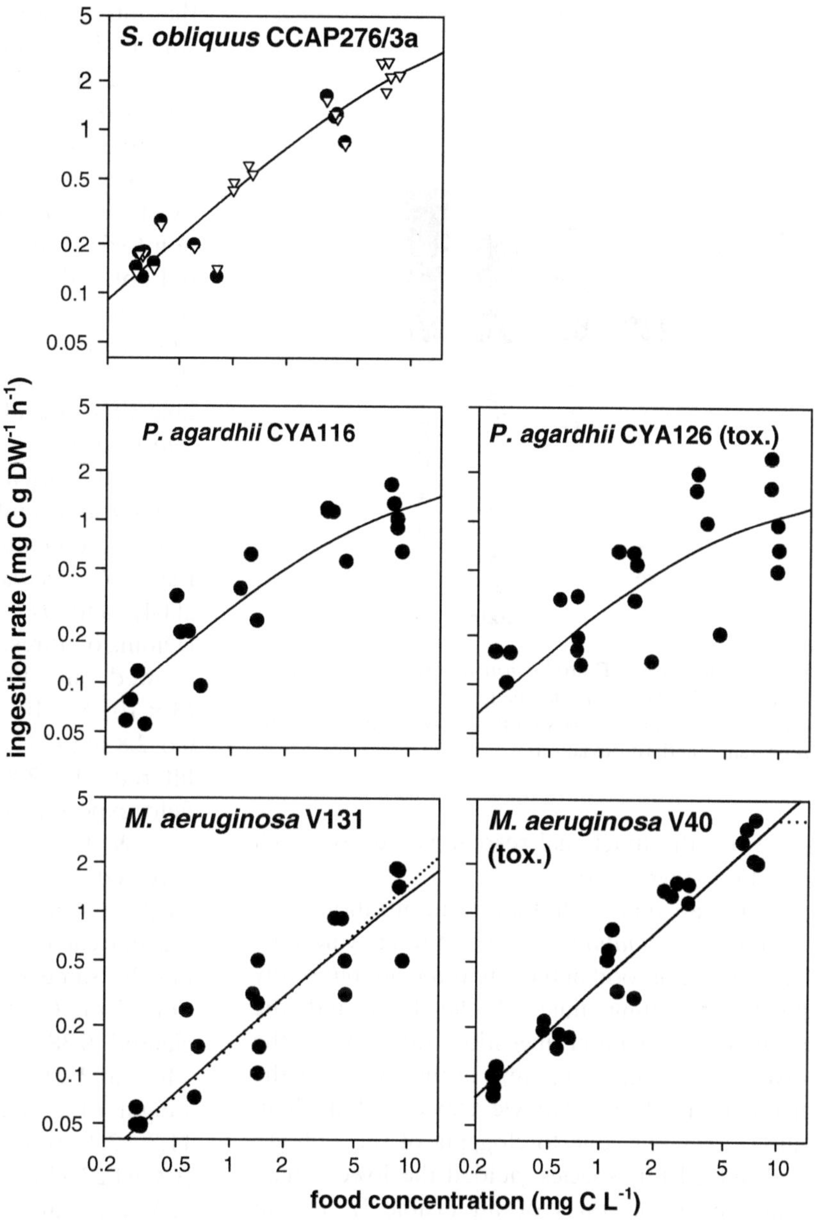

Fig. 3 Mass-specific ingestion rates (IR) of *Anodonta anatina* grazing on different concentrations of five different phytoplankton strains. Food concentrations are geometric means of initial and final concentrations. Dots represent measured ingestion rates, white triangles are estimated ingestion rates (see Materials and methods), solid lines represent Type II functional response model regressions, dotted lines represent Type I functional response model regressions

filamentous *P. agardhii* than on non-toxic *M. aeruginosa*, and equal to toxic *M. aeruginosa*, showing that *P. agardhii* is filtered at least equally efficient from the water column at low food densities. On the other hand, maximum ingestion rates were higher for both non-toxic and toxic *M. aeruginosa*, which indicates that *M. aeruginosa* is preferred over *Planktothrix* at high food densities. Probably the higher maximum clearance rates on both *P. agardhii* strains are artefacts caused by the larger size of these algae, which makes it more likely that they are filtered by chance. The high PF production on non-toxic and toxic *P. agardhii* species suggests that they are not very suitable food sources. At higher concentrations, long handling times of the filaments rapidly becomes a serious limiting factor in the ingestion process, which is visible in the curviness of the functional response (Figs. 2, 3, low *K* values in Table 3). With *M. aeruginosa*,

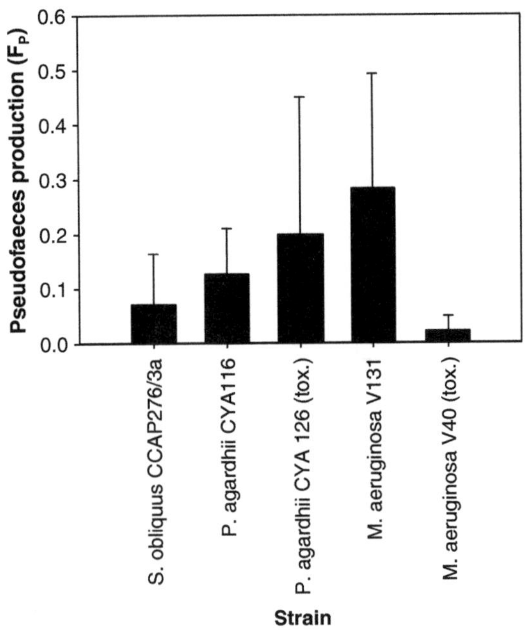

Fig. 4 Fraction of PF production relative to gross ingestion (F_P) of *Anodonta anatina* grazing on five different phytoplankton strains. Bars represent averages, error bars represent standard deviations

however, the functional responses did not reach satiation (Figs. 2, 3).

Thus, at very high food concentrations, the feeding of *A. anatina* was not limited anymore by the amount of food in the environment, but solely by the handling time of the food and the estimated maximum ingestion rates were the inverse of the handling time of the food. For the filamentous *P. agardhii* we expected long handling times (e.g. due to clogging of the siphon), and indeed this species yielded the lowest maximum ingestion rates of all (Table 3). In our study, biovolume (Table 1) was inversely related to the maximum ingestion rate (Table 3), which shows that food particle size can be used as a predictor of feeding rates.

Contrary to our hypothesis 5, toxicity did seem not seem to affect the grazing behaviour of *A. anatina.* For *M. aeruginosa*, the toxic strain is preferred over the non-toxic strain. Moreover, the higher maximum ingestion rate of the toxic *M. aeruginosa*, as compared to *S. obliquus*, suggests that toxin production did not interfere with filtration and ingestion. Previous findings describe that mussels are relatively insensitive to microcystin producing cyanobacteria (e.g. Dionisio Pires et al., 2004). We are unable to distinguish the effects of size (toxic *P. agardhii* twice as long as non-toxic *P. agardhii* (Table 1)) and toxicity of the toxic *P. agardhii* on the lower ingestion rates by *A. anatina* on this strain as compared to the non-toxic *P. agardhii* strain. However, because we did not find a significant effect of microcystin on the feeding behaviour of *A. anatina* in the *Microcystis* treatments, it seems more plausible that size rather than toxin caused the lower ingestion rate of *A. anatina* on the toxic *Planktothrix* strain.

For the discussion on the applicability of *A. anatina* as a biofilter, we compare the filtration rates of *A. anatina* (mean dry weight: 3.5 g) with that of *Corbicula leana* in the field (Hwang et al., 2004) and *D. polymorpha* in the laboratory (Dionisio Pires, 2005). Hwang et al. (2004) showed in the hypertrophic Lake Ilgam, (3×10^5–8×10^5 cells ml^{-1} of *M. aeruginosa* and 0.6–2.8 mg C l^{-1} of *P. agardhii*) that *C. leana* filtered 1.6–7.8 l mussel^{-1} day^{-1}. Furthermore, field experiments with *D. polymorpha* by Reeders & Bij de Vaate (1990) in a Dutch lake revealed a filtration rate of 1.2 l mussel^{-1} day^{-1} (with a shell length of 18 mm). Comparing grazing capacities per unit biomass when offering the same phytoplankton species (concentration 2 mg C l^{-1}), we found that *A. anatina* cleared 0.38; 0.22; 0.34; 0.22 and 0.42 ml mg^{-1} h^{-1}, and *D. polymorpha* (mean individual dry weight of 9 mg; Dionisio Pires, 2005) 4.0; 4.6; 11.3; 3.4 and 3.1 ml mg^{-1} h^{-1} of *S. obliquus* (CCAP276/3a), non-toxic- and toxic *P. agardhii* (CYA116 and CYA126) and non-toxic- and toxic *M. aeruginosa* (<60 μm) (V131 and V40), respectively. This means that for the phytoplankton strains used in this study, the grazing capacity, on a weight-specific basis, of *D. polymorpha* is ~7–33 times higher than that of *A. anatina.* Despite the low clearance per unit of biomass, the grazing capacities per individual of *A. anatina* per day (CR_{max} of 42.6; 33.5; 37.1; 21.3 and 33.1 l^{-1} day^{-1} of *S. obliquus*, non-toxic *P. agardhii*; toxic *P. agardhii*, non-toxic *M. aeruginosa* and toxic *M. aeruginosa*, respectively) were comparably high with those of marine bivalves (*M. edulis* 35.3 l $mussel^{-1}$ day^{-1} and *Cerastoderma edule* (Linnaeus, 1758) 31.2 l $mussel^{-1}$ day^{-1}; Ward & Shumway, 2004). On a weight-specific basis, the grazing capacity of *Anodonta* is low compared to mussel species as *D. polymorpha*. However, per individual, *Anodonta's* grazing capacity is equal or higher than other species.

The current abundance of Unionid bivalves in the field (0–5 mussels or 17,500 mg DW m^{-2}, as found in a field study in four shallow lakes in The Netherlands during 2005, unpublished data) however, is too low for *A. anatina* to have a strong effect on algal biomass. In contrast, in high density areas of Lake IJsselmeer and Lake Markermeer, *D. polymorpha* has been described to remove ±70% of the algal population (Lammens, 1999). When we take into account the number of 1,000 individuals per m^{-2} of *Dreissena* observed in Lake IJsselmeer (Bij de Vaate, 1991), an average individual DW for *D. polymorpha* and *A. anatina* of 9 and 3,600 mg respectively, and the grazing rates for *Anodonta* found in this study and for *Dreissena* found in Dionisio Pires (2005), we would need 83 *Anodonta* mussels per m^{-2} to clear the same volume of water from e.g. the toxic *P. agardhii* (CYA126) (at a food density of 2 mg C l^{-1}) as would 1,000 *Dreissena* mussels. However, densities of Unionid mussels in the field appears to be low (unpublished data). On the other hand, the observed high densities of *D. polymorpha* (individuals per m^{-2}) 500–1,000 and 400–1,000 individuals per m^{-2} in Lake IJsselmeer and Lake Markermeer, respectively (Bij de Vaate, 1991); 30,000 individuals per m^{-2} in Lake Zürich; 20,000 individuals per m^{-2} in Lake Garda and 21,000 individuals per m^{-2} in Lake Constance (Burla & Lubini-Ferlin, 1976; Franchini, 1978 and Suter, 1982 in Bij de Vaate, 1991) make it more likely that these smaller mussels have a larger effect on the primary productivity and a more realistic achievement for biomanipulation. If *A. anatina* is to be used as a biofilter in shallow Dutch lakes, its densities must be promoted. Current densities may be low because after years of eutrophication the stability of the sediments has changed, providing unsuitable habitat for these mussels.

Summary and conclusion

From this study we conclude that *A. anatina* mussels can filter and ingest colonial as well as larger filamentous cyanobacteria. This observation indicates that *A. anatina* may be useful as a biofilter even when cyanobacteria blooms are toxic, provided that densities of these mussels in the field are increased through suitable restoration measures (to be developed).

Acknowledgements Publication 3951 Netherlands Institute of Ecology (NIOO-KNAW). We thank the members of the Plant Animal Interactions group of the NIOO in Nieuwersluis, for providing us fresh mussels. This is a contribution of OBN (Restoration and management of Dutch peat lakes), funded by the Ministry of Agriculture, Nature and Fisheries, The Netherlands.

References

Bij de Vaate, A., 1991. Distribution and aspects of population dynamics of zebra mussel, *Dreissena polymorpha* (Pallas, 1771), in the Lake IJsselmeer area (The Netherlands). Oecologia 86: 40–50.

Bishop, Y. M., S. E. Fienberg & P. W. Holland, 1975. Formal goodness of fit: summary statistics and model selection. In Discrete Multivariate Analysis: Theory and Practice. MIT Press, Cambridge, 123–175.

Coughlan, J., 1969. The estimation of filtering rate from the clearance of suspensions. Marine Biology 2: 356–358.

Dionisio Pires, L. M., 2005. Grazing for clarity. PhD Thesis, Radboud University Nijmegen, The Netherlands.

Dionisio Pires, L. M., K. Karlsson, J. A. O. Meriluoto, E. Kardinaal, P. M. Visser, K. Siewertsen, E. Van Donk & B. W. Ibelings, 2004. Assimilation and depuration of microcystin-LR by the zebra mussel, *Dreissena polymorpha*. Aquatic Toxicology 69: 385–396.

Dionisio Pires, L. M., B. M. Bontes, E. Van Donk & B. W. Ibelings, 2005. Grazing on colonial and filamentous, toxic and non-toxic cyanobacteria by the zebra mussel *Dreissena polymorpha*. Journal of Plankton Research 27: 331–339.

Fahnenstiel, G. L., T. B. Bridgeman, G. A. Lang, M. J. McCormick & T. F. Nalepa, 1995. Phytoplankton productivity in Saginaw Bay, Lake Huron: effects of zebra mussels (*Dreissena polymorpha*) colonization. Journal of Great Lakes Research 21: 465–475.

Hassell, M. P., 1978. The Dynamics of Arthropod Predator-Prey Systems. Princeton University Press, Princeton, NJ.

Holling, C. S., 1959. Some characteristics of simple types of predation and parasitism. The Canadian Entomologist 91: 339–347.

Hwang, S., H. Kim, J. Shin, J. Oh & D. Kong, 2004. Grazing effects of a freshwater bivalve (*Corbicula leana* Prime) and large zooplankton on phytoplankton in two Korean lakes. Hydrobiologia 515: 161–179.

Lammens, E., 1999. Het voedselweb van IJsselmeer en Markermeer: Veldgegevens, hypotheses, modellen en scenario's (in Dutch). RIZA report 99.008. Rijksinstituut voor Integraal Zoetwaterbeheer en Afvalwaterbehandeling, Lelystad, The Netherlands.

Lürling, M. & A. M. Verschoor, 2003. F_0-spectra of chlorophyll fluorescence for the determination of zooplankton grazing. Hydrobiologia 491: 145–157.

Prins, T. C., A. C. Smaal & R. F. Dame, 1998. A review of the feedbacks between bivalve grazing and ecosystem processes. Aquatic Ecology 31: 349–359.

Reeders, H. H. & A. Bij de Vaate, 1990. Zebra mussels (*Dreissena polymorpha*): a new perspective for water quality management. Hydrobiologia 200/201: 437–450.

Rigler, F. H., 1961. Relation between concentration of food and feeding rate of *Daphnia Magna* Straus. Canadian Journal of Zoology 39: 857–1961.

Scheffer, M., 1998. Ecology of Shallow Lakes. Chapman & Hall, London.

Sprung, M. & U. Rose, 1988. Influence of food size and food quantity on the feeding of the mussel *Dreissena polymorpha*. Oecologia 77: 526–532.

Vanderploeg, H. A., J. R. Liebig, W. W. Carmichael, M. A. Agy, T. H. Johengen, G. A. Fahnenstiel & T. F. Nalepa, 2001. Zebra mussel (*Dreissena polymorpha*) selective filtration promoted toxic *Microcystis* blooms in Saginaw Bay (Lake Huron) and Lake Erie. Canadian Journal of Fisheries and Aquatic Sciences 58: 1208–1221.

Ward, J. E. & S. E. Shumway, 2004. Separating the grain from the chaff: particle selection in suspension- and deposit feeding bivalves. Journal of Experimental Marine Biology and Ecology 300: 83–130.

Hydrobiologia (2007) 584:205–213
DOI 10.1007/s10750-007-0598-5

SHALLOW LAKES

The role of herbivorous water birds in aquatic systems through interactions with aquatic macrophytes, with special reference to the Bewick's Swan – Fennel Pondweed system

Marcel Klaassen · Bart A. Nolet

Abstract The role of aquatic macrophytes in stimulating biodiversity and maintaining clear waters is currently undisputed. The management of (eutrophic) shallow waters is therefore often directed at (re-)establishing macrophyte domination. In contrast, the role of water birds has long been considered of minor importance for the functioning of fresh water ecosystems. Indeed, in terms of biomass and production, water birds constitute only a minor part of these systems. However, water birds may graze heavily on water plants under certain circumstances, and the question arises whether herbivorous water birds have an important indirect effect on shallow fresh water systems. Mainly illustrated with the interaction between Bewick's Swans and Fennel Pondweed, we present data on the role that water plants may play in the life of water birds and how water birds may impact water plants' fitness in terms of survival, production, dispersal and competitive ability. It appears that water plants may be crucial for water birds during periods of high-energy requirements, such as migration. Despite the plants' costs associated with water bird grazing, the interaction between water birds and water plants varies in nature from an apparent predator–prey relationship to a mutually beneficial interaction depending on the context and the perspective. For the case of the Bewick's Swan–Fennel Pondweed interaction, regular bird grazing is sustainable and may actually favour the plant's dispersal. Thus, Bewick's Swans themselves may in fact play a crucial role in establishing and maintaining the Fennel Pondweed rich staging sites between the swans' wintering and breeding grounds, which are vital for the swans' successful migration.

Keywords Bewick's Swan · *Cygnus columbianus* · Dispersal · Grazing · Plant competition · Plant–herbivore interaction · *Potamogeton pectinatus*

Guest editors: R. D. Gulati, E. Lammens, N. De Pauw & E. Van Donk
Shallow lakes in a changing world

M. Klaassen (✉) · B. A. Nolet
Department of Plant-Animal Interactions, Centre for Limnology, Netherlands Institute of Ecology (NIOO-KNAW), P. O. Box 1299, Maarssen 3600 BG, The Netherlands
e-mail: m.klaassen@nioo.knaw.nl

Introduction

The structuring role of aquatic macrophytes in lakes, and in particular their importance as a food source, has long been underrated (Lodge, 1991). Less than a century ago, Shelford (1918) for instance wrote: "One could probably remove all the larger plants and substitute glass structures of the same form and surface structure without greatly affecting the immediate food relations". A more recent example is a statement like "macrophytes enter the food web only as detritus" (Polunin, 1984). The main argument was that losses due to herbivore grazing were usually small, between 0.5% and 8% of the total annual macrophyte production (Wetzel, 1983, p. 543). We now know that aquatic macrophytes in fact play important roles in freshwaters such as promoting a clear water state through the competition with algae and the stabilisation of the bottom sediment, enhancing biodiversity by offering shelter and substrate, and providing food for herbivorous water birds (Scheffer, 1998; Knapton & Petrie, 1999; Engelhardt & Ritchie, 2002; Noordhuis et al., 2002; Declerck et al., 2005). Nowadays, water managers strive for freshwater bodies dominated by aquatic macrophytes to fulfil these structuring functions and to support herbivorous water birds, which are often target species for management (Coops et al., 2002).

Conversely, effects of herbivores on aquatic macrophytes are generally still thought to be of minor importance. Only during the establishment phase, for instance after lake restoration, herbivorous water birds have been shown to delay colonization by aquatic macrophytes (Lauridsen et al., 1993; Søndergaard et al., 1996; Perrow et al., 1997; Lodge et al., 1998). Nevertheless, the lives of many aquatic macrophytes and their consumers may be importantly intertwined. Rather than trying to review the recent literature on the interaction between aquatic macrophytes and waterfowl, we present an overview of the possible facets of this fascinating interplay. We exemplify these various aspects of the interaction between grazing birds and macrophytes by the interaction between Fennel Pondweed (*Potamogeton pectinatus*) and Bewick's Swan (*Cygnus columbianus bewickii*), which we have studied in detail over the past decade.

Bewick's Swans and their reliance on Fennel Pondweed

Herbivorous waterfowl can be found all over the world but their breeding presence in the Arctic is particularly evident (Bellrose, 1980; Scott & Rose, 1996; Miyabayashi & Mundkur, 1999). There they make up a large part of the avifauna having a considerable impact on the tundra vegetation (Person et al., 2003) and the functioning of the ecosystem as a whole (Kerbes et al., 1990; Jefferies et al., 1994, 2006). Each autumn these birds leave the Arctic in search of more benign environments to spend the winter. Most of the waterfowl breeding in the European Arctic and partly those breeding in Siberia and Greenland, aggregate in north-western Europe (Scott & Rose, 1996). Tundra Swans (*Cygnus columbianus*) are circumpolar breeding, with Bewick's Swans (*C. c. bewickii*) breeding in northern Russia. The Netherlands appear to be very attractive to Bewick's Swans, with up to 70% of the population spending the winter there (Beekman, 1997). The attractiveness to these swans, and in fact many other species of waterfowl, probably lays in the country's relatively mild marine climate, the large availability of fresh water bodies and its agricultural production (Van Eerden et al., 2005). That The Netherlands is special in this latter respect can be exemplified by its fertiliser input, which is by far the highest in Europe with nearly 500 kg of N/ha of agricultural land per year (http://epp.eurostat.cec.eu.int).

If the attraction of The Netherlands for waterfowl partly lays in the readily available agricultural products and their harvest remains, what may then possibly be the role of water plants in the life of these birds and in return the birds' impact on water plants? Historic data on habitat use of Bewick's Swans in the 1930s reveals that at this time the swans in fact exclusively foraged on aquatic vegetation and notably Fennel Pondweed while wintering in The Netherlands (Brouwer & Tinbergen, 1939). Fennel Pondweed is a pseudo-annual highly relying on the production of asexual, starch-rich tubers to survive the winter. Bewick's Swans forage on these tubers by uprooting them from the sediment by trampling, subsequently sieving them from the sediment with their beak. Only after the explosion of agricultural production and accompanying decrease of submerged macrophytes in the 1960s and 1970s, Bewick's Swans were seen feeding on agricultural lands (Merne, 1972; Mullié & Poorter, 1977). Being subsidised in this way, Bewick's Swans, like many other waterfowl species using agricultural resources, increased in population size (Abraham et al., 2005; Van Eerden et al., 2005). Despite the current great dependency on agricultural products relative to aquatic resources, the first food source to be selected by the swans after arrival in The Netherlands on their autumn migration are still *Potamogeton* tubers (Dirksen et al., 1991). To determine the diet choice of Bewick's Swans away from the wintering grounds we equipped 18 individuals with satellite transmitters. It appeared that both during spring and autumn migration Bewick's Swans spend most of their stop-over time in wetlands suggesting a great use of aquatic vegetation, notably Fennel Pondweed (Nolet et al., 2001; Beekman et al., 2002) (Fig. 1).

Using heart rate telemetry we estimated the energetic costs of aquatic and terrestrial foraging modes. It appeared that foraging energy expenditure alone could not explain the preference for aquatic habitats, since aquatic foraging was actually more expensive (38.9 Watt) than foraging on land (28.2 Watt). Next we estimated daily energy intake rates in free-living animals on a range of food sources. To this end we used a number of methods that mainly relied on measurements of dropping rates and estimates of grazing intensity in combination with food availability. We further measured energy assimilation in captive swans, which appeared to be highest (90%) for pondweed tubers (Nolet et al., 2002). Depending on food availability, estimates of daily intake rates on grassland ranged up to 40 Watt, on crops (and their remains) up to 86 Watt, and on Fennel Pondweed tubers up to 136 Watt (Fig. 2). Daily energy intake rates on Fennel Pondweed are among the highest ever recorded in free-living birds. Not the rate of intake while feeding is so exceptional in tuber-foraging swans, but rather the time they can spend feeding per day. While on land, these heavy birds with their poor take-off abilities are potential prey for land predators such as foxes. Therefore, their safe foraging on land is precluded to the daylight hours. On water, however, they forage around the clock (Nolet & Klaassen, 2005). As a result, the swans only leave the pondweed beds after these have been depleted down to the level that daytime beet-feeding becomes equally profitable (Beekman et al., 1991; Nolet et al., 2002).

We conclude that Bewick's Swans heavily feed on water plants during periods of high-energy demand, such as during migration, and that the availability of these resources is probably crucial in fulfilling their annual cycle. Whether swans, with their presumably high flight costs and their poor take-off abilities, and their need for high-energy food such as tubers, are exceptional in their reliance on aquatic resources requires further attention.

Changing perspectives: Fennel Pondweed and its consumers

Above we have argued that Bewick's Swans (and possibly other waterfowl) may greatly benefit from consuming water plants. The question is whether this consumer-resource relationship is purely antagonistic or whether there are also

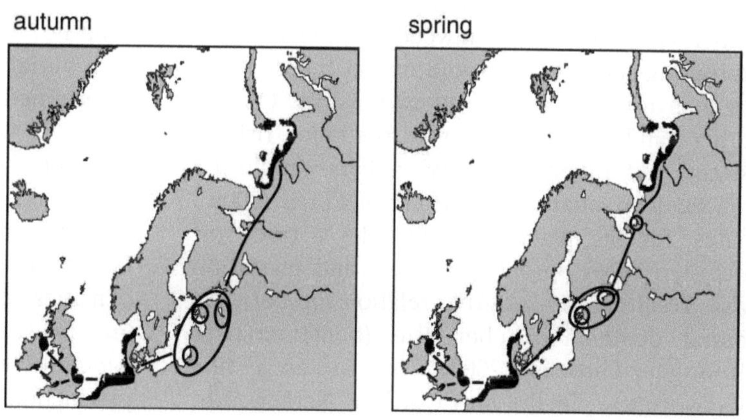

Fig. 1 Breeding an wintering distribution of Bewick's Swan and the major stop-over sites used during migration in both spring and autumn. During migration Bewick's Swans heavily rely on wetlands where they preferentially forage on aquatic vegetation

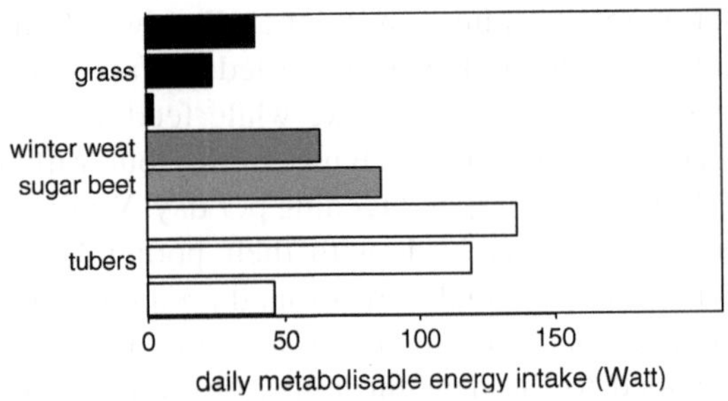

Fig. 2 Average daily metabolisable energy intake of free-living Bewick's Swans feeding on grass (data from three different years/study sites), winter weat, sugar beet harvest remains and Fennel Pondweed tubers (data from three different years/study sites). Collated from data presented in Beekman et al. (2002)

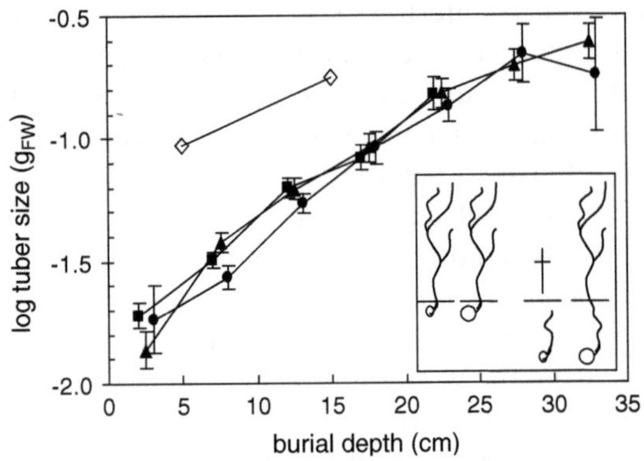

Fig. 3 Tuber size of Fennel Pondweed versus burial depth. Measurements are depicted from Lauwersmeer 1980 (open diamonds; Beekman et al., 1991), 1995 (closed squares ± SD), 1996 (closed triangles) and 2003 (closed circles). The inset explains the tuber-size dependent mortality costs associated with burial depth. Tuber size and burial depth are explained by a trade-off between sprouting mortality costs, the resources needed to make large tubers and burial-depth dependent predation risk

aspects to it that are neutral and possibly mutualistic by nature?

Plant defences against herbivores

Plants may defend themselves in various ways against herbivore attack. Such defences may include chemical and structural defences, but also hiding in time (by being out of phase with the predator) and space. It is the latter that apparently happens in Fennel Pondweed when they are faced with increased predation pressure by tuber digging swans.

Fennel Pondweed tubers may easily vary one order of magnitude in volume (Santamaría & Rodríguez-Gironés, 2002). For small tubers to sprout successfully they have to be in the upper sediment layers. However, given their larger stores, big tubers may successfully sprout from deep sediment layers (Fig. 3). Indeed, field data indicate that burial depth is positively related to tuber size in Fennel Pondweed. Still, to increase sprouting success one would expect tubers to be as high up in the sediment as possible. On the other hand, however, to avoid predation risk by swans and other waterfowl, tubers should be placed deep into the sediment (Nolet et al., 2006). Thus, for the plant there may be a trade off between tuber sprouting success and predation risk, resulting in a positive relationship between burial depth and tuber size (Santamaría & Rodríguez-Gironés, 2002). In an area like Lauwersmeer, in the north of The Netherlands, this relationship is particularly strong. However, the outcome of the trade off between sprouting success and predation risk has not always been the same. Comparing our recent Lauwersmeer measurements on tuber burial depth with those of Beekman et al. (1991), shortly after the area was taken into use by Bewick's Swans, indicate that tubers of a given size tended to be buried approximately 10 cm higher up in the sediment (Fig. 3). Thus in recent times, predation pressure apparently weighs heavier in the trade off than sprouting success. There is a genetic component to tuber size and burial depth (Hangelbroek & Santamaría, 2004), and genotypic selection can therefore be held responsible for the change in tuber burial depth within 20 years. However, this need not necessarily be the case, since tuber size and burial depth are also strongly maternally determined (Hangelbroek & Santamaría, 2004).

Plant production in the face of herbivory

Despite sometimes intensive grazing on tubers, Fennel Pondweed stands regain similar tuber densities as when grazing is absent within one growing season. Experiments have shown that *per capita* tuber production (i.e., the annual total production of tubers divided by initial tuber density) is negatively related to initial tuber density at the start of the growing season, self-shading being one of the most likely candidates explaining the lowered production at high plant densities (Rodríguez-Gironés et al., 2003). Modelling studies have indicated that this effect may easily lead to compensation if swans graze the tuber stocks down to a certain level (Jonzén et al., 2002). Indeed, field work confirmed that grazing levels by swans are generally such that compensation occurs and tuber yield to the swans is large (Nolet 2004). Also in the absence of subsidiary (agricultural) food sources this will probably remain the case. The swans are predicted to graze down the tuber bank to a lower level, namely that at which foraging costs would exceed food intake rather than to a level at which pondweed foraging yields as much as foraging on the surrounding beet fields (Nolet et al., 2002). However, at this lower level, compensatory growth of the pondweed will also be higher. Hence, Fennel Pondweed is very well capable of withstanding grazing of the tuber stocks by swans.

Herbivores as vectors for dispersal

"But the wide distribution of fresh-water plants (...) apparently depends in main part on the wide dispersal of their seeds (...) by animals, more especially by fresh-water birds, which have great powers of flight, and naturally travel from one piece of water to another." (Darwin, 1859). Indeed, of many water plant species the seeds may stick to the feathers or may be transported in the guts of water birds (see Clausen et al., 2002). Most ingested seeds are digested wholly or partly, and are no longer viable after passing the digestive tract. Still, for Fennel Pondweed it has been shown that some seeds may remain in the digestive tract of ducks for as long as 54 h while retaining viability (Charalambidou et al., 2005). This is more than the calculated maximum flight durations of long-distance flights, so this would mean that such seeds might potentially travel 3000 3500 km (i.e., the maximum flight distance) on board of a migrating duck (Clausen et al., 2002). Hence, there is a small chance for a seed to survive such a long-distance dispersal and to successfully germinate. The next question is what potential such a seed has to compete with locally adapted conspecifics in a foreign environment. For Fennel Pondweed Santamaría et al. (2003) conducted reciprocal transplant experiments to investigate in how far clones from various regions in Europe thrived at distant sites. They showed that northern European clones tended to be much better tuber producers than Mediterranean clones, which relied almost exclusively on sexual reproduction through seeds. Although seeds have a much higher survival to desiccation than tubers, with their large nutrient stores tubers probably have a much better competitive ability to seeds. This led Santamaría et al. (2003) to hypothesise that northern clones could successfully compete with southern clones as long as no drought would strike, allowing a steady gene flow from northern to southern Europe along the flyways of migratory waterfowl. Gene flow in the opposite direction is thought to be less likely.

Whether this hypothesis holds still remains to be tested, yet, evidence for migratory waterfowl being an important vector for dispersal now comes from a number of population genetic studies. Mader et al. (1998) showed that for Fennel Pondweed at sites not visited by Bewick's Swans a clear correlation exists between genetic distance and geographic distance, indicating isolation by distance. However, for Fennel Pondweed from populations situated along the main NE-SW migratory flyway of waterfowl in Europe, this was not the case (Fig. 4). Furthermore, individual clones from nearby populations were more diverse along the flyway than outside the flyway. These data indicate more mixing among populations along the flyway, possibly as a result of migratory waterfowl connecting these populations. Similarly, genetic differentiation among *P. pectinatus* populations around the Baltic Sea basin increased with geographical distance, but this effect was smaller along the southeastern Swedish coast, where most water bird traffic took place. Presumably these birds acted as dispersal vectors for the plant (King et al., 2002).

Plant competition under herbivory

For terrestrial herbivores it is well established that they can affect the composition of the

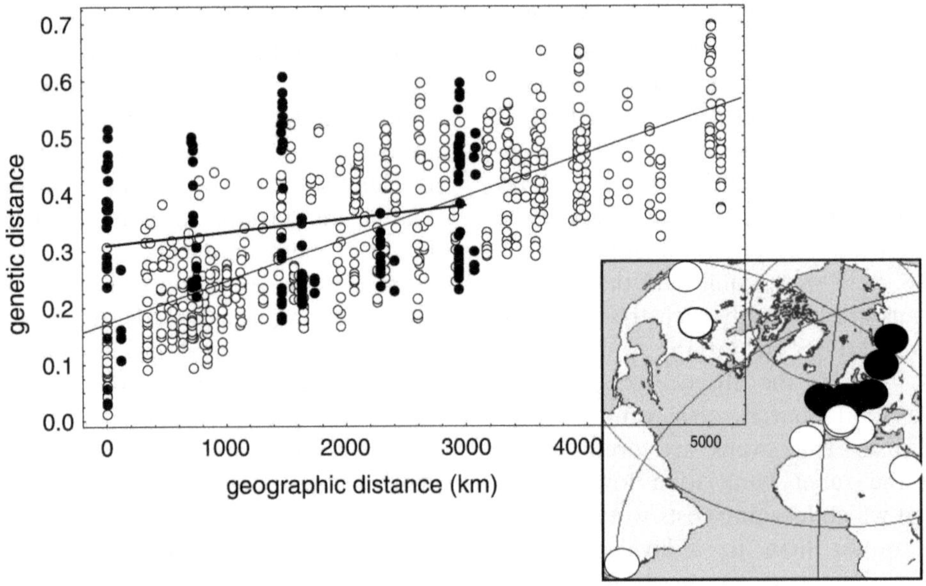

Fig. 4 Correlation of genetic distances of Fennel Pondweed (calculated from 0/1 matrix of RAPD data) versus geographic distances between the sites where pondweed samples were collected. Open symbols represent samples from non-swan-visited populations and closed symbols represent samples from swan-visited populations. The heavy and thin line represent correlations for swan visited and non-swan visited populations, respectively (redrawn from Mader et al., 1998)

vegetation, especially by removing the competitive plant dominant (Pacala & Crawley, 1992; Olff & Ritchie, 1998). It is well established for terrestrial systems that herbivores may maintain specific vegetations, grassland systems being the most renown example. Among water birds, geese are known to create so-called grazing lawns through repeated grazing. These grazing lawns have a specific vegetation composition of plants tolerating grazing, at the same time providing a high and sustainable yield to the geese (Person et al., 2003). Similar phenomena may also apply to aquatic vegetations (Santamaría, 2002).

Fennel Pondweed and Perfoliate Pondweed (*P. perfoliatus*) can often be found in close proximity, Perfoliate Pondweed often occuring in somewhat deeper water than Fennel Pondweed. However, there are several areas where Perfoliate Pondweed occurs in shallow as well as deep waters, while Fennel Pondweed is predominant at intermediate depths. This is the case in the Pechora Delta, an important staging site for Bewick's swans in northern Russia, and in southern Sweden. We therefore conducted an experiment to learn if this typical distribution is related to the high tuber grazing intensity by Bewick's Swans in both areas (Sandsten & Klaassen, 2002). This hypothesis originated through the observation that swans prefer intermediate water depths, where the water is deep enough to uproot the tubers out of the sediment by trampling but where at the same time the water is shallow enough for the swans to reach to the uprooted tubers on the bottom with their beak. In early spring Perfoliate Pondweed stolons were planted in exclosures at intermediate water depths and let to grow. Just before the swans arrived some of these exclosures were removed. The subsequent grazing by notably Bewick's Swans had a dramatic effect. Nearly all the Perfoliate Pondweed was wiped out in the unprotected sites whereas in the protected sites the Perfoliate Pondweed continued to proliferate (Sandsten & Klaassen, 2002). Fennel Pondweed tubers that escape predation have a high chance of sprouting in the following spring. However, because of their size and structure the stolons and roots of Perfoliate Pondweed are probably more easily detected and, if not eaten, damaged.

Swans, however, not always promote their main food plant *P. pectinatus*, especially not when it is the competitive dominant. This may explain why grazing by Trumpeter Swans (*Cygnus buccinator*) was demonstrated to increase submerged plant species richness in pools dominated by *P. pectinatus* (LaMontagne et al., 2003). It may also explain why charophytes can displace canopy-forming *P. pectinatus* in shallow turbid lakes after oligotrophication. Due to eutrophication many shallow lakes like the Veluwemeer lost many of its submerged macrophytes during the 1960s and 1970s, until only *P. pectinatus* was remaining (Van Vierssen et al., 1994). During the recovery phase *P. pectinatus* was largely displaced by charophytes, notably *Chara aspera* (Van den Berg et al., 2003). Competition for inorganic carbon may lead to alternative states in which either species is dominant (Van Nes et al., 2003), but why charophytes can displace canopy-forming *P. pectinatus* in such shallow systems when turbidity is reduced is not understood (Van den Berg et al., 1998). Possibly herbivory by water birds, which presumably mainly affects *P. pectinatus*, may explain this switch (Van den Berg et al., 2003). In theory, Bewick's Swans additionally feeding on *Chara* bulbils (Noordhuis et al., 2002) would be able to deplete the *P. pectinatus* tubers down to very low levels. *Chara* would then affect *P. pectinatus* negatively in an indirect way through supporting their shared enemy (i.e., apparent competition). The hypothesis that remains to be tested is whether bird grazing tips the balance from a competitive advantage of *P. pectinatus* to that of *C. aspera*. In that way water birds would indirectly promote the clear water state of a shallow lake.

Final caveat

One of the points we wanted to make is the importance of aquatic plants for waterfowl. Despite the increased reliance on agricultural crops and their remains, aquatic vegetations form an important food source for many birds. Several species, like Bewick's Swan, cannot do without aquatic plants during migration, which is among the most energy craving periods during the annual cycle. In making our point on the importance of aquatic macrophytes for waterfowl, the focus has been mainly on the need to fulfil their energy requirements. Although of prime importance, foraging is not exclusively a matter of fulfilling energy needs. Also the requirements for (specific) proteins, essential fatty acids and other essential nutrients elements should be met. For waterfowl, it might be easier to maintain a balanced dietary intake while foraging in (semi-) natural environments than while foraging on agricultural fields (cf. Prop & Black, 1998). Fitness consequences of dietary composition and the importance of natural wetlands to meet these demands require further study.

Water birds may have a range of effects on water plants. They may impact their population dynamics, meta-population dynamics, as well as vegetation composition. The consequences of these interactions for the plants involved may be labelled negative to positive. In part this depends on the species of plants involved, but it tends to depend mostly on the perspective one takes in viewing these interactions. By grazing, waterfowl may change the competitive ability of plant species or increase their dispersal rate. Thus the nature of these plant–animal interactions are often perspective dependent. The ultimate outcome of these interactions for the plants, i.e. in terms of plant fitness, depends on the balance between these various interactive processes, which are probably context dependent (Hay et al., 2004). Migrating Bewick's swans might in fact regularly visit the same foraging sites where they harvest the re-growth of the plants time and again. One could thus potentially view Bewick's swans as "nomadic farmers". Therefore, for the case of the interaction between Bewick's Swans and Fennel Pondweed the picture starts to emerge that this often may be a mutualistic relationship were ultimately both the swan and the plant benefit from their interactions.

Acknowledgements A warm thanks to all members of the department of Plant–Animal Interactions for the many stimulating discussions, and their enthusiasm and efforts in conducting the research that formed the basis of this publication. Also the views of Ruurd Noordhuis and Luc de Meester on a previous version of this manuscript are greatly appreciated. This is publication 3995, NIOO-KNAW, Centre for Limnology, Nieuwersluis, The Netherlands and publication 479 of the Centre for Wetland Ecology.

References

Abraham, K. F., R. L. Jefferies & R. T. Alisauskas, 2005. The dynamics of landscape change and snow geese in mid-continent North America. Global Change Biology 11: 841–855.

Beekman, J. H., 1997. International censuses of the NW-European Bewick's Swan population, January 1990 and 1995. Swan Specialist Group Newsletter 6: 7–9.

Beekman, J. H., B. A. Nolet & M. Klaassen, 2002. Skipping swans: fuelling rates and wind conditions determine differential use of migratory stopover sites of Bewick's Swans *Cygnus bewickii*. Ardea 90: 437–460.

Beekman, J. H., M. R. Van Eerden & S. Dirksen, 1991. Bewick's Swans *Cygnus columbianus bewickii* utilising the changing resource of *Potamogeton pectinatus* during autumn in the Netherlands. Wildfowl Suppl. 1: 238–248.

Bellrose, F. C., 1980. Ducks, Geese and Swans in North America, 3 edition. Stackpole Books, Harrisburg, PA, USA.

Brouwer, G. A. & L. Tinbergen, 1939. De verspreiding der kleine zwanen, *Cygnus b. bewickii* Yarr., in de Zuiderzee, vóór en na de verzoeting. Limosa 12: 1–18.

Charalambidou, I., L. Santamaría, C. Janssen & B. A. Nolet, 2005. Digestive plasticity in Mallard ducks modulates dispersal probabilities of aquatic plants and crustaceans. Functional Ecology 19: 513–519.

Clausen, P., B. A. Nolet, A. D. Fox & M. Klaassen, 2002. Long-distance endozoochorus dispersal of submerged macrophyte seeds by migratory waterbirds in northern Europe – a critical review of possibilities and limitations. Acta Oecologica 23: 191–203.

Coops, H., E. H. Van Nes, M. S. Van den Berg & G. D. Butijn, 2002. Promoting low-canopy macrophytes to compromise conservation and recreational navigation in a shallow lake. Aquatic Ecology 36: 483–492.

Darwin, C., 1859. On the origin of species by means of natural selection, or the preservation of favoured races in the struggle for life. John Murray: London.

Declerck, S., J. Vandekerkhove, L. Johansson, K. Muylaert, J. M. Conde-Porcuna, K. Van der Gucht, C. Perez-Martinez, T. Lauridsen, K. Schwenk, G. Zwart, W. Rommens, J. Lopez-Ramos, E. Jeppesen, W. Vyverman, L. Brendonck & L. De Meester, 2005. Multi-group biodiversity in shallow lakes along gradients of phosphorus and water plant cover. Ecology 86: 1905–1915.

Dirksen, S., J. H. Beekman & T. H. Slagboom, 1991. Bewick's Swans *Cygnus columbianus bewickii* in the Netherlands: numbers, distribution and food choice during the wintering season. Wildfowl Suppl. 1: 228–237.

Engelhardt, K. A. M. & M. E. Ritchie, 2002. The effect of aquatic plant species richness on wetland ecosystem processes. Ecology 83: 2911–2924.

Hangelbroek, H. H. & L. Santamaría, 2004. Regulation of propagule size in the aquatic pseudo-annual *Potamogeton pectinatus*: are genetic and maternal non-genetic effects additive? Evolutionary Ecology Research 6: 147–161.

Hay, M. E., J. D. Parker, D. E. Burkepile, C. C. Caudill, A. E. Wilson, Z. P. Hallinan, & A. D. Chequer, 2004. Mutualisms and aquatic community structure: the enemy of my enemy is my friend. Annual Review of Ecology Evolution and Systematics 35: 175–197.

Jefferies, R. L., A. P. Jano & K. F. Abraham, 2006. A biotic agent promotes large-scale catastrophic change in the coastal marshes of Hudson Bay. Journal of Ecology 94: 234–242.

Jefferies, R. L., D. R. Klein & G. R. Shaver, 1994. Vertebrate herbivores and northern plant communities: reciprocal influences and responses. Oikos 71: 193–206.

Jonzén, N., B. A. Nolet, L. Santamaría & M. G. E. Svensson, 2002. Seasonal herbivory and mortality compensation in a swan-pondweed system. Ecological Modelling 147: 209–219.

Kerbes, R. H., P. M. Kotanen & R. L. Jefferies, 1990. Destruction of wetland habitats by lesser snow geese: a keystone species on the west coast of Hudson Bay. Journal of Applied Ecology 27: 242–258.

King, R. A., R. J. Gornall C. D. Preston & J. M. Croft, 2002. Population differentiation of *Potamogeton pectinatus* in the Baltic Sea with reference to waterfowl dispersal. Molecular Ecology 11: 1947–1956.

Knapton, R. W. & S. A. Petrie, 1999. Changes in distribution and abundance of submerged macrophytes in the Inner Bay at Long Point, Lake Erie: implications for foraging waterfowl. Journal of Great Lakes Research 25: 783–798.

LaMontagne, J., L. J. Jackson & R. M. R. Barclay, 2003. Compensatory growth responses of *Potamogeton pectinatus* to foraging by migrating trumpeter swans in spring stop over areas. Aquatic Botany 76: 235–244.

Lauridsen, T. L., E. Jeppesen & F. Østergaard Andersen, 1993. Colonization of submerged macrophytes in shallow fish manipulated Lake Væng: impact of sediment composition and waterfowl grazing. Aquatic Botany 46: 1–15.

Lodge, D. M., 1991. Herbivory on freshwater macrophytes. Aquatic Botany 41: 195–224.

Lodge, D. M., G. Cronin, E. van Donk & A. J. Froelich, 1998. Impact of herbivory on plant standing crop: comparisons among biomes, between vascular and nonvascular plants, and among freshwater herbivore taxa. In Jeppesen E., M. Søndergaard, M. Søndergaard, & K. Christoffersen (eds), The Structuring Role of Submerged Macrophytes in Lakes. Springer, New York: 149–174.

Mader, E., W. Van Vierssen & K. Schwenk, 1998. Clonal diversity in the submerged macrophyte *Potamogeton pectinatus* L. inferred from nuclear and cytoplasmic variation. Aquatic Botany 62: 147–160.

Merne, O. J., 1972. Bewick's Swans feeding on waste potatoes and other agricultural crops. British Birds 65: 394–395.

Miyabayashi, Y. & T. Mundkur, 1999. Atlas of Key Sites for Anatidae in the East Asian Flyway. Wetland

International. http://www.jawgp.org/anet/aaa1999/aaaendx.htm.

Mullié, W. C. & E. P. R. Poorter, 1977. Aantallen, verspreiding en terreinkeus van de kleine zwaan bij vijf landelijke tellingen in 1976 en 1977. Watervogels 2: 85–96.

Nolet, B. A., 2004. Overcompensation and grazing optimisation in a swan–pondweed system? Freshwater Biology 49: 1391–1399.

Nolet, B. A., V. A. Andreev, P. Clausen, M. J. M. Poot & E. G. J. Wessel, 2001. Significance of the White Sea as a stopover for Bewick's Swans *Cygnus columbianus bewickii* in spring. Ibis 143: 63–71.

Nolet, B. A., R. M. Bevan, M. Klaassen, O. Langevoord & Y. G. J. T. Van der Heijden, 2002. Habitat switching by Bewick's swans: maximisation of average long-term energy gain? Journal of Animal Ecology 71: 979–993.

Nolet, B. A., V. N. Fuld & M. E. C. Van Rijswijk, 2006. Foraging costs and accessibility as determinants of giving-up densities in a swan-pondweed system. Oikos 112: 353–362.

Nolet, B. A. & M. Klaassen, 2005. Time and energy constraints in demanding phases of the annual cycle: an example of time limitation in refuelling migratory swans. Oikos 111: 302–310.

Noordhuis, R., D. T. Van der Molen & M. S. Van den Berg, 2002. Response of herbivorous water-birds to the return of *Chara* in Lake Veluwe, The Netherlands. Aquatic Botany 72: 349–367.

Olff, H. & M. E. Ritchie, 1998. Effects of herbivores on grassland plant diversity. Trends in Ecology and Evolution 13: 261–265.

Pacala, S. W. & M. J. Crawley, 1992. Herbivores and plant diversity. American Naturalist 140: 243–260.

Perrow, M. R., J. H. Schutten, J. R. Howes, T. Holzer, F. J. Madgwick & A. J. D. Jowitt, 1997. Interactions between coot (*Fulica atra*) and submerged macrophytes: the role of birds in the restoration process. Hydrobiologia 342/343: 241–255.

Person, B. T., M. P. Herzog, R. W. Ruess, J. S. Sedinger, R. M. Anthony & C. A. Babcock, 2003. Feedback dynamics of grazing lawns: coupling vegetation change with animal growth. Oecologia 135: 583–592.

Polunin, N. V. C., 1984. The decomposition of emergent macrophytes in fresh-water. Advances in Ecological Research 14: 115–166.

Prop, J. & J. M. Black, 1998. Food intake, body reserves and reproductive success of barnacle geese *Branta leucopsis* staging in different habitats. Norsk Polarinstitutt Skrifter 200: 175–193.

Rodríguez-Gironés, M. A., H. Sandsten & L. Santamaría, 2003. Asymmetric competition and the evolution of propagule size. Journal of Ecology 91: 554–562.

Sandsten, H. & M. Klaassen, 2002. Waterfowl foraging affects competitive ability and distribution of *Potamogeton pectinatus* and *P. perfoliatus*. In Sandsten H. (ed), Waterfowl Herbivory on Submerged Macrophytes in Eutrophic Lakes. Lund University, Lund.

Santamaría, L., 2002. Selective waterfowl herbivory affects species dominance in a submerged plant community. Archiv für Hydrobiologie 153: 353–365.

Santamaría, L., J. Figuerola, J. J. Pilon, M. Mjelde, A. J. Green, T. de Boer, R. A. King & R. J. Gornall, 2003. Plant performance across latitude: the role of plasticity and local adaptation in an aquatic plant. Ecology 84: 2454–2461.

Santamaría, L. & M. A. Rodríguez-Gironés, 2002. Hiding from swans: optimal burial depth of sago pondweed tubers foraged by Bewick's swans. Journal of Ecology 90: 303–315.

Scheffer, M., 1998. Ecology of Shallow Lakes. Chapman & Hall, London [etc.].

Scott, D. A. & P. M. Rose, 1996. Atlas of Anatidae Populations in Africa and Western Eurasia. Wetlands International Publication No. 41, Wageningen, The Netherlands.

Shelford, V. E., 1918. Conditions of coexistence. In Ward H. B. & G. C. Whipple (eds), Freshwater Biology. John Wiley, New York: 21–60.

Søndergaard, M., L. Bruun, T. Lauridsen, E. Jeppesen & T. V. Madsen, 1996. The impact of grazing waterfowl on submerged macrophytes: in situ experiments in a shallow eutrophic lake. Aquatic Botany 53: 73–84.

Van den Berg, M. S., H. Coops, J. Simons & A. De Keizer, 1998. Competition between *Chara aspera* and *Potamogeton pectinatus* as a function of temperature and light. Aquatic Botany 60: 241–250.

Van den Berg, M. S., W. Joosse & H. Coops, 2003. A statistical model predicting the occurrence and dynamics of submerged macrophytes in shallow lakes in the Netherlands. Hydrobiologia 506–509: 611–623.

Van Eerden, M. R., R. H. Drent, J. Stahl & J. P. Bakker, 2005. Connecting seas: western Palaearctic continental flyway for water birds in the perspective of changing land use and climate. Global Change Biology 11: 894–908.

Van Nes, E. H., M. Scheffer, M. S. Van den Berg & H. Coops, 2003. Charisma: a spatial explicit simulation model of submerged macrophytes. Ecological Modelling 159: 103–116.

Van Vierssen, W., M. Hootsmans & J. Vermaat, 1994. Lake Veluwe, a Macrophyte-dominated System under Eutrophication Stress. Kluwer Academic Publishers, Dordrecht [etc.].

Wetzel, R. G., 1983. Limnology, 2nd edition. W.B. Saunders, Philadelphia, PA, USA.

Hydrobiologia (2007) 584:215–222
DOI 10.1007/s10750-007-0581-1

SHALLOW LAKES

Cascading trophic effects in pampean shallow lakes: results of a mesocosm experiment using two coexisting fish species with different feeding strategies

María B. Boveri · Rolando Quirós

Abstract Planktivorous fish, both visual predators and filter feeders, enhance eutrophication processes in lakes. In pampean shallow lakes several planktivorous species may coexist but often two species dominate: silverside (*Odontesthes bonariensis*), a visual planktivorous fish when young adult, and sabalito (*Cyphocharax voga*), an omnivorous filter feeder. To assess the relative strength of the cascading trophic effects of the two species, a mesocosm experiment was conducted using different proportions of both species as treatments. Differences were found in water transparency, phytoplankton biomass, crustacean abundance, individual size and biomass. Our results suggest that visual predators intensify eutrophication effects more strongly than filter feeders do.

Keywords Planktivorous fish · Trophic interactions · Pampean shallow lakes · Mesocosm experiments

Guest editors: R. D. Gulati, E. Lammens, N. De Pauw & E. Van Donk
Shallow lakes in a changing world

M. B. Boveri (✉) · R. Quirós
Sistemas de Producción Acuática, Departamento de Producción Animal Facultad de Agronomía, Universidad de Buenos Aires, Av. San Martín 4453, CP1417 Ciudad de Buenos Aires, Argentina
e-mail: boveri@agro.uba.ar

Introduction

Planktivorous fish enhance eutrophication processes in lakes. Feeding by visual predators or filter feeders usually cascades through zooplankton to the base of the food web and thus promote algal growth via suppression of large herbivorous zooplankters (Hurlbert et al., 1972; Carpenter et al., 1985), though the strength of the cascade may depend on various lake ecosystem characteristics (Jeppesen et al., 2003; Borer et al., 2005).

The Pampa plain in South America covers approximately 500,000 km^2 and is sprinkled with several thousand shallow lakes whose state range from eutrophic to highly hypertrophic (Quirós & Drago, 1999). Phytoplankton biomass is usually high, with chlorophyll levels often exceeding 50 mg m^{-3} during summer (Quirós et al., 1988). Because of their shallowness, the lakes are strongly influenced by wind and therefore polymictic (Rennella & Quirós, 2002). Several planktivorous fish species may coexist, though in different proportions depending on the condition of the aquatic environment. However, two species tend to dominate: silverside (*Odontesthes bonariensis*, Valenciennes 1835, Atherinidae) and "sabalito" (*Cyphocharax voga*, Hensel 1870, Curimatidae). As young adult silverside is a visual planktivore (Boveri & Quirós, 2002). Sabalito is described as a planktivore, feeding on benthos by Destefanis & Freyre (1972), as a planktivore and periphyton sucker by Grosman et al. (1996), and as a detritivore by Brassesco et al. (2004). Sabalito dominance in shallow lakes occurs most frequently in eutrophicated lakes and lakes impacted by siltation (Rosso, personal communication). As silverside is an important angling fish, there is some concern to understand which factors may affect the development of its populations, particularly if other fish species outcompete silverside populations.

To obtain more insight into the trophic relationships between silverside and sabalito, and the relative strength of their cascading trophic effects, a mesocosm experiment was conducted using different combinations of the two species as treatments. Previous studies suggest that the two planktivores have different trophic cascade pathways (Drenner et al., 1986), but comparisons between the strength of this effects are only few (Lazzaro et al., 1992; Williams & Moss, 2003). Although mesocosm experiments with visually and filter-feeding planktivorous fish are not new (Drenner et al., 1990), our experimental design using similar abundances but different species proportions is unique.

Materials and methods

The experiment was run for a 5-week period in late summer—early fall 2000 (March 15th–April 17th) in ten outdoor mesocosms located at the Agronomy Faculty of Buenos Aires University. All ten tanks hosted 16 young individuals of pampean silverside and sabalito in five different densities of each species.

The mesocosms consisted of large white fibreglass tanks, 2 m in diameter and 2 m high, holding approximately 5500 l. To prevent stratification and simulate the wind-induced mixing of shallow lake systems, each mesocosm was continually mixed with air that was softly pumped into the tank near the bottom.

In a 70 m^3 reservoir we prepared the experimental medium with groundwater enriched with 4750 g NH_4Cl and 230.5 g KH_2PO_4 and salt (120 kg NaCl) which dissolution were stimulated by a vigorous bubbling of air from eight 1/2″ hoses distributed near the reservoir bottom. Thus, the following average concentrations in the tanks on the first sampling day were achieved: 17.83 mg TN l^{-1}, 1.21 mg TP l^{-1}, achieving a N:P ratio of 31.75 on molar base, and a salinity of 0.28%. The experimental medium mimicked the average conditions of salinity, nutrient contents and ratios of a typical pampean shallow lake and was inoculated with natural phytoplankton and zooplankton imported from de Gomez shallow lake, a typical pampean hypertrophic shallow lake (Quirós et al., 2002). All tanks were filled by pumping at the same time during a 48-h period prior to the onset of the experiment.

Fish were caught by seining in Lake de Gomez. On average, stocked silversides (sil) weighed 3.5 ± 0.8 g (mean weight ± SD) and were 86 ± 7 mm long (mean total length ± SD), while the sabalitos (sab) weighed 6.5 ± 2.1 g and were 73 ± 0.8 mm long.

Silversides and sabalitos differed in individual body weight, and the resulting overall fish biomass was distinct between fish treatment combinations. However, we found that, in our study, fish number was more critical than fish biomass, because we focused at differences between the impacts of two species differing in foraging modes and activity levels. Indeed, silverside is an active open water swimmer, while sabalito is less mobile, and it is well-known that fish with different activity levels have different food demands (Margalef, 1983).

The experimental design comprised five treatment combinations with duplicate replication. Each tank received 16 young individual fishes. The five treatments consisted of five different proportions of Pampean silversides and sabalitos: 0 sil + 16 sab, 4 sil + 12 sab, 8 sil + 8 sab, 12 sil + 4 sab, and 16 sil + 0 sab. Treatments were randomly assigned to tanks, by stocking fish on 15 March (day 1).

Because fish mortality may potentially result in release of nutrients, potentially affecting the outcome of tank mesocosm experiments (Threlkeld, 1988), sick and dead fish detected at the surface or at the bottom were removed during daily inspections of the tanks and replaced by healthy specimens. Due to high undetected mortality rates common of experiments involving silversides (Boveri & Quirós, 2002), we assumed a weekly undetected mortality rate for silverside of 10%. Then, 8 and 24 days after the onset of the experiment we added 0/1/2/3/4 silversides to the tanks with 0 sil + 16 sab, 4 sil + 12 sab, 8 sil + 8 sab, 12 sil + 4 sab, and 16 sil + 0 sab, respectively.

Tanks were monitored daily at noon for pH, electrical conductivity (mS cm^{-1}), turbidity (NTU), dissolved oxygen content (mg l^{-1}), and water temperature (°C) with a HORIBA-U10 limnological probe, and for Secchi disk depth. Water and plankton samples were taken on five dates (days 1, 6, 17, 27, and 34). The first sampling was done prior to fish release on day 1.

Zooplankton was sampled with four vertical tows using a plankton net (mesh size: 69 μm, diameter: 14 cm), pooled, and preserved in 5% sucrose—formalin. The zooplankton was then counted and measured under a dissecting microscope in a Bogorov chamber. Zooplankton biomass was calculated using length-weight regressions (Bottrell et al., 1976).

Integrated water-column samples were taken for analyses of chlorophyll-a (Chl), total phosphorus (TP) and total nitrogen (TN). Samples were collected with a 1.5 cm diameter PVC hose which was lowered to 10 cm above the tank bottom and then slowly moved upwards while water was siphoned into the can. To determine Chl, water was filtered through Whatman GF/F filters, which were subsequently wrapped in aluminium foil and cooled. Chl was extracted in chloroform—methanol in the dark at 4°C for 48 h and absorbance was then read at 665 nm (Wood, 1985).

TP was analysed by the ascorbic acid method corrected for turbidity following current APHA procedures (American Public Health Association, 1995). With a sub-sample of collected water, particles were retained on a Whatman GF/C filter to allow distinction between the soluble and particulate fraction.

Total organic nitrogen was determined by the Kjeldahl method and ammonia in the digested samples was determined using an ORION specific electrode (American Public Health Association, 1995). The concentration of NO_3^- was read with an ORION specific electrode following standard procedures (American Public Health Association, 1995). TN is considered as the sum of both fractions of nitrogen.

At the end of the experiment the water of the tanks was carefully drained through a small mesh seine and all remaining live fish were recovered, weighed and measured. Fish unaccounted for at the time of the tank emptying were assumed dead and decomposed in the tanks during the experiment (on average: 45% and 60%, respectively, of the silversides and sabalitos introduced). Both detected and undetected mortality turned out to be relatively high in our experiment. Thus, the number of live silversides recovered by the end of the experiment was 13 and 11 in the 16 sil tanks, 12 and 8 in the 12 sil tanks, 7 and 5 in the 8 sil tanks, and 4 and 4 in the 4 sil tanks, allowing us to interpret the treatments as they were planned.

Data were analysed by one-way ANOVA with NCSS 2000 (Hintze, 1998). Because of low replication and statistical power, we chose a probability level of $\alpha < 0.10$ to reduce the risk of making the type II error of failing to reject a false null hypothesis. We present here the probability values for the comparison between the 0 sil + 16 sab and the 16 sil + 0 sab treatments.

ANOVA of response variables did not reveal any effects due to tank location or pre-treatment variation, except for a few zooplankton variables. Thus, cladoceran abundance ($P = 0.02$), cladoceran biomass ($P = 0.02$), macrozooplankton abundance ($P = 0.02$), macrozooplankton biomass ($P = 0.02$), and *Daphnia* biomass ($P = 0.02$), were significantly different between the tanks before the start of the experiment. However, this may be ignored because the association between these differences and the treatments applied did not coincide with the differences expected to occur as a result of treatments. On the other hand, the differences disappeared within the first 5 days of the experiment ($P = 0.32$, 0.56, 0.42, 0.57, and 0.57 on the second sampling date, respectively).

Results

Growth of each fish species was inversely proportional to its density in tanks (Fig. 1). At the end of the experiment, silversides had grew by 150 ± 48% above their initial weight in the 4 sil

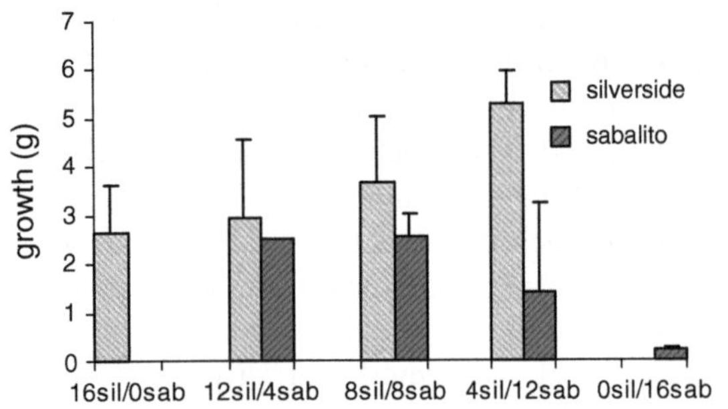

Fig. 1 Average fish growth of silverside and sabalito (final weight – initial weight) recorded at the end of the experiment in each treatment. Vertical lines represent 1 SD

tanks, while sabalitos had enhanced their weight by approx. 38 ± 14% in the 4 sab tanks.

There were no significant differences in macrozooplankton biomass between the treatments ($P = 0.46$, Fig. 2). However, the effects of filter feeding and visual planktivorous fish on the zooplankton community differed.

In the tanks where visual planktivores were most abundant, cladoceran abundance was continuously low and by the end of the experiment they had completely disappeared. In contrast, cyclopoid abundance increased during the first week of the experiment after which it declined (Fig. 3a). In the tanks containing only filter feeders, cladocerans and cyclopoids seemed to be equally controlled by fish (Fig. 3e). Tanks with both visual planktivores and filter feeders exhibited intermediate levels (Fig. 3b–d).

During the experiment, cladoceran size decreased markedly ($P = 0.01$ in comparison

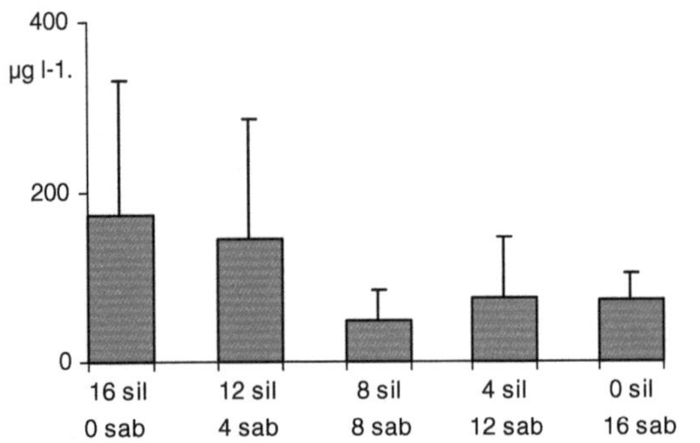

Fig. 2 Effect of the different proportions of visually and filter-feeding planktivores on macrozooplankton biomass (µg DW l^{-1}). Vertical lines represent 1 SD

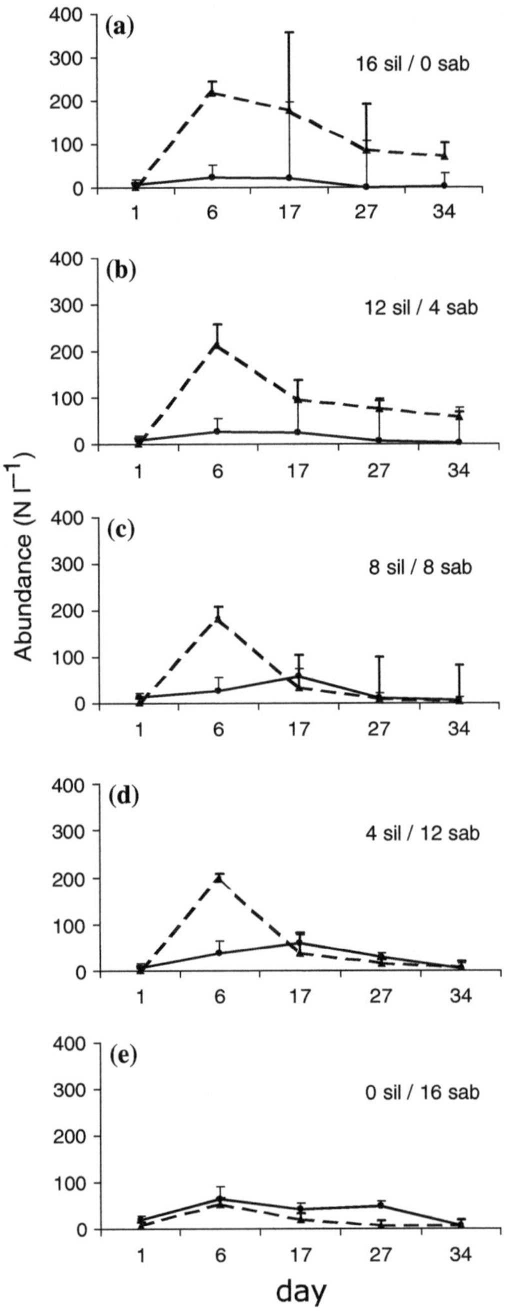

Fig. 3 Average cladoceran (full line) and cyclopoid (broken line) abundance (number of individuals l^{-1}) at all sampling dates for the five treatments. Vertical lines represent ±1SD

between the last and the first day of the experiment), and so, though less pronouncedly, did that of cyclopoids ($P = 0.20$) in the silverside-only tanks (Fig. 4a). However, neither cladoceran

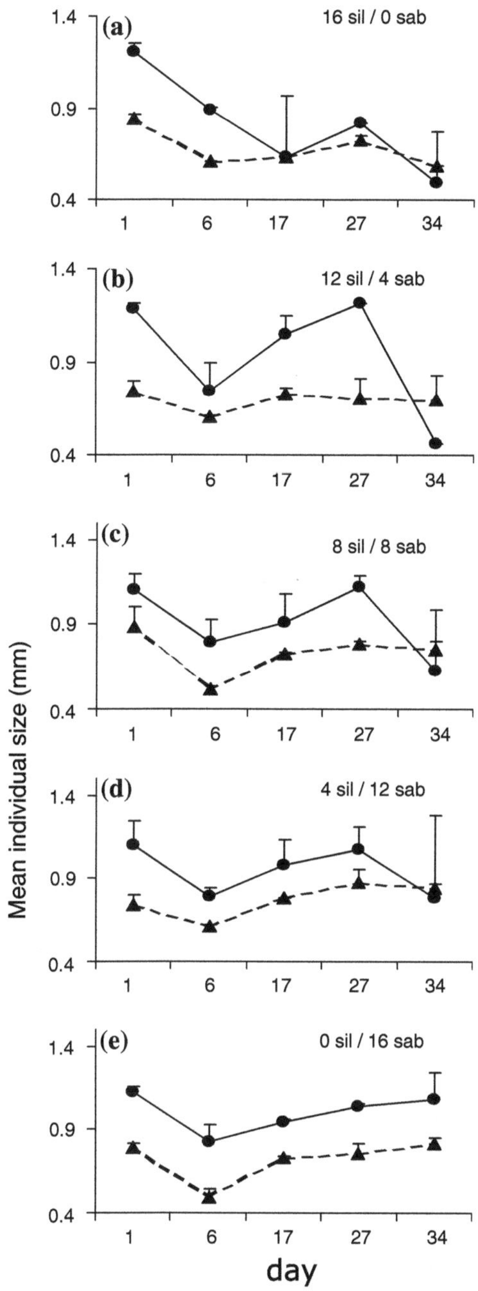

Fig. 4 Average cladoceran (full line) and cyclopoid (broken line) size (mm) at all sampling dates for the five treatments. Vertical lines represent ±1 SD

(P = 0.77) nor cyclopoid size (P = 0.42) decreased in the tanks containing only sabalitos (Fig. 4e).

At the end of the experiment, the sabalito-dominated mesocosms exhibited larger cladocerans (P = 0.05), insignificant differences in cladoceran abundance (P = 0.49), and an enhanced, though not significantly so, cladoceran biomass (P = 0.13) compared to the silverside-mesocosms (Table 1). Despite limited abundance, the cladoceran community was dominated by *Daphnia* in the sabalito tanks, *Daphnia* constituting <20% in the silverside-tanks (P = 0.04, Table 1). Moreover, cladocerans larger than 1 mm were present only in the tanks with few or no silversides (P = 0.02, Table 1).

The fish effects on cyclopoids differed from those on cladocerans. At the end of the experiment, cyclopoids were more abundant in tanks dominated by silversides (P = 0.06, Table 2). In tanks with only filter feeders, cyclopoids were controlled in the same manner as the cladocerans. Silverside prompted only a small reduction of cyclopoid size (P = 0.25, Table 2), especially when compared with the large reduction observed in cladoceran size. Cyclopoid biomass was not affected particularly by silverside density (P = 0.33, Table 2. However, cyclopoids dominated the macrozooplankton community in the silverside-dominated tanks (P = 0.08, Table 2).

During the first week of the experiment, water turbidity increased in all the tanks, after which it declined steadily. In tanks with only visual planktivores (16 sil/0 sab), turbidity increased consistently from the 12th day until the end of the experiment (excepting the effect of a clearwater state due to 100 mm precipitation on the 24th day after the onset of the experiment). For intermediate treatments, turbidity increased as from the 18th day (12 sil/4 sab), 23th day (8 sil/8 sab), and 30th day (4 sil/16 sab). In tanks with only filter feeders (0 sil/16 sab), water turbidity remained low from the third day until the end of the experiment.

At the end of the experiment, tanks with visual planktivores had more turbid water (P = 0.09, Fig. 5a). Moreover, tanks holding only silversides had more than 200 times the chlorophyll concentration than tanks with only filter feeders (P = 0.0008, Fig. 5b). The more visual planktivores, the more turbid the water and the more abundant the phytoplankton, compared with tanks dominated by filter feeders.

During the experiment, nutrient concentrations tended to decrease (P = 0.0001 for TN; P = 0.0001 for TP in a comparison between the last and the first day for all the tanks). The fish stock composition did not affect nitrogen or phosphorus concentrations.

Table 1 Effect of the different proportions of visual predators and filter-feeding planktivores on cladoceran variables at the end of the experiment: abundance (number l^{-1}); mean individual size (mm); biomass (μg l^{-1}); proportion of Daphnia of total cladoceran number (%); and proportion of cladocerans larger than 1 mm (%). In brackets: 1 SD

Cladoceran	16 sil 0 sab	12 sil 4 sab	8 sil 8 sab	4 sil 12 sab	0 sil 16 sab
Abundance (number l^{-1})	2.1 (1.1)	1.1 (1.6)	6.0 (7.1)	5.5 (6.4)	6.6 (6.2)
Mean individual size (mm)	0.5 (0.1)	0.5 (–)	0.6 (0.4)	0.8 (0.5)	1.1 (0.2)
Biomass (μg l^{-1})	2.2 (0.8)	0.8 (1.2)	27.6 (38.5)	45.8 (63.9)	43.9 (24.0)
% of Daphnia of total clad.	17 (23)	0 (–)	45 (64)	78 (67)	100 (0)
% of clad. larger than 1 mm	0 (0)	0 (–)	27 (37)	35 (49)	36 (8)

Table 2 Effect of the different proportions of visual predators and filter-feeding planktivores on cyclopoid variables at the end of the experiment: abundance (number l^{-1}); mean individual size (mm); biomass (μg l^{-1}); and proportion of cyclopoids of total macrozooplankton number (%). In brackets: 1 SD

Cyclopoids	16 sil 0 sab	12 sil 4 sab	8 sil 8 sab	4 sil 12 sab	0 sil 16 sab
Abundance (number l^{-1})	70.4 (23.9)	58.7 (69.1)	5.2 (0.3)	5.9 (2.2)	6.0 (2.3)
Mean individual size (mm)	0.6 (0.2)	0.7 (0.1)	0.8 (0.1)	0.8 (0.0)	0.8 (0.0)
Biomass (μg l^{-1})	170 (158)	143 (141)	20.5 (3.4)	28.4 (8.8)	27.2 (7.8)
% of cyc. of macrozooplank.	97 (0)	91 (13)	58 (39)	62 (28)	54 (18)

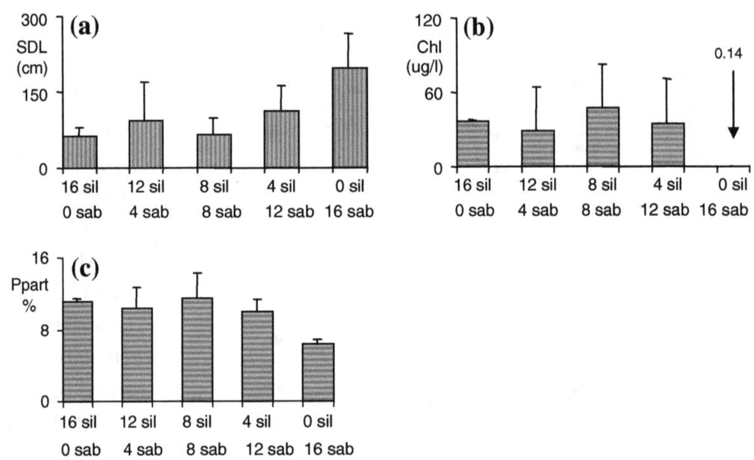

Fig. 5 Effect of the different proportions of visually and filter-feeding planktivores on (**a**) water transparency (SDL, cm); (**b**) phytoplankton biomass (chlorophyll concentration, μg l^{-1}); and (**c**) proportion of particulate phosphorus of total phosphorus (%). Vertical lines represent ±1 SD

However, there were significant differences for phosphorus fractions between treatments. Tanks with more visual planktivores had a higher proportion of particulate phosphorus than tanks dominated by filter feeders ($P = 0.09$, Fig. 5c).

Discussion

Our results show that the feeding strategy of planktivorous fish is an important factor for the outcome of cascading trophic effects.

While filter feeders controlled both cladoceran and copepod size and abundance, visual planktivores selected strongly for cladocerans, cyclopoid abundance thus remaining high. These results agree with those of Janssen (1976) who experimentally showed that a suction feeding device is less effective at catching calanoid copepods than a filtering device. We found that this was true for cyclopoids as well. Moreover, our observations on the behaviour of visual planktivores in the mesocosms also agree with those of Hairston et al.

(1982) concerning the attack of visual planktivores upon their selected prey items.

Our results show that visual planktivores strongly suppressed the largest prey items, the opposite being true for the filter feeders. In consequence, even though the final communities of the silverside- and sabalito-dominated mesocosms had the same total crustacean biomass, the types of macrozooplankton communities differed widely. Thus, the macrozooplankton in the silverside tanks was dominated by cyclopoids and a few small cladocerans, while that of the sabalito tanks was composed of equal proportions of cyclopoids and cladocerans.

The silverside mesocosms were turbid due to a higher phytoplankton biomass, while the sabalito tanks were clearer. This was to be expected from the different composition of the grazer communities (Brooks & Dodson, 1965; Carpenter et al., 1985; Bergquist et al., 1985; Declerck et al., 1997). The grazer community dominated by cyclopoids and small cladocerans should be, and was, less efficient at removing algae than the other dominated by large-sized cladocerans. However, the observed effect of zooplankton on phytoplankton may be the consequence not only of grazing, but the result of direct as well as indirect impacts, such as the recycling of nutrients or removal of competitors (Sommer et al., 2001).

We conclude that through both processes—the preference of visual planktivores for large-sized cladocerans, and the filter feeders' reliance on both cladocerans and cyclopoids as prey items—visual predators enhance eutrophication effects more than filter feeders. Visual planktivorous fish have stronger cascading effects on phytoplankton biomass. Fish directly act on the herbivorous zooplankton via predation, with indirect effects on the phytoplankton. The effects of the visual planktivores on the most efficient herbivores, large-sized cladocerans, cascade strongly down the food web and lead to a marked deterioration of water transparency via the growth of the algal standing crop.

Our results are not in agreement with those of Lazzaro et al. (1992), who found that cyclopoid density was higher in tanks with the filter feeder gizzard shad (*Dorosoma cepedianum*) than with the visual feeder bluegill (*Lepomis macrochirus*), and our results on phytoplankton growth are also the opposites.

The results obtained allow us to elucidate some of the trophic relations occurring in the complex pelagic system of pampean shallow lakes of relevance for managers.

Our results do not support the hypothesis proposing that the replacement of visual planktivores by filter feeders in hypertrophic shallow lakes is a result of exploitative competition. There are important differences in many characteristics of the zooplankton community resulting from the feeding behaviour of both fish species. In fact, where filter feeders and visual planktivores coexist, the size reduction of cladocerans is similar to that observed in tanks inhabited only by visual planktivores. This important effect on the largest cladocerans is hardly attributable to a cladoceran enlargement in the filter feeder tanks but rather to the feeding activities of the visual predators, as concluded also in previous studies (Boveri & Quirós, 2002). We therefore hypothesize that the two species only have a few prey items in common, which contradicts the argument of competitive displacement of visual planktivores by filter feeders.

We suspect that changes in the environment of pampean shallow lakes induced by human activities may affect the survival of silverside. The current intensification of agriculture, stockbreeding activities and urbanization have triggered intense eutrophication processes (Quirós et al., 2002) that may potentially lead to dominance by other fish species than those dominant today. Filter feeders, detritivores, and omnivores often substitute visual predators at increasing eutrophication (Horne & Goldman, 1994). Because silverside and sabalito often dominate the fish communities in pampean shallow lakes, the ecological relationships among fish species, plankton communities and water quality may have important implications for lake management, and the intensified land use in the lake catchments calls for the implementation of new management measures in our shallow lakes.

Acknowledgements R. Quirós acknowledges research support from the Consejo Nacional de Investigaciones Científicas y Tecnológicas (CONICET). We thank J. J. Rosso, C. A. Petracchi, A. M. Rennella, D. Blanco Bello, V. Asikian and H. T. von Bernard for their assistance, and Miss Anne Mette Poulsen for language assistance, and are grateful to Alejandro Sosnovsky and Armando Martín Rennella for their insightful comments. Thanks go also to Erik Jeppesen for his valuable comments and the improvement of an early version of the manuscript. We also appreciate the critics, comments and ideas of two reviewers: Xavier Lazzaro and an anonymous person. This work was supported by Agencia Nacional de Promoción Científica y Técnica grant PMT-PICT 4698.

References

American Public Health Association, American Water Works Association, Water Environment Federation, 1995. Standard methods for examination of water and wastewater. 19th edn.

Bergquist, A. M., S. R. Carpenter & J. C. Latino, 1985. Shifts in phytoplankton size structure and community composition during grazing by contrasting zooplankton assemblages. Limnology and Oceanography 30: 1037–1045.

Bottrell, H. H., A. Duncan, Z. M. Gliwicz, E. Grygierek, A. Herzig, A. Hillbricht-Ilkowska, H. Kurasawa, P. Larson & T. Weglenska, 1976. A review of some problems in zooplankton production studies. Norwegian Journal of Zoology 24: 419–456.

Borer, E. T., E. W. Seabloom, J. B. Shurin, K. E. Anderson, C. A. Blanchette, B. Broitman, S. D. Cooper & B. S. Halpern, 2005. What determines the strength of a trophic cascade?. Ecology 86: 528–537.

Boveri, M. B. & R. Quirós, 2002. Trophic interactions in pampean shallow lakes: evaluation of silverside predatory effects in mesocosm experiments. Verhandlungen Internationale Vereinigung für Theoretische und Angewandte Limnologie 28: 1274–1278.

Brassesco, M. S., M. C. Pastori, H. A. Roncati & A. S. Fenocchio, 2004. Comparative cytogenetic studies of Curimatidae (Pisces, Characiformes) from the middle Paraná River (Argentina). Genetics and Molecular Research 3: 293–301.

Brooks, J. & S. Dodson, 1965. Predation, body size, and composition of Plankton. Science 150: 28–35.

Carpenter, S., J. Kitchell & J. Hodgson, 1985. Cascading trophic interactions and lake productivity. Bioscience 35: 634–639.

Declerck, S., L. De Meester, N. Podoor & J. M. Conde Porcuna, 1997. The relevance of size efficiency to biomanipulation theory: a field test under hypertrophic conditions. Hydrobiologia 360: 265–275.

Destefanis, S. & L. Freyre, 1972. Relaciones tróficas de los peces de la laguna de Chascomús con un intento de referenciación ecológica y tratamiento bioestadístico del espectro trófico. Acta Zoológica Lilloana 29: 17–33.

Drenner, R. J., J. D. Smith, J. R. Mummert & H. F. Lancaster, 1990. Responses of a eutrophic lake community to separate and combined effects of N:P supply and planktivorous fish: a mesocosm experiment. Hydrobiologia 208: 161–167.

Drenner, R. W., S. D. Threlkeld & M. D. McCracken, 1986. Experimental analysis of the direct and indirect effects of an omnivorous filter-feeding clupeid on plankton community structure. Canadian Journal of Fisheries and Aquatic Sciences 43: 1935–1945.

Grosman, M. F., J. R. González & E. J. Usunoff, 1996. Trophic niches in an Argentine pond as a way to assess functional relationships between fishes and other communities. Water SA 22: 345–350.

Hairston, N. G. Jr, K. T. Li & S. S. Easter Jr, 1982. Fish vision and the detection of planktonic prey. Science 218: 1240–1242.

Hintze, J. L., 1998. Number Cruncher Statistical System (NCSS). Version 2000. User's Guide.

Horne, A. & C. Goldman, 1994. Limnology. 2nd edn, McGrawHill, New York.

Hurlbert, S. H., J. Zedler & D. Fairbanks, 1972. Ecosystem alteration by mosquitofish (Gambusia affinis) predation. Science 175: 639–641.

Janssen, J., 1976. Selectivity of an artificial filter feeder and suction feeders on calanoid copepods. American Midland Naturalist 95: 491–493.

Jeppesen, E., J. P. Jensen, C. Jensen, B. Faafeng, P. Brettum, D. Hessen, M. Søndergaard, T. Lauridsen & K. Christoffersen, 2003. The impact of nutrient state and lake depth on topdown control in the pelagic zone of lakes: study of 466 lakes from the temperate zone to the Arctic. Ecosystems 6: 313–325.

Lazzaro, X., R. W. Drenner, R. A. Stein & J. D. Smith, 1992. Planktivores and plankton dynamics: effects of fish biomass and planktivore type. Canadian Journal of Fisheries and Aquatic Sciences 49: 1466–1473.

Margalef, R., 1983. Limnología. Editorial Omega, Barcelona.

Quirós, R. & E. Drago, 1999. The environmental state of Argentinean lakes: An overview. Lakes & Reservoirs. Research and Management 4: 55–64.

Quirós, R., J. J. Rosso, A. Rennella, A. Sosnovsky & M. Boveri, 2002. Análisis del estado trófico de las lagunas pampeanas (Argentina). Interciencia 27: 584–591.

Quirós, R., C. R. M. Baigun, S. Cuch, R. Delfino, A. de Nichilo, C. Guerrero, M. C. Marinone, S. Menu Marque, S. & M. C. Scapini, 1988. Evaluación del rendimiento pesquero potencial de la republica Argentina: I. Datos. Instituto Nacional de Investigación y Desarrollo Pesquero. Informes Técnicos del Departamento de Aguas Continentales N°7.

Rennella, A. & R. Quirós, 2002. Relations between planktivorous fish and zooplankton in two very shallow lakes of the Pampa Plain. Verhandlungen Internationale Vereinigung für Theoretische und Angewandte Limnologie 28: 887–891.

Sommer, U., F. Sommer, B. Santer, C. Jamieson, M. Boersma, C. Becker & T. Hansen, 2001. Complementary impact of copepods and cladocerans on phytoplankton. Ecology Letters 4: 545–550.

Threlkeld, S. T., 1988. Planktivory and planktivore biomass effect on zooplankton, phytoplankton and the trophic cascade. Limnology and Oceanography 33: 1364–1377.

Williams, A. & B. Moss, 2003. Effects of different fish species and biomass on plankton interactions in a shallow lake. Hydrobiologia 491: 331–346.

Wood, L. W., 1985. Chloroform—methanol extraction of chlorophyll-a. Canadian Journal of Fisheries and Aquatic Sciences 42: 38–43.

Hydrobiologia (2007) 584:223–237
DOI 10.1007/s10750-007-0587-8

SHALLOW LAKES

Impact of hydrology on aquatic communities of floodplain lakes along the Daugava River (Latvia)

Davis Gruberts · Ivars Druvietis · Elga Parele · Jana Paidere · Arkadijs Poppels · Janis Prieditis · Arturs Skute

Abstract During July 2004, various limnological characteristics of 24 floodplain lakes and reservoirs have been explored along the Middle Daugava for the first time in order to reveal possible impact of the long-term mean annual flooding frequency on their phytoplankton, zooplankton, macrozoobenthos and macrophyta communities. Obtained data series were analysed by Spearman's rank correlation method, Principal Component Analysis (PCA) method and Renkonen's similarity test. UPMGA method was used for single linkage clustering of the lakes based on the abundance of phyto- and zooplankton taxa. Low similarity between the obtained cluster trees and hydrological grouping was stated indicating minor impact of the flooding hydrology on summer plankton communities of these lakes. Significant correlation between the flooding frequency and several physicochemical and biological parameters was found. Six main factors, which explain observed variations, were extracted by PCA. Significant negative impact of hydrological connectivity on zooplankton species diversity as well as positive impact on Oligochaeta density was stated, whereas other biotic parameters were affected by local factors, such as lake morphology, internal loading of nutrients from sediments, throphic interactions as well as local source of dissolved organic matter.

Keywords Floodplain lakes · Seasonal flooding · Long-term impact · Daugava River

Guest editors: R. D. Gulati, E. Lammens, N. De Pauw & E. Van Donk
Shallow lakes in a changing world

D. Gruberts (✉) · J. Paidere · J. Prieditis · A. Skute
Daugavpils University, Daugavpils, Latvia
e-mail: davis@dau.lv

I. Druvietis · E. Parele
University of Latvia, Institute of Biology, Salaspils, Latvia

A. Poppels
Agency of Latvia's Fish Resources, Riga, Latvia

Introduction

The "flood pulse" is commonly perceived as the main factor that controls the existence and productivity of floodplain lake ecosystems of major world rivers (Junk et al., 1989; Junk, 1997; Lewis et al., 2000). Seasonal flooding stimulates an elemental transfer between sediments and water column, simplifies ecosystem's structure, rejuvenates resources and stimulates phytoplankton productivity in large Amazonian floodplain lakes (Furch & Junk, 1993; Huszar & Reynolds, 1997). On the other hand, despite their widespread distribution and comparatively easy access, smaller lakes and ponds are much less explored to permit

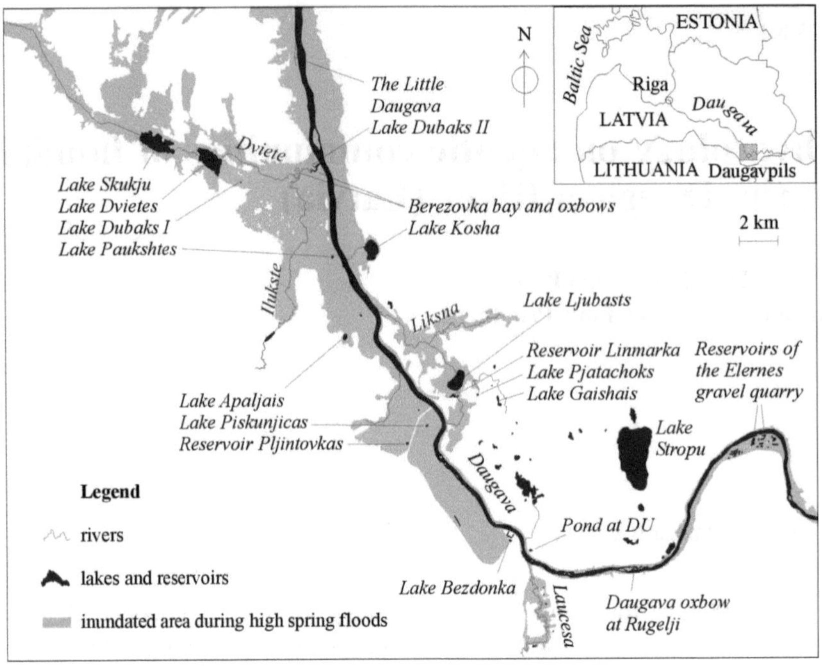

Fig. 1 Location of selected floodplain lakes and reservoirs along the Middle Daugava, South-East Latvia

any generalized assessment of their plankton ecology (Reynolds, 1993).

Still, there were some comprehensive studies in recent years focused on the role of hydrology in plankton communities of floodplain lakes of highly eutrophic rivers of Central Europe. Hydrology was found to exercise a major impact on water chemistry and phyto- and zooplankton community composition in floodplain lakes distributed along the lower Rhine and Meuse, in Netherlands (Van den Brink et al., 1994). However, in-lake processes rather than river dynamics seem to be driving the turbidity of floodplain lakes along the lower Rhine (Roozen et al., 2003).

The Daugava River (Zapadnaja Dvina) is among the largest East-European watercourses. It is about 1,000 km long, flows through Russia, Byelorussia and Latvia and drains to the Baltic Sea. Since the second part of 20th century, dams of three huge hydroelectric power stations have completely blocked river transport as well as migratory paths for salmon, eel and lamprey in the Lower Daugava, in Latvia. Construction of the Cascade also created other serious environmental problems, such as an erosion of the shores of reservoirs, formation of ice jams during the spring floods at Jekabpils City and so on. Despite these obvious faults, during the next 10–12 years construction of four new dams is proposed on the Upper Daugava, in Byelorussia. Therefore, ecosystems of floodplain lakes along the entire Daugava River are under threat of anthropogenic impact today.

The largest concentration of such small, shallow water bodies in the Daugava floodplain can be found in Daugavpils region, South-east Latvia (Fig. 1).

In total, there are more than 40 floodplain lakes and reservoirs of glacial, fluvial and artificial origin on both sides of the river in South-east Latvia. So far, their hydrology remains rather "natural", without any impact from the existing hydroelectric power stations located further downstream. Most of them are connected to the Daugava during the spring floods and become more or less isolated in other seasons. Therefore, their water levels are significantly influenced by such natural factors, as the amount of snow accumulated in the Daugava's drainage area during the winter, the air temperature increase in spring and formation of ice jams in the Daugava's valley during the floods (Gruberts, 2003).

Until now, only one study has focused on the ecological role of seasonal flooding of the Daugava River in this region. In 1999, an impact of the Daugava flood pulse on its largest floodplain lake ecosystem (Lake Skukju or Grivas) located in the Dvietes Floodplain Nature Park (Fig. 1) has been explored. During the spring floods of 1999, considerable water level rise and dilution in this shallow lake was observed. Impact on phytoplankton composition and diversity was also stated during the floods. Three distinct phases in hydrological regime (isolation, filling, drainage) as well as compositional and structural shift from planktonic to epiphytic (benthic) algae groups controlled by water level change were observed in the phytoplankton community of the lake (Gruberts & Druvietis, 2001; Gruberts et al., 2005).

In recent years, a comparative research was carried out in order to evaluate the ecological effect of the long-term artificial pumping of the Daugava's water into the Lake Stropu near Daugavpils City for industrial needs (Fig. 1). This study revealed an increased eutrophication of the lake's water column and significant alternations of the phytoplankton community due to this impact (Katelo, 1995).

These two examples suggested that short-term (seasonal) as well as long-term impacts of the Daugava's hydrology on its floodplain lakes are possible. The aim of this study was to search for an impact of long-term annual flooding conditions (flooding frequency) on phytoplankton, zooplankton, macrozoobenthos and macrophyta communities of various floodplain lakes and reservoirs located along the Middle Daugava, in South-east Latvia.

Materials and methods

In order to select sampling sites for this study, relative height above the mean summer water level in the Daugava of more than 40 lakes and reservoirs was calculated by applying high-resolution topographic maps (scale 1:10,000). Besides, their long-term annual flooding frequency was determined by using hydrological data from six posts, which were located along the Middle Daugava and operated for more than 100 years (Gosudarstvennij vodnij kadastr 1987). After this preliminary research, 24 floodplain lakes and reservoirs of the Middle Daugava characterised by different flooding conditions were selected for further investigations (Table 1). In most cases, physicochemical and biological parameters of selected water bodies were explored for the first time.

Field measurements and sampling in selected water bodies were performed during July 18–28, 2004. The largest depth was measured from boat by echo sounding. Water temperature, pH, conductivity (CND), total diluted solids (TDS), dissolved oxygen, red-ox potential (ORP) and turbidity was measured at 0.5 m depth and down

Table 1 Hydrological grouping of floodplain lakes and reservoirs of the Middle Daugava according to the long-term annual flooding frequency

Long-term annual flooding frequency	Lakes (reservoirs)	Points[a]
Not flooding at all even at the highest flood level because of artificial isolation	Lake Gaishais, pond at Daugavpils University (DU)	0
Flooding rarely (once or twice in a century), only during the highest spring floods	Lakes Pjatachoks, Apaljais	1
Flooding regularly (once or twice in every 10 years), at the mean perennial flood level	Lakes Paukshtes, Bezdonka, Piskunjicas, Kosha, reservoir Linmarka	2
Flooding repeatedly (once or twice per year), not only in spring but also at the highest summer (autumn) water level caused by heavy rainstorms	Lakes Skukju, Dvietes, Ljubasts, Dubaks I, reservoir Pljintovkas, reservoirs of the Elerne's gravel quarry	3
Flooding frequently (several times per year), even at the mean perennial summer (autumn) water level	Lake Dubaks II, Berezovka oxbow II, The Little Daugava	4
Flooding very frequently (many times per year), even at the mean lowest summer water level	Berezovka oxbow I, Berezovka bay, Daugava oxbow at Rugelji	5

[a] Relative numeric scale of flooding frequency created for statistical analysis

to the bottom at 1 m interval in the deepest parts of the lakes by *Hydrolab Surveyor 4* miniprobe fitted with adequate sensors. Water transparency was measured by *Secchi* disc, 30 cm in diameter.

Water samples were collected at the 0.5 m depth in the deepest parts of the lakes by Ruttner type sampler and analysed for total nitrogen (total N) and total phosphorus (total P) at Ecological Laboratory of Daugavpils Regional Environment Board on the same or the next day after sampling according to standard methods accredited in Latvia (LVS 340:2001 & LVS EN 1189:2000).

Phytoplankton samples were collected at 0.5 m depth in the deepest parts and littoral zones of the lakes by Ruttner type sampler and immediately fixed with the Lugol's iodine solution. Phytoplankton taxa were identified and enumerated some time after sample collection, following the inverted microscope method of Utermöh (1958). Algal biomass was established by means of the volume measure method comparing algal shapes of particular species and their individual volume. Concentrated samples of zooplankton (100 l in volume) were collected from the epilimnion of the deepest parts and littoral zones by 65 μm plankton mesh and fixed in the field by 4% formaldehyde solution (Wetzel & Likens, 1991). The light microscope was used for the zooplankton species identification. Macrozobenthos samples (three at each sampling site) were collected in the deepest parts and littoral zones by the Ekman type grab with a 1/40 m^2 surface. A sieve with a mesh size 0.5 mm and 4% formaldehyde solution was used for the zoobenthos samples concentration and fixation. Composition and distribution of aquatic vegetation was observed in the field.

In order to evaluate accuracy of the long-term annual flooding frequency data obtained before, seasonal amplitude of water level fluctuation was determined for each lake by instrumental levelling of the highest position of the spring flood debris to actual water level in the lake on the day of sampling.

To test an impact of the long-term annual flooding conditions and other environmental factors on the biotic communities of the lakes, Spearman's rank correlation coefficients were calculated between the all measured parameters. In order to extract main factors, which explain the observed variance in biotic communities, a Kaiser–Meyer–Olkin test of sampling adequacy and Principal Component (Factor) Analysis was applied to selected independent parameters using *SPSS for Windows* statistic software package. Several biotic parameters, which demonstrated significant correlation to the mean annual flooding frequency and/or an amplitude of the water level fluctuation, were used for the principal component extraction. Some other environmental variables, which correlate with the selected biotic parameters, were also included in this procedure. After the initial extraction, the principal components were rotated in order to increase their scores in relation of the variables. For this, the Varimax method with the Kaiser Normalization was applied.

To determine if the selected water bodies can be classified together or need to be separated, calculations of the Renkonen's coefficients (percentage similarity) were performed after the phyto- and zooplankton community data standardization in terms of percentages. They were used further for an average linkage clustering of the selected lakes and reservoirs according to the unweighted pair-group method using arithmetic averages (UPGMA method) (Krebs, 1989). The obtained cluster trees were used in order to illustrate large differences between phyto- and zooplankton communities of these water bodies and for a qualitative comparison with the hydrological grouping presented above.

Results

During the field measurements, traces of considerable water level fluctuation (>0.3 m $year^{-1}$) were observed on the shore of 21 lakes out of 24 selected for this study. Remaining three water bodies were not inundated because of isolation by the flood protection dams (Lake Gaishais, pond at Daugavpils University), or because of the average magnitude of the flood pulse (Lake Pjatachoks). Seasonal amplitude of water level fluctuation varied among the lakes depending on the long-term annual flooding frequency and relative height of the lakes above the mean summer water level in the river (Table 2, Fig. 2). The largest amplitude of water level fluctuation (6–7 m $year^{-1}$) was stated for the water bodies characterised by low position within the floodplain and high frequency of flooding (Lake Rugelju, Berzovka Bay, etc.) (Table 1), but, in most cases, it varied between 2 and 4 m per year.

In spite of similar flooding amplitude, significant variation in physicochemical parameters of the selected water bodies was stated during this study (Fig. 3). Large differences were stated in the water turbidity, temperature, dissolved O_2, total N and other physicochemical parameters, even for morphologically and hydrologically similar lakes like Dvietes and Skukju. Except for four reservoirs located within the Elernes gravel quarry at Daugavpils City and characterised by lower pH values (6.6–7.2), any other group of physically or chemically similar water bodies was not found. Meanwhile, pH values had the least variability between the measured parameters.

Distinct thermal stratification was found in some of the lakes with the maximum depth >4 m (Fig. 4) characterised by a sharp decrease in water temperature and dissolved oxygen concentrations with the depth. In most cases thermocline was located at 1–3 m depth, but its scale was different, indicating a significant impact of the lakes' morphometry on the water column stability. For example, in the lake Pjatachoks, which is about 9 m deep and has a small surface area (0.5 ha), differences in water temperature between the upper and deeper layers reached 20°C whereas in the lake Kosha, which has similar maximum depth but much larger surface area (60 ha), it was only about 5°C (Fig. 4).

Although 125 taxa (species) of phytoplankton, 50 taxa of zooplankton, 77 taxa of macrozoobenthos as well as 85 taxa of macrophyta were stated during this study, only 29 taxa could be considered as common, found in more than 50% of the sampled sites (Table 3). Dominance of rare species (taxa), each found in only one lake, was obvious, especially in the composition of macrozoobenthos community, where it exceeded 50% of total number of taxa.

The rare species formed also the highest biomass (abundance) within these communities rather than common ones. For example, phytoplankton biomass of the Kosha Lake was completely dominated by the Cyanophyta *Oscillatoria* sp., which forms the summer algal blooms in many shallow, eutrophic lakes and reservoirs in Latvia (Druvietis, 1998) (Fig. 5). On the contrary, phytoplankton biomass of the Lake Pjatachoks was dominated by the Chloromonadophyta *Gonyostomum semen* (Ehrenberg) Diesing, a typical taxa for the peat bog lakes and freshwater marshes (Fott, 1971).

After the Spearman's rank correlation analysis, significant correlations were found between several biotic parameters (mainly the zooplankton community characteristics), the long-term mean annual flooding frequency and amplitude of water level fluctuation (Table 4). For example, both hydrological variables significantly correlated with the number of zooplankton taxa as well as the number of Cladocera, Copepoda and, especially, *Cyclopus* sp. individuals per m^3. On the other hand, these biotic parameters significantly correlated also to some environmental variables, such as the water temperature, red-ox potential, biomass of some algae taxa and so on.

After the Kaiser–Meyer–Olkin test of sampling adequacy, some biotic parameters were unsuitable for the first step of factor analyses, like the number of Oligochaeta individuals, which was known for the less than 1/2 of the selected water bodies. On the other hand, it was possible to unite the surface area and maximum depth data into a single surface/depth ratio variable. By initial application of Principal Component Analysis (PCA) to these data, six components (factors) were extracted. They explained 81% of total variance. After their rotation by Varimax method, the first three components indicated significant impact on the most of variables (Table 5).

According to the results of PCA, the first component explains variation in flooding frequency, water level amplitude, conductivity and number of zooplankton taxa. Component 2 explains variation in the area/depth ratio, temperature, dissolved O_2 and number of Cladocera individuals; component 3—in pH, red-ox potential, total P, *Anabaena* sp. biomass and number of Copepoda individuals; component 4—in total N and *Cryptomonas* sp. biomass; component 5—in

Table 2 Selected morphometrical, hydrological and biological parameters

Sampling site	Surface area/ maximum depth ratio	Relative height (m)	Level amplitude (m year^{-1})	Transparency (m)	Phytoplankton biomass (mg l^{-1})	Zooplankton abundance (×10^6 ind. m^{-3})	Zoobenthos biomass (×10^2 ind. m^{-2})	Vegetation cover (%)	Number of taxa			
									Phyto-plankton	Zoo-plankton	Macrozoo-benthos	Macro-phyta
Lake Skukju (Grivas)	100	3.0	3.6	1.1[a]	0.0	0.4	0.1	80	18	11	21	31
Lake Dvietes	89	2.9	3.6	0.9[a]	0.0	1.5	0.4	95	4	6	27	27
Lake Dubaks I	0.0	4.3	3.1	1.5	0.6	2.2	–	10	18	9	–	26
Berezovka oxbow I	1.0	0.8	5.8	0.5	0.0	0.3	2.0	20	11	7	10	20
Lake Dubaks II	0.0	2.1	4.5	0.8	0.3	–	–	10	4	–	–	20
Berezovka Bay	0.2	0.7	5.9	1.3	1.3	0.0	0.4	8	20	6	4	15
Berezovka oxbow II	0.2	2.3	4.3	0.8	0.3	4.9	0.4	95	19	9	5	20
The Little Daugava	1.6	1.7	5.3	0.8	0.8	0.7	0.4	70	15	5	9	21
Lake Paukshtes	0.4	6.0	1.1	1.6	0.1	13.8	0.8	35	43	13	4	20
Lake Apaljais	10	7.9	1.0	0.5[a]	1.0	4.8	0.3	100	26	16	4	15
Reservoir Pljintovkas	0.5	6.8	3.0	2.5	0.2	1.0	47	15	28	11	9	23
Lake Kosha	13	4.5	1.7	0.4	5.0	8.3	1.3	8	5	15	4	28
Reservoir Linmarka	1.3	6.0	2.9	0.8	0.2	15.1	3.1	20	11	15	2	18
Lake Ljubasts	60	5.6	3.3	1.0[a]	0.3	2.3	1.4	95	48	20	5	24
Pond at DU	0.4	5.4	0.3	2.2	0.1	3.6	0.0	10	25	20	2	24
Lake Gaishais	1.3	9.9	0.1	3.2	0.3	1.8	0.3	5	19	12	8	22
Lake Piskunjicas	1.3	5.7	2.4	1.2	0.8	12.9	–	25	24	18	–	28
Lake Bezdonka	0.1	6.5	2.0	1.6	0.1	4.3	–	10	12	13	–	16
Lake Pjatachoks	0.1	9.5	0.1	1.8	0.5	1.3	–	1	9	8	–	14
Lake Rugelju	0.1	2.5	6.9	1.3	4.5	1.2	0.4	40	28	7	3	15
Reservoir Elernes I	0.3	5.5	3.1	1.2	0.2	–	–	30	29	–	–	9
Reservoir Elernes II	0.3	5.2	3.3	1.5[a]	0.1	4.7	–	5	16	13	–	14
Reservoir Elernes III	0.5	5.2	3.4	1.5[a]	0.1	0.5	–	5	17	10	–	3
Reservoir Elernes IV	0.4	5.5	3.1	1.3	0.1	9.1	–	10	10	10	–	4

[a] Transparency down to bottom

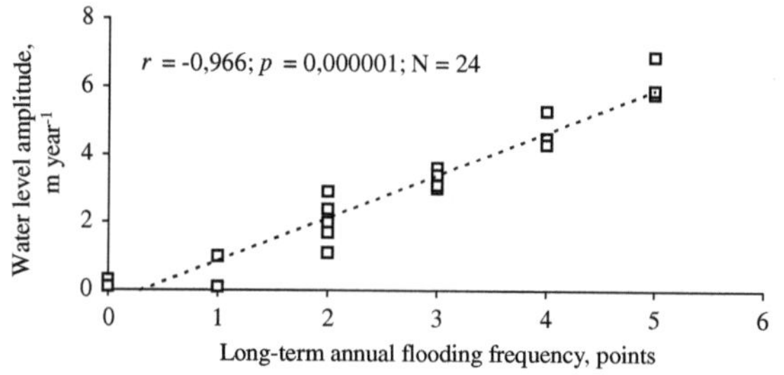

Fig. 2 Correlation between the amplitude of water level fluctuation and the long-term annual flooding frequency

number of Rotatoria and *Cyclopus* sp. individuals; component 6—in *Euglena* sp. biomass. After some additional analyses, number of Oligochaeta individuals tended to be most closely related to component 1, vegetation cover—to component 2.

Strong differences between the explored lake communities were found after the application of similarity and cluster analysis. Single linkage clustering based on the Renkonen's similarity coefficients, which were calculated from the phytoplankton and zooplankton species abundance data, resulted in two quite different cluster tree diagrams (Fig. 6).

In the cluster tree A, selected water bodies resembled each other with respect to their phytoplankton communities. From this cluster, five groups were identified (Fig. 6a). Group I included nine lakes and reservoirs, which were characterised by high percentage of Cryptophyta taxa *Cryptomonas* sp. in total biomass of their phytoplankton communities. The highest percentage of this taxa was stated in three most similar water bodies: the Little Daugava, Lake Ljubasts and reservoir Pljintovkas (Renkonen's similarity coefficients > 95%), which were united in one subgroup (I a). Group II was composed of three other floodplain lakes characterised by higher biomass of Chrysophyta species *Dynobrion divergens* in phytoplankton biomass. Group III was composed by two water bodies, which were similar to each other by higher biomass of Chlorophyta species *Volvox aureus*. Group IV was also composed by 3 lakes characterised by higher biomass of Chlorophyta species *Ankistrodesmus falcatus*, *Botryococcus braunii* and *Staurastrum* sp. Group V was composed by four lakes characterised by higher biomass of Chlorophyta species *Scenedesmus quadricauda* and Dinophyta species *Peridinium cinctum* and *Ceratium hirudinella*.

The four other water bodies (reservoir Elernes 4, Berezovka bay, Lake Kosha and Pjatachoks) were quite different from each other and the rest of the lakes (Renkonen's similarity coefficients <5%) and therefore were not included in any group mentioned above (Fig. 6a). Reservoir Elernes 4 is characterised by higher biomass of Cyanophyta species *Gomphosphaenia lacustris* and Bacillariophyta species *Melosira varians*; Berezovka bay—by higher biomass of Bacillariophyta species *Fragilaria capucina*, *Melosira*

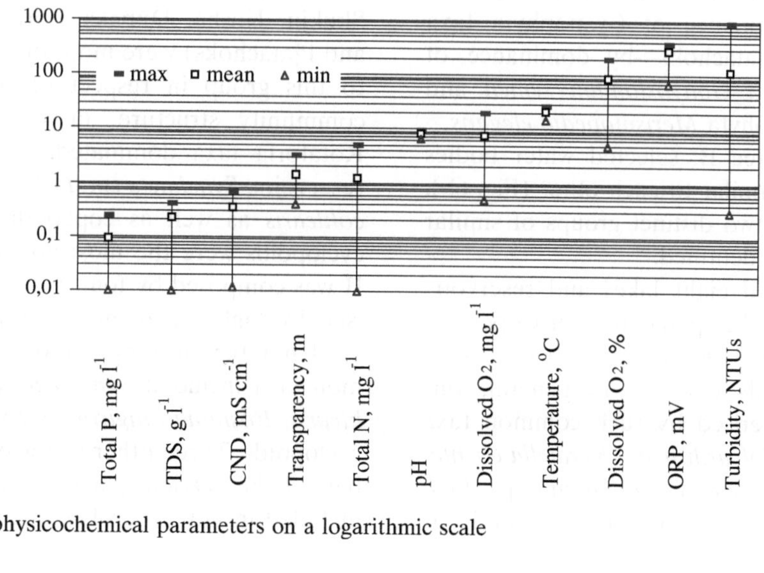

Fig. 3 Variation in physicochemical parameters on a logarithmic scale

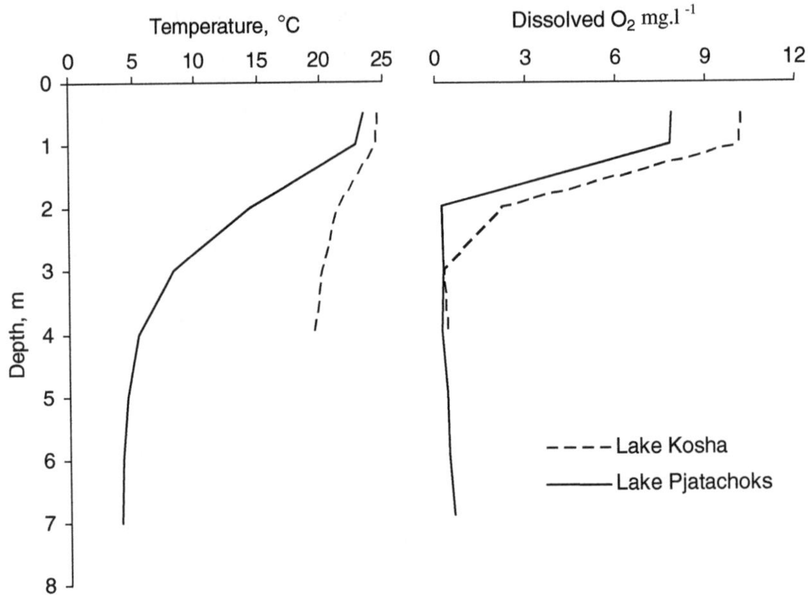

Fig. 4 Thermal stratification in the lakes Kosha and Pjatachoks, July 2004

Table 3 Common species (taxa)* of the Daugava's floodplain lakes and reservoirs, July 2004

Phytoplankton	*Anabaena* sp., *Euglena* sp., *Cryptomonas* sp., *Gymnodinium* sp., *Synedra acus* Kütz.
Zooplankton	*Synchaeta* sp., *Polyarthra* sp., *Keratella cochlearis* Gosse, *Keratella quadrata* Müll., *Bosmina longirostris* Müll., *Cyclops strennus* Fischer
Macrozoobenthos	*Limnodrilus hoffmeisteri* Claparede, *Chaoborus flavicans* Meigen, Chironomidae, Culicoidae Gen. sp.
Macrophyta	*Agrostis stolonifera* L., *Alisma plantago-aquatica* L., *Carex acuta* L., *Elodea canadensis* Michx., *Equisetum fluviatile* L., *Hydrocharis morsus-ranae* L., *Lemna minor* L., *Lemna trisulca* L., *Nuphar lutea* (L.) Sm., *Potamogeton* sp., *Rorripa amphibia* (L.), Besser *Sagittaria sagittifolia* L., *Sparganium erectum* L., *Spirodela polyrhiza* (L.) Schleid.

* Found in more than 50% of the sampled sites

varians and *Rhoicosphaenia curvata* as well as Euglenophyta species *Euglena acus*; Lake Kosha—by total dominance of Cyanophyta taxa *Oscillatoria* sp.; Pjatachoks—by dominance of Chloromonadophyta *Gonyostomum semen* and presence of Cyanophyta *Merismopedia elegans*.

In the cluster tree B, selected water bodies were much more similar to each other (Fig. 6b). From this cluster, two distinct groups of similar water bodies were identified.

Group I included eight lakes and reservoirs characterised by higher percentage of Copepoda taxa and relatively high total number of zooplankton individuals per m^3. In general, this group was characterised by such common taxa as *Synchaeta* sp., *Polyarthra* sp., *Keratella cochlearis*, *Keratella quadrata*, *Ceriodaphnia* sp., *Bosmina longirostris*, *Cyclops* sp. as well as copepodite of cyclopoids. Seven other water bodies (reservoir Elernes 3, Lake Apaljais, Skukju, Kosha, Dubaks 1, reservoir Elernes 4 and Pjatachoks) were more or less closely related to this group in respect of their zooplankton community structure, in which Copepoda or Rotatoria taxa dominated. In these lakes and reservoirs, *Synchaeta* sp., *Polyarthra* sp., *Keratella cohlearis* as well as copepodite and nauplii of cyclopoids were the most abundant taxa. Group II was composed by four water bodies characterised by higher percentage of Rotatoria taxa in total number in zooplankton individuals. Common taxa included *Synchaeta* sp., *Keratella cochlearis, Bosmina longirostris* and, also, nauplii of cyclopoids. Three other water bodies (Berezovka bay, Lake Dvietes and Gaishais) were not included in any group because of quite different

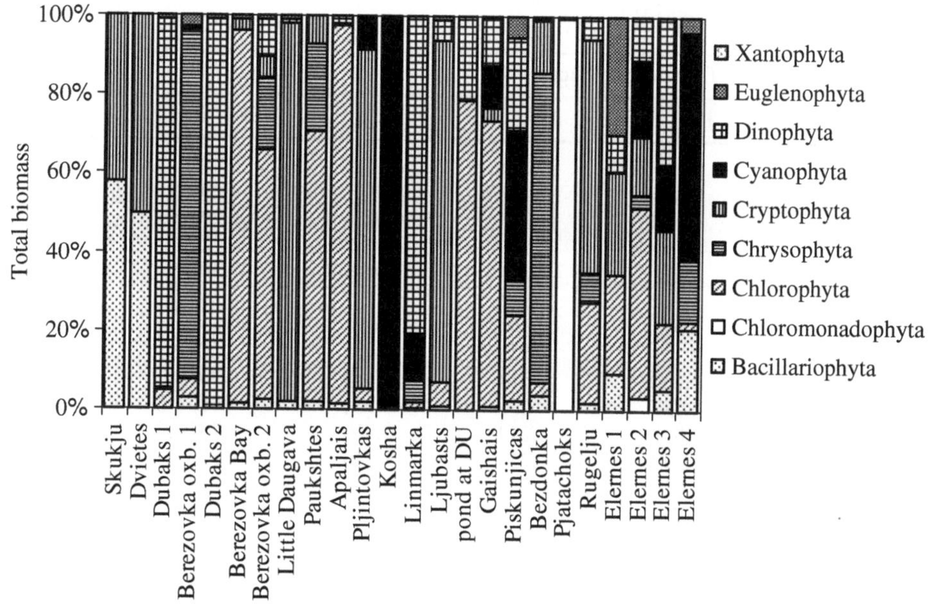

Fig. 5 Percentage composition of total biomass in phytoplankton communities

Table 4 Spearman's rank correlation coefficients between selected morphometrical, hydrological and physicochemical and biotic parameters

	Cryptomonas sp. biomass	Zooplankton taxa	*Cyclopus* sp. (ind./m^3)	Rotatoria (ind./m^3)	Cladocera (ind./m^3)	Copepoda (ind./m^3)	Zooplankton (ind./m^3)	Oligochaeta (ind./m^3)
Surface area	–0.29	0.38	0.47	–0.09	0.60*	0.14	0.12	–0.40
Maximum depth	–0.23	0.05	0.21	0.44**	–0.05	0.15	0.20	–0.53
Flooding frequency	0.66*	–0.68*	–0.62**	–0.38	–0.48**	–0.54*	–0.52**	0.75*
Water level amplitude	0.57**	–0.64*	–0.62**	–0.46**	–0.38	–0.54*	–0.53**	0.74*
Temperature	0.05	0.59*	–0.13	0.63*	0.01	0.43**	0.45**	–0.59**
Conductivity	0.24	–0.44**	0.13	–0.30	–0.33	–0.39	–0.41	0.73*
pH	–0.05	0.51**	0.39	0.20	0.43	0.24	0.23	–0.45
Dissolved O_2	0.40	0.02	–0.58**	0.24	–0.13	–0.06	–0.05	–0.19
Red-ox potential	0.60*	–0.44**	–0.68*	–0.04	–0.66*	–0.31	–0.32	0.43
N total	–0.32	0.16	0.86*	0.17	0.18	0.23	0.34	–0.20
P total	–0.46**	0.05	0.64**	0.00	0.42	0.30	0.29	–0.17
Anabaena sp. biomass	–0.31	0.13	0.18	–0.14	0.31	0.59**	0.40	0.29
Cryptomonas sp. biom.	1.00	–0.33	–0.67**	–0.20	–0.34	–0.35	–0.43	0.29
Euglena sp. biomass	0.47	0.36	0.70	0.85*	–0.47	0.36	0.60	–0.76
Vegetation cover (%)	0.11	0.03	0.50	–0.15	0.20	0.06	0.11	0.10

* Correlation is significant at the 0.01 level (two-tailed)

** Correlation is significant at the 0.05 level (two-tailed)

Table 5 Rotated component matrix

	Component (factor)					
	1	2	3	4	5	6
Area/depth	0.006	–0.712	–0.012	0.052	–0.353	0.101
Flooding frequency	0.886	0.062	–0.141	0.205	–0.092	0.045
Water level amplitude	0.879	0.080	–0.056	0.296	–0.204	0.073
Temperature	–0.480	0.696	–0.015	0.128	–0.055	0.359
Conductivity	0.622	–0.204	0.094	0.309	–0.222	–0.052
pH	–0.289	–0.082	0.722	0.355	–0.210	0.050
Dissolved O_2	–0.005	0.860	0.091	0.210	–0.171	–0.007
Red-ox potential	0.361	0.536	–0.626	–0.010	–0.152	0.193
N total	0.204	–0.018	0.114	0.907	0.136	–0.029
P total	0.057	–0.256	0.784	–0.010	0.284	–0.095
Anabaena sp. biomass	–0.021	0.299	0.810	–0.137	0.060	0.076
Cryptomonas sp. biomass	0.298	0.223	–0.140	0.862	–0.072	–0.069
Euglena sp. biomass	–0.023	0.030	–0.023	–0.067	0.007	0.892
Zooplankton taxa number	–0.765	–0.012	0.240	0.059	0.047	0.373
Rotatoria, individuals	–0.198	0.251	0.145	–0.017	0.659	0.524
Cladocera, individuals	–0.435	–0.618	0.272	0.113	0.126	–0.261
Copepoda, individuals	–0.172	0.196	0.594	–0.114	0.552	0.361
Cyclopus sp., individuals	–0.310	–0.239	0.123	0.117	0.802	–0.150

zooplankton community composition and structure they has in relation to the rest of the lakes and reservoirs explored during this study.

Discussion

Based on characteristics of six components, which were extracted by the PCA, it is possible to identify them with certain environmental factors. With respect to the aim of this study, it is important to discuss these factors in more detail, because they might explain the results described in previous section.

Both hydrological parameters were related most closely to the *component 1* in the Rotated Component Matrix (Table 5). Increasing in its value resulted in higher frequency of flooding and larger amplitude of water level fluctuation. Obviously, it could be identified with the *hydrological connectivity* of the lakes to the Daugava River. According to the Flood Pulse Concept proposed by Junk et al. (1989) and expanded later by Tockner et al. (2000), hydrological connectivity via surface waters is the key factor for the ecological functioning of the large river floodplain ecosystems in tropics (Junk, 1997; Lewis et al., 2000). Its significant role in plankton ecology has been stated also in the studies of floodplain lakes along other European rivers. For example, many physicochemical parameters of the water and sediment as well as the species composition and biodiversity of flora and fauna shows significant correlation with the annual flood duration (frequency) in floodplain lakes along the highly eutrophic lower reaches of the Rhine and Meuse, in the Netherlands (Van den Brink & Van der Velde, 1994; Van den Brink et al., 1994). Increased flood duration and frequency results in an increase of salinity, levels of total N and P, heavy metals and organic micropollutants in the water and sediments, resulting in eutrophication, decrease in biological diversity, blooms of Cyanophyta, shift from the macrophyta-dominated community towards a phytoplankton-dominated one and so on. Clear correlation between hydrological connectivity and plankton characteristics has been found also in ecological studies of the River Danube floodplain segments near Vienna, Austria (Hein et al., 2001). The biomass ratio of phyto- and bacterioplankton declined as hydrological connectivity decreased. The highest total biomasses of phyto- and bacterioplankton were related to the open connection site, whereas biomass of

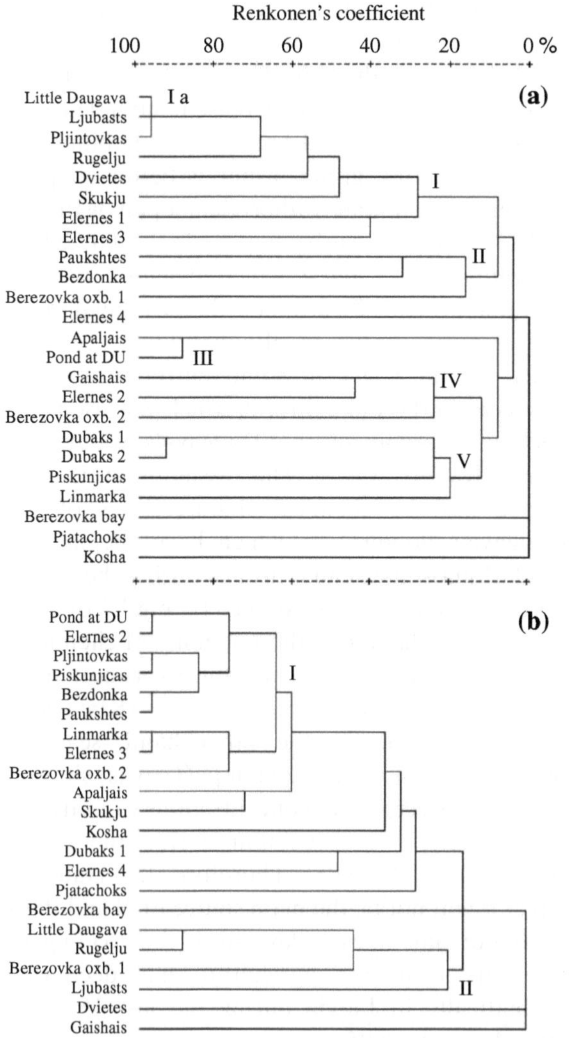

Fig. 6 Tree diagrams resulting from the single linkage clustering of the selected lakes and based on phytoplankton (**a**) and zooplankton (**b**) community data

zooplankton was higher in the isolated one, explained by the higher role of biotic interactions and availability of nutrients in lentic conditions.

According to the Spearman's correlation and PCA, decrease in the number of zooplankton taxa (mainly Cladocera and Rotatoria) as well as zooplankton abundance was associated with an increasing hydrological connectivity (Tables 4, 5; Fig. 7) according to the results obtained in the studies of floodplain lakes located along the Rhine, Meuse and Danube (Van den Brink et al., 1994). Obviously, it was not related to a decrease in abundance of aquatic vegetation as observed in

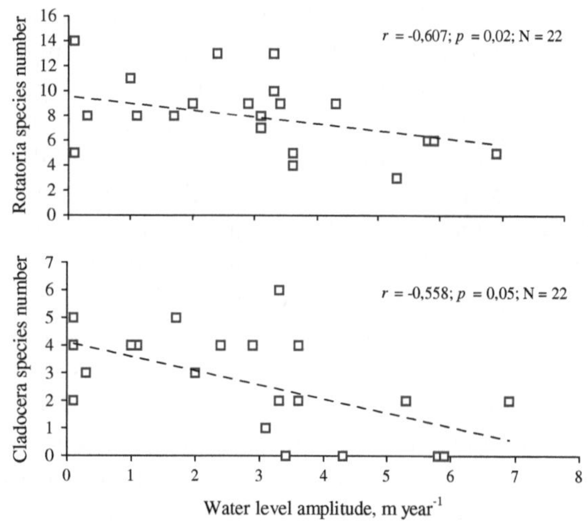

Fig. 7 Relation between amplitude of water level fluctuation and Rotatoria and Cladocera species number

floodplain lakes of the Lower Rhine and Meuse (Van den Brink et al., 1994), since rank correlations between these variables were not significant (Table 4). Since high concentrations of suspended particles strongly reduce the population growth of cladocerans (Kirk & Gilbert, 1990), observed relationship might be partly explained by the lower transparency in frequently flooded lakes (Spearman's correlation significant at 0.05 level; $r = -0.51$). Strong negative effect of the flood event was also found on Rotatoria in the Schelde Estuary, Belgium (Muylaert & Vyverman, 2006).

Higher number of Oligochaeta individuals in frequently flooded lakes of the Daugava River could also be related to the short retention time of water masses and an increased sedimentation of suspended organic matter during the floods. Oligochaeta worms are vulnerable for O_2 saturation in the water and are feeding on a particulate organic matter and decomposing aquatic vegetation, which is a common substance for the alluvial sediments and floodwaters of the Daugava River (Kumsare, 1967). Therefore, frequent influx of fresh water and suspended organic matter via the floodwaters could explain variance in the abundance of these animals in the floodplain lakes of the Middle Daugava.

Significant correlation of electric conductivity (CND) to the annual flooding frequency (Spearman's correlation coefficient significant at 0.01 level; $r = 0.53$) also might be related to some local factors instead of the Daugava's flood pulse. On the one hand, influx of turbulent floodwater might cause a resuspension of the upper sediment layers and promote an elemental transfer between the sediments and water columns of floodplain lakes (Furch & Junk, 1993). Since the duration of such direct influence largely depends on hydrological connectivity, higher CND might be observed in the lakes characterised by larger flooding amplitude and frequency. On the other hand, floodwaters of the Daugava River are characterised by low level of mineralization (Gruberts et al., 2005), and sediment resuspension in shallow lakes are also caused by the action of wind, waterbirds or fish populations (Scheffer, 2004). In fact, many lakes explored during this study are large and shallow (max. depth <2 m), almost completely covered by aquatic vegetation (U. Suško, unpublished) and rich in bird fauna (Račinskis, 2004). Therefore, sediment resuspension by wind or waterbirds could probably account for the variation in CND of the selected water bodies. In addition, higher CND could also be related to the higher role of the groundwater in the water balance of these lakes. As stated by Lesack & Melack (1995), local rainfall and runoff, groundwater exchange and evaporation might be more important for certain floodplain lakes than the seasonal flooding. In South-east Latvia, the Daugava's valley acts as a local output base for the groundwater, which is characterised by the relatively high mineralization (200–500 mg l^{-1}) in comparison to the surface waters. It plays also the most important role in the river runoff during the summer isolation period (Klavinš & Cimdinš, 2005).

According to PCA, *component 2* shows close connection to the surface area/depth ratio, water temperature and dissolved O_2 (Table 5) and therefore could be identified by *morphology and thermal stratification* of the lakes. It was also associated with a decrease in the number of Cladocera individuals per m^3. Larger maximum depth resulted in lower area/depth ratio and higher temperature and concentration of dissolved O_2 in the epilimnion of the lakes. These variances, obviously, were associated with the development of distinct summer thermal stratification, which was characteristic for the smaller and deeper lakes and less typical for the large and shallow ones (Fig. 4). Decrease in the number of Cladocera individuals in deeper lakes and reservoirs might arise from a decrease in abundance of aquatic vegetation and less suitable habitats for this zooplankton group, which is better adapted to complex habitats rather than open water in shallow lakes during the daytime (Scheffer, 2004).

Components 3 and 4, which explain variations in total N and P, are not directly related to hydrological conditions, although higher connectivity should result in some eutrophication as observed in the Lake Stropu, which was used as a backup reservoir of the Daugava's waters for years (Katelo, 1995). This assumption is supported also by chemical characteristics of the Daugava's floodwaters, which have the highest level of phosphates and nitrates during the spring flood maximum (for example, 1.5 mg l^{-1} of NO_3 in April) explained by the reloading of chemicals from the sediments and the surface runoff from agricultural areas covered by snow (Klavinš & Cimdinš, 2005). On the other hand, seasonal observations of the Lake Skukju (Grivas) showed significant decrease in concentrations of nutrients in the lakes' water column during the floods (Gruberts et al., 2005). Therefore, the Daugava River might not be the main source of N and P for its floodplain lakes today, since the loading of nutrients from the agriculture has decreased dramatically in Latvia during the last 15 years (Briede et al., 2005).

Obviously, these components are related to some other sources of nutrients, probably, the sediments of the lakes. On the other hand, the flooding might be responsible for accumulation of inorganic P and N in these sediments in the past, when using of artificial fertilizers in this region was quite intensive, concentration of population and industry in Daugavpils City relatively high and water treatment facilities few (Kumsare, 1967; Katelo, 1995). Therefore, observed variation in total P and N could be explained by the *internal loading of nutrients from the sediments*, especially during the summer stagnation, as observed in floodplain lakes of other eutrophic European rivers. For example, the phosphate-laden sediments of frequently flooded lakes along the Lower Rhine and Meuse have been shown regeneration of P after depletion from the water column (Van den Brink et al., 1994).

Association of an increase of the biomass of *Anabaena* sp. with the higher loading of phosphorus is self-evident, since availability of P is the most critical factor for its growth rate (Reynolds, 1993). Summer bloom of *Anabaena* usually occurs in warm, stratified euthropic lakes. During this study, maximum biomass of *Anabaena* sp. (0.19 mg l^{-1}) was observed in the reservoir Linmarka, which has distinct thermal stratification and the highest level of total P (2.4 mg l^{-1}) among the lakes, where this species has been found. Less obvious is relation of Copepoda abundance to this factor. Probably, it could be explained by the presence of a large number of smaller protozoans and Rotatoria, who are feeding on the decomposing *Anabaena* filaments and act as the food source for copepods. Unfortunately, they were not counted during this study because of a relatively large mesh size of the plankton net (65 μm). Therefore, this is only an assumption rather than fact.

Increased biomass of *Cryptomonas* sp. in the lakes rich in total N (Component 4; Table 5) could also be related to their tolerance to critical limiting factors. As illustrated by Reynolds (1993), availability of nutrients (especially N) might be a limiting factor for their summer growing in temperate lakes, if other factors are subcritical and there is no severe effect of zooplankton grazing.

Component 5 explains variation in the number of Rotatoria and *Cyclopus* sp. individuals, therefore indicating a significant role of throphic interactions between these zooplankton groups. It might be indentified with the *Cyclopus predation on Rotatoria*, as observed in other lakes too. For example, according to Stemberger & Evans (1984), predation by *Cyclopus* and food resource levels has been important factors affecting the Rotatoria, specially *Synchaeta* genus, in the Lake Michigan. The density of *Synchaeta* was negatively correlated with the population density of *Cyclops* and positively correlated with standing stocks of flagellates and small, non-flagellate (<30 μm on one dimension) phytoplankton. In our case, correlation between the number of *Synchaeta* and *Cyclopus* sp. was positive, therefore suggesting bottom-up control of zooplankton density.

Finally, *component 6* could be identified with the *local source of dissolved organic matter*, since it explains the variance in *Euglena* sp. biomass, which assimilates a paramylum and oil instead of nitrogen (Reynolds, 1993).

Taking into account these relationships, dissimilarities between the cluster tree diagrams are better understandable. Only some water bodies are grouped in the same clusters on both diagrams (Fig. 6). For instance, Lake Paukshtes and Bezdonka converged to the same clusters characterised by regular flooding (once or twice in every 10 years), higher biomass of Chrysophyta *Dynobrion divergens* and relatively high percentage of Copepoda individuals per m^3. Little Daugava, Ljubasts and Lake Rugleju, which are flooding more frequently and characterised by high percentage of *Cryptomonas* sp. in phytoplankton biomass as well as higher percentage of Rotatoria in the number of zooplankton individuals, formed similar clusters also. On the other hand, oxbows of the Berezovka River and reservoirs of the Elernes gravel quarry were distributed between different clusters. Obviously, flooding frequency has minor impact on summer plankton communities of these floodplain lakes, influenced by many local factors discussed in this section. In general, this statement coincides with the conclusions of Roozen (2005) in his doctoral thesis based on the research of shallow floodplain lakes along the Rhine, in the Netherlands, where plankton community characteristics are depending largely on slow ecosystem components.

Conclusions

The first ecological studies of floodplain lakes and reservoirs along the Middle Daugava in July 2004 showed considerable diversity of summer aquatic communities, which were characterised by high percentage of rare taxa in their species composition and abundance (Table 2). Even morphologically and hydrologically similar water bodies were different according to their physicochemical parameters and dominant taxa of phytoplankton and zooplankton communities (Table 3; Fig. 5). Single linkage clustering of selected water bodies resulted in quite different cluster trees (Fig. 6) suggesting that most of them could not be classified together with respect to their plankton communities. On both cluster diagrams, only five water bodies were grouped in similar clusters, which revealed clear correlation to hydrological grouping based of the long-term annual flooding frequency (Table 1).

After the Spearman's rank correlation analysis, significant correlations between several biotic parameters, long-term annual flooding frequency and amplitude of water level fluctuation were found (Table 4). On the other hand, PCA showed, that most of them could be explained by other environmental factors instead of hydrological connectivity (Table 5). Hydrological connectivity was found to explain only variation in the number of zooplankton taxa (negative impact) and Oligochaeta density (positive impact), probably, caused by an increased turbidity and loading of suspended organic matter in more frequently flooding lakes. Variations in other biological and physicochemical parameters were explained by local factors, such as the lake morphology and thermal stratification (Fig. 4), internal loading of nutrients from the phosphate and nitrate rich sediments, throphic interactions between the plankton groups as well as the local source of dissolved organic matter (see previous section). In fact, hydrological connectivity was also considered as local factor, since it is largely influenced by the relative height of the lakes above the mean water level of the Daugava River in summer (Fig. 2). Therefore, all observed variations in the physicochemical and biological parameters are determined, at first, by floodplain geomorphology, as stated by Lewis et al. (2000) in their study of the Orinoco river floodplain.

In contrast to floodplain lakes along the highly eutrophic Rhine and Meuse, an impact of the long-term flooding hydrology is minor and/or selective in this case, possibly, because of lower level of pollution and more natural hydrology of the Middle and Upper Daugava today. Creation and operation of the proposed dams on the Upper Daugava would, possibly, increase an eutrophication of the river and destroy the natural flood pulse and seasonal runoff variation therefore reducing the capacity of these floodplain lake ecosystems to resist an intermediate disturbance caused by the seasonal flooding.

Acknowledgements This study has been supported by VPD1/ESF/PIAA/04/NP/3.2.3/0003/0065 project. The authors would like to thank Ecological Laboratory of Daugavpils Regional Environment Board for preparing chemical analysis of water samples.

References

Briede, A., I. Kokorīte, M. Klavinš & I. Druvietis, 2005. Daugava River case study – Latvia. In Modelling Nutrient Loads and Response in River and Estuary Systems. NATO/CCMS Pilot study. Final Report. North Atlantic Treaty Organization, 223–256.

Druvietis, I., 1998. Observations on cyanobacteria blooms in Latvia's inland waters. In Reguera, B., I., Blanco, M. L., Fernández & T. Wyatt (eds), Harmful Algae. Xunta de Galicia, Intergovernmental Oceanographic Commission of UNESCO, 35–36.

Fott, B., 1971. Algenkunde. 2. auflage. Veb Gustav Fisher Verlag, Jena.

Furch, K. & W. J. Junk, 1993. Seasonal nutrient dynamics in an Amazonian floodplain lake. Archiv für Hydrobiologie 128: 277–285.

Gruberts, D., 2003. The four largest floodplain lakes in Latvia: hydrology, hydrochemistry and hydrobiology. In Järvet, A. & E. Lode (eds), Ecohydrological Processes in Northern Wetlands. Selected papers. Tartu University Press, Tallinn, 196–202.

Gruberts, D. & I. Druvietis, 2001. Phytoplankton periodicity in floodplain lake Grivas, Latvia – Baltic State. In Partnerships for Sustainable Life in Lake Ecosystems. Proceedings of the 9th International Conference On The Conservation And Management Of Lakes. BIWAKO, Shiga, 54–57.

Gruberts, D., I. Druvietis & M. Klavinš, 2005. Seasonal variability of aquatic chemistry and phytoplankton communities in a shallow floodplain lake of the Daugava River, Latvia. Ecohydrology and Hydrobiology 5(2): 155–164.

Hein, T., C. Baranyi, G. Heiler, C. Holarek, P. Riedler & F. Schiemer, 2001. Plankton communities controlled by hydrology in two Danubian floodplain segments and the River Danube, Austria. Verhandlungen International Verein für Limnologie 27, 1–2.

Huszar, V. L. D. & C. S. Reynolds, 1997. Phytoplankton periodicity and sequences of dominance in an Amazonian floodplain lake (Lago-Batata, Para, Brazil) – responses for gradual environmental change. Hydrobiologia 346: 169–181.

Junk, W. J., 1997. The Central Amazon Floodplain: Ecology of a Pulsing System. Springer-Verlag, New York.

Junk, W. J., P. B. Bayley & R. E. Sparks, 1989. The flood pulse concept in river-floodplain systems. In Dodge, D. P. (ed.), Proceedings of the International Large River Symposium. Canadian Special Publication on Fisheries and Aquatic Sciences, 106: 110–127.

Katelo, I., 1995. Eutrophication process in Lake Stropu, Latvia. Proceedings of the Latvian Academy of Sciences 573: 101–104.

Kirk, K. L. & J. J. Gilbert, 1990. Suspended clay and the population dynamics of planktonic rotifers and cladocerans. Ecology 71: 1741–1755.

Klavinš, M. & P. Cimdinš, 2005. Water Quality and its Protection. University of Latvia, Riga (in Latvian).

Krebs, C. J., 1989. Ecological Methodology. Harper Collins Publishers.

Kumsare, A., 1967. Gidrobiologija reki Daugava [Hydrobiology of the Daugava River]. Zinatne, Riga.

Lesack, L. F. W. & J. M. Melack, 1995. Flooding hydrology and mixture dynamics of lake water derived from multiple sources in an Amazon floodplain lake. Water Resources Research 31: 329–345.

Lewis, W. M., S. J. Hamilton, M. A. Lasi, M. Rodríguez & J. F. Saunders III, 2000. Ecological determinism on the Orinoco floodplain. BioScience 50: 681–692.

Muylaert, K. & W. Vyverman, 2006. Impact of a flood event on the planktonic food web of the Schelde estuary (Belgium) in spring 1998. Hydrobiologia 559(1): 385–394.

Račinskis, E., 2004. Important Bird Areas of European Union Importance in Latvia. Latvian Ornithological Society, Riga.

Reynolds, C. S., 1993. The Ecology of Freshwater Phytoplankton. University Press, Cambridge.

Roozen, F. C. J. M., G. J. Geest, B. W. Ibelings, R. Roijackers, M. Scheffer & A. D. Buijse, 2003. Lake age and water level affect the turbidity of floodplain lakes along the lower Rhine. Freshwater Biology 48: 519–531.

Roozen, F. C. J. M., 2005. Transparency of floodplain lakes; a study of plankton and suspended mater along the lower Rhine. Doctoral thesis. Wageningen Universiteit.

Scheffer, M., 2004. Ecology of Shallow Lakes. Kluwer Academic Publishers, Dordrecht.

Stemberger, R. S. & M. S. Evans, 1984. Rotifer seasonal succession and copepod predation in Lake Michigan. Journal of Great Lakes Research 10(4): 417–428.

Tockner, K., F. Malard & J. V. Ward, 2000. An extension of the flood pulse concept. Hydrological Processes 14: 2861–2883.

Utermöh, H., 1958. Zur Vervollkomnung der quantitativen Phytoplankton-Methodik. Mitt International Ver Theoretical and Applied Limnology 19: 100–124.

Van den Brink, F. W. B. & G. Van der Velde, 1994. Impact of hydrology and water chemistry on floodplain lake communities along the lower Rhine and Meuse. Water Science and Technology 29: 57–60.

Van den Brink, F. W. B., M. M. Van Katwijk & G. Van der Velde, 1994. Impact of hydrology on phyto- and zooplankton community composition in floodplain lakes along the lower Rhine and Meuse. Journal of Plankton Research 16: 351–373.

Wetzel, R. G. & G. E. Likens, 1991. Limnological Analyses, 2nd edn. Springer-Verlag, New York.

Hydrobiologia (2007) 584:239–252
DOI 10.1007/s10750-007-0596-7

SHALLOW LAKES

Shallow lake restoration by nutrient loading reduction—some recent findings and challenges ahead

Erik Jeppesen · Martin Søndergaard · Mariana Meerhoff · Torben L. Lauridsen · Jens Peder Jensen

Abstract Shallow lakes respond to nutrient loading reductions. Major findings in a recent multi-lake comparison of data from lakes with long time series revealed: that a new state of equilibrium was typically reached for phosphorus (P) after 10–15 years and for nitrogen (N) after <5–10 years; that the in-lake Total N:Total P and inorganic N:P ratios increased; that the phytoplankton and fish biomass often decreased; that the percentage of piscivores often increased as did the zooplankton:phytoplankton biomass ratio, the contribution of *Daphnia* to zooplankton biomass, and cladoceran size. This indicates that enhanced resource and predator control often interact during recovery from eutrophication. So far, focus has been directed at reducing external loading of P. However, one experimental study and cross-system analyses of data from many lakes in north temperate lakes indicate that nitrogen may play a more significant role for abundance and species richness of submerged plants than usually anticipated when total phosphorus is moderate high. According to the alternative states hypothesis we should expect ecological resistance to nutrient loading reduction and P hysteresis. We present results suggesting that the two alternative states are less stable than originally anticipated. How global warming affects the water clarity of shallow lakes is debatable. We suggest that water clarity often will decrease due to either enhanced growth of phytoplankton or, if submerged macrophytes are stimulated, by reduced capacity of these plants to maintain clear-water conditions. The latter is supported by a cross-system comparison of lakes in Florida and Denmark. The proportion of small fish might increase and we might see higher aggregation of fish within the vegetation (leading to loss of zooplankton refuges), more annual fish cohorts, more omnivorous feeding by fish and less specialist piscivory. Moreover, lakes may have prolonged growth seasons with a higher risk of long-lasting algal blooms and at places dense floating plant communities. The effects of global

Guest editors: R. D. Gulati, E. Lammens, N. De Pauw & E. Van Donk
Shallow lakes in a changing world

E. Jeppesen (✉) · M. Søndergaard · M. Meerhoff · T. L. Lauridsen · J. P. Jensen
Department of Freshwater Ecology, National Environmental Research Institute, University of Aarhus, Vejlsøvej 25, 8600 Silkeborg, Denmark
e-mail: ej@dmu.dk

E. Jeppesen · M. Meerhoff
Department of Plant Biology, University of Aarhus, Ole Worms Allé, Building 135, 8000 Aarhus C, Denmark

M. Meerhoff
Departmento de Ecología, Facultad de Ciencias, Universidad de la República, Iguá 4225, C.P. 11400 Montevideo, Uruguay

warming need to be taken into consideration by lake managers when setting future targets for critical loading, as these may well have to be adjusted in the future. Finally, we highlight some of the future challenges we see in lake restoration research.

Keywords Recovery · Lake restoration · Nitrogen · Climate change · Resilience · Regime shifts

Introduction

During the past 20–30 years much effort has been directed, not least in Europe and North America, at combating the eutrophication of shallow lakes by reducing external phosphorus (P) loading. Loading with sewage and industrial waste has declined, but the nutrient input from diffuse sources has remained high, particularly in intensively cultivated countries (Van der Molen & Portielje, 1999; Kronvang et al., 2005). Many lakes have responded positively to nutrient loading reductions; however, delayed recovery has often been observed, the causes being internal P loading or biological resistance (Sas, 1989; Marsden, 1989; Jeppesen et al., 2005). In the present paper we first describe the highlights of a recent cross-system analysis of 22 long-term studies of shallow lakes recovering from eutrophication. Then we discuss the role of nitrogen (N) for recovery and resilience and how climate change might influence lake ecosystem state and the expected nutrient targets for shifts to good or high ecological state. Finally, we identify some important research areas for the future within the field of lake recovery and restoration.

A recent analysis of 22 case studies on re-oligotrophication

A recent a cross-system analysis was conducted on long term data series from one North American and 21 shallow European lakes (mostly north temperate lakes) in recovery from eutrophication (Jeppesen et al., 2005). It was found that internal P loading as expected delayed recovery. However, a new equilibrium with respect to total phosphorus (TP) was typically reached after <10–15 years, and the response time was only marginally dependent on hydraulic retention time. The response time to N loading reduction was typically <5–10 years. As loading reductions have mainly focused on P, the TN:TP and inorganic N:ortho-P ratios have increased (Fig. 1), often markedly. Lake TP typically declined during all seasons, but particularly in winter, followed by spring and autumn. Changes in chlorophyll *a* followed those of TP, indicating that nutrient constraint (resource control) has been a key factor for the changes in seasonal dynamics of phytoplankton biomass (Fig. 1). However, enhanced top-down control by zooplankton in the recovery process may also play a role. Thus in most lakes for which fish data were available the catch-per-unit-effort (test-fishing or commercial catches) by weight declined, often substantially, while the proportion of piscivores increased. In accordance with the expected cascading effects of such changes in the fish community, the zooplankton:phytoplankton biomass ratio (and thus probably also phytoplankton grazing) rose in many lakes, while the contribution of *Daphnia* to the total biomass (Fig. 1) and mean individual biomass of cladocerans increased. It must be emphasised, though, that the zooplankton data mainly derived from north temperate lakes. A different scenario is likely for warm temperate and (sub)tropical lakes where fish abundance, and likely then also the predation pressure on large-bodied zooplankton, is higher (Meerhoff et al., 2003; Romo et al., 2005; Jeppesen et al., 2007).

Nitrogen reduction—do we have to consider it?

In most studies, recovery follows a reduction in P loading, as P is generally considered the most important limiting nutrient in lakes (Schindler, 1977). However, some recent studies in shallow lakes emphasise also the role of N (Moss, 2001; González Sagrario et al., 2005; James et al., 2005). Enclosure experiments run at various combinations of P and N in a shallow Danish lake indicated that the risk of loss of submerged macrophytes was high when N was above 1.2–2 mg l^{-1} and P higher than 0.1–0.2 mg l^{-1} (González Sagrario

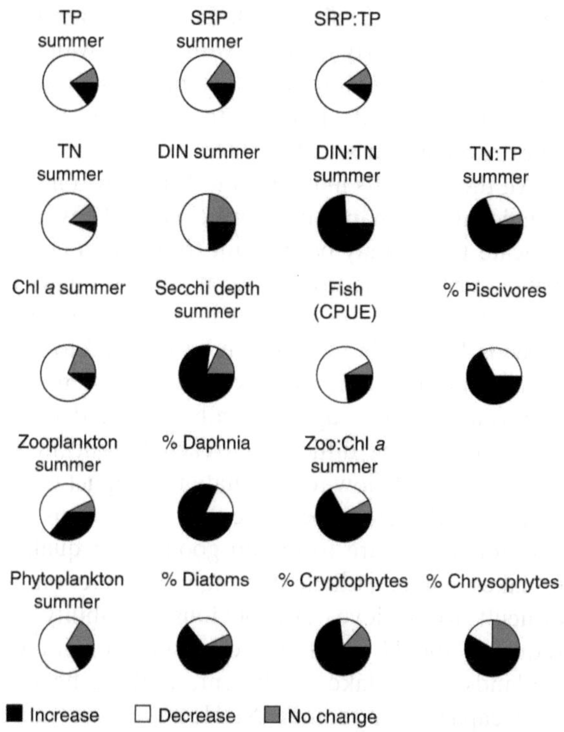

Fig. 1 Some major results of a multi-lake (22 lakes) comparison of responses to nutrient loading reduction. The pies show direction of changes (% of lakes) during recovery 5–25 years) of a series of biological and chemical variables, all summer means (May 1–October 1 in cold temperate lakes and May 1–November 1 in warm temperate and subtropical lakes). DIN is dissolved inorganic N, SRP is soluble reactive phosphorus, CPUE is catch-per-unit-effort based on test-fishing or in a few cases commercial catches. Zooplankton and phytoplankton are total biomass and total biovolume, respectively, % *Daphnia* is contribution of *Daphnia* to the total zooplankton biomass and % of algae taxa their contribution to the total biovolume. Zoo is zooplankton biomass. (Modified from Jeppesen et al., 2005)

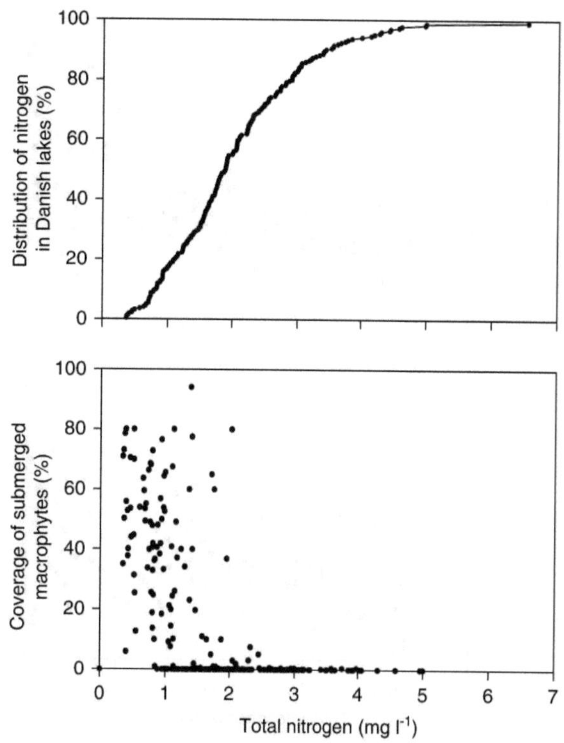

Fig. 2 The distribution of summer averages of nitrogen in 204 Danish lakes (left, upper) and the coverage of submerged macrophytes in late summer in 44 Danish lakes (246 lake years) (left, lower) along a nitrogen gradient. All lakes have a mean depth lower <5 m and a lake area >5 ha (from González Sagrario et al., 2005)

et al., 2005). However, at low P concentrations, variations in N loading had no impact. In accordance with these results, data from 44 Danish lakes (and 246 lake years) suggest that submerged plants typically disappear at concentrations of 1–2 mg N l^{-1} when TP is moderately high (González Sagrario et al., 2005; Fig. 2). These N concentrations, which apparently trigger the loss of macrophytes, are low compared to the present concentrations in Danish lakes (Fig. 2).

But why should these plants be lost at high N loadings? In eutrophic shallow lakes, nitrate is often low during late summer due to high denitrification and low summer input of nitrogen, and TP is high due to internal loading (Søndergaard et al., 2005). Nitrogen limitation of periphyton or filamentous algae on plant surface and phytoplankton is therefore likely to occur during mid-summer. Thus when N loading is high growth of these algae may be stimulated and potentially then outshade the submerged macrophytes. To gain further insight, we depicted, in an N:P diagram, summer data from the 44 lakes with an inorganic N:P ratio below 7, where algae are likely N limited if the inorganic N concentrations are low (Smith, 1983), and focused on the P concentrations at which a shift to a clear-water state may typically occur, i.e. <0.15 mg P l^{-1} (Jeppesen et al., 1990) (Fig. 3). Lakes with inorganic N:P <7 are detected within the 0.5–2 mg TN l^{-1} interval (Fig. 3), i.e. where submerged macrophytes actually disappear (Fig. 2). Moreover, the concentration of inorganic N in these lakes usually

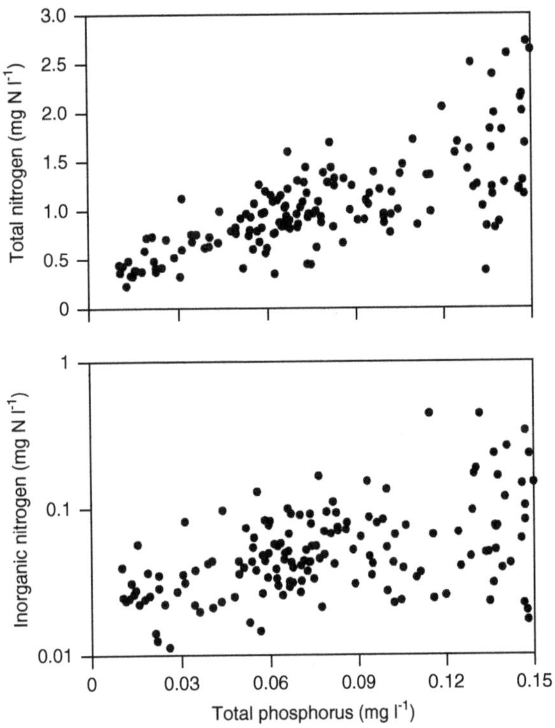

Fig. 3 The concentration of nitrogen (total nitrogen upper and inorganic nitrogen lower) along a phosphorus gradient. The analysis comprises only lakes with a summer average inorganic N:P ratio in the lake water <7 (56 lakes in the analysis (152 lake years)). All lakes have an area >5 ha, a mean depth <5 m and a total phosphorus concentration <0.15 mg l^{-1}

remains <0.1 mg l^{-1} when TP is between 0.015 mg l^{-1} and 0.12 mg l^{-1}, meaning that N limitation of algae is likely within this TP range. If, instead, we used data from late summer (15 July–15 August) when submerged macrophyte biomass typically reaches a seasonal maximum and nitrate its seasonal minimum, the evidence of N limitation would be even stronger.

N not only seems to affect the abundance of submerged plants. Studies conducted in Polish and British lakes have further shown that plant species richness declines with increasing winter nitrate concentrations (James et al., 2005). Development of a particular or few species may cause large interannual variations in plant abundance, as indicated by results from Danish lakes dominated by a single species, such as rigid hornwort (*Ceratophyllum demersum*) or Canadian pondweed (*Elodea canadensis*) (Søndergaaard et al., 1997; Lauridsen et al., 2003). In some years the lakes may be nearly filled up with plants, while in other years plants may be completely absent (Søndergaard et al., 1997), with great risk of the lakes reverting to a turbid state.

All these results may indicate N to play a more significant role than usually anticipated. They also indicate that it may be difficult to obtain a clear-water state in shallow lakes at medium-high P concentrations if the N concentration in summer exceeds 1–2 mg N l^{-1}. If P can be reduced to low concentrations, N is not likely to be of major importance, but in agriculturally impacted landscapes it may be difficult to reach P concentrations sufficiently low to eliminate the impact of N. One may therefore perceive N reduction as an additional measure to obtain good water quality in lakes where sufficiently high P reduction is difficult to achieve. An obvious possibility of reducing the N input is (re-) establishment of wetlands in the lake catchment, as these have a high capacity to remove N. However, we must emphasise that the results presented only allow preliminary conclusions and need further testing. Moreover, they only cover north temperate shallow lakes of the size 5–3,900 ha and intermediate to high P concentrations. In ponds and small lakes (<5 ha) submerged macrophytes may well be abundant at higher N concentrations (authors' unpublished data). This may also be true in lakes—not only ponds but also larger lakes—in other geographical areas (see discussion in González Sagrario et al., 2005).

P hysteresis in the recovery phase—do we need to push the systems?

According to the alternative state hypothesis (Scheffer et al., 1993), we should expect not only chemical resistance to nutrient loading reduction due to internal P loading but also ecological resistance and P hysteresis, i.e. the shift from turbid to clear occurs at a lower P concentration when P is decreasing than from clear to turbid when P is increasing. When approaching the P range of alternative states after nutrient loading reduction, a push (e.g. biomanipulation) may help shifting the lakes to the clear state (Moss, 1990;

Jeppesen et al., 1990; Scheffer et al., 1993). An example of P hysteresis comes from Lake Veluwe (Meijer et al., 1999). One of the factors that, according to the theory, is expected to contribute to ecological resistance and P hysteresis is homeostasis among fish due to high longevity and persistence of some of the key species dominating in the turbid state, such as bream (*Abramis brama*) and various carp species. They then continue to control zooplankton and stir up suspended sediment a long time after the external nutrient loading and P concentration has been reduced. However, analyses of data from Danish lakes (Jeppesen et al., 2002) and the 22 lakes (see above) showed relatively fast responses towards reduced abundance of plankti-benthivorous fish (including bream) and a higher proportion of predatory fish.

Besides enhancing internal loading, resuspension of sediment detritus accumulated during the eutrophication period is also considered to contribute to P hysteresis as water clarity is reduced. However, a study of 15 Danish shallow lakes showed major reduction in suspended matter during recovery and an almost proportional reduction in chlorophyll *a*, detritus and inorganic suspended matter. This indicates that resuspension of old detritus, accumulated during the eutrophication period, does not hinder an improvement of water transparency in Danish lakes <3,900 ha. (Jeppesen et al., 2003), though it is debatable whether the same holds true for very large lakes (Bachmann et al., 1999; Coveney et al., 2005).

Delayed response of submerged macrophytes due to lack of propagules, waterfowl and fish grazing, poor substrate conditions and fish control of snails (indirectly contributing to periphyton-induced out-shading of the plants) are also considered to create hysteresis (Moss, 2001; Hilt et al., 2006). The multi-lake comparison of the 22 lakes (Jeppesen et al., 2005) suggested different response patterns. In some lakes no colonisation was observed despite raised water clarity, while in others a sudden increase was seen, supporting the hypothesis. However, in some cases abundance, coverage, plant volume inhabited (PVI) or depth distribution of submerged macrophytes increased gradually concurrently with the reduction in in-lake P.

In view of the above, it can be discussed whether P hysteresis is as important for lake recovery as suggested in the alternative state theory (Scheffer et al., 1993). Some of the arguments for P hysteresis seem less strong than originally thought. To gain further insight into the impact of P hysteresis, we analysed a series of Danish lakes with long-term data during the recovery phase (Fig. 4a). We found major reductions in chlorophyll *a* (Chla), which largely followed TP in lakes subjected only to external nutrient loading control, indicating that P hysteresis is of minor importance. In biomanipulated (Fig. 4b) lakes an abrupt drop in Chla and TP was seen after biomanipulation, followed in some cases by an increase some years later. In these cases there was evidence that the Chla:TP ratio was lower immediately upon biomanipulation when Chla reached low concentrations, which is in accordance with expectations in the P hysteresis hypothesis. However, Lake Arreskov became unstable shifting between the clear and turbid state as seen in several other Danish case studies of biomanipulation (Lauridsen et al., 2003). Large year-to-year variations between low and high Chla and TP are also seen in some of the lakes with no or minor changes in external loading, such as Lake Stigsholm and Lake Hornum (Fig. 5c), indicating that the "stability" of the clear-water and turbid states may not be as strong as predicted for the TP range where alternative states are to be expected. We further plotted Chla against TP for the lakes presented in Fig. 4a, c and seven other Danish non-biomanipulated shallow lakes with long time series (Fig. 5). While one lake showed a tendency to a stepwise decrease of the Chla:TP ratio at a specific P threshold, the overall impression is that Chla follows TP, though the year-to-year variation could be high in the P range of alternative states as seen also in several lakes elsewhere (Mitchell, 1989; Moss et al., 1990; Perrow et al., 1994; Blindow et al., 1993).

Our results therefore indicate that resistance to changes with decreasing TP concentrations is less strong than predicted by the alternative state hypothesis and the stability of the two states within the window of alternative states weaker. In northern temperate lakes a shift to the clear state

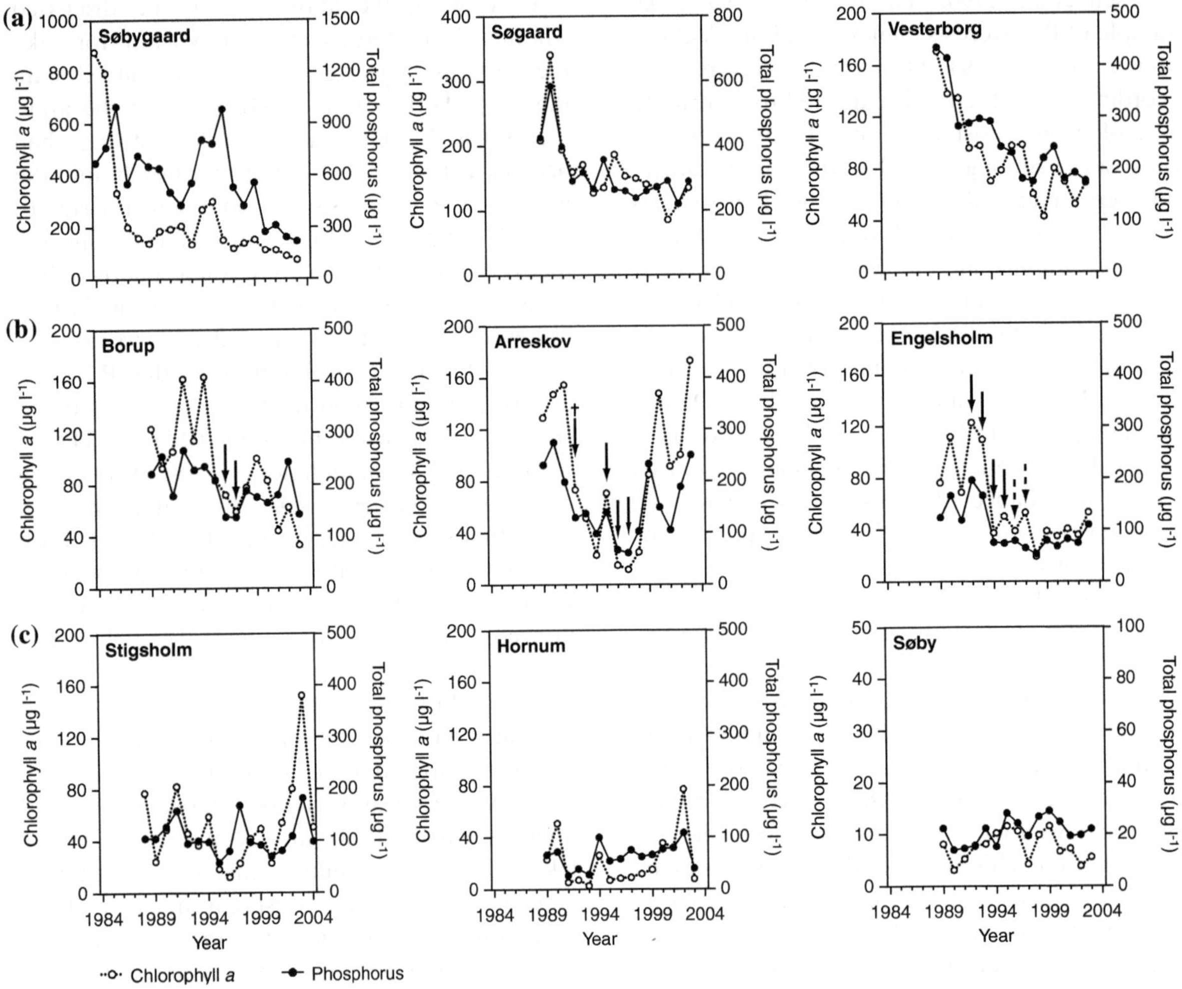

Fig. 4 Time series of summer mean (1 May–1 October fortnightly samplings) concentration of chlorophyll *a* (Chla) and total phosphorus in the lake water of nine shallow Danish lakes following: (**a**) external nutrient loading reduction only, (**b**) external loading reduction and biomanipulation conducted in the years indicated by arrows, and (**c**) lakes with minor or no reduction in external loading. Note the different scale on axes. The arrows indicate years of fish removal or fish kills (+), if broken then only little effort was used to remove fish that particular year

may occur gradually, concurrently with the reduction in internal P loading and the in-lake P. Whether additional tools such as biomanipulation are needed to push the system from turbid to clear has therefore to be evaluated carefully in each case. Certainly, many experiments involving a reduction in fish biomass have led to abrupt changes from turbid to clear (Hansson et al., 1998; Meijer et al., 1999; Mehner et al., 2002) (Fig. 4b) and to reduced internal P loading, having thus reinforced the recovery, at least in the short term. So far, the long-term perspectives are less promising when based on a single event to change the system from turbid to clear (Meijer et al., 1999; Olin et al., 2005; authors' unpublished results). More focus should be on how to stabilise the lake ecosystem to get long term success of biomanipulation (Mehner et al., 2002), perhaps by less intensive but more frequent manipulation of the fish stock. Methods to improve establishment and growth of submerged macrophytes may be needed if plants do not emerge after a rise in water clarity, as has been the case for several lakes (Köhler et al., 2005, Jeppesen et al., 2005). However, a reduction of the external nutrient load appears as the most

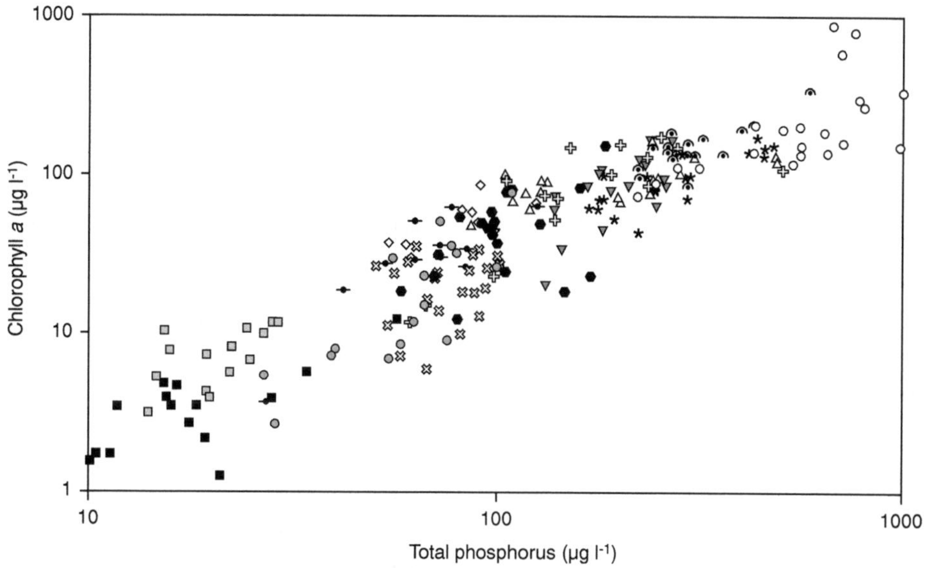

Fig. 5 Summer mean (1 May–1 October fortnightly samplings) chlorophyll *a* versus total phosphorus in non-biomanipulated shallow lakes with long time series. Each symbol represents one lake and each point the summer average for one of the years included

straightforward strategy to improve the ecological status of shallow eutrophic lakes.

Climate change—impact on the future ecological state?

How changes in climate will affect the response to loading reduction and the new ecological state at steady state is debatable. Few studies have tested the effects of increased temperature and nutrient-enrichment using an ecosystem approach in controlled experiments (McKee et al., 2003, Liboriussen et al., 2005). Mckee et al. (2003) showed that warming did not promote phytoplankton blooming but increased P concentrations and deoxygenation. Long-term field data show that effects of warming on water quality can be contradictory and the net effect therefore difficult to predict (Carvalho & Kirika, 2003). Here, we first discuss how climate change, mostly warming, may affect the shallow lake ecosystem and, secondly, the potential consequences for the future ecological state of recovering lakes and possible adaptation strategies for managers.

A possible scenario is that macrophyte growth is stimulated by climate warming (Scheffer et al., 2001) due to the higher temperature, and in the Mediterranean region due to reduced water table as well (Beklioglu et al., 2006). Moreover, unless compensated by higher precipitation induced N loading, nitrate in summer will probably be lower in the future due to enhanced denitrification (Weyhenmeyer et al., in press), which potentially may be beneficial for the macrophytes as outlined above. Therefore, shallow lakes may potentially shift from the phytoplankton to the macrophyte dominated state. However, the role of submerged macrophytes for water clarity differs along the climate gradient. While high abundance of plants is synonymous with clear water in most north temperate non-humic lakes, the effect is much weaker in warm temperate (Romo et al., 2005) and subtropical lakes (Bachmann et al., 2002). Thus, Bachmann et al. (2002) found no difference in the chlorophyll:TP or Secchi depth:TP relationships in shallow lakes in Florida with low, medium high or high plant coverage (COV) or plant volume infested (PVI).

We conducted a comparative study of relationships between Chla, nutrients and submerged plant abundance in lakes from Florida and Denmark, and the results showed Chla at a given nutrient concentration to be overall higher in Florida lakes than in Danish lakes when macrophytes are abundant (Fig. 6). Accordingly, the

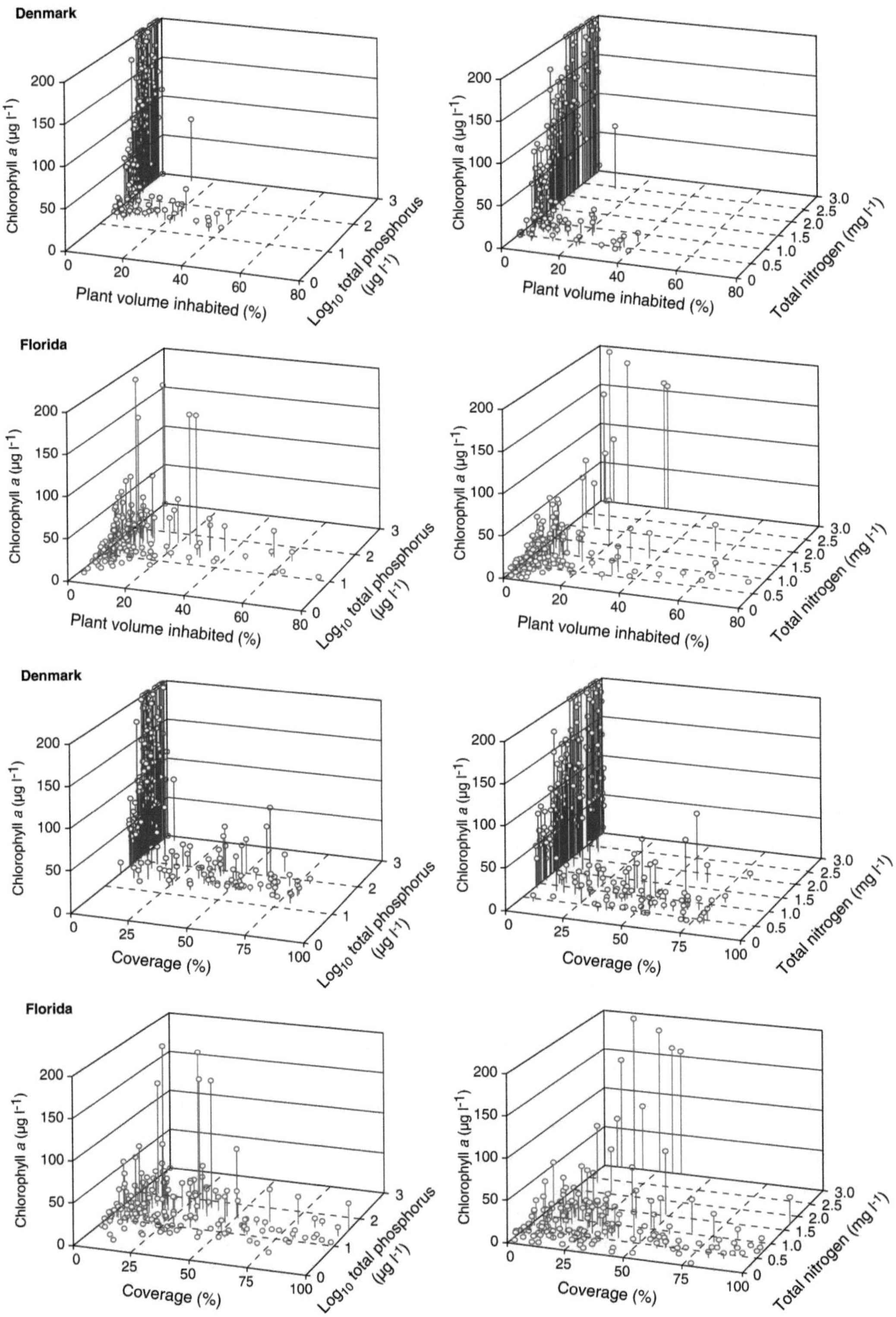
Denmark
Florida
Denmark
Florida
Chlorophyll a (µg l-1)
Plant volume inhabited (%)
Coverage (%)
Log10 total phosphorus (µg l-1)
Total nitrogen (mg l-1)

Table 1 Relationship between $\log_e$ of chlorophyll *a* (1g l^{-1}) and $\log_{10}$ of total nitrogen (TN, mg l^{-1}), total phosphorus (TP, 1g l^{-1}), coverage (COV, %+1) and plant volume infested (PVI, % +1) in a series of lakes from Denmark (DK) and Florida (FL) in summer

	Intercept	Log_{10} (TP)	Log_{10} (TN)	Log_{10} (COV)	Log_{10} (PVI)	r^2	$P <$	n
COV								
DK	1.23 ± 0.33	1.47 ± 0.14		-0.73 ± 0.08^a		0.78	0.0001	217
FL	−0.63 ± 0.26	2.54 ± 0.16		-0.25 ± 0.12^b		0.61	0.0001	175
DK	4.03 ± 0.10		2.01 ± 0.25	-0.88 ± 0.09^a		0.75	0.0001	227
FL	3.20 ± 0.18		2.50 ± 0.18	-0.38 ± 0.13^b		0.53	0.0001	175
DK	1.58 ± 0.33	1.17 ± 0.15	1.07 ± 0.25	-0.58 ± 0.09^a		0.80	0.0001	217
FL	0.71 ± 0.31	1.71 ± 0.19	1.31 ± 0.20	-0.29 ± 0.11^b		0.68	0.0001	175
PVI								
DK	$-0.24 \pm 0.37^{n.s.}$	2.07 ± 0.16			-0.69 ± 0.15^a	0.78	0.0001	179
FL	−0.86 ± 0.22	2.58 ± 0.16			-0.22 ± 0.12^b	0.60	0.0001	175
DK	3.59 ± 0.09		3.42 ± 0.26		-0.77 ± 0.15^a	0.77	0.0001	189
FL	2.97 ± 0.13		2.57 ± 0.18		-0.36 ± 0.14^b	0.53	0.0001	175
DK	1.06 ± 0.40	1.28 ± 0.20	1.90 ± 0.33		-0.45 ± 0.14^a	0.81	0.0001	179
FL	$0.53 \pm 0.28^{n.s.}$	1.74 ± 0.19	1.35 ± 0.20		-0.32 ± 0.11^a	0.69	0.0001	175

All slopes (±SE) on COV and PVI were significantly different among the two countries except for PVI together with nitrogen and phosphorus. Tests for differences in slope on COV and PVI between DK and FL are shown (a, b different; a, a not)

slopes for COV and PVI in multiple regressions, including also N and/or P as independent variables and Chla as dependent variable, were much lower for Florida lakes (Table 1). One reason for these lower slopes may be that macrophytes are not acting as proper refuges for zooplankton in warm lakes. Several studies in subtropical and tropical lakes have shown that fish, particularly the smallest species and individuals, aggregate in high numbers in the vegetation in subtropical and tropical lakes (Conrow et al., 1990; Meerhoff et al., 2003) and some are even confined to this biotope (Sazima & Zamprogno, 1985; Delariva et al., 1994). Fish aggregation in the vegetation and lack of zooplankton aggregation among plants have also been observed in brackish lakes dominated by small-sized sticklebacks (*Gasterosteus aculeatus*) in Denmark and eutrophic brackish lakes thus remain turbid even when macrophyte coverage is high (Jeppesen et al., 1997). Besides, in shallow lakes in the tropics and subtropics, fish zooplanktivory is stronger due to multiple or continuous reproductive events (Paugy & Lévêque, 1999), lower densities of large specialist piscivores with sit-and-wait hunting behaviour more frequent among them (Quirós, 1998), widespread omnivory (Branco et al., 1997, Yafe et al., 2002), and high population densities (but not necessarily high biomass) of fish, particularly of small and juvenile individuals (Mazzeo et al., 2003). Accordingly, (sub)tropical lakes are usually dominated by small forms of zooplankton (Crisman & Beaver, 1990; Meerhoff et al., 2003) that exert low grazing pressure on phytoplankton. Although more studies are needed to elucidate the differences in predation on zooplankton, it may be concluded that this community structure and common habitat use of fish may be one of the key factors determining the weaker effect of submerged plants on water clarity in warm lakes. We may therefore expect that north temperate lakes move in the direction of a decreased capacity of the submerged plants to create and maintain clear-water conditions and of a higher predation on large-bodied zooplankton and lower grazer control of phytoplankton. Supporting this view, Gyllström et al. (2005) found the fish CPUE (multi-mesh sized gillnets) to zooplankton biomass ratio to increase and the zooplankton:phytoplankton biomass ratio to decrease from cold to warm shallow temperate lakes in Europe.

◀**Fig. 6** Chlorophyll *a* in the lake water at contrasting concentrations of total nitrogen and total phosphorus and coverage (A) and percentage of water occupied by plants (PVI, sensu Canfield et al., 1984) of submerged macrophytes in Florida and Danish shallow lakes in summer. Abbreviation at 200 µg l^{-1} chlorophyll *a* in some of the Danish lakes. Only one lake (grey circle) shows a clear threshold shift in chl *a*: TP with declining TP

An alternative scenario to increasing macrophyte growth is stronger dominance of phytoplankton, as phytoplankton growth may be stimulated by higher internal P loading, which is highly sensitive to temperature in shallow lakes (Jensen & Andersen, 1992, Søndergaard et al., 2003; Jensen et al., 2006). Moreover, the growing season is extended and the risk of prolonged periods with blooming of noxious cyanobacteria most likely increases. Phytoplankton dominance may be further stimulated as the above outlined changes in fish community composition, size distributions and habitat choice may enhance predation on zooplankton and thus release phytoplankton from grazing. Moreover, in north temperate lakes higher disturbance of sediment is to be expected if cyprinids as predicted (Lehtonen, 1996) and likely also carp (*Cyprinus carpio*) become more abundant in the future warmer climate. In north-west Europe, higher precipitation combined with reduced winter frost enhance nutrient loss from cultivated fields, not least during autumn and winter (Andersen et al., submitted; JRC, 2005). In Southern Europe the risk of drought or low water concentrations in inland waters will increase and salinisation become a more widespread phenomenon due to increased evaporation and likely also enhanced consumption of water for irrigation purposes (Williams, 1998; Zalidis et al., 2002). Less precipitation in these areas means lower nutrient loading to lakes, but this is not expected to compensate for the negative consequences of water loss. Nutrients will likely be up-concentrated and eutrophication increase (Zalidis et al., 2002). Moreover, at intermediate nutrient concentrations saline lakes tend to be more turbid than freshwater lakes at similar nutrient concentrations (Jeppesen et al., 1994).

Apart from the effects of warming, both phytoplankton and macrophyte production are expected to increase with increased CO_2 concentrations (Schippers et al., 2004a, b), at different rates according to variations in external conditions, such as nutrient status and alkalinity. Besides, local extinctions and the arrival of higher-temperature tolerant species will further complicate the final state and function of shallow lakes. Complications may also occur if changes in seasonality affect key interactions among different trophic levels (e.g. among fish, zooplankton and phytoplankton) leading to a mismatch that may cascade through the food-web and ultimately affect the ecological state of the lake (Boersma et al., 1996; Benndorf et al., 2001; Winder & Schindler, 2004).

However, we suggest that the most probable outcome of the effects of climate change, at least in Europe, is that the eutrophic lakes, whether phytoplankton or macrophyte dominated, will reach a more turbid state and, in southern Europe, become more saline as well. If true, it may therefore be more difficult to fulfil the present-day objectives set for the ecological state of the lakes without undertaking additional efforts to reduce nutrient loading to levels lower than the present-day expectations. This demands better management practise, including optimised fertiliser use and less intensive farming and improved sewage treatment. Moreover, re-establishment of lost wetlands will enhance N loss (particularly) and the P retention capacity, leading to reduced eutrophication. Re-establishment of buffer zones along streams may also reduce the nutrient input. To reduce the risk of salinisation, species loss, eutrophication and lowering of the water table of shallow lakes in Southern Europe, restrictions on human use of water are needed, which may imply less intensive agriculture in sensitive areas, improved recycling of water within catchments, increased efficiency with which water is allocated among different uses, and drought control.

Some challenges ahead

Although much new knowledge of how shallow lakes respond to nutrient loading reduction has been gained recently, a large number of questions are unresolved and need further consideration. Examples are:

Role of nitrogen. How important is nitrogen loading for the presence and absence of submerged macrophytes? If nitrogen is of importance, as recent studies suggest, then what are the causal mechanisms? How does the threshold concentration for plant disappearance vary with lake size, depth, trophic state, plant composition and climate?

Resistance, P hysteresis, alternative states. When do we expect P hysteresis and when not? Can P hysteresis sometimes be attributed to the fact that P loading reduction is usually not accompanied by a strong (if any) N loading reduction (apparent P hysteresis)? How is hysteresis influenced by climatic changes? Is it less strong in subtropical and tropical lakes with natural fish populations?

What is the role of the benthic-pelagic coupling and benthic facilitation for resistance to changes in loading, P hysteresis and existence of alternative states?

Fish manipulation: When is fish manipulation required to shift the lakes from turbid to clear? What are the long-term perspectives? Do we need repeated manipulations in the early phase to avoid strong perturbations that may eventually shift them back to the turbid state? How important is the risk of accumulating a ticking P bomb after biomanipulation (higher P accumulation in the sediment after manipulation) and how can we prevent its explosion?

Macrophyte growth: Which factors are most important for delaying the colonisation of plants after loading reduction—sediment content, waterfowl and fish grazing, limited amount of propagules and seeds, shading by periphyton and phytoplankton (including lack of grazing due to predation on snails and zooplankton, respectively)? How can we reinforce growth of macrophytes—sediment removal, protection against grazers, addition of seeds, change of water level or biomanipulation of fish? How can macrophytes be maintained, once established, and are some macrophyte communities more stable than others?

Role of climate: For northern shallow lakes more research is needed on the role of climate change on lake ecosystems not least on the effect of changes in seasonality on nutrient dynamics, trophic interactions and the potential for shifts to a clear state after nutrient loading reduction. Little is known about how lakes in the subtropics and the tropics respond to nutrient loading reduction and about additional methods to reinforce the improvement of the ecological state. To what extent can we transfer knowledge about recovery from eutrophication from temperate lakes to the (sub)tropics and vice versa? How realistic are the present-day objectives set for the ecological state of shallow lakes under the global warming scenario?

Acknowledgements This study was supported by the Danish Natural Science Research Council (research project "CONWOY" on the effects of climate changes on aquatic ecosystems), the EU EUROLIMPACS project (www.eurolimpacs.ucl.ac.uk) on the effects of climate changes on freshwater, and the Finnish CARE and Danish CLEAR research projects. Mariana Meerhoff was supported by the Danish Research Agency. We wish to thank the Danish counties for access to some of the data included in our analyses. Finally thanks to Mark V. Hoyer, Roger W. Bachmann and Daniel E. Canfield for most inspiring discussions on similarities and differences in structure and function of shallow lakes in the temperate zone and the subtropics and for providing the Florida data used in Fig. 6 and Table 1.

References

Andersen, H. E., B. Kronvang, S. E. Larsen, C. C. Hoffmann, T. S. Jensen & E. K. Rasmussen, 2006. Climate change impacts on hydrology and nutrients in a Danish lowland river basin. Science of the Total Environment 365: 223–237.

Bachmann, R. W., M. V. Hoyer & D. E. Canfield Jr., 1999. The restoration of Lake Apopka in relation to alternative stable states. Hydrobiologia 394: 219–232.

Bachmann, R. W., C. A. Horsburgh, M. V. Hoyer, L. K. Mataraza & D. E. Canfield Jr., 2002. Relations between trophic state indicators and plant biomass in Florida lakes. Hydrobiologia 470: 219–234.

Beklioglu, M., G. Altinaya & C. O. Tan, 2006. Water level control over submerged macrophyte development in five shallow lakes of Mediterranean Turkey. Archiv für Hydrobiologie 166: 535–556.

Benndorf, J., J. Kranich, T. Mehner & A. Wagner, 2001. Temperature impact on midsummer decline of *Daphnia galeata*: an analysis of long-term data from biomanipulated Bautzen Reservoir (Germany). Freshwater Biology 46: 199–211.

Blindow, I., G. Andersson, A. Hargeby & S. Johansson, 1993. Long-term pattern of alternative stable states in two shallow eutrophic lakes. Freshwater Biology 30: 159–167.

Boersma, M., O. F. R van Tongeren & W. M. Mooij, 1996. Seasonal patterns in the mortality of *Daphnia* species in a shallow lake. Canadian Journal of Fisheries and Aquatic Science 53: 18–28.

Branco, C. W. C., T. Aguiaro, F. A. Esteves & E. P. Caramaschi, 1997. Food sources of the Teleost *Eucinostomus argenteus* in two coastal lagoons of Brazil. Studies on Neotropical Fauna & Environment 32: 33–40.

Canfield, D. E. Jr., J. V. Shireman, D. E. Colle, W. T. Haller, C. E. I. I. Watkins & M. J. Maceina, 1984.

Prediction of chlorophyll *a* concentrations in Florida Lakes: Importance of aquatic macrophytes. Canadian Journal of Fisheries and Aquatic Science 41: 497–501.
Carvalho, L. & A. Kirika, 2003. Changes in lake functioning: response to climate change and nutrient reduction. Hydrobiologia 506/509: 789–796.
Conrow, R., A. V. Zale & R. W. Gregory, 1990. Distributions and abundances of early stages of fishes in a Florida lake dominated by aquatic macrophytes. Transactions of the American Fisheries Society 119: 521–528.
Coveney, M. F., E. F. Lowe, L. E. Battoe, E. R. Marzolf & R. Conrow, 2005. Response of a eutrophic, shallow subtropical lake to reduced nutrient loading. Freshwater Biology 50: 1718–1730.
Crisman, T. L. & J. R. Beaver, 1990. Applicability of planktonic biomanipulation for managing eutrophication in the subtropics. Hydrobiologia 200: 177–186.
Delariva, R., A. Agostinho, K. Nakatani & G. Baumgartner, 1994. Ichthyofauna associated to aquatic macrophytes in the upper Paraná River floodplain. Unimar 16: 41–60.
González, Sagrario, M. A., E. Jeppesen, J. Gomà, M. Søndergaard, T. Lauridsen & F. Landkildehus, 2005. Does high nitrogen loading prevents clear-water conditions in shallow lakes at intermediate high phosphorus concentrations. Freshwater Biology 50: 27–41.
Gyllström, M., L.-A. Hansson, E. Jeppesen, F. García-Criado, E. Gross, K. Irvine, T. Kairesalo, R. Kornijow, M. R. Miracle, M. Nykänen, T. Nõges, S. Romo, D. Stephen, E. Van Donk & B. Moss, 2005. Zooplankton community structure in shallow lakes: interaction between climate and productivity. Limnology and Oceanography 50: 2008–2021.
Hansson, L.-A., H. Annadotter, E. Bergman, S. F. Hamrin, E. Jeppesen, T. Kairesalo, E. Luokkanen, P.-Å. Nilsson, M. Søndergaard & J. Strand, 1998. Biomanipulation as an application of food chain theory: constraints, synthesis and recommendations for temperate lakes. Ecosystems 1: 558–574.
Hilt, S., E. M. Gross, M. Hupfer, H. Morscheid, J. Mählmann, A. Melzer, J. Poltz, S. Sandrock, E.-M. Scharf, S. Schneider & K. van de Weyer, 2006. Restoration of submerged vegetation in shallow eutrophic lakes – A guideline and state of the art in Germany. Limnologica 36: 155–171.
James, C., J. Fisher, V. Russell, S. Collings & B. Moss, 2005. Nitrate availability and plant species richness: implications for management of freshwater lakes. Freshwater Biology 50: 49–63.
Jensen, H. S. & F. Ø. Andersen, 1992. Importance of temperature, nitrate and pH for phosphorus from aerobic sediments of four shallow, eutrophic lakes. Limnology and Oceanography 37: 577–589.
Jensen, J. P., A. R. Pedersen, E. Jeppesen & M. Søndergaard, 2006. An empirical model describing the seasonal dynamics of phosphorus in 16 shallow eutrophic lakes after external loading reduction. Limnology and Oceanography 51: 791–800.
Jeppesen, E., J. P. Jensen, P. Kristensen, M. Søndergaard, E. Mortensen, O. Sortkjær & K. Olrik, 1990. Fish manipulation as a lake restoration tool in shallow, eutrophic, temperate lakes 2: threshold levels, long-term stability and conclusions. Hydrobiologia 200/201: 219–227.
Jeppesen, E., M. Søndergaard, E. Kanstrup, B. Petersen, R. B. Henriksen, M. Hammershøj, E. Mortensen, J. P. Jensen & A. Have, 1994. Does the impact of nutrients on the biological structure and function of brackish and freshwater lakes differ?. Hydrobiologia 275/276: 15–30.
Jeppesen, E., T. L. Lauridsen, T. Kairesalo & M. Perrow, 1997. Impact of submerged macrophytes on fish-zooplankton relationships in lakes. In Jeppesen, E., Ma. Søndergaard, Mo. Søndergaard & K. Christoffersen (eds), The Structuring Role of Submerged Macrophytes in Lakes. Ecological Studies, Vol. 131. Springer-Verlag, New York, 91–115.
Jeppesen, E., J. P. Jensen & M. Søndergaard, 2002. Response of phytoplankton, zooplankton and fish to re-oligotrophication: an 11-year study of 23 Danish lakes. Aquatic Ecosystems and Health Management 5: 31–43.
Jeppesen, E., J. P. Jensen, M. Søndergaard, K. Sandby Hansen, P. H. Møller, H. U. Rasmussen & V. Norby, 2003. Does resuspension prevent a shift to a clearwater state in shallow lakes during re-oligotrophication? Limnology and Oceanography 48: 1913–1919.
Jeppesen, E., M. Søndergaard, J. P. Jensen, K. Havens, O. Anneville, L. Carvalho, M. F. Coveney, R. Deneke, M. Dokulil, B. Foy, D. Gerdeaux, S. E. Hampton, K. Kangur, J. Köhler, S. Körner, E. Lammens, T. L. Lauridsen, M. Manca, R. Miracle, B. Moss, P. Nõges, G. Persson, G. Phillips, R. Portielje, S. Romo, C. L. Schelske, D. Straile, I. Tatrai, E. Willén & M. Winder, 2005. Lake responses to reduced nutrient loading – an analysis of contemporary long-term data from 35 case studies. Freshwater Biology 50: 1747–1771.
Jeppesen, E., M. Meerhoff, B. A. Jakobsen, R. S. Hansen, M. Søndergaard, J. P. Jensen, T. L. Lauridsen, N. Mazzeo & C. W. Branco, 2007. Restoration of shallow lakes by nutrient control and biomanipulation – the successful strategy varies with lake size and climate. Hydrobiologia 581: 269–285.
JRC (Joint Research Centre), European Commission, 2005. Climate change and the European Water Dimension. S. J. Eisenreich, JRC, Ispra, Italy (Ed). EUR 21553, 253 pp.
Kronvang, B., E. Jeppesen, D. J. Conley, M. Søndergaard, S. E. Larsen, N. B. Ovesen & J. Carstensen, 2005. Nutrient pressures and ecological responses to nutrient loading reductions in Danish streams, lakes and coastal waters. Journal of Hydrology 304: 274–288.
Köhler, J., S. Hilt, R. Adrian, A. Nicklisch, H. P. Kozerski & N. Walz, 2005. Long-term response of shallow, flushed Müggelsee (Berlin, Germany) to reduced external P and N loading. Freshwater Biology 50: 1639–1650.

Lauridsen, T. L., H. Sandsten & P. H. Møller, 2003. The restoration of a shallow lake by introducing Potamogeton spp.: the impact of waterfowl grazing. Lakes & Reservoirs: Research Management 8: 177–187.

Lehtonen, H., 1996. Potential effects of global warming on northern European freshwater fish and fisheries. Fisheries Management and Ecology 3: 59–71.

Liboriussen, L., F. Landkildehus, M. Meerhoff, M. Søndergaard, K. Christoffersen, K. Richardson, M. Søndergaard, T. L. Lauridsen & E. Jeppesen, 2005. Global warming: design of a flow-through shallow lake mesocosm climate experiment. Limnology and Oceanography: Methods 3: 1–9.

Marsden, S., 1989. Lake restoration by reducing external phosphorus loading: the influence of sediment phosphorus release. Freshwater Biology 21: 139–162.

Mazzeo, N., L. Rodríguez-Gallego, C. Kruk, M. Meerhoff, J. Gorga, C. Lacerot, F. Quintans, M. Loureiro, D. Larrea & F. García-Rodrígues, 2003. Effect of *Egeria densa* Planch. beds on a shallow lake without piscivorous fish. Hydrobiologia 506/509: 591–602.

McKee, D., D. Atkinson, S. E. Colling, J. W. Eaton, A. B. Gill, I. Harvey, K. Hatton, T. Heyes, D. Wilson & B. Moss, 2003. Response of freshwater microcosm communities to nutrients, fish, and elevated temperature during winter and summer. Limnology and Oceanography 48: 707–722.

Meerhoff, M., N. Mazzeo, B. Moss & L. Rodríguez-Gallego, 2003. The structuring role of free-floating versus submerged plants in a shallow subtropical lake. Aquatic Ecology 37: 377–391.

Mehner, T., J. Benndorf, P. Kasprzak & R. Koschel, 2002. Biomanipulation of lake ecosystems: successful applications and expanding complexity in the underlying science. Freshwater Biology 47: 2453–2465.

Meijer, M.-L., I. de Boois, M. Scheffer, R. Portielje & H. Hosper, 1999. Biomanipulation in the Netherlands: an evaluation of 18 case studies in shallow lakes. Hydrobiologia 408/409: 13–30.

Mitchell, S. F., 1989. Primary production in a shallow eutrophic lake dominated alternately by phytoplankton and by submerged macrophytes. Aquatic Botany 33: 101–110.

Moss, B., 1990. Engineering and biological approaches to the restoration from eutrophication of shallow lakes in which aquatic plant communities are important components. Hydrobiologia 200/201: 367–377.

Moss, B., 2001. The Broads. The People's Wetland. The New Naturalist. Harper Collins Publishers, London.

Moss, B., J. Stansfield & K. Irvine, 1990. Problems in the restoration of a hypertrophic lake by diversion of nutrient-rich inflow. Verhandlungen International Vereinigung für Theoretische und Angewandte Limnologie 24: 568–572.

Olin, M., M. Rask, J. Ruuhijärvi, J. Keskitalo, J. Horppila, P. Tallberg, T. Taponen, A. Lehtovaara & I. Sammalkorpi, 2005. Effects of biomanipulation on fish and plankton communities in ten eutrophic lakes of southern Finland. Hydrobiologia 553: 67–88.

Paugy, D. & C. Lévêque, 1999. La Reproduction. In Lévêque, C. & D. Paugy (eds), Less poisons des eaux continentales africaines: Diversité, écologie, utilisation par ĺhomme. IRD Editions, Paris, 129–152.

Perrow, M. R., B. Moss & J. Stansfield, 1994. Trophic interactions in a shallow lake following a reduction in nutrient loading – a long-term study. Hydrobiologia 276: 43–57.

Quirós, R., 1998. Fish effects on trophic relationships in the pelagic zone of lakes. Hydrobiologia 361: 101–111.

Romo, S., M.-J. Villena, M. Sahuquillo, J. M. Soria, M. Giménez, T. Alfonso, E. Vicente & M. R. Miracle, 2005. Response of a shallow Mediterranean lake to nutrient diversion: does it follow similar patterns as in northern shallow lakes? Freshwater Biology, 50: 1706–1717.

Sas, H. (ed.), 1989. Lake restoration by reduction of nutrient loading. Expectation, experiences, extrapolation. Acad. Ver. Richardz Gmbh.

Sazima, I. & C. Zamprogno, 1985. Use of water hyacinths as shelter, foraging place, and transport by young piranhas, *Serrasalmus spilopleura*. Environmental Biology of Fishes 12: 237–240.

Scheffer, M., S. H. Hosper, M.-L. Meijer, B. Moss & E. Jeppesen, 1993. Alternative equilibria in shallow lakes. Trends in Ecology and Evolution 8: 275–279.

Scheffer, M., D. Straile, E. H. Van Nes & H. Hosper, 2001. Climatic warming causes regime shifts in lake food webs. Limnology and Oceanography 46: 1780–1783.

Schindler, D. W., 1977. Evolution of phosphorus limitation in lakes. Science 195: 260–262.

Schippers, P., M. Luring & M. Scheffer, 2004a. Increase of atmospheric CO2 promotes phytoplankton productivity. Ecology Letters 7: 446–451.

Schippers, P., J. E. Vermaat, J. de Klein & W. M. Mooij, 2004b. The effect of atmospheric carbon dioxide elevation on plant growth in freshwater ecosystems. Ecosystems 7: 63–74.

Smith, V. H., 1983. Low nitrogen to phosphorus ratios favour dominance by blue-green algae in lake phytoplankton. Canadian Journal of Fisheries and Aquatic Sciences 43: 1101–1112.

Søndergaard, M., T. L. Lauridsen, E. Jeppesen & L. Bruun, 1997. Macrophyte–waterfowl interactions: tracking a variable resource and the impact of herbivory on plant growth. In Jeppesen, E., Ma. Søndergaard, Mo. Søndergaard & K. Christoffersen (eds), The Structuring Role of Submerged Macrophytes in Lakes. Ecological Studies, Vol. 131. Springer-Verlag, New York, 298–307.

Søndergaard, M., J. P. Jensen & E. Jeppesen, 2003. Role of sediment and internal loading of phosphorus in shallow lakes. Hydrobiologia 506/509: 135–145.

Søndergaard, M., J. P. Jensen & E. Jeppesen, 2005. Seasonal response of nutrients to reduced phosphorus loading in 12 Danish lakes. Freshwater Biology 50: 1605–1615.

Van der Molen, D. T. & R. Portielje, 1999. Multi-lake studies in The Netherlands: trends in eutrophication. Hydrobiologia 409: 359–365.

Weyhenmeyer, G., E. Jeppesen, R. Adrian, L. Arvola, T. Blenckner, T. Jankowski, E. Jennings, P. Nõges, T. Nõges & D. Straile, in press. Nitrate depleted conditions on the increase in shallow northern European lakes. Limnology and Oceanography.

Williams, W. D., 1998. Salinity as determinant of the structure of biological communities in salt lakes. Hydrobiologia 381: 191–201.

Winder, M. & D. E. Schindler, 2004. Climate change uncouples trophic interactions in an aquatic ecosystem. Ecology 85: 2100–2106.

Yafe, A., M. Loureiro, F. Scasso & F. Quintans, 2002. Feeding of two Cichlidae species in a hypertrophic urban lake. Iheringia Séries Zoologica, Porto Alegre 92: 73–79.

Zalidis, G., S. Stamatiadis, V. Takavakoglou, K. Eskridge & N. Misopolinos, 2002. Impacts of agricultural practices on soil and water quality in the Mediterranean region and proposed assessment methodology. Agriculture Ecosystems and Environment 88: 137–146.

Hydrobiologia (2007) 584:253–264
DOI 10.1007/s10750-007-0603-z

SHALLOW LAKES

Reaction of large and shallow lakes Peipsi and Võrtsjärv to the changes of nutrient loading

Tiina Nõges · Arvo Järvet · Anu Kisand · Reet Laugaste · Enn Loigu · Boris Skakalski · Peeter Nõges

Abstract More than 20-year monitoring of Estonian rivers reveals that the loading of nitrogen to large shallow lakes Peipsi (3,555 km^2, mean depth 7.1 m) and Võrtsjärv (270 km^2, mean depth 2.8 m) decreased substantially in the 1990s. Phosphorus loading decreased to a much smaller extent than nitrogen loading. In L. Võrtsjärv both N and P concentrations followed the decreasing trends of loading, which show the high sensitivity of large shallow lakes to catchment processes. Our study showed a positive relationship between P content in sediments and the relative depth of the lake. Assumingly the resilience of a lake in responding to the reduction of nutrient loading decreases together with the decrease of its relative depth. In L. Peipsi the concentration of P has not decreased since the 1990s. Our data show indirectly that P loading from Russia to L. Peipsi may have increased. The N/P ratio has decreased in both lakes. Cyanobacterial blooms have been common in both lakes already at the beginning of the 20th century. The blooms disappeared during heavy nitrogen loading in the 1980s but started again in L. Peipsi in recent years together with the drop of the N/P ratio. In L. Võrtsjärv the N/P ratio is higher and the ecosystem is more stable although the share of N_2-fixing cyanobacteria increased from the 1990s. Reappearing cyanobacterial blooms in L. Peipsi have caused fish-kills in recent years. In L. Peipsi summer/autumn fish-kills during water-blooms are a straightforward consequence of reduced nitrogen level at remaining high phosphorus level while in L. Võrtsjärv the climatic factors affecting water level are more critical—at low water level winter fish-kills may occur. In L. Võrtsjärv nutrient loading has

Guest editors: R. D. Gulati, E. Lammens, N. De Pauw & E. Van Donk
Shallow lakes in a changing world

T. Nõges (✉) · A. Kisand · R. Laugaste · P. Nõges
Centre for Limnology, Institute of Agricultural and Environmental Sciences, Estonian University of Life Sciences, Rannu, Tartumaa 61101, Estonia
e-mail: tiina.noges@emu.ee

A. Järvet
Institute of Geography, University of Tartu, Vanemuise 46, Tartu 51014, Estonia

E. Loigu
Department of Environmental Engineering, Tallinn Technical University, Ehitajate tee 5, Tallinn 19086, Estonia

B. Skakalski
Russian Hydrometeorological University, Malookhtinsky avenue 98, Saint Petersburg 195196, Russian Federation

P. Nõges
Institute for Environment and Sustainability, European Commission – DG Joint Research Centre, Ispra, VA 21020, Italy

decreased and water quality has improved, present ecological status seems to be mostly controlled by climatic factors through changes of water level. The most important measure to improve water quality in L. Peipsi would be the reduction of phosphorus loading from both Estonian and Russian subcatchments.

Keywords Nutrient loading · Nitrogen reduction · Large shallow lakes · Sediment phosphorus · Relative depth · Cyanobacteria

Introduction

In the past 40–50 years, many European lakes have been subjected to anthropogenic eutrophication caused by the increased supply of nutrients. A decrease in loadings occurring in the late 1980s or early 1990s has up to now resulted in considerable improvement in the state of the ecosystems of many lakes (Jeppesen et al., 2005b). The structure of changes in the nutrient supply is rather different in the Western European countries compared to the countries of the former Soviet Union. In western countries, e.g. in Denmark, extensive efforts have been made to improve wastewater purification and to reduce phosphorus concentration in the effluent (Jeppesen et al., 2005a; Kronvang et al., 2005). In Eastern Europe, e.g. in Baltic States, a major nitrogen reduction occurred as a result of collapse of the soviet-type agriculture (Blinova, 2001; Järvet, 2001; Juhna & Klavins, 2001).

In the present paper we review the changes of riverine loadings of nutrients into two large shallow lakes in North-East Europe, Peipsi and Võrtsjärv, and address the question, how these changes are reflected in the ecological status of these lakes. The changes of water chemistry and phytoplankton composition are discussed on the background of sediment chemistry.

Study site

By its surface area Lake Peipsi (3,555 km^2, mean depth 7.1 m) is the fourth largest lake in Europe after lakes Ladoga, Onega, and Vänern. Being shared between Estonia and Russia, L. Peipsi is the largest transboundary lake in Europe. The lake consists of three parts: L. Peipsi *sensu stricto*, L. Pihkva, and the river-shaped L. Lämmijärv connecting the formers. The outflowing River Narva runs into the Gulf of Finland of the Baltic Sea. L. Võrtsjärv (270 km^2, mean depth 2.8 m) is a large shallow lake located in the watershed of L. Peipsi (Fig. 1, Table 1). The water residence time is about two years in L. Peipsi and about one year in L. Võrtsjärv. Both lakes are ice-covered from mid-November to mid-April.

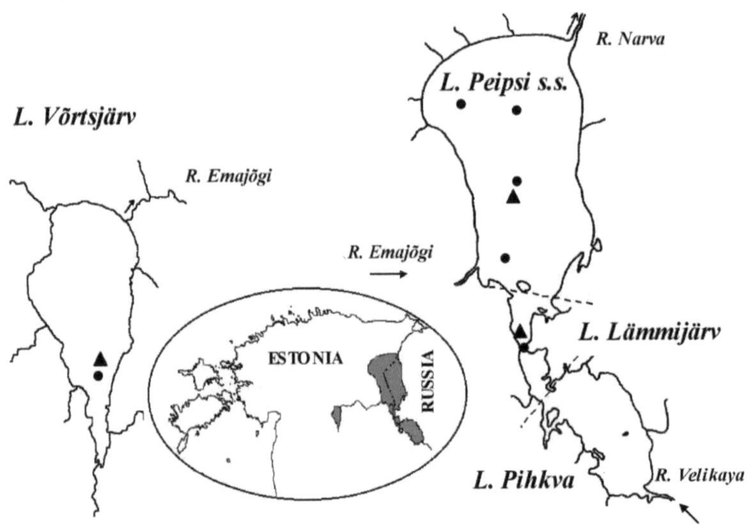

Fig. 1 Location map and sampling sites of lakes Peipsi and Võrtsjärv. Closed circles—hydrochemistry and plankton, triangles—sediment chemistry

Table 1 Morphometric and trophic state characteristics of L. Peipsi with its subbasins, and L. Võrtsjärv (Haberman et al., 1998; Jaani, 2001; Laugaste et al., 2001; Starast et al., 2001; Nõges P. et al., 2003; Nõges T. et al., 2003; Järvet et al., 2004)

Parameter		Whole Peipsi	Pihkva	Lämmijärv	Peipsi s.s.	Võrtsjärv
Catchment area (without lake area)	(km^2)	44245	28130	2957	13158	3104
Area	(km^2)	3555	708	236	2611	270
Mean depth	(m)	7.1	3.8	2.5	8.3	2.8
Volume	(km^3)	25.07	2.68	0.60	21.79	0.75
TP	(mg l^{-1})	0.042	0.063	0.053	0.035	0.054
TN	(mg l^{-1})	0.768	1.010	0.923	0.678	1.600
Chlorophyll *a*	(mg m^{-3})	18	26	25	14	24
Secchi depth	(m)	1.8	1.3	1.4	1.8	1

Riverine transport is the most important pathway for the input of nutrients into both lakes. More than 80% of phosphorous and nitrogen compounds enter L. Peipsi by the rivers Velikaya carrying biologically treated wastewaters from the Russian town Pskov (ca. 200,000 inhabitants), and Emajõgi (outflow from L. Võrtsjärv), which receives wastewater from the Estonian town Tartu (ca. 100,000 inhabitants). The R. Emajõgi contributes about 17% of all nutrient loading into L. Peipsi and 70% of the total nitrogen (TN) and phosphorus (TP) loading discharged to L. Peipsi from Estonian territory. The Russian R. Velikaya contributes about 65% of all nutrient loading into L. Peipsi (Stålnacke et al., 2002). Seventy Percent of L. Peipsi drainage area belongs to lakes Pihkva and Lämmijärv and 81% of this area is accounted for the R. Velikaya; 74% of the drainage area of L. Peipsi s.s. is accounted for the R. Emajõgi (Nõges P. et al., 2003; Nõges T. et al., 2003).

L. Peipsi s.s. and L. Võrtsjärv are unstratified eutrophic lakes, L. Lämmijärv has some dyseutrophic features, while L. Pihkva is a eutrophic, or even hypertrophic water body (Table 1). During the last two decades, both lakes have been subjected to big changes in nutrient discharges (Blinova, 2001; Järvet, 2001). Compared to the level of the end of the 1980s, only 5–10% of N-, P- and K- mineral fertilizers, and 30% of the manure were applied in the agricultural lands at the end of the 1990s (Järvet et al., 2002).

Methods

We used the data on riverine discharges and total nutrient concentrations gathered in Estonia within different monitoring programs and research projects since the 1980s. We used daily water discharge data and monthly measured concentrations to calculate the annual riverine loadings (L):

$$L = \sum_{i=1}^{12} Wki \times C_{ki}$$

W_{ki}: volume of monthly discharge, C_{ki}: monthly concentration.

Loadings from non-monitored areas (L_n) of the Estonian watershed were calculated as:

$$L_n = L_m \frac{A_n}{A_m}$$

L_m: loading from monitored area, A_n: area of non-monitored part of the watershed, A_m: area of monitored part of the watershed

TN and TP were analysed according to the methods described by Grasshoff et al. (1982). From the Russian part of the catchment nutrient concentrations in the R. Velikaya in 1980–2003 were available for present study. Up to 1999 only mineral forms of N and P were measured, TP measurements started in 2000 and TN measurements in 2001. As monthly measurements started only in 2003 and earlier the measurement frequency was mainly four times per year (varying from 1–8 times), the loading calculation for the Russian catchment was not possible.

Phytoplankton samples were preserved with formaldehyde (1962–1995) and Lugol's iodine (since 1996). Microscopic counting was performed at 400× magnification in a Fux-Rosenthal chamber (1962–1999), since 2000 the Utermöhl

(1958) method was used. For chlorophyll analysis, plankton was filtered on Whatman GF/C filters, pigments were extracted with 96% ethanol and analysed spectrophotometrically. The equation of Jeffrey & Humphrey (1975) was applied in the calculation of chlorophyll *a* (Chl *a*).

The content and vertical distribution of different phosphorus forms in the sediments were analysed in L. Võrtsjärv in June 1996, in the central part of L. Peipsi s.s, in L. Lämmijärv and, for comparison in a much smaller but also shallow L. Kaiavere (area 2.5 km^2, mean depth 2.8 m) in July 2001. The water depth at these sites was 5, 10, 15 and 4 m, respectively. The sediment cores were taken with a plexiglas corer (0.5 m long, 8 cm in diameter) keeping at least 10 cm of water above the top of the sediment to gain the sediment surface undisturbed. The sediment cores were sliced into sublayers with the thickness of 1 cm along the upper 5 cm, then 2 cm (5–7 cm), 3 cm (7–10 cm) and 5 cm (deeper than 10 cm). The content of the sediment dry matter was determined after drying at 105°C during 24 h. Four phosphorus fractions were extracted using 0.1 M NH_4Cl, 0.11 M bicarbonate-dithionite, 0.1 M NaOH and 0.5 M HCl as extracting agents (Psenner et al., 1988, with modification in the concentration of NaOH). The sediment of Lake Võrtsjärv was analysed according to the fractionation scheme of Hieltjes and Lijklema (1980). It lacks the bicarbonate-ditionite extraction, and the phosphorus of this fraction is included in NaOH-extractable fraction. The fraction extractable with ammonium chloride (NH_4Cl-RP) was assumed to represent immediately available labile P in sediments, bicarbonate-dithionite (BD-RP) extracts redox-sensitive (mainly Fe-bound) P (Psenner et al., 1988) and the sum of all fractions was taken for the total storage of sediment P.

Relative depth (z_r) of lakes was calculated as the percent ratio of the maximum depth (z_m) to the mean diameter of the circle equivalent to the lake surface area (Wetzel, 1983):

$$z_r = \frac{50 \cdot z_m \cdot \sqrt{\pi}}{\sqrt{A_o}}$$

A_o: surface area

We used the program Statistica for Windows version 6.0 for statistical analysis. To avoid the effects of the non-normal distribution of data, Spearman rank order correlation analysis was applied.

Results

In the 1980s, the discharge of nutrients by Estonian rivers into lakes Peipsi and Võrtsjärv increased while in the early 1990s a sharp decrease occurred, first of all in TN loadings (Fig. 2a). As TN loading decreased faster than TP loading (Fig. 2b), the TN/TP ratio in the loadings decreased (Fig. 2c). Similar tendencies were evident in nutrient concentrations and in the N/P ratio in the R. Velikaya (Fig. 2d–f). From the 1980s to 1991s (P1—period before breakdown of Soviet Union type of economics) the annual loading of TN, and the TN/TP ratio were significantly higher than in 1992–2004 (P2) in case of both lakes. In the loadings to L. Võrtsjärv also TP was lower in P2 while in case of L. Peipsi this difference was not significant. The median annual in-lake concentrations of TN and TP and the TN/TP ratio did not differ significantly between P1 and P2 (Table 2). In L. Võrtsjärv the concentration of TN and the TN/TP ratio decreased significantly during the whole period of 1983–2004 ($r = 0.2$, $P < 0.001$) with the strongest decrease since the beginning of the 1990s (Fig. 3g and i). The decreasing trend of TP concentration was not statistically significant. However, since 1997 TP concentrations have not exceeded the value of 0.14 mg l^{-1} (Fig. 3h). In L. Peipsi s.s. TN and TP concentrations increased in 1985–1987 (Fig. 3a, b) followed by a significant decrease of TN and TN/TP ratio in 1992–2002 (Fig. 3a and c, Fig. 4). In last years TN, TP and the TN/TP ratio started to increase again (Fig. 3a–c), and the linear trends for the period of 1992–2004 were mostly not significant (Fig. 4). In L. Lämmijärv TP increased and TN/TP decreased substantially in 1992–2002, after that the direction of the changes has reversed (Fig. 3e and f, Fig. 4). In all studied lakes the ratio of chlorophyll *a* (Chl *a*) to TP has increased and the Secchi depth has decreased (Fig. 4).

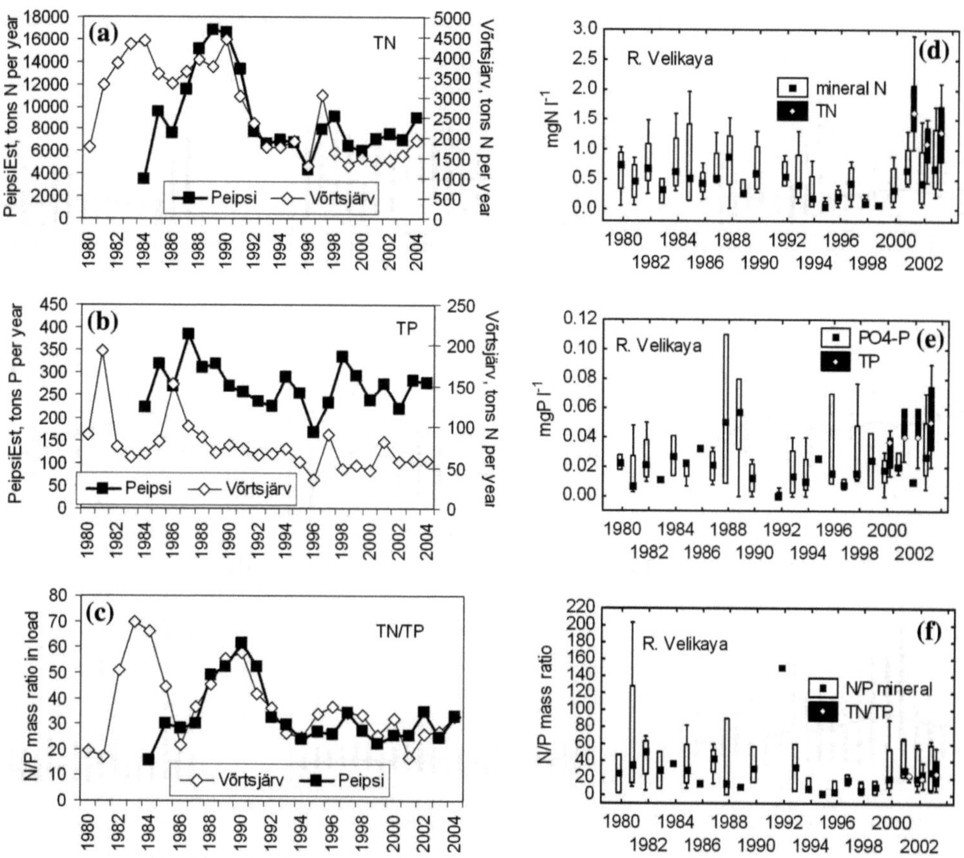

Fig. 2 Annual runoff of total nitrogen (TN) and phosphorus (TP), and the TN/TP mass ratio in loadings of L. Võrtsjärv and of Estonian catchment of L. Peipsi (**a**, **b** & **c**, correspondingly), and the concentration of mineral and total N and P and N/P ratio in the River Velikaya (**d**, **e** & **f**, respectively) in 1980–2004

The content of labile P, redox sensitive P and TP in sediments was inversely related to the relative depth of the lake (Table 3, Fig. 5).

In L. Võrtsjärv the mean phytoplankton wet biomass in the ice-free period (May–October) of the investigated years was 20 ± 14 g m^{-3} (± standard deviation) and the median 17, in L. Peipsi s.s. the corresponding values were 7 ± 12, and 5 g m^{-3}.

In L. Peipsi the N_2-fixing *Gloeotrichia echinulata* (J. E. Smith) Richter, *Aphanizomenon*

Table 2 Comparison (unequal n HSD test) of yearly riverine loadings of TN, TP and their mass ratio (TN/TP), and the median annual values of in-lake concentrations of the same indices in 1980–1991 and 1992–2001

Period	Peipsi loading from Estonia			Võrtsjärv loading		
	TN, t y^{-1}	TP, t y^{-1}	TN/TP	TN, t y^{-1}	TP, t y^{-1}	TN/TP
1980–1991	**11770**	294	**40**	**3621**	**94**	**44**
1992–2004	**7124**	256	**28**	**1762**	**61**	**30**
P	**0.006**	0.11	**0.032**	**0.000**	**0.011**	**0.012**
	Peipsi in-lake			Võrtsjärv in-lake		
	TN, mg l^{-1}	TP, mg l^{-1}	TN/TP	TN, mg l^{-1}	TP, mg l^{-1}	TN/TP
1980–1991	0.67	0.037	21	1.6	0.048	42
1992–2004	0.67	0.043	17	1.3	0.047	29
P	0.97	0.38	0.40	0.15	0.85	0.17

Significantly different ($P < 0.05$) values are marked bold

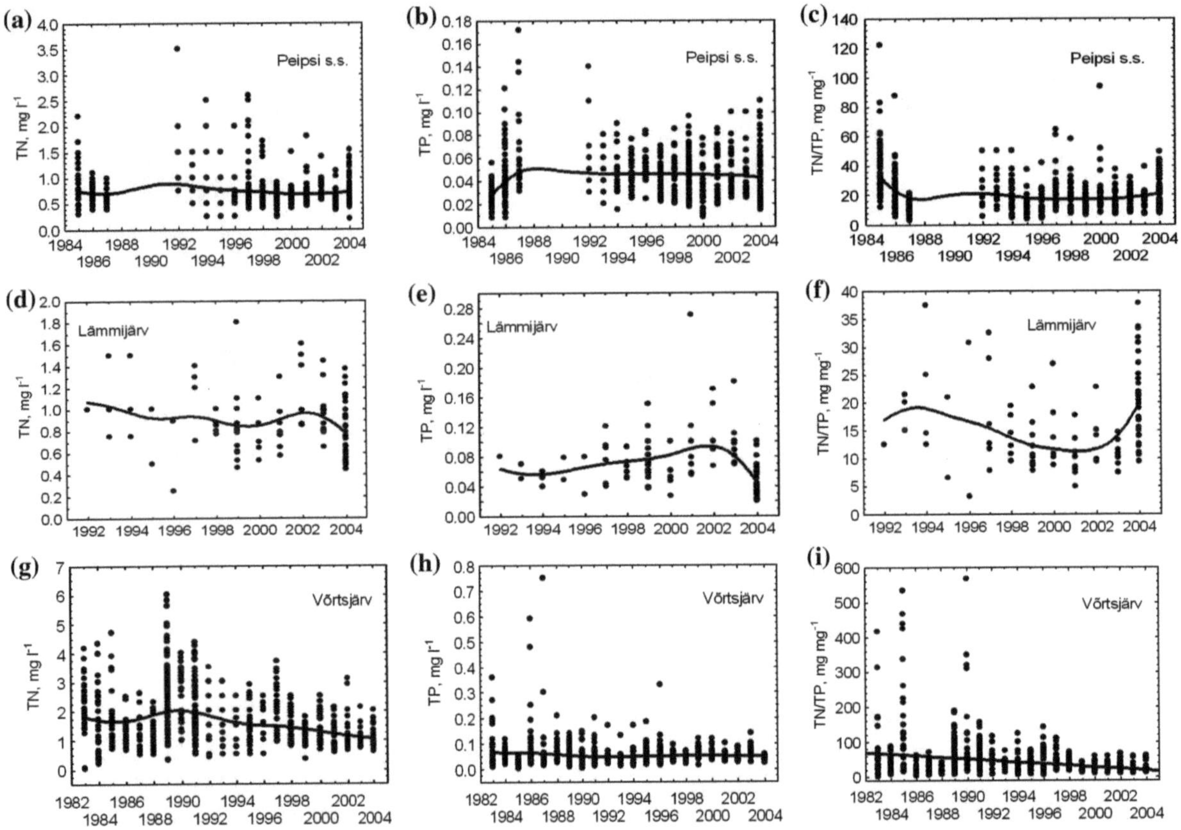

Fig. 3 Long-term changes in the concentration of total nitrogen (TN), total phosphorus (TP) and the TN/TP ratio in lakes Peipsi s.s. (**a**, **b** & **c**, correspondingly), Lämmijärv (**d**, **e** & **f**, respectively), and Võrtsjärv (**g**, **h** & **i**, correspondingly). Distance weighed least square fitting lines reflect changes in concentrations

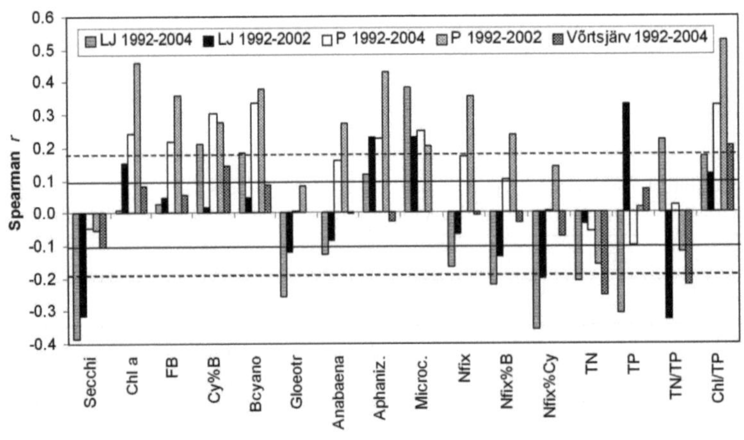

Fig. 4 Trends of changes of water chemistry and phytoplankton indices in lakes Peipsi s.s. (P), Lämmijärv (LJ) in 1992–2002 and 1992–2004, and in L. Võrtsjärv in 1992–2004. Solid line indicates P = 0.05 level for lakes Peipsi s.s. and Võrtsjärv, and dashed line indicates P = 0.05 for L. Lämmijärv. Chl *a*—concentration of chlorophyll *a*; FB—biomass of phytoplankton; CY%B—share of cyanobacteria in phytoplankton biomass; Bcyano—biomass of cyanobacteria; Nfix—biomass of N_2-fixing cyanobacteria; Nfix%B—share of N_2-fixing cyanobacteria in total phytoplankton biomass; Nfix%Cy—share of N_2-fixing species in total biomass of cyanobacteria

Table 3 Morphological characteristics, relative depth and concentration of different phosphorus fractions in sediments of lakes in present study and from literature sources

Lake	Area km^2	Max depth, m	Mean depth, m	Relative depth, z_r	Total P mg kg DW^{-1}	Labile P, mg l^{-1} pore water	Redox-sensitive P mg kg DW^{-1}	Studied sediment layer	
Dongting	2625	30.8	6.7	0.05	862			0–10 cm	Table 2 in Wang et al., 2006
Poyang	3210	25	8.4	0.04	403			0–10 cm	Table 2 in Wang et al., 2006
Peipsi s.s	2611	12.9	8.3	0.02	794	3	134	0–10 cm	Present study
Võrtsjärv	270	6	2.8	0.03	787	2		0–10 cm	Present study
Lämmijärv	236	15.3	2.5	0.09	899	11	246	0–10 cm	Present study
Kaiavere	2.51	5	2.8	0.28	1210	26	334	0–10 cm	Present study
Loch Leven	13.3	25	3.9	0.61	2000	110	600	0–3 cm	Fig. 5h in Spears et al., 2006
Müggelsee	7.18	8	4.9	0.26	2800			0–10 cm	Kleeberg & Kohl, 1999
Ringsjon Sätofta	4.2	17	3	0.73	1890			Bulk sediment	Graneli, 1999
Ringsjon Eastern	20.5	16.4	6.1	0.32	2320			Bulk sediment	Graneli, 1999
Ringsjon Western	14.8	5.4	3.1	0.12	1740			Bulk sediment	Graneli, 1999
Baptiste	5.07	15.5	5.6	0.61	3300			0–4 cm	Burley et al., 2001
Crooked	3.38	9	4.1	0.43	1600			0–4 cm	Burley et al., 2001
SpembergReservoir	6.75	9	3.5	0.31	3000		800	Surface sediment	Fig. 6 in Heidenreich & Kleeberg, 2003
Sønderby	0.08	5.7	2.8	1.79	3000		1000	0–10 cm	Fig. 2 in Reitzel et al., 2005

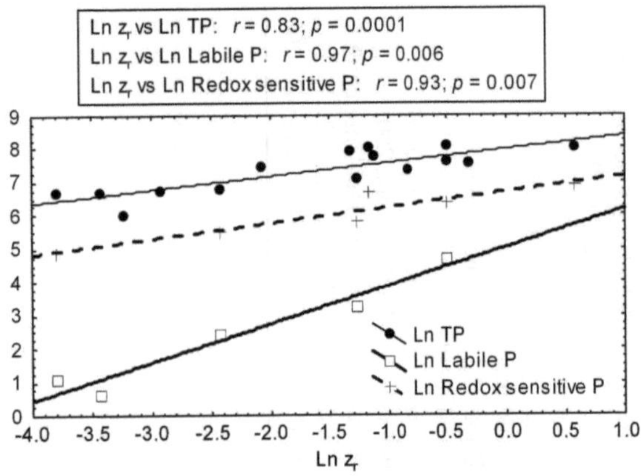

Fig. 5 Relation of the concentration of labile P fraction in pore water (mg P l^{-1}), redox sensitive P fraction (mg P kg DW^{-1}) and total phosphorus (mg P kg DW^{-1}) in sediments, and the relative depth (z_r) of the lakes referred in Table 3

flos-aquae (L.) Ralfs and *Anabaena* spp., and the N_2-non-fixing *Microcystis* spp. prevail in summer phytoplankton. In Lake Võrtsjärv, the dominant cyanobacterial species *Limnothrix planktonica* (Wolosz.) Meffert, *L. redekei* (Van Goor) Meffert and *Planktolyngbya limnetica* (Lemm.) Kom.-Legn. are not able to fix N_2; the main N_2-fixing species, *Aphanizomenon skujae* Kom.-Legn. & Cronb., *Anabaena* spp. commonly do not achieve the dominant status. In both lakes the biomass of cyanobacteria and N_2-fixing species has increased in the 1990s (Figs. 4 and 6).

In L. Peipsi the biomass of N_2-fixing cyanobacteria in July–August was inversely correlated with TN/TP (r = -0.54, P = 0.04). If the biomass of *Microcystis* was also added, the correlation with TN/TP turned non-significant (r = -0.49, P = 0.06) while a negative correlation with the average wind speed was revealed (r = -0.61, P = 0.03). In L. Võrtsjärv we did not find significant correlations with the corresponding indices.

Discussion

According to paleolimnological evidences, the eutrophication of L. Peipsi was escalated in 1970–1980s (Heinsalu et al., 2007). The eutrophication history is similar also for L. Võrtsjärv (Nõges & Järvalt, 2004). In both lakes cyanobacterial blooms have been a common phenomenon already in the beginning of the 20th century (Mühlen & Schneider, 1920; Laugaste et al., 2001). The blooms disappeared in the conditions of heavy nitrogen loading in the 1980s (Fig. 5) and started again in recent years together with the decrease of the TN/TP ratio (Figs. 3, 4 and 6). The increase of the Chl-*a*/TP ratio in lakes Peipsi s.s. and Võrtsjärv (Fig. 4) shows that the phytoplankton community has increased the efficiency of utilizing phosphorus. In general, cyanobacteria are rather poor competitors for P, however, some of them including *Planktolyngbya*, *Anabaena* and *Aphanizomenon species*, which can produce alkaline phosphatase, have an access to phosphate bound to organic compounds (Huisman & Hulot, 2005). As *Planktolyngbya*, *Anabaena* and *Aphanizomenon* are among the dominating cyanobacteria in both lakes Peipsi and Võrtsjärv, an increasing efficiency of P utilisation together with the increase of the biomass of these species could be assumed that explains the increasing trend of Chl-*a*/TP (Fig. 4). Reappearing cyanobacterial blooms in *L. Peipsi* have caused serious fish-kills in recent years (Kangur et al., 2005). These fish-kills seem to be a straightforward consequence of reduced nitrogen level at remaining high phosphorus level and, thus, the changed N/P ratio. In L. Võrtsjärv where recurrent winter fish-kills have been observed, the climatic factors affecting the water level are more crucial (Nõges P. et al., 2003; Nõges T. et al., 2003, Fig. 7).

A well-known phenomenon, which often interferes the improvement of water quality in the lakes where the external P-loading has been reduced (Marsden, 1989; Madgwick, 1999), is the internal loading based on high phosphorus content of sediments. However, in large and shallow polymictic lakes the bottom sediments are permanently stirred up and washed during the whole ice-free period not allowing the labile P to accumulate. Our results and published data showed a strong and significant decrease of the concentrations of labile P, redox sensitive P and TP together with the decreasing z_r of the lake (Table 3, Fig. 5). Therefore we assume that in large and shallow lakes with a small z_r and sediments in close contact with water, the potential of internal P loading and the resilience of ecosystem recovery after the removal of external P sources is much less than in smaller

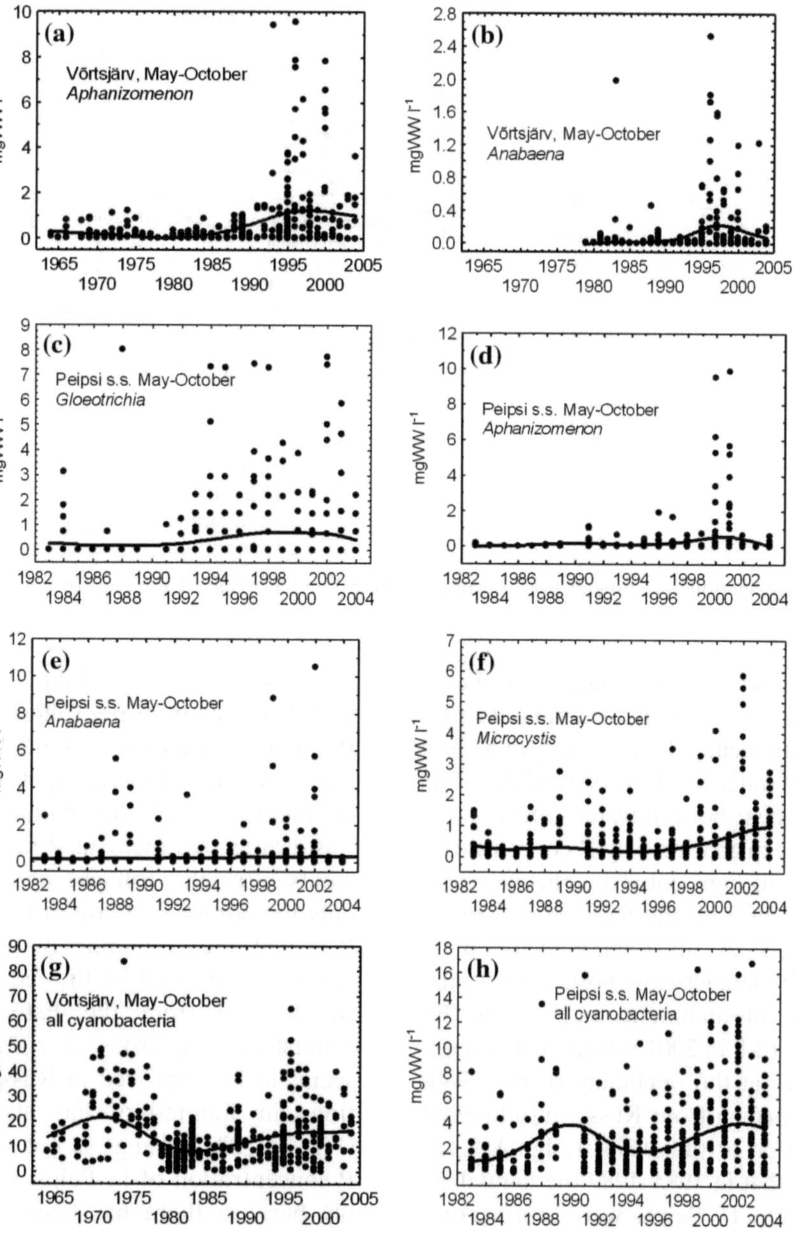

Fig. 6 Long-term changes of the biomass of all cyanobacteria, *Microcystis* and N_2-fixing species in lakes Võrtsjärv (**a**, **b** & **g**, correspondinly) and Peipsi s.s. (**c**, **d**, **e**, **f** & **h**, respectively). Distance weighed least square fitting lines are reflecting changes in concentrations

and deeper lakes with greater z_r. Basing on this assumption the subject for further comparative studies would be to investigate weather large shallow lakes exhibit a lower nutrient retention capacity than smaller and less shallow lakes.

Compared to L. Peipsi s.s., the concentration of labile sediment P was somewhat higher in L. Lämmijärv (Table 3), the narrow river-shaped part of the lake through which the water from the southern basin enters L. Peipsi s.s. and which includes also the deepest area (15.3 m) of the whole basin (Fig. 1). Anoxic conditions that likely occur in this area in wintertime may cause a release of redox sensitive P and influence to some

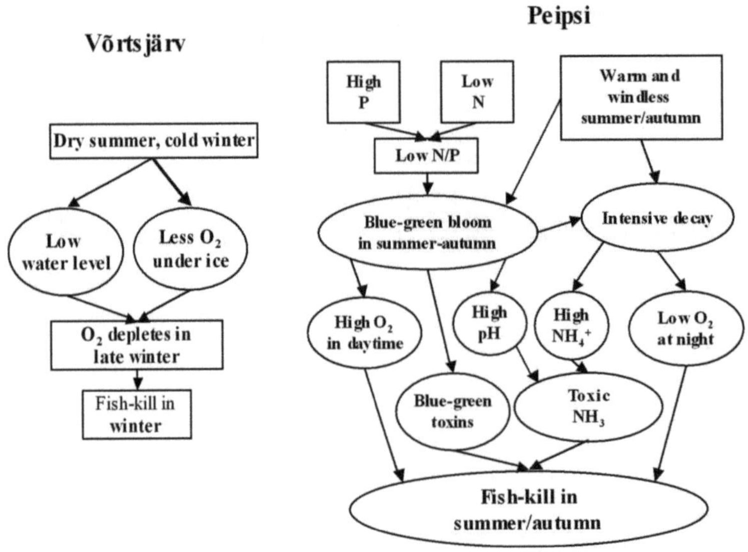

Fig. 7 Schematic explanation of causes of fish-kills in lakes Peipsi and Võrtsjärv (redrawn from Nõges & Nõges, 2004)

extent also the larger part of the lake. In L. Peipsi s.s. winter anoxia has not been documented and in L. Võrtsjärv it has occurred quite rarely in winters of extremely low water level (in 1996 whole water column, in 2002 near-bottom water layer).

One possible explanation why P concentration in L. Peipsi is still high could be the increased phosphorus loading from the Russian part of the catchment. Nõges et al. (2005) analysed all available data on loadings from both the Estonian and the Russian subcatchments and showed that the total loading of P in 2001 was almost equal to the high loadings at the beginning of the 1980s, and that the P loading from Russia in mid 1990s even exceeded that of the 1980s. Loigu & Leisk (1996) reported that in 1985–1989 the annual TN and TP loads into L. Peipsi were, correspondingly, 55,350 and 1,163 tonnes. According to Nõges P. et al. (2003) and Nõges T. et al. (2003), the respective values in 1998 were 23,800 and 1,300 tonnes, and 60% of the total TP loading entered the southernmost basin (L. Pihkva) from the R. Velikaya (Fig. 1). The information on loadings from Russia is extremely scarce and inaccessible but as the major part of this loading reaches L. Peipsi s.s. through L. Lämmijärv, we can indirectly judge on the dynamics of Russian loadings on the basis of the situation in L. Lämmijärv. As TP concentration increased in this basin until 2002 (Figs. 3e and 6), a corresponding growth of phosphorus loading from the Russian subcatchment could be supposed. The reason for that, however quite speculative, could be the failure of the maintenance of efficient wastewater treatment in Pskov and other settlements after the collapse of Soviet Union. The only temporary decrease of P concentration in L. Peipsi at the beginning of the 1990s (Fig. 3b) could be explained by the reduction of phosphorus loading from the Estonian part of the watershed (Fig. 2b). After 2002, the situation seems to be improved in Russia as P concentrations in Lämmijärv has started to decrease (Fig. 3e). This change is not yet reflected in the P concentration of L. Peipsi s.s. (Fig. 3b) probably because there has been no reduction of P loading from the Estonian subcatchment in recent years (Fig. 2b). In L. Võrtsjärv the fact that both N and P concentrations (Fig. 3g and h) follow the decreasing trends of loading (Fig. 2a and b), show the high sensitivity of large shallow lakes to changes of human activity in the catchment.

As to conclude, the nutrient loading to L. Võrtsjärv has decreased, the water quality has improved, and the present ecological situation is mostly controlled by climatic factors, which are mediated to the ecosystem through water level fluctuations (Nõges P. et al., 2003; Nõges T. et al.,

2003). Lake Peipsi is as well very sensitive to external nutrient loading and the destabilisation of the ecosystem is the result of unbalanced changes: nitrogen loading has decreased while phosphorus loading has remained unchanged or has even increased. The most important measure to achieve the improvement of the water quality in L. Peipsi would be the reduction of phosphorus loading from both the Estonian and the Russian parts of the catchment.

Acknowledgements The funding was provided by Estonian Ministry of Education (0362480s03), by Estonian Science Foundation grant 5738, and by the EC project CLIME (EVK1-CT-2002–00121). Data collection in state monitoring program was supported by Estonian Ministry of Environment. We are grateful to the two referees for their valuable comments and suggestions.

References

Blinova, I., 2001. Water quality and pollution load of the rivers of L. Peipsi basin. Riverine load into the L. Peipsi. In Nõges, T. (ed.), Lake Peipsi. Meteorology, Hydrology, Hydrochemistry. Sulemees Publishers, Tallinn, 94–96.

Burley, K. L., E. E. Prepas & P. A. Chambers, 2001. Phosphorus release from sediments in hardwater eutrophic lakes: the effects of redox-sensitive and -insensitive chemical treatments. Freshwater Biology 46: 1061–1074.

Graneli, W., 1999. Internal phosphorus loading in Lake Ringsjön. Hydrobiologia 404: 19–26.

Grasshoff, K., M. Ehrhardt & K. Kremling (eds), 1982. Methods of Seawater Analysis. Verlag Chemie, 125–187.

Haberman, J., P. Nõges, E. Pihu, T. Nõges, K. Kangur & V. Kisand, 1998. Characterization of L. Võrtsjärv. Limnologica 28: 3–11.

Heidenreich, M. & A. Kleeberg, 2003. Phosphorus-binding in ironrich sediments of a shallow revervoir: spatial characterization based on sonar data. Hydrobiologia 506(509): 147–153.

Heinsalu, A., T. Alliksaar, A. Leeben & T. Nõges, 2007. Sediment diatom assemblages and composition of pore-water dissolved organic matter as indicators of recent eutrophication history of Lake Peipsi (Estonia/Russia). Hydrobiologia 584: 133–143.

Hieltjes, A. H. M. & L. Lijklema, 1980. Fractionation of inorganic phosphates in calcareous sediments. Journal of Environmental Quality 9: 405–407.

Huisman, J. & F. D. Hulot, 2005. Population dynamics of harmful cyanobacteria. Factors affecting species composition. In Huisman, J., H. C. P. Mattehijs & P. M. Visser (eds), Harmful Cyanobacteria. Springer, 143–176.

Jaani, A., 2001. The location, size and general characterization of Lake Peipsi. In Nõges, T. (ed.), Lake Peipsi. Meteorology, Hydrology, Hydrochemistry. Sulemees Publishers, Tallinn, 10–17.

Järvet, A., 2001. Long-term changes of nutrient runoff in a Lake Võrtsjärv catchment area. In Seepõld, M. (ed.), Environmental Impact and Water Management in a Catchment Area Perspective. Tallinn Technical University, Tallinn, 54–62.

Järvet, A., R. Karukäpp & I. Arold, 2004. Location and physico-geographical conditions of the catchment area. In Haberman, J., E. Pihu & A. Raukas (eds), Lake Võrtsjärv. Estonian Encyclopedia Publishers, Tallinn, 11–26.

Järvet, A., Ü. Mander, A. Kull & V. Kuusemets, 2002. Nutrient runoff change in a rural catchment in South Estonia. Large Rivers. Archiv für Hydrobiology, Supplement 141: 305–319.

Jeffrey, S. W. & G. F. Humphrey, 1975. New spectrophotometric equations for determining chlorophylls a, b, c1 and c2 in higher plants, algae and natural phytoplankton. Biochem Physiol Pflanzen 167: 191–194.

Jeppesen, E., J.-P. Jensen, M. Søndergaard & T. L. Lauridsen, 2005a. Response of fish and plankton to nutrient loading reduction in eight shallow Danish lakes with special emphasis on seasonal dynamics. Freshwater Biology 50: 1616–1627.

Jeppesen, E., M. Søndergaard, J.-P. Jensen, K. E. Havens, O. Anneville, L. Carvalho, M. F. Coveney, R. Deneke, M. T. Dokulil, B. Foy, D. Gerdeaux, S. E. Hampton, K. Kangur, J. Köhler, S. Hilt, E. H. H. R. Lammens, T. L. Lauridsen, M. Manca, M. R. Miracle, B. Moss, P. Nõges, G. Persson, G. Phillips, R. Portielje, S. Romo, C. L. Schelske, D. Straile, I. Tatrai, E. Willen & M. Winder, 2005. Lake responses to reduced nutrient loading – an analysis of contemporary long-term data from 35 case studies. Freshwater Biology 50: 1747–1771.

Juhna, T. & M. Klavins, 2001. Water-Quality changes in Latvia and Riga 1980–2000: Possibilities and problems. Ambio 30: 306–314.

Kangur, K., A. Kangur, P. Kangur & R. Laugaste, 2005. Fish kill in Lake Peipsi in summer 2002 as a synergistic effect of a cyanobacterial bloom, high temperature and low water level. Proceedings of the Estonian Academy of Sciences. Biology, Ecology 54: 67–80.

Kleeberg, A. & J.-G. Kohl, 1999. Assessment of the long-term effectiveness of sediment dredging to reduce benthic phosphorus release in shallow Lake Müggelsee (Germany). Hydrobiologia 394: 153–161.

Kronvang, B., E. Jeppesen, D. J. Conley, M. Søndergaard, S. E. Larsen, N. B. Ovesen & J. Carstensen, 2005. Nutrient pressures and ecological responses to nutrient loading reductions in Danish streams, lakes and coastal waters. Journal of Hydrology 304: 274–288.

Laugaste, R., T. Nõges, P. Nõges, V. V. Jastremskij, A. Milius & I. Ott, 2001. Algae. In Pihu, E. & J. Haberman (eds), Lake Peipsi. Flora and Fauna. Sulemees Publishers, Tallinn, 31–49.

Loigu, E. & Ü. Leisk, 1996. Water quality of rivers in the drainage basin of Lake Peipsi. Hydrobiologia 338: 25–35.

Madgwick, F. J., 1999. Restoring nutrient-enriched shallow lakes: Integrating of theory and practice in the Broads, UK. Hydrobiologia 408(409): 1–12.

Marsden, M. W., 1989. Lake restoration by reducing external phosphorus loading: The influence of sediment phosphorus release. Freshwater Biology 21: 139–162.

von zur Mühlen M. & G. Schneider, 1920. Der See Wirzjerw in Livland. Archiv für die Naturkunde des Ostbaltikums 14: 1–156.

Nõges, P. & A. Järvalt, 2004. Long-term changes in the ecosystem of Lake Võrtsjärv and their causes. In Haberman, J., E. Pihu & A. Raukas (eds) Lake Võrtsjärv. Estonian Encyclopaedia Publishers, Tallinn, 347–355.

Nõges, P., Ü. Leisk, E. Loigu, A. Reihan, B. Skakalski. & T. Nõges, 2003. Nutrient budget of Lake Peipsi in 1998. Proceedings of the Estonian Academy of Sciences. Biology. Ecology. 4: 407–422.

Nõges, T. & P. Nõges, 2004. Consequences of catchment processes and climate changes on the ecosystems of large shallow temperate lakes. In Rupp, G. L. & M. D. White (eds), Proceedings of the 7th International Symposium on Fish Physiology, Toxicology, and Water Quality, Tallinn, Estonia, May 12–15, 2003, U.S. Environmental Protection Agency, Ecosystems Research Division, Athens, Georgia, USA. EPA 600/R-04/049, 341–350.

Nõges, T., P. Nõges & R. Laugaste, 2003. Water level as the mediator between climate change and phytoplankton composition in a large shallow temperate lake. Hydrobiologia 506: 257–263.

Nõges, T., R. Laugaste, E. Loigu, I. Nedogarko, B. Skakalski & P. Nõges, 2005. Is the destabilisation of Lake Peipsi ecosystem caused by increased phosphorus loading or decreased nitrogen loading? Water Science and Technology 51: 267–274.

Psenner, R. B., M. Boström, K. Dinka, K. Pettersson, R. Pucsko & M. Sager, 1988. Fractionation of phosphorus in suspended matter and sediment. Archiv für Hydrobiologie – Beiheft Ergebnisse der Limnologie. 30: 98–110.

Reitzel, K., J. Hansen, F. O. Andersen, K. S. Hansen & H. S. Jensen, 2005. Lake restoration by dosing aluminum relative to mobile phosphorus in the sediment. Environmental Science & Technology 39: 4134–4140.

Spears, B. M., L. Carvalho, R. Perkins, A. Kirika & D. M. Paterson, 2006. Spatial and historical variation in sediment phosphorus fractions and mobility in a large shallow lake. Water Research 40: 383–391.

Stålnacke, P., Ü. Sults, A. Vasiliev, B. Skakalsky, A. Botina, G. Roll, K. Pachel & T. Maltsman, 2002. An assessment of riverine loads of nutrients to Lake Peipsi, 1995–1998. Archiv für Hydrobiologie, Supplement 141: 437–457.

Starast, H., A. Milius, T. Möls & A. Lindpere, 2001. Hydrochemistry. In Nõges, T. (ed.), Lake Peipsi. Meteorology, Hydrology, Hydrochemistry. Sulemees Publishers, Tallinn, 97–130.

Utermöhl, H., 1958. Zur Vervollkomnung der quantitativen Phytoplanktonmethodik. Mitteilungen Internationale Vereinigung für Theoretische und Angewandte Limnologie 9: 1–38.

Wang, S., X. Jin, H. Zhao & F. Wu, 2006. Phosphorus fractions and its release in the sediments from the shallow lakes in the middle and lower reaches of Yangtze River area in China. Colloids and Surfaces A: Physicochem Eng Aspects 273: 109–116.

Wetzel, R. G., 1983. Limnology. 2nd edn. Saunders College Publishing, Philadelphia.

Hydrobiologia (2007) 584:265–275
DOI 10.1007/s10750-007-0607-8

SHALLOW LAKES

Spatial and temporal diversity of small shallow waters in river Lužnice floodplain

David Pithart · Radka Pichlová · Michal Bílý · Jaroslav Hrbáček · Kateřina Novotná · Libor Pechar

Abstract To propose a concept of their mutual diversity, twenty-nine permanent shallow floodplain pools and oxbows in the river Lužnice floodplain were analysed for area, depth, shape, flooding, and shading by terrestrial vegetation, and sampled in all seasons for their water chemistry, phytoplankton composition and biomass, and zooplankton composition. The sites are regularly flooded, eutrophic, and often shaded by surrounding vegetation. Cryptophyceae, Chrysophyceae and Euglenophyceae dominated the phytoplankton, while Cyanophytes were rare. Within the rich zooplankton assemblage (63 species), cladocerans and rotifers dominated. Correlation matrices and multivariate analyses indicated that shaded and relatively deeper sites had lower oxygen saturation and higher concentrations of PO_4–P and NH_4–N. Shade and relative depth correlated negatively with phytoplankton biomass and number of phytoplankton taxa, and positively with Cryptophytes and large cladocerans—thus indicating poor mixing, poor light availability and low fish pressure on herbivores. Decomposition of leaf litter increased oxygen consumption, while shade from terrestrial vegetation restricted photosynthesis and decreased oxygen production. Larger sites were more species-rich in phytoplankton and supported Euglenophyceae, green algae and rotifers.

Keywords Floodplain · Terrestrial–aquatic interactions · Phytoplankton · Zooplankton · Shading · Flooding

Guest editors: R. D. Gulati, E. Lammens, N. De Pauw & E. Van Donk
Shallow lakes in a changing world

D. Pithart (✉) · L. Pechar
Institute of Systems Biology and Ecology, Academy of Sciences of the Czech Republic, Dukelská 145, 379 01 Třeboň, Czech Republic
e-mail: d.pithart@usbe.trebon.cz

K. Novotná · L. Pechar
Faculty of Agriculture, Applied Ecology Laboratory, University of South Bohemia, Studentská 13, 370 05 České Budějovice, Czech Republic

R. Pichlová
Cooperative Institute for Limnology and Ecosystem Research, University of Michigan, 2205 Commonwealth Blvd., Ann Arbor, MI 48105, USA

M. Bílý
T. G. Masaryk Water Research Institute, Podbabská 30, 160 62 Praha 6, Czech Republic

J. Hrbáček
Hekrova 820, Praha 4, Czech Republic

Introduction

In a floodplain with a natural hydrological regime, aquatic ecosystems are characterised by their transience between being either marshy, lotic or lentic in character. Their mutual connectivity remains relatively high because water flowing through the underlaying sediments influences the lotic subsystems biogeochemically and ecologically from the adjacent river, and from the soil and bedrock sediments (Jones & Mulholland, 2000). Moreover, flood events enhance the connectivity, accelerating the exchange of plankton communities, plant seeds and fish among the various pools and oxbows. After each flood event, a process of diversification begins again as the water level declines, the water current diminishes, and finally the floodplain lakes and oxbows become disconnected (Pithart, 1999).

Water level fluctuations and the transient character of the many aquatic parts of the floodplain intensify the interactions between the floodplain's terrestrial and aquatic environments (Prach, 1996). As the volume of each water body decreases, the ratio between its surface and volume increases, enhancing the intensity of exchange between the sediment and water column. Similarly, the relative impact of terrestrial vegetation—with its effect of shading and leaf litter—increases with decreasing area. The small, shallow lakes and pools can be relatively deep, with a tendency towards intense stratification and oxygen deficits, influencing the whole aquatic community, and in particular the fish (Pithart & Pechar, 1995; Pechar et al., 1996).

The aim of this study is to describe and explain the spatial and temporal diversity of the permanent lotic ecosystems of the hydrologically well-preserved Lužnice river floodplain. We focus on the environmental factors specific to these pools and oxbows, and try to maintain a multi-dimensional viewpoint that incorporates both spatial as well as temporal changes (Ward, 1989). We use the term diversity sensu lato as the overall heterogeneity of our studied subsystems; one that includes environmental variables as well as species occurence, composition and biomass.

Long-term limnological monitoring of two floodplain pools has been carried out for about 20 years, resulting in several papers being published: about the plankton of the Lužnice floodplain, describing the plankton dynamics (Hrbácek et al., 1994; Pechar et al., 1996); flagellate vertical migrations (Pithart, 1997) or and phytoplankton composition (Kylbergerová et al., 2002). This more synthetic paper is based on repeated synoptical samplings of a larger set of different localities.

Sampling sites

The Lužnice floodplain within the study area remains one of the last floodplains in the Czech Republic whose natural hydrological regime has not been altered substantially by any massive river regulation. The river channel is still meandering and changing its course after every major flood event, including the cutting of new meanders. The studied area is flooded for several weeks every year, mostly in March and April, with occasional floods in winter or summer. The area, about 25 km long and 1–2 km wide, lies to the south of the Czech Republic between the Austrian border and the town of Suchdol nad Lužnicí (Fig. 1). The average (50 years) river discharge is 5.8 m^3s^{-1}. The floodplain area consists of: the main stream and standing water bodies (15% of area, about 200 permanent pools and oxbows); meadows, both maintained and abandoned (the latter prevailing, mainly overgrown by *Phallaris arundinacea and Urtica dioica*); pastures; and floodplain softwood forest (15% dominated by willow and white poplar with an admixture of oak and alder). In winter, the pools regularly freeze over, the ice being mostly covered by snow. Water levels fluctuate within a range of 1.5 m.

The 29 permanent oxbows and pools chosen for this study represent a range of parameters typical of the studied area in terms of their surface area, depth, shape, level of shading by shore vegetation, distance from the river, and macrophyte vegetation.

In this paper, we use the term "oxbow lake" for water bodies with a length many times exceeding their width; they originate in former meanders that have been cut off from the active river channel. The term "pools" we use for circular or oval-shaped water bodies that originated from oxbows but were reshaped by heterogeneous silting-up and erosion caused by floods.

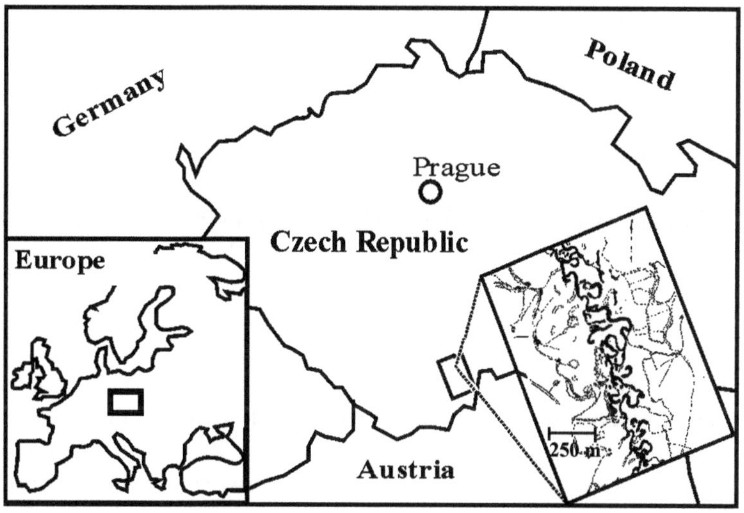

Fig. 1 Location of studied sites and landscape character of the river Lužnice floodplain

Methods

Sampling was carried out five times to cover different parts of the seasonal development: September 6–7 1994, October 25–26 1994, January 24–25 1995, May 2–3 1995, and August 19–20 1997.

During each of these samplings, water temperature, dissolved oxygen concentration, pH and conductivity were measured in situ by WTW field probes and Multi-line P4 measuring set.

Total alkalinity (TA) was determined by end-point (pH = 4.5) titration with 0.1 M HCl. For nutrient and phytoplankton analysis, an integrated water sample (3 l) was taken with a Plexiglas tube from the first 1–1.5 m of the water column. Concentrations of NH_4-N, NO_3-N, and PO_4-P were determined using a Tecator flow injection analyser (FIA Star 5010, Tecator, Sweden; Ruzicka & Hansen, 1981) in samples filtered through Whatman GF/C filters. Total nitrogen (TN) and total phosphorus (TP) were determined in samples filtered through 100 μm mesh after mineralisation by persulphate as NO_3-N and PO_4-P (for details, see Pithart & Pechar, 1995). Chlorophyll-*a* concentration was estimated spectrophotometrically or fluorometrically after extraction in an acetone–methanol mixture (Pechar, 1987).

For enumeration, 100 ml of the sample was preserved by Lugol solution, sedimented in chambers and analysed under the inverted microscope. Algal species were counted in random fields at a magnification of 200× or 400× up to the number of 300 individuals for each dominant species; cell volume was calculated as the volume of corresponding geometrical body (Javornický, 1978). Macrophytes in the pools were determined to species, and their coverage was estimated.

Zooplankton was sampled in a qualitative way. One or two mixed samples were taken from each of the pools, two samples in the case of a separate free-water area and nearshore macrophyte area. From free-water areas (without macrophytes), several bottom-up tows by a conical plankton net (mesh size 100 μm) were taken and mixed together. Zooplankton from macrophyte beds was sampled from throughout the water column using a plastic vessel, and filtering through a plankton net (mesh size 100 μm). All samples were preserved by formalin. Later, 300–400 individuals were determined and counted. Zooplankton was not sampled in August 1997. In zooplankton samples, cladocerans and rotifers were determined mostly to species, or at least to genus level. Copepods were determined only to groups: nauplii, copepodids, diaptomids and harpacticids (Straškraba, 1964).

The dimensions of pools were measured during a period of average water level to calculate their area and shape. Maximum depth was measured from the ice cover. Relative depth was calculated according to the equation $z_r = 50 \cdot z_{max} \cdot \sqrt{\pi} / \sqrt{A}$ (without dimension, usually <2 for shallow lakes; Hutchinson, 1957) and Landscape Shape Index (equalling 1 for a circle) according to the formula LSI = p/(2$\sqrt{\pi \cdot A}$) (McGarigal & Marks, 1995); where A—area, z_{max}—maximum depth, z_r—relative depth, and p—perimeter. Position along the longitudinal river course (river km) and distance from river was estimated from aerial photographs.

Shading was measured during clear-sky days by simultaneous measurement of relative PhAR by Li-Cor 150 with SA 190—sensors at an unshaded site outside the locality, and on the water surface of the water body, at each point of a virtual network consisting of 2×2, 3×3 or 5×5 m squares (according to pool area; total number of points per water body ranged from 20 to 40). Average values from all points of the network and two measurements (morning and afternoon) were used.

An index of flooding was developed as an estimate of probability of flooding for each of the pools, i.e., direct surface contact of a pool with main stream water. We had observed that pools underwent more flooding depending on their proximity to the river, and the more they were upstream. Also, we found that NO_3-N concentration strongly correlated linearly in the same way, i.e., negatively with distance from river and positively with river kilometre. Thus, we used NO_3-N concentration as an indicator of flooding. Using our dataset (distance from river, river kilometre and average NO_3-N concentration) we developed a linear function for calculating the flood index:

$$\text{FI} (\sim NO_3 - N) = -3610.65 + 32.67 * \text{river km} + (-5.06 * \text{river distance})$$

Correlation between the measured and predicted NO_3-N concentration was 0.30, $P = 0.009$. From this analysis 6 pools were excluded, as these pools clearly obtained most of their water supply from terrace runoff and hence their NO_3-N values were very high.

For data analyses, correlation matrices were calculated for all parameters for each pool using all sampling data (total), from the average value of parameters, and finally for particular seasonal samplings separately. For multivariate statistics we used CANOCO software; DCA method was tried first and finally RDA or CCA. Only the most relevant environmental factors were used for models. For phytoplankton, only relative contributions of selected important groups to the total biovolume were used; for zooplankton, only relative contributions of large pelagial cladocerans and pelagial rotifers to total zooplankton biomass. To make the biplots more understandable, the environmental factors were divided into two groups: (1) morphology, flooding, shading and distance to the river; and (2) water chemistry parameters.

Results

Morphology, distance to the river, flooding and shading

The studied set of water bodies are small, shallow lakes (Table 1) of different shapes, from almost circular pools to elongated oxbows (LSI = 3.7). Some of them were permanently connected to the river (parapotamon-type side arms). Despite most pools maximum or average depth being rather shallow, relative depths can reach rather extreme values (maximum 17). Due to their small area and bankside woody vegetation, pools can be substantially shaded (up to 79%). Some morphological parameters were correlated (see Fig. 2); shape index was correlated significantly with area because the elongated oxbows were also the largest water bodies. All oxbows were also close to the river, therefore their relative depth correlated with distance from river.

Table 1 Mean, minimum and maximum values of morphometric characteristics and shading of studied localities

	Maximum depth (m)	Relative depth (%)	Area (m^2)	LSI	Shade (%)	Distance to main stream (m)
Mean	1.1	8.6	753	1.6	36	44
Min–Max	0.2–3.3	1.5–17.0	58–5,800	1.03–3.74	3–79	0–135

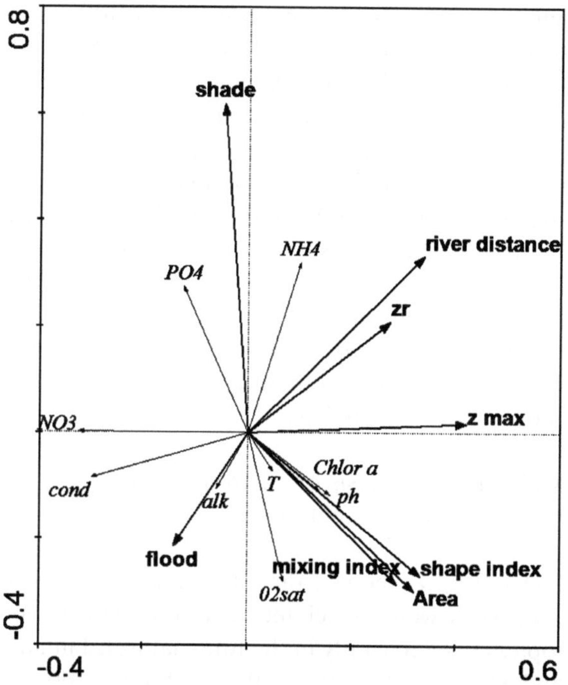

Fig. 2 RDA analysis of morphometry, shading, flooding and distance from river (explaining variables, thick vectors) and water chemistry as explained parameters. First axis explains 10.3% of variability, second 0.2%. Monte Carlo permutation test for first axis: F-ratio = 16.03, P = 0.058

Water chemistry

According to total nitrogen (TN, mean value 3.7 mg l^{-1}), total phosphorus (TP, 130 μg l^{-1}), and chlorophyll a (67 μg l^{-1}) concentrations, the studied sites would be classified as eutrophic or hypertrophic. Ranges between minimum and maximum values showed extreme variations in most parameters regardless of sampling date and site. Conductivity fluctuated mostly between 200 and 300 μS $cm^{-1.}$ Alkalinity below 1 meq l^{-1} was found in 70% of cases; values higher than 2.00 meq l^{-1} were recorded in less then 5% of measurements. Values of pH ranged mostly from 6.5 to 7.

Some chemical parameters were associated with shading (Fig. 2), which was positively and significantly correlated with PO_4 and NH_4 and negatively with oxygen saturation. Area was correlated with chlorophyll *a* concentration, pH, temperature and shape index and associated with oxygen. Flooding was associated with NO_3-N, alkalinity and conductivity.

NO_3-N was the prevailing component in the nitrogen budget for most sites: on average nitrate represented about 60% of TN. No significant correlations between NO_3-N and PO_4-P or NH_4-N were found. Also ratios of dissolved inorganic nitrogen to phosphate (DIN/PO_4-P) and TN/TP were determined mainly by the nitrate concentrations (not shown).

At some sites, NO_3-N concentration exceeded the six-year river maximum (monitored monthly). These pools (6 in all) were considered to have an important source of water from some origin other than the river. The highest values of conductivity, sulphates and high alkalinity as well as lower temperatures in summer in these 6 pools indicate an intensive throughflow of underground water. The remaining 23 water bodies were assumed to be sourced by infiltrating river water. The flood index of these 23 pools correlated with biomass of diatoms (0.58, P = 0.004) abundant in the main stream plankton, and lotic species of diatoms (*Achnanthes lanceolata, Navicula lanceolata, and Navicula gregaria*; Poulíčková, 1997) were found there.

Some parameters also showed seasonal aspects (Fig. 3). High NO_3-N and oxygen were associated with spring (time of flooding), whereas high NH_4-N concentrations were associated with winter. Summer was characterised by higher chlorophyll *a*, PO_4-P and alkalinity.

Plankton composition

Phytoplankton composition was very different compared to that of other types of shallow or deep lakes (Kylbergerová et al., 2002). Typical was the year-round prevalence of Cryptophyceae, (70% of algal biomass in the whole set of localities in September, 67% in January), namely species such as: *Cryptomonas curvata, C. marssonii, C. reflexa* and *Rhodomonas minuta*; abundance of Euglenophyceae (*Trachelomonas volvocina, T. volvocinopsis* and a variety of species of *Euglena* and *Phacus*); Dinophyceae; and Chrysophyceae (namely *Synura* spp., *Mallomonas* spp. and *Chrysococcus* spp.) In contrast, except for diatoms, the phytoplankton groups

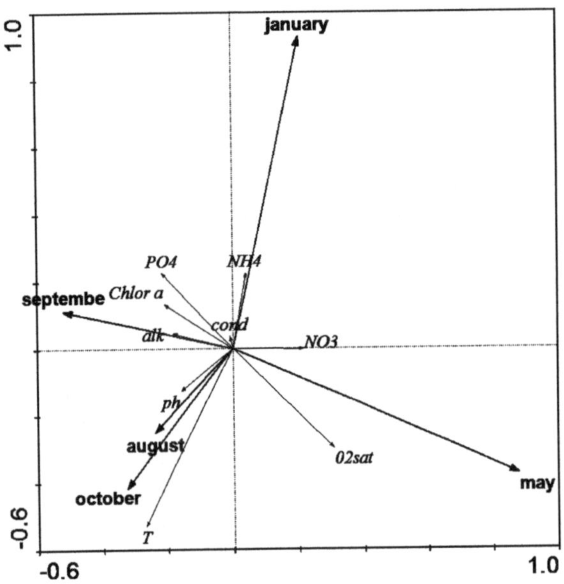

Fig. 3 RDA analysis of season as explaining variable (thick lines) and water chemistry as explained variable. First axis explains 4.5% of variability, second 0.1%. Monte Carlo permutation test for first axis: *F*-ratio = 13.144, $P = 0.004$

typical of eutrophic shallow lakes were either quite rare (Cyanophyceae contributed up to 2%) or were remarkably less abundant (green chlorococcal algae contributed up to 10%). The species richness for the whole set of water bodies was high, especially for green algae (66 species) and the flagellate groups—Euglenophytes (34 species) and Chrysophytes (16 species). Species richness per particular locality varied considerably from 4 to 42 species.

Zooplankton was collected in four sets of samples; 33 taxa of cladocerans and 30 taxa of rotifers were determined (copepods were not identified to species level). For the purpose of this analysis, the species were grouped in several distinguishable categories to simplify the dataset. In general the most relatively abundant were pelagial rotifers: species of genera *Anureopsis*, *Brachionus*, *Conochilus*, *Filinia*, *Gastropus*, *Hexarthra*, *Keratella*, *Notholca*, *Polyarthra*, *Pompholyx*, *Asplanchna*, *Synchaeta*, nauplii and copepodids, contributing together 75–90% to total individuals in zooplankton. Rotifers dominated in late summer (September, 39% of total count of zooplankton) and autumn (October 32%), whereas copepods including nauplii were most numerous in winter and spring (about 40%). Large pelagial cladocerans (*Daphnia longispina*, *Daphnia pulicaria*, and *Daphnia pulex*), littoral and surface cladocerans (*Sida* sp., *Simocephalus* sp., *Ceriodaphnia megops*, *C. reticulata*, *C. rotunda*, *Megafenestra* sp., and Chydoridae) were virtually absent in January, whereas small pelagic cladocerans (*Bosmina* sp., *Ceriodaphnia affinis*, *C. pulchella*, *Daphnia cucullata*, *Daphnia parvula*, *Diaphanosoma sp. and Moina* sp.) were at least present in May and otherwise held a stable share in the zooplankton assemblage.

Multivariate analysis of environmental parameters and plankton

RDA (Fig. 4) showed that shaded and relatively deep sites were associated with abundant Cryptophytes, significantly in the total dataset, January and May ($r = 0.49$; $P = 0.01$) and large cladocerans, whereas large sites were associated with chlorophyll *a*. Area significantly correlated with chlorophyll *a* in August and September ($r = 0.44$; $P = 0.008$), with euglenophytes (August, $r = 0.67$; $P = 0.00$), and was positively associated with green algae and Cyanophytes (significantly in total dataset, $r = 0.79$; $P = 0.00$), and rotifers.

CCA (not shown) tested the impact of season on phytoplankton groups and showed that Euglenophytes, diatoms and green algae were typical for summer, whereas Cryptophyceae and large cladocera were more typical in autumn.

RDA used to model the relations of water chemistry parameters and plankton is shown in Fig. 5. A more detailed seasonal analysis (not shown) showed that surface oxygen saturation correlated in May with chlorophyll *a* ($r = 0.43$; $P = 0.02$) and with pH ($r = 0.58$; $P = 0.001$) but not with shade (at the beginning of May most leaves had not yet developed), indicating that the rapid growth of the algal spring bloom was the main source of oxygen in the water bodies. The correlation between the biomass of Chrysophytes (0.44; $P = 0.02$) and Cryptophytes ($r = 0.62$; $P = 0.017$) as they reached their spring peak, supports this explanation. On the other hand, in summer, oxygen saturation correlated significantly with pH, chlorophyll *a* and negatively

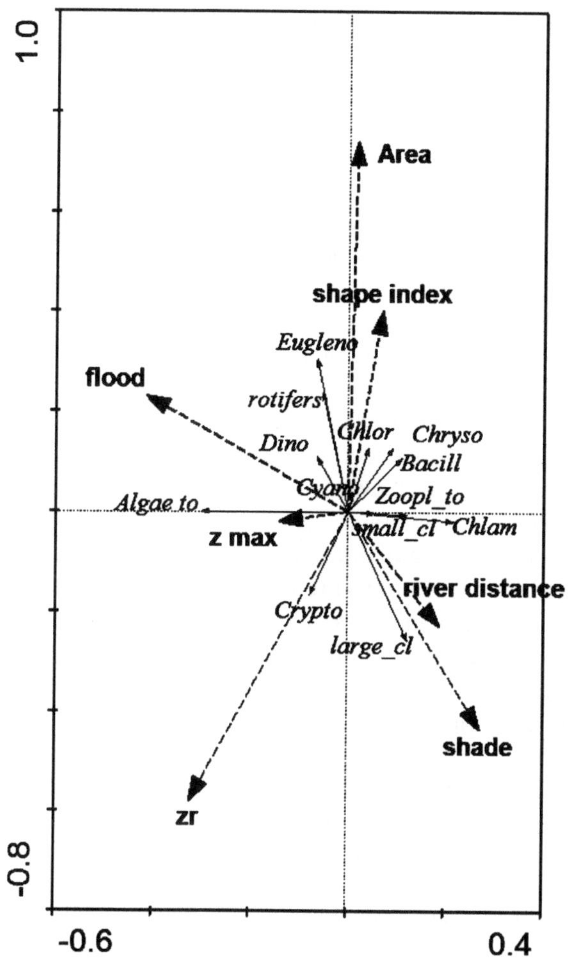

Fig. 4 RDA analysis of morphometry, shading, flooding and distance from river (explaining variables, dashed vectors) and plankton variables (relative contribution of groups to total biomass (phytoplankton) or count (zooplankton) as explained parameters. First axis explains 12% of variability, second 5%. Monte Carlo permutation test for first axis: *F*-ratio = 13.821, *P* = 0.102

with shade, whereas in October it correlated only with shade ($r = -0.66$; $P = 0.00$), indicating the reduction of photosynthesis as the day shortened and enhancing the role of sediment decomposition in the oxygen dynamics.

Concentrations of NH_4-N and PO_4-P (mostly mutually correlated, Fig. 2) were apparently bound to the oxygen regime as well. These variables were also associated with shade, a stronger correlation for PO_4-P (significantly correlated in September, January and October, and also for all seasons) than that for NH_4-N (significant only in September).

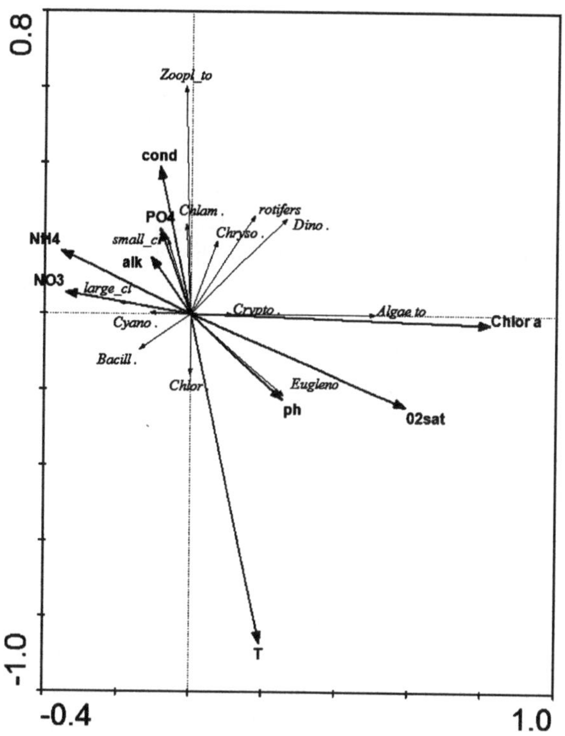

Fig. 5 RDA analysis of water chemistry and temperature (explaining variables, thick lines) and plankton variables—relative contribution of groups to total biomass (phytoplankton) or count (zooplankton)—as explained variables. First axis explains 25% of variability, second 0.4%. Monte Carlo permutation test for first axis: *F*-ratio = 45.5, *P* = 0.002

Also a correlation between large pelagic daphniids and shade was repeatedly significant (in September $r = 0.66$, $P = 0.02$), indicating weak fish predation at these sites.

Cryptophyceae was the only algal group positively correlated with shade (significantly in September, $r = 0.58$; $P = 0.008$, and in the total and average dataset); the other algal groups, both in terms of biomass and proportion, showed a tendency to correlate negatively with shade (mostly not significantly). Not surprisingly, shade restricted algal species number ($r = -0.53$; $P = 0.015$ in September).

Discussion

Concept of spatial and temporal heterogeneity

The data analysis indicated the mutual interplay of the different mechanisms that result in the ecosystem diversity of the studied water bodies. Even if it is possible to explain the observed phenomena through the action of only one mechanism, it is more likely that other different mechanisms share a contribution to every phenomena, in different proportions from case to case or from time to time. To explain such a network of relations, we divide the factors driving the diversity among the studied localities into three groups (Fig. 6): (i) source of water: infiltration and flooding; (ii) morphology: area, shape and depth; and (iii) terrestrial vegetation: shading and leaf litter.

Source of water: infiltration and flooding

Ground water seeping to the floodplain water bodies through the permeable sandy alluvial sediments may flow in from either the main river stream or laterally from the elevated river terrace. The nutrient load of these two water sources is markedly different and thus their relative proportion in each case influences the alluvial water bodies substantially. Due to the adjoining intensive agriculture, the water seeping laterally has a much higher conductivity and concentration of HCO_3^-, NO_3^-, and a relatively lower concentration of PO_4-P.

Perhaps somewhat surprisingly the river water is not the most important source of N and P for the studied sites. Correlation between water chemistry in the pools and the river is low most of the year, which indicates the prevalence of some other autonomous processes (Pechar et al., 1996). The high level of nutrients (N, P) in the pools can be considered the result of natural input and accumulation of allochtonous, mainly organic, matter from terrestrial vegetation.

An increasing discharge results in surface flow from the river channel to the adjacent ditches, oxbows and pools; later, the flooding of the whole floodplain connects the subsystems into one water body. The flood events influence the basic chemistry of floodplain waters, especially the concentrations of major ions. During a high discharge, their concentrations are usually diluted—with the exception of nitrates. With decreasing discharge and water levels, and even before the surface separation of water subsystems, a horizontal heterogeneity of water chemistry and plankton already appears, being accelerated by the subsequent surface separation of water bodies. Thus, the different spatial and temporal course of each flood causes a time shift in water chemistry and plankton development in different parts of the floodplain (Pithart, 1999).

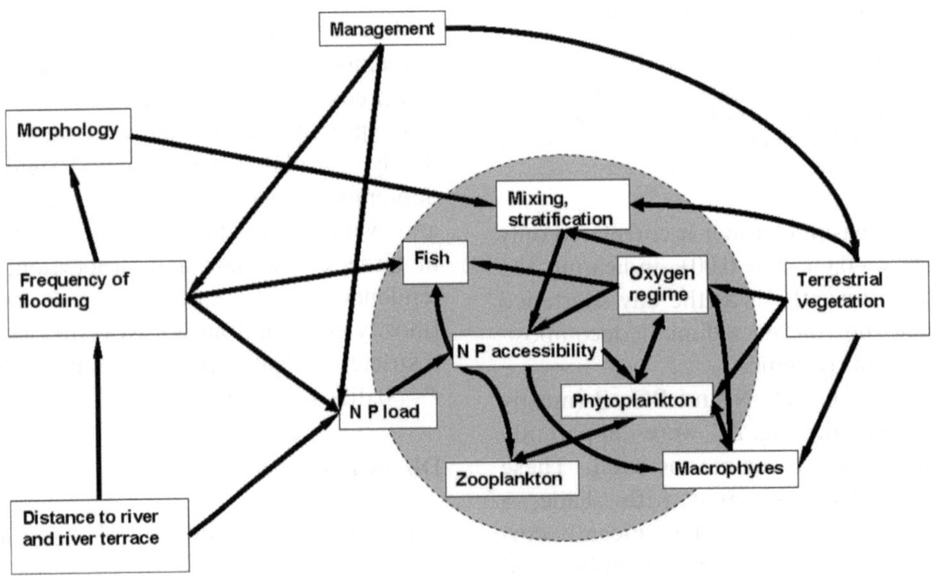

Fig. 6 Scheme of terrestrial–aquatic interactions in floodplain ecosystem complex. Grey circle indicates water bodies, dashed circle symbolizes the open character of the aquatic system

Morphometry: area, shape and relative depth

The morphometry of any water body has a crucial impact on the mixing of the water column. Mixing was generally supposed to be poor throughout the entire set of studied localities. Even the relatively large oxbows have an elongated shape, curved character and elevated banks, preventing wind action. Larger and/or relatively shallower water bodies were better mixed -enhancing the growth of a wider spectrum of algal species.

Cryptophyceae, being well adapted to poor light conditions, also prospered well in the very stabilized water column, there occurring no water mixing to move the phytoplankton upwards to the light (Reynolds, 1984). As efficient swimmers among flagellates, they are able to locate themselves according to light conditions and nutrient gradients (Jones, 1988). Cryptophyceae dominate throughout most of the year in waters whose trophic status are very different, but which have a poorly mixed (and often shaded) water column (e.g., in Finish forest lakes: Smolander & Arvola, 1988; Lužnice river floodplain: Pithart, 1997; karstic lakes in Catalonia: Pedrós-Alió & Guerrero, 1993; and lakes covered with macrophytes: Blinn & Green, 1986, Izaguirre et al., 2004).

Planktic cyanobacteria frequently predominate in shallow eutrophic lakes all over the world (Scheffer, 1998), just as in the Třeboň basin (Pechar et al., 2002). They are well adapted to low light conditions and, similarly as with *Cryptophycea*, they are able to regulate their position in the water column. However, unlike *Cryptophyceae* and other phytoplankton groups, *Cyanophyceae* were much less abundant in the alluvial pools, a fact that can be explained by the low pH of the alluvial waters. A similar disappearance of blue-greens as a result of a decrease in pH has been described by Shapiro (1984). Moreover, the decomposition of different plant materials, as for example barley straw (Barret et al., 1999), can restrict the development of *Cyanophyceae* and may favour the mixotrophic flagellates (Jones, 1988).

The impact of fish should also differ substantially between larger and smaller water bodies. Oxygen in large oxbows is never depleted throughout the whole water column; a better oxygen saturation in winter and more frequent flooding (long oxbows have a closer connectivity with the river) can enhance the fish population. The distinctive zooplankton structure in large pools may serve as indirect evidence of the impact of fish: large pelagic cladocerans are virtually absent in the pelagial and small cladocerans contribute very little. In the littoral zone, however, small pelagial cladocerans significantly correlate positively with area, indicating the growing fish impact that results in the change of spatial distribution of zooplankton, as they seek refuge among macrophytes (Timms & Moss, 1984).

Pelagic rotifers most contribute to the zooplankton in large pools, correlating significantly and consistently with chlorophyll *a* (for the whole set of localities). This phenomenon seems to be the effect of both a release from competition, when large filtrators are not present (Lampert & Rothhaupt, 1991, Urabe, 1992) and simultaneously the elevated concentration of suitable food (Merriman & Kirk, 2000).

Similarly, the absence of any significant correlation between those nutrients occurring in high concentrations and phytoplankton biomass may also indicate the frequent light limitation (Scheffer, 1998) and therefore the importance of such mechanisms as mixing.

Terrestrial vegetation: shading and leaf litter

Terrestrial vegetation has a crucial impact on the chemistry and life in these water bodies. Permanent shading and nutrient-rich leaf litter affect the oxygen regime and water chemistry as a whole, macrophyte development, fish survival, and consequently the whole structure of the plankton community (Fig. 6). Height and density of terrestrial vegetation on the banks and a pool's total area, determine the relative proportion of shaded area.

Shade restricts algal photosynthesis, resulting in lower oxygen production. At the same time, oxygen is consumed by decomposition: the amount of organic rich leaf litter in the sediment would be expected to enhance this process. These two mechanisms have not been experimentally separated, but analysis of the seasonal aspects can help towards understanding which one is more important in which situation for determining the final oxygen concentration.

Oxygen tends to be depleted under the ice cover in shaded small pools; therefore fish cannot survive in these localities and large filtrators are not consumed. On the other hand, larger pools with a longer, more twisted shoreline may support more fish, which would relieve the pressure of food competition for smaller cladocerans as fish graze on larger species. Poor mixing, shade, and humic substances leaching from the leaf litter, are all different mechanisms that again result in poor light conditions in the water column, to which most algal species are unable to adapt.

Acknowledgements This study was financially supported by Research Projects AV0Z60870520 and MSM 6007665806 of the Ministry of Education, Youth and Sport and projects of the Czech Grant agency GAČR 206/01/1113 and Ministry of Environment VaV - SL/1/6/04. We are obliged to Martina Kylbergerova for help with phytoplankton counting and Steve Ridgill for language corrections.

References

Barrett, P. R. F., J. W. Littlejohn & J. Curnow, 1999. Long-term algal control in a reservoir using barley straw. Hydrobiologia 415: 309–313.

Blinn, D. W. & J. Green, 1986. A pump sampler study of microdistribution in Walker Lake, Arizona, U.S.A.: a senescent crater lake. Freshwater Biology 16: 175–185.

Hrbácek, J., L. Pechar & V. Dufková, 1994. Anaerobic conditions in winter shape of the seasonal succession of Copepoda and Cladocera in pools in forested inundations. Verhandlungen Internationale Verein Limnologie 25: 1335–1336.

Hutchinson, G. E., 1957. A treatise on limnology v.1. Geography, Physics and Chemistry, Wiley, 1015 pp.

Izaguirrre, I., I. O'Farrell, F. Urein, R. Sinitro, M. dos Santos Afonso & G. Tell, 2004. Algal assemblages across a wetland, from a shallow lake to relictual oxbow lakes (Lower Paraná River, South America). Hydrobiologia 511: 25–36.

Javornický, P., 1978. Metódy kvantitatívného štúdia fytoplanktónu (Methods of phytoplankton quantitative study). In Hindák, F. (ed.), Sladkovodné riasy (Freshwater algae). ŠPN, Bratislava, Czechoslovakia, 35–57.

Jones, J. B. & P. J. Mulholland (eds), 2000. Streams and ground waters. Academic Press, San Diego, 425 pp.

Jones, R. I., 1988. Vertical distribution and diel migration of flagellated phytoplankton in a small humic lake. Hydrobiologia 161: 75–87.

Kylbergerová, M., D. Pithart & M. Rulík, 2002. Algological survey of small floodplain backwaters. Algological Studies 104: 169–187.

Lampert, W. & K. O. Rothhaupt, 1991. Alternating Dynamics of Rotifers and Daphnia Magna in a Shallow Lake. Archiv für Hydrobiologie 120: 447–456.

McGarigal, K. & B. J. Marks, 1995. FRAGSTATS: spatial pattern analysis program for quantifying landscape structure. Program documentation. Gen Tech Rep US Dep Agric For Serv Pac Northwest Res Stn 351, Portland.

Merriman, J. L. & K. L. Kirk, 2000. Temporal patterns of resource limitation in natural populations of rotifers. Ecology 8: 141–149.

Pedros-Alió, C. & R. Guerrero, 1993. Microbial ecology of Lake Cisó. In Jones, J. G. (ed.), Advances in microbial ecology, Volume 13. Plenum Press, New York, 155–207.

Pechar, L., 1987. Use of an acetone-methanol mixture for the extraction and spectrophotometric determination of chlorophyll a in phytoplankton. Archiv für Hydrobiologie, Suppl. (Algological Studies 46) 78: 99–117.

Pechar, L., J. Hrbáček, D. Pithart & J. Dvořák, 1996. Ecology of pools in the floodplain. In Prach, K., J. Jeník & A. R. G. Large (eds), Floodplain ecology and management. The Lužnice River in the Třeboň Biosphere Reserve, Central Europe., Backhuys Publishers Amsterdam, 209–226.

Pechar, L., I. Přikryl & R. Faina, (2002). Hydrobiological evaluation of Třeboň fishponds since the end of 19th century. In Květ, J., J. Jeník & L. Soukupová (eds), Freshwater wetlands and their sustainable future: A case study of the Třeboň basin biosphere reserve, Czech Republic. Man and the Biosphere Series 28, UNESCO & The Parthenon Paris, 31–62.

Pithart, D., 1997. Diurnal vertical migration study during a winter bloom of Cryptophyceae in a floodplain pool. Internationale Revue der Gesamten Hydrobiologie 82: 33–46.

Pithart, D., 1999. Phytoplankton and water chemistry of several alluvial pools and oxbows after the flood event–a process of diversification. Algological Studies 95: 93–113.

Pithart, D. & L. Pechar, 1995. The stratification of pools in the alluvium of the river Lužnice. Internationale Revue der Gesamten Hydrobiologie 80: 61–75.

Poulíčková, A., 1997. Rozsivková flóra povodi Lužnice. (Diatom flora of Lužnice River catchment). Preslia 68: 257–264.

Prach, K., 1996. The river Lužnice and its floodplain. In Prach, K., J. Jeník & A. Large (eds), Floodplain Ecology and Management. The Lužnice River in the Třeboň Biosphere Reserve, Central Europe, SPB Academic Publishing, Amsterdam, 11–18.

Reynolds, C. S., 1984. The ecology of freshwater phytoplankton. Cambridge University Press, Cambridge, New York, Melbourne, 384 pp.

Ruzicka, J. & E. H. Hansen, 1981. Flow injection analysis. John Wiley, New York, 207 pp.

Shapiro, J., 1984. Blue-green dominance in lakes: The role and management significance of pH and CO_2. Inter-

nationale Revue der Gesamten Hydrobiologie 69: 765–780.
Scheffer, M. (ed.), 1998. Ecology of shallow lakes. Chapman & Hall London, Weinheim, 357 pp.
Straškraba, M., 1964. Preliminary results of a new method for the quantitative sorting of freshwater net plankton into main groups. Limnology and Oceanography 9: 268–270.
Smolander, U. & L. Arvola, 1988. Seasonal variation in the diel vertical distribution of the migratory alga Cryptomonas marssonii (Cryptophyceae) in a small, highly humic lake. Hydrobiologia 161: 89-98.
Timms, R. M. & B. Moss, 1984. Prevention of growth of potentially dense phytoplankton populations by zooplankton grazing in the presence of zooplanktivorous fish in a shallow wetland ecosystem. Limnology and Oceanography 29: 472–486.
Urabe, J., 1992. Midsummer succession of rotifer plankton in a shallow eutrophic pond. Journal of Plankton Research 14: 851–866.
Ward, J. V., 1989. The four dimensional nature of lotic ecosystems. Journal of North American Benthological Society 8: 2–8.

Hydrobiologia (2007) 584:277–290
DOI 10.1007/s10750-007-0593-x

SHALLOW LAKES

Potential for the development of submerged macrophytes in eutrophicated shallow peaty lakes after restoration measures

R. J. W. van de Haterd · G. N. J. Ter Heerdt

Abstract Biomanipulation of eutropicated peaty lakes has rarely been successful; clear water with dense macrophyte stands fails to develop in most cases. It was unclear whether (1) high turbidity due to resuspension by benthivorous fish or wind is the major cause of low macrophyte density or whether (2) the establishment of submerged macrophyte stands is prevented by a lack of propagules, low cohesive strength of the lake sediment, high concentrations of phytotoxics, grazing by waterfowl and/or shading by periphyton growth. These hypotheses were tested in an experiment in a shallow peat lake in the Netherlands (Terra Nova). Removal of fish from a 0.5 ha experimental site resulted in clear water and the development of a dense (90% coverage) and species-rich (10 species) submerged vegetation. At a fish-stocked site and a control site the water remained turbid and dense macrophyte stands did not develop. The establishment of submerged macrophytes appeared not to be limited by a lack of propagules. Introduced plants grew poorly in turbid water, but very well in clear water. Exclosures showed that bird grazing reduced the plant biomass. In clear water grazing seemed to enhance the vegetation diversity. Periphyton development did not prevent plant growth in clear water. After the experiment, the fish stock was greatly reduced in the whole lake (85 ha), to test if (3) in a large lake, submerged macrophyte stands will not develop after biomanipulation. In the first season after fish reduction, transparency increased and species-rich submerged macrophyte stands developed, covering 60% of the shallow parts of the lake. Most of the species known to have occurred in the past re-established. The results indicate that high turbidity caused by benthivorous fish in combination with bird grazing were the major causes of the absence of submerged macrophyte stands in this lake. Abiotic conditions after the clearing of the lake were suitable for the growth of macrophytes. We infer that the restoration potential of submerged macrophyte stands in eutrophicated peaty lakes can be high, and results can be obtained quickly.

Keywords Shallow peaty lake · Restoration · Biomanipulation · Macrophytes · Herbivory · Periphyton

Guest Editors: R. D. Gulati, E. Lammens, N. De Pauw & E. Van Donk
Shallow lakes in a changing world

R. J. W. van de Haterd (✉)
Bureau Waardenburg, P.O. Box 365, 4100 AJ Culemborg, The Netherlands
e-mail: r.vandehaterd@buwa.nl

G. N. J. Ter Heerdt
Waternet, P.O. Box 8169, 1005 AD Amsterdam, The Netherlands
e-mail: gerard.ter.heerdt@waternet.nl

Introduction

Many of the freshwater lakes in the Netherlands originate from peat digging and are very shallow (<2 m). Once the water was oligotrophic and clear, but during the 20th century external nitrogen (N) and phosphorus (P) loading caused eutrophication. At first, this resulted in increased macrophyte growth and increased biodiversity. Since the 1950s, cyanobacterial blooms have resulted in high turbidity, poor underwater light climate and loss of submerged vegetation in most lakes. Restoration measures have been employed to reduce the external loading, but the clear water and macrophyte-dominated state generally did not return. For the history of the eutrophication and restoration of shallow lakes, see Van Liere & Gulati (1992), Scheffer (1998) and Gulati & van Donk (2002).

One of the reasons why nutrient reduction does not result in the return of a clear water state is the presence of large numbers of planktivorous and benthivorous fish. Planktivorous fish prevent zooplankton from becoming abundant and grazing down phytoplankton blooms. Benthivorous fish stir up the sediment, resuspending large amounts of detritus. A drastic reduction of fish stocks, a form of biomanipulation or biological control, often results in a shift from a turbid state towards one with clear water (Jeppesen et al., 1997; Scheffer, 1998; Gulati & van Donk, 2002). However, unlike results obtained in lakes with mineral sediment, biomanipulation in peaty lakes in the Netherlands has rarely been successful (Van Liere & Gulati, 1992; Meijer et al., 1999; Gulati & Van Donk, 2002).

The failure of restoration measures might be caused by poor colonisation of the lake sediment by submerged macrophytes (Hansson et al., 1998; Körner, 2002). Macrophytes are able to absorb large quantities of nutrients, thereby suppressing algal blooms, and they stabilise the sediment surface, reducing resuspension of detritus. Macrophytes also function as a refuge for zooplankton and are the habitat of predatory fish (Scheffer, 1998; Hansson et al., 1998; Gulati & Van Donk, 2002). A sustainable clear water situation requires the development of a dense submerged macrophyte cover (Jeppesen et al., 1997; Scheffer, 1998; Gulati & Van Donk, 2002). The process of colonisation by submerged macrophytes can be hampered by factors like a lack of propagules, shading by abundant periphyton growth, insufficient light conditions in deeper water and grazing by waterfowl (Jeppesen et al., 1997; Hansson et al., 1998; Scheffer, 1998; Van Donk & Otte, 1996). In peaty lakes, the development of submerged macrophytes may also be limited by high turbidity, due to resuspension of sediment by wind and wave action (Meijer et al., 1999), low cohesive strength of the lake sediment in combination with wave action (Schutten & Davy, 2000; Schutten et al., 2005) or high concentrations of phytotoxics such as sulphide (Smolders & Roelofs, 1993; Lamers et al., 1998). Hence, biomanipulation in peaty lakes may be less successful than in lakes with a mineral sediment. Meijer et al. (1999) showed that biomanipulation results for peaty lakes were indeed less favourable than those for other lakes, but did not clarify whether this was caused by insufficient fish removal or by the factors and conditions mentioned above.

Waternet is the water quality manager of many shallow peaty lakes in the region south of Amsterdam and restoration projects are going on or planned for several of these turbid lakes to restore the clear-water and macrophyte-dominated state. Before starting drastic and expensive measures, we wanted to make sure that macrophytes would be able to colonise peaty lakes rapidly. As part of a large-scale biomanipulation experiment, we started a pilot study of the colonisation by submerged macrophytes in two small (0.5 ha) and shallow (0.7 m) experimental sites. In this part of the experiment we tested the following hypotheses:

(1) High turbidity is the major cause of low macrophyte density.
(2) The establishment of submerged macrophyte stands is prevented by a lack of propagules, low cohesive strength of the lake sediment, high concentrations of phytotoxics, grazing by waterfowl and/or shading by periphyton growth.

Even if the colonisation by macrophytes is not inhibited in the enclosed sites, it would still not be certain that biomanipulation would work in larger peaty lakes, where waves and poorer light conditions due to greater depth (1.75 m) might prevent macrophyte establishment. Therefore we also performed a whole-lake biomanipulation experiment in an 85 ha lake, testing the following hypothesis:

(3) In a larger lake, submerged macrophyte stands will not develop after biomanipulation.

Materials and methods

Study area

This study was performed at Lake Terra Nova, also called Lake West Loenderveen (LWL). Lake Terra Nova is an 85 ha shallow peaty lake in the centre of the Netherlands, originating from peat excavation and subsequent erosion, and forming part of the Loosdrecht lake system (Hofstra & Van Liere, 1992). In 1941, more than 50% of the lake bottom was covered by submerged macrophytes, dominated by Characeae. Water transparency was 2–2.5 m Secchi depth (Leentvaar & Mörzer Bruijns, 1962; Best et al., 1984). Between 1941 and 1980, several observations indicated that species-rich macrophyte stands were present in Lake Terra Nova. In 1980–1982, dense stands of Characeae, *Elodea* sp., *Najas marina* and *Potamogeton* spp. were found, totalling 26 submerged macrophyte species (Table 1, Best et al., 1984). Secchi depth in summer was 1.2–1.9 m. Between 1977 and 1983, the average total phosphorus content in summer gradually increased from 0.04 to 0.08 mg/l, and remained around that level since.

Blooms of cyanobacteria appeared in 1987 (Dekker et al., 1992), while transparency decreased to 0.4 m Secchi depth from that date onwards. In 1994, a quick scan revealed very sparse growth of *Ceratophyllum demersum*, *Potamogeton lucens*, *Potamogeton obtusifolius*, *Potamogeton pectinatus*, *Myriophyllum spicatum* and some specimens of *Stratiotes aloides*. In 2003, we found only a very few specimens of *C. demersum*, *Elodea nuttallii*, *P. obtusifolius* and *Chara globularis*. *Nymphaea alba* and *Nuphar lutea* fields were present in shallow parts of Lake Terra Nova.

Experimental setup

Two experimental sites of 0.5 ha each were constructed by building dams between parallel banks in the western part of Lake Terra Nova (Fig. 1). Behind these dams, waves were greatly reduced, and water exchange and fish migration were no longer possible. Water depth at these sites ranged between 30 and 90 cm. The bottom was covered with a 0.9 m organic mud layer. At both sites, we greatly reduced the fish stock, using fykes, seine nets, gill nets and electro fishery. Petersen's mark and recapture method was used to estimate the fish population before and after fishing (Klinge et al., 2003). Fishing was repeated until the stock was below 25 kg fresh weight (FW) of benthivorous fish ha^{-1} and below 15 kg FW of planktivorous fish ha^{-1}. One site—referred to below as "fish-stocked"—was restocked with 180 kg FW cyprinids ha^{-1} (the average density in the lake) in April 2003. In December 2003, the fish population at the fish-stocked site had increased to 320–400 kg FW of cyprinids ha^{-1}. At the other site, which was not restocked—referred to as "fish-less", the fish stock had increased to 82.4 kg FW of cyprinids ha^{-1} in December 2003. Next to the experimental sites, a third, morphologically identical but unenclosed site served as control. The fish stock in the control site was estimated at 180 kg FW of cyprinids ha^{-1}. In the winter of 2003–2004, the fish stock had increased to 244 kg FW of cyprinids ha^{-1}.

We expected that the water of the "fish-less" site would become clear, and that the restocked and control sites would remain turbid. If our first hypothesis should be true, submerged macrophytes should develop much better at the clear "fish-less" site. If significant submerged macrophytes should develop, this would mean that lack of propagules, low cohesive strength of the lake sediment and high concentrations of phytotoxics do not prevent macrophyte development.

To test the effect of bird grazing, eight 4 m^2 exclosures were constructed at each of the three sites, with nets to keep birds out (Fig. 1). Next to each exclosure an unprotected 4 m^2 plot was

Table 1 Submerged macrophyte species found in Lake Terra Nova between 1949 and 1983 (Best et al., 1984; unpublished data), at the experimental sites in 2003 and in the lake in 2004, with indications of abundance

Species	Lake 1943–1983	2003 control	2003 fish-stocked	2003 fish-less	Lake 2004
Ceratophyllum demersum	xx	xx	xx	xx	Very common
Elodea nuttallii	xx	xx	xx	xx	Common
Potamogeton obtusifolius	xx	x	xx	xx	Common
Nitella mucronata				xx	Common
Najas marina	xx			x	Common
Utricularia vulgaris	xx				Common
Potamogeton crispus	x			xx	Fairly common
Potamogeton lucens	xx				Fairly common
Chara globularis	xx		x	xx	Fairly common
Stratiotus aloides	xx			x	Less common
Elodea canadensis	xx				Less common
Potamogeton pectinatus	xx	x			Fairly rare
Potamogeton pusillus	xx				Fairly rare
Potamogeton mucronatus	+			xx	Fairly rare
Myriophyllum spicatum	xx				Fairly rare
Potamogeton compressus	+				Rare
Nitella flexilis				xx	Rare
Nitellopsis obtusa	xx				1 individual
Potamogeton perfoliatus	x				1 individual
Chara vulgaris	xx				
Fontinalis antipyretica	xx				
Myriophyllum verticillatum	xx				
Potamogeton trichoides	xx				
Ranunculus circinatus	xx				
Utricularia minor	x				
Ranunculus aquatilis	+				

'+' means present with unknown abundance; 'x' means found once or twice; 'xx' means found more often

marked, where birds had free access. Vegetation developments in the exclosures and unprotected plots were assessed and compared; any significant difference was attributed to bird grazing. In view of the possibility that plants would not establish due to lack of propagules, we introduced plants at each site, viz. in four of the exclosures and in four of the adjacent unprotected plots (Fig. 1). The species introduced had all been found growing in Lake Terra Nova during the past 10 years and had different growth forms: *P. lucens*, *C. demersum* and *E. nuttallii*. Ten young plants of each species were individually planted in half-open pots, mounted in a 50 × 50 cm metal frame that was sunk into the sediment, one frame per species per plot.

Light conditions

Water transparency and vertical light attenuation were both measured each week from April to September and once every 2 weeks in winter. Transparency was measured with a Secchi disk. The vertical light attenuation coefficient (K_d) was calculated as: $K_d = (\ln(I_0/I_Z))/Z$, where I_0 and I_Z are the intensities of the Photosynthetically Active Radiation (PAR) just under the water surface and at a depth $Z = 0.5$ m (Scheffer, 1998). I_0 and I_Z were measured with two LI-192SA underwater quantum sensors, mounted in a frame, 0.5 m apart. Measurements were taken every second for 30 s, using a LI-1400 data logger. K_d was calculated as the average of 3–10 of such 30-s series, 1 min apart.

Light attenuation by periphyton

Thin 1.5 cm^2 glass discs were used to measure shading by periphyton (Van Dijk, 1993). Ten glass discs were placed at a depth of 10 cm below the water level, and ten other plates at a depth of 50 cm. The incubation time was 14 days, as this is

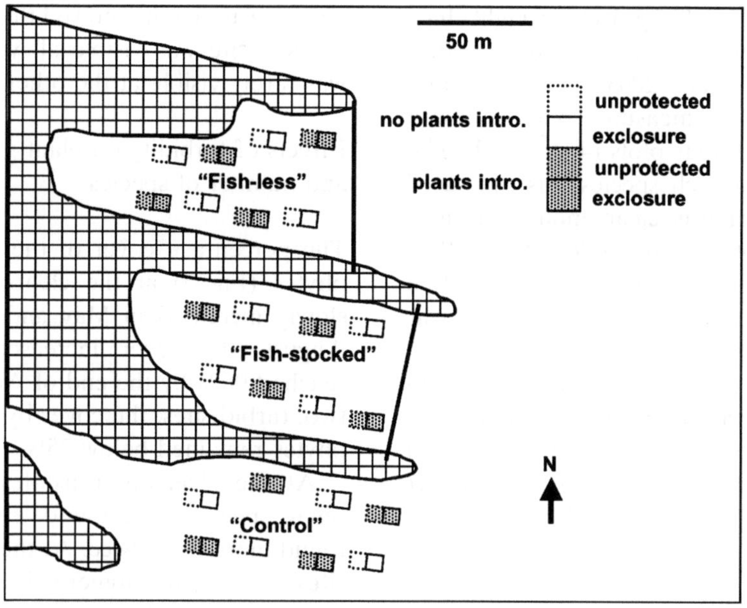

Fig. 1 Overview of the experimental setup

approximately the lifespan of leaves of submerged plants (Hootsmans, 1994). After 14 days, when the discs had become covered with a periphyton layer, light attenuation (% PAR) was compared with that of a clean glass disc, using two LI-COR underwater sensors.

Vegetation sampling

The percentage cover of the submerged macrophytes and of the floating-leaved species at each site was estimated on a decimal scale (adapted from Londo, 1975). Measurements were made weekly from April to September 2003 and fortnightly in the winter of 2003–2004. Estimations were made visually, using a hydroscope when needed, and with a rake when transparency was poor. On 23 June and 11 August 2003, the vegetation was surveyed in detail, by estimating the plant cover at the three experimental sites in 35 evenly distributed plots of approximately 0.25 m^2.

To determine plant biomass, all 4 m^2 plots were harvested in August 2003. If the volume of the plants collected from a plot was more than approximately 10 l, a representative subsample (1/2, 1/4 or 1/8) was taken. Species were separated, washed to remove detritus and stored in paper bags, and the dry weight of the plants (70°C, 48 h) was determined.

Statistical analysis

Differences in species biomass were tested with multiple regression analysis using Genstat 7.1. The data were log-transformed and model checking was applied. Post-hoc contra *t*-tests were carried out according to *t*-values. Least significant differences at a 5% level were used to test differences between factor classes of the exclosure/unprotected plot treatments. Blocks were used for the fish-less, fish-stocked and control sites. Differences in total biomass were log-transformed and tested with a *t*-test. Differences in light attenuation by periphyton layers at different depths were also tested with a *t*-test. The average number of species per treatment was tested with a Mann–Whitney *U*-test, because they were not normally distributed.

Whole-lake biomanipulation experiment

In the winter of 2003/2004, the fish stock in the whole Lake Terra Nova was reduced from 244 kg fresh weight ha^{-1} to 48 kg fresh weight ha^{-1} (estimated with Petersen's mark and recapture technique; Klinge et al., 2003). Turbidity and vegetation cover were measured once a month at six locations, with four plots (4 m^2) each. The percentage cover of each species was estimated visually if the water was clear enough, using a decimal scale (adapted from Londo, 1975). Afterwards, all plants in the plots were collected with a rake. In August, total dry biomass in the plots was determined as described above.

The vegetation of the lake was mapped in August by identifying homogeneous areas, surveying the species (visually and with a rake) in each area and estimating their percentage cover on a decimal scale (adapted from Londo, 1975).

Results

Light conditions at the experimental sites

After biomanipulation in the winter of 2002–2003, the water in the fish-less site became clear in April 2003, while the water in the fish-stocked site and in the control site remained turbid (Fig. 2). In summer, light conditions decreased at all three sites, but light conditions at the fish-less site remained better than at the other sites. Secchi-disc measurements showed that the bottom of the fish-less site was visible all year round, while it was rarely visible at the other sites.

Effects of turbidity on plant growth and number of species

The percentage cover of the submerged macrophytes (Fig. 3) at the clear-water fish-less site slowly increased in May and June. In July and August, the cover increased rapidly until it reached 90% by the end of August. At the other two, turbid sites, the cover by submerged macrophytes remained below 25%.

At the clear-water fish-less site, ten macrophyte species were found, eight of which were found more than twice. At the persistently turbid sites, only four submerged macrophyte species were found (Table 1).

There was no significant difference in total plant biomass between the turbid control and the fish-stocked sites, nor between the plots with or without introduced plants (*t*-tests, see Table 2). The total biomass at the clear-water fish-less site was significantly higher than at the two turbid sites, both with and without introduced plants (*t*-tests, see Table 2). At species level, the same effect was seen for the plots without introduced species (Fig. 4): the dominant *E. nuttallii* and the

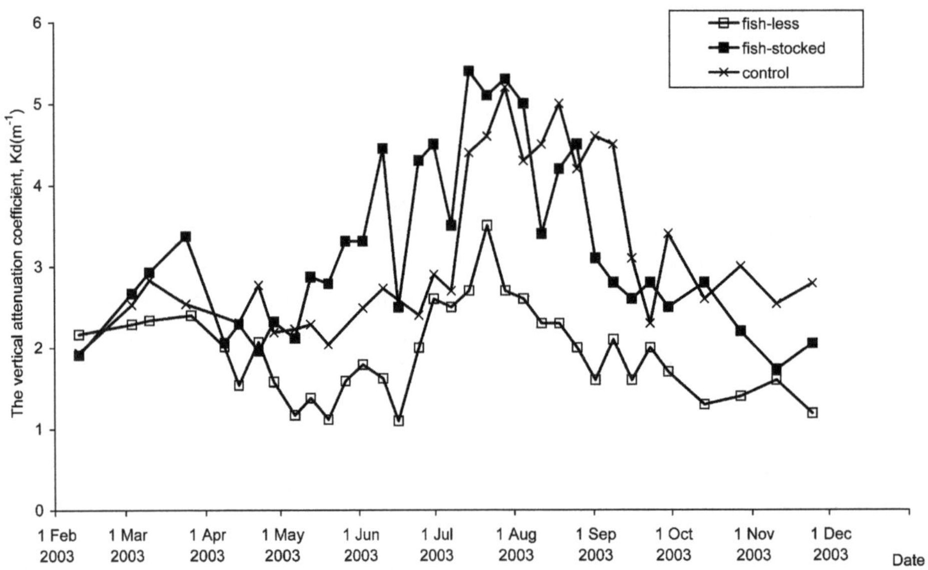

Fig. 2 Light conditions under water at the experimental sites in 2003

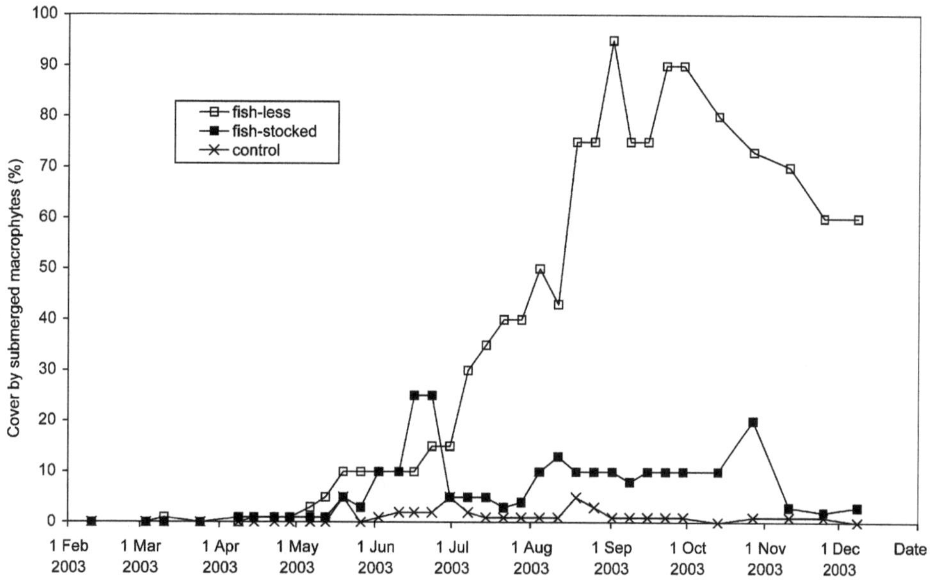

Fig. 3 Development of cover by submerged macrophytes during 2003 at the three experimental sites

less abundant Characeae both had significantly higher biomass at the fish-less site, whereas the control site and the fish-stocked site did not differ significantly (regression post-hoc *t*-tests, see Table 3). For the plots with introduced species, the pattern seemed similar (Fig. 5), but a significant effect was only found for *E. nuttallii*, and not for the dominant *P. lucens* (regression post-hoc *t*-tests, Table 3).

Light attenuation by periphyton

The light attenuation by periphyton, measured as the periphyton layer on glass discs after 14 days in the water, was not significantly different between the two depths of 10 and 50 cm (*t*-test, $t = 0.629$, $n = 830$, $P = 0.53$). Light attenuation by this periphyton layer generally remained below 20% (Fig. 6).

Table 2 Differences in total biomass at the sites, as tested with a *t*-test

With introduced species	
Control–fish-stocked	$t = -.918$, $n = 16$, $P = .374$
Control–fish-less	$t = -3.546$, $n = 15$, $P = .006$
Fish-stocked–fish-less	$t = -2.826$, $n = 15$, $P = .019$
Without introduced species	
Control–fish-stocked	$t = -.676$, $n = 15$, $P = .511$
Control–fish-less	$t = -8.478$, $n = 16$, $P < .001$
Fish-stocked–fish-less	$t = -6.527$, $n = 15$, $P < .001$

Effects of bird herbivory

At the fish-less site, *E. nuttallii* grew up to the water surface inside the exclosures. Outside the exclosures, it rarely did, as it was "clipped" a few decimetres below the water surface, a pattern which is consistent with grazing by herbivorous waterbirds (Common coot *Fulica atra* was actually observed eating *E. nuttallii*). At the unprotected plots without introduced plants in clear water, all common species (*E. nuttallii*, *C. demersum*, *P. obtusifolius* and *C. globularis*) established, and *E. nuttallii* became dominant. At the sites with turbid water, however, hardly any biomass developed in the unprotected plots (Fig. 4).

The biomass of the dominant species (*P. lucens* with and *E. nuttallii* without introduction) was significantly higher inside than outside the exclosures (regression post-hoc *t*-tests, Table 3). This was true for all three sites, both with and without introduction, with only one exception, viz. the fish-less site without introduction.

In the plots without introduced plants at the fish-less site, the average number of species outside the exclosures was 4.3, whilst the number

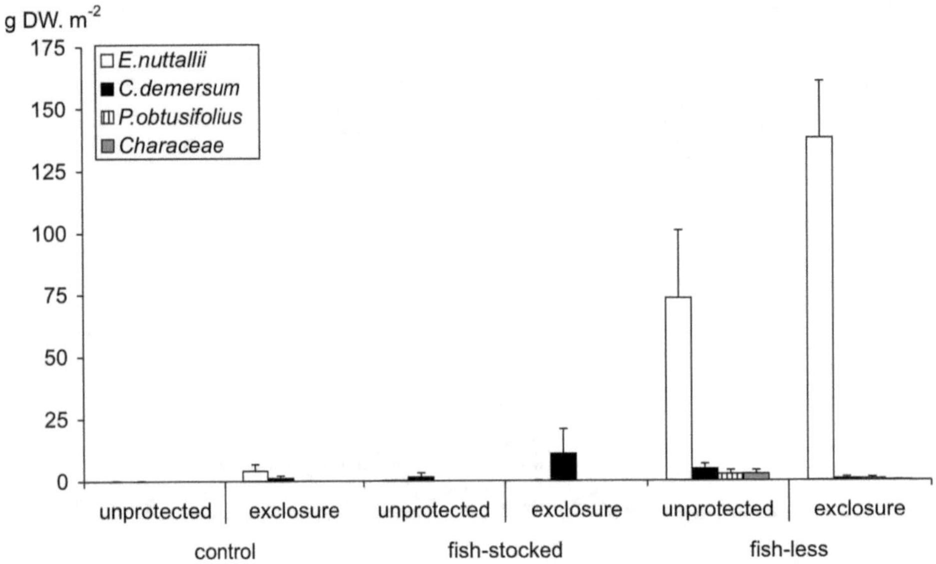

Fig. 4 Average biomass of the plots without introduced plants. For each treatment condition, the biomass in the plots open to bird grazing and in the exclosures is presented separately. *Potamogeton crispus* and *Najas marina* (not shown) occurred only in the unprotected plots of the fish-less site, with a very low biomass (too low to be visible). Error bars represent standard error of the average biomass

of species inside the exclosures was 2.3. However, this difference was significant only at the 10% level (Mann–Whitney $U = 2$, $n = 8$, $P = 0.074$).

Whole lake biomanipulation experiment

Biomanipulation of the whole Lake Terra Nova in the winter of 2003–2004 resulted in clear water (Fig. 7; for details see Ter Heerdt & Hootsmans, 2007). Submerged macrophytes quickly developed in the shallow parts (<1.4 m) of the lake. Plants were occasionally present in the deeper parts, but average cover remained below 1%. In August, submerged macrophytes covered about 30% of the whole lake, and 60% of the shallow parts (<1.4 m), although they remained absent from a few shallows. A total of 19 submerged macrophyte species, including four species of Characeae, were found in 2004 (Table 1). Whereas the vegetation at the fish-less experimental site was dominated by *E. nuttallii* (with almost 100% coverage), the vegetation in the lake as a whole was more diverse (Fig. 8). The shallowest parts (<1.1 m) were co-dominated by *C. demersum, Najas marina* and *P. obtusifolius*, in a patchy pattern. The cover of individual species hardly ever exceeded 70%. Characeae were frequently observed in the shallowest parts; high cover by Characeae only occurred at a few locations. At relatively sheltered, 1.1–1.4 m deep locations, only *C. demersum* covered more than 10%. At the more exposed places, *Potamogeton lucens* formed scattered stands at depths between 1.3 and 1.7 m.

Discussion

Effect of turbidity on colonisation by macrophytes

After the water had cleared, many submerged macrophyte species established, both at the experimental site and in the lake as a whole. Macrophytes reached a high cover, which did not happen at the turbid fish-stocked experimental site. We therefore conclude that the high turbidity was the major factor limiting the development of submerged macrophytes in Lake Terra Nova. The negative effect of high turbidity on the biomass of the dominant species was significant at the plots without introduced species, but not at the plots with introduced species. This difference suggests that high turbidity has a negative effect on the establishment of plants, rather than on their growth rate.

Table 3 Results of the regression analysis

Fitted terms: constant + blocks + grazing		With introduced plants			Without introduced plants	
		ln(*E. nuttallii*)	ln(*C. demersum*)	ln(*P. lucens*)	ln(*E. nuttallii*)	ln(characaea)
Regression summary						
N		22	22	22	23	23
F		5.3	4.7	9.0	38.5	4.1
P		0.007	0.012	<.001	<.001	0.021
R^2		0.36	0.33	0.51	0.84	0.30
Parameter estimates						
Constant		0.745	0.768	3.707	–0.611	–2.619
t, P		$t = 1.04$, $P = 0.312$	$t = 1.72$, $P = 0.101$	$t = 6.58$, $P = <.001$	$t = –1.16$, $P = 0.259$	$t = –6.38$, $P = <0.001$
Blocks fish-stocked		0.833	1.192	0.284	-0.973	0.045
t, P		$t = 0.95$, $P = 0.355$	$t = 2.18$, $P = 0.041$	$t = 0.41$, $P = 0.685$	$t = –1.47$, $P = 0.158$	$t = 0.09$, $P = 0.931$
Blocks fish-less		2.769	1.864	1.124	5.456	1.39
t, P		$t = 3.15$, $P = 0.005$	$t = 3.41$, $P = 0.003$	$t = 1.63$, $P = 0.119$	$t = 8.54$, $P = <0.001$	$t = 2.78$, $P = 0.012$
Grazing		–1.689	–0.666	–2.774	–0.966	0.633
t, P		$t = –2.35$, $P = 0.029$	$t = –1.49$, $P = 0.151$	$t = –4.92$, $P = <.001$	$t = –1.81$, $P = 0.087$	$t = 1.51$, $P = 0.147$
Post-hoc t-tests						
Effect of fish in unprotected plots						
Control unprot.	Fish-stocked unprot	ns	$P < 0.05$	ns	ns	ns
Control unprot.	Fish-less unprot	$P < 0.05$	$P < 0.05$	ns	$P < 0.05$	$P < 0.05$
Fish-stocked unprot	Fish-less unprot	$P < 0.05$	ns	ns	$P < 0.05$	$P < 0.05$
Effect of fish in exclosures						
Control excl.	Fish-stocked excl	ns	$P < 0.05$	ns	ns	ns
Control excl.	Fish-less excl	$P < 0.05$	$P < 0.05$	ns	$P < 0.05$	$P < 0.05$
Fish-stocked excl	Fish-less excl	$P < 0.05$	ns	ns	$P < 0.05$	$P < 0.05$
Effect of exclosures (grazing)						
Control unprot.	Control excl	$P < 0.05$	ns	$P < 0.05$	$P < 0.05$	ns
Fish-stocked unprot	Fish-stocked excl	$P < 0.05$	ns	$P < 0.05$	$P < 0.05$	ns
Fish-less unprot	Fish-less excl	$P < 0.05$	ns	$P < 0.05$	ns	ns

Only significant regressions are shown (regression of *P. obtusifolius* and *C. demersum* at the plots without introduction were not significant)

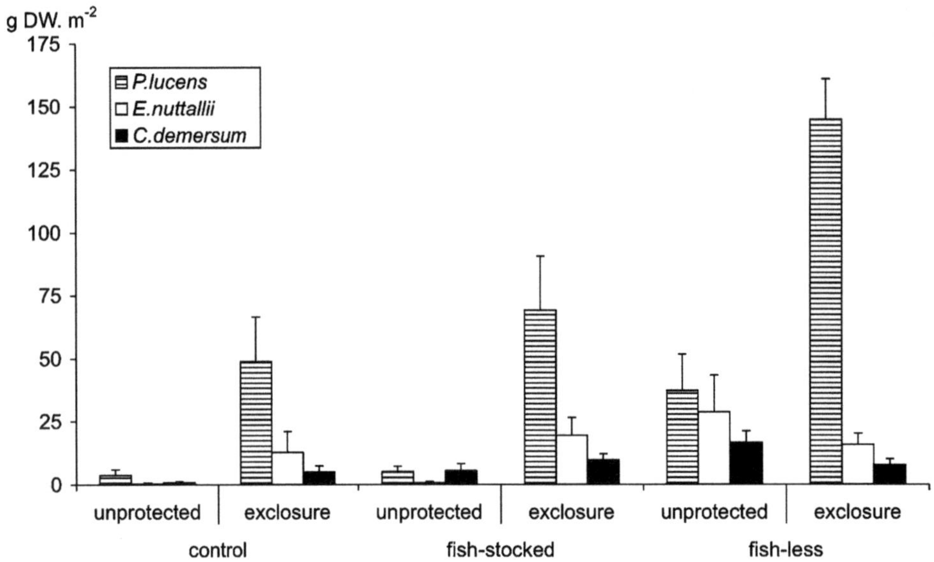

Fig. 5 Average biomass of the plots with introduced plants. For each treatment condition, the biomass in the plots open to bird grazing and the exclosures is presented separately. Error bars represent standard error of the average biomass

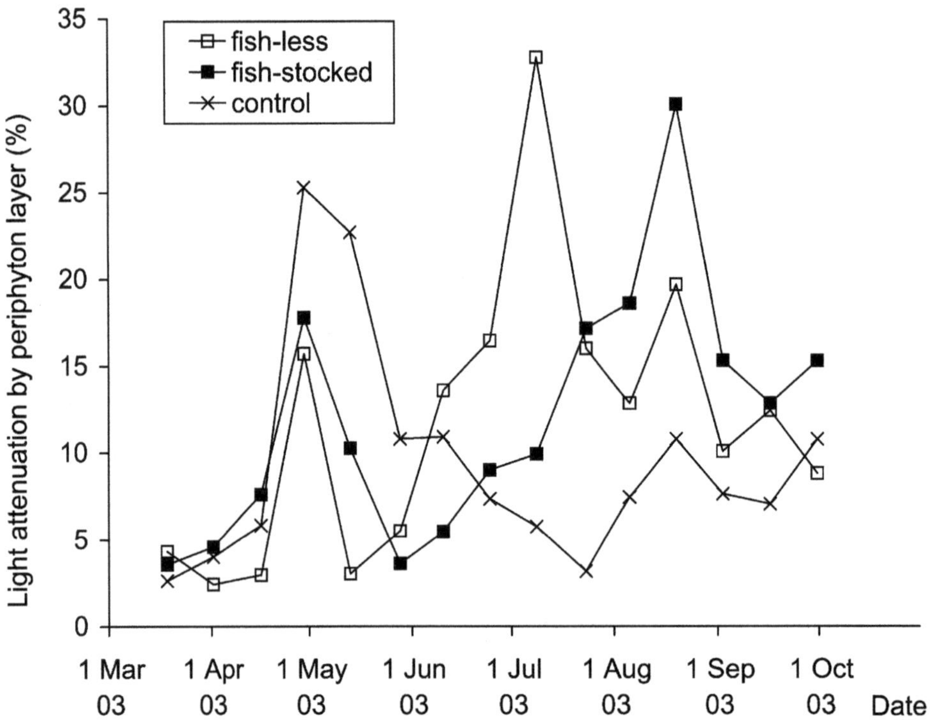

Fig. 6 Light attenuation by a periphyton layer growing on a glass disc submerged for 14 days at the experimental sites

Colonisation

The large number of species found and the rapid colonisation of the lake indicate there was no lack of propagules in Lake Terra Nova. Most species known to have occurred in the past recovered in 2004, including some rare species. Although some of the species found in 2004 had been encoun-

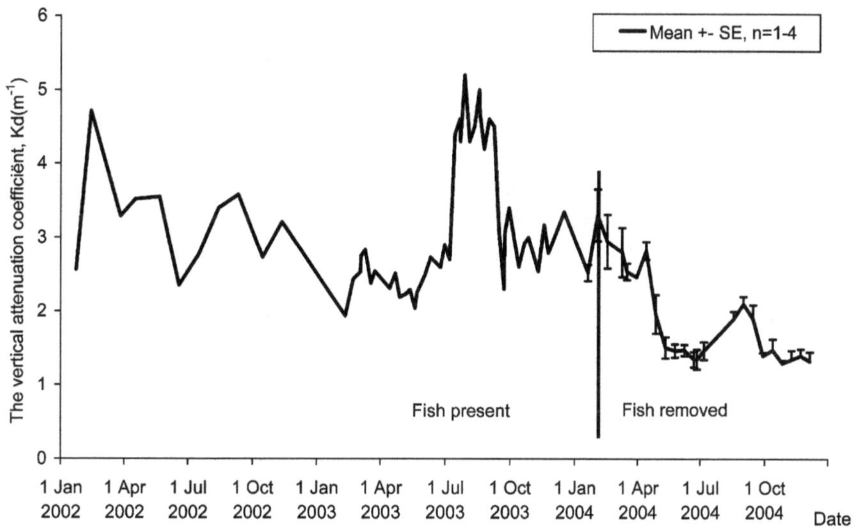

Fig. 7 Development of the vertical light attenuation coefficient in Lake Terra Nova (excluding the dammed off experimental sites)

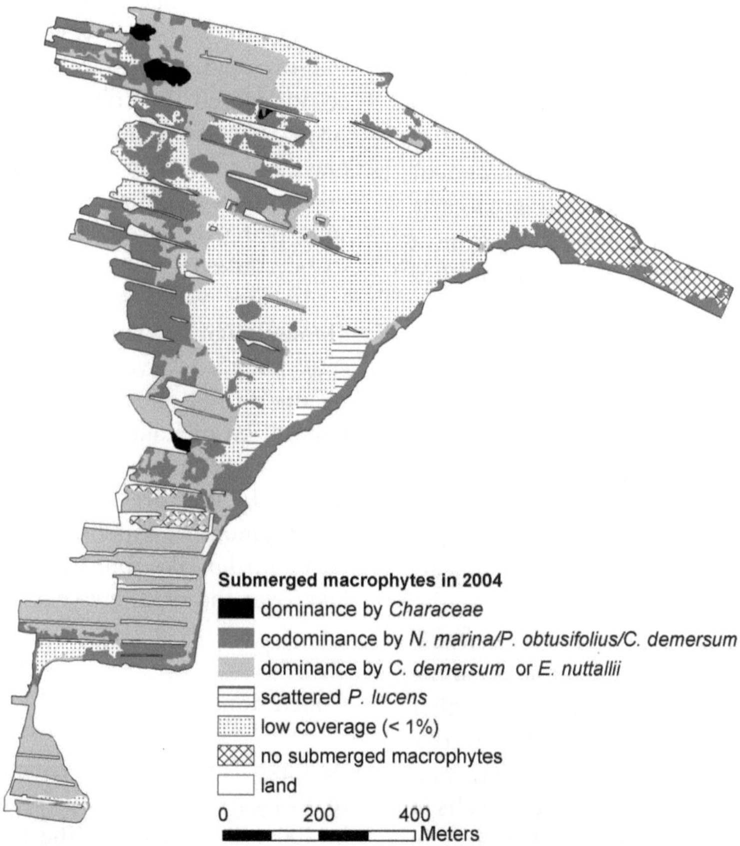

Fig. 8 Map of the submerged macrophytes in Lake Terra Nova in 2004

tered regularly in the lake since 1984, most had not been observed. Some of these species probably re-established from soil seed banks, such as Characeae, which are known to have persistent seed banks (Thompson et al., 1997). Transport of diaspores by water is also known to play an important role in the establishment of submerged macrophytes, especially *E. nuttallii* and *C. demersum* (Boedeltje et al., 2003, 2004). Birds are known to transport propagules of aquatic plants frequently, at least at a local scale, both by endozoochory (internal transport of seeds, e.g. those of *Potamogeton* species) and ectozoochory (external transport of vegetative parts, e.g. those of *Elodea* species) (Figuerola & Green, 2002). Large numbers of geese, ducks and swans visit Terra Nova and thus may import propagules from a wide area. Vegetation development in the lake started with only a few individuals, but expanded exponentially (Fig. 3). As the processes summarised above are quite common, it seems reasonable to assume that propagules will be present in many other lakes too. This implies that the restoration potential may also be high in similar lakes.

The low cohesive strength of the lake sediment and the presence of phytotoxics were apparently no major constraints in Lake Terra Nova, as most shallow parts became covered by submerged macrophytes.

Grazing by waterbirds

Exclosures yielded higher biomasses than unprotected plots, where grazing and uprooting by birds could occur. Because the effect was observed not only at the fish-stocked and control sites, but also at the fish-less site, we conclude that grazing by birds reduces the total biomass. In turbid water, grazing reduced biomass to almost zero, whereas in clear water, the reduction of the biomass of the dominant species seemed to result in a higher number of species. It is well-known from terrestrial grasslands (e.g. Olff & Ritchie, 1998), that grazing (at least extensive grazing) often results in higher plant diversity. We suggest that also in macrophyte stand, a reduction of the competition between species by a reduction of the dominant species results in higher vegetation diversity.

Light attenuation by periphyton

Although plant growth is reduced even by moderate shading, we do not consider 20% light attenuation by a periphyton layer (grown in 14 days) to be a major obstacle to plant growth in clear and shallow water. Van Dijk & Van Vierssen (1991) found that 26% shading did not have a significant effect on the growth of *P. pectinatus* in Lake Veluwe, whereas 45% and 73% shading did have significant effects on the growth of this species.

Restoration experiment in the whole lake

The vegetation that developed at the (shallow) experimental fish-less site differed from the vegetation in shallow parts of the lake as a whole. Whilst the experimental fish-less site was dominated by one species (*E. nuttallii*; covering nearly 100%), the vegetation in the other shallow parts was less dense and mostly co-dominated by two or three species. We suspect that the difference in wind dynamics played a role. Why some parts of the lake were not colonised by submerged macrophytes, despite being sheltered, shallow and clear, remains to be studied.

A clear water state in the first year after biomanipulation sometimes returns to turbid state after a few years (Meijer et al., 1999). Our results do not allow us to conclude whether the restoration of Lake Terra Nova is sustainable or whether supplementary measures will be required. We are optimistic because of the development of a dense submerged macrophyte cover in the lake, but will continue to study the lake over the next few years.

Conclusions

At Lake Terra Nova, a reduction of the fish stock to 48 kg ha^{-1} of cyprinids resulted in clear water in 2004. Turbidity was the major factor limiting the growth of submerged macrophytes in Terra Nova. In this lake, which has been turbid since 1988, the density of propagules enabled colonisation of shallow parts of the lake within 1 year. Grazing by birds reduced the total biomass, and hardly any biomass remained in turbid water. In clear water, the reduction of the biomass of the dominant species seemed to enhance vegetation diversity. Periphyton growth generally did not limit light conditions to such an extent that plant growth was prevented in the shallow parts.

This study has shown that the potential for restoration of species-rich submerged macrophyte stands in peaty lakes is high, even after two decades in a turbid state. Whether the restoration of Lake Terra Nova is sustainable or whether supplementary measures will be required, cannot be concluded from the results.

Acknowledgements We would like to thank our colleagues at the Water Laboratory for field assistance and lab analysis, Victoria Correa de la Torre for constructing the exclosures and introducing the plants, M. Hootsmans (Waternet), A. Bak and G. Bonhof (Bureau Waardenburg) for their contributions to the project and their useful suggestions. M. Poot and K. Krijgsveld (Bureau Waardenburg) performed the statistical analysis. Jan Klerkx from Beta translations improved the English. Last but not least, we would like to thank two anonymous referees for their constructive comments. Construction of the experimental sites was made possible by a financial contribution from the Utrecht provincial authorities. The biomanipulation work at Lake Terra Nova was funded by the Dutch Ministry of Agriculture, Nature and Food Quality.

References

Best, E. P. H., D. De Vries & A. Reins, 1984. The macrophytes in the Loosdrecht Lakes: a story of their decline in the course of eutrophication. Verhandlung Internationale Vereinigung Limnologie 22: 868–875.

Boedeltje, G., J. P. Bakker, R. M. Bekker, J. M. van Groenendael & M. Soesbergen, 2003. Plant dispersal in a lowland stream in relation to occurrence and three specific life-history traits of the species in the species pool. Journal of Ecology 91: 855–866.

Boedeltje, G., J. P. Bakker, A. ten Brinke, J. M. van Groenendael & M. Soesbergen, 2004. Dispersal phenology of hydrochorous plants in relation to discharge, seed release time and buoyancy of seeds: the flood pulse concept supported. Journal of Ecology 92: 786–796.

Dekker, A. G., T. J. Malthus, M. M. Wijnen & E. Seyhan, 1992. Remote sensing as a tool for assessing water quality in Loosdrecht Lakes. Hydrobiologia 233: 137–159.

Figuerola, J. & A. J. Green, 2002. Dispersal of aquatic organisms by waterbirds: a review of past research and priorities for future studies. Freshwater Biology 47: 483–494.

Gulati, R. D. & E. van Donk, 2002. Lakes in the Netherlands, their origin, eutrophication and restoration: state-of-the-art review. Hydrobiologia 478: 73–106.

Hansson, L.-A., H. Annadotter, E. Bergman, S. F. Hamrin, E. Jeppesen, T. Kairesalo, E. Luokkanen, P.-Å. Nilsson, M. Søndergaard & J. Strand, 1998. Biomanipulation as an application of food-chain theory: constraints, synthesis, and recommendations for temperate lakes. Ecosystems 1: 558–574.

Hofstra, J. J. & L. van Liere, 1992. The state of the environment of the Loosdrecht lakes. Hydrobiologia 233: 11–20.

Hootsmans, M. J. M., 1994. A growth analysis model for Potamogeton pectinatus L. Geobotany 21: 250–286.

Jeppesen, E., J. P. Jensen, M. Sondergaard, T. Lauridsen, L. J. Pedersen & L. Jensen, 1997. Top-down control in freshwater lakes: the role of nutrient state, submerged macrophytes and water depth. Hydrobiologia 342: 151–164.

Klinge, M., G. Hensens, A. Brennikmeijer & L. Nagelkerke, 2003. Handboek visstandbemonstering, voorbereiding, bemonstering, beoordeling (Manual for Fish stock monitoring, in Dutch). STOWA 2002-07, Utrecht.

Körner, S., 2002. Loss of submerged macrophytes in shallow lakes in North-Eastern Germany. International Review of Hydrobiology 87: 375–384.

Leentvaar, P. & M. F. Mörzer-Bruijns, 1962. The pollution of the Loosdrecht Lakes and its consequences. De Levende Natuur 65: 45–48 (in Dutch).

Lamers, L. P. M., H. B. M. Tomassen & J. G. M. Roelofs, 1998. Sulphate-induced eutrophication and phytotoxicity in freshwater wetlands. Environmental Science & Technology 32: 199–205.

Londo, G., 1975. De decimale schaal voor vegetatiekundige opnamen van permanente kwadraten (The decimal scale for vegetation survey of permanent plots; in Dutch). Gorteria 7: 101–105.

Meijer, M. L., I. de Boois, M. Scheffer, R. Portielje & H. Hosper, 1999. Biomanipulation in shallow lakes in the Netherlands. An evaluation of 18 case studies. Hydrobiologia 408(409): 13–30.

Olff, H. & M. E. Ritchie, 1998. Effects of herbivores on grassland plant diversity (review). Trends in Ecology and Evolution 13: 261–265.

Scheffer, M., 1998. Ecology of Shallow Lakes. Population and community biology series 22. Chapman and Hall, London.

Schutten, J. & A. J. Davy, 2000. Predicting the hydraulic forces on submerged macrophytes from current velocity, biomass and morphology. Oecologia 123: 445–452.

Schutten, J., J. Dainty, A. J. Davy, 2005. Root anchorage and its significance for submerged plants in shallow lakes. Journal of Ecology 93: 556–571.

Smolders, A. J. P. & J. G. M. Roelofs, 1993. Sulphate mediated iron limitation and eutrophication in aquatic ecosystems. Aquatic Botany 39: 267–293.

Ter Heerdt, G. N. J. & M. Hootsmans, 2007. Why biomanipulation can be effective in peaty lakes. Hydrobiologia 584: 305–316.

Thompson, K., J. P. Bakker & R. M. Bekker, 1997. The soil seed banks of north west Europe. Journal of Applied Ecology 34: 834–835.

Van Dijk, G. M., 1993. Dynamics and attenuation characteristics of periphyton upon artificial substratum under various light conditions and some additional observations on periphyton upon *Potamogeton pectinatus* L. Hydrobiologia 252: 143–161.

Van Dijk, G. M. & W. Van Vierssen, 1991. Survival of a *Potamogeton pectinatus* L. population under various light conditions in a shallow eutrophic lake Lake Veluwe in The Netherlands. Aquatic botany 39: 121–130.

Van Donk, E. & A. Otte, 1996. Effects of grazing by fish and waterfowl on the biomass and species composition of submerged macrophytes. Hydrobiologia 340: 285–290.

Van Liere, L. & R. D. Gulati, 1992. Restoration and recovery of shallow eutrophic lake ecosystems in The Netherlands. Kluwer Academic Publishers, Dordrecht.

Hydrobiologia (2007) 584:291–303
DOI 10.1007/s10750-007-0611-z

SHALLOW LAKES

The importance of drawdown and sediment removal for the restoration of the eutrophied shallow Lake Kraenepoel (Belgium)

Jeroen Van Wichelen · Steven Declerck · Koenraad Muylaert · Ivan Hoste · Vanessa Geenens · Jochen Vandekerkhove · Erik Michels · Niels De Pauw · Maurice Hoffmann · Luc De Meester · Wim Vyverman

Abstract Lake Kraenepoel (Belgium) is a shallow lake (22 ha), divided in two basins since 1957 by a shallow dike. The lake was used for fish farming until World War II and was drawn down about every 5 years to harvest fish. Despite its dense historical carp population, it had clear water and a rich Littorelletea vegetation. During the course of the 20th century, the lake became eutrophic and the Littorelletea vegetation degraded. The northern basin, which was still drawn down about every decade after 1957, retained its clear water and had a dense submerged macrophyte vegetation. The southern basin, which was never drawn down after 1957 and which received direct surface water inputs, had become a turbid shallow lake with phytoplankton blooms in summer. In 2000, efforts were taken to restore the lake: the entire lake was drawn down, the fish community was biomanipulated, nutrient-rich surface water inputs were diverted from the southern basin and sediments were removed (only in the northern basin). Fish biomanipulation and sediment removal were successful in the northern basin, as nutrient levels declined and the Littorelletea vegetation recovered. In the southern basin, sediment analyses indicated that drawdown resulted in sediments with a lower water and organic matter content and water column turbidity decreased after the drawdown. But pH in the southern basin declined to <4, probably because sulphides in the sediment were oxidized during drawdown and sediment desiccation. In contrast, desiccated sediments were removed from the northern basin and pH did not decline below 6 after restoration. In spite of the still high dissolved nutrient concentrations, phytoplankton biomass declined significantly in the southern basin, probably due to acidification. However, no Littorelletea species colonised the lake bottom in the southern basin. Thus, lake drawdown may be a useful

Guest editors: R. D. Gulati, E. Lammens, N. De Pauw & E. Van Donk
Shallow lakes in a changing world

J. Van Wichelen (✉) · K. Muylaert · V. Geenens · W. Vyverman
Laboratory of Protistology and Aquatic Ecology, Ghent University, Ghent, Belgium
e-mail: jeroen.vanwichelen@UGent.be

S. Declerck · J. Vandekerkhove · E. Michels · L. De Meester
Laboratory of Aquatic Ecology, Catholic University Leuven, Leuven, Belgium

I. Hoste
National Botanic Garden of Belgium, Meise, Belgium

N. De Pauw
Laboratory of Environmental Toxicology and Aquatic Ecology, Ghent University, Ghent, Belgium

M. Hoffmann
Laboratory of Terrestrial Ecology, Ghent University, Ghent, Belgium

management technique to promote clear water conditions in shallow lakes. However, acidification due to sulphide oxidation may be an undesirable outcome and should be considered in drawdown and sediment desiccation manipulations.

Keywords Shallow lake · Restoration · Drawdown · Macrophyte · Littorelletea · Sediment

Introduction

Due to intensification of agriculture and domestic wastewater discharges, many shallow lakes have undergone eutrophication during the 20th century. In non-disturbed, oligotrophic shallow lakes, the water is usually clear and primary production tends to be dominated by benthic algae and small isoetid macrophytes (Murphy, 2002; Vadeboncoeur et al., 2003). The first sign of eutrophication in shallow lakes is often the profuse development of submerged macrophytes in the water column. Upon further eutrophication, turbidity of the water increases and submerged macrophytes are replaced by phytoplankton. Ecological lake management often aims at restoring a clear water state with a submerged or benthic vegetation of macrophytes because it harbours a more diverse fauna and flora (Scheffer, 1998; Declerck et al., 2005). Although external nutrient inputs tend to be the main cause of turbid water in shallow lakes, reduction of these external inputs is rarely sufficient to restore water clarity. Additional measures are therefore needed, like biomanipulation of the fish community, removal of lake sediments or a reduction of the water level (e.g. Gulati & Van Donk, 2002; Jeppesen et al., 2005).

A complete drawdown has been referred to as a potentially powerful method to restore water clarity in shallow lakes. Complete drawdown of shallow lakes has been shown to result in a consolidation of the lake sediment (e.g. James et al., 2001), which may reduce sediment resuspension after refilling the lake. Complete drawdown of shallow lakes also facilitates control of fish biomass as well as removal of accumulated sediments (Scheffer, 1998). On the other hand, remineralization of organic matter is accelerated in dried-out lake sediments and this may result in increased nutrient levels in the water column when the lake is refilled (e.g. Qiu & McComb, 1996; James et al., 2001, 2004). Although complete drawdown of lakes may be a relatively cost-efficient management method in shallow lakes, relatively few studies have reported on the impact of drawdown events on the chemistry and ecology of shallow lakes.

Lake drawdown is used frequently in the management of fish culture ponds and in

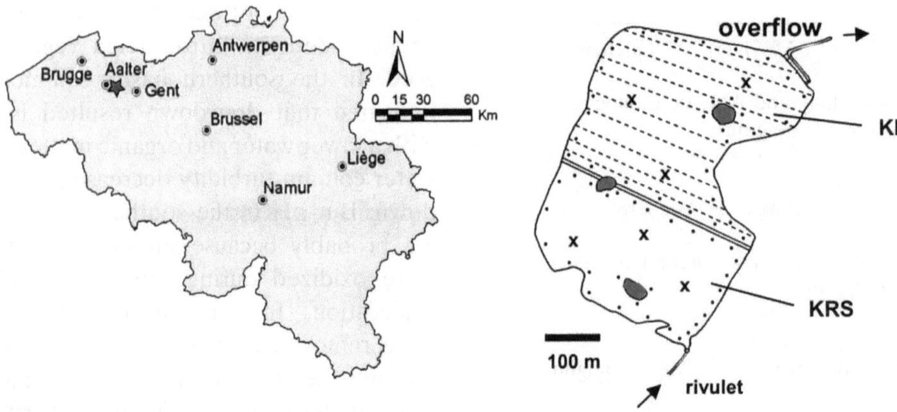

Fig. 1 Location of Lake Kraenepoel (indicated with a *star*) in Belgium (left) and detailed map (right) of the lake divided in a northern (KRN) and southern (KRS) basin by a dike in the middle. The lake is fed by a small rivulet (Bloembeeksken) in the southern basin and the water level is regulated by an outlet structure in the northern basin. *Dark areas* indicate the presence of small islands and the *crosses* indicate the locations in each lake part where water was sampled regularly during the project. *Dashed lines* represent transects used for vegetation mapping and the *black dots* indicate the places were electro-fishing was performed

artificial reservoirs. An example of a lake that was used for fish culture and that was regularly drawn down is Lake Kraenepoel (Belgium) (Fig. 1). Lake Kraenepoel is a relatively large (22 ha), shallow (mean depth 1.0 m, maximum depth 1.5 m) lake that most probably originated in a peat or sandstone extraction site in the 16th century. In the early 19th century, a shallow dike was constructed around the lake to raise the water level for carp breeding purposes. The lake was drawn down about every 5 years to harvest fish (up to 15–20 ton in the 20th century). In the 19th century, the lake was well-known among botanists for having one of the richest Littorelletea vegetations in Belgium. The shallow areas of the lake harboured a diverse community with many rare isoetid plants while the deeper zones were colonized by *Myriophyllum alterniflorum* (Vander Meersch, 1874). The lake also harboured an exceptionally diverse desmid community, which included 177 species (Van Oye, 1941). From the end of the 19th century onwards, however, the vegetation degraded and the Littorelletea vegetation lost many of its characteristic species (Table 1). After World War II, fish farming was given up and the lake was divided in two basins (north and south) through the construction of a shallow dike in 1957. The southern basin received nutrient-rich surface water inputs through a rivulet and was never completely drawn down after 1957. The northern basin only received nutrient-rich water indirectly when the shallow dike separating the two basins flooded in winter. Complete water level drawdown for several months (generally in summer) and subsequent fish removal were still conducted almost every decade in the northern basin after 1957. Despite these drawdowns, the Littorelletea vegetation continued to degrade and, in 1976, the diversity of desmids had dropped to only 21 species (Hoozee, 1978). Eutrophication resulted in a large accumulation of silt on the bottom of both parts of the lake (<1 cm thick in wind-exposed areas but up to 60 cm thick in the deeper middle part of both basins).

Table 1 Characteristic plant species observed in Lake Kraenpoel, before (1999) and after (2001, 2002) restoration. The last year of observation is indicated for disappeared species. Littorelletea species are marked in bold

	Last observation	1999	2001	2002
Apium inundatum	1984	–	–	–
Baldellia ranunculoides	1894	–	–	–
Carex lasiocarpa	1924	–	–	–
Carex rostrata	1924	–	–	–
Chara fragifera	1873	–	–	–
Deschampsia setacea	1874	–	–	–
Elatine hexandra	Not disappeared	+	+	+
Eleocharis acicularis	Not disappeared	+	+	+
Eleocharis multicaulis	1924	–	–	–
Eleocharis ovata	1994	–	–	–
Eleocharis quinqueflora	1874	–	–	–
Eriophorum latifolium	1874	–	–	–
Eriophorum polystachion	1894	–	–	–
Gnaphalium luteoalbum	Not disappeared	+	+	–
Hypericum elodes	1994	–	+	+
Juncus bulbosus	Not disappeared	+	+	+
Littorella uniflora	1985	–	–	–
Lobelia dortmanna	1924	–	–	–
Luronium natans	1985	–	–	–
Lythrum portula	Not disappeared	+	+	–
Myriophyllum alterniflorum	1985	–	–	–
Nitella translucens	1873	–	–	+
Potamogeton berchtoldii	Not disappeared	+	–	–
Potamogeton natans	1995	–	+	+
Potamogeton obtusifolius	Not disappeared	+	–	+
Potamogeton polygonifolius	1874	–	–	–
Scirpus fluitans	1853	–	–	–
Scirpus setaceus	Not disappeared	+	+	–

In 1998, a LIFE-project funded by the European Union was initiated to restore the lake and several restoration measures were taken.

(1) The rivulet flowing into the southern basin was diverted to stop further external nutrient inputs.
(2) The lake was completely drawn down between July and December 2000. This was done by draining the lake through a rivulet connected to the northern basin. In addition, pumps were used to lower the ground water level down to the mineral substrate below the lake.
(3) After several weeks of drying, the lake sediment layer was removed down to the mineral substrate using hydraulic excavators. Due to financial and time constraints, sediments (24,600 m^3) were removed only in the northern basin and not in the southern basin.
(4) A biomanipulation of the fish community was carried out. When the lake was drawn down, the entire fish population was removed, and, after refilling, the lake was stocked with 7000 juvenile pike.
(5) Eroded parts of the dike separating the two basins were raised to avoid exchange of water between the two basins when the water level was high.

Starting 1 year before and up to 2 years after the restoration measures, water chemistry and the major components of the pelagic food web were monitored in both basins of the lake. In this paper, we discuss the effect of the management activities on the ecology of the lake and compare them with the historical management regime in the lake, focusing on the impact of water level drawdown.

Materials and methods

Sampling

The lake was monitored during three consecutive years (June 1999–June 2002), starting 1 year before and ending 2 years after the start of the restoration measures. Water samples were taken fortnightly between April and October and monthly between November and February in both basins. Conductivity (WTW LF 191) and pH (WTW pH 197) were measured at a fixed location in each basin. In each basin, water was collected at three randomly chosen locations (Fig. 1) using Lund tubes and mixed to yield a spatially integrated water sample from each basin. These pooled samples were refrigerated, transported to the lab, and sub-sampled within 3 h for the analysis of suspended particulate matter (SPM), dissolved nutrients, phytoplankton and rotifers. A 50 ml sub-sample for enumeration of ultra-phytoplankton (<5 μm) was fixed with 0.2 μm-filtered formaldehyde to a final concentration of about 3%. A 250 ml sub-sample for enumeration and identification of phytoplankton was fixed according to the lugol–formaline–thiosulphate method (Sherr & Sherr, 1993). For enumeration of rotifers, a sub-sample was filtered over a 30 μm mesh. Samples for macro-zooplankton (cladocerans and copepods) were collected at the same three locations in each basin using a Schindler-Patalas sampler (mesh size 64 μm). Macro-zooplankton and rotifer samples were fixed with sugar-saturated formaldehyde to a final concentration of 4% (Haney & Hall, 1973). To investigate the biomass and composition of the fish community present in the lake before the restoration measures, all fish removed during drawdown of the lake were identified and weighed. For the most abundant fish species, a representative sub-sample (n = 30) was weighed. It should be noted that, especially in the southern basin, this census was probably an underestimate as many fish were buried in the mud during the drawdown. In addition, to investigate the composition of the fish community before and after the restoration, fish populations in the littoral zone were sampled by electro-shocking at 30 locations (Fig. 1) in each basin in the autumn of 1999 and 2001. Submerged macrophyte cover was estimated in July 2000, 2001 and 2002. For each macrophyte species, total cover and vegetation height were measured visually in about 120 1.0 m^2 quadrants that were evenly distributed along 10 parallel transects across the lake (Fig. 1). Vegetation mapping was only carried out in the northern basin as, apart from some isolated

Nymphaea alba specimens, macrophytes were absent in the southern basin during the study period. In each basin before restoration, the sediment layer (up to the mineral substrate) was sampled in January 1999 at 4–5 randomly chosen sites with a core sampler. After the drawdown, sediment cores from 24 randomly distributed sites in the southern basin were gathered in February 2001. Sediment cores were refrigerated during transport to the lab.

Laboratory analyses

For determination of dissolved nutrient concentrations (nitrate, ammonium and orthophosphate), water was filtered over a GF/F glass fibre filter and stored frozen until analysis using a Skalar auto-analyser (Koroleff, 1976; Grasshoff, 1976). SPM was measured gravimetrically after filtration of a known volume of water on pre-weighed GF/F filters. The water content of the sediment was measured gravimetrically after heating (105°C) a sub-sample (50 g) from each homogenized sediment core. The organic matter content was determined by sulfochromic oxidation (ISO 14235).

Ultra-phytoplankton (<5 μm) was enumerated using epifluorescence microscopy. Two size classes were discerned (<2 and 2–5 μm) and a minimum of 100 cells were counted per sample. For eukaryotic and prokaryotic cells of each size class, 100 individuals from different samples were measured and biovolume was converted to biomass using a conversion factor of 0.22 pg C μm^{-3} (Reynolds, 1984). Phytoplankton >5 μm was identified up to genus level and enumerated using inverted microscopy. For each genus, 50 individuals were measured and biovolume was converted to biomass using the equations given by Menden-Duer & Lessard (2000). Macro-zooplankton and rotifers were enumerated using a dissection microscope. Cladocerans were identified up to species level while copepods were identified to the order level. For each macro-zooplankton taxon in each sample, 30 individuals were measured to convert abundances to biomass using published length–weight regressions (Bottrell et al., 1976). Rotifers were identified up to species level (where possible) or to genus level. Abundances of rotifers were converted to biomass using published data on biomass content of the taxa encountered (Ruttner-Kolisko, 1974; Dumont et al., 1975; Pontin, 1978). Zooplankton grazing rates were calculated according to Jeppesen et al. (1994) and Blindow et al. (2000).

Results

Temporal variation of abiotic factors in the two basins of Lake Kraenepoel is illustrated in Fig. 2 while variation in biotic variables is shown in Fig. 3. Averages for both basins before and after the restoration are compared in Table 2. In 1999, the year before the restoration, no significant differences in the concentration of nitrate and orthophosphate were observed between the northern and southern basin. Nitrate reached maximal concentrations (up to 3000 μg NO_3–N l^{-1}) in winter, while phosphate concentrations were maximal during summer (up to 1000 μg PO_4–P l^{-1}). Nitrate and ammonium concentrations were often below the detection limit (1 μg l^{-1}) during summer. pH was generally neutral in both basins, except in summer when the lake water became more alkaline. Conductivity did not differ significantly between the two basins and oscillated around 400 μS cm^{-1}. SPM concentrations were higher in the southern than in the northern basin. Sediments in the northern basin had a significant lower water (*P*-level *t*-test: 0.003) and organic matter (*P*-level *t*-test: 0.032) content than those in the southern basin (Table 3). Phytoplankton biomass was higher in the southern than in the northern basin, although this difference was not significant. Phytoplankton blooms were dominated by cyanobacteria and chlorophytes in the northern basin and by cyanobacteria, euglenophytes and chlorophytes in the southern basin. Only three desmid species were found in the lake. The zooplankton biomass was comparable in both basins. The zooplankton community was dominated by small taxa in the southern basin (mainly rotifers, cyclopoid copepods and *Bosmina longirostris*). In the northern basin, rotifers were less important and zooplankton biomass

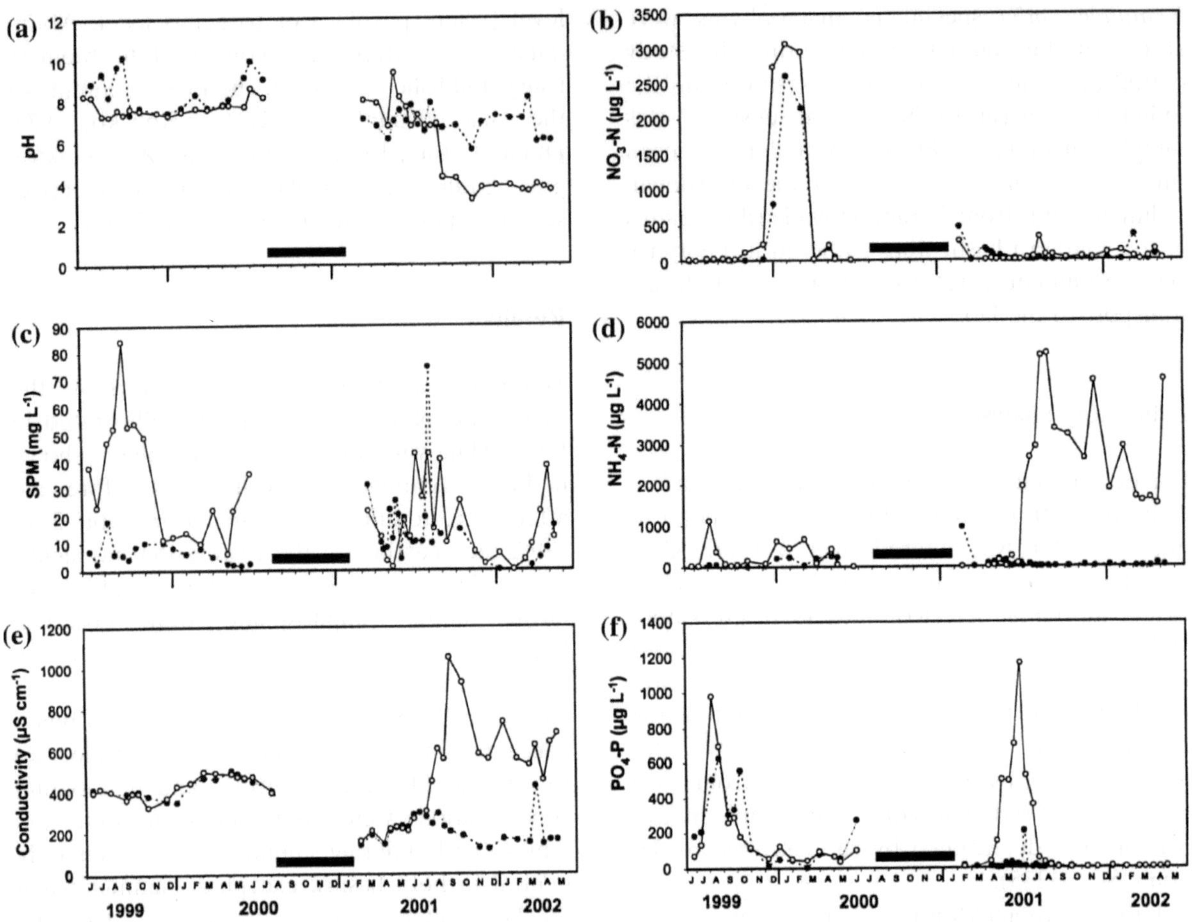

Fig. 2 Seasonal fluctuation for some important physical and chemical parameters for the northern (*black circles*) and southern (*white circles*) basin of Lake Kraenepoel during the monitoring. The *black bar* indicates the period of water level withdrawal. (SPM = Suspended Particulate Matter)

was dominated by cyclopoid copepods and *Bosmina longirostris*. In both sections, large cladocerans (>1 mm) like *Daphnia ambigua* and *D. galeata* were restricted to the spring period. Although estimated zooplankton grazing pressure on phytoplankton did not differ significantly between the two basins, a high grazing pressure from daphnids was observed in the northern basin in spring while this did not occur in the southern basin. Fish biomass assessed during the drawdown in August 2000 (Table 4) revealed an order of magnitude difference between the southern (385 kg ha^{-1}) and the northern basin (37 kg ha^{-1}). In the northern basin, fish biomass was dominated by large carp (*Cyprinus carpio*, 37%) and pike (*Esox lucius*, 35%) while small perch (*Perca fluviatilis*) dominated numerically (84%). In the southern basin, fish biomass and abundance was dominated by large bream (*Abramis brama*, >60%). Electrofishing in the littoral zone of the northern basin revealed a numerical dominance of small perch while tench (*Tinca tinca*) was also important in terms of biomass (Fig. 4). In the southern basin, bream and perch dominated both in terms of abundance and biomass. Macrophytes were virtually absent in the southern basin, except for a few dispersed individuals of *Nymphaea alba* and some sparse mats of filamentous algae (*Spyrogyra* sp.) in shallow areas. In contrast, in the northern basin, submerged macrophytes covered almost 40% of the lake bottom and 35% of the water volume (Fig. 5). Dominant species were the narrow-leaved *Potamogeton* species

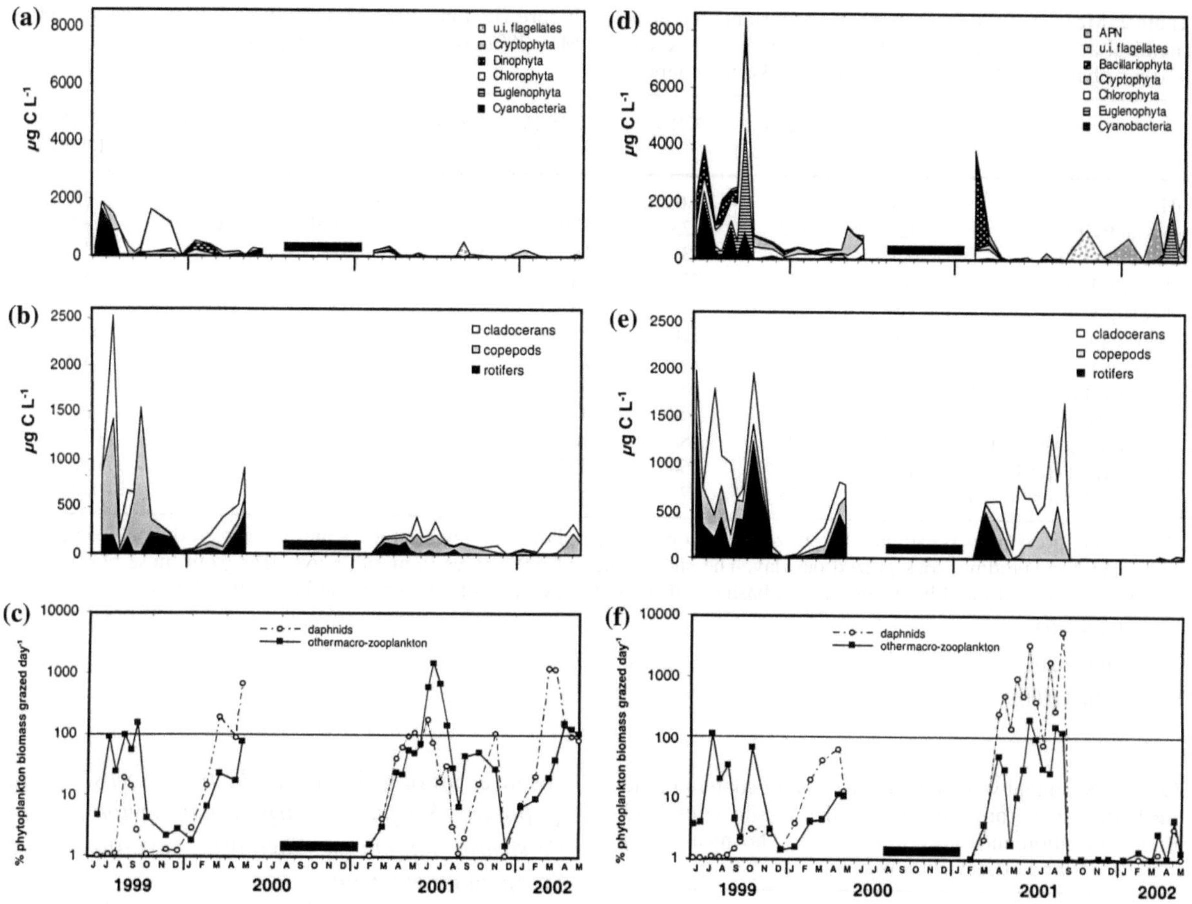

Fig. 3 Seasonal fluctuation in carbon biomass for the most important representatives of the pelagic foodweb in the northern (**a–c**) and southern (**d–f**) basin of Lake Kraenepoel during the study period. The *black bar* indicates the period of water level withdrawal. (**a**) and (**d**): fluctuation in C-biomass of phytoplankton, divided in taxonomical groups, (**b**) and (**e**): fluctuations in C-biomass of zooplankton, (**c**) and (**f**): fluctuations in the estimated grazing rates of macro-zooplankton (cladocerans and copepods)

P. pectinatus and *P. pusillus*. The only remaining representatives of the Littorelletae community, *Elatine hexandra*, *Juncus bulbosus* and *Eleocharis acicularis*, occurred mainly in the shallow parts of the lake and contributed only to 16% of total macrophyte cover.

After the restoration in the northern basin, ammonium and phosphate concentrations in the water column decreased significantly relative to pre-restoration values. Pre- versus post-restoration differences were only marginally significant for nitrate. Peak concentrations of all nutrients in 2001 and 2002 were much lower than peak concentrations in 1999–2000. pH declined significantly by about 1 unit. Conductivity also declined significantly by about 50%. Phytoplankton biomass decreased significantly and biomass maxima were about three times lower than before the restoration. Large cyanobacteria almost completely disappeared, while cryptophytes became more important. The diversity of desmids increased up to 30 species. Biomass of cyclopoid copepods and rotifers decreased substantially after the restoration, while cladoceran biomass remained constant. In spring, *Daphnia pulex* was the dominant cladoceran while *Diaphanosoma brachyurum* dominated during summer. The estimated grazing pressure of macro-zooplankton on phytoplankton increased significantly after the restoration measures. Electro-fishing in the littoral zone yielded only a few individuals of tench, pike, perch and eel (*Anguilla anguilla*). Canopy-forming

Table 2 Mean values for the most important physical, chemical and biological parameters for both basins (KRN: northern basin, KRS: southern basin) of Lake Kraenepoel before (B: June 1999–May 2000) and after (A: June 2001–May 2002) restoration measures were taken

	KRN B (n = 16)	KRN A (n = 23)	KRS B (n = 16)	KRS A (n = 18)	KRN B – KRN A	KRS B – KRS A	KRN B – KRS B	KRN A – KRS A
pH	8.3	6.91	7.6	5.1	>***	>***	>*	>***
Conductivity	422	212	423	570	>***	0.060	n.s.	<***
SPM (mg l^{-1})	6.5	12	33	18	n.s.	>**	<***	<**
NH_4^+ (µg l^{-1})	90	17	277	2653	>**	<***	<*	<***
NO_3^- (µg l^{-1})	392	24	632	63	0,063	n.s.	n.s.	<*
PO_4^{3-} (µg l^{-1})	207	13	210	158	>***	>**	n.s.	n.s.
SiO_2 (µg l^{-1})	919	307	1320	1044	>*	>*	<*	0.062
Phytoplankton (µg C l^{-1})	737	85	1874	521	>***	0.058	0.083	<***
Rotifers (µg C l^{-1})	135	8.8	368	4.4	>***	>***	n.s.	n.s.
Copepods (µg C l^{-1})	339	82	188	106	>**	>***	n.s.	<**
Cladocerans (µg C l^{-1})	187	63	254	237	n.s.	>*	n.s.	n.s.
Grazing pressure (%)	108	374	27	673	<*	n.s.	n.s.	0.088

The grazing pressure was estimated from zooplankton to phytoplankton biomass ratio's and is expressed as percentage of phytoplankton standing stock grazed per day. The significance of differences in monthly averages from these periods was tested for each basin and between the two basins with paired t-tests (results shown in the last four columns)

* P-value <0.05

** P-value <0.01

*** P-value<0.001

Table 3 Sediment water and organic matter content in the northern basin before and in the southern basin before and after restoration measures were taken. The mean of n samples ± standard deviation is shown

	North before	South before	South after
Water (%)	61.9 ± 4.8	81.6 ± 6.4	65.8 ± 16.6
Organic matter (%)	8.8 ± 3.3	18.7 ± 6.7	13.3 ± 5.5
N	4	5	24

Potamogeton species disappeared almost completely after the restoration and were replaced by meadow-forming plant species that covered a larger part of the lake surface but a similar or lower fraction of the lake volume. The charophyte *Nitella flexilis* appeared in spring but by July 2001, an *Oedogonium* sp. dominated the vegetation. However, Littorelletea species like *Elatine hexandra*, *Juncus bulbosus* and *Eleocharis acicularis* expanded compared to 2000. In 2002, the same Littorelletea species expanded further, another Littorelletea species (*Hypericum elodes*) reappeared in the lake while filamentous algae (now mainly *Zygnema* sp.) declined. Several other macrophyte species reappeared in the northern basin of the lake, some of them after a long time of absence, among them *Nitella translucens*, *Chara globularis*, *Scirpus lacustris*, *Potamogeton natans*, *Persicaria amphibia*, *Sagittaria sagittifolia* and *Sparganium emersum*.

In the southern basin after the drawdown, pH was initially circum-neutral but dropped below 5 in September 2001 and remained low until the end of the study period. pH even reached a minimum of 3.2 (November 2001). Phosphate concentrations displayed a peak in June 2001 that was comparable to the pre-restoration year (up to 1200 µg P l^{-1}) but dropped dramatically to levels below 10 µg l^{-1} in September 2001 to remain low until the end of the study period. During summer 2001, ammonium concentrations rose to 5200 µg N l^{-1} while nitrate concentrations decreased to levels below 100 µg N l^{-1}. Conductivity was initially lower than before the drawdown but increased in July 2001 up to a maximum above 1000 µS cm^{-1} in September. Significantly lower SPM concentrations were measured after the drawdown and very low values were observed in winter 2002. The water and organic matter content of the sediment had declined substantially, although the difference was not significant due to a large variability among the replicate

Table 4 Number of individuals and biomass categorized by size for each fish species captured during the water level withdrawal in each basin of Lake Kraenepoel (August 2000)

	Size (cm)	KRS				KRN			
		# Captured	Kg Captured	# ha^{-1}	Kg ha^{-1}	# Captured	Kg captured	# ha^{-1}	Kg ha^{-1}
Perca fluviatilis	<10	41300	284	4640	32	3895	27	302	2
(Perch)	>10	3003	125	337	10	376	24	29	2
Rutilus rutilus	<10	48	0	5	0	269	2	21	<1
(Roach)	>10	752	48	84	5	5	2	<1	<1
Abramis brama	<10	78813	355	8855	40	34	0	3	<1
(Bream)	10–30	1895	130	213	15	0	0	0	0
Cyprinus carpio	>30	831	2076	93	233	4	16	<1	1
(Carp)	>30	18	134	2	15	19	176	2	14
Anguilla anguilla (Eel)	>30	144	103	16	12	16	8	1	<1
Esox lucius	<30	103	17	12	2	214	36	17	3
(Pike)	>30	66	124	7	14	107	131	8	10
Leucaspius delineatus (Belica)	<10	0	0	0	0	61	<1	5	<1
Tinca tinca	<10	4	<1	<1	<1	96	3	7	<1
(Tench)	>10	28	33	3	4	49	46	4	4
Total		127,005	3429	14,270	385	5145	471	399	37

samples collected in 2001 (*P*-level *t*-test: 0.088). Phytoplankton biomass declined significantly and small peaks in phytoplankton biomass occurred only in spring and autumn after the drawdown. Large cyanobacteria and euglenophytes were replaced by ultra-phytoplankton (<5 μm) and unidentified phytoflagellates. In the first months after the drawdown, a high biomass of large cladocera (*Daphnia pulex* and *D. magna*) was observed, coinciding with a strong estimated grazing pressure of macro-zooplankton on phytoplankton. In September 2001, however, zooplankton virtually disappeared from the southern basin until the end of the study period. Only a few eels were captured in the southern basin in 2001. A sudden fish kill (about 20 eels) coincided with the strong pH decrease in September 2001. Filamentous green algae (*Ulothrix* spp. and *Stigeoclonium* spp.) expanded but no other macrophytes colonized the lake bottom in the southern basin.

Discussion

The presence of a rich Littorelletea vegetation and a diverse desmid community indicate that, in the early 19th century up to World War II, Lake Kraenepoel must have been a clear water lake with relatively low nutrient concentrations (Murphy, 2002). The low nutrient levels can be ascribed to extensive (low intensity) agricultural

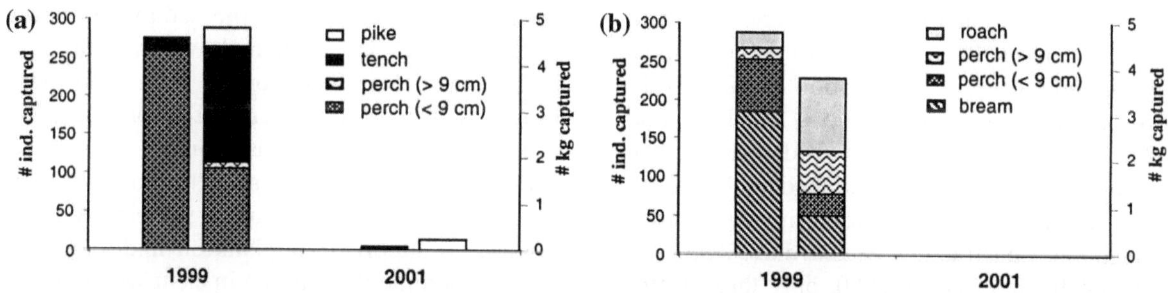

Fig. 4 Number of individuals (left columns) and biomass (right columns) for the main fish species captured with electro-shocking in 30 littoral sites in the northern (**a**) and southern (**b**) basin of Lake Kraenepoel before (1999) and after (2001) fish removal

Fig. 5 Relative contribution of the main taxa of submerged vegetation to total coverage in the northern basin of Lake Kraenepoel before (2000), one (2001) and two years (2002) after restoration measures took place. Note that the sum can be higher than 100 % due to lamination of the vegetation

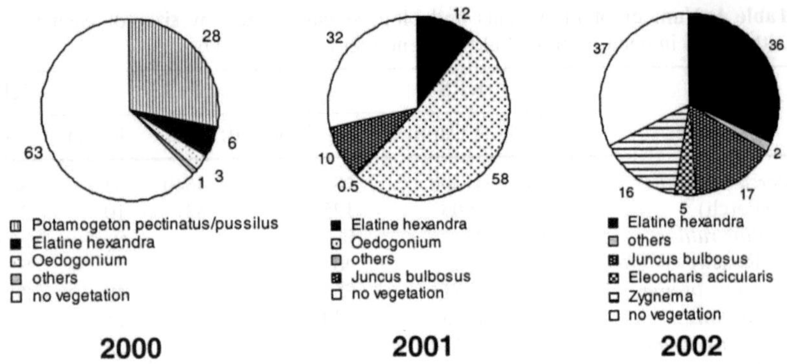

farming and limited wastewater discharges in those days. The low turbidity, however, was surprising given the high biomass of benthivorous fish like carp present in the lake. Benthivorous fish are an important cause of turbidity in shallow lakes because they disturb the sediment through their feeding activities (Scheffer et al., 2003). The lack of a negative effect of carp on water clarity in Lake Kraenepoel before World War II may have been due to the regular drawdown of the lake in those days. Lake drawdown may result in a consolidation and respiration of organic matter in the sediment and may therefore reduce sediment resuspension (James et al., 2001). Also, the low nutrient levels may have resulted in a lower accumulation of fine organic material on the lake bottom, which may also have reduced the influence of benthivorous fish on turbidity.

From the end of the 19th century, the Littorelletea vegetation in the lake started to lose several of its indicative plant species. Moreover, desmid diversity declined from 177 species to 22 species in 1976. It is likely that this degradation of the vegetation was due to increased nutrient inputs due to an intensification of agriculture and waste discharges. Intensive fish farming may have contributed to this eutrophication through addition of supplementary food to the lake (ca. 1000 kg corn year^{-1}). After World War II, fish farming activities were abandoned and in 1957 Lake Kraenepoel was divided in two basins through the construction of the dike. The lake degraded further and, in 2000, only three Littorelletea species and three desmid species were left. By then, both basins of Lake Kraenepoel had developed into typical eutrophic shallow lakes with high concentrations of dissolved inorganic N (up to >2500 μg l^{-1}) and dissolved inorganic P (up to >500 μg l^{-1}). The northern basin had retained a clear water state, with low SPM concentrations, low phytoplankton biomass and a dense submerged macrophyte vegetation. The dominant macrophytes in the northern basin, however, were narrow-leaved *Potamogeton* species typical for eutrophic conditions (e.g. Van den Berg et al., 2003). The southern basin had developed into a turbid lake with high SPM concentrations, a high phytoplankton biomass and an important contribution of cyanobacteria to the phytoplankton community. In agreement with the alternative stable states theory for shallow lakes (Scheffer et al., 1993), the southern basin had a higher biomass of fish than the northern basin and had a lower grazing pressure of zooplankton on phytoplankton.

The difference in water clarity between the two basins after the subdivision of the lake may have several causes. First, in contrast to the southern basin, the northern basin continued to be drawn down about every decade since World War II. This probably explains the lower moisture and organic matter content of the sediments in the northern basin. The regular drawdown may also explain the lower fish biomass in the northern basin, as fish stocks were probably severely reduced when the lake was drained. As a result of the lower fish biomass, planktivorous or benthivorous fish never attained sufficiently high biomass to suppress zooplankton and/or re-suspend the sediment, resulting in clear water conditions. The order of magnitude of difference in fish biomass between the two basins could probably be maintained because the dike separating both basins flooded only in winter, when most fish are dormant. Second, the southern basin received continuous inputs of nutrients through the rivulet feeding the lake while the northern basin received nutrients only in winter, when the dike separating both basins flooded. This continuous input of nutrients may have weakened the competitive advantage of submerged macrophytes over phytoplankton. The southern basin may have acted as a nutrient buffer for the northern basin and, in that way, may have prevented degradation of the northern basin towards a turbid system. The separation between the two basins in 1957 may therefore have allowed several Littorelletea species to survive in the lake.

The management activities carried out in 2000 had a strong effect in both the northern and southern basin. In the northern basin, the water quality clearly improved: SPM, ammonium and phosphate concentrations and biomass of phytoplankton decreased. Probably as result of increased water clarity and reduced nutrient levels (Roelofs, 1983, 1996; Brouwer et al., 2002) the abundant submerged, narrow-leaved *Potamogeton* species were replaced by a benthic Littorelletea vegetation. One Littorelletea species reappeared in the lake (*Hypericum elodes*) and desmid diversity increased from 3 to 30 species. The reduced nutrient levels and phytoplankton biomass can probably be ascribed to removal of both external (through the diversion of the rivulet) and internal (through sediment removal) sources of nutrients. The reduced turbidity can be ascribed to the removal of fine sediments in the northern basin.

In the southern basin, where the drawdown of the lake was not accompanied by sediment removal, the management activities had an entirely different outcome. After refilling the lake, concentrations of P and N increased to the same levels as before the drawdown, despite the removal of external nutrient inputs. These nutrients probably originated from the sediment. In dried-out lake sediments, mineralization of organic compounds is often accelerated (De Groot & Van Wijck, 1993) and nutrients produced during this process may diffuse into the water after refilling the lake (Qiu & McComb, 1996; James et al., 2001, 2004). In September 2001, a few months after refilling, the lake acidified and pH dropped to around 4. This may have been due to oxidation of sulphides in dried-out lake sediments (Holmer & Storkholm, 2001). Acidification has also been observed in rich fens or fen soil cores that were dried-out (van Dam & Buskens, 1993; Van Haesebroeck et al., 1997). The diversion of the rivulet feeding the southern basin resulted in a higher contribution of rainwater to the lake and probably resulted in a reduced buffering capacity, as is suggested by the decrease in conductivity of the water. This probably made the lake more prone to acidification. The acidification of the lake in turn resulted in immobilisation of phosphate at low pH (Ulrich and Pöthig, 2000; Kopacek et al., 2001) and prevented conversion of ammonium to nitrate by nitrifying bacteria (Rudd et al., 1988). The low pH probably also had a negative influence on fish (Rahel and Magnuson, 1983) and macrophytes (Maessen, et al., 1992) in the lake. The low phosphate levels resulting from the low pH also had a negative effect on phytoplankton biomass, which in turn resulted in a low zooplankton biomass. The management activities had a positive effect on SPM concentrations, which decreased by about 50%. This can be ascribed to the disappearance of benthivorous fish and the compaction of the sediment due to the drawdown of the lake. This observation is in line with a survey of floodplain lakes of the river Rhine, which revealed that clear water conditions often occurred in lakes that dried out regularly (Van Geest et al., 2005).

Conclusions

In the northern basin of Lake Kraenepoel, the combination of lake drawdown, diversion of surface water inputs, sediment removal and fish biomanipulation resulted in reduced nutrient levels and allowed the Littorelletea community to expand. In the southern basin, the same management strategy without sediment removal resulted in acidification of the basin. Although this acidification resulted in low nutrient levels, it appeared to have prevented Littorelletea species from re-colonising the basin. Historical observations in Lake Kraenepoel suggest that lake drawdown may favour a diverse macrophyte vegetation in shallow lakes. Observations before and after the restoration measures in the lake, point to a positive effect of drawdown on lake sediment composition. However, the dramatic pH drop observed in the southern basin indicates that drawdown without sediment removal should be avoided when large amounts of organically rich sediments are present. Possibly, a pH drop might be avoided by refilling the lake after drawdown with water with a sufficiently high buffering capacity.

Acknowledgements The research presented in this paper was carried out in the framework of a LIFE-project (LIFE98NAT/B/5172): 'Restoration and management of Lake Kraenepoel (Aalter)' financially supported by the EC and the Flemish (AMINAL/NATUUR) and local (Aalter) governments. Special thanks to the Pettiaux family, owners of the northern section of the lake, for their agreement to study and restore their property, Koen Himpe (Belconsulting NV) for supervising the restoration efforts, Nele Nuyten for the help during sampling and zooplankton analysis, Sara Denayer for the counting of several phytoplankton samples, Dirk Libbrecht (Geolab BVBA) and Wim Kerstens (Belconsulting NV) for gathering and analyzing data concerning sediment characteristics, Pieter Vanormelingen for some statistical questions and identifications of desmids, Dirk Van Gansbeke for the analysis of dissolved nutrients and the head (Viviane Vandenbil, AMINAL) and members of the supervising committee for their stimulating discussions. SD is a post-doctoral fellow with the Fund for Scientific Research, Flanders (FWO-Vlaanderen).

References

Blindow, I., A. Hargeby, B. M. A. Wagner & G. Andersson, 2000. How important is the crustacean zooplankton for maintenance of water clarity in shallow lakes with abundant submerged vegetation? Freshwater Biology 44: 185–197.

Bottrell, H., A. Duncan, Z. M. Gliwicz, E. Grygierek, A. Herzig, A. Hillbricht-Ilkowska, H. Kurasawa, P. Larson & T. Weglenska, 1976. A review of some problems in zooplankton production studies. Norwegian Journal of Zoology 24: 419–456.

Brouwer, E., R. Bobbink & J. G. M. Roelofs, 2002. Restoration of aquatic macrophyte vegetation in acidified and eutrofied softwater lakes: an overview. Aquatic Botany 73: 405–431.

Declerck, S., J. Vandekerkhove, L. Johansson, K. Muylaert, J. M. Conde-Porcuna, K. Van der Gucht, C. Pérez-Martínez, T. Lauridsen, K. Schwenk, G. Zwart, W. Rommens, J. López-Ramos, E. Jeppesen, W. Vyverman, L. Brendonck & L. De Meester, 2005. Multi-group biodiversity in shallow lakes along gradients of phosphorus and water plant cover. Ecology 86: 1905–1915.

De Groot, C. J. & C. Van Wijck, 1993. The impact of desiccation of a fresh-water marsh (Garcines Nord, Camargue, France) on sediment water vegetation interactions. 1. The sediment chemistry. Hydrobiologia 252: 83–94.

Dumont, H. T., I. van de Velde & S. Dumont, 1975. The dry weight estimate of biomass in a selection of Cladocera, Copepoda and Rotifera from the plankton, periphyton, and benthos of continental waters. Oecologia 19: 225–246.

Grasshoff, K., 1976. Determination of nitrate and nitrite. In Grasshoff, K. (ed.), Methods of Seawater Analysis. Verlag Chemie, Weinheim, New York, 317 pp.

Gulati, R. D. & E. Van Donk, 2002. Lakes in the Netherlands, their origin, eutrophication and restoration: state-of-the-art review. Hydrobiologia 478: 73–106.

Haney, J. & D. Hall, 1973. Sugar-coated Daphnia: A preservation technique for Cladocera. Limnology and Oceanography 18: 331–333.

Holmer, M. & P. Storkholm, 2001. Sulphate reduction and sulphur cycling in lake sediments: a review. Freshwater Biology 46: 431–451.

James, W. F., J. W. Barko, H. L. Eakin & D. R. Helsel, 2001. Changes in sediment characteristics following drawdown of Big Muskego Lake, Wisconsin. Archiv für Hydrobiologie 151: 459–474.

James, W. F., J. W. Barko & H. L. Eakin, 2004. Impacts of sediment dewatering and rehydration on sediment nitrogen concentration and macrophyte growth. Canadian Journal of Fisheries and Aquatic Sciences 61: 538–546.

Jeppesen, E., M. Søndergaard, E. Kanstrup, B. Petersen, R. B. Eriksen, M. Hammerhøjj, M. Mortensen, J. P. Jensen & A. Have, 1994. Does the impact of nutrients on the biological structure and function of freshwater and brackish lakes differ? Hydrobiologia 275/276: 15–30.

Jeppesen, E., M. Søndergaard, J. P. Jensen, K. E. Havens, O. Anneville, L. Carvalho, M. F. Coveney, R. Deneke, M. T. Dokulil, B. Foy, D. Gerdeaux, S. E. Hampton, S. Hilt, K. Kangur, J. Köhler, E. H. H. R. Lammens, T. L. Lauridsen, M. Manca, M. R. Miracle, B. Moss, P. Nõges, G. Persson, G. Phillips, R. Portielje, S. Romo, C. L. Schelske, D. Straile, I. Tatrai, E. Willén & M. Winder, 2005. Lake responses to reduced nutrient loading – an analysis of contemporary long-term data from 35 case studies. Freshwater Biology 50: 1747–1771.

Hoozee, J., 1978. Ekologisch-systematische studie van het "Kraenepoel"-fytoplankton (O.-Vl.). Unpublished masters thesis, Ghent University, Gent, 160 pp.

Kopacek, J., K. U. Ulrich, J. Hejzlar, J. Borovec & E. Stuchlik, 2001. Natural inactivation of phosphorus by aluminium in atmospherically acidified water bodies. Water Research 35: 3783–3790.

Koroleff, F., 1976. Determination of ammonia, phosphorus and silicon. In Grashoff, K. (ed.), Methods of Seawater Analysis. Verlag Chemie, Weinheim, New York, 317 pp.

Maessen, M., J. G. M. Roelofs, M. J. S. Bellemakers & G. M. Verheggen, 1992. The effects of aluminium, aluminium/calcium ratios and pH on aquatic plants from poorly buffered environments. Aquatic Botany 43: 115–127.

Menden-Deuer, S. & E. J. Lessard, 2000. Carbon to volume relationships for dinoflagellates, diatoms, and other protist plankton. Limnology and Oceanography 45: 569–579.

Murphy, K. J., 2002. Plant communities and plant diversity in softwater lakes of northern Europe. Aquatic Botany 73: 287–324.

Pontin, R. M., 1978. A Key to British Freshwater Planktonic Rotifera. Freshwater Biological Association. Scientific Publication 38, 178 pp.

Qiu, S. & A. J. McComb, 1996. Drying-induced stimulation of ammonium release and nitrification in reflooded lake sediment. Marine and freshwater research 47: 531–536.

Rahel, F. J. & J. J. Magnuson, 1983. Low pH and the absence of fish species in naturally acidic Wisconsin lakes – interferences for cultural acidification. Canadian Journal of Fisheries and Aquatic Sciences 40: 3–9.

Reynolds, C. S., 1984. Freshwater Phytoplankton. Cambridge University Press, Cambridge, 384 pp.

Roelofs, J. G. M., 1983. Impact of acidification and eutrophication on macrophyte communities in soft waters in the Netherlands. I. Field Observations. Aquatic Botany 17: 139–155.

Roelofs, J. G. M., 1996. Restoration of eutrophied shallow softwater lakes based upon carbon and phosphorus limitation. Netherlands Journal of Aquatic Ecology 30: 197–202.

Rudd, J. W. M., C. A. Kelly, D. W. Schindler & M. A. Turner, 1988. Disruption of the nitrogen cycle in acidified lakes. Science 240: 1515–1517.

Ruttner-Kolisko, A, 1974. Plankton rotifers biology and taxonomy. Binnengewasser supplement 36: 1–146.

Scheffer, M., S. H. Hosper, M.-L. Meijer, B. Moss & J. Jeppesen, 1993. Alternative equilibria in shallow lakes. Trends in Ecology and Evolution 8: 275–278.

Scheffer, M., 1998. Ecology of shallow lakes. Population and Community Biology Series 22. Chapman and Hall, London, 357 pp.

Scheffer, M, R. Portielje & L. Zambrano, 2003. Fish facilitate wave resuspension of sediment. Limnology & Oceanography 48: 1920–1926.

Sherr, E. B. & B. F. Sherr, 1993. Preservation and storage of samples for enumeration of heterotrophic protists. In Kemp, P. F., B. F. Sherr, E. B. Sherr & J. J. Cole (eds), Handbook of Methods in Aquatic Microbial Ecology. CRC Press, Boca Raton, 207–212.

Ulrich, K. U. & R. Pöthig, 2000. Precipitation of aluminium and phosphate affected by acidification. Acta Hydrochimica et Hydrobiologica 28: 313–322.

Vadeboncoeur, Y., E. Jeppesen, M. J. Van der Zanden, H.-H. Schierup, K. Christoffersen & D. M. Lodge, 2003. From Greenland to green lakes: cultural eutrophication and the loss of benthic energy pathways in lakes. Limnology & Oceanography 48: 1408–1418.

van Dam, H. & R. F. M. Buskens, 1993. Ecology and management of moorland pools – balancing acidification and eutrophication. Hydrobiologia 265: 225–263.

Van den Berg, M. S. W. Joosse & H. Coops, 2003. A statistical model predicting the occurrence and dynamics of submerged macrophytes in shallow lakes in the Netherlands. Hydrobiologia 506: 611–623.

Van der Meersch, E., 1874. Notice sur la florule du Kraene-Poel. Bulletin de la société Royale de botanique de Belgique 13: 224–241.

Van Geest, G. J., H. Coops, R. M. M. Roijackers, A. D. Buijse & M. Scheffer, 2005. Succession of aquatic vegetation driven by reduced water-level fluctuations in floodplain lakes. Journal of Applied Ecology 42: 251–260.

Van Haesebroeck, V., D. Boeye, B. Verhagen & R. F. Verheyen, 1997. Experimental investigation of drought-induced acidification in a rich fen soil. Biogeochemistry 37: 15–32.

Van Oye, P., 1941. Etude biologique des Desmidiées de l'étang du Kraenepoel (Belgique). Biologisch Jaarboek Dodonaea 8: 173–293.

Hydrobiologia (2007) 584:305–316
DOI 10.1007/s10750-007-0594-9

SHALLOW LAKES

Why biomanipulation can be effective in peaty lakes

Gerard ter Heerdt · Michiel Hootsmans

Abstract The effects of fish stock reduction (biomanipulation) was studied in an 85 ha shallow peaty turbid lake. The lake cleared in a 4-week period in April–May 2004, which demonstrated that biomanipulation can be effective in peaty lakes. We demonstrated that it is possible to reduce the fish stock to <25 kg ha^{-1} benthivorous fish and <15 kg ha^{-1} planktivorous fish, sufficiently low to switch the lake from a turbid to a clear state. Knowledge of lake morphology, fish stock, fish behaviour, and a variety of fishing methods was necessary to achieve this goal. It is expected that continuation of fisheries to remove young of the year planktivorous species is needed for several years, until macrophytes provide sufficient cover for zooplankton and can compete with phytoplankton. Cladocerans developed strongly after fish removal. The clearing of the lake coincided with a sudden decrease of filamentous cyanobacteria and suspended detritus, and a strong increase of *Bosmina*. We assume that *Bosmina* was able to reduce filamentous prokaryotes and detritus. After the disappearance of the cyanobacteria, *Bosmina* disappeared too. After the clearing of the lake *Daphnia* dominated in zooplankton and apparently was able to keep phytoplankton levels low. In our case, wind resuspension did not prevent biomanipulation from being successful. No correlation between windspeed and turbidity was found, neither in an 85 ha nor in a 230 ha shallow peaty lake. Regression analysis showed that on average 50% of the amount of suspended detritus can be explained by resuspension by fish and 50% by phytoplankton decomposition. The main goal of this biomanipulation experiment, clear water and increased submerged plant cover in a shallow peaty lake, was reached.

Keywords Biomanipulation · Peaty lakes · Phytoplankton · Zooplankton · Fish · Wind · Resuspension · Detritus · Macrophytes

Guest editors: R. D. Gulati, E. Lammens, N. De Pauw & E. Van Donk
Shallow lakes in a changing world

G. ter Heerdt (✉) · M. Hootsmans
Waternet, Post Box 8169, 1005 AD Amsterdam, The Netherlands
e-mail: gerard.ter.heerdt@waternet.nl

M. Hootsmans
e-mail: michiel.hootsmans@waternet.nl

Introduction

Fish stock reduction, or 'biomanipulation', can switch a turbid lake to a clear state, provided that nutrient levels are not too high (Jeppesen et al. 1990a, 1990b, 1997; Meijer et al., 1994, 1999; Perrow et al., 1997; Scheffer, 2001; Gulati & Van Donk, 2002). By reducing benthivorous fish, such as bream (*Abramis brama* L.), sediment resuspension is reduced. Reducing planktivorous fish, such as roach (*Rutilus rutilus* L.) and small bream, reduces predation pressure on zooplankton, leading to top–down control of phytoplankton. The reduction of suspended solids and algae will increase water clarity and promotes increase of submerged aquatic plants that help maintaining the new phase (Jeppesen et al., 1997; Scheffer, 2001; Gulati & Van Donk, 2002).

However, compared with results in mineral bottom lakes, biomanipulation in peaty lakes in the Netherlands has not been successful (Van Liere & Gulati, 1992; Meijer et al., 1999; Gulati & Van Donk, 2002). In this paper we will discuss the possible reasons for this failure, illustrated with a whole lake biomanipulation experiment in the shallow peaty Lake Terra Nova in The Netherlands.

Fish stock reduction

Many failures of biomanipulation measures in peaty lakes are due to insufficient fish removal (Van Der Vlugt et al., 1992; Van Donk et al., 1994; Meijer et al., 1999). One explanation for this failure can be the large amount of canals, ditches and banks where fish can hide. Young cyprinids concentrate in small waters during winter and can escape netting (Meijer et al., 1999; Klinge et al., 2002). Therefore lake managers often suppose that sufficient fish removal is impossible in peaty lakes. On the other hand, when one knows where fish is hiding and how it migrates, it should be possible to catch sufficient amounts of fish to make biomanipulation work.

According to Meijer et al. (1999) fish stock reduction should be at least 75% of the cyprinids. This reduction percentage does not, however, take the original stock into account, nor the stock to be achieved (Gulati & Van Donk, 2002). In the Netherlands a more absolute threshold to be reached is considered to be <25 kg ha^{-1} benthivorous fish and <15 kg ha^{-1} planktivorous fish (Meijer et al., 1998). Other thresholds mentioned in literature are <50 kg ha^{-1} planktivore fish (Gulati & Van Donk, 2002), and 10–40 kg ha^{-1} (Moss et al., 1996).

Recruitment of young fish can lead to a rapid recolonisation of the lake (Meijer et al., 1994, 1995; Gulati & Van Donk, 2002), and fish removal may need to be repeated several times.

In this study we want to demonstrate that it is possible to reduce the fish stock in an 85 ha peaty lake sufficiently to switch the lake from a turbid to a clear state, and keep it that low the years after.

Zooplankton grazing

Zooplankton grazing, mainly by large bodied cladocerans such as *Daphnia*, is considered a key factor in the clearing of a lake after biomanipulation (Perrow et al., 1997; Meijer et al., 1999; Scheffer, 2001). Generally, large bodied cladocerans are not abundant in eutrophic peaty lakes. This may be because of high predation pressure by fish (Gulati et al., 1992; Van Donk et al., 1994), or unsuitability of the phytoplankton dominated by filamentous cyanobacteria (Blindow et al., 2000; DeMott et al., 2001a, b). The latter could be one reason why *Daphnia* does not develop in peaty lakes after biomanipulation (Van Donk et al., 1990; Lammens et al., 1992).

In this paper we will test the hypothesis that cladocerans will not develop after fish removal in a peaty lake.

Resuspension by wind and fish

In shallow lakes, turbidity or light attenuation is caused, generally, by phytoplankton, suspended detritus and dissolved coloured organic matter (Scheffer, 2001). The major sources of suspended detritus in a peaty lake are phytoplankton decomposition and resuspension of sediment (Gons et al., 1992). Major causes of resuspension are wind (Gons et al., 1986, 1991, 1992), and benthivorous fish (Meijer et al., 1990). Benthivorous fish, however, have not received adequate attention in studies of lake restoration (Gulati & Van Donk, 2002).

Unsuccessful lake biomanipulation is often explained by wind induced resuspension (Van Donk et al., 1990; Van Der Vlugt et al., 1992), and biomanipulation in peaty lakes, with their loose bottom sediments, might not work due to wind resuspension (Van Liere & Gulati, 1992; Van Liere & Janse, 1992; Scheffer, 2001; Gulati & Van Donk, 2002). In none of these studies, however, the role of wind resuspension and its interaction with fish stock removal was quantified. In general, resuspension by wind is estimated from windspeed, wind fetch, lake depth and sediment type (Meijer et al., 1999; Scheffer, 2001). Gons et al (1986, 1991) and Scheffer (2001) showed that it is surprisingly difficult to estimate wind resuspension, leading to a reliance on empirical relationships between turbidity, suspended matter and windspeed (Gons et al., 1991; Scheffer, 2001).

In this study we want to quantify both wind and fish induced resuspension in peaty lakes. We hypothesise that resuspension by wind and fish are major factors causing turbidity.

Materials and methods

Study area

Lake Terra Nova is an 85 ha shallow (0.5–2 m) peaty lake. At the west-side of lake lies a series of banks in a west–east direction. The eastern side of the lake is open water. The fetch depends largely on the wind angle and can be up to 0.75 km. A map of the area is shown in van de Haterd and Ter Heerdt (2007).

As a consequence of eutrophication, algal blooms started in 1988. Turbidity increased from 1.2–1.9 to 0.3–0.4 m Secchi-depth. The lake lost its varied, mesotrophic, macrophyte vegetation completely (see van de Haterd and Ter Heerdt, 2007). Part of Lake Terra Nova is covered by *Nymphaea alba* L. and *Nuphar lutea* (L.) Sm. Phytoplankton is dominated by filamentous cyanobacteria species that originally have been termed "*Oscillatoria limnetica*-like". This is a group of several related species, co-occurring in the lake, which belong to the *Limnothrix/Pseudanabaena* group, and *Prochlorothrix hollandica* (Zwart et al., 2005). Zooplankton is dominated by copepods, rotifers and *Bosmina* spp.

Total phosphorus concentration (TP) in 2001–2003 was 0.071 ± 0.004 mg l^{-1} (Mean ± SE, $n = 31$). The fish population is typical for a eutrophicated peaty lake. Based on size and feeding habit we estimate that in 2000 there was 134 kg ha^{-1} benthivorous fish (mainly bream, *A. brama*) and 80 kg ha^{-1} planktivorous fish (mainly roach, *R. rutilus*) in the lake.

Lake Loenderveen is a 230 ha shallow peaty lake without banks and cuts. The fetch lies between 1.25 km and 2.0 km. Data from this lake were used to test the effect of wind on turbidity.

Fish stock reduction

Fish was removed in October 2003–June 2004 and September 2004–June 2005. This time of the year was chosen because it is before spawning (Moss et al., 1996), before a spring increase in cladocerans (Perrow et al., 1997), and during the migration period of small cyprinids (Meijer et al., 1999; Klinge et al., 2002). From mid-September onwards, many fyke nets were placed in the canals to detect migration routes. Ditches were electrofished to remove winter concentrations of small fish. Open water was fished with 375 m seine nets during winter. Cuts were fished with 1,000 m gill nets, mesh 120–140 mm.

All larger fish were counted, weighted, and their length measured. Representative subsamples of smaller fish were measured. Weights are given in kg ha^{-1} fresh weight. A Capture Mark Recapture method was used to estimate fish stock development during biomanipulation (Klinge et al., 2002).

Light under water

Vertical attenuation coefficient (K_d) was measured as $K_d = (\ln(I_0/I_Z))/Z$ where I_0 and I_Z are the intensities of Photosynthetically Active Radiation (PAR) just under the water surface and at depth z (Scheffer, 2001). PAR was measured with two LI-192SA underwater quantum sensors, mounted in a frame, 0.5 m apart. Measurements were taken every second during 30 s and averaged, using a LI-1400 data logger. Average K_d was measured over a 3- to 10-min period.

Hydrochemical and biological data

Two sites were sampled, one open-water site (since 2001) and one *Nymphaea* site (since 2003). Each site was 50 × 50 m^2. Fortnightly to monthly samples were taken.

Water samples were taken with a tube which contained one (at 0.5 m water depth) to two litre water (at >1 m water depth). At least 20 subs-amples were taken and combined in a tub, until a volume of 30–40 l was gathered. From this volume samples were drawn for analysis. Per zooplankton sample 20 l was used.

Chlorophyll *a* and phaeopigment were measured after extraction in hot ethanol. Phytoplankton and zooplankton numbers were measured microscopically (minimal 200 individuals). Image analysis (Leica QWin v.3) was used to measure the body size dimensions for calculating zooplankton biovolume.

Suspended matter dry weight was measured by filtering up to 3 l of water over a weighted 1.5 μ filter. Phytoplankton dry weight concentration was estimated as 46*CHL (under light limitation, turbid water, $K_d \geq 2$ m^{-1}) or 70*CHL (less light limitation, clear water, $K_d < 2$ m^{-1}) adapted from Gons et al. (1992) and Scheffer (2001). Suspended detritus dry weight was estimated as suspended matter minus phytoplankton dry weight.

Means are given as mean ± SE. Data between 1st January and 15th April 2004, when fish stock was rapidly decreasing, were left out of statistical analysis.

Wind and fish induced resuspension

The contribution of wind and fish to the amount of suspended matter and turbidity was estimated from their empirical relationship. Though this relationship should be roughly sigmoidal, a linear model is sufficient if concentrations of suspended matter stay below saturation level (Gons et al., 1991; Scheffer, 2001; Scheffer et al., 2003).

Wind data between January 2001 and May 2005 were measured at Schiphol Amsterdam Airport, at a height of 10 m. Schiphol is located in a open landscape, comparable with the study area at a distance of 25 km.

Results

Fish stock reduction

We found several locations were fish concentrated and a few migration routes. This eased removal of most of the small fish with fykes and electrofishery (Table 1). Large fishes were mostly caught in open water with seine nets and in the cuts with gill nets (Table 1). Our goal of <25 kg ha^{-1} benthivorous fish and <15 kg ha^{-1} planktivorous fish was reached both in 2004 and 2005 (Table 2).

The recruitment in 2004 of perch (*Perca fluviatilis* L.) was 22.2 kg ha^{-1}, of roach it was 14.1 kg ha^{-1}. Hardly any young of the year bream were found.

Clearing of the lake

The clear water state reestablished in 2004 between April 14 and May 10 (Fig. 1). The mean attenuation coefficient after fish removal dropped from 3.10 ± 0.12 m^{-1} to 1.58 ± 0.09 m^{-1} (open-water site) and from 3.16 ± 0.18 m^{-1} to 1.49 ± 0.07 m^{-1} (*Nymphaea* site). This decrease was highly significant (Table 3). Lake bottom was visible all over the lake after fish removal.

Table 1 Catch per fishery method in Lake Terra Nova, winter 2003–2004 (kg ha^{-1})

	Fish <25 cm	Fish >25 cm
Fykes	140.3	28.2
Electro fishery	14.5	0.6
Seine nets	0.3	27.6
Gill nets	0.0	51.0
Total	155.1	107.3

Table 2 Lake Terra Nova fish stock per year in kg ha^{-1}, based on Capture Mark Recapture method (*estimation based on catch)

	2003	2004	2005
A. brama benthivorous	119.5	8.7	9.3
R. rutilus planktivorous	41.6	1.3	4.1
P. fluvialilis planktivorous	6.8	0.4	Low*
Tinca tinca L.	32.1	11.9	10*
Esox lucius L.	28	22.7	20*
Others	15.8	2.8	Low*
Total	243.8	47.8	–

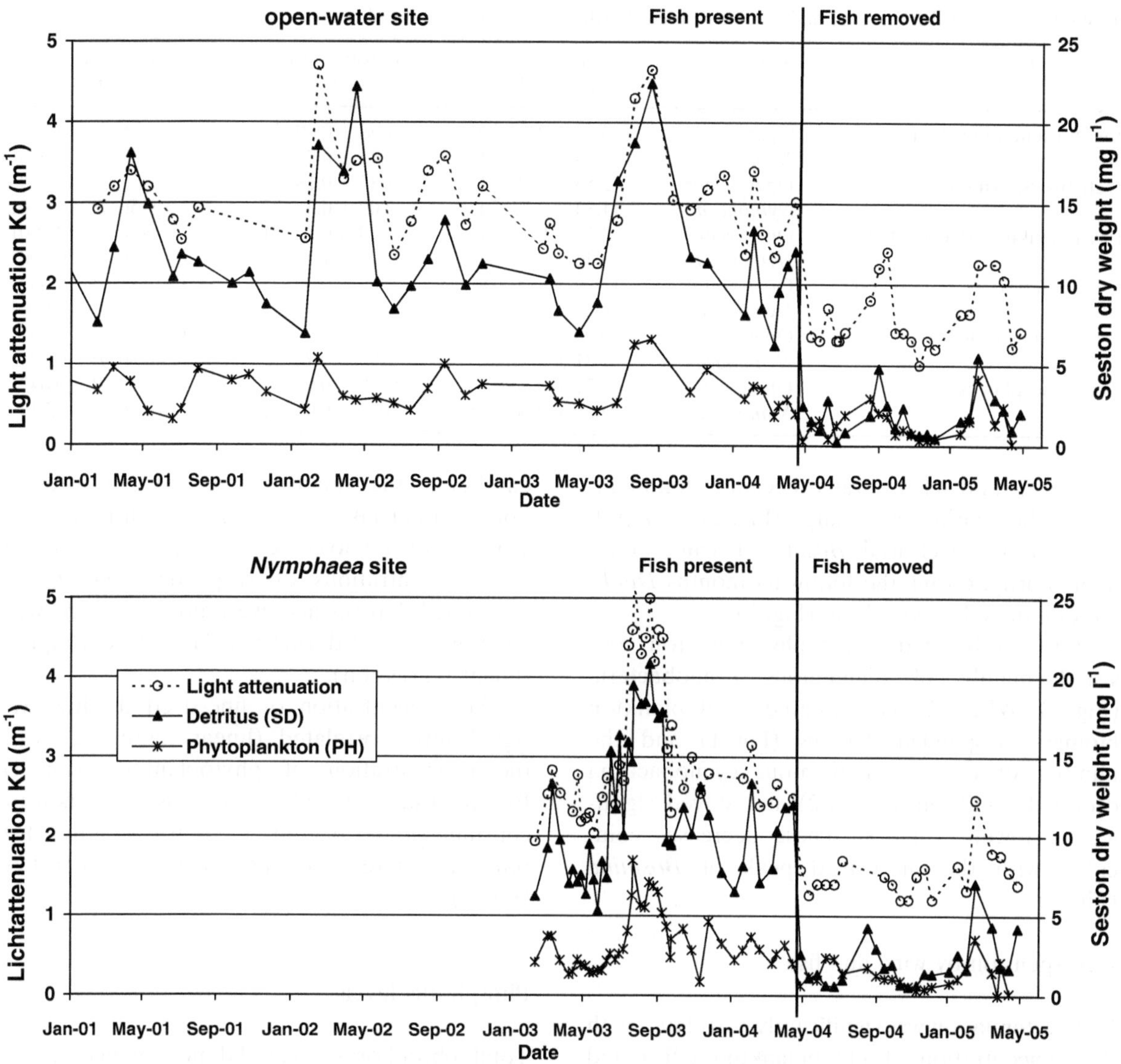

Fig. 1 Light attenuation K_d (m^{-1}), phytoplankton biomass (mg l^{-1}) and suspended detritus concentration (mg l^{-1}) before and after fish removal in two stations in Lake Terra Nova in the period January 2001–May 2005

The lake transparency increased suddenly, not gradually. Between April 14 and May 10, the concentration of phytoplankton (PH) and suspended detritus (SD) dropped sharply (Fig. 1). The difference in mean PH and SD before and after fish removal (Table 3) was highly significant.

The submerged vegetation cover increased from 0% to 30%. The vegetation was characterised by *Ceratophyllum demersum* L., *Najas marina* L., *Nitella mucronata* A. Braun, *Elodea nuttallii* Planch., *Potamogeton obtusifolius* L., *Utricularia vulgaris* L., *P. crispus* L., *Lemna trisulca* L., *P. lucens* L. and *Chara globularis* Thuillier. (in order of decreasing cover, van de Haterd & Ter Heerdt, 2007).

Zooplankton grazing

Before fish removal copepods and rotifers dominated zooplankton (Fig. 2). Cladoceran biovolume increased significantly after fish removal

Table 3 Effects of fish removal on light, suspended detritus, plankton and phosphorus

Parameter	Site	Statistics before (mean ± SE, *n*)	Statistics after (mean ± SE, *n*)	Comparison of means (*t*-test)
Light attenuation, K_d (m^{-1})	Open water	3.10 ± 0.12, 30	1.58 ± 0.09, 22	$t = 9.86$, $P < 0.0001$
	Nymphaea	3.16 ± 0.18, 31	1.49 ± 0.07, 21	$t = 7.48$, $P < 0.0001$
Detritus, SD (mg l^{-1})	Open water	12.55 ± 0.79, 31	1.85 ± 0.28, 22	$t = 11.10$, $P < 0.0001$
	Nymphaea	11.84 ± 0.77, 34	2.09 ± 0.34, 22	$t = 9.56$, $P < 0.0001$
Phytoplankton, PH (mg l^{-1})	Open water	3.57 ± 0.23, 31	1.29 ± 0.22, 21	$t = 6.84$, $P < 0.0001$
	Nymphaea	3.46 ± 0.34, 34	1.22 ± 1.22, 22	$t = 4.99$, $P < 0.0001$
Cyanobacteria (n l^{-1}) × 1,000	Open water	174.9 ± 71.0, 18	0.63 ± 0.30, 23	$t = 2.78$, $P = 0.008$
	Nymphaea	209.4 ± 38.8, 35	0.48 ± 0.26, 23	$t = 4.29$, $P < 0.0001$
Cladocerans (mg l^{-1})	Open water	0.06 ± 0.00, 5	4.84 ± 1.28, 24	$t = 3.75$, $P = 0.001$
	Nymphaea	1.21 ± 0.21, 35	2.79 ± 0.76, 25	$t = 2.30$, $P = 0.025$
Total phosphorus (mg l^{-1})	Open water	0.071 ± 0.004, 31	0.079 ± 0.020, 24	$t = 0.47$, $P = 0.642$
	Nymphaea	0.068 ± 0.005, 35	0.073 ± 0.019, 22	$t = 0.36$, $P = 0.718$

(Table 3) and cladocerans became dominant, though fluctuations were large (Fig. 2). In April, when the water cleared, *Bosmina* became abundant shortly, during the following months *Daphnia* dominated zooplankton (Fig. 2).

Before fish removal the phytoplankton consisted mainly of filamentous cyanobacteria (Fig. 2). When the lake cleared, phytoplankton biomass, suspended detritus (Fig. 1) and the number of cells of filamentous cyanobacteria dropped significantly (Fig. 2) and stayed significantly lower (Table 3). These decreases coincided with the short-lived peak of *Bosmina* (Fig. 2).

Resuspension by wind and fish

Light attenuation was significantly correlated with the concentration of phytoplankton (PH) and suspended detritus (SD) (Table 4). These two parameters explained about 80% of the variation in light attenuation.

When the regression formulas (Table 4) are applied on the mean PH and SD (Table 3), the percentage contribution to light attenuation by constant, PH and SD can be estimated. Before fish removal, phytoplankton and suspended detritus contributed largely to turbidity (Table 5). After fish removal their combined contribution dropped to less than 35%. Light attenuation after fish removal is mainly explained by the constant.

No significant correlation (linear regression) was found between mean windspeed during the day before the sampling date and light attenuation in either lake (Fig. 3). After fish removal in Terra Nova still no correlation was found (Fig. 3). The concentrations of suspended matter and suspended detritus also were not correlated with mean windspeed during the day before sampling (linear regression).

The concentration of fine-sized detritus was significantly correlated (linear regression) with the concentration of phytoplankton (Fig. 4). Before fish removal there was a significant constant of 5.5–6.39 mg l^{-1} detritus. After fish removal there was no significant constant (Fig. 4).

Phosphorus levels

Total phosphorus level did not change significantly after fish removal (Table 3). However, a remarkable peak of 0.45–0.50 mg P l^{-1} was recorded in 2004 between April 27 and May 10, during the sudden clearing of the lake. Orthophospate levels on these dates (0.01 and 0.07 mg P l^{-1}, respectively) were only slightly higher than the usual <0.01 mg P l^{-1}.

Discussion

Fish stock reduction

Fish stock can be reduced in peaty lakes with many cuts and banks when there is sufficient

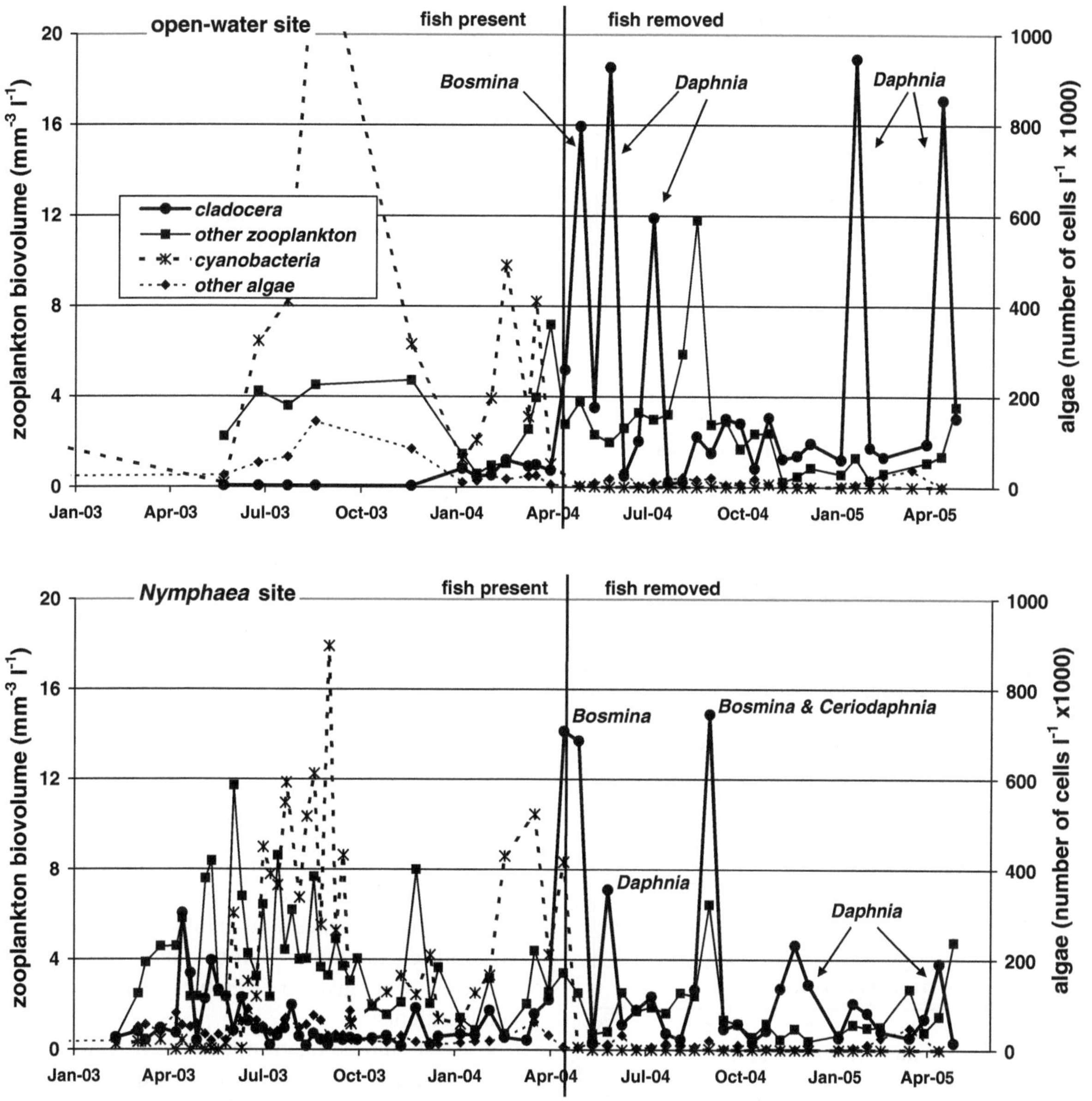

Fig. 2 Changes in zooplankton (biovolume, left axis) and in cyanobacteria (number of cells, right axis), before and after fish removal in Lake Terra Nova

knowledge of fish behaviour and habits. Different groups of fish and different areas require a variety of fishing methods. A stock of <25 kg ha^{-1} benthivorous fish and <15 kg ha^{-1} planktivorous fish can be reached and is sufficiently low to let a peaty lake return to a clear state. Even when the fish stock is very low, however, recruitment can be strong and the fish stock can grow above the threshold within 1 year. Meijer et al. (1994) also conclude that recruitment of planktivorous cyprinids in productive lakes can be strong. The large amount of cladocerans after the clearing of the lake provides a perfect food source. Until the cover with submerged macrophytes is large enough to compete effectively with phytoplankton (Blindow et al., 2000; Gulati & Van Donk, 2002), high amounts of cladocerans are needed to maintain clear water (Meijer et al., 1994). There-

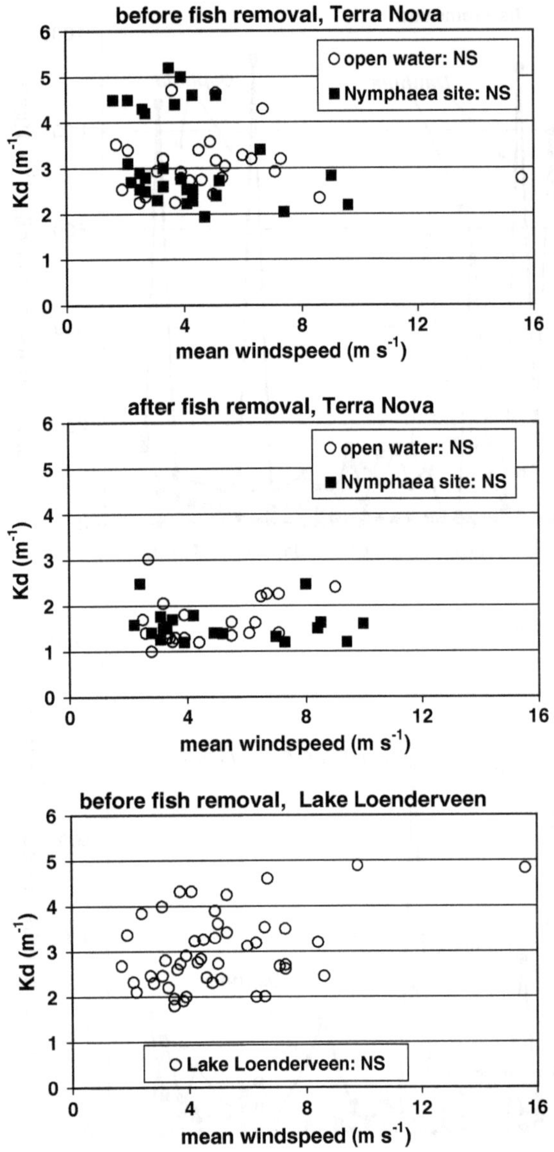

Fig. 3 Effect of mean wind speed (m s^{-1}) the day before sampling on light attenuation K_d (m^{-1}) between January 2001 and May 2005 in Lake Terra Nova and Lake Loenderveen. Linear regression never was significant

Table 4 Linear regression analysis effect of phytoplankton [PH, mg l^{-1}] and detritus (SD, mg l^{-1}) on light attenuation K_d (m^{-1}) K_d = constant + B_1 [PH] + B_2 [SD]

	Constant	B_1 (PH)	B_2 (SD)	R^2
Open-water site, before fish removal	1.28^{***}	0.25^{***}	0.08^{***}	0.80
Nymphaea site, before fish removal	1.00^{***}	0.27^{***}	0.10^{***}	0.79
Open-water site, after fish removal	1.14^{***}	0.15^{*}	0.14^{***}	0.88
Nymphaea site, after fish removal	1.21^{***}	0.09^{NS}	0.10^{***}	0.81

* $P < 0.05$; ***$P < 0.001$; NS, not Significant

fore fisheries have to be continued for several years to keep planktivorous fish stock low. We presume that after a few years the macrophytes can provide sufficient cover for zooplankton and the fishery effort can be reduced.

In temperate peaty lakes the winter period is the most suitable period for fish removal as the fish concentrates during this time. Throughout the summer the fish are more homogeneously distributed (Klinge et al., 2002). For the same reason, however, the winter season is unsuitable for accurate fish stock monitoring (Klinge et al., 2002). Fish stock sampling with trawl or seine nets is difficult in peaty lakes due to the large area of cuts, ditches and banks. Capture Mark Recapture methods before and during the fisheries will give a reliable insight in the achievements.

Care should be taken not to use fish stock data of more than 1 or 2 years old to determine the amounts of fish to be caught, as annual differentiation in growth, recruitment or death might change population sizes rapidly.

Table 5 Percentage contribution to light attenuation of constant, phytoplankton and suspended detritus. Linear regression analysis

	Constant	Phytoplankton	Detritus
Open-water site, before fish removal	43	28	30
Nymphaea site, before fish removal	34	30	37
Open-water site, after fish removal	66	12	22
Nymphaea site, after fish removal	74	8	18

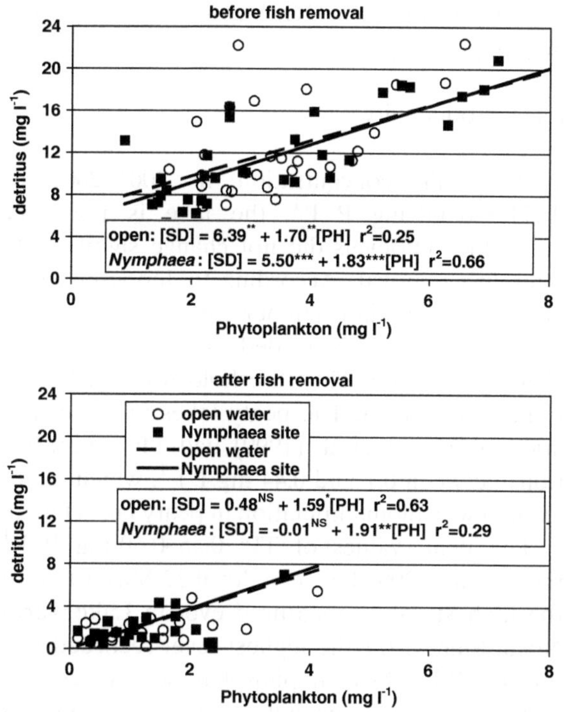

Fig. 4 Effect of phytoplankton concentration (PH) on suspended detritus concentration (SD) in Lake Terra Nova. Significant linear regression equations are shown also

Zooplankton grazing

In contrast with our expectations cladocerans, such as *Bosmina*, *Daphnia* and *Ceriodaphnia* became abundant after fish removal. The sudden appearance of large numbers of *Bosmina* coincided with the sudden decrease of cyanobacteria and detritus between April 14 and May 10. We cannot imagine any other explanation for this decrease than that *Bosmina* was able to consume filamentous cyanobacteria and maybe detritus as well. We found some indications in literature that support this thought. In Lake Vogelenzang in The Netherlands the cyanobacteria disappeared after a high peak of *Bosmina* also (Van Der Vlugt et al., 1992). *Bosmina* species are well adapted to food dominated by cyanobacteria, can be the most important grazers in a peaty lake and often appear in early spring (Gulati et al., 1992). On the other hand Meijer et al. (1999) consider potential grazing of *Bosmina* negligible compared to that of Daphnia.

After an initial 'bloom' *Bosmina* decreases again in Lake Terra Nova. *Daphnia* takes over when the amount of cyanobacteria is negligible. The remaining phytoplankton will be of good quality for cladocerans (DeMott et al., 2001a, b). This corresponds to the observation of Kurmayer (2001) that larger cladocerans have a competitive advantage over smaller cladocerans in exploiting limited food resources.

DeMelo et al. (1990) debated the whole concept of top–down control after biomanipulation. Our study shows that top–down control by cladocerans likely will be one of the factors determining whether a lake is turbid or clear.

Resuspension by wind and fish

We conclude that resuspension by wind is not a major factor determining detritus concentration and light attenuation, as we did not find any correlation between windspeed and suspended detritus in the 85 ha Lake Terra Nova, nor in the 230 ha Lake Loenderveen. In our shallow peaty lakes, though the sediment is very loose, resuspension by wind will not prevent biomanipulation being successful.

This is not as unexpected as it appears. Gons et al. (1986) measured a strong resuspension when windspeed exceeded 7 m s^{-1} in Lake Loosdrecht (980 ha). One day later, however, concentration of suspended matter returned closely to its previous level. Short-term variations in turbidity, between 16 and 30 NTU, followed windspeed and an averaged hourly windspeed of 5 m s^{-1} was the threshold for wind resuspension in Lake Loosdrecht (Gons et al., 1991). The lakes we studied are much smaller and therefore the threshold will be larger and the wind induced fluctuations in turbidity less frequent and smaller.

We conclude that the amount of detritus resuspended by fish on average was 5.50–6.39 mg l^{-1}, which is about 50% of the mean detritus concentration, as the constant in our linear regression equation describing the relation between detritus and phytoplankton became not significantly different from zero after fish removal (Fig. 4). Gons et al. (1991) postulate from their regression analysis that resuspension by fish was quantitatively unimportant. As the constant of about 16 FTU in their equation remained unexplained, however, the activity of benthic fish still might be a major cause of resuspension.

Sinking rates of fine sized detritus can be very low. Gons et al. (1992) concluded that about 14% of the fine sized detritus might be present 1 year later. In our study, however, the concentration of suspended detritus showed rapid changes, following phytoplankton concentration. This implies not only that significant amounts of detritus can be 'produced' rapidly, but also that detritus disappears rapidly from the water column. During the clearing of the lake, detritus concentration dropped to near zero within a few weeks. This suggests that sinking rates of detritus are higher than expected. One other explanation might be that *Bosmina* consumed detritus as well as algae.

In several other studies, resuspension by wind, even in large lakes, did not prevent lakes to become clear after biomanipulation in the Netherlands (Meijer et al., 1999; Scheffer, 2001). Benthivorous fish can facilitate wind resuspension largely by reducing the erosion resistance of the sediment and fish removal thus can lead to a decrease of wind resuspension (Scheffer et al., 2003). Wind resuspension will play a role in a lake ecosystem, however, as it might stir up the bottom temporarily and sinking sediment can cover plants. But we conclude that it will not prevent biomanipulation from being successful.

As expected, most of the variation in light attenuation can be explained by variation in phytoplankton and suspended detritus (Gons et al., 1992; Scheffer, 2001). Like in Lake Loosdrecht (Gons et al., 1992), in Lake Terra Nova the concentrations of detritus and phytoplankton are coupled. Apparently phytoplankton decomposition will be an important source of detritus.

The constant of 1.00–1.28 found in the regression between light attenuation and detritus and phytoplankton (Table 3) might be caused by the presence of Coloured Dissolved Organic Matter (CDOM). Concentrations of CDOM can be quite high in peaty lakes (Rijkeboer et al., 1997). In Lake Terra Nova CDOM concentration is around 13 mg l^{-1}. The major source of CDOM could be the sediment; the concentration in peat pore water in neighbouring Lake Loenderveen was 30–66 mg l^{-1}. When the bottom is less disturbed by fish and wind the CDOM concentration in the lake might drop.

Lake productivity

With a TP concentration in 2001–2003 of 0.071 ± 0.004 mg P l^{-1}, the lake is not very eutrophic. TP-levels did not change significantly after fish removal. According to Jeppesen et al. (1990b) a minimum summer concentration below 0.08–0.15 mg P l^{-1} is needed to establish a stable clear water state. This threshold was reached in Lake Terra Nova. The peaty lakes in the evaluation of Meijer et al. (1999) had TP concentrations between 0.1 and 0.44 mg P l^{-1}, which might explain why biomanipulation in peaty lakes failed.

The peak values of TP; 0.45–0.50 mg P l^{-1} measured in 2004 between April 27/May 10, were the highest values measured ever in Lake Terra Nova. Since 1977 the highest level recorded was 0.15 mg P l^{-1}. As orthophosphate levels were not raised much, we conclude that this P-load was present in suspended matter: zooplankton, phytoplankton or detritus. The source of this sudden increase remains unknown, but it coincided with the decrease of phytoplankton and detritus and the short term increase of *Bosmina*. We assume that there is a relation between the clearing of the lake and the peak TP-values. As we cannot offer any reliable explanation for this phenomenon, we will not speculate about it.

Conclusions

With sufficient knowledge of lake morphology, fish stock, fish behaviour, and a variety of fishing methods the fish stock in a peaty lake can be reduced to very low levels. Continuation of fisheries to remove young of the year of planktivorous species is needed for several years until macrophytes provide sufficient cover to zooplankton.

The main goal of this biomanipulation experiment, clear water and increased submerged plant cove was reached.

Most likely *Bosmina* is able to reduce filamentous cyanobacteria and detritus, while *Daphnia* is able to keep phytoplankton levels low.

In our case, resuspension by wind is not a major factor determining light attenuation and detritus concentration and did not prevent biomanipulation from being successful. On average, resuspension of detritus by fish explained about half of the amount of detritus, the other half can be explained by phytoplankton.

Acknowledgements The biomanipulation of Lake Terra Nova was funded by the Dutch Ministry of Agriculture, Nature and Food Quality. We want to thank our colleagues from The Water Laboratory for the analysis of the numerous samples. Dr. K. Irvine helped us to improve the English text.

References

Blindow, I., A. Hargeby, B. M. A. Wagner & G. Andersson, 2000. How important is the crustacean plankton for the maintenance of water clarity in shallow lakes with abundant submerged vegetation? Freshwater Biology 44: 185–197.

DeMelo, R., R. France & D. J. McQueen, 1990. Biomanipulation: Hit or Myth? Limnology and Oceanography 37: 192–207.

DeMott, W. R., R. D. Gulati & E. Van Donk, 2001a. Effects of dietary phosphorus deficiency on the abundance, phosphorus balance, and growth of *Daphnia cucullata* in three hypertrophic Dutch lakes. Limnology and Oceanography 8: 1871–1880.

DeMott, W. R., R. D. Gulati & E. Van Donk, 2001b. *Daphnia* food limitation in three hypereutrophic Dutch lakes: evidence for exclusion of large bodied species by interfering filaments of Cyanobacteria. Limnology and Oceanography 46: 2054–2060.

Gons, H. J., T. Burger-Wiersma, J. H. Otten & M. Rijkeboer, 1992. Coupling of phytoplankton and detritus in a shallow, eutrophic lake (Lake Loosdrecht, The Netherlands). Hydrobiologia 233: 51–59.

Gons, H. J., J. H. Otten & M. Rijkeboer, 1991. The significance of wind resuspension for the predominance of filamentous cyanobacteria in a shallow, eutrophic lake. Memorie Dell'Instituto Italiano De Idrobiologia Dott. Marco De Marchi 48: 233–249.

Gons, H. J., R. Veeningen & R. Van Keulen, 1986. Effects of wind on a shallow lake ecosystem: resuspension of particles in the Loosdrecht lakes. Hydrobiological Bulletin 20: 109–120.

Gulati, R. D., A. L. Ooms-Wilms, O. F. R. Van Tongeren, G. Postema & K. Siewertsen, 1992. The dynamics and role of limnic zooplankton in Loosdrecht Lakes (The Netherlands). Hydrobiologia 233: 69–86.

Gulati, R. D. & E. Van Donk, 2002. Lakes in the Netherlands, their origin, eutrophication and restoration: state-of-the-art review. Hydrobiologia 478: 73–106.

Jeppesen, E., J. P. Jensen, P. Kristensen, M. Sondergaard, E. Mortensen, O. Sortkjaer & O. K. Olrik, 1990b. Fish manipulation as a lake restoration tool in shallow, eutrophic, temperate lakes 2: threshold levels, long-term stability and conclusions. Hydrobiologia 200/201: 219–227.

Jeppesen, E., J. P. Jensen, M. Sondergaard, T. Lauridsen, L. J. Pedersen & L. Jensen, 1997. Top–down control in freshwater lakes: the role of nutrient state, submerged macrophytes and water depth. Hydrobiologia 342: 151–164.

Jeppesen, E., J. P. Jensen, M. Sondergaard, E. Mortensen, P. Kristensen, B. Rieman & P. J. Jensen, 1990a. Fish manipulation as a lake restoration tool in shallow, eutrophic, temperate lakes 1: cross-analysis of three Danish case-studies. Hydrobiologia 200/201: 205–218.

Klinge, M., G. Hensens, A. Brennikmeijer & L. Nagelkerke, 2002. Handboek visstandbemonstering, voorbereiding, bemonstering, beoordeling (Manual for Fish stock monitoring, in Dutch). STOWA, Utrecht.

Kurmayer, R., 2001. Competitive ability of *Daphnia* under dominance of non-toxic filamentous cyanobacteria. Hydrobiologia 442: 279–289.

Lammens, E. H. R. R., N. Boesewinkel-De Bruyn, H. Hoogveld & E. Van Donk, 1992. P-load, phytoplankton, zooplankton and fish stock in Loosdrecht Lake and Tjeukemeer: confounding affects of predation and food availability. Hydrobiologia 233: 87–94.

Meijer, M. L., I. De Boois & R. Portielje, 1998. Actief Biologisch Beheer: de meest effectieve maatregel? (Active Biological Management: the most effective measure? In Dutch) H2O, 23–26.

Meijer, M. L., I. De Boois, M. Scheffer, R. Portielje & H. Hosper, 1999. Biomanipulation in shallow lakes in The Netherlands: an evaluation of 18 case studies. Hydrobiologia 409: 13–30.

Meijer, M. L., M. W. De Haan, A. W. Breukelaar & H. Buitenveld, 1990. Is reduction of the benthivorous fish an important cause of high transparency following biomanipulation in shallow lakes? Hydrobiologia 200/201: 303–315.

Meijer, M. L., E. H. R. R. Lammens, A. J. P. Raat, J. G. P. KleinBreteler & M. P. Grimm, 1995. Development of fish communities in lakes after biomanipulation. Netherlands Journal of Aquatic Ecology 29: 91–101.

Meijer, M. L., E. Jeppesen, E. Van Donk, B. Moss, M. Scheffer, E. H. R. R. Lammens, E. Van Nes, J. A. Van Berkum, G. J. De Jong, B. A. Faafeng & J. P. Jensen, 1994. Long-term responses to fish-stock reduction in small shallow lakes – interpretation of 5-year results of 4 biomanipulation cases in The Netherlands and Denmark. Hydrobiologia 276: 457–466.

Moss, B., J. Madgwick & G. Phillips, 1996. A Guide to the Restoration of Nutrient-enriched Shallow Lakes. The Broads Authority, Norwich.

Perrow, M. R., M. L. Meijer, P. Dawidowicz & H. Coops, 1997. Biomanipulation in shallow lakes: state of the art. Hydrobiologia 342: 355–365.

Rijkeboer, M., A. G. Dekker & H. Gons, 1997. Subsurface irradiance reflectance spectra of inland waters differing in morphology and hydrology. Aquatic Ecology 31: 313–323.

Scheffer, M., 2001. Ecology of shallow lakes. Kluwer Academic Publishers, Dordrecht, Boston, London.
Scheffer, M., R. Portielje & L. Zambrano, 2003. Fish faciliate wave resuspension of sediment. Limnology and Oceanography 48: 1920–1926.
van de Haterd, R. J. W. & G. N. J. Ter Heerdt, 2007. Potential for the development of submerged macrophytes in eutrophicated shallow peaty lakes after restoration measures. Hydrobiologia 584: 277–290.
Van Der Vlugt, J. C., P. A. Walker, J. Van Der Does & A. J. Raat, 1992. Fisheries management as an additional lake restoration measure: biomanipulation scaling-up problems. In van Liere, L. & R. D. Gulati (eds), Restoration and Recovery of Shallow Eutrophic Lake Ecosystems in the Netherlands. Kluwer Academic Publishers, Dordrecht/Boston/London, 213–224.
Van Donk, E., M. P. Grimm, R. D. Gulati, P. G. M. Heuts, W. A. De Kloet & L. Van Liere, 1990. First attempt to apply whole-lake foodweb manipulation on a large scale in The Netherlands. Hydrobiologia 200/201: 291–301.
Van Donk, E., M. P. Grimm, P. G. M. Heuts, G. Blom, K. Everards & O. F. R. Van Tongeren, 1994. Use of mesocosms in a shallow eutrophic lake to study the effects of different restoration measures. Archiv für Hydrobiologie Beiheft 40: 283–294.
Van Liere, L. & R. D. Gulati, 1992. Restoration and recovery of shallow eutrophic lake ecosystems in The Netherlands: epilogue. Hydrobiologia 233: 283–287.
Van Liere, L. & J. H. Janse, 1992. Restoration and resilience to recovery of the Lake Loosdrecht ecosystem in relation to its phosphorus flow. Hydrobiologia 233: 95–104.
Zwart, G., M. P. Kamst-van Agterveld, I. Van der Werff-Stagerman, F. Hagen, H. Hoogveld & H. J. Gons, 2005. Molecular characterization of cyanobacterial diversity in a shallow eutrophic lake. Environmental Microbiology 7: 365–377.

Hydrobiologia (2007) 584:317–326
DOI 10.1007/s10750-007-0577-x

SHALLOW LAKES

State of the art in the functioning of shallow Mediterranean lakes: workshop conclusions

Meryem Beklioglu · Susana Romo · Ifigenia Kagalou · Xavier Quintana · Eloy Bécares

Abstract Studies on shallow lakes from the north temperate zone show that they alternate between clear and turbid water states in response to control factors. However, the ecology of semi-arid to arid shallow Mediterranean lakes is less explored. Hydrological effects (e.g. water level fluctuations, water residence time) on major ions and nutrient dynamics and processes, and ecology of submerged macrophytes appear to have a crucial role for food webs in shallow Mediterranean lakes. Nutrient control may be of greater priority in eutrophicated warm shallow lakes than in similar lakes at higher latitudes. This will be relevant for the implementation of the European Water Framework Directive, and conservation and management of these ecosystems. Strong trophic cascading effects of fish resulting from dominance of omnivorous and benthivorous fish species, whose diversity is usually high, together with frequent spawning and absence of efficient piscivores, seem to be the reason for the lack of large-bodied grazers that could control phytoplankton. However, such effects may vary within the region depending on fish distribution and community. These factors need elaboration in order to allow shallow lake ecologists and managers to develop better restoration strategies for eutrophicated shallow Mediterranean lakes. Consequently, modifications for the implementation of the European Water Framework Directive for determining ecological status in shallow Mediterranean lakes appear to be necessary. Furthermore, the implications of climate warming may be even more challenging than in high latitude lakes since shallow lakes in the Mediterranean region are among the most sensitive to extreme climate changes. There is an urgent need for data on the ecology of shallow lakes in the region. An appeal is made for international

Guest editors: R. D. Gulati, E. Lammens, N. De Pauw & E. Van Donk
Shallow lakes in a changing world

M. Beklioglu (✉)
Biology Department, Middle East Technical University, Ankara 06531, Turkey
e-mail: meryem@metu.edu.tr

S. Romo
Área de Ecología, Facultad Biología, Edificio Investigación, Campus Burjasot, Valencia 46100, Spain

I. Kagalou
Department of Biological Applications and Technology, University of Ioannina, Ioannina 45110, Greece

X. Quintana
Institut d'Ecologia Aquàtica, Universitat de Girona, Campus Montilivi, Girona 17071, Spain

E. Bécares
Departmento de Ecología, Facultad de Biología, Universidad de León, León 24071, Spain

cooperation, development of large-scale research and information exchange to facilitate this and a web-based discussion list has been implemented.

Keywords Climate change · Top-down control · Nutrients · Submerged plants · Water level · Water Framework Directive

Introduction

Mediterranean climate, with dry sub-tropical summers, dominates regions from 32° to 40° north and south of the Equator. It covers five significant regions among which that bordering the Mediterranean Sea is the focal point of this paper. In Mediterranean regions there is a segregation of two well-separated seasons, a wet winter, during which most of the precipitation is concentrated, and an arid summer with no precipitation (Bolle, 2003; Naselli-Flores & Barone, 2005; Álvarez-Cobelas et al., 2005). Further, in Mediterranean areas groundwater hydrology may greatly affect the persistence and functioning of aquatic ecosystems since groundwater discharge is largely controlled by rainfall seasonality (Álvarez-Cobelas et al., 2005). Summer drought places a great deal of stress on the hydrology of freshwaters, especially shallow lakes. Changes in hydrological regimes may especially affect growth rates, timing of reproduction, balance of photosynthesis and respiration, rates of mineralization, with local to large-scale regional effects (Coops et al., 2003; Beklioglu et al., 2006).

During the last three decades, increased urbanization, sewage disposal, regulation of wetlands and more intensive farming practices have increased the nutrient loading to many shallow lakes world-wide. Such deterioration causing water quality and amenity problems has led to the initiation of many studies on north temperate shallow lakes that have contributed greatly to a general understanding of the functioning of these ecosystems (Scheffer et al., 1993; Moss et al., 1996; Moss, 1998; Jeppesen et al., 1998, 2005a). Comparable information on tropical and subtropical lakes including semi-arid Mediterranean areas, is limited (Lazzaro, 1997; Talling and Lamolle, 1998; Jeppesen et al., 2005b; Beklioglu et al., 2003; Romo et al., 2004; Romo et al., 2005, Beklioglu et al., 2006).

This paper is based on a mini-workshop entitled "Shallow Mediterranean Lakes: Variation in the Theme" held during the Shallow Lakes 2005 Conference, in The Netherlands. Our primary aim is to compile and evaluate existing knowledge and experience on the ecology of shallow Mediterranean lakes and to compare this compilation with that of north-temperate shallow lakes. The fact that current understanding of shallow lake ecology has mainly been formed in the north-temperate region, makes this approach necessary. This contribution explores the role of hydrology on major ions and nutrients (N and P), trophic interactions in shallow Mediterranean lakes (SMLs) and postulates some hypotheses to be tested. Problems associated with the implementation of the European Water Framework Directive (WFD) on SMLs are also analysed. We also include possible changes in the ecology of SMLs in response to global warming and climate change.

Alternative stable states in shallow Mediterranean lakes

Recent observations on shallow lakes have led to the wide acceptance of two alternative stable states in northern temperate lakes, a turbid and a clear one (Scheffer et al., 1993; Moss et al., 1996; Jeppesen et al., 2003). Several mechanisms that cause these two alternative states have been reported (Blindow, 1992; Scheffer et al., 1993; Moss et al., 1996; Jeppesen et al., 1998; Jeppesen et al., 2003) and recently the influence of the effect of water level fluctuations on the stable states has been explored for floodplain lakes (Van Geest et al., 2005). Most of the differences depend on the extent of growth of submerged plants.

For shallow Mediterranean lakes, submerged macrophytes may be even more effective in maintaining water clarity, because of their potential persistence all year round resulting in advantages in the competition for nutrients and light between plants and algae (Álvarez-Cobelas et al., 2005). Dense macrophyte beds significantly increased water clarity at low and intermediate nutrient levels in mesocosm experiments carried out in Spanish lakes (Romo et al., 2004; Fernández-Aláez et al., 2004). However, at high nutrient levels (TP > 0.25 mg L^{-1}) submerged plants did not prevent deterioration in water transparency in some SMLs (Romo et al., 2004; Tan & Beklioglu, 2005; Karapinar, 2005). A survey of 319 shallow and polymictic subtropical Florida lakes (Bachmann et al., 2002) concluded that there was no clear and direct relationship between plant coverage and water transparency. In tropical and subtropical shallow lakes, aggregation of fish inside macrophyte beds may weaken the role of plants as refuges for zooplankton and therefore the significant effect of grazers on water transparency (Meerhoff et al., 2003). Furthermore, Meerhoff et al., (2006) reported from a laboratory experiment using (sub)tropical species that *Daphnia* did not take refuge among the plants but rather swam away from them when exposed simultaneously to plants and alarm signals of fish and conspecific. They concluded that in the (sub)tropics, therefore, aquatic plants may not promote cascading effects via large-bodied grazers on phytoplankton as seen in temperate lakes. Further investigations are needed in order to develop criteria for lake water quality in these climatic zones or construct general models on the ecology of shallow lakes. Control factors and their thresholds that may induce state shifts in SMLs need to be explored thoroughly.

Hydrology

Functioning of water bodies depends on water balance, which depends on groundwater and surface water hydrology. Water level fluctuation emerges as a decisive hydrological element in SMLs, with severe reductions experienced in summer owing to high evaporation. This often makes difficult the management of water for diverse uses (e.g. agriculture, domestic purposes, conservation etc.). Many SMLs experience progressive decrease in water level and are seriously threatened (Mitraki et al., 2004; Beklioglu et al., 2006). Pronounced water level changes may act as catastrophic disturbances leading to state shifts and affecting the functioning of the biotic components (Chow-Fraser et al., 1998; Gafny & Gasith, 1999; Fernández-Aláez et al., 1999; Coops et al., 2003; Naselli-Flores & Barone, 2003; Romo et al., 2004; Beklioglu et al., 2004; Tan & Beklioglu, 2005, 2006; Van Geest et al., 2005; Beklioglu et al., 2006). Low water levels may enhance submerged plant biomass and coverage in SMLs (Romo et al., 2004; Fernández-Aláez et al., 2004; Mitraki et al., 2004; Tan & Beklioglu, 2006). The impact of water depth in determining the extent of the littoral zone is dependent also on lake morphometry and bathymetry (Beklioglu et al., 2006).

Concentrations of the major ions and nutrients (nitrogen and phosphorus) vary with water levels (Talling, 2001; Nöges et al., 2003) and significantly increase during drought periods in the Mediterranean region (Quintana et al., 1998; Hambright et al., 2000; Tan, 2002; Tan & Beklioglu, 2005). Drought-induced decrease in water level and increase in water residence time may provide longer contact with sediment that may enhance internal release of nutrients, such as phosphorus (Quintana et al., 1998; Romo et al., 2005; Karapınar, 2005). Biota of several SMLs have become adapted to rapid and alternative internal and external input of nutrients in relation to hydrological water changes (Comín et al., 1992; Quintana et al., 1998; Quintana et al., 1999). Drop in water level favoured cyanobacteria over other algal taxa in many SMLs (Moustaka-Gouni, 1993; Tryfon & Moustaka-Gouni, 1997; Romero et al., 2002; Kagalou et al., 2003; Beklioglu et al., 2003; Romo et al., 2004).

Hydrological changes may exert a strong control over plankton and fish community structure. Plant-associated rotifers and salinity-tolerant calanoid copepods can predominate during low water level periods taking advantage of the structure offered by macrophytes (Tan, 2002; Romo et al., 2004; Fernández-Aláez et al., 2004). Salinity may limit the presence of Cladocera, copepods gaining competitive advantage at salinities above 0.5‰ (Jeppesen et al., 1994). However, some large daphnids, such as *Daphnia magna* and *D. mediterranea*, tolerate changes in salt concentration in temporary and SMLs (Alonso, 1996; Boronat et al., 2001). The strong structuring role of fish predation on zooplankton communities in these systems (see top-down control) may preclude their persistence (Romo et al., 2004; Tan, 2002). Furthermore, drop in water level may cause deterioration of spawning grounds, and change spawning behaviour and migration patterns of fish, as observed in Lake Pamvotis, Greece (Theocharis, 1989). Many endemic fish species have also become extinct due to reduced water quantity and quality in Greece (Economidis et al., 1998). Water-level fluctuations affect fish spawning success and fish migration patterns, thus ultimately the level of zooplanktivory and predictability of fish effects on food webs and plankton dynamics.

Water-level fluctuations in SMLs may be one of the important reasons for the hypothesis that the rate of change of fish and planktonic-benthic communities can be high and variable and therefore more unpredictable with implications for management and conservation of these systems.

Nutrients

The relative influence of nutrients on water quality of shallow lakes appears to increase southward in Europe (Moss et al., 2004). Nutrient availability in SMLs may shift dominance rapidly from submerged macrophytes, to periphyton and phytoplankton (Romo et al., 2004). There is some evidence that nutrient concentrations for oligotrophication and the recovery of submerged macrophytes can be lower than in northern temperate shallow lakes (Romo et al., 2004; Romo et al., 2005; Karapinar, 2005). A stable clear-water vegetated state is unlikely to occur in northern temperate shallow lakes when total phosphorus (TP) level of the lake is between 0.05 and 0.15 mg l^{-1} (Jeppesen et al., 2003). These values were unable to prevent the decline of vegetation in a shallow Mediterranean lake, which rapidly reached turbid states with TP > 0.1 mg l^{-1} (Romo et al., 2004). Thus, phosphate loadings giving concentrations ≤ 0.05 mg l^{-1} were suggested as a possible critical threshold to avoid algal turbid states and maintain submerged macrophytes in SMLs (Romo et al., 2004). Critical nutrient thresholds allowing SMLs to maintain vegetated states need further research. Nutrient control may be a greater priority in the restoration of eutrophicated warm shallow lakes than in similar lakes at higher latitudes (Moss et al., 2004).

Lately, it has been suggested for temperate lakes that low nitrogen may have an important role in stabilizing submerged plants, especially at relatively high phosphorus levels (Moss, 2001). Nitrogen availability may be lower in shallow lakes of warmer regions, owing to higher growth rates and faster nitrogen losses by denitrification (Talling & Lamolle, 1998). Additionally, external loading of nitrogen to lakes could be also lower owing to lesser runoff of rainfall, although this can be compensated by intensive agriculture. Fernández-Aláez et al. (1999) reported nitrogen-limited macrophyte development in northwest Spanish lakes and Beklioglu et al., (2003) recorded that low availability of nitrogen can be crucial in maintaining macrophyte dominance at intermediate to high TP levels in a Turkish shallow Mediterranean lake. This agrees with González-Sagrario et al. (2005), who showed that nitrogen negatively affected submerged macrophytes at moderately high TP levels in mesocosm experiments carried out in a shallow Danish lake. Consequently, more stringent phosphorus control, but also nitrogen control, might be needed for long-term recovery of SMLs with rich submerged plant diversity.

Effects of nutrient enrichment on the food web have been shown for temperate shallow lakes (Moss et al., 1996; Jeppesen et al., 2005a). Some of these patterns have been also observed in SMLs and nutrient enrichment often results in dominance and persistence of cyanobacteria and green algae (Lowe et al., 2001; Scasso et al., 2001; Romero et al., 2002; Romo et al., 2005), although cyanobacteria can remain dominant in SMLs over a wider range of nutrient levels (Romo et al., 2004). Implications of their presence for water users should be taken into account when assessing toxicity risks. Low oxygen concentrations owing to nutrient inputs, and also food-limitation affect macroinvertebrate community composition in SMLs (Mihaljevic et al., 2004; García-Criado et al. 2005; Kagalou et al., 2006). In nutrient-enriched SMLs, cyprinid fish are favoured over percids as they are in temperate lakes (Blanco et al., 2003; Beklioglu et al., 2003; Romo et al., 2005; Kagalou et al., 2006). Direct fish effects on water chemistry were insignificant in the presence of submerged macrophytes (Romo et al., 2004; Fernández-Aláez et al., 2004), but disturbance of lake sediment by fish was severe in unvegetated SMLs (Blanco et al., 2003; Beklioglu et al., 2003; Romo et al., 2005). Presence of bottom dwelling carp and dominance of "invasive" cyprinid species (*Carrasius gibelio*), as happens in almost all Greek lakes, affected water turbidity and chemistry (Economidis et al., 1998; Theocharis, 1989). Indirect effects of fish on nutrient balance and light climate will depend on the size of the fish and feeding modes of the fish communities.

Top-down control

In permanent SMLs, zooplankton community seems mostly dominated by small-sized species (Beklioglu et al., 2003; Romo et al., 2004; Fernández-Aláez et al., 2004; Romo et al., 2005). Dominance of omnivorous and benthivorus fish species, whose diversity is usually high, together with frequent spawning and absence of piscivores seem responsible for the lack of large-bodied zooplankton grazers (Blanco et al., 2003; Fernández-Aláez et al., 2004; Romo et al., 2005). Variation in fish communities along a north-south gradient may create a pattern with lack of prominent predators fish and increased omnivory and spawning frequency (Blanco et al., 2003), similarly to tropical systems (Lazzaro, 1997; Lazzaro et al., 2003). Furthermore, the ratio of zooplankton to phytoplankton biomass was significantly smaller in southern Europe (Gyllström et al., 2005; Jeppesen et al., 2005a). In temporary Mediterranean water bodies dominated by macrozooplankton and almost absence of fish, species richness can be low but highly diverse in size (Brucet et al., 2006). SMLs can thus remain in a turbid state with zooplankton exerting only a weak control over algal biomass, as a consequence of fish predation being far more intense than in temperate lakes for a given nutrient level (Moss et al., 2004; Romo et al., 2004). Nonetheless, the size-structure of zooplankton in relation to the fish community needs further investigation for the Mediterranean zone. The effects of omnivorous fish on food webs of SMLs also deserve attention.

The distribution and composition of the fish community can also be influenced by altitude. Comparison among low altitude lakes in Greece and Spain and high altitude lakes in Turkey all located at latitude 39° N, denoted a lack of piscivorous fishes in the former with communities mostly composed by omnivores, whereas in high altitude shallow Turkish lakes piscivorous (e.g. pike, perch) and omnivorous species were present (Blanco et al., 2003; Beklioglu et al., 2003; Tan, 2002; Tan & Beklioglu, 2005). The differences were partly due to the effect of winter ice-cover and the restricted biogeographical distribution of fish at higher altitudes, but can be also related to other factors, such as water temperature and water visibility for visual predators (Vijverberg et al., 1990; Liao et al., 2004). Fish composition and ontogenetic changes in diet may cause remarkable differences in the outcome of biomanipulation (Moss et al., 1996; Jeppesen et al., 2005b). Jeppesen et al. (2005b) concluded that biomanipulation as a restorative measure for subtropical and tropical lakes can be less successful owing to a very diverse fish fauna characterized by high biomass of omnivorous fish. However, the response of the Turkish Lake Eymir to about 50% removal of carp and tench improved trophic conditions for pike growth in the lake and a shift in transparency sufficient for macrophytes to develop coverages of 40–90% (Tan, 2002; Beklioglu et al., 2003; Tan & Beklioglu, 2005). The potential for biomanipulation in SMLs is still to be finally determined.

Furthermore, there is also a variation in theme of top-down control at lower trophic levels of SMLs. Weaker top-down effects on bacteria and ciliates in SMLs than in northern shallow temperate lakes was recorded, probably due to the stronger bottom-up control and higher growth rates of bacteria at southern latitudes (A. Conty, pers. Comm.). Also mixotrophic dinoflagellates can exert top-down control under certain conditions in some Mediterranean saltmarshes (López-Flores et al., 2006). Thus, complexity of food webs in SMLs needs further studies and new insights.

European Union Water Framework Directive and shallow Mediterranean lakes

The European Water Framework Directive (Directive 2000/60/EC, WFD 2000) aims to establish a framework for the community action in the field of water policy. Difficulties in the implementation of the WFD in shallow lakes are discussed by Moss et al. (2003). We here explore the implementation of the WFD in SMLs by taking into account the hydrological changes and unpredictability characteristic of the Mediterranean climate and differences in food web processes.

Making of an inventory of shallow Mediterranean lakes is one of the first difficulties in the implementation of the WFD, since some shallow lakes forming part of wetlands or temporary lakes are not clearly included and defined in the Directive. According to the WFD, a continental water body should be categorized as a lake, river or transitional water. Coastal shallow lakes and wetlands may be included as transitional waters, but inland wetlands can be considered as littoral zones associated with deeper lakes or as surface springs of aquifers. In the WFD lake surface of 50 ha is regarded as a possible cut-off for categorizing lakes under monitoring and conservation, but in some Mediterranean countries most shallow lakes are smaller (e.g. in Spain usually less than 10 ha Casado & Montes (1995)). Furthermore, there are numerous small and isolated ponds of high ecological and biodiversity interest. From a management perspective, the ecological interest should prevail because of conservation considerations. Water depth per se is crucial in SMLs, due to water level changes and importance in the zone of temporary and endorheic systems.

The determination of reference sites imposes additional problems, because of the intensive anthropogenic alteration of water bodies (Moss et al., 2003; Moss, 2007). Criteria to be considered for identifying reference sites in SMLs are difficult to determine. Criteria for assessment of the ecological status of shallow lakes have focused mainly on shallow lakes from central and northern Europe (Moss et al., 2003). For SMLs with their salinity and water level fluctuations, criteria are still undetermined and there is a crucial need for comparative methodological studies. Macroinvertebrate indices for biomonitoring the ecological status have been more commonly used for rivers (Davis & Simon, 1995; Wright et al., 2000), but their use is debatable for shallow lakes (Moss et al., 2003). A recent study has explicitly shown that some widely used macroinvertebrate indices are problematic for use in SMLs (Garcia-Criado et al., 2005), because variable hydrological conditions could have a negative synergistic effect on benthic biodiversity. Consequently, the concept of macroinvertebrate species indicators for SMLs needs to be revised. Alternatively, body size distribution of benthic organisms may be a better indicator of environmental conditions in SMLs (Basset et al., 2004). There is evidence that plant architecture and species composition can be more important for structuring macroinvertebrate communities than the lake trophic status (García-Criado et al., 2005). The use of microcrustaceans as determinants of water quality have shown satisfactory results when composition was identified at least to genus level in both SMLs (Boix et al., 2005) and shallow north temperate lakes (King & Richardson, 2002). Tools for assessments of water quality and conservation status have been successfully developed at a regional scale in northeast Spain, including the use of aquatic invertebrates (Sala et al., 2004; Boix et al., 2005). Diatoms have given good results as bioindicators in freshwater shallow lakes (Blanco et al., 2004), but their use is less satisfactory in brackish waters, where some species may change their shape in response to changing environmental conditions (Trobajo et al., 2004). Unfortunately, there is no inter-calibration research currently being conducted in the implementation of WFD, in which Mediterranean lakes are involved.

Climate change in shallow Mediterranean lakes

The Mediterranean region is considered as one of the most sensitive areas regarding global warming and extreme climate changes (Sánchez et al., 2004). The predicted trend of mean precipitation for the Mediterranean zone is of significantly decreasing seasonal precipitation with important fluctuations (Sánchez et al.,

2004). Furthermore, the daily temperature range is anticipated to be larger than at present, and extreme cold and hot days are expected to increase in frequency with a higher overall increase of 0.1–0.4°C decade^{-1}, which is twice that of northern Europe (IUCN, 2002). According to general circulation models, the mean annual precipitation is expected to decrease south of 45° N. Winter precipitation is expected to fall, due to less rain and snow, and earlier snowmelt (e.g. shifting to January, February and March) (Harvey, 2000). Consequently, shortage of rainfall is expected to accentuate water level fluctuations in SMLs.

Prolonged drought periods in SMLs may emphasise the issue of increased salinity and conductivity, enhancing internal nutrient recycling and affecting communities as mentioned above. This would favour eutrophication and particularly outbreaks of inedible and possibly toxic algae. Concerning nutrient dynamics, especially nitrogen, climate change in the Mediterranean region might be critical. Decrease in nitrate-nitrogen availability is anticipated as a result of reduced leaching, probably a prolonged terrestrial growing season and increased terrestrial nitrogen assimilation, and the elevated in-lake assimilation of nitrogen by increased biological activity (Sommaruga-Wograth et al., 1997; Kopacek et al., 2005). Possible alterations in major ions and nutrient dynamics are likely to affect distinctly the state of shallow lakes and food webs, maybe in some of the ways aforementioned. Some of us believe that many of our SMLs will become temporary lakes with increased salinity level in the future.

Acknowledgements To enhance cooperation between shallow lake scientists in the Mediterranean climate regions, an e-mail based discussion group, named "Warm lakes discussion group" with the address warmlakes@sulakalan.org has begun following the workshop. We wish to express our gratitude to Nuray Akbulut, Carles Alcaraz, Javier Armengol, Douglas Branch Mc Nair, Luciano Caputo, Ana Conty, Kemal Çelik, Cristina Trigal Domínguez, Máximo Florin, Marija Gligora, Mariana Meerhoff, Sarian Korsten, Erik Jeppesen, Marten Scheffer, Paola Lombardo, Dani Boix, Nestor Mazzeo, Carla Micheli, David Da Motta Marques, Biel Obrador, Didem Oğuzkurt, Tugba Ongun, Aleksandar Popijač, Kostas Stefanidis, Ivancica Ternjej, Ana Cristina Cardoso and Luis Zambrano. We would like also to thank Maria José Boavida and the organizing Committee of the Shallow Lakes Conference in Dalfsen (The Netherlands) for encouraging the organization of this workshop on SMLs. We specially thank Brian Moss for kindly revising the manuscript and the two referees for their valuable comments.

References

Alonso, M., 1996. Crustacea, Branchiopoda. In Ramos, M. A. et al. (eds), Fauna Ibérica, Vol. 7. Museo Nacional de Ciencias Naturales. CSIC. Madrid, 486 pp.

Álvarez-Cobelas, M., C. Rojo & D. Angeler, 2005. Mediterranean limnology: current status, gaps and future. Journal of Limnology 64(1): 13–29.

Bachmann, R., C. A. Horsburgh, M. V. Hoyer, L. K. Mataraza & D. E. Canfield, 2002. Relationship between trophic state indicators and plant biomass in Florida Lakes. Hydrobiologia 470: 219–234.

Basset, A., F. Sangiorgio & M. Pinna, 2004. Monitoring with benthic macroinvertebrates: advantages and disadvantages of body size descriptors. Aquatic Conservation: Marine and Freshwater Ecosystems 14(1): 43–58.

Beklioglu, M., Ö. Ince & I. Tüzün, 2003. Restoration of Eutrophic Lake Eymir, Turkey, by biomanipulation undertaken following a major external nutrient control I. Hydrobiologia 489: 93–105.

Beklioglu, M., G. Altınayar & C. O. Tan, 2004. Ülkemiz sığ göllerinin suiçi bitki gelişiminin belirlenmesinde, su seviyesi değişimi, besin tuzları ve bentik balık beslenmesinin değişen rolleri. In Beklioglu, M. (ed.), Sığ Göl Sulakalanlar: Ekoloji, Ötrofikasyon ve Restorasyon. Pozitif Matba, Ankara, 23–29.

Beklioglu, M., G. Altinayar & C. O. Tan, 2006. Water level control over submerged macrophyte development in five shallow lakes of Mediterranean Turkey. Archive für Hydrobiology 166(4): 535–556.

Blanco, S., L. Ector & E. Bécares, 2004. Epiphytic diatoms as water quality indicators in Spanish shallow lakes. Vie Milieu 54: 71–79.

Blanco, S., S. Romo, M. J. Villena & S. Martínez, 2003. Fish communities and food web interactions in some shallow Mediterranean lakes. Hydrobiologia, 506(509): 473–480.

Blindow, I., 1992. Long- and short-term dynamics of submerged macrophytes in two shallow eutrophic lakes. Freshwater Biology 28: 15–27.

Boix, D., S. Gascon, J. Sala, M. Martinoy, J. Gifre & X. D. Quintana, 2005. A new index of water quality assessment in Mediterranean wetlands based on crustacean and insect assemblages: the case of Catalunya (NE Iberian Peninsula). Aquatic Conservation: Marine and Freshwater Ecosystems 15: 635–651.

Bolle, H. J., 2003. Mediterranean climate. variability and trends. Springer Verlag, Berlin, 320 pp.

Boronat, L., R. Miracle & X. Armengol, 2001. Cladoceran assemblages in a mineralization gradient. Hydrobiologia 442: 75–88.

Brucet, S., D. Boix, R. López-Flores, A. Badosa, R. Moreno-Amich & X. D. Quintana, 2006. Size diversity and species diversity of zooplankton communities in fluctuant Mediterranean salt marshes. Estuarine Coastal and Shelf Science 67: 424–432.

Casado, S. & C. Montes, 1995. Guía de los lagos y humedales de España. Ed. J. M. Reyero, Madrid, 255 pp.

Chow-Frase, P., V. Lougheed, V. Thiec, B. Grosbie, L. Simser & J. Lord, 1998. Long-term response of the biotoc community to fluctuating water levels and changes in water quality in Cootes Paradise Marsh, a degraded costal wetland of Lake Ontario. Wetland Ecology and Management 6: 19–42.

Comín, F. A., X. Rodo & P. Comín, 1992. Lake Gallocanta (Aragon, NE Spain): a paradigm of fluctuations at different scales of time. Limnetica 8: 79–86.

Coops, H., M. Beklioglu & T. L. Crisman, 2003. The role of water-level fluctuation in shallow lake ecosystems-workshop conclusions. Hydrobiologia 506(509): 23–27.

Coveney, M. F., E. F. Lowe, L. E. Battoe, E. R. Marzolf & R. Conrow, 2005. Response of a eutrophic, shallow subtropical lake to reduced nutrient loading. Freshwater Biology 50: 1718–1731.

Davis, W. S. & T. P Simon, 1995. Biological Assessment and Criteria: tools for water resource planning and decision making. Lewis Publishers, Inc., 432 pp.

Economidis, P. S., A. I. Sinis & G. Stamou, 1998. Spectral analysis of exploited fish populations in Lake Koronia (Macedonia-Greece) during the years 1947–1983. Cybium 12: 151–159.

Fernández-Aláez, C., M. Fernández-Aláez & E. Bécares, 1999. Influence of water level fluctuations on the structure and composition of the macrophyte vegetation in two small temporary lakes in the northwest of Spain. Hydrobiologia 415: 155–162.

Fernández-Aláez, M., C. Fernández-Aláez, E. Bécares, M. Valentín, J. Goma & P. Castrillo, 2004. A two-year experimental study on nutrient and predator influences on food web constituents in a shallow lake of north-west Spain. Freshwater Biology 49: 1574–1592.

Gafny, S. & A. Gasith, 1999. Spatially and temporally sporadic appearance of macrophytes in the littoral zone of Lake Kinneret, Israel: taking advantage of a window of opportunity. Aquatic Botany 62: 249–267.

García-Criado, F., E. Bécares, C. Fernández-Aláez & M. Fernández-Aláez, 2005. Plant-associated invertebrates and ecological quality in some Mediterranean shallow lakes: implications fort he application of the EC Water Framework Directive. Aquatic Conservation: Marine and Freshwater Ecosystems 15: 31–50.

González-Sagrario, M. A., E. Jeppesen, J. Gomà, M. Søndergaard, J. P. Jensen, T. Lauridsen & F. Landkildehus, 2005. Does high nitrogen loading prevent clear-water conditions in shallow lakes at moderately high phosphorus concentrations? Freshwater Biology 47(3): 343–365.

Gyllström, M., L. A. Hansson, E. Jeppesen, F. García-Criado, E. Gross, K. Irvine, T. Kairesalo, R. Kornijow, M. R. Miracle, M. Nykänen, T. Noges, S. Romo, D. Stephen, E. Van Donk & B. Moss, 2005. The role of climate in shaping zooplankton communities of shallow lakes. Limnology & Oceanography 50(6): 2008–2021.

Hambright, K. D., A. Parparov & T. Berman, 2000. Indices of water quality for sustainable management and conservation of an arid region lake, Lake Kinneret (Sea of Galilee), Israel. Aquatic Conservation: Marine And Freshwater Ecosystems 10: 393–406.

Havey, L. D. D., 2000. Climate and global environmental change. Prentice Hall, 240 pp.

IUCN (the World Conservation Union), 2002. Report on climate change, water and wetlands: building linkages for their integrated management, 4–9.

Jeppesen, E., J. P. Jensen, M. Søndergaard, T. Lauridsen & F. Landikildehus, 2000. Trophic structure, species richness and biodiversity in Danish lakes: changes along a phosphorus gradient. Freshwater Biology 45: 201–213.

Jeppesen, E., M. Søndergaard, E. Kanstrup, B. Petersen, R. B. Eriksen, M. Hammershøj, E. Mortensen, J. P. Jensen & A. Have, 1994. Does the impact of nutrients on the biological structure and function of brackish and freshwater lakes differ? Hydrobiologia 275/276: 15–30.

Jeppesen, E., M. Sondergaard, J. P. Jensen, E. Mortensen, A. M. Hansen & T. Jorgensen, 1998. Cascading trophic interactions from fish to bacteria and nutrients after reduced sewage loading: an 18-year study of a shallow hypertrophic lake. Ecosystems 3: 250–267.

Jeppesen, E., M. Søndergaard & J. P. Jensen, 2003. Recovery from Eutrophication: Global perspectives. In Kumagai, M. & W. F. Vincent (eds), Freshwater management–global versus local perspectives. Springer, Tokyo, 135–152.

Jeppesen, E., M. Søndergaard, J. P. Jensen, K. E. Havens, O. Anneville, L. Carvalho, M. F. Coveney, R. Deneke, M. T. Dokulil, B. Foy, D. Gerdeaux, S. E. Hampton, K. Kangur, J. Köhler, S. Hilt, E. H. H. R. Lammens, T. L. Lauridsen, M. Manca, M. R. Miracle, B. Moss, P. Nõges, G. Persson, G. Phillips, R. Portielje, S. Romo, C. L. Schelske, D. Straile, I. Tatrai, E. Willén & M. Winder, 2005a. Lake responses to reduced nutrient loading—an analysis of contemporary long-term data from 35 case studies. Freshwater Biology 50: 1747–1772.

Jeppesen, E., M. Sondergaard, N. Mazzeo, M. Meerhoff, C. C. Branco, V. Huszar & F. Scasso, 2005b. Lake restoration and biomanipulation in temperate lakes: relevance for subtropical and tropical lakes. In Reddy M. V. (ed.), Restoration and management of tropical Eutrophic lakes. Oxford & IBH Publishing Co., Pvt, Ltd., New Delhi.

Kagalou, I., E. Papastergiadou, G. Tsimarakis & D. Petridis, 2003. Evaluation of the trophic state of Lake Pamvotis, Greece. A shallow urban lake. Hydrobiologia 506–509: 745–752.

Kagalou, I., G. Economidis, I. Leonardos & C. Papaloukas, 2006. Assessment of a Mediterranean shallow lentic ecosystem (Lake Pamvotis) using benthic community diversity: response to environmental parameters. Limnologica 36: 269–278.

Karapinar, B. 2005. Role of hydrology, sewage effluent diversion and fish on mass balance of nutrients in a system of shallow Lakes Eymir and Mogan. Msc. Thesis, Middle East Technical University, Ankara.

King, R. S. & C. J. Richardson, 2002. Evaluating subsampling approaches and macroinvertebrate taxonomic resolution for wetland bioassessment. Journal of North American Benthological Society 21(1): 150–171.

Kopacek, J., E. Stuchlik & R. Wright, 2005. Long term trends and spatial variability in nitrate leaching from alpine catchment lake ecosystems in the Tartra Mountains (Slovakia, Poland). Environmental Pollution 136: 89–101.

Lazzaro, X., 1997. Do the trophic cascade hypothesis and classical biomanipulation approaches apply to tropical lakes and reservoirs? Verhandlungen Internationale Vereinigung für Theoretiche und Angewandte Limnologie 26: 719–730.

Lazzaro, X., M. Bouvy, R. Ribeiro-Filho, V. Oliviere, L. Sales, A. Vasconcelos & M. Mata, 2003. Do fish regulate phytoplankton in shallow Eutrophic Northeast Brazilian Reservoirs. Freshwater Biology 48: 649–668.

Liao, H. S., C. L. Pierce & J. G. Larscheid, 2004. Consumption dynamics of the adult piscivorous fish community in Spirit Lake, Iowa. North American Journal of Fisheries Management 24(3): 890–902.

López-Flores, R., D. Boix, A. Badosa, S. Brucet & X. D. Quintana, 2006. Phytoplankton composition and size distribution in a confined Mediterranean salt marsh ecosystem. Marine Biology 149: 1313–1324.

Lowe, E. F., L. E. Battoe, M. F. Coveney, C. L. Schelske, K. E. Havens, E. R. Marzolf & K. Reddy, 2001. The restoration of Lake Apopka in relation to alternative stable states: an alternative view. Hydrobiologia 448: 11–18.

Meerhoff, M., C. Fosalba, C. Bruzzone, N. Mezzeo, N. Noordoven & E. Jeppesen, 2006. An experimental study of habitat choice by *Daphnia*: plants signal danger more than refuge in subtropical lakes. Freshwater Biology 51: 1320–1330.

Meerhoff, M., N. Mazzeo, B. Moss & L. Rodriguez-Gallego, 2003. The structuring role of free-floating versus submerged plants in a subtropical shallow lake. Aquatic Ecology 37(4): 377–391.

Mihaljevic, Z., M. Kerovec, I. Ternjej & A. Popijač, 2004. Long-term changes in the macroinvertebrate community structure of a shallow Mediterranean lake. Ekologia (Bratislava) 23: 421–429.

Mitraki, C, T. L. Crisman & G. Zalidis, 2004. Lake Koronia, Greece: Shift from autotrophy to heterotrophy with cultural eutrophication and progressive water-level reduction. Limnologica 34(1–2): 110–116.

Moss, B., 1998. The ecology of freshwaters, man and medium, past to future 3rd edn. Blackwell Science, Oxford, 557 pp.

Moss, B., 2001. The broads. The people's wetland. The new naturalist. Harper Collins Publishers, London, 392 pp.

Moss, B., 2007. Shallow lakes, the water framework directive and life: What should it all be about? Hydrobiologia 584: 381–394.

Moss, B., J. Madwewick & G. Phillips, 1996. A guide to the restoration of nutrient-enriched shallow lakes. W.W. Hawes, UK, 177 pp.

Moss, B., D. Stephen, C. Álvarez, E. Bécares, W. van de Bund, S. E. Collings, E. van Donk, E. de Eyto, T. Feldmann, C. Fernández-Aláez, M. Fernández-Aláez, R. J. M. Franken, F. García-Criado, E. M. Gross, M. Gyllström, L.-A. Hansson, K. Irvine, A. Järvalt, J.-P. Jensen, E. Jeppesen, T. Kairesalo, R. Kornijów, T. Krause, H. Künnap, A. Laas, E. Lill, B. Lorens, H. Luup, M. R. Miracle, P. Nõges, T. Nõges, M. Nykänen, I. Ott, W. Peczula, E. T. H. M. Peeters, G. Phillips, S. Romo, V. Russell, J. Salujõe, M. Scheffer, K. Siewertsen, H. Smal, C. Tesch, H. Timm, L. Tuvikene, I. Tonno, T. Virro, E. Vicente & D. Wilson, 2003. The determination of ecological status in shallow lakes—a tested system (ECOFRAME) for implementation of the European Water Framework Directive. Aquatic Conservation: Marine and Freshwater Ecosystems 13: 507–549.

Moss, B., D. Stephen, D. Balayla, E. Bécares, S.E. Collings, C. Fernández-Aláez, M. Fernández-Aláez, C. Ferriol, P. García, J. Gomá, M. Gyllström, L.-A. Hansson, J. Hietala, T. Kairesalo, M. R. Miracle, S. Romo, J. Rueda, V. Russell, A. Ståhl-Delbanco, M. Svennson, K. Vakkilainen, M. Valentin, W. J. Van de Bund, E. Van Donk, E. Vicente & M. J. Villena, 2004. Continental-scale patterns of nutrient and fish effects on shallow wetland lakes: synthesis of a pan-European mesocosm experiment. Freshwater Biology 49: 1633–1650.

Moustaka-Gouni, M., 1993. Phytoplankton succession and diversity in a warm monomictic, relatively shallow lake: Lake Volvi, Macedonia, Greece. Hydrobiologia 249: 33–42.

Naselli-Flores, L. & R. Barone, 2003. Steady-state assemblages in a Mediterranean hypertrophic reservoir. The role of *Microcystis* ecomorphological variability in maintaining an apparent equilibrium. Hydrobiologia 502: 133–143.

Naselli-Flores, L. & R. Barone, 2005. Water-level fluctuations in Mediterranean reservoirs: setting a dewatering threshold as a management tool to improve water quality. Hydrobiologia 548: 85–99.

Nõges, P., T. Nõges, L. Tuvikene, H. Smal, L. Sawomir, R. Kornijow, P. Woiciech, E. Bécares, F. García-Criado, C. Alvarez-Carrera, C. Fernández-Aláez, C. Ferriol, M. R. Miracle, E. Vicente, S. Romo, E. van Donk, W. Van de Bund, J. P. Jensen, E. Gross, L. A. Hansson, M. Gyllström, M. Nykänen, E. de Eyto, K. Irvine, D. Stephen, S. Collings & B. Moss, 2003. Factors controlling hydrochemical and trophic state variables in 86 shallow lakes in Europe. Hydrobiologia 506-509 (509): 51–58.

Osborne, P., 2005. Eutrophication of shallow tropical lakes. In ÓSullivan, P. & C. S. Reynolds (eds), The lakes handbook. Vol. II. Blackwell Science, Oxford, 278–299.

Quintana, X. D., R. Moreno-Amich & F. A. Comin, 1998. Nutrient and plankton dynamics in a Mediterranean salt marsh dominated by incidents of flooding. Part 1: differential confinement of nutrients. Journal of Plankton Research 20(11): 2089–2107.

Quintana, X. D., L. Suner, R. Trobajo, S. R. Gascon & R. Moreno-Amich, 1999. Nutrientes y renovación del agua en Aiguamolls de l'Empordà (NE de la Península Ibérica). Uso potencial de agua residual para la inundación de zonas húmedas. Limnetica 16: 55–65.

Romero, J., I. Kagalou, J. Imberger, D. Hela, M. Kotti, A. Bartzokas, T. Albanis, N. Evmiridis, S. Karkabounas, I. Papagiannis & A. Bithava, 2002. Seasonal water quality of shallow and eutrophic lake Pamvotis, Greece: implications for restoration. Hydrobiologia 474: 91–105.

Romo, S., M. R. Miracle, M. J. Villena, J. Rueda, C. Ferriol & E. Vicente, 2004. Mesocosm experiments on nutrient and fish effects on shallow lake food webs in a Mediterranean climate. Freshwater Biology 49: 1593–1607.

Romo, S., M. J. Villena, M. Sahuquillo, J. M. Soria, M. Gimenez, T. Alfonso, E. Vicente & M. R. Miracle, 2005. Response of a shallow Mediterranean lake to nutrient diversion: does it follow similar patterns as in northern shallow lakes? Freshwater Biology 50(10): 1706–1717.

Sala, J., S. Gascón, D. Boix, J. Gesti & X. D. Quintana, (2004). Proposal of a rapid methodology to assess the conservation status of Mediterranean wetlands and its application in Catalunya (NE Iberian Peninsula). Archiv des Sciences 57: 141–152.

Sánchez, E., C. Gallardo, M. A. Gaertner, A. Arribas & M. Castro, 2004. Future climate extreme events in the Mediterranean simulated by a regional climate model: a first approach. Global and Planetary Change 44: 183–180.

Scasso, F., N. Mazzeo, J. Gorga, C. Kruk, G. Lacerot, J. Clemente, D. Fabián & S. Bonilla, 2001. Limnological changes in a sub-tropical shallow hypertrophic lake during its restoration: two years of a whole-lake experiment. Aquatic Conservation: Marine & Freshwater Ecosystems 11: 31–44.

Scheffer, M., S. H. Hosper, M.-L. Meijer, B. Moss & E. Jeppesen, 1993. Alternative equilibria in shallow lakes. Trends in Ecology & Evolution 8: 275–279.

Sommaruga-Worgrath, S., K. Koinig, R. Schmidt, R. Sommaruga, R. Tessandri & R. Psenner, 1997. Temperature effects on the acidity of remote alpine lakes. Nature 387: 64–67.

Talling, J. F., 2001. Environmental controls on the functioning of shallow tropical lakes. Hydrobiologia 458: 1–8.

Talling, J. F. & R. J. Lamolle, 1998. Ecological dynamics of tropical inland waters. Cambridge University Press, Cambridge, 441 pp.

Tan, C. O. 2002. The roles of hydrology and nutrients in alternative equilibria of two shallow lakes of Anatolia, Lake Eymir and Lake Mogan: Using monitoring and modeling approaches. Msc. Thesis, Middle East Technical University, Ankara.

Tan, C. O. & M. Beklioglu, 2005. Catastrophic-like shifts in two Turkish lakes: a modeling approach. Ecological Modelling 183: 425–434.

Tan, C. O. & M. Beklioglu, 2006. Modelling complex nonlinear responses of shallow lakes to fish and hydrology using artificail neural networks. Ecological Modelling 196: 183–194.

Theocharis, V. S., 1989. Observations on the ecology and trophic habitat of fish populations in a eutrophic Ioannina Lake. In Armantrout, N. & R. Wolotira (eds), Proceedings of the World Fisheries Congress on: Condition of the World's Aquatic Habitats, 398–403 pp.

Trobajo, R., E. J. Cox & X. D. Quintana, 2004. The effects of some environmental variables on the morphology of *Nitzschia frustulum* (Bacillariophyta), in relation to its use as bioindicator. Nova Hedwigia 79: 433–445.

Tryfon, E. & M. Moustaka-Gouni, 1997. Species composition and seasonal cycles of phytoplankton with special reference to the nanoplankton of Lake Mikri Prespa. Hydrobiologia 351: 61–75.

Van Geest, G. J., H. Wolters, F. C. J. M. Roozen, H. Coops, R. M. M. Roijackers, A. D. Buijse & M. Scheffer, 2005. Water-level fluctuations affect macrophyte richness in floodplain lakes. Hydrobiologia 539: 239–248.

Vijverberg, J., M. Boersma, W. Vandensen, W. Hoogenboezem, E. Lammens & W. Mooıj, 1990. Seasonalvariatıon in the interactıons between piscivorous fısh, planktivorous fısh and zooplankton in a shallow eutrophıc lake. Hyrobiologia 207: 279–286.

WFD European Communities, 2000. Directive 2000/60/EC of the European Parliament and of the Council of 23 October 2000 establishing a framework for community action in the field of water policy. Official journal of the European Communities L327 (43): 1–72.

Wright, J. F., D. W. Sutcliffe & M. T. Furse, 2000. Assessing the biological quality of fresh waters. RIVPACS and other techniques. Freshwater Biological Association, 373 pp.

Hydrobiologia (2007) 584:327–335
DOI 10.1007/s10750-007-0585-x

SHALLOW LAKES

Changes in bacterial and ciliate densities with trophic status in Mediterranean shallow lakes

Ana Conty · Francisco García-Criado · Eloy Bécares

Abstract Ciliate and bacterial densities and their link with eutrophication were studied in fourteen shallow lakes in northwest Spain. Total phosphorus (TP) in these lakes varied between 30 μg l^{-1} and 925 μg l^{-1} and chlorophyll a concentration (chla) between 0.5 μg l^{-1} and 107 μg l^{-1}. Bacterial abundance ranged from 1×10^6 to 14×10^6 cells ml^{-1}, while ciliate abundance ranged from 0.6 cells ml^{-1} to 229 cells ml^{-1}. Lakes were classified into three trophic types from their TP and chla concentrations. Bacterial abundance was significantly correlated with trophic type, as well as with TP and with chla separately, whereas ciliate abundance was only correlated with chla. No significant relationship could be established between bacterial and ciliate abundance across the trophic gradient. A general pattern was observed in the ratios of bacterial abundance to TP and chla concentrations, of decreasing ratios with increases in the nutrient loading. This pattern was not found for ciliates. The dominant zooplankton group in 13 of the 14 lakes studied was Rotifera, which accounted for a mean of 71% of total zooplankton abundance (41% of zooplankton biomass). The positive correlation between bacteria and ciliates with this group, and the absence of any relationship with Cladocera suggest that top down control by cladocerans was weaker in our lakes than previously shown in northern European shallow lakes. Rotifers could be important predators of bacteria in the high-nutrient lakes of our study. Higher slopes of regressions on bacterial abundance towards the hypertrophic range indicate that top-down control was weaker in our lakes than in northern European shallow lakes.

Keywords Bacteria · Ciliates · Zooplankton · Nutrients · Shallow lakes · Mediterranean region

Guest editors: R. D. Gulati, E. Lammens, N. De Pauw & E. Van Donk
Shallow lakes in a changing world

A. Conty (✉) · F. García-Criado · E. Bécares
Faculty of Biology and Environmental Sciences, Area of Ecology, University of León, 24071 Leon, Spain
e-mail: degacf@unileon.es

Introduction

Since the introduction of the microbial food web concept (Pomeroy, 1974) and the microbial loop (Azam et al., 1983) were proposed, there have been many studies on the functioning of the pelagic food web that include the importance of bacteria and protozoans. Many of these studies pointed out the essential role of the microbial loop as a carbon source or sink and in the energy flow, and for recycling of nutrients to higher trophic levels. However, although microbial components of the food web have important implications in biogeochemical fluxes (Cotner & Biddanda, 2002) both in oceans and freshwaters (Pomeroy & Wiebe, 1988; Weisse et al., 1990; Berninger et al., 1991; Porter, 1996; Burns & Schallenberg, 1996; Šimek et al., 1998; Jürgens & Jeppesen, 2000), their composition, structure and regulation are largely unknown in Mediterranean shallow lakes.

Shallow lakes of the north of Spain differ from their northern European counterparts in both structure and function (Fernández-Aláez et al., 1999). Mediterranean shallow lakes have stronger water level fluctuations and they tend to be isolated from other lakes or rivers, particularly during the summer dry period, when bodies of water have smaller area and depth. These features might set them apart from the larger, less fluctuating and highly connected shallow lakes in the north of Europe. Moreover, the intensity of light and the high water temperature in the Mediterranean lakes could allow for a higher productivity and a higher nutrient turnover rate. In shallow fertile lakes, where nutrient constraints are less important and food supply can support large microbial population sizes, predation pressure may be the dominating structuring force for the microbial community (Jürgens et al., 1999). Jeppesen et al. (1997) observed that top-down control of large cladocerans on bacteria is weak at high nutrient concentrations because of the disappearance of these cladocerans due to higher predation on macrozooplankton with increasing trophic state and decreasing mean depth.

In this study we examined the relationships between ciliates and bacterioplankton, and other biological groups (e.g. phyto and zooplankton) to identify potential factors involved in regulating their trends in abundance across a trophic gradient. Our main hypothesis was that patterns of the top-down and bottom-up control of microbial components in Mediterranean lakes would be different from those found in northern lakes. It was also important to test whether evidence that bacterial abundance does not increase as rapidly as algal abundance along an increasing nutrient gradient (Gasol & Duarte, 2000; Biddanda & Cotner, 2001; Cotner & Biddanda, 2002), could be extrapolated to Mediterranean lakes.

Materials and methods

The study was conducted in 14 lakes situated in the northwest plateau of Spain (Castilla-León and Asturias regions, 880 m a.s.l. mean altitude) (Fig. 1). The lakes were small, with a mean area of 5 ha, (from 0.6 ha to 24 ha), and shallow, with a mean depth average of 1 m (from 0.33 m to 3 m). The mean water temperature during summer for those lakes above 1000 m a.s.l. was 15°C, and 20°C for those lakes below that altitude.

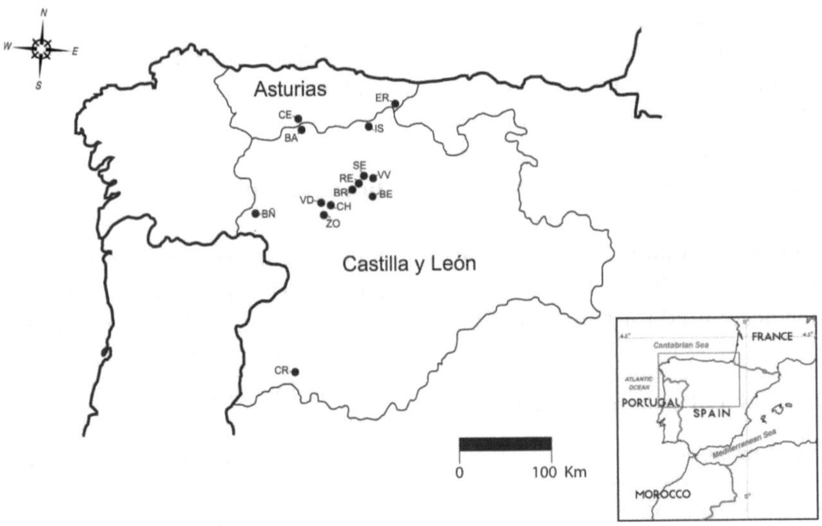

Fig. 1 Location of the lakes studied in Spain. (BA) Babia, (BE) Bercianos, (BÑ) Baña, (BR) Burgo, (CE) Cerveriz, (CH) Chozas, (CR) Cristo, (ER) Ercina, (IS) Isoba, (RE) Redos, (SE) Sentiz, (VD) Villadangos, (VV) Villaverde, (ZO) Zotes

Conductivities were from 10 μS cm^{-1} to 489 μS cm^{-1}, well within the exorheic range and thus comparable to northern, temperate lakes. Further data on morphological and physico-chemical variables are presented elsewhere (Moss et al., 2003; Fernández-Aláez et al., 2004; García-Criado et al., 2005).

Lakes were sampled in July and August 2000 and in July 2001, with a total of 30 samples. An integrated water sample of 20 l was collected in each lake with a one metre long tube sampler (6 cm in diameter) along transects from the shore to the centre of the lake, taking samples randomly of free water areas in the pelagic zone. Routine measurements were made of lake surface, mean depth, pH, oxygen, conductivity, temperature and Secchi disc depth, by standard instruments. Subsamples were taken from the integrated water sample for chemical and biological variables. Total phosphorus (TP) and total nitrogen (TN) subsamples were preserved in mercuric chloride (0.1% v/v final concentration) until further processing in the laboratory. TP concentrations were obtained using the molybdenum-blue method (Mackereth et al., 1978). TN concentrations were estimated by oxidation to nitrate (Grasshoff et al., 1983). For estimating the planktonic chlorophyll *a* (chla) concentration, subsamples were filtered through GF/F glass fibre filters and extracted in hot ethanol (Jeffrey & Humphrey, 1975; Arvola, 1981). The concentration of total (TSS) and volatile (VSS) suspended solids was estimated on GF/C glass fibre filters (Standard Methods, 1985). Subsamples for bacteria were fixed with formalin (4% final concentration), filtered onto black polycarbonate filters (Millipore) with a pore size of 0.2 μm and stained with the fluorescent dye 4′, 6′-diamidino-2-phenylindole (DAPI) (final concentration 0.4 μg ml^{-1}; modified from Porter & Feig, 1980). On each sample, between 400 and 500 bacteria were counted using a Zeiss Axioskop 2 epifluorescence microscope. Ciliates were counted by the Utermöhl sedimentation method (Utermöhl, 1958) on water subsamples preserved in lugol's iodine, using a Nikon Eclipse TE300 inverted microscope. Other biological variables (Percentage Volume Infested of macrophytes or PVI, total phytoplankton and zooplankton abundances) were sampled and counted using the methodology described in Moss et al. (2003).

Owing to the absence of normality, even after transformation, data were analysed with Spearman rank correlation and Kruskal–Wallis ANOVA. Chla and TP concentrations were used to classify samples according to their trophic status by means of joining clustering analysis (Euclidean distances and unweighted pair-group method using arithmetic averages or UPGMA). All statistical analyses were performed with STATISTICA 6 software (StatSoft, 2001).

Results

The trophic level of each sample was evaluated using TP and chla concentrations. These two variables were strongly correlated (Spearman $R = 0.7$, $n = 30$, $P = 0.000016$). Results of the cluster analysis applied to these two variables allowed us to distinguish three groups of samples or types of lakes. TP and chla values for each type are presented in Fig. 2. The least nutrient-rich group (type 1, 13 samples) was characterized by a mean TP concentration of 43 μg l^{-1}, (from 30 μg l^{-1} to 54 μg l^{-1}) and mean chla value of 8 μg l^{-1} (from 0.5 μg l^{-1} to 34 μg l^{-1}). The most fertile group (type 3, 7 samples) had a mean TP value of 426 μg l^{-1} (from 300 μg l^{-1} to 925 μg l^{-1}) and mean chla concentration of 64 μg l^{-1} (from 12 μg l^{-1} to 107 μg l^{-1}). The intermediate group

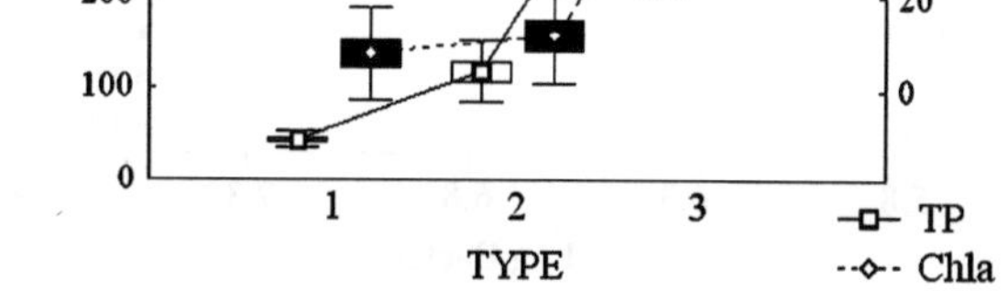

Fig. 2 Box plots of total phosphorus (TP) and chlorophyll a (chla) concentrations for each trophic type

(type 2, 10 samples) had a mean TP and chla concentration of 117 μg l^{-1} (from 77 μg l^{-1} to 176 μg l^{-1}) and 12 μg l^{-1} (from 1 μg l^{-1} to 30 μg l^{-1}), respectively. Differences between the three types of lakes were statistically significant for both TP (KW-H = 25.2, $P = 0.000003$) and chla (KW-H = 13.5, $P = 0.001$). (Kruskal–Wallis (KW-H) variance analysis test, Fig. 2).

Bacterial abundance in the lakes ranged from 1.1×10^6 to 14.6×10^6 cells ml^{-1}, (mean of 7.3×10^6 cells ml^{-1}). Mean ciliate abundance was 43 cells ml^{-1}, ranging from 1 cells ml^{-1} to 229 cells ml^{-1}. Bacteria and ciliate abundances were not correlated, neither for the whole set of data or separately in each trophic type (Fig. 3). Regression analysis between bacterial abundance and chla and TP gave the following significant relationships: $\log_{10}$ bacteria = 5.58 + 0.56 $\log_{10}$ TP ($r^2 = 0.31$, $P < 0.01$) and $\log_{10}$ bacteria = 6.35 + 0.39 $\log_{10}$ chla ($r^2 = 0.341$; $P < 0.01$). Ciliate abundances were significantly related only to chla: $\log_{10}$ ciliates = 0.07 + 0.97 $\log_{10}$ chla ($r^2 = 0.45$, $P < 0.0001$).

A general pattern was observed in the ratios of bacterial abundance to TP and chla concentrations, of decreasing ratios with increases in either TP or chla (Fig. 4). While the relationship between bacteria/TP and TP was almost linear ($\log_{10}$ bacteria/$\log_{10}$ TP = 6.03 – 1.29 $\log_{10}$ TP, $r^2 = 0.86$, $P < 0.001$, Fig. 4B), bacteria/chla vs. chla followed an almost perfect negative exponential ($\log_{10}$ bacteria/$\log_{10}$ chla = 5.4 + 7.79 exp (–5.07 $\log_{10}$ chla), $r^2 = 0.984$, Fig. 4A). A summary of significant Spearman correlations between main physico-chemical parameters and microbial components are shown in Table 1.

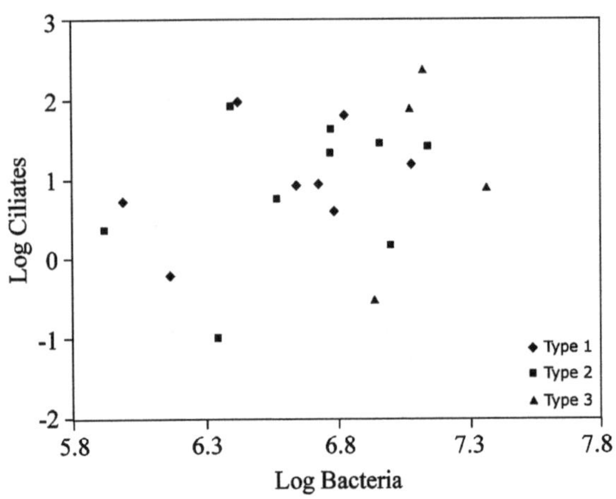

Fig. 3 Relationships between bacterial and ciliate abundances (cells ml^{-1}) for each trophic type

Bacteria and ciliate densities increased with the trophic status of the lakes. Highest densities were found in the type 3 lakes although differences were only significant for bacteria (KW-H = 5.56, $P = 0.05$; Fig. 5). Correlations between the abundances of these microbial loop components and other biological taxa were significant and positive for total zooplankton, rotifers and phytoplankton (Table 2), being the last two groups also correlated ($R = 0.65$, $P < 0.01$). Ciliates were also positively correlated with nauplii larvae and copepods.

With regard to zooplankton density, rotifers were the dominant group with a mean abundance of 71% of total zooplankton (ranging from 34% to 99%). Regarding zooplankton biomass, rotifers and cladocerans accounted for a mean of 41% and 35% of total zooplankton, respectively. Those lakes in which the percentage of cladocerans was higher showed less ciliate abundance ($R = -0.6$, $P < 0.01$) and phytoplankton biomass ($R = -0.5$, $P < 0.05$).

Discussion

The bacteria to phytoplankton ratio tends to be greater in oligotrophic waters than in eutrophic waters. Bacterial abundance increases more slowly than phytoplankton numbers as nutrient concentrations increase along a trophic gradient (see Cotner & Biddanda, 2002 for a review). The underlying cause for this relation is unclear, but there are some possible explanations regarding to our data. One of them argues that bacterial abundance is relatively greater than that of algae in oligotrophic systems owing to their greater ability for acquiring nutrients at very low ambient concentrations (Cotner & Wetzel, 1992). However, we did not find very low nutrient concentrations in our lakes as the trophic scale studied ranged from moderate to very high inorganic nutrient concentrations. Therefore, the higher bacterial abundances found in the mesotrophic

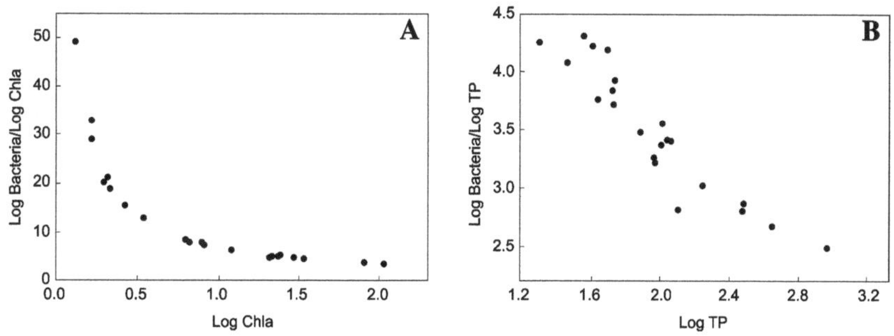

Fig. 4 Relationships between the $\log_{10}$ bacteria/$\log_{10}$ chla ratio and $\log_{10}$ chlorophyll a (**A**) and the $\log_{10}$ bacteria/$\log_{10}$ TP ratio and $\log_{10}$ TP (**B**)

Table 1 Spearman rank order correlation coefficients between ciliates and bacterial abundances and physico-chemical variables

	T[a]	pH	Mean depth	Conductivity	Secchi	TP	TN	Chla
Bacteria	0.56**		–0.5*	0.77**	–0.53*	0.51*	0.68**	0.59**
Ciliates		0.57**						0.68**
Bacteria/TP						–0.51*		
Ciliates/TP								
Bacteria/chla								–0.67**
Ciliates/chla								

TP: total phosphorous; TN: total nitrogen; Chla: chlorophyll a. Only significant values are shown (* $P < 0.05$; ** $P < 0.01$)

lakes relative to the hypertrophic ones might not be properly explained by a higher ability of bacteria in taking up inorganic nutrients. There is another bottom-up hypothesis based on carbon limitation. It is expected that the more fertile the systems are, the less proportion dissolved/particulate organic C (DOC:POC) they have (Biddanda et al., 2001, Cotner et al., 2002). This could mean that bacteria would have less readily consumable carbon in the more eutrophic lakes.

Another explanation about the observed pattern of decreasing bacterial abundances with nutrient loading, is the increasing predation pressure by bacterivores along the trophic gradient (Sanders et al., 1992). In shallow fertile lakes, where nutrient constraints are less important and food supply can support large microbial population sizes, predation pressure may be the dominating structuring force for the microbial community (Jürgens et al., 1999). We could then hypothesize that predation, as opposed to nutrient availability, is the main forcing factor in the more fertile of our lakes. Higher predation levels with increasing trophy can be caused by zooplankton (Jürgens et al., 1999), protozoa (Berninger et al., 1991; Šimek et al., 1995; Jezbera et al., 2003), or viruses (Weinbauer & Peduzzi, 1995; Bettarel et al., 2003). Experiments in northern shallow lakes (Jeppesen et al., 1992, 1996) have argued for an impact of zooplankton grazing on bacterioplankton, which is highest at intermediate phosphorus levels, when *Daphnia* is the dominant cladoceran. Studies of cladocerans have shown that they can consume a large size spectrum of resources from the smallest bacteria to many of the larger ciliates and rotifers (Pace & Vaqué, 1994) causing strong predation effects on all microbial components (Jürgens & Jeppesen, 2000). Although not statistically significant, large cladocerans tended to be more abundant in the lakes with the highest trophy, coinciding with the highest bacterial abundances of all lakes in our set. A number of shallow lakes in our region have higher cladoceran biomass than expected from their TP concentrations. Natural fish mortality in summer in some lakes may be caused by strong oxygen depletion. In other lakes, large water level fluctuations leading to isolation of the lake may end in their complete dry out. In these lakes, fish communities may be naturally absent, unless the lake has been stocked by anglers.

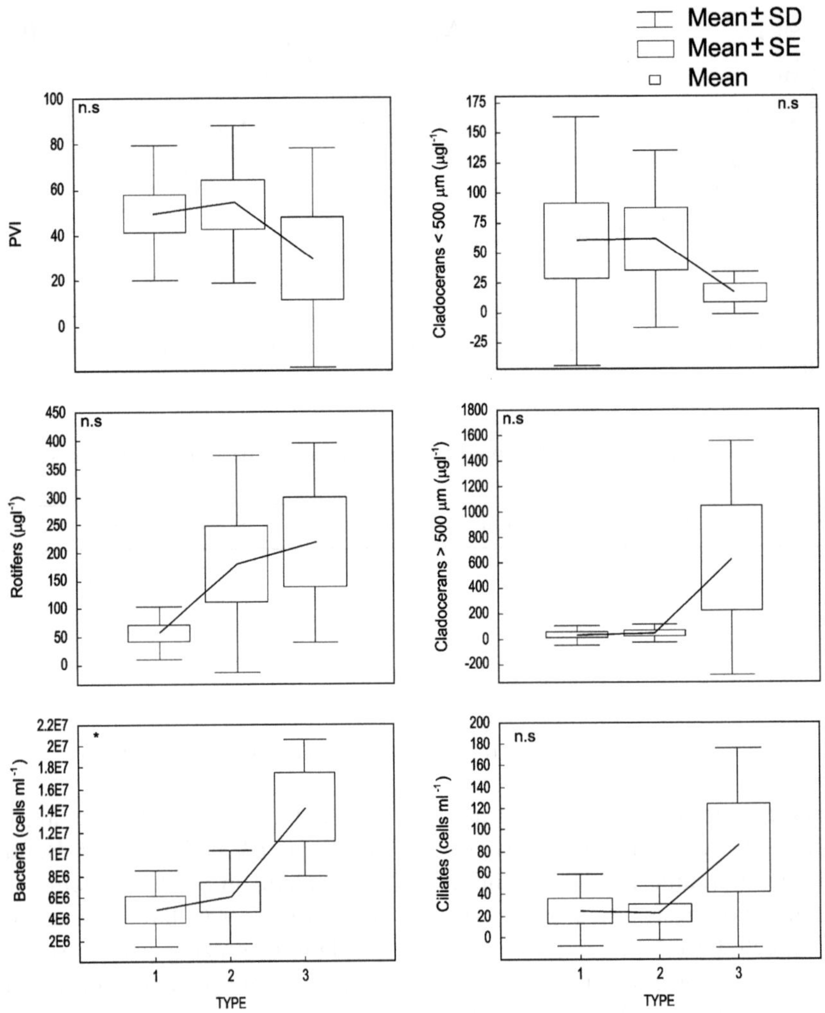

Fig. 5 Box plots of plant volume infested (PVI), bacteria, ciliates and other zooplankton groups for each trophic type. (n.s) Kruskal–Wallis test (KW-H) not significant. (*) KW-H = 5.96, $P < 0.05$

Rotifers may have been important predators of bacteria in the high-nutrient lakes in our study. The significant correlation between bacteria and rotifers suggests that bacteria can also be an important food resource for the rotifer species, together with phytoplankton. Due to the fact that microbial components may have reproductive rates which are considerably higher than mortality rates due to rotifer grazing (Arndt, 1993), rotifers should not be able to depress the microbial groups

Table 2 Spearman rank order correlation coefficients between ciliates and bacterial abundances and other biological groups depending on their abundance (A) or biomass (B)

	PVI	Phytoplankton	Zooplankton	*Cladocerans*		*Nauplii*	*Copepods*	*Rotifers*
				<500 µm	>500 µm			
Bacteria	n.s	0.59 (B)	0.515 (A)	n.s	n.s	n.s	n.s	0.52 (A)
Ciliates	n.s	0.57 (B) 0.59 (A)	0.564 (A)	n.s	n.s	0.59 (B)	0.487 (A)	0.57 (A) 0.63 (B)

Significant coefficients are shown ($P < 0.05$). n.s = not significant

via direct grazing impact, but they could be one of the responsible groups for the relative less abundance of bacteria along the trophic gradient.

Rotifers and ciliates have a considerably overlap in their food niches and in addition to direct predation and interference effects, exploitative competition might occur. In our study both groups were positively correlated with phytoplankton, which may suggest that they used the same resource. The fact that rotifers can serve as alternate food sources for crustacean zooplankton (Adrian, 1991; Jack & Gilbert, 1997) is another indirect positive effect on ciliates. This could explain why ciliates and rotifers showed a positive correlation. Cladocerans might exert a top-down control on ciliates as there was a negative relationship between the percentage of cladocerans and the ciliate abundance.

Regarding bacteria–ciliate relationship, some work demonstrates that the two groups are not mutually dependent (Riemann & Christoffersen, 1993; Gasol & Duarte, 2000). Results from shallow lakes in northern latitudes are not consistent, as some suggest a clear control of bacteria by ciliates (Auer et al., 2004), whereas others do not (Riemann & Christoffersen, 1993; Jeppesen et al., 1998). In our case no any relationship was found between these two groups. The highly positive correlations of ciliates with phytoplankton but the absence of any relationship with bacteria could indicate that ciliates are using the algae as their main food source in our lakes. Nevertheless, further studies about the ciliate community composition must be carried out in the future, in order to elucidate if there are changes in the proportion of bacterivorous/algivorous ciliate species that could lead to different predation pressure on bacterioplankton along the trophic gradient.

On the other hand, it is not known whether top-down control is stronger in Mediterranean than in north-Europe lakes. The slope of the relationship between bacteria and chla (bact/chla) may be used as a measure of algal grazing pressure, with higher slopes meaning lower predator pressure (Pace & Cole, 1994; see Gasol & Duarte, 2000 for a review). In our lakes, the slope was 0.39, higher than the slopes found in lakes in northern Europe (0.167 and 0.29 in Auer et al., 2004 and in Jeppesen et al., 1997, respectively). This difference between slopes in northern Europe and Mediterranean lakes is also consistent with other evidences suggesting that zooplankton predation pressure is lower in southern than in northern lakes (Moss et al., 2004). On the other hand, the slope could have been even higher in our lakes because of sediment resuspension. The studied lakes are strongly affected by wind and fish bioturbation, especially when plants are lost at the higher nutrient levels. Higher turbidity levels in the generally more exposed southern lakes could favour bacterial populations (higher bact/chla ratio). If the coupling with benthos was higher in the Mediterranean lakes than in northern lakes, intensified as plants are lost, bacteria would be relatively more important in our lakes even as chla concentrations increase, leading to the higher slopes observed for this relationship in Mediterranean than in northern shallow lakes.

No clear patterns of the relative importance of top-down and bottom-up control of bacterial production have been established for northern European shallow lakes. This makes comparisons with Mediterranean lakes difficult. The difficulty of this comparison is increased by the absence of data from the southern lakes. Further studies, also covering the potential roles of viruses and protozoan flagellates (Berninger et al., 1991; Wommack & Colwell, 2000), and data from a large number of lakes with wide ranges of TP and biological conditions are needed to give a conclusive test of the relative roles of top-down and bottom-up effects on the microbial loop in Mediterranean lakes.

Acknowledgements This work was funded by the European Community Project ECOFRAME (EVK1-CT-1999-00039). Authors want to thank Margarita Fernández-Aláez, Camino Fernández-Aláez, Cristina Álvarez, Saúl Blanco Lanza, Carlos F. Rodríguez and Benito Fuertes for their valuable assistance in the field and laboratory.

References

Adrian, R., 1991. The feeding behaviour of *Cyclops kolensis* and *C. vicinus* (Crustacea, Copepoda). Verhandlungen Internationale Vereinigung für Theoretische und Angewandte Limnologie 24: 2852–2863.

Arndt, H., 1993. Rotifers as predators on components of the microbial web (bacteria, heterotrophic flagellates, ciliates) – a review. Hydrobiologia 255/256: 231–246.

Arvola, L., 1981. Spectrophotometric determination of chlorophyll a and phaeopigments in ethanol extraction. Annales Botanici Fennici 18: 221–227.

Auer, B., U. Elzer & H. Arndt, 2004. Comparison of pelagic food webs in lakes along a trophic gradient and with seasonal aspects: influence of resource and predation. Journal of Plankton Research 26: 697–709.

Azam, F., T. Fenchel, J. G. Field, J. S. Gray, L. A. Meyer-Reil & T. F. Thingstad, 1983. The ecological role of water-column microbes in the sea. Marine Ecology Progress Series 10: 257–263.

Berninger, U. G., B. J. Finlay & P. Kuuppo-Leinikki, 1991. Protozoan control of bacterial abundances in freshwaters. Limnology and Oceanography 36: 139–147.

Bettarel, Y., T. Sime-Ngando, C. Amblard, J. F. Carrias & C. Portelli, 2003. Virioplankton and microbial communities in aquatic systems: a seasonal study in two lakes of differing trophy. Limnology and Oceanography 48: 810–822.

Biddanda, B., M. Ogdahl & J. Cotner, 2001. Dominance of bacterial metabolism in oligotrophic relative to eutrophic waters. Limnology and Oceanography 46: 730–739.

Burns, C. W. & M. Schallenberg, 1996. Relative impacts of copepods, cladocerans and nutrients on the microbial food web of a mesotrophic lake. Journal of Plankton Research 18: 683–714.

Cotner, J. B. & R. G. Wetzel, 1992. Uptake of dissolved inorganic and organic phosphorus compounds by phytoplankton and bacteria. Limnology and Oceanography 37: 232–243.

Cotner, J. B. & B. A. Biddanda, 2002. Small players, large role: microbial influence on biogeochemical processes in pelagic aquatic ecosystems. Ecosystems 2: 105–121.

Fernández-Aláez, C., M. Fernández-Aláez & E. Bécares, 1999. Influence of water level fluctuation on the structure and composition of the macrophyte vegetation in two small temporary lakes in the northwest of Spain. Hydrobiologia 415: 155–162.

Fernández-Aláez, M., C. Fernández-Aláez, E. Bécares, M. Valentín, J. Gomá & P. Castrillo, 2004. A two-year experimental study on nutrient and predator influences on food webs constituents in a shallow lake of north west Spain. Freshwater Biology 49: 1574–1592.

García-Criado, F., E. Bécares, C. Fernández-Aláez & M. Fernández-Aláez, 2005. Plant associated invertebrates and ecological quality in some Mediterranean shallow lakes: implications for the application of the EC Water Framework Directive. Aquatic Conservation: Marine and Freshwater Ecosystems 15: 31–50.

Gasol, J. M. & C. M. Duarte, 2000. Comparative analyses in aquatic microbial ecology: how far do they go?. FEMS Microbiology Ecology 31: 99–106.

Grasshoff, K., M. Ehrhardt & K. Kremling, 1983. Methods of seawater analysis. Verlag Chemie, Weinheim.

Jack, J. D. & J. J. Gilbert, 1997. Effects of metazoan predators on ciliates in freshwater plankton communities. Journal of Eukaryotic Microbiology 44: 194–199.

Jeffrey, S. W. & G. C. Humphrey, 1975. New spectrophotometric equations for the determination of chlorophylls-a, b, c1 and c2 in higher plants, algae and natural phytoplankton. Biochemie und Physiologie der Pflanzen 167: 191–194.

Jeppesen, E., O. Sotkjaer, M. Sondergaard & M. Erlandsen, 1992. Impact of a trophic cascade on heterotrophic bacterioplankton production in two shallow fish-manipulated lakes. Archiv fur Hydrobiologie 37: 219–231.

Jeppesen, E., M. Sondergaard, J. P. Jensen, E. Mortensen & O. Sortkjaer, 1996. Fish-induced changes in zooplankton grazing on phytoplankton and bacterioplankton: a long term study in shallow hypertrophic Lake Sobygaard. Journal of Plankton Research 19: 1605–1625.

Jeppesen, E., M. Erlandsen & M. Sondergaard, 1997. Can simple empirical equations describe the seasonal dynamics of bacterioplankton in lakes: an eight-year study in shallow hypertrophic and biologically highly dynamic lake Sobygard, Denmark. Microbial Ecology 34: 11–26.

Jeppesen, E., M. Sondergaard, J. P. Jensen, E. Mortensen, A. M. Hansen & T. Jorgensen, 1998. Cascading trophic interactions from fish to bacteria and nutrients alter reduced sewage loading: an 18-year study of a shallow hypertrophic lake. Ecosystems 1: 250–267.

Jezbera, J., J. Nedoma & K. Šimek, 2003. Longitudinal changes in protistan bacterivory and bacterial production in two canyon-shaped reservoirs of different trophic status. Hydrobiologia 504: 115–130.

Jürgens, K., O. Skibbe & E. Jeppesen, 1999. Impact of metazooplankton on the composition and population dynamics of planktonic ciliates in a shallow, hypertrophic lake. Aquatic Microbial Ecology 17: 61–75.

Jürgens, K. & E. Jeppesen, 2000. The impact of metazooplankton on the structure of the microbial food web in a shallow, hypertrophic lake. Journal of Plankton Research 22: 1047–1070.

Mackereth, F. J. H., J. Heron & J. F. Talling, 1978. Water analysis: some revised methods for limnologists. Freshwater Biological Association. Scientific publication No. 36.

Moss B., D. Stephen, C. Álvarez, E. Bécares, W. J. Van de Bund, S. E. Collings, E. Van Donk, E. De Eyto, T. Feldmann, C. Fernández-Aláez, M. Fernández-Aláez, R. J. M. Franken, F. García-Criado, E. M. Gross, M. Gyllström, L.-A. Hansson, K. Irvine, A. Järvalt, J.-P. Jensen, E. Jeppesen, T. Kairesalo, R. Kornijów, T. Krause, H. Künnap & A. Laas, et al., (49 authors), 2003. The determination of ecological status in shallow lakes – a tested system (ECO-FRAME) for implementation of the European Water Framework Directive. Aquatic Conservation: Marine and Freshwater Ecosystems 13: 507–549.

Moss, B., D. Stephen, D. M. Balayla, E. Bécares, C. Fernández-Aláez, M. Fernández-Aláez, C. Ferriol, P. García, J. Goma, M. Gyllström, L.-A. Hansson, J. Hietala, T. Kairesalo, M. R. Miracle, S. Romo, J. Rueda, V. Russell, A. Stahl-Delbanco, M. Svensson, K. Vakkilainen, M. Valentin, W. J. Van de Bund, E. Van Donk, E. Vicente & M. J. Villena, 2004.

Continental-scale patterns of nutrient and fish effects on shallow lakes: synthesis of a pan-European mesocosm experiment. Freshwater Biology 49: 1633–1649.

Pace, M. L. & J. J. Cole, 1994. Comparative and experimental approaches to top-down and bottom-up regulation of bacteria. Microbial Ecology 28: 181–193.

Pace, M. L. & D. Vaqué, 1994. The importance of *Daphnia* in determining mortality rates of protozoans and rotifers in lakes. Limnology and Oceanography 39: 985–996.

Pomeroy, L. R., 1974. The ocean's food web, a changing paradigm. BioScience 24: 499–504.

Pomeroy, L. R. & W. J. Wiebe, 1988. Energetics of microbial food webs. Hydrobiologia 159: 7–18.

Porter, K. G. & Y. S. Feig, 1980. The use of DAPI for identifying and counting aquatic microflora. Limnology and Oceanography 25: 943–948.

Porter, K. G., 1996. Integrating the microbial loop and the classic food chain into a realistic planktonic food web. In Polis, G. A. & K. O. Winemiller (eds), Food webs. Integration of Patterns & Dynamics. Ed Chapman & Hall, NY, 51–59.

Riemann, B. & K. Christoffersen, 1993. Microbial trophodynamics in temperate lakes. Marine Microbial Food Webs 7: 69–100.

Sanders, R. W., D. A. Caron & U. G. Berninger, 1992. Relationships between bacteria and heterotrophic nanoplankton in marine and fresh waters – an inter-ecosystem comparison. Marine Ecology Progress Series 86: 1–14.

Šimek, K., J. Bobková, M. Macek, J. Nedoma & R. Psenner, 1995. Ciliate grazing on picoplankton in a eutrophic reservoir during the summer phytoplankton maximum: a study at the species and community level. Limnology and Oceanography 40: 1077–1090.

Šimek, K., J. Armengol, M. Comerma, J. C. García, T. H. Chrzanowski, M. Macek, J. Nedoma & V. Straškrabová, 1998. Characteristics of protistan control of bacterial production in three reservoirs of different trophy. International Review of Hydrobiology 83: 485–494.

Standard Methods for the examination of water, wastewater. 1985. Ed. APHA, Washington, D.C. 16th edn.

Utermöhl, H, 1958. Zur vervollkommung der quantitativen phytoplankton-methodik. Mitteilung Internationale Vereinigung fuer Theoretische unde Angewandte Limnologie 9: 1–38.

Weinbauer, M. G. & P. Peduzzi, 1995. Significance of viruses versus heterotrophic nanoflagellates for controlling bacterial abundance in the northern Adriatic Sea. Journal of Plankton Research 17: 1851–1856.

Weisse, T., H. Müller, R. M. Pinto-Coelho, A. Schweizer, D. Springmann & G. Baldringer, 1990. Response of the microbial loop to the phytoplankton spring bloom in a large prealpine lake. Limnology and Oceanography 35: 781–794.

Wommack, K. E. & R. R. Colwell, 2000. Virioplankton: viruses in aquatic ecosystems. Microbial and Molecular Biology Reviews 64: 69–114.

Hydrobiologia (2007) 584:337–346
DOI 10.1007/s10750-007-0590-0

SHALLOW LAKES

The relationship between phytoplankton species dominance and environmental variables in a shallow lake (Lake Vrana, Croatia)

Marija Gligora · Anđelka Plenković-Moraj · Koraljka Kralj · István Grigorszky · Danijela Peroš-Pucar

Abstract The shallow Lake Vrana was studied over a 1-year period, special attention being paid to the phytoplankton. Phytoplankton was investigated monthly with respect to temporal variability of selected environmental factors. The regular annual development observed was in species contribution to total biomass rather than in seasonal changes in species composition. The assemblage was dominated by *Cosmarium tenue* Arch. and *Synedra* sp. In winter and in spring the phytoplankton assemblage was dominated by *Cosmarium tenue* and high contribution of *Synedra* sp. was observed during the summer and autumn. Results suggest that concentrations of inorganic nitrogen and phosphorus were critical in regulating phytoplankton biomass and species dominance.

Guest editors: R. D. Gulati, E. Lammens, N. De Pauw & E. Van Donk
Shallow lakes in a changing world

M. Gligora (✉) · A. Plenković-Moraj · K. Kralj
Division of Biology, Department of Botany, Faculty of Science, University of Zagreb, Rooseveltov trg 6, HR-10000 Zagreb, Croatia
e-mail: mgligora@zg.biol.pmf.hr

I. Grigorszky
Botanical Department, Debrecen University, P.O. Box 14, Debrecen 4010, Hungary

D. Peroš-Pucar
Public Health Institute Zadar, Kolovare 2, Zadar, Croatia

Keywords Phytoplankton · Shallow lake · Species dominance

Introduction

One of the two states in shallow lakes is a clear state rich in submerged vegetation (Scheffer, 1998). Macrophytes provide refuge for pelagic grazers (Stephen et al., 1998), support a diverse fish population, prevent sediment resuspension and are also one of several mechanisms that can suppress phytoplankton growth by reducing nutrient concentration in the lake water column (Jeppesen et al., 1997; Søndergaard et al., 2003; Fernández-Aláez et al., 2004). The rates at which organisms consume resources depend on the availability of such resources. Phytoplankton utilizes nutrients and becomes a component of the pelagic food web (Lampert & Sommer, 1997; Gliwicz, 2002; Stephen et al., 2004; Auer et al., 2004). Interannual variability in nutrient resources can play an important role in the determination of phytoplankton distribution and abundance (Sommer, 1987; Reynolds, 1997; Naselli-Flores, 2000). Phytoplankton is affected by external factors but is also influenced by the outcome of the competition itself (Kilham &

Tilman, 1979; Tilman et al., 1982; Huisman & Weissing, 1999). There were two working hypotheses during this investigation. First, that environmental variables influence phytoplankton species composition and biomass. Second, that phytoplankton seasonality and changes in species dominance are correlated with nutrient availability in a shallow vegetated lake.

Study site

Lake Vrana (Vransko jezero) is situated on the East Adriatic coast. It is a freshwater (salinity range 0.7–1.2‰), karstic cryptodepression connected at the south west to the Adriatic Sea by a narrow artificial channel. Lake Vrana is polymictic, shallow waterbody with a mean depth of 2 m. Catchments are characterized mainly by karst. Because of the karst morphology and the agricultural nature of the surrounding area, in conjunction with the presence of a bird sanctuary in the north west part, the lake receives considerable external nutrient loads. The lake is dominated by macrophytes, especially during the warm summer months. Macrophyte coverage, as visually estimated, exceeded 50% of the surface area during summer (dominants: *Najas marina* L., *Potamogeton pectinatus* L. and *Potamogeton perfoliatus* L.). The macrophyte association also includes *Phragmites australis* (Cav.) Trin. ex Steud., *Scirpus triqueter* L. Tavlen, *Myriophyllum spicatum* L., *Utricularia australis* R. Br, *Chara* sp., *Typha angustifolia* L. and *Potamogeton lucens* L., as accompanying species (M. Mrakovčić, pers. comm.).

Materials and methods

Phytoplankton was sampled monthly at four selected sites, in the period from January to December 2004. Samples for phytoplankton analysis were collected at 0.5 m depth, preserved in 2% formaldehyde (final concentration) and stored at 4°C. After 24 h sedimentation, 10 ml subsamples were analyzed. Cell counts were obtained with the inverted microscope following Uthermöhl's method (1958). A minimum of 400 settling units were counted, nanophytoplankton (<20 μm) at 1000× magnification in 15 randomly selected fields and microphytoplankton (>20 μm) cells in a transect at 400× magnification, providing a counting error of <10% (Lund et al., 1958). Measurements of the 20 randomly chosen cells were obtained by AxioVision software and biomass was calculated according to Rott (1981). Transparency was measured with a Secchi disc. Samples for water chemistry were taken simultaneously with phytoplankton samples. They were stored in refrigerated boxes and analyzed in the laboratory for nitrate, nitrite, ammonium, soluble reactive phosphorus, total phosphorus, chlorides, and pH, alkalinity, conductivity, oxygen and its saturation (APHA, 1995). Chlorophyll *a* concentrations were analyzed fluorometrically according to Method 445.0 after filtration onto Whitman GF/F glass filters and acetone extraction (Arar & Collins, 1997).

For statistical evaluation PRIMER 5 software package was employed (Clarke & Warwick, 2001). Principal-component analysis (PCA) of all physical and chemical variables (Table 1, excepted chl *a*) was used to identify the main environmental variables in the dataset. Distances between samples on the ordination attempt to match the corresponding dissimilarities in the environmental data. The correlation between the individual and combined environmental parameters and the phytoplankton biomass was analyzed using PRIMER 5 submodule BIO-ENV. A standard Spearman's rank correlation was used in this procedure. Only taxa contributing more than 5% to the total phytoplankton biomass were included in the analysis. All physical and chemical data as well as phytoplankton biomass were normalized for PCA and BIO-ENV. To reduce the influence of absolute biomass, the data were standardized to a 0–1 range (Jackson, 1993). The same transformation was used for all physical and chemical parameters since data were not normally distributed (except conductivity and chloride values). The draftsman plot of all pairwise combinations suggested that linearity was satisfied under this transformation, which is appropriate for these variables. The analysis of differences between samples and groups was restricted to non-parametric tests, Kruskal–Wallis and Mann–Whitney

Table 1 Variables average and *P*-values of Kruskal–Wallis and Mann–Whitney *U* tests on four groups identified by statistical analyses

	Unit	Group I avrg	Group II avrg	Group III avrg	Group IV avrg	Kruskal–Wallis test	Mann–Whitney test					
							Group I vs. Group II	Group I vs. Group III	Group I vs. Group IV	Group II vs. Group III	Group II vs. Group IV	Group III vs. Group IV
Temperature	°C	3.41	19.13	21.25	7.56	***<0.001	***<0.001	***<0.001	***<0.001	ns	***<0.001	***<0.001
Secchi depth	m	1.6	1.4	1.7	1.4	ns	ns	ns	ns	ns	ns	ns
pH		8	8.3	8.3	8.0	ns	ns	ns	ns	ns	ns	ns
Alkalinity	mg l^{-1}	157.66	140.10	89.58	134.40	***<0.001	ns	***<0.001	***<0.001	***<0.001	ns	***<0.001
Conductivity	(μS cm^{-1})	3006.25	2293.33	2402.33	2433.75	ns	**0.002	ns	ns	ns	ns	ns
O_2	mg l^{-1}	12.79	10.82	9.44	11.65	***<0.001	***0.001	***<0.001	***<0.001	**0.003	ns	***<0.001
Saturation	%	95.5	116.2	112.6	93.0	***<0.001	**0.003	***<0.001	ns	ns	***<0.001	***<0.001
NH_4–N (NH_4^+)	mg l^{-1}	0.18	0.03	0.018	0.17	***<0.001	***<0.001	***<0.001	ns	ns	***<0.001	***<0.001
NO_2–N (NO_2^-)	mg l^{-1}	0.007	0.018	0.005	0.006	***<0.001	***<0.001	ns	ns	***<0.001	***<0.001	ns
NO_3–N (NO_3^-)	mg l^{-1}	1.7	1.77	0.66	1.02	***<0.001	ns	***<0.001	ns	***<0.001	***0.001	**0.009
Cl^-	mg l^{-1}	887.5	639.17	657.08	620.63	**0.003	***<0.001	**0.003	**0.002	ns	ns	ns
Total P	mg l^{-1}	0.13	0.03	0.03	0.04	ns	ns	ns	ns	ns	ns	ns
PO_4–P (PO_4^{3-})	mg l^{-1}	0.022	0.017	0.007	0.012	ns	ns	ns	ns	***<0.001	ns	ns
Chl *a*	μg l^{-1}	1.65	4.90	1.31	1.15	***<0.001	***0.001	ns	ns	***<0.001	***0.001	ns

*** $P \leq 0.001$, ** $P < 0.01$, ns = non significant

U tests. Pearson's correlation coefficients, as well as Kruskal–Wallis and Mann–Whitney *U* tests, were calculated using Statistica 6.0 software.

Results

In the course of the year water temperature in Lake Vrana ranged between 1.6°C and 26.2°C and was greatly influenced by the Mediterranean climatic conditions. It was typically higher during the summer (Table 1, group III) and lower in the winter period (Table 1, group I). The water was alkaline (7.7–9.4), with very high conductivity (1638–3960 μS cm^{-1}). Nitrate concentrations varied between 0.5 mg l^{-1} and 2.34 mg l^{-1} reaching, a maximum in March and decreasing towards summer. The lowest nitrate concentration was recorded in September. Ammonium ranged from 0.004 mg l^{-1} to 0.3 mg l^{-1} with a decreasing trend from winter to spring-summer. Mean concentration of ammonium during summer (from July to September, Group III) was 0.018 mg l^{-1}. Following lower summer values, an increase in ammonium concentration was recorded. Nitrite concentrations ranged from 0.0013 mg l^{-1} to 0.023 mg l^{-1}. Nitrite concentrations were higher (mean 0.018 mg l^{-1}) during the spring period (from April to June, Group II) and decreased in the lake during the summer and autumn. Monthly variation of total phosphorus concentrations changed between 0.01 mg l^{-1} and 0.059 mg l^{-1} except in January, when a mean value of 0.25 mg l^{-1} was recorded. Soluble phosphorus ranged from 0.001 mg l^{-1} to 0.025 mg l^{-1} during the study period except for a peak in January (January mean value 0.04 mg l^{-1}). Dissolved inorganic N (as NH_4–N + NO_2–N + NO_3–N) to P (PO_4–P) ratio was higher than 30:1 during whole research period.

Among the 55 species identified during 2004, only 15 species contributed more than 5% to the total phytoplankton biomass. The assemblage was dominated by *Cosmarium tenue* Arch. and *Syneda* sp. The winter and spring assemblages were dominated by *Cosmarium tenue*. The assemblage was determined by a high contribution of *Synedra* sp. during summer and autumn (Fig. 1). The high species diversity in January was caused by the highest species number (24). Diversity significantly decreased from January to June and it was lowest in June, due to the great predominance of *Cosmarium tenue*.

The two axes in PCA analyses accounted for 60.3% of the cumulative variance in physicochemical data set with eigenvalues of 5.29 and 2.55, respectively (Table 2). Axes 3, 4 and 5 accounted for 29.5% and are not discussed further. PCA 1 was presented by temperature, in the positive direction ($r = 0.395$) and ammonium in the negative ($r = -0.382$), explaining 40.7% of the variance. PCA axis 2 was positively influenced by NO_2–N ($r = 0.490$) and NO_3–N ($r = 0.575$) (Fig. 2).

The BIO-ENV procedure presented phytoplankton assemblages and biomass in correlation with concentrations of inorganic nitrogen compounds, temperature and chlorides ($\rho_w = 0.389$).

According to the analyses, winter samples (January/March) were joined in Group I and correlated with high ammonium concentrations (Table 1) but also with high oxygen, chlorides and alkalinity values (Fig. 2). The dominant species during this period was *Cosmarium tenue*, accompanied by *Gonatozygon* sp. (Table 3). There was significant difference in phosphorus concentration between January and March ($p < 0.05$). Group II joined the spring samples (April/May/Jun) with the clear dominance of *Cosmarium tenue* (Table 3) and high concentrations of nitrites and nitrates, but lower ammonium values (Table 1). The temporal distribution of *Synedra* sp. mostly affected groups III and IV. According to the statistical analysis, Group III represents summer (July/August/September) assemblages dominated by *Synedra* sp. associated with low nitrogen concentrations, oxygen and chlorides (Fig. 2). In terms of species dominance two subgroups of Group III can be recognized. The first subgroup consists of July samples where *Synedra* sp. was a codominant species due to biomass, and *Gomphosphaeria* sp. was the dominant species accompanied by *Planktolyngbya contorta* (Lemmermann) Anagnostidis & Komárek and *Cosmarium tenue*. The second subgroup was represented by August and September assemblages with the clear dominance of *Synedra* sp.,

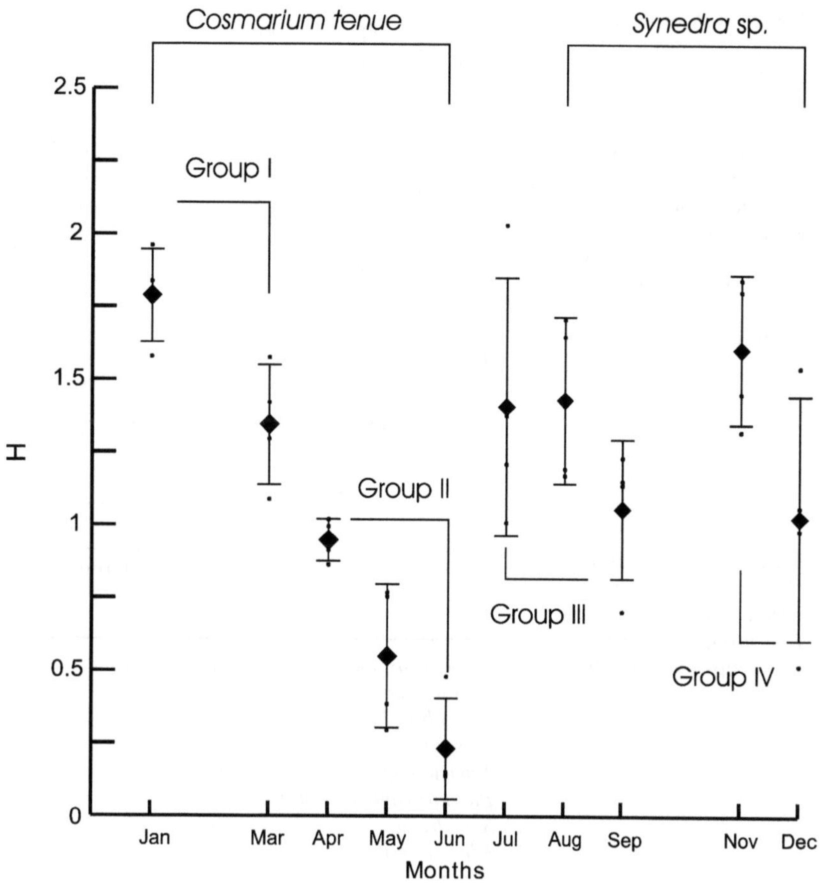

Fig. 1 Changes in Shannon–Wiener Diversity Index (H) through the investigation period (mean values and standard deviation on four sampling points) indicating groups identified by statistical analyses and dominant species

Table 2 Summary statistics for the first five axes of PCA performed on the environmental data set during the research period

PCA axis	1	2	3	4	5
Eigen values	5.29	2.55	1.76	1.27	0.80
% Variation	40.7	19.6	13.5	9.8	6.2
Cumulative % variation	40.7	60.3	73.5	83.5	89.7

and *Gomphospaheria* sp. as accompanying species (Table 3). There was a statistically significant difference in levels of nitrites and nitrates between subgroups within Group III ($p < 0.01$). During autumn, nitrogen compounds again reached higher values (Table 1) determining the assemblage in Group IV with the dominance of *Synedra* sp. and codominance of *Cosmarium tenue*.

Other species, *Pseudanabaena catenata* Lauterborn, *Chroococcus* sp., *Crucigenia tetrapedia* (Kirchner) W. & G. S. West, *Monoraphidium minutum* (Nägeli) Komárková-Legenerová, *Ankistrodesmus densus* Korshikov, *Koliella tenuis* (Nygaard) Hindák and *Navicula* sp. also typified the assemblages in Lake Vrana but contributed less to the total phytoplankton biomass during most of the investigation period, with no distinct seasonal differences in the assemblages.

Kruskal–Wallis and Mann–Whitney *U* tests performed on the four statistically identified groups confirmed significant differences in the nitrogen concentrations and also temperature, alkalinity, oxygen, concentrations of chlorides and chlorophyll *a* (Table 1).

Concentrations of chlorophyll *a* ranged from 0.05 μg l^{-1} to 9.54 μg l^{-1} and clearly correlated with concentration of nitrites ($r = 0.66$, $p < 0.001$). Biomass ranged from 0.94 mg l^{-1} to 51.60 mg l^{-1} and reached a higher concentration during winter period due to the high contribution of large benthic diatoms to the phytoplankton biomass. Total phytoplankton biomass showed a

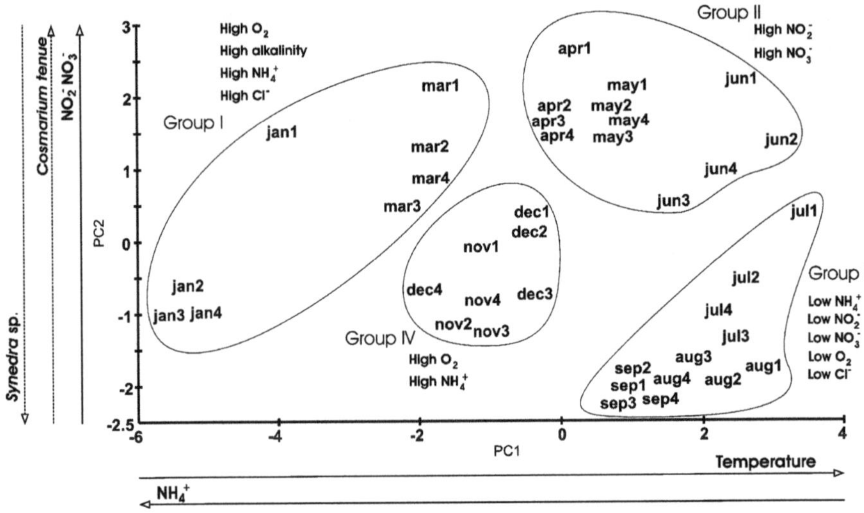

Fig. 2 PCA ordination of physico-chemical parameters showing sample grouping and general trends of those parameters for each group. Full line arrows indicate tendencies of main parameters associated with first and second PC axis. Dotted arrows are projection of phytoplankton species dominance on axis 2

Table 3 Contribution (%) of dominant and accompanying species in four groups and two subgroups identified by statistical analyses

	Group I	Group II	Group III	Subgroups III		Group IV
				Jul	Aug Sep	
Cosmarium tenue	50	87	10	15	7	29
Diatoma tenue	7					
Gomphosphaeria sp.			30	35	22	15
Gonatozygon sp.	14					
Navicula sp.	6				6	
Planktolyngbya contorta				10		
Synedra sp.	6		40	20	54	35
Ulnaria ulna	8					

great positive correlation with phosphates ($r = 0.54$, $p < 0.001$) but also a very high correlation with all nitrogen compounds, especially with nitrites ($r = 0.33$, $p < 0.05$).

Discussion

The present study indicates that nutrients are significant variables influencing phytoplankton biomass. The regular annual development observed was in species contribution to total biomass rather than in seasonal changes in species composition. According to the statistical analysis, it was the inorganic nitrogen compounds in the water column that mostly affected species dominance. PCA showed seasonality of nitrogen compounds, which was followed by changes in species dominance. High nitrogen concentrations during winter and especially during spring period coincided with the dominance and high biomass of *Cosmarium tenue*. Nitrogen compounds reached their lowest values during summer, and this coincided with change in species dominance. In this time of the year, a low concentration of inorganic nitrogen was followed by a high contribution of *Synedra* sp. to the phytoplankton biomass. This relationship was further supported by the species biomass–environmental correlations in statistical analyses.

According to the catchments morphology and the characteristics of the surrounding area it can be assumed that Lake Vrana receives a high nutrient input. Despite the high external nutrient loads, phosphorus was present in much lower concentrations than nitrogen and might have a limited effect on phytoplankton total biomass. The Redfield ratio is one of the methods frequently used to identify potentially limiting nutrients, and phosphorus limitation in Lake Vrana is evident considering the high concentration values of nitrogen in relation to phosphorus in water column (Redfield, 1958). The influence of phosphorus on the phytoplankton assemblage is shown by the high correlation between the total phytoplankton biomass and phosphates. Although not supported by statistical analyses, Table 1 suggests that the lowest values of phosphates were recorded in group III when the concentration of soluble reactive phosphorus was on average below 0.01 mg l^{-1} during three summer months. Apart from the external nutrient load in lake, the internal load, especially of phosphorus, is very important in shallow lakes. Although the phosphorus and nitrogen content in sediment were not determined in this study, the phosphorus exchange process over the sediment-water interface can play important role in the phosphorus budget in shallow lakes. There is considerable evidence for phosphorus sediment immobilization resulting in the phosphorus limitation condition in the water column (Scheffer, 1998). Chemical adsorption of phosphorus in sediment may be caused by a number of processes, which can be critical to phosphorus release and can apparently cause phosphorus limitation for phytoplankton. Water that is rich in calcium and carbonate, such as karstic Lake Vrana, may buffer the pH and lead to a decrease in phosphorus release from the sediment (Scheffer, 1998). In Lake Vrana, a high concentration of nitrate was detected. It has been suggested that a high concentration of nitrate may buffer the redox potential of the sediment surface, preventing a release of iron bound phosphorus in the same way as oxygen (Scheffer, 1998). Also, at high turbulence, which is frequently present in the shallow Lake Vrana, the sediment surface becomes oxygenated and phosphorus can be immobilized by iron again. The great accumulation of phosphorus in sediment found in shallow lakes does not occur with nitrogen (Jensen et al., 1991). Apart from chemical absorption macrophytes can change nutrient retention and its effect on the nitrogen is more consistent than on phosphorus. Spring and summer concentration of nitrogen decreased in the presence of high macrophyte coverage. The high uptake of nitrogen by macrophytes apparently caused a nitrogen limitation for the phytoplankton (Van Donk et al., 1993). The low concentration of nutrients during the summer period in Lake Vrana followed the dynamics of macrophyte coverage. It can be assumed that aquatic macrophytes in Lake Vrana, due to high coverage during summer, affect nutrient dynamics in the lake and have an impact on phytoplankton biomass (Scheffer et al., 1993; Jeppesen et al., 1997) and changes in the species dominance. *Cosmarium tenue* was dominated as long as high concentrations of inorganic nitrogen were recorded in water column. The summer dominant, *Synedra* sp., was present in the lake with high biomass when nitrogen was exhausted from the water column and it is evident that phosphorus limitation during the summer period for *Cosmarium tenue* was not limiting for *Synedra* sp. (Sommer, 1987).

Results showed a range of diversity values in this stable, well-mixed system, inconstant with respect to chemical features, and temporarily disturbed. Low diversity values were strongly affected by the most abundant species. Individual species adaptations to an environment suitable to host steady-state assemblages (Mischke & Nixdorf, 2003) resulted in low diversity values, especially during June. From January to June, diversity decreased steeply, affected by the predominance of only one species (Padisák, 1993), *Cosmarium tenue*. Diversity was reduced to minimal levels by competitive exclusion or some other biotic interaction that can result in steady state assemblages (Naselli-Flores et al., 2003; Rojo & Álvarez-Cobelas, 2003). Overlooking the fact that steady state phases are not frequently attained in phytoplankton succession (Padisák et al., 2003), and the monthly (instead of weekly) sampling dynamics during this investigation (Rojo & Álvarez-Cobelas, 2003), theoretically, this phytoplankton assemblage, is close to establishing equilibrium (Sommer et al., 1993), as was manifested by the monodominance of *Cosmarium tenue* (Padisák et al., 2003), with more than 80% in total biomass over 3 months. However, external factors may prevent the establishment of equilibrium by favoring the dominance of a new species (Reynolds, 1993). Changes in species dominance and high contribution of *Synedra* sp. coincided with decreases in concentrations of inorganic nitrogen. The same values of nitrogen yielded greater diversity values. Considering diversity and disturbance as indirectly linked (Reynolds et al., 1993) and disturbance as a force that can be measured as a reaction, it could be suggested that the sufficient intensity of disturbance (Sommer et al., 1993) was caused by a sudden event of low concentration of nitrogen in the lake water.

According to BIO-ENV analysis, assemblages were also affected by temperature and chloride concentrations. High concentrations of chlorides due to the artificial connection between the Lake and Adriatic Sea were recorded in winter, while lower and stable concentrations were observed in summer. During stable summer temperature conditions, changes in environmental variables, such as nitrogen availability, produced noticeable shifts in species dominance.

The phytoplankton assemblage in Lake Vrana can be characterized in general as nanoplankton. Cell size was found to be important in the determination of predominant species but could not be considered the decisive factor for species selection and their growth kinetics (Suttle et al., 1987; Sommer, 1989). Species with small cell-sizes, because of volume/surface area ratio and cell shape, benefit from low nutrient concentration and from phosphorus limited conditions and outcompete larger-sized species (Smith & Kalff, 1982; Grover, 1989). The species *Cosmarium tenue* in Lake Vrana is characterized by a large mucous cell envelope. The presence of an extracellular mucous envelope is considered evidence indicative of several functions suggested by Coesel (1994), Decho (1990) and Whitton (1967). According to Coesel (1994) mucilage sheets of desmids might be indicatively related to the capture of scarce nutrients. Apart from such results there are contradictory experimental data with no clear indication of the storage role of the extracellular mucous envelope (Spijkerman & Coesel, 1998). It can be deduced from Table 1 and statistical analyses that species assemblage in Lake Vrana is naturally selected according to the temperature, salinity conditions, high conductivity, pH values and availability of CO_2 (Reynolds, 1997) but also affinity for the resource and species storage capacity (Spijkerman & Coesel, 1996a, b). According to Sommer et al. (1993) phosphorus limitation of species with high requirements becomes possible if SRP concentration falls below 10 $\mu g\ l^{-1}$ and intracellular stores have been depleted. Such conditions in Lake Vrana were evident occasionally during the year and for a period of 3 months during summer. Species with low phosphorus demand, such as pennate diatoms, would only become limited at undetectable SRP concentrations (Sommer et al., 1993). It can be assumed that the species dominance in a given lake is determined by competition for nutrients (Sommer, 1989). The species *Synedra* sp. was competitively superior for phosphorus at a low concentration of nitrogen and it became a more productive summer species. This ability gives a significant advantage to these organisms in summer and *Synedra* sp. developed a larger population than *Cosmarium tenue*. It was able to compete for phosphorus and reach dominance as long as *Cosmarium tenue* was excluded by nitrogen depletion (Lampert & Sommer, 1997).

In conclusion, the present study shows that in the vegetated shallow Lake Vrana the concentration of nitrogen, like that of phosphorus, may be important in the determination of phytoplankton dominants. The phytoplankton annual succession in Lake Vrana seems to be mainly controlled by nutrients. Nutrient resources seem to be critical in regulating phytoplankton species dominance and total species biomass in Lake Vrana. In this way, the phytoplankton community might be regulated by external factors and also by competitive interactions between the dominant species. The influence of nutrients on the phytoplankton in Lake Vrana was discussed in the specific lake environment and should be considered a factor controlling phytoplankton assemblages, biomass and changes in dominance overlooking other environmental features of the Lake.

Acknowledgments We would specially like to thank Professor M. Mrakovčić, who initiated the study of Lake Vrana. We, are also we are grateful to the Croatian Ministry of Science, Education and Sport, and the Hungarian Scholarship Board for financial support. Special gratitude goes to the reviewers whose comments helped us improve this paper.

References

American Public Health Association (APHA), 1995. Standard Methods for the Examination of Water and Waster Water, 19th edn. APHA, Washington.

Arar, E. J. & G. B. Collins, 1997. In Vitro Determination of Chlorophyll *a* and Pheophytin a in Marine and Freshwater Algae by Fluorescence. National

Exposure Research Laboratory, Office of Research and Development, US Environmental Protection Agency, Cincinnati, Ohio.
Auer, B., U. Elzer & H. Arndt, 2004. Comparison of pelagic food webs in lakes along a trophic gradient and with seasonal aspects: influence of resource and predation. Journal of Plankton Research 26: 637–709.
Clarke, K. R. & R. M. Warwick, 2001. Change in Marine Communities: An Approach to Statistical Analysis and Interpretation. Plymouth Marine Laboratory, Plymouth.
Coesel, P. F. M., 1994. On the ecological significance of a cellular mucilaginous envelope in planktic desmids. Algological Studies 73: 65–74.
Decho, A. W., 1990. Microbial exopolymer secretions in ocean environments: their role(s) in food webs and marine processes. Oceanography and Marine Biology – An Annual Review 28:73–153.
Fernández-Aláez, M., C. Fernández-Aláez, E. Bécares, M. Valentín, J. Goma & P. Castrill, 2004. A 2-year experimental study on nutrient and predator influences on food web constituents in a shallow lake of north-west Spain. Freshwater Biology 49: 1574–1592.
Gliwicz, Z. M., 2002. On the different nature of top–down and bottom–up effects in pelagic food webs. Freshwater Biology 47: 2296–2312.
Grover, J. P., 1989. Influence of cell shape and size on algal competitive ability. Journal of Phycology 25: 402–405.
Huisman, J. & F. J. Weissing, 1999. Biodiversity of plankton by species oscillations and chaos. Nature 402: 407–410.
Jackson, D. A., 1993. Multivariate analysis of benthic invertebrate communities: the implication of choosing particular data standardizations, measures of associations, and ordination methods. Hydrobiologia 268: 9–26.
Jensen, J. P., P. Kristensen & E. Jeppensen, 1991. Relationships between nitrogen loading and in-lake nitrogen concentrations in shallow Danish lakes. Verhandlungen Internationale Vereinigung für theoretische und angewandte Limnologie 24: 201–204.
Jeppesen, E., J. P. Jensen, M. Søndergaard, T. Laurisden, L. J. Pedersen & L. Jensen, 1997. Top–down control in freshwater lake: the role of nutrient state, submerged macrophytes and water depth. Hydrobiologia 342/343: 151–164.
Kilham, P. & D. Tilman, 1979. The importance of resource competition and nutrient gradients for phytoplankton ecology. Ergebnisse der Limnologie 13: 110–111.
Lampert, W. & U. Sommer, 1997. Limnoecology: The Ecology of Lakes and Streams. Oxford University Press, New York.
Lund, J. W. G., C. Kipling & E. D. Le Cren, 1958. The inverted microscope method of estimating algal numbers and statistical basis of estimations by counting. Hydrobiologia 11: 6–21.
Mischke, U. & B. Nixdorf, 2003. Equilibrium phase conditions in shallow German lakes: how cyanoprokaryota species establish a steady state phase in late summer. Hydrobiologia 502: 123–132.
Naselli-Flores, L., 2000. Phytoplankton assemblages in twenty-one Sicilian reservoirs: relationships between species composition and environmental factors. Hydrobiologia 424: 1–11.
Naselli-Flores, L., J. Padisák, M. T. Dokulil & I. Chorus, 2003. Equilibrium/steady-state concept in phytoplankton ecology. Hydrobiologia 502: 395–403.
Padisák, J., 1993. The influence of different disturbance frequencies on the species richness, diversity and equitability of phytoplankton in shallow lakes. Hydrobiologia 249: 135–156.
Padisák, J., G. Borics, G. Fehér, I. Grigorszky, I. Oldal, A. Schmidt & Z. Zámbóné-Doma, 2003. Dominant species, functional assemblages and frequency of equilibrium phases in late summer phytoplankton assemblages in Hungarian small shallow lakes. Hydrobiologia 502: 157–168.
Redfield, A. C., 1958. The biological control of chemical factors in the environment. American Scientist 46: 205–221.
Reynolds, C. S., 1993. Scales of disturbance and their role in plankton ecology. Hydrobiologia 249: 157–171.
Reynolds, C. S., 1997. Vegetation processes in the pelagic: a model for ecosystem theory. Oldendorf. Ecology Institute.
Reynolds, C. S., J. Padisák & U. Sommer, 1993. Intermediate disturbance in the ecology of phytoplankton and the maintenance of species diversity a synthesis. Hydrobiologia 249: 183–188.
Rojo, C. & M. Álvarez-Cobelas, 2003. Are there steady state phytoplankton assemblages in the field? Hydrobiologia 502: 3–12.
Rott, E., 1981. Some results from phytoplankton counting intercalibration. Schweizerische Zeitschrift für Hydrologie 43: 35–62.
Scheffer, M., 1998. Ecology of Shallow Lakes. Chapman & Hall, London.
Scheffer, M., S. H. Hosper, M. L. Meijer, B. Moss & E. Jeppesen, 1993. Alternative equilibria in shallow lakes. Trends in Ecology and Evolution 8: 275–279.
Smith, R. E. H. & J. Kalff, 1982. Size-dependent phosphorus uptake kinetics and cell quota in phytoplankton. Journal of Phycology 18: 275–284.
Sommer, U., 1987. Factors controlling the seasonal variation in phytoplankton species composition. A case study for a deep, nutrient rich lake (Lake Constance). Progress in Phycological Research 5: 122–178.
Sommer, U., 1989. The role of competition for resources in phytoplankton succession. In Sommer, U. (ed.), Plankton Ecology: Succession in Plankton Communities. Springer-Verlag, New York: 57–107.
Sommer U., A. Gaedeke & A. Schweizer, 1993. The first decade of oligotrophication in Lake Constance. II: the response of phytoplankton taxonomic composition. Oecologia 93: 276–284.
Søndergaard, M., J. P. Jensen & E. Jeppesen, 2003. Role of sediment and internal loading of phosphorus in shallow lakes. Hydrobiologia 506–509: 135–145.

Spijkerman, E. & P. F. M. Coesel, 1996a. Competition for phosphorus among planktonic desmid species in continuous flow culture. Journal of Phycology 32: 939–948.

Spijkerman, E. & P. F. M. Coesel, 1996b. Phosphorus uptake and growth kinetics of two planktonic desmid species. European Journal of Phycology 31: 53–60.

Spijkerman, E. & P. F. M. Coesel, 1998. Alkaline phosphatase activity in two planktonic desmid species and the possible role of an extracellular envelope. Freshwater Biology 39: 503–513.

Stephen, D., D. M. Balayla, S. E. Collings & B. Moss, 2004. Two mesocosm experiments investigating the control of summer phytoplankton growth in a small shallow lake. Freshwater Biology 49: 1551–1564.

Stephen, D., B. Moss & G. Phillips, 1998. The relative importance of top–down and bottom–up control of phytoplankton in a shallow macrophyte-dominated lake. Freshwater Biology 39: 699–713.

Suttle, C. A., J. G. Stockner & P. J. Harrison, 1987. Effects of nutrient pulses on community structure and cell size of a freshwater phytoplankton assemblage in culture. Canadian Journal of Fisheries and Aquatic Sciences 44: 1768–1774.

Tilman, D., S. S. Kilham & P. Kilham, 1982. Phytoplankton community ecology: the role of limiting nutrients. Annual Review of Ecology and Systematics 13: 349–372.

Utermöhl, H., 1958. Zur Vervollkommnung der quantitativen Phytoplankton Methodik. Mitteilungen. Internationale Vereiningung fuer Theoretische und Angewandte Limnologie 9: 1–38.

Van Donk, E., R. D. Gulati, A. Iedema & J. Meulemans, 1993. Macrophyte-related shifts in the nitrogen and phosphorus contents of the different trophic levels in a biomanipulated shallow lake. Hydrobiologia 251: 19–26.

Whitton, B. A., 1967. Phosphate accumulation by colonies of *Nostoc*. Plant and Cell Physiology 8: 293–296.

Hydrobiologia (2007) 584:347–359
DOI 10.1007/s10750-007-0579-8

SHALLOW LAKES

Comparative biodiversity of crustaceans and aquatic insects from various water body types in coastal Mediterranean wetlands

Dani Boix · Jordi Sala · Stéphanie Gascón · Mònica Martinoy · Jaume Gifre · Sandra Brucet · Anna Badosa · Rocío López-Flores · Xavier D. Quintana

Abstract Coastal wetlands are characterized by a high biodiversity. At the same time, biodiversity is one of the main criteria used to establish protection policy priorities, or to propose management actions. In this study, crustacean and aquatic insect species richness in the Empordà wetlands was investigated. These two groups contribute in an important way to the total biodiversity, and still they are seldom taken into account in the management of natural areas. Representative samples (38 points) of all aquatic water body types in the Empordà wetlands were taken monthly (dip net with 250 μm mesh). Sampling was carried out between 1996 and 2000, but until present, only qualitative data have been extracted. A rich fauna of 125 crustacean taxa and 295 aquatic insect taxa were found. Some environments were characterized by low richness and high singularity (isolated artesian freshwater springs), some by high richness and high singularity (estuarine waters, brackish and meso-eutrophic freshwater wetlands), and others by low richness and low singularity (hypertrophic freshwater wetlands and hyperhaline wetlands). Factors determining singularity and richness are discussed. Comparison with crustacean richness of other western Mediterranean wetlands showed a similar high species richness in our study sites, probably due to high spatial heterogeneity of these areas.

Keywords Empordà wetlands · Species richness · Singularity · Similarity · Shannon-Wiener diversity · Evenness

Introduction

Wetlands, marshes and other permanent or temporary shallow waters have experienced an increasing interest from conservation biologists during recent years. Several studies stress their ecological importance, based on the fact that they contain species or environments that are becoming rarer, but also because they show a high species richness (Giudicelli & Thiéry, 1998; Gopal & Junk, 2000; Boix et al., 2001). The high species richness is due to the transitional nature of wetlands, characterized by the presence of both resident and transient species, the latter coming from adjacent

Guest editors: R. D. Gulati, E. Lammens, N. De Pauw & E. Van Donk
Shallow lakes in a changing world

D. Boix (✉) · J. Sala · S. Gascón · S. Brucet · A. Badosa · R. López-Flores · X. D. Quintana
Institute of Aquatic Ecology and Department of Environmental Sciences, Faculty of Sciences, campus Montilivi, University of Girona, 17071 Girona, Spain
e-mail: dani.boix@udg.es

M. Martinoy · J. Gifre
Mosquito Control Service of Badia de Roses and Baix Ter, Girona, Spain

terrestrial or aquatic environments or migrating from distant environments (Lefeuvre et al., 2003). However, although the importance of these ecosystems has been recognized in the European Directive, which considers them as priority habitats (92/43/CEE), they are still under the threat of degradation and loss (e.g., Brown, 1998; Gibbs, 2000).

Biodiversity is one of the main criteria used in the protection of wetlands (Ramsar Convention Bureau, 2005). However, the knowledge of the composition of aquatic invertebrate species in these environments is relatively poor (Britton & Crivelli, 1993), and most of the management efforts are focused on the conservation of a small number of species, mainly water birds (Britton, 1982; Mocci, 1983). Nevertheless, at present other criteria are taken into account (Ramsar Convention Bureau, 2005). On the other hand, recent trends in water management have been focused on integrated catchment management (Everard, 1999; Verdonschot, 2000), which requires a broadening of biodiversity knowledge to promote its conservation. From this perspective, comparing the biodiversity characteristics of the different water body types found in the catchment area, it is necessary to assess the contribution of each type to the total biodiversity (Williams et al., 2003). This is especially important in places where several water body types are present in small areas, such as in the Mediterranean wetlands (Britton & Crivelli, 1993). However, such an approach is not common, and comparative studies of biodiversity of several wetland types are scarce (e.g., Forès et al., 1986; Valladares et al., 2002).

The Empordà wetlands present a wide range of water body types, due to gradients of salinity, water permanence and nutrients. The aim of this paper is to analyze the relationship between environmental gradients and biodiversity in a group of Mediterranean coastal wetlands by comparing species richness and faunal singularity among different water body types. Biodiversity was assessed in terms of crustacean and aquatic insect richness and singularity. The crustacean biodiversity of the Empordà wetlands is compared with that of other western Mediterranean wetlands to assess the importance of their biodiversity on a wider geographical scale.

Study area

The Empordà wetlands are located in the northeastern part of the Iberian Peninsula. They are situated in a deltaic plain originated by quaternary depositions from the Muga, Fluvià, Ter and Daró rivers. The Montgrí Massif divides the deltaic plain into two halves near the coast, which are connected further inland (Fig. 1). The wetlands of the northern part of the deltaic plain are more extended (4,731 ha) and protected by the Empordà Wetlands Natural Park. Protection of the wetlands in the southern part of the plain is weaker and restricted to the largest lagoons. The littoral wetlands are free from tidal influence and the hydrology is characterized by sudden and irregular flooding (caused by sea storms, rainfall and inputs of water from rivers or channels), followed by dry periods when most of the basins become isolated and gradually dry out (Quintana, 2002). The hydrology of the inland wetlands is influenced by rainfall and river inputs (superficial or groundwater), and the marine influence is scarce or nonexistent. Additional information about the physical, chemical and biological characteristics of each water body type are shown in Table 1.

Methods

Quantitative sampling was carried out monthly during two different survey periods (1996–1997 and 1999–2000). For detailed information about the sampling surveys see Martinoy et al. (2006). The location of all the water bodies sampled is shown in Fig. 1. The number of sites sampled for each water body type depended on the abundance of the type in the studied area (Table 2). To assess the regional biodiversity of the Empordà wetlands, data from qualitative samplings (carried out during the last 15 years) and bibliographical data (Martinoy et al. (2006) and references therein) were taken into account.

Captures in lentic ecosystems were done using a dip net with a diameter of 20 cm and a mesh size of 250 μm. In each water body 20 dip net "sweeps" in a rapid sequence per visit were carried out. In estuarine ecosystems, additional captures were done using a Surber net with a

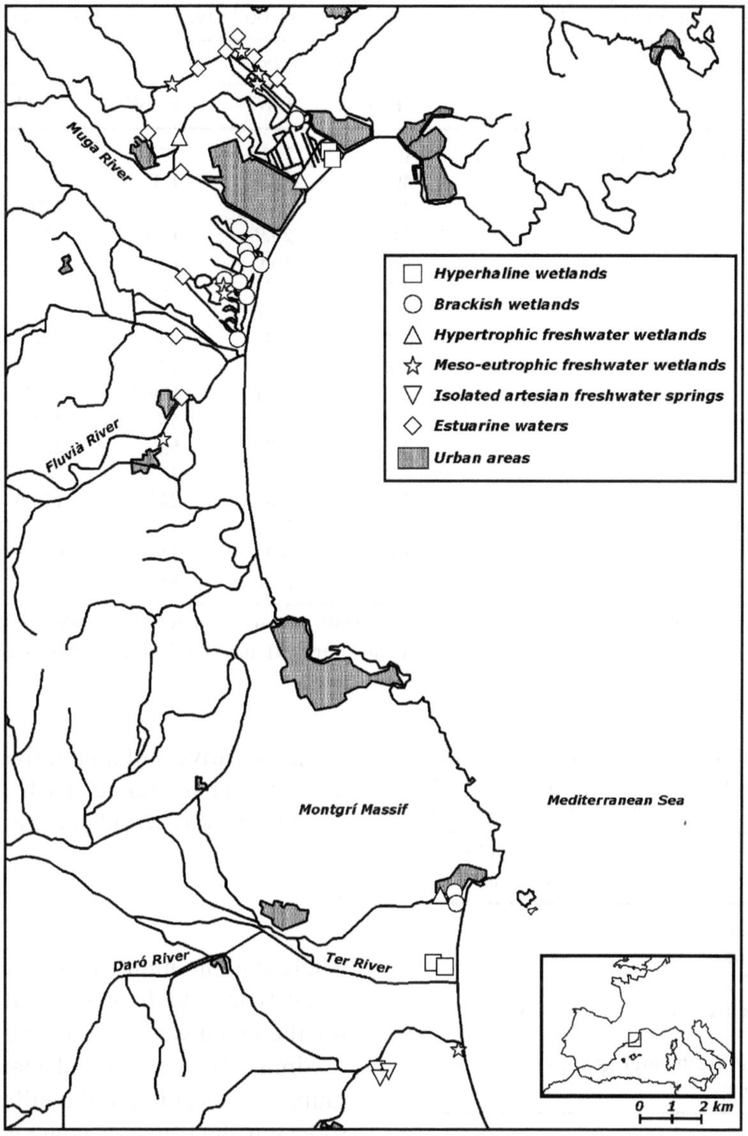

Fig. 1 Map of the study site with the locations of the 38 water bodies studied

30 × 30 cm opening and a 1 mm mesh size, when there was water flow.

Temperature, dissolved oxygen, conductivity and pH were measured in situ. Filtered (125 ml) and unfiltered (250 ml) water samples were collected and frozen upon arrival to the laboratory. Analyses of dissolved inorganic nutrients (ammonium, nitrite, nitrate, phosphate) from filtered samples and total nutrients (total nitrogen and phosphorus) from unfiltered samples were carried out following Grasshoff et al. (1983). Water chlorophyll-*a* content was extracted using 80% methanol, after filtering the water samples (Whatman GF/C filters), and measured according to Talling and Driver (1963).

Following Trobajo et al. (2002) all water bodies were grouped into five different types found in the study area (Table 2): hyperhaline wetlands (HHW), brackish wetlands (BW), hypertrophic freshwater wetlands (HTW), meso-eutrophic freshwater wetlands (MFW) and isolated artesian freshwater springs, known as "ullals" in the Iberian Mediterranean coast, hereafter freshwater springs (FS). This classification is based on physical,

Table 1 Physical, chemical and biological characteristics of each water body type in the Empordà wetlands

	Waterbody types					
	HHW	BW	HTW	MFW	FS	EW
Approximated hydroperiod length (months·year^{-1})	12	7–12	12	7–12	12	12
Conductivity (mS·cm^{-1})	**49.6**	**18.9**	**1.3**	**3.9**	**1.1**	**0.8**
	48	*66*	*136*	*179*	*99*	*76*
Chlorophyll-*a* (μg·l^{-1})	**21.5**	**20.9**	**12.1**	**14.3**	**18.1**	**6.5**
	186	*183*	*107*	*150*	*143*	*103*
Ammonium (μM)	**9.1**	**4.3**	**34.1**	**8.6**	**2.6**	**29.8**
	196	*260*	*109*	*245*	*351*	*214*
Nitrite (μM)	**0.7**	**0.5**	**6.9**	**2.0**	**5.5**	**10.9**
	569	*234*	*156*	*289*	*148*	*193*
Nitrate (μM)	**2.1**	**11.1**	**109.9**	**53.7**	**705.9**	**205.6**
	253	*312*	*58*	*151*	*25*	*100*
Phosphate (μM)	**4.3**	**3.7**	**5.8**	**2.9**	**0.93**	**8.5**
	298	*253*	*173*	*174*	*368*	*159*
Total N (μM)	**219.3**	**164.8**	**212.7**	**125.0**	**851.9**	–
	160	*94*	*49*	*75*	*27*	
Total P (μM)	**10.5**	**9.2**	**11.0**	**6.1**	**2.0**	–
	248	*213*	*83*	*121*	*197*	

Mean values in bold and coefficients of variation in italics. HHW: hyperhaline wetlands; BW: brackish wetlands; HTW: hypertrophic freshwater wetlands; MFW: meso-eutrophic freshwater wetlands; FS: isolated artesian freshwater springs; EW: estuarine waters

Table 2 Water body types studied in the Empordà wetlands, classified following Trobajo et al. (2002). The number of water bodies sampled in each type is shown

Code	Waterbody type	Number of sampled sites
HHW	Hyperhaline wetlands	4
BW	Brackish wetlands	12
HTW	Hypertrophic freshwater wetlands	3
MFW	Meso-eutrophic freshwater wetlands	8
FS	Isolated artesian freshwater springs	3
EW	Estuarine waters	8

chemical and biological characteristics of 22 water bodies from the study area. Since the classification of Trobajo et al. (2002) did not encompass all types of water bodies in the area, one more type was included: estuarine waters (EW).

In order to compare the community structure of the different water body types, the following community parameters were calculated using quantitative data: species richness per visit, Shannon-Wiener diversity and evenness (Pielou, 1969). Calculation of Shannon-Wiener diversity and evenness was based on $\log_2$. Cumulative species richness, singularity and similarity were calculated using presence/absence data obtained by quantitative and qualitative samples, and were used to characterize each water body type. Singularity was calculated as:

$$s = \frac{e}{E}$$

where e is the number of species only found in a given type, and E is the total number of species for the given water body type.

Similarity was also calculated to compare the faunal composition of the different types. In order to lessen the effect of comparing two inventories with different number of taxa, the following similarity index was selected:

$$S = \frac{a}{A}$$

where a is the number of coincident species from both inventories, and A is the total number of taxa from the inventory with less taxa.

A canonical correspondence analysis (CCA) using CANOCO 4.5 (ter Braak and Šmilauer, 2002) was performed to relate the faunal composition of each water body to major environmental gradients. The species matrix (153 taxa) was converted to presence/absence data for each water body (in the matrix bibliographical and qualitative samples were not included). Abundance data were not used because they are not useful to compare faunal compositions among water body types due to the higher abundance of microcrustaceans in comparison with insects. The environmental parameters used were temperature, conductivity, pH, percentage of dissolved oxygen, ammonium, nitrite, nitrate, phosphate, and chlorophyll-*a*. For each environmental parameter, two variables were included in the CCA. The first one corresponded to the median, as an estimator of the magnitude, and the second one corresponded to the variation coefficient, as an estimator of the variability. All medians were log transformed [$\log_{10}(x + 1)$], except that of pH. The environmental parameters retained for the analysis were those with an inflation factor <20 (ter Braak & Šmilauer, 2002) to avoid correlation between variables. The significance of the canonical axes represented in the solution was tested according to ter Braak and Šmilauer (2002) by means of a Monte Carlo test (499 permutations). The Monte Carlo tests show if species variation is explained by each canonical axis.

In order to compare the species richness of the Empordà wetlands with other western Mediterranean wetlands, those wetlands which had enough faunal data of different water body types were selected. Information on aquatic insect composition of these wetlands is scarce, so only the crustacean richness was compared. The wetlands selected were the Camargue (Hertzog, 1935; Schachter & Conat, 1951; Hartmann, 1953; Petit & Schachter, 1954; Aguesse, 1955, 1956, 1957, 1959, 1960; Aguesse & Dussart 1956; Aguesse & Bigot, 1960; Schachter, 1960; Nourisson & Aguesse, 1961; Marazanof, 1963, 1964, 1965; Verhoeven, 1975; Pont, 1977; Crivelli, 1982; Henry & Magniez, 1983), the Ebre Delta (Chinchilla & Comín, 1977; Sabater, 1986; Forès et al., 1986; Forès, 1988; Balada, 1989; España et al., 1993; Boix et al., 2005), the Albufera of Mallorca (Jaume, 1995), the Albufera of València (Oltra & Miracle, 1984, 1992; Alfonso & Miracle, 1987, 1990; Alfonso, 1996), and Doñana (Margalef, 1953; Dussart, 1964, 1967; Bigot & Marazanof, 1965; Marazanof, 1967; Estrada, 1973; Armengol, 1976; Furest & Toja, 1981, 1987; López et al. 1991; Galindo et al. 1994; Ruiz et al., 1996; Cuesta et al., 1996; Serrano & Toja, 1998; Fahd et al., 2000; León et al., 2004; Arechederra et al., 2004; Cuesta et al., 2006).

Results

The remarkably rich fauna found in the studied area included 420 taxa, 125 of which were crustaceans and 295 were aquatic insects. Within the crustaceans, Copepoda and Branchiopoda were the best represented classes (41 and 42 taxa, respectively), followed by Ostracoda (24 taxa) and Malacostraca (18 taxa). Within the aquatic insects, Coleoptera (129 taxa), Diptera (94 taxa), Odonata (36 taxa) and Heteroptera (25 taxa) were the richest orders, representing together 97% of the insect taxa found in the area. Following these orders were Ephemeroptera (7 taxa), Trichoptera (2 taxa), Neuroptera (1 taxon) and Plecoptera (1 taxon). The complete faunal list appears in Martinoy et al. (2006).

Six variables were retained for the CCA analysis: conductivity median and coefficient of variation (condMD and condCV), chlorophyll-*a*

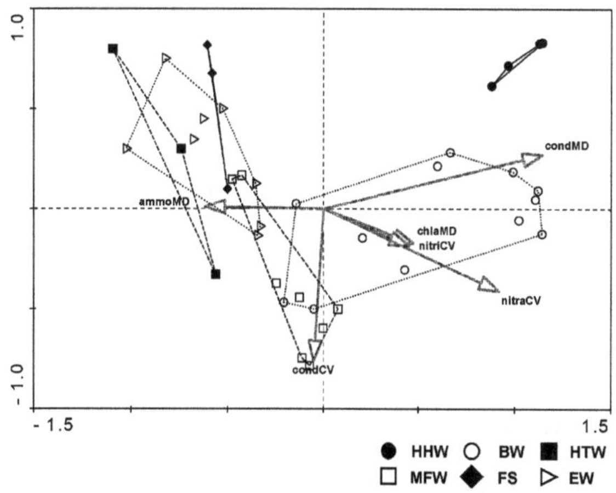

Fig. 2 CCA ordination diagram. Biplot showing sample position in relation to environmental variables in the space created by the first two axes. HHW: hyperhaline wetlands; BW: brackish wetlands; HTW: hypertrophic freshwater wetlands; MFW: meso-eutrophic freshwater wetlands; FS: isolated artesian freshwater springs; EW: estuarine waters; condMD: conductivity median; condCV: conductivity coefficient of variation; chlaMD: chlorophyll-*a* median; nitriCV: nitrite coefficient of variation; nitraCV: nitrate coefficient of variation; ammoMD: ammonium median

median (chlaMD), nitrite coefficient of variation (nitriCV), nitrate coefficient of variation (nitraCV), and ammonium median (ammoMD). Both axes were significant (F-ratio: 2.966 and $P < 0.01$ for the 1st axis; F-ratio: 1.831 and $P < 0.01$ for the 2nd axis). The first axis separated the water body type assemblages according to their conductivity (median): HHW and BW had positive scores, while MFW, FS, EW and HTW had negative scores (Fig. 2). Other environmental characteristics were associated with the conductivity gradient. Thus, water bodies with high conductivity had a higher variation in nitrite and nitrate concentration, higher chlorophyll-*a* concentration, and lower ammonium concentration than water bodies with low conductivity. The second axis distinguished the assemblages of water bodies with high conductivity variations (variation coefficient) from the rest, according to their faunal composition. These water bodies, belonging to the BW or MFW types, are located very near to the coast and are characterized by a change of the water input origin throughout the year (river floods or sea storms) and a high confinement (important evaporation rates).

Three water body types were characterized by higher cumulative species richness (BW, MFW and EW) than the other three (HHW, HTW and FS) (Table 3). However, this may be due to the fact that water body types with more cumulative species richness had been sampled more extensively. Nevertheless, when species richness per visit (the mean number of taxa captured during each visit, previously log-transformed) was analysed, significant differences related to water body types also appeared ($F_{5,432} = 11.89$; $P < 0.0005$). HHW and HTW were the two types with the lowest species richness per visit, whereas FS, BW and MFW were the types with highest richness per visit (Fig. 3a). An ANOVA could not be performed to test differences in Shannon-Wiener diversity and evenness among water body types due to heterogeneity of the variances. Although it was not possible to apply statistical tests, the lowest values of Shannon-Wiener diversity were observed in the two saline water body types:

Table 3 Cumulative species richness of different crustacean and insect groups in each water body type

	Waterbody types					
	HHW	BW	HTW	MFW	FS	EW
Crustacea	35	62	48	76	24	47
Branchiopoda	6	14	24	26	8	14
Copepoda	19	24	12	29	10	14
Ostracoda	5	11	9	12	3	8
Malacostraca	5	13	3	9	3	11
Insecta	18	47	24	95	47	81
Ephemeroptera	0	1	1	2	1	7
Plecoptera	0	0	0	0	0	1
Odonata	0	0	1	4	1	3
Heteroptera	2	7	1	15	8	8
Neuroptera	0	0	0	0	1	0
Coleoptera	9	24	13	55	25	45
Trichoptera	0	1	0	2	0	2
Diptera	7	14	8	17	11	15
Cumulative species richness	53	109	72	171	70	128
Singularity	0.04	0.19	0.04	0.20	0.13	0.29
Within Crustacea	0.00	0.21	0.06	0.11	0.04	0.06
Within Insecta	0.11	0.35	0.09	0.34	0.17	0.51
Ratio						
Crustacea/Insecta	1.9	1.3	2.0	0.8	0.5	0.6
Branchiopoda/Copepoda	0.3	0.6	2.0	0.9	0.8	1.0
Coleoptera/Diptera	1.3	1.7	1.6	3.2	2.3	3.0

Singularity and several faunal ratios are also shown. HHW: hyperhaline wetlands; BW: brackish wetlands; HTW: hypertrophic freshwater wetlands; MFW: meso-eutrophic freshwater wetlands; FS: isolated artesian freshwater springs; EW: estuarine waters

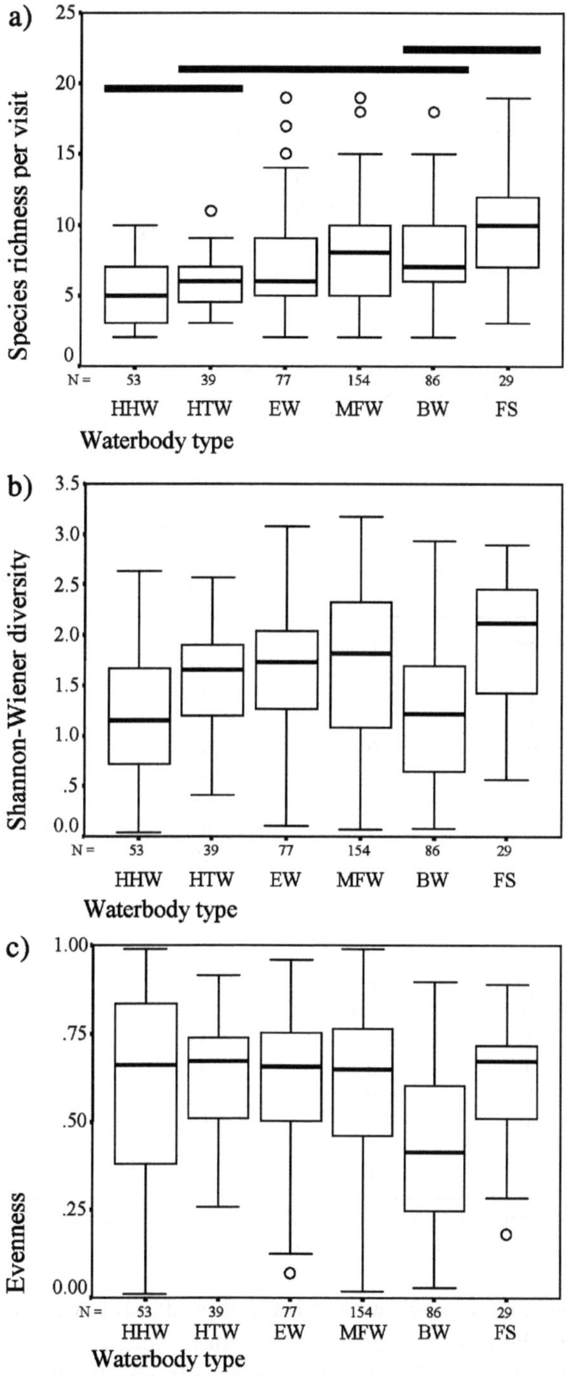

Fig. 3 Box-plots of (**a**) species richness per visit, (**b**) Shannon-Wiener diversity, and (**c**) evenness for each water body type. In figure **a**, water body types under the same bar do not differ significantly ($P > 0.05$, Scheffé post-hoc test). Below the abscissa axis for each figure appears the number of cases for each water body type. HHW: hyperhaline wetlands; BW: brackish wetlands; HTW: hypertrophic freshwater wetlands; MFW: meso-eutrophic freshwater wetlands; FS: isolated artesian freshwater springs; EW: estuarine waters

HHW and BW (Fig. 3b). In HHW these lower diversity values were explained by the low values of species richness, whereas in BW the explanation was the high dominances observed (lowest values of evenness; Fig. 3c).

In HHW and BW, the cumulative crustacean richness was higher than the cumulative insect richness, but the pattern was reversed in the remaining water body types, which were characterized by having lower values of conductivity (Table 3). Among the crustaceans, Branchiopoda had higher richness values in freshwater water body types, while Malacostraca presented more species in BW, followed by EW, due to the mix of fluvial, brackish and marine faunas in this water body type. The aquatic insect orders that presented more species were the Coleoptera and Diptera, having maximum values of species richness in MFW and EW. Heteropterans also had a high number of species in freshwater types, with the exception of HTW.

The highest singularity value was found in EW, whereas in water body types highly constrained by salinity or high nutrient concentration (HHW and HTW), the values of singularity were the lowest (Table 3). These water body types also presented the lowest values of similarity, while the maximum values were observed between MFW and the rest of the water body types (Table 4).

A comparison of the crustacean fauna of the Empordà wetlands with that of several western Mediterranean wetlands showed that the species richness observed in the Empordà wetlands was lower to those reported from Doñana or the Camargue and higher than those reported from the Albufera of València, the Albufera of Mallorca and the Ebre Delta (Table 5). However, the studies carried out in the Albufera of València were focused on the lagoon and the rice fields, and therefore not all the types of aquatic environments were analyzed, meaning that gradients like salinity or temporality of water were smaller than those analyzed in the Camargue, Doñana, the Ebre Delta or the Empordà wetlands. In the case of the Albufera of Mallorca, the low species richness found can be explained by the low number of sampling stations and by the non-extensive sampling in time. The case of the low

Table 4 Assemblage similarities among different water body types

		Waterbody types					
		HHW	BW	HTW	MFW	FS	EW
Waterbody types	HHW	–	**0.78**	*0.27*	**0.78**	0.37	0.49
	BW		–	0.52	**0.69**	0.44	0.43
	HTW			–	**0.86**	0.41	0.36
	MFW				–	**0.79**	**0.65**
	FS					–	0.57
	EW						–

minimum value in italics, values higher than 0.6 in bold

HHW: hyperhaline wetlands; BW: brackish wetlands; HTW: hypertrophic freshwater wetlands; MFW: meso-eutrophic freshwater wetlands; FS: isolated artesian freshwater springs; EW: estuarine waters

Table 5 Cumulative species richness of crustacean groups in several western Mediterranean wetlands

	Western Mediterranean wetlands					
	CA	EW	ED	AM	AV	DO
Crustacea	134	124	89	(33)	(42)	156
Branchiopoda	40	40	25	(11)	19	54
Anostraca	4	2	1	0	0	5
Notostraca	1	1	1	0	1	1
Spinicaudata	1	0	0	0	0	1
Ctenopoda	1	0	0	0	0	1
Anomopoda	33	36	21	11	18	46
Onychopoda	0	1	2	0	0	0
Copepoda	47	42	30	(11)	16	45
Poecilostomatoida	0	1	1	0	0	0
Calanoida	9	6	5	2	1	14
Cyclopoida	25	18	13	6	12	24
Harpacticoida	13	17	11	3	3	7
Ostracoda	23	23	13	(4)	–	26
Malacostraca	24	18	29	(7)	(7)	31
Mysidacea	4	2	2	1	0	2
Cumacea	1	0	0	0	0	0
Tanaidacea	0	1	1	0	0	0
Isopoda	6	3	4	2	1	2
Amphipoda	7	7	14	3	3	1
Decapoda	6	5	8	1	3	26

CA: Camargue; EW: Empordà wetlands; ED: Ebre Delta; AM: Albufera of Mallorca; AV: Albufera of València; DO: Doñana. In parentheses, taxa that have a subestimation of their species richness due to lack of information

species richness found in the Ebre Delta cannot be explained by any methodological reason, as it is comparable to the present study. It is worth pointing out that the Ebre Delta presents the maximum value of Malacostracan richness in all the wetlands compared in this study.

Several resemblances were found between the faunal composition of the wetlands compared. Branchiopoda and Copepoda, the faunal groups best represented, have a similar number of species in each wetland. Cyclopoida and Amphipoda were the groups with more species within the Copepoda and Malacostraca, respectively. In contrast, the ratio Harpacticoida/Cyclopoida differed between wetlands. Thus, there was a similar number of species of both groups in the Empordà wetlands and the Ebre Delta, while in the rest of the wetlands the number of Cyclopoida species was twice as high, or more, than the number of Harpacticoida species.

Discussion

The Empordà wetlands present a high spatial heterogeneity due to three main environmental gradients: salinity, water permanence and eutrophy (Trobajo et al., 2002; Gascón et al., 2005). These environmental gradients are typical for the Mediterranean region and are accentuated by the high variability caused by the lack of tidal influence, seasonality, rainfall/evaporation deficit, unpredictable sea storms and freshwater flood events, the high interannual variation, and the anthropogenic influence to which they are subjected (Guelorget & Perthuisot, 1983; Quintana et al., 1998). Furthermore, the existence of several river mouths (Muga, Fluvià, Ter and Daró rivers) in the study area also favours this high spatial heterogeneity (Britton & Crivelli, 1993), which is also typical of other western Mediterranean wetlands, such as the Camargue, the Ebre Delta or Doñana. Such high spatial heterogeneity favours the high value of crustacean and aquatic insect biodiversity found in the Empordà wetlands, which is comparable to other western Mediterranean wetlands.

Highly constrained environments, such as hyperhaline (HHW) and hypertrophic wetlands (HTW), presented the lowest values of species richness in the Empordà wetlands. Salinity and hypertrophy have been widely recognized as constraints on species richness in aquatic environments (Wiederholm, 1984; Hammer et al., 1990; Williams et al., 1990; Petridis, 1993). In the present study, FS also presented low species richness values, but this can be explained by the small size of this type of water bodies and the scarce presence of these environments in the Empordà wetlands (only three small springs are present in the study area, they occur close to each other, but they are not connected). The high species richness per visit of these environments implies that the low cumulative richness can be explained by their low presence in the Empordà wetlands.

The fauna of the more constrained environments (HHW and HTW) was characterized not only by lower species richness, but also by their species composition. The lowest singularity obtained for these environments showed that the faunal composition was characterized by species also present in other water body types. In contrast, the presence of several faunal groups in EW, like Ephemeroptera and Plecoptera, explained the highest singularity of this water body type. On the other hand, the high similarity values found between the fauna of MFW and that of other water body types could be explained by the fact that MFW act as faunal reservoirs of aquatic biodiversity in the Empordà wetlands because of their wide range in conductivity and nutrient concentration. In contrast, the assemblages that inhabit HTW and HHW showed the most different faunal composition among all water body types, which were reflected in the lowest similarity values obtained for them. Thus, the aquatic fauna is not adapted to survive both extreme conditions: high salinity and high nutrient concentration.

The environments constrained by salinity or trophic state present a higher proportion of crustacean species than insect species (Timms, 1993). Insect species richness is known to have lower values in saline environments (Timms et al., 1986; Williams & Williams, 1998) or in eutrophic situations (Wiederholm, 1984) as can be observed in the Empordà wetlands. The decline of species richness due to salinity increases has been reported often, whereas the effect of eutrophication has been more related to changes in composition or abundance than to a decline in species richness (e.g., McCormick et al., 2004; García-Criado et al., 2005). A similar pattern was not clearly observed for crustaceans in the present study, as HHW and HTW presented a species richness similar to other water body types. In estuarine environments where there is a higher influence of the sea, there is also no clear relation between crustacean species richness and salinity (Josefson & Hansen, 2004).

Each constraint affects the various faunal groups differently. Thus, HHW and BW present a higher proportion of Copepoda than Branchiopoda, while HTW present the inverse pattern. On the other hand, Branchiopoda species prefer freshwater environments, although there are some species adapted to brackish waters, or even hyperhaline waters (Frey, 1993). In HTW, the number of Branchiopoda species exceeds that of copepods, but in other types of freshwater environments (MFW, FS and EW), there are similar values of Branchiopoda and Copepoda species abundance. However, the dominance of Branchiopoda in eutrophic environments has been found in many studies (e.g., McNaught, 1975; Pace, 1986; George et al., 1990).

In the case of insects, Coleoptera were more affected by the constraints of salinity and nutrient inputs than Diptera. Both groups were less represented in HHW and HTW water body types, but higher proportion of Coleoptera species avoided these constrained water body types than Dipterans. A partial explanation could be the different life stages in which Coleoptera and Diptera species richness was estimated. Identification of coleopterans to species level was performed on adults, while this was not possible for dipterans, whose adults are not aquatic. Thus, adult dipterans choose the water body without entering it (during reproduction), whereas coleopterans have two chances of selection, before and after entering the water body. Several studies have shown the active movement of coleopterans between water bodies (e.g., de Szalay & Resh, 2000; Davy-Bowker, 2002).

The factors determining species richness and its relative importance vary among faunal groups (Eitam et al., 2004). When using biodiversity as the criteria for environmental protection and/or management, it has to be estimated in several faunal groups and these groups have to represent an important proportion of the total biodiversity. Species richness is a good parameter, but others like rarity also have to be taken into account (Nijboer & Verdonschot, 2004). Thus, several environment types in one area need to be evaluated using, at least, measures of species richness and rarity (i.e., singularity). As an example, the low Coleoptera richness observed in Mediterranean saline environments is not a reason to neglect the importance of their fauna, because the singularity and the number of endemic species present in them is high (Sánchez-Fernández et al., 2004).

Acknowledgements This work was supported by a grant from the Comisión de Investigación Científica y Técnica (CICYT), Programa de Recursos Naturales (ref. CGL2004-05433/BOS). N. L. Arroyo and E. Olafsson helped to improve the English and provided useful comments.

References

Aguesse, P., 1955. Liste des cladocères nouveaux pour la Camargue. Terre et Vie 9: 313–314.

Aguesse, P., 1956. Quelques considérations sur les copépodes de Camargue. Vie et Milieu 7: 38–42.

Aguesse, P., 1957. Complement a l'inventaire de la faune invertébrée des eaux camarguaises. Terre et Vie 11: 241–252.

Aguesse, P., 1959. Complement a l'inventaire de la faune invertébrée des eaux camarguaises (3me note). Terre et Vie 13: 158–161.

Aguesse, P., 1960. Complement a l'inventaire de la faune invertébrée des eaux camarguaises (4e note). Terre et Vie 14: 132–135.

Aguesse, P. & L. Bigot, 1960. Observations floristiques et faunistiques sur un étang de moyenne Camargue: la Baisse-salée de la Tour du Valat. Vie et Milieu 8: 248–307.

Aguesse, P. & B. H. Dussart, 1956. Sur quelques crustacés de Camargue et leur écologie. Vie et Milieu 7(4): 481–520.

Alfonso, M. T., 1996. Zooplancton del Parque Natural de la Albufera de Valencia. Ph.D Thesis, Universitat de València.

Alfonso, M. T. & M. R. Miracle, 1987. Estudio comparativo del zooplancton en tres ullales del Parque Natural de la Albufera de Valencia. Limnetica 3: 263–272.

Alfonso, M. T. & M. R. Miracle, 1990. Distribución espacial de las comunidades zooplanctonicas de la Albufera de Valencia. Scientia gerundensis 16: 11–25.

Arechederra, A., L. Serrano, M. Reina, E. Reyes, G. Martín & J. Toja, 2004. Descripción de la comunidad de zooplancton del lucio "El Cangrejo Grande", Parque Natural de Doñana (S. O. España). Abstract book of IV Iberian Conference of Limnology, 60 pp.

Armengol, J., 1976. Crustáceos del Coto de Doñana. Oecologia acuatica 2: 93–97.

Balada, R., 1989. El Delta del Ebro. INCAFO, Madrid.

Bigot, L. & F. Marazanof, 1965. Considérations sur l'écologie des invertébrés terrestres et aquatiques des Marismas du Guadalquivir. Vie et Milieu 16: 411–473.

Boix, D., J. Sala & R. Moreno-Amich, 2001. The faunal composition of Espolla pond (NE Iberian peninsula): the neglected biodiversity of temporary waters. Wetlands 21: 577–592.

Boix, D., S. Gascón, J. Sala, M. Martinoy, J. Gifre & X. D. Quintana, 2005. A new index of water quality assessment in Mediterranean wetlands based on crustacean and insect assemblages: the case of Catalunya (NE Iberian peninsula). Aquatic Conservation: Marine and Freshwater Ecosystems 15: 635–651.

Britton, R. H., 1982. Managing the prey fauna. In Scott, D. A. (ed.), Managing wetlands and their birds. Proceedings of the Third Technical Meeting on Western Palearctic Migratory Bird Management, October 12–15, 1982, Münster, Germany: IWRB. Slimbridge, 92–98.

Britton, R. H. & A. J. Crivelli, 1993. Wetlands of southern Europe and North Africa: Mediterranean wetlands. In Whigham, D. F., D. Dykyjová & S. Hejný (eds), Wetlands of the world I: Inventory, ecology and management. Kluwer Academic Publishers, Dordrecht, 129–194.

Brown, K. S., 1998. Vanishing pools taking species with them. Science 281: 626.

Chinchilla, M. & F. A. Comín, 1977. Contribució al coneixement dels crustacis del delta de l'Ebre. Treballs de la Institució Catalana d'Història Natural 8: 119–144.

Crivelli, A. J., 1982. Biology of three Malacostraca (Decapoda) in a Mediterranean lagoon with particular emphasis on the effect of rapid environmental changes on the activity (catchability) of the species. Estuarine, Coastal and Shelf Science 15: 591–604.

Cuesta, J. A., L. Serrano, M. R. Bravo & J. Toja, 1996. Four new crustaceans in the Guadalquivir river estuary (SW Spain), including an introduced species. Limnetica 12: 41–45.

Cuesta, J. A., E. González-Ortegón, A. Rodríguez, F. Baldó, C. Vilas & P. Drake, 2006. The decapod crustacean community of the Guadalquivir Estuary (SW Spain): seasonal and inter-year changes in community structure. Hydrobiologia 557: 85–95.

Davy-Bowker, J., 2002. A mark and recapture study of water beetle (Coleoptera: Dytiscidae) in a group of semi-permanent and temporary ponds. Aquatic Ecology 36: 435–446.

de Szalay, F. A. & V. H. Resh, 2000. Factors influenting macroinvertebrate colonization of seasonal wetlands: responses to emergent plant cover. Freshwater Biology 45: 295–308.

Dussart, B. H., 1964. Copépodes d'Espagne. Bulletin de la Société Zoologique de France 89: 117–125.

Dussart, B. H., 1967. Contribution à l'étude des Copépodes d'Espagne. Publicaciones del Instituto de Biología Aplicada 42: 87–105.

Eitam, A., L. Blaustein, K. Van Damme, H. J. Dumont & K. Martens, 2004. Crustacean species richness in temporary pools: relationships with habitat traits. Hydrobiologia 525: 125–130.

España, A., C. Ibañez, N. Prat & A. Rodrigues-Capítulo, 1993. Caracterització limnològica dels ullals de Baltasar. Butlletí del Parc Natural Delta de l'Ebre 7: 38–41.

Estrada, M., 1973. Nota sobre diaptòmids del Coto de Doñana. Treballs de la Institució Catalana d'Història Natural 32: 127–134.

Everard, M., 1999. Towards sustainable development of still water resources. Hydrobiologia 396: 29–38.

Fahd, K., L. Serrano & J. Toja, 2000. Crustacean and rotifer composition of temporary ponds in the Doñana National Park (SW Spain) during floods. Hydrobiologia 436: 41–49.

Forès, E., 1988. Els ostràcodes dels arrossars del delta de l'Ebre: sitemàtica, ecologia i distribució geogràfica. Treballs de la Institució Catalana d'Història Natural 55: 47–57.

Forès, E., M. Menéndez & F. A. Comín, 1986. Contribución al conocimiento de crustáceos y rotíferos del delta del Ebro. Miscel·lània Zoològica 10: 105–111.

Frey, D. G., 1993. The penetration of cladocerans into saline waters. Hydrobiologia 267: 233–248.

Furest, A. & J. Toja, 1981. Ecosistemas acuáticos del Parque Nacional de Doñana: distribución del zooplancton. In Actas del Simposio del Agua en Andalucía 1, March 23–26, 1981, Granada, Spain: Universidad de Granada, 151–165.

Furest, A. & J. Toja, 1987. Tipificación de lagunas andaluzas según sus comunidades de crustáceos. Oxyura 4: 89–100.

Galindo, M. D., A. J. Mata, N. Mazuelos & L. Serrano, 1994. Microcrustacean and rotifer diversity and richness relating to water temporality in dune ponds of the Doñana National Park. Verhandlungen / Internationale Vereinigung für theoretische und angewandte Limnologie 25: 1350–1356.

García-Criado, F., E. Bécares, C. Fernández-Aláez & M. Fernández-Aláez, 2005. Plant-associated invertebrates and ecological quality in some Mediterranean shallow lakes: implications for the application of the EC Water Framework Directive. Aquatic Conservation: Marine and Freshwater Ecosystems 15: 31–50.

Gascón, S., D. Boix, J. Sala & X. D. Quintana, 2005. Variability of benthic assemblages in relation to the hydrological pattern in Mediterranean salt marshes (Empordà wetlands, NE Iberian Peninsula). Archiv für Hydrobiologie 163: 163–181.

George, D. G., D. P. Hewitt, J. W. G. Lund & W. J. P. Smyly, 1990. The relative effects of enrichment and climate change on the long-term dynamics of *Daphnia* in Esthwaite Water, Cumbria. Freshwater Biology 23: 55–70.

Gibbs, J. P., 2000. Wetland loss and biodiversity conservation. Conservation Biology 14: 314–317.

Giudicelli, J. & A. Thiéry, 1998. La faune des mares temporaires, son originalité et son intérêt pour la biodiversité des eaux continentales méditerranéennes. Ecologia mediterránea 24: 135–143.

Gopal, B. & W. J. Junk, 2000. Biodiversity in wetlands: an introduction. In Gopal, B., W. J. Junk & J. A. Davis (eds) Biodiversity in wetlands: assessment, function and conservation. Backhuys Publishers, Leiden, 1–10.

Grasshoff, K., M. Ehrhardt & K. Kremling, 1983. Methods of seawater analysis. Verlag Chemie, Weinheim.

Guelorget, O. & J. P. Perthuisot, 1983. Le domaine paralique: expressions géologiques, biologiques et économiques du confinement. Travaux du laboratoire de Géologie 16. Presses de l'École Normale Supérieure, Paris.

Hammer, U. T., J. S. Sheard & J. Kranabetter, 1990. Distribution and abundance of littoral benthic fauna in Canadian prairie saline lakes. Hydrobiologia 197: 173–192.

Hartmann, G., 1953. Ostracodes des étangs méditerranées. Vie et Milieu 4: 707–712.

Henry, J. P. & G. Magniez, 1983. Crustaces Isopodes (Principalment asellotes). In Ginet, R. (ed.), Introduction pratique a la systematique des organismes des eaux continentales françaises, 4. Association Française de Limnologie, Lyon.

Hertzog, L., 1935. Notes faunistiques de Camargue. Bulletin de la Société Zoologique de France 60: 265–282.

Jaume, D., 1995. Una llista dels crustacis de S'Albufera. In Martínez, A. & J. Mayol (eds), S'Albufera de Mallorca. Editorial Moll, Palma de Mallorca, 119–124.

Josefson, A. B. & J. L. S. Hansen, 2004. Species richness of benthic macrofauna in Danish estuaries and coastal areas. Global Ecology and Biogeography 13: 273–288.

Lefeuvre, J.-C., P. Laffaille, E. Feunteun, V. Bouchard & A. Radureau, 2003. Biodiversity in salt marshes: from patrimonial value to ecosystem functioning. The case study of the Mont-Saint-Michel bay. Comptes rendues biologies 326: 125–131.

León, D., L. Serrano & J. Toja, 2004. Zooplancton en las lagunas del P.N. de Doñana: revisión bibliográfica y nuevas citas. In IV Iberian Conference of Limnology, July 5–9, 2004, Porto, Portugal: AEL, 60.

López, T., J. Toja & A. Gabellone, 1991. Limnological comparison of two peridunar ponds in the Doñana National Park (Spain). Archiv für Hydrobiologie 120: 357–378.

Marazanof, F., 1963. Cycle annuel des populations de cladocères et copépodes du Saint-Seren, de la Baisse-salée des Relongues et Cerisières de la Tour du Valat. Terre et la Vie 3: 335–356.

Marazanof, F., 1964. Cladocères nouveaux pour la Camargue. Terre et la Vie 18: 380–382.

Marazanof, F., 1965. Ostracodes de Camargue. Annales de Limnologie 1: 95–102.

Marazanof, F., 1967. Ostracodes, Cladocères, Hétéroptères et hydracariens noveaux pour les Marismas du Guadalquivir (Andalousie). Données écologiques. Annales de Limnologie 3: 47–64.

Margalef, R., 1953. Los crustáceos de las aguas continentales ibéricas. In Biología de las aguas continentales, 10. Ministerio de Agricultura, Instituto Forestal de Investigaciones y Experiencias, Madrid.

Martinoy, M., D. Boix, J. Sala, S. Gascón, J. Gifre, A. Argerich, R. Barrera, S. Brucet, A. Badosa, R. López-Flores, M. Méndez, J. M. Utgé & X. D. Quintana, 2006. Crustacean and aquatic insect assemblages in the Mediterranean coastal ecosystems of Empordà Wetlands (NE Iberian Peninsula). Limnetica 25: 665–682.

McCormick, P. V., R. B. E. Shuford III & P. S. Rawlik, 2004. Changes in macroinvertebrate community structure and function along a phosphorus gradient in the Florida Everglades. Hydrobiologia 529: 113–132.

McNaught, D. C., 1975. A hypothesis to explain the succession from calanoids to cladocerans during eutrophication. Verhandlungen/Internationale Vereinigung für theoretische und angewandte Limnologie 19: 724–731.

Mocci, A., 1983. Rapporto e risultati della "Conferenze sulla convenzione per le zone umide d'importanza internazionale" (Cagliari 24/29 novembre 1980) e riflessioni sugli stagni della Sardegna. Alytes 1: 21–40.

Nijboer, R. C. & P. F. M. Verdonschot, 2004. Rare and common macroinvertebrates: definition of distribution classes and their boundaries. Archiv für Hydrobiologie 161: 45–64.

Nourisson, M. & P. Aguesse, 1961. Cycle annuel des phyllopodes d'une mare temporaire de Camargue. Bulletin de la Société Zoologique de France 86: 754–762.

Oltra, R. & M. R. Miracle, 1984. Comunidades zooplanctónicas de la Albufera de Valencia. Limnetica 1: 51–61.

Oltra, R. & M. R. Miracle, 1992. Seasonal succession of zooplankton populations in the hypertrophic lagoon Albufera of Valencia (Spain). Archiv für Hydrobiologie 124: 187–204.

Pace, M. L., 1986. An empirical analysis of zooplankton community size structure across lake trophic gradients. Limnology and Oceanography 31: 45–55.

Petit, G. & D. Schachter, 1954. La Camargue: étude écologique et faunistique. Annales Biologiques 30: 194–252.

Petridis, D., 1993. Macroinvertebrate distribution along an organic pollution gradient in Lake Lysimachia (Western Greece). Archiv für Hydrobiologie 128: 367–384.

Pielou, E. C., 1969. An introduction to mathematical ecology. Wiley-Interscience, New York.

Pont, D., 1977. Structure et évolution saisonnière des populations de copépodes, cladocères et ostracodes des rizières de Camargue. Annales de Limnologie 13: 15–28.

Quintana, X. D., 2002. Estimation of water circulation in a Mediterranean salt marsh and its relationship with flooding causes. Limnetica 21: 25–35.

Quintana, X. D., R. Moreno-Amich & F. A. Comín, 1998. Nutrient and plankton dynamics in a Mediterranean salt marsh dominated by incidents of flooding. Part 1: differential confinement of nutrients. Journal of Plankton research 20: 2089–2107.

Ramsar Convention Bureau. 2005. Strategic framework and guidelines for the future development of the List of Wetlands of International Importance of the Convention on Wetlands (Ramsar, Iran, 1971). Available: http://www.ramsar.org/key_guide_list2006_e.htm.

Ruiz, F., M. L. González-Regalado, L. Serrano & J. Toja, 1996. Ostrácodos de las lagunas temporales del Parque Nacional de Doñana. Aestuaria 4: 125–140.

Sabater, F., 1986. Some intersticial species of the crustacean communities of the Ter and Ebre rivermouths (NE Spain). Miscel·lània Zoològica 10: 113–119.

Sánchez-Fernández, D., P. Abellán, J. Velasco & A. Millán, 2004. Selecting areas to protect the biodiversity of aquatic ecosystems in a semiarid Mediterranean region using water beetles. Aquatic Conservation: Marine and Freshwater Ecosystems 14: 465–479.

Schachter, D., 1960. Données nouvelles sur l'évolution de quelques étangs du delta du Rhône. Bulletin de la Commission Internationale pour l'Exploration Scientifique de la mer Méditerrannée, Monaco 15: 155–162.

Schachter, D. & M. Conat, 1951. Note préliminaire sur la faune des rizières. Bulletin de la Société Zoologique de France 76: 365–370.

Serrano, L. & J. Toja, 1998. Interannual variability in the zooplankton community of a shallow temporary pond. Verhandlungen/Internationale Vereinigung für theoretische und angewandte Limnologie 26: 1575–1581.

Talling, J. F. & D. Driver, 1963. Some problems in the estimation of chlorophyll a in phytoplankton. In Proceedings of the Conference on primary productivity measurement, marine and freshwater, 1961, Hawaii: University of Hawaii, USA Atomic Energy Commission, Division of Technical Information TID-7633: 142–146.

ter Braak, C. J. F. & P. Šmilauer, 2002. CANOCO Reference manual and CanoDraw for Windows user's guide. Biometrics, Wageningen.

Timms, B. V., 1993. Saline lakes of the Paroo, inland New South Wales, Australia. Hydrobiologia 267: 269–289.

Timms, B. V., U. T. Hammer & J. W. Sheard, 1986. A study of benthic communities in some saline lakes in Saskatchewan and Alberta, Canada. Internationale Revue der gesamten Hydrobiologie 71: 759–777.

Trobajo, R., X. D. Quintana & R. Moreno-Amich, 2002. Model of alternative predominance of phytoplankton-periphyton-macrophytes in lentic waters of Mediterranean coastal wetlands. Archiv für Hydrobiologie 154: 19–40.

Valladares, L. F., F. J. Vega, R. A. Mazé, J. A. Régil & F. García-Criado, 2002. Biodiversidad de los macroinvertebrados acuáticos del Parque Natural de Valderejo (Álava): implicaciones en conservación. Boletín de la Asociación española de Entomología 26: 37–55.

Verdonschot, P. F. M., 2000. Integrated ecological assessment methods as a basis for sustainable catchment management. Hydrobiologia 422: 389–412.

Verhoeven, J. T. A., 1975. *Ruppia*-communities in the Camargue, France. Distribution and structure in relation to salinity and salinity fluctuations. Aquatic Botany 1: 217–241.

Wiederholm, T., 1984. Responses of aquatic insects to environmental pollution. In Resh, V. H. & D. M. Rosenberg (eds) The ecology of aquatic insects. Praeger, New York.

Williams, D. D. & N. E. Williams, 1998. Aquatic insects in an estuarine environment: densities, distribution and salinity tolerance. Freshwater Biology 39: 411–421.

Williams, W. D., A. J. Boulton & R. G. Taaffe, 1990. Salinity as a determinant of salt lake fauna: a question of scale. Hydrobiologia 197: 257–266.

Williams, P., M. Whitfield, J. Biggs, S. Bray, G. Fox, P. Nicolet & D. Sear, 2003. Comparative biodiversity of rivers, streams, ditches and ponds in an agricultural landscape in Southern England. Biological Conservation 115: 329–341.

Hydrobiologia (2007) 584:361–372
DOI 10.1007/s10750-007-0606-9

SHALLOW LAKES

Land use changes and associated environmental impacts on the Mediterranean shallow Lake Stymfalia, Greece

E. S. Papastergiadou · A. Retalis · P. Kalliris · Th. Georgiadis

Abstract Land cover and land use changes affect ecological landscape functions and processes. Land use changes mainly caused by human activities, is a common reason for wetlands degradation worldwide. Lake Stymfalia, located at Peloponnese, southern Greece, is an ancient wetland with a great ecological value. Lake Stymfalia has been severely degraded and transformed during the past 60 years due to agricultural activities in the surrounding areas and watercourses alterations. In this context, we investigated the land cover/ use changes and the role of the reed beds in the terrestrialization process of this shallow wetland. This particular effort utilized remotely sensed data and Geographical Information Systems (GIS) techniques to estimate land use alterations for the period 1945–1996. Patch related landscape indices were generated to analyze impacts on landscape features. Spatial and thematic information concerning the surface area and the major land cover types of the lake for years 1945, 1960, 1972, 1987, 1992, and 1996 was obtained from aerial photographs and land surveys of the area, and was stored in the GIS database. The 1996 map was ground verified, corrected and updated to 2004 conditions. From the spatio-temporal analysis of the stored data, a permanent decrease of the open water surface has been observed between the years 1945 and 1996. The results indicated that the reed beds expanded dramatically, increasing by 89.3%, and is the predominant aquatic vegetation of the whole wetland. Open water areas and wet meadows decreased by 53.7 and 96.5% respectively. Landscape analyses and, in particular, the use of selected landscape metrics, proved useful for detecting and quantitatively characterising dynamic ecological processes. As land cover/use analysis of the wetland has shown much serious environmental degradation, conservation measures should be undertaken urgently.

Keywords Aerial photography · Remote sensing · GIS · Land cover/use · Landscape metrics · Reed beds

Guest editors: R. D. Gulati, E. Lammens, N. De Pauw & E. Van Donk
Shallow lakes in a changing world

E. S. Papastergiadou (✉) · Th. Georgiadis
Department of Biology, Section of Plant Biology, University of Patras, Patras 265 00, Greece
e-mail: evapap@upatras.gr

A. Retalis
Institute for Environmental & Research Sustainable Development, National Observatory of Athens, Palaia Pendeli, Athens 152 36, Greece
e-mail: adrianr@meteo.noa.gr

P. Kalliris
Forest Management Service of Korinthos, Korinthos, Korinthian Prefecture, Greece

Introduction

Natural ecosystems are exposed to growing pressure due to intensification of agricultural land use, tourism development and other activities. In most of the EU member states conservation and sustainable land use of such areas have a high priority in environmental policy and administration (Bromley, 1997). On the European scale the legal and operational framework has been set up with the NATURA 2000 network linking the habitats directive (92/43/EEC) and bird's directive (79/409/EEC).

Wetland ecosystems currently receive much attention in environmental science and policy. European legislation has increasingly recognized the importance of preserving wetland ecosystems such as shallow lakes. The Water Framework Directive (WFD, 2000/60/EC) embodies many of the existing directives that have implications for wetlands. Over time shallow lakes are susceptible to anthropogenic alteration, eutrophication by increased nutrient and sediment loads and altered hydrology. Aquatic macrophyte communities can be influenced by these changes in nutrient levels and physical conditions. Macrophyte species richness has been observed to decrease as lakes eutrophy due to light limitation to submerged vegetation (Sand- Jensen, 1997; Scheffer et al., 1992). As water transparency decreases, macrophyte communities can shift in composition from a dominance of submerged to floating leaves and to emergent vegetation (Sand- Jensen, 1997; van den Berg, 1999). Another factor that can mediate the dominance and structure of macrophyte communities is varying water levels resulting from anthropogenic modulation of hydrology. As water levels decrease macrophytes may overcome light limitation. Because anthropogenic eutrophication has occurred over a great time period, long-term data are essential to an understanding of the trajectory of changes in macrophyte communities. Only long-term data can reveal trends that exceed the sub decade scale (Magnuson et al., 1997). Long-term data on lake ecosystems, especially those predating anthropogenic influences, are very scarce, making such records particularly valuable (Sand- Jensen, 1997).

Interpretation of spatial relationships among land cover/use categories, comparisons between areas and their surroundings, monitoring of dynamic phenomena, such as vegetative succession and /or degradation, past changes in land cover/use and future trends, can only be accomplished by analyzing existing aerial photographs because they are the only source of information on past conditions for the area. Geographic Information Systems (GIS) offer excellent capabilities as tools for analysis of wetland changes caused by long-term lake level fluctuations (Burrough, 1998; Williams et al., 1997). When accurate historical information on wetland areas and wetland characteristics are available in digital form, GIS change detection algorithms (Rembold et al., 2000; Kalivas et al., 2003) can be employed to rapidly provide information on changes in wetland area and changes in wetland vegetation class associated with water level fluctuations (Stalmans et al., 2001).

Our study is focused on land cover changes occurring as a result of human activities in the second half of 20th century in a small ancient freshwater wetland of Southern Greece, Lake Stymfalia that is nowadays a protected area. The purpose of this analysis was to examine changes in macrophyte community composition and species occurrence with water level changes over the past century, using aerial photos and GIS techniques. A long-term study of this nature can further elucidate and document the impact of human activities on aquatic macrophytes in shallow lake ecosystems.

The objectives of this study were: to detect and measure the land cover/use changes which occurred during the last 60 years; to ecologically evaluate the environmental impact of these changes; to determine present and probable future trends of land cover/use and finally, to assess conflicts and designate areas more susceptible to human alterations that affect the wetlands values.

Materials and methods

Site description

The study was conducted in Lake Stymfalia, Southern Greece (Prefecture of Korinthos,

22°28′ E, 37°51′ N). This lake is the largest mountainous lake of the whole Peloponnese (catchments area around 153 km^2). It is an ancient Mediterranean, shallow lake (mean depth, 0.8 m, maximum depth of 1.2 m, about 600 m above sea level) and occupies an area of 3.8 km^2 (Fig. 1). It is well known from Greek mythology as the kingdom of the ancient God Artemis, where Hercules killed the Stymphalides ornithes (Kalliris & Spinthakis, 1997). The study area contains 1309 ha mountainous catchments of Lake Stymfalia basin, included to the protected areas of the NATURA 2000 network (Dafis et al., 1996). This site is a very important refuge for migratory birds and it is home of the endemic fish species *Pseudophoxinus s. stymphalicus* and the subspecies *Leuciscus cephalus peloponnensis* (Georgiadis et al., 1999).

The region incorporates significant water resources since it includes a shallow freshwater body and many karstic springs. The examined area mainly comprises of calcareous rocks, which belongs to the Gavrovo – Tripoli geological zone. There are many karstic wells surrounding the lake suppling fresh water. The area belongs to the Greek Public and the local communities. From time to time the lake is even totally drained. During the droughts of the 1990s, the lake was completely dry and was used for agricultural purposes by the inhabitants of the nearby villages (Georgiadis et al., 1999).

Natural karstic sewers reduce the water of the lake. Artificial drainage is also a common method used to acquire land for cultivation. Another more or less natural threat to the lake is the deposition of silt and the following development of reed bed vegetation and hence the reduction of the surface area of the lake (Georgiadis et al., 1999). Agricultural pressures includes over-pumping of the water for irrigation, which has a detrimental effect on the hydrological balance; the application of high amounts of fertilizers and agrochemicals in adjacent fields causes water quality degradation, waste disposal; and the expansion of cultivations in the wetland area.

Efforts to make the water of the lake worthy for irrigation of the dry coastal areas of Korinthos have been made from the times of the Emperor Adrianos (128 AD) and are even partially applied today. The Lake is supplied with waters from the surface flow of the hydrologic basin and from springs in the Velatsouri region. In this area exists a characteristic ditch, which is evident mainly from the place "Siouri". The ditch is constructed in the north-eastern end of the lake, where the waters from surrounding springs are concentrated. A few sewers are used for the natural drainage of lake. The main important ones are

Fig. 1 Location of Lake Stymfalia in Southern Greece. Circles denoted the most important natural karstic sewers

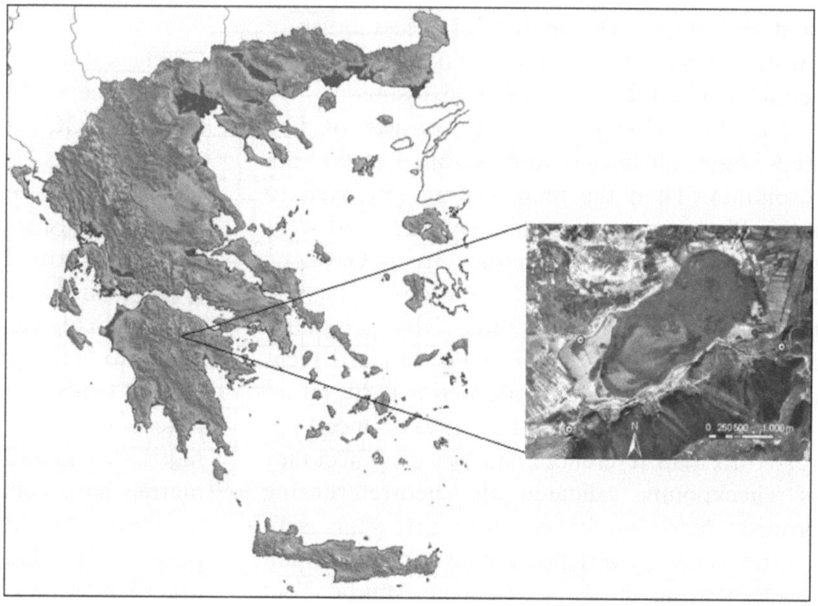

those of "Gidomantra" and "Forsta". "Gidomantra" sewer is located in the south -western end of lake. During the most part of the year the sewer remains inactive and is activated in periods of intense rainfalls (Kalliris & Spinthakis, 1997). "Forsta" sewer is located in the north–western end of lake. In June when the waters start to withdraw, a small flow from the lake to the sewer is developed. The drainage operation of this sewer is inverted during the winter months and supplies the lake with water. The drainage of lake is also supplemented with small sinks located near the spring of Velatsouri.

Remotely sensed data elaboration

Remote sensing techniques have been applied to identify and quantify land cover alterations during the examined period. For the purposes of this scientific effort remote sensing techniques have been used to acquire the land cover data of the area and Arc View 8.3 GIS program has enhanced their processing and change detection analysis. In particular a data-set of black and white aerial imagery acquired from the Hellenic Army Geographical Survey (HAGS) for the years 1945, 1960, 1972, 1987, 1992, and 1996 were used in order to acquire land cover data of the area. The aerial imagery data set was acquired during the summer period (mainly August at scale of 1:30 000–1:40 000) allowing a reliable comparison of the specific land cover types. The absence of excess moisture surplus or humidity during the period of photo acquisition and season secured this issue.

The methodology followed consists of two steps. First, all images were scanned at 600 dpi resolution. Then the images were imported to Leica Photogrammetric Suite software and were rectified and projected to the Greek Geodetic Reference System (EGSA 87) using the nearest neighbour resampling method. The geometric correction was carried out using standard techniques proposed in the international literature (Schowengerdt, 1997; Mather, 1999). Once converted to map references, assessing the accuracy of checkpoints validated the geo-referencing process. Next, all images for each year were geo-referenced, overlapping images were merged together using the Erdas Imagine Mosaic Tool (ERDAS, 2003) to provide continuous coverage of each target area, in the form of orthophoto maps for each year of analysis, respectively (Weng, 2001).

The second step involves the recognition and mapping of the presented land cover types from the available aerial photographs. Stereoscopic observation and photo-interpretation were used and the borders of these areas were defined. Next, the geo-referenced aerial photographs were stored in a GIS spatial database. Conducted land surveys of the area were used for ground truthing of the photo interpretation results, with the aid of ArcGIS 8.3 software (ESRI, 2002). Thus, land cover maps were derived showing the relative cover type areas for each year, respectively. The outer perimeter boundary of the lake's catchment's area polygon was used to identify the protected area proposed to be included to the European network NATURA 2000 network. The prepared thematic map from the 1996 aerial photographs was ground verified using GPS, so that each area represented most closely to the conditions occurring in 2004. Focus was mainly given to the cover of reed beds, open water areas (lake) and periodically flooded regions. Moreover, change detection analysis has been applied to derive land cover maps and respective changes on spatial and temporal basis.

Landscape Analysis

Changes at the landscape level were assessed using the landscape structure analysis program FRAGSTATS*ARC (Mc Garigal & Marks, 1995). Patch characteristics can be used to describe landscape features (Forman, 1995). Thus, several patch-related landscape indices were generated and analysed across the study area based on classified image data. The classification maps were exported to the Grid Module of ArcInfo GIS program (ArcInfo is the product of ESRI Inc.) and patch-related landscape indices listed below were computed. We used a series of landscape pattern metrics to quantify change in the landscape. Such metrics have been widely applied in similar studies, and allow the description of the temporal pattern of landscape change and comparison with other landscapes (Turner et al., 2001).

The indices chosen to describe the changes in landscape structure were number of patches (NP), patch density (PD number per 100 ha), largest patch index (LPI, %), total edge lengths (TE, meters), landscape shape index (LSI), and patch area mean (AREA_MN).

Total landscape area (TA, ha), equals the total area (m^2) of the landscape, divided by 10,000 (to convert to hectares). Number of patches (NP), equals the number of patches in the landscape. Patch density (PD) equals the number of patches in the landscape, divided by total landscape area (m^2), multiplied by 10,000 and 100 (to convert to 100 ha). Patch density (PD) as an index has the same basic utility as the number of patches, except that it expresses the number of patches per unit area. Total edge lengths (TE), equals the sum of the lengths (m) of all edge segments in the landscape. Largest patch index (LPI, %), equals the area (m^2) of the largest patch in the landscape divided by total landscape area (m^2), multiplied by 100 (to convert to a percentage); in other words, LPI equals the percent of the landscape that the largest patch comprises. Landscape shape index (LSI), equals the total length of edge in the landscape, given in number of cell surfaces, divided by the minimum total length of edge possible, also given in number of cell surfaces, which is achieved when the landscape consists of a single patch. Mean patch area (AREA_MN), equals the sum, across all patches of the corresponding patch type (or all patches in the landscape), of the corresponding patch metric values, divided by the number of patches of the same type (or total number of patches). MN is given in the same units as the corresponding patch metric.

A thorough explanation of these metrics is given by Mc Garigal & Marks (1995) and Turner et al., (2001).

Results

Land cover types analyses

A spatio-temporal quantification of the changes of Lake Stymfalia between the years 1945–1996 was performed. Over the period covered by the six surveys the corresponding thematic maps were produced. A classification system of land cover/use types was developed for the purposes of the study (Table 1) and the identified and delineated boundaries of the polygons of the categories were transferred from the aerial photographs on to the base maps. The descriptions of these types are shown in Table 1.

For the entire NATURA 2000 proposed area, we classified gross habitat types (open water, aquatic emergent's, seasonally flooded areas, woody vegetation, other), mapped their distribution and calculated their areas. Extensive field's observations had been used to correct ambiguities in land cover/use mapping. In addition, indices of patchiness' and edge vegetation (open water - reed beds) were calculated for each region and each image.

Table 1 The lands cover/use types classification system used for Stymfalia wetland, Greece (followed EUNIS habitat classification system, Davies et al., 2004)

Land cover/use types	Description
Periodically flooded areas	Periodically flooded areas mainly during winter months
Arable land	Agricultural land presently cultivated by crops, mainly irrigated from the lake.
Garrigues	Garrigues of the eastern Mediterranean areas (dominated by *Quercus coccifera, Pistacia lentiscus, Cistus spp.*)
Open water areas / lakes	Lakes open water area possibly containing aquatic plants of Potametea or Lemnetea class close to the littoral zone.
Wet meadows / marshes	Areas periodically inundated which are characterized by vegetation that requires saturated soils for growth. Usually communities of Molinio- Holoschoenion, Magnocaricetea or in some wetter places different associations of Phragmitetalia.
Reed beds	Usually reed- dominates associations of Phragmitetalia.
Scrubs and trees	Wet forested land of *Salix alba* and *Populus alba* species.
Forests	Greek forests of the endemic species *Abies cephalonica*

The frequency and the quality of the aerial photographs available for the analysis necessarily limit the study. The assumption is made that the changes have occurred at the time that they were recorded in the database. This approach overlooks the changes that take place between the snapshots and considers that the environmental and human activities affecting the changes in the lake were instantaneous (Kalivas et al., 2003). The changes that occur between the time overlays must be determined from the spatial patterns between two successive overlays.

The comparative study of natural conditions in Stymfalia wetland between 1945 and 1996 presented the changes in land cover/use that occurred during this period. The changes since 1945 extended mainly throughout the eastern and western part of the wetland. They involved decreases in wet-meadows (96.5%, Table 2) and open water areas (53.7%). The land cover/use categories that increased were the reed beds (89.3%) and irrigated agricultural land (3.96%, Table 2, Fig. 2, 3). Long-term variability in aggregate characteristics of the wetland vegetation was linked with environmental changes and human impacts on the catchments area.

The central and eastern part of the lake that is today covered by dense mat of reed beds does not appear in the 1945 air photos in its present formation. There was much more open water area, with only small patches of reed beds (Fig. 3, 4).

The total area of reed beds increased mainly during the 1945–1960, 1960–1972, and 1992–1996 periods (Table 2). The emergent vegetation mainly occupied by reed increases from 10.1% in 1945 to 19.2% in 1996 (Fig. 2, 3). Reed bed development was almost two times faster in the last than in the first period (6.64 ha yr^{-1}, compared to 2.81 ha yr^{-1}, Table 2). The reed *Phragmites australis* is the dominant species in macrophyte communities of the lake. The open water area of the wetland has decreased since 1945, as 104.48 ha have been lost, with a comparable increase in the extent of reed beds (Table 2, Fig. 3, 4). Aerial photo-interpretation shows that new reed beds are extending along both the eastern and even the central part of the lake, with great changes between the different years. From the quantitative analysis (Table 2) open water areas fell 53.7% of the total, through 39.8% in 1960 and 26.4% in 1996, as the water level was lowered. Of particular interest is that in 1987 an increase was observed mainly due to increased water level and the subsequent decrease of periodically flooded areas. In 1987 the surface of the lake covered by the periodically flooded areas subsequently decreased by 70.2% (Table 2). The wet meadows have been subjected to considerable losses, 35.5 ha since 1945 (96.51%, Table 2, Fig. 3).

Only minor changes (4.29 ha and 6.12 ha) occurred on forests and garrigues (Table 2, Fig. 3). All such changes are located on the northern and southern slopes and were mainly into pastures and secondarily into irrigated agricultural lands. Although the general structure of the forested and scrub areas did not change between 1945 and 2004, some forest areas in the 1945 aerial photos seem to have been less dense than in 1996. The small increase in forest area attributed to the rehabilitation works of the local forest service. Also during the last years some plantations of *Populus alba* at the eastern part of the lake increase the total area of scrubs and trees (Table 2).

Landscape changes

The landscape metrics for the years 1945, 1960, 1972, 1987, 1992, and 1996 in landscape level are presented in Table 3.

As indicated by landscape indices (Table 3), the intensification of agriculture and human induced activities have modified landscape characteristics and this process varies according to land cover types. Patch density (PD), landscape shape index (LSI), and total edge length (TE) decreased at the landscape level over the period concerned (Table 3). PD decreased from 2.6 patches /100 ha to 1.9 patches /100 ha. LSI decreased from value 4.86 to 4.33 and TE decreased from 48.33 to 40.62 km.

When total landscape area remains constant, patch density and the number of patches conveys the same information. PD is the index of heterogeneity, because a landscape with a higher patch density shows greater spatial heterogeneity (Zhao

Table 2 Changes in the total area (ha) and the area of different land use/ cover extent (%) for the period 1945–1996, in the study area

Land use	Area (ha)						% change						Annual change (ha/year)				
	1945	1960	1972	1987	1992	1996	1945–1960	1960–1972	1972–1987	1987–1992	1992–1996	1945–1996	1945–1960	1960–1972	1972–1987	1987–1992	1992–1996
Arable land	752.35	768.29	768.42	778.58	791.46	782.17	2.12	0.02	1.32	1.65	-1.17	3.96	1.06	0.01	0.68	2.57	-2.32
Forest	26.51	27.61	28.38	30.80	30.80	30.80	4.17	2.80	8.51	0.00	0.00	16.21	0.07	0.06	0.16	0.00	0.00
Garrigues	122.07	121.77	121.93	115.10	116.06	115.95	-0.25	0.13	-5.60	0.84	-0.09	-5.01	-0.02	0.01	-0.46	0.19	-0.03
Open water areas/ lake	194.64	117.21	94.17	138.05	101.62	90.16	-39.78	-19.66	46.60	-26.39	-11.28	-53.68	-5.16	-1.92	2.93	-7.29	-2.87
Periodically flooded areas	44.00	82.32	66.87	19.92	38.23	34.61	87.11	-18.77	-70.22	91.96	-9.46	-21.33	2.56	-1.29	-3.13	3.66	-0.90
Reed beds	132.55	174.67	203.42	217.98	224.32	250.90	31.77	16.46	7.16	2.91	11.85	89.28	2.81	2.40	0.97	1.27	6.64
Scrubs and trees		0.51	1.08	2.86	2.94	3.03		111.21	165.39	2.73	3.37		0.03	0.05	0.12	0.02	0.02
Wet meadows / marshes	36.80	16.54	24.64	5.64	3.49	1.28	-55.06	49.02	-77.11	-38.09	-63.27	-96.51	-1.35	0.68	-1.27	-0.43	-0.55

et al., 2003). Like patch density total edge length (TE) is also related to the degree of spatial heterogeneity in the landscape, but serves as a measure of the shapes of patches.

Additionally and at the same time periods, Mean Patch Area (MN) values increased from 38.50 ha to 52.36 ha and the Largest Patch Index (LPI) percentages increased 38.7–59.7% at the landscape level in Lake Stymfalia (Table 3). These results suggest a tendency for decreased spatial heterogeneity in the study area. The main reason for that may be the increased abstraction of water from the surroundings streams and the subsurface drains in the fields as part of the intensification of agriculture (see also site description). The inclination noted for the year 1987 (Table 3) could be due to the increase of open water areas and reed beds and the significant decrease of flooded areas in comparison with the other examined periods (see Table 2).

Discussion

Archive maps contain much information about past environments and if viewed in a historical series, can depict environmental change and response in some detail. As one element of a complex study of wetland dynamics this archive cartography can be an important resource.

The proposed methodology, for deriving baseline data covered the needs for spatial information of a monitoring protocol. In particular, baseline data for land use, wetland habitats, and infrastructures were derived at the basin level. The operational remote sensing and GIS methods outlined here, provided spatially distributed information of high quality. This information was presented in the form of thematic maps (Fig. 3, 4), which described the state of ecosystem as a result of human pressures.

Furthermore, data limitations are a major problem for quantifying past and present lake - water level variations with high precision. The results of this study show the value and efficiency of using remotely sensed data and GIS utilities in a low cost and time basis providing reliable results in environmental management. The importance of these baseline data can be directly derived by

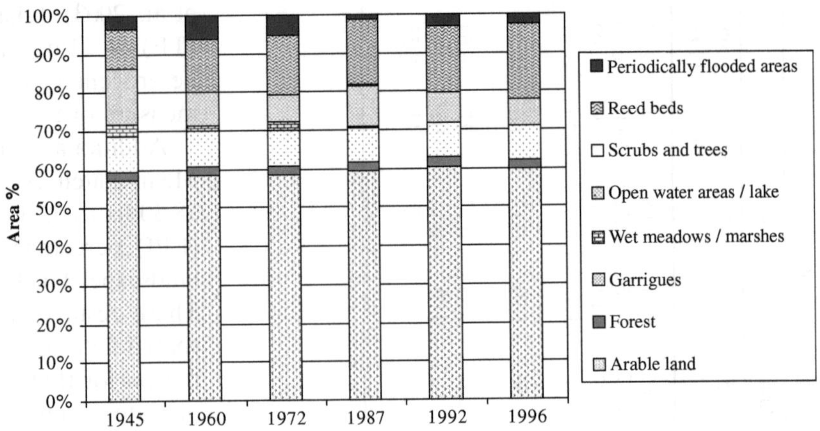

Fig. 2 Area extent (%) of the main land cover classes as derived from the classification of black and white aerial photographs of Lake Stymfalia

the obligations that are clearly defined in Directives 92/43/EEC and 2000/60/EC (WFD, 2000).

Aerial photographs have proved to be an efficient tool in providing reliable, low-cost, classification data on land cover/use of the Stymfalia wetland. Displays of these data, in the form of digital maps will permit resource managers and planners to acquire the basic information that is necessary for analyzing and understanding environmental problems and developing suitable management plans. Need for fieldwork was kept to a minimum, as automated image analysis techniques were preferred against traditional surveys, therefore reducing the total costs.

GIS-based analysis of six images of the Stymfalia wetland, taken between 1945 and 1996, demonstrated variability of habitat features in response to environmental changes and human impact (Fig. 2). Indeed, the change detection maps provided useful information regarding changes of certain land cover classes, which described the dominant pressures in the study area.

Reeds are widespread in the lake because *Phragmites australis* is the most competitive species. It forms dense monospecific stands all over the littoral zone of the lake. The recent appearance of the extended dense reed formations (Fig. 2, Table 2) covering more than 60% of the open water areas (which was observed in all studied aerial photos) may be a result of man-made alterations to the bed of the perennial stream and spring wells (Kalliris & Spinthakis, 1997). These wells brought large quantities of water to the area (particularly during snow-melt season), causing an increase in local water volume and in the lake's depth, and at the same time preventing the spread of reeds on the lakeshores. The water quality of the lake was improved as the highly oxygenated springs water enters the lake. After the disappearance of the main springs for irrigation purposes, local water depth decreased and the existing sediments enhanced the formation of the reed beds mats. The reed in Lake Stymfalia have increased in the last 50 years (Fig. 3, 4), most likely due to eutrophication, low-water periods and declination of cattle breeding in the shore areas (Kalliris & Spinthakis, 1997). Today less dynamic conditions prevail (Table 3) and the wetland has a different morphological structure and vegetation composition than in the past. The expansion of reeds has changed significantly the habitat for fauna communities in the littoral zone of the lake.

Another serious problem of this shallow lake is the disappearance of submerged macrophytes that plays an important role in the structuring of the ecosystem. Due to the lack of complete species lists of the lake as well as historical data on plant composition, information on species diversity is missing. A serious decline in species diversity of the lake during the 90's was found by Georgiadis et al. (1999) and Papastergiadou et al. (unpublished data), the latter reporting a decrease of submerged macrophytes, and the presence of some free floating species such as *Lemna minor* and *Spirodella polyrhiza*, with low abundances. Also a well-developed structure during

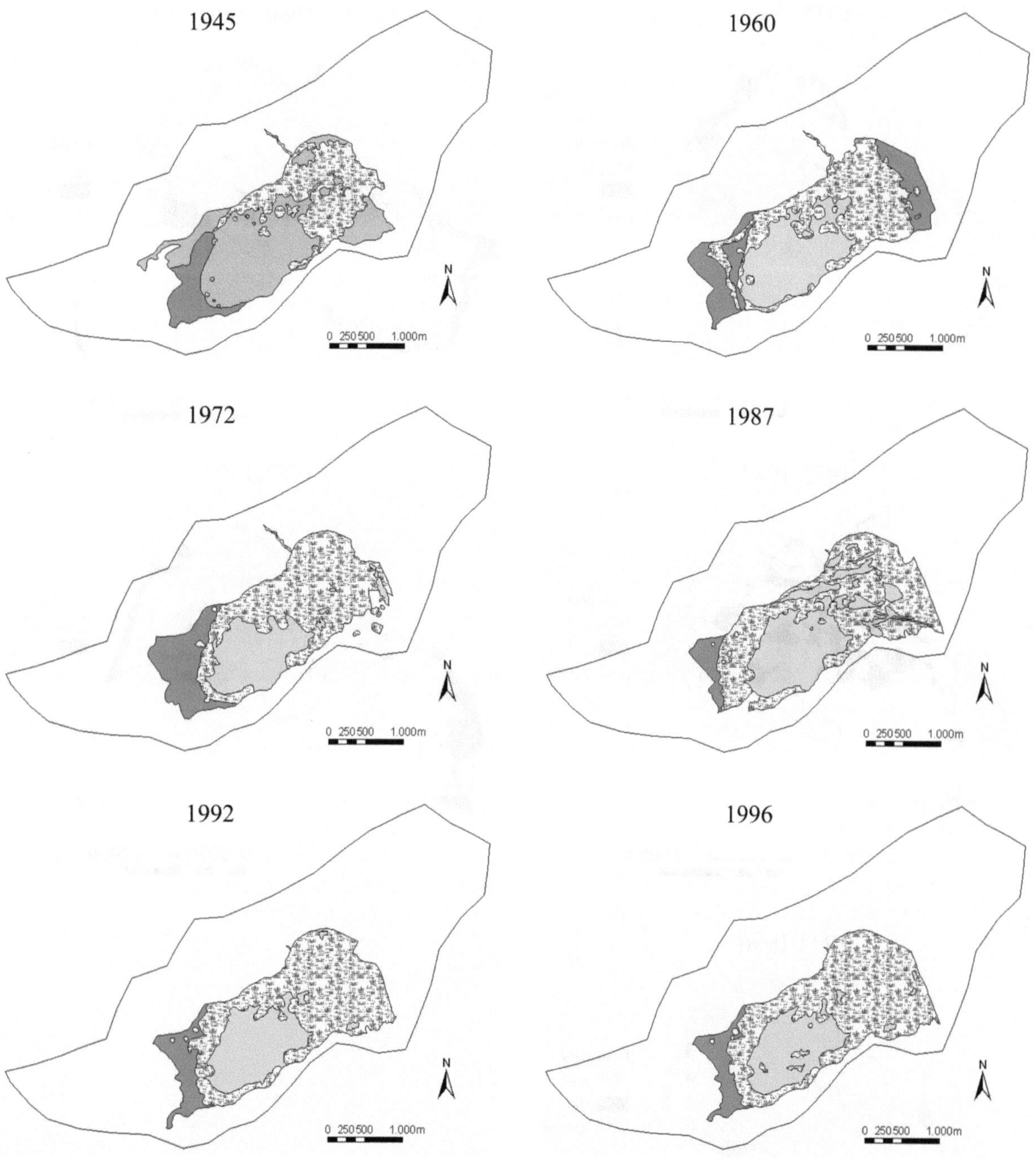

Fig. 3 Temporal cover changes for the types of open water areas – lake (light gray), periodically flooded areas (dark gray) and reed beds (dashed gray) for the period 1945–1996; the outer line corresponds to the Nature zone

spring months was recorded, with *Polygonum amphibium*, covering relatively small areas. Besides eutrophication, the intense lowering of the water table during recent years seems to have contributed considerably to the decline of many species of submerged macrophytes and disappearances of the most intolerant species. The loss of species diversity and biomass production results in loss of habitats for fauna species and changes in food- webs, carbon and nutrient cycles

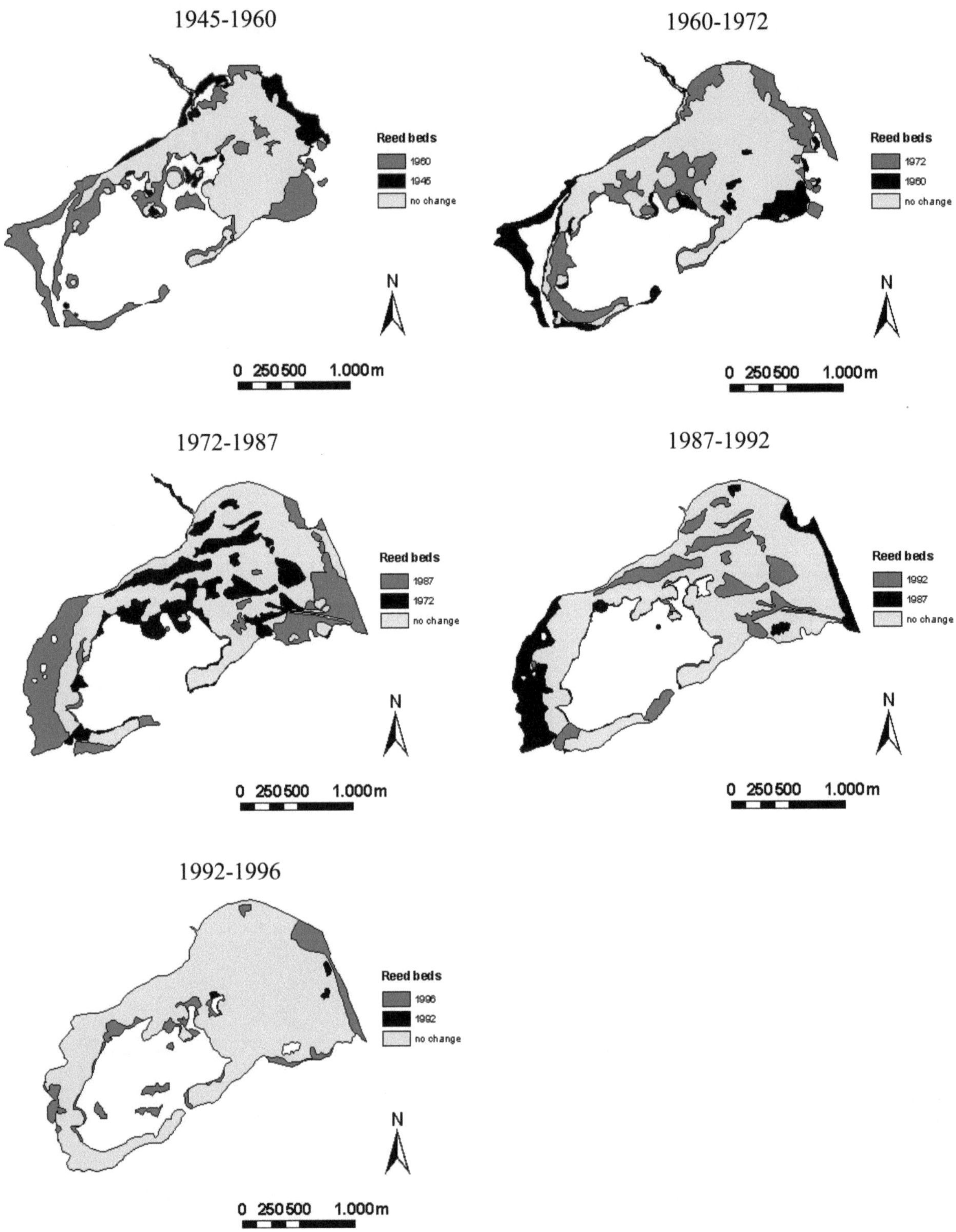

Fig. 4 Spatial-temporal changes in the cover area of reed beds for the period 1945–1996

Table 3 Landscape metrics for different periods in Lake Stymfalia, Greece

	1945	1960	1972	1987	1992	1996
Total landscape area (TA) ha	1308.9	1308.90	1308.90	1308.90	1308.90	1308.90
Number of Patches (NP)	34	25	27	35	21	25
Patch Density (PD) n/100 ha	2.60	1.91	2.06	2.67	1.60	1.91
Largest Patch Index (LPI) %	38.71	40.00	40.38	59.48	60.47	59.76
Total Edge length (TE) km	48.33	48.86	40.09	53.44	40.75	40.62
Landscape Shape Index (LSI)	4.86	4.90	4.29	5.22	4.34	4.33
Mean Patch Area (AREA_MN) ha	38.50	52.36	48.48	37.40	62.33	52.36

(Scheffer et al., 1992; Sand-Jensen et al., 2000).

The lake's surface and depth are restricted by the deposition of sediments from their drainage basin. The suspended material considerably reduces depth of light penetration, which is an essential factor for macrophytes growth, leading to water quality deterioration, while also negatively affecting the fish population of the lake (Scheffer et al., 1992).

The extent of emergent vegetation in the Stymfalia wetland, although variable between images, remained 50% or higher of the total area (Fig. 2, Table 2). At present there is no management of the reed beds and decomposition of these large quantities of reed biomass, which accumulate on the lake bottom, causes accelerated ageing and succession to more terrestrial habitats of the lake (Papastergiadou et al., 2002).

A significant part of the periodically flooded areas and wet meadows have been lost during this period, leading to direct environmental loss and allowing non-environmental friendly agricultural activities to be developed very close to the lake. Almost all areas covered by wet meadows has been drained and transformed to farmland (Fig. 2, 3, Table 2). All of the lost wet meadows are located within the irrigation/ drainage network mainly at the eastern part of the lake. Wet meadows still remain today in very few places, mainly at the naturally flooded areas only during the spring and early summer period at the eastern part of the lake.

As expected the intensification of agriculture has meant that the landscape has become more homogeneous, less diverse at the landscape level (Table 3), and lost the habitat fragmentation of the past years (Forman, 1995). The results (Table 3), suggest a tendency for decreased spatial heterogeneity in the studied area. This problem is enhanced by the fact that the particular area is still under agricultural development and therefore, future water demands are expected to increase.

Based on these considerations, a detailed watershed analysis including water abstractions, evaporation, groundwater discharge, and future irrigation demands will be conducted in the area, for the implementation of the appropriate restoration measures.

Since the irrigation/drainage network was improved, there has been a general trend of concentrating land use in the lowlands, close to the lake. In this area land use activities are continuously becoming more concentrated and more intensive. Modification of the natural environment is quite serious in certain cases, and the current land uses are not compatible with the objectives of the NATURA 2000 protected area. Concentration of agricultural activities close to or within the wetland is eroding the soil and destroying the natural vegetation and moreover affecting negatively the populations of endangered fauna species. Present agricultural methods demand increased inputs of chemicals and fertilizers, which flow into the lake with a consequent impact on the water quality and fisheries.

In general, the studied area shows increasing environmental degradation and resource depletion, while present conservation measures are not adequate to meet the environmental standards of a protected NATURA 2000 area. Therefore, a management plan is urgently needed, which will regulate all of the areas resources, in an integrated ecologically sound manner. The usefulness of such a survey is widespread, since environmental managers, policy-makers, scientists and local governments can utilize them in the planning process of an area's developmental phase as well as in designing protection and preservation measures and in generally, increase stakeholders' understanding.

Acknowledgements Aerial photo's used with kind permission of the Greek Ministry of Agriculture and HAGS. We would like to thank two anonymous reviewers, whose comments on a previous draft of the paper helped us substantially improve the manuscript.

References

Bromley, P., 1997. Nature Conservation in Europe: Policy And Practice. E & FN Spon, London.

Burrough, P. A., 1998. Principles of Geographical Information Systems. Oxford University Press, New York.

Dafis, S., E. Papastergiadou, K. Georghiou, D. Babalonas, Th. Georgiadis, M. Papageorgiou, E. Lazaridou & B. Tsiaoussi (eds), 1996. The Greek Habitat Project: NATURA 2000. An Overview. Commission of the European Communities, Goulandris Natural History Museum, 917 pp.

Davies, C. E., D. Moss & M. Hill, 2004. EUNIS Habitat Classification, revised 2004. Final Report to the European Topic Center on Nature Conservation, European Environment Agency, 307 pp.

ERDAS, 2003. ERDAS Imagine 8.7 Field Guide, ERDAS Inc., Atlanta, Georgia, USA.

ESRI, 2002. ArcGIS 8.3, Environmental Systems Research Institute, Redlands, California, USA.

Forman, R. T. T., 1995. Land Mosaic: The Ecology of Landscapes and Regions. Cambridge University Press, Cambridge, 43–48.

Georgiadis, Th., G. Iatrou, V. Chondropoulos, S. Fragedaki-Tsolis, G. Mitsainas & G. Tryfonopoylos, 1999. Identification and description of habitat types in NATURA 2000 sites, of southern Greece. GR 2530002 Limni Stymfalia. Ministry of Environment and Public Works (in Greek).

Kalliris, P. & E. Spinthakis, 1997. Lake Stymfalia: Specific Study of Ecological Management. Stymfalia Development Association, 237 p. (in Greek).

Kalivas, D. P., V. J. Kollias, G. Karantounias, 2003. A GIS for the assessment of the spatio-temporal changes of the Kotychi Lagoon, Western Peloponnese, Greece. Water Resources Management 17: 19–36.

Magnuson, J. J., K. E. Webster, R. A. Assel, C. J. Bowser, P. J. Dillon, J. G. Eaton, H. E. Evans, E. J. Fee, R. I. Hall, L. R. Mortsch, D. W. Schindler & F. H. Quin, 1997. Potential effects of climate changes on aquatic systems: Laurentian Great lakes and Precambrian Shield Regional. Hydrological Proceedings 11: 825–871.

Mather, P., 1999. Computer Processing of Remotely – Sensed Images, John Wiley.

McGarigal, K. & B. J. Marks, 1995. FRAGSTATS: spatial pattern analysis program for quantifying landscape structure. US Forest Service General Technical Report PNW 351: 103–104.

Papastergiadou, E., M. Agami & Y. Waisel, 2002. Restoration of aquatic vegetation in mediterranean wetlands. In: G. Zalidis, T. Crisman, P. A. Gerakis (eds), Restoration of Mediterranean wetlands, Medwet, 47–69.

Rembold, F., S. Carnicelli, M. Nori & G. A. Ferrari, 2000. Use of aerial photographs, Landsat TM imagery and multidisciplinary field survey for land-cover change analysis in the lakes region (Ethiopia). International Journal of Applied Earth Observation and Geoinformation 2(3–4): 181–189.

Sand-Jensen, K., 1997. Eutrophication and plant communities in Lake Fure during 100 years. In: K. Sand-Jensen & O. Pedersen (eds), Freshwater Biology: Priorities and Development in Danish Research. Gad Publishers, Copenhagen. 26–53.

Sand-Jensen, K., T. Riis, O. Vestergaard & S. E. Larsen, 2000. Macrophyte decline in Danish lakes and streams over the past 100 years. Journal of Ecology 88: 1030–1040.

Scheffer, M., M. R. de Redelijkheid & F. Noppert, 1992. Distribution and dynamics of submerged vegetation in a chain of shallow eutrophic lakes. Aquatic Botany 42: 199–216.

Schowengerdt, R. A., 1997. Remote Sensing: Models and Methods for Image Processing. Second Edition, Academic Press, New York.

Stalmans, M., K. Balkwill, E. Witkowski & K. H. Rogers, 2001. A landscape ecological approach to address scaling problems in conservation management and monitoring. Environmental Management 28: 389–401.

Turner, M. G., R. H. Gardner & R. V. O'Neil, 2001. Landscape Ecology in Theory and Practice: Pattern and Process. Springer – Verlag, New York.

Van den Berg, M. S., 1999. Charophyte Colonization in Shallow Lakes. Processes, ecological effects and implications for lake management. Ph.D. Thesis, University of Amsterdam, 138 p.

Weng, Q., 2001. Modelling urban growth effects on surface runoff with the integration of remote sensing and GIS. Environmental Management 28: 737–748.

WFD 2000/60/EC., 2000. Council Directive of 23 October 2000. On establishing a framework for community action in the field of water policy. Council of Europe 2000.

Williams, D. C. & J. G. Lyon, 1997. Historical aerial photographs and a geographic information system (GIS) to determine effects of long-term water level fluctuations on wetlands along the St. Marys River, Michigan, USA. Aquatic Botany 58: 363–378.

Zhao, S., J. Fang, W. Ji & Z. Tang, 2003. Lake restoration from impoldering: impact of land conversion on riparian landscape in Honghu Lake area, Central Yangtze. Agric. Ecosys. Environm. 95: 111–118.

Hydrobiologia (2007) 584:373–379
DOI 10.1007/s10750-007-0604-y

SHALLOW LAKES

Impact of climatic variability on parameters used in typology and ecological quality assessment of surface waters—implications on the Water Framework Directive

Peeter Nõges · Wouter Van de Bund · Ana Cristina Cardoso · Anna-Stiina Heiskanen

Abstract In most cases the negative impacts of climate change to aquatic ecosystems cannot be mitigated by measures in the river basin management. Ignoring climate change by the Water Framework Directive may have strong implications for the typology and quality assessment systems used for water bodies. As a result of climate change, water bodies, especially those located near the type boundaries may change their type. Compared to typology characteristics, water quality parameters are even more labile and may be easily affected by climate change. The paper exemplifies that the anticipated deterioration of water quality within the time frame relevant for WFD implementation may be large enough to endanger the fulfillment of the set water quality objectives. The review of the river basin characterization every six years, as required by the WFD, might also include re-evaluation of reference conditions according to the changes observed at pristine reference sites. As a consequence, the restoration targets (i.e., the good ecological status) would also need to be evaluated periodically.

Keywords Climate change · Water Framework Directive · Typology of water bodies · Ecological water quality · Reference conditions

Guest editors: R. D. Gulati, E. Lammens, N. De Pauw & E. Van Donk
Shallow lakes in a changing world

P. Nõges (✉) · W. Van de Bund · A. C. Cardoso · A.-S. Heiskanen
DG Joint Research Centre, Institute for Environment and Sustainability, Rural, Water and Ecosystem Resources Unit, TP 290, European Commission, 21020 Ispra, VA, Italy
e-mail: peeter.noges@jrc.it

Introduction

The Water Framework Directive (Directive, 2000/60/EC; WFD) creates a legislative framework to manage, use, protect, and restore surface water and groundwater resources in the European Union. The WFD approaches water management at the scale of major river catchments (river basins) that in many cases include several countries. The WFD requires the establishment of a 'river basin management plan' (RBMP) for each of the river basins. The RBMP is a detailed account of how the environmental objectives (i.e., good ecological status of natural water bodies and good ecological potential of heavily modified and artificial water bodies) are to be achieved. The environmental objectives of the WFD should be reached by 2015. For those countries, which can demonstrate that this is not feasible without disproportionate economic and social costs, the Directive allows the possibility for a delay until 2030. This sets a time scale for restoration of the water bodies during which a considerable change in climate can be expected. Although the WFD does not explicitly mention risks posed by climate change to the achievement of environmental objectives, there is capacity within the WFD to mitigate some of the climate change consequences and adapt to the others through the river basin planning process.

The WFD defines ecological status as "... an expression of the quality of the structure and functioning of aquatic ecosystems associated with surface waters ..." (Directive, 2000: Article 2: 21). The assessment of the ecological status in the WFD requires classification of water bodies in five quality classes using the Ecological Quality Ratio (EQR), which is the ratio between reference conditions and measured status of the biological quality elements. Reference conditions should equal almost natural conditions. The anticipated effects of the climate change will affect most of the physical, chemical and biological parameters of water bodies used for ecological quality assessment although, it may be often hard to disentangle the climate change effects on surface waters from the local direct human impact (IPCC, 2001; Eisenreich, 2005). In some cases, climate change may bring a relief to environmental problems and help to achieve the restoration goals set by the Directive. For example, shorter duration of the ice-cover and higher water level will diminish the risk of winter fish-kills in northern shallow lakes (Fang & Stefan, 2000; Järvalt et al., 2005), a phenomenon, which may be considered natural but has been aggravated by anthropogenic eutrophication. Shortening of water retention time in lakes and reservoirs in areas with a prospected increase in precipitation will suppress water bloom development (UNEP/UNESCO, 2002). However, in most cases pressures to aquatic ecosystems will increase as a result of climate change imposing effects that cannot be mitigated by measures taken in river basin management.

In this paper we discuss the conceptual impacts of the climate change on the process and requirements of the ecological status assessment in the WFD. Using some examples we demonstrate the potential extent of changes that climate change may bring about and consider the various steps in the classification and assessment where these effects should be considered. We argue that reference conditions cannot be considered as static but will change as a consequence of climate change impacts on physical and chemical conditions of water bodies. The review of the river basin characterization every six years, as required by the WFD, might also include re-evaluation of reference conditions. As a consequence, the restoration targets (i.e., the good ecological status) would also need to be evaluated periodically.

Mitigation of climate change effects or adaptation to them?

Although the influence of human activities on climate represents an adverse environmental impact, the management of that impact is global rather than only a European-scale issue. Considering that climate change may affect the natural range of values of the WFD quality elements, Owen et al. (2001) showed that if reference conditions are not updated in these circumstances, a deterioration in status could be recorded under the Directive's classification scheme. The concept of reference conditions is the anchor point of the WFD methodology for water quality assessment and any change in these conditions can be justified only if it

(1) does not counteract to the WFD's main goal of protecting and improving the quality of European waters and
(2) would avoid setting unachievable goals and making unreasonable investments into measures that cannot be effective in the changed climatic boundary conditions.

It should be carefully analysed which aspects of water quality might be given some flexibility and which standards are so essential to our use and appreciation of water bodies that they should be fixed, despite the impacts of climate change. Certainly no alleviations can be made for example to the environmental quality standards of priority substances in water although the mobility of these substances, which controls their release from secondary sources like soils and lake sediments, is likely to be affected by climate change (Eisenreich, 2005).

In single cases, measures applied in the river basin management can indeed mitigate the adverse effects of climate impact. For example, the climate-induced acceleration of eutrophication resulting from altered soil processes in the catchment (McDowell et al., 2001; Randall & Mulla, 2001; Justic et al., 2003), increased nutrient release from the sediments (Liikainen, 2002), or enhanced microbial decomposition of organic matter (Beznosov & Suzdaleva, 2004), can to a certain extent be remediated by appropriate practices in agriculture and forestry. However, there will be adverse effects of the climate change that cannot be avoided, even with coordinated action at a European level. As a consequence, classification scales need to be adapted taking into account the effects of climate change. Reference conditions would need to be reviewed also in cases when climate-induced changes in timings of biological key events like spawning, hatching or growth phases of different biological groups cause a mismatch in the food web leading to altered functioning of the entire ecosystem (Winder & Schindler, 2004). Species forced to migrate because of climate change may have a similar effect on ecosystems in which they act as alien invaders and may turn the established quality assessment scales inapplicable. The following two examples illustrate climate change impacts on biological quality elements which cannot be remediated within river basins: the habitat loss for benthic invertebrates and fish with developing anoxia in lakes, and with increasing effect of droughts in intermittent streams.

A number of examples demonstrate that climate change may have a strong effect on hypolimnetic dissolved oxygen concentration in lakes that affect habitats for benthic macroinvertebrates (Quinlan et al., 1998, 2002; Stendera & Johnson, 2005) and fish (Nurnberg, 2004). The decrease of hypolimnetic oxygen levels is a dual effect of temperature caused, on one hand, by the lower solubility of oxygen in warmer water but mostly by the decrease of mixing intensity due to stronger thermal stratification (Salmaso et al., 2003; Salmaso, 2005). Among climatic factors, land use changes and urban impacts, climate (winter temperature) explained the largest part of community variance and significantly influenced the relative abundance of littoral taxa in a 73-year time-series on chironomids from eight Canadian lakes (Quinlan et al., 2002).

As shown by the ECHAM4 and HadCM3 model simulations for the 2020s, in some parts of Europe the today's 100-year droughts will return every ten years or even more frequently (Lehner & Döll, 2001). During droughts river flow falls to extremely low levels for extended periods of time. The most frequently demonstrated effects of drought were population declines, loss of habitat, changes in the community, negative effects from changes in water quality, loss of stream connectivity, alteration of food resources, changes in the strength and structure of interspecific interactions and crowding of animals in reduced microhabitats (Lake, 2003; Matthews & Marsh-Matthews, 2004). After drought the recruitment by taxa that lack desiccation-resistant stages or have limited mobility is delayed (Boulton, 2003). One of the few restoration measures for intermittent streams would be the creation of drought refugia (Lake, 2003).

Typology and type specific reference conditions of water bodies

Depending on geographic region, catchment geology and size, water bodies obtain naturally different chemical and physical features and offer different habitats for biota. Water bodies located in areas of fertile soils with rich mineral composition show naturally higher nutrient concentrations and higher productivity compared, e.g., to water bodies in silicious catchments. Trophic scales like that worked out by the OECD (1982) are useful tools for describing these differences but do not allow distinguishing between natural and anthropogenic eutrophication. To enable the distinction between natural variability and anthropogenic changes, the WFD implementation is based on a detailed typology of water bodies and type specific reference conditions (RC) that could serve as the basis for water quality assessment. As the principle issues for the WFD implementation related to typology of water bodies, Owen et al. (2001) pointed out the following questions:

- how much natural variation can be accommodated within types?
- how can we differentiate between natural variation and impact?
- should we update the natural range of values to accommodate "natural" changes, such as climate change?

The factors used to build up the typology are summarised in Table 1. Several of the typology factors suggested by the WFD are actually climatic variables (mean air temperature, precipitation), have an intimate linkage to them (altitude, latitude, longitude, river flow, water level, lake mixing) or may be influenced by climate in a longer perspective (morphometric characteristics).

Table 1 Obligatory and optional factors for characterisation of surface water body types given in Annex II of the WFD

	Rivers	Lakes
Obligatory factors	Altitude Latitude Longitude Geology Size	Altitude Latitude Longitude Depth Geology Size
Optional factors	Distance from river source	Mean water depth
	Energy of flow	Lake shape
	Mean water width	Residence time
	Mean water depth	Mean air temperature
	Mean water slope	Air temperature range
	Form and shape of main river bed	Mixing characteristics
	River flow category	Acid neutralising capacity
	Valley shape	Background nutrient status
	Transport of solids	Mean substratum composition
	Acid neutralising capacity	Water level fluctuation
	Mean substratum composition	
	Chloride	
	Air temperature range	
	Mean air temperature	
	Precipitation	

The following examples illustrate the extent of the observed and anticipated changes in inland waters related to climate change. As a result of climate change, water bodies, especially those located near the type boundaries, may change their type. The most probable changes, already observed in some cases, are the changes in lake mixing type. In warmer climate cold monomictic lakes may stratify in summer and become dimictic (Sorvari, 2001). The disappearance of ice-cover will cause a continuous mixing in winter turning previously dimictic lakes into warm monomictic lakes (Croley et al., 1998). Higher stability of the water column may prevent full mixing of deep lakes changing them from holomictic to meromictic (Ambrosetti et al., 2003). An increase in annual mean temperature by 1°C corresponds approximately to a latitudinal habitat shift of 100–200 kilometers (Richardson et al., 2004; Wheeler et al., 2003) or to an altitude change of nearly 200 meters (Körner, 1999; Agustí-Panareda & Thompson, 2002). Hence, even the geographic coordinates may prove not to be sufficiently good type descriptors in changing climate. Even the morphometry of water bodies that generally can be considered rather conservative, may change. In more than 10,000 large lakes in rapidly warming Siberia a widespread decline in lake abundance and area was observed since 1973 caused by rapid lake draining into the subsurface after the thaw and breaching of permafrost (Smith et al., 2005). Decrease in precipitation often accompanied by increased summer temperatures in the future climate scenarios for warm regions will consistently change river discharges, increase evaporation and cause a shift from permanent to temporary water bodies. All these changes will have a major impact on the aquatic ecosystems.

Drift in water quality parameters

Compared to the more conservative typology characteristics, water quality parameters within each type are supposed to change according to pressure gradients. As being labile by their nature, water quality parameters may be easily affected by climate change. The quality class boundaries within the WFD are set in relation to the reference conditions. Six principal methods were suggested by the REFCOND Working Group (REFCOND, 2003) for setting type-specific reference conditions. Four of them, based on using spatial networks, modelling, curve fitting, and expert judgement, are able to consider also climate change impacts, while methods based on paleo-ecology and historical data are more static.

By introducing the category of heavily modified water bodies (HMWB), the WFD acknowledges that due to major physical alterations of some water bodies restoration to good ecological status may not be achievable even in the long-term. The inability to achieve good ecological status may also be caused by irreversible environmental changes. In order to incorporate those in the WFD assessment systems, some flexibility should be introduced to RC making them capable of responding to natural changes.

There are several ways to make the RC flexible. This could be based on periodical monitoring of a set of reference sites and adjusting the values of reference conditions taking into account the effects of long-term natural processes (de Wilde et al., 2002) including climate change. Periodic updating of RC would be in line with the concept of adaptive management which acknowledges that policies must be continually modified to adapt to new emerging uncertainties. Natural variation can be included within the RC also by relating the latter to the natural causes of variability. In this case the procedure of setting reference conditions should, instead of a single RC value, yield a table or a nomogramm showing the relationship of the quality parameter to the most important natural factors. For example, the growth of phytobenthos and submerged macrophytes in lakes depends on the illumination of the lake bottom. In Lake Võrtsjärv the proportion of aphotic bottom area (receiving <1% of the incident light at the water surface) may change from less than 5% to 90% of the total lake area as a function of variable water transparency and fluctuating water level (Fig. 1). Hence, the RC for macrophyte abundance in this lake for high- and low-water years should be different.

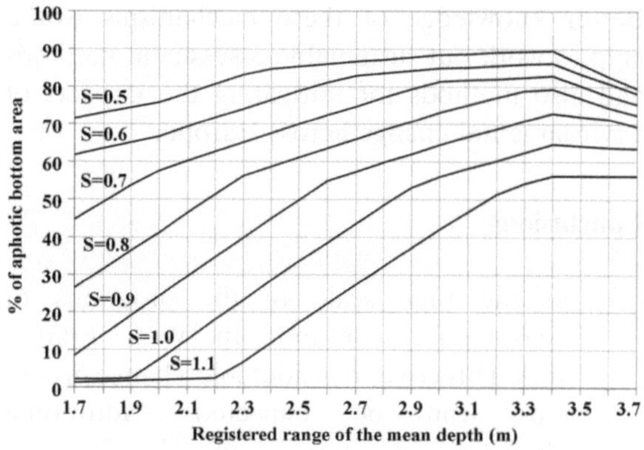

Fig. 1 Percentage of aphotic bottom area (receiving <1% of the incident light at the surface) in L. Võrtsjärv as a function of the mean depth of the lake and the Secchi depth (S). The light attenuation coefficient was calculated as $k = (1.48 \pm 0.05)/S$ (Nõges & Nõges, 1998)

Climate change sets new challenges for ecosystem research and monitoring

Most of the methods currently used to monitor and model water bodies are based on a deterministic rather than a probabilistic approach and do not properly represent the 'cascade of uncertainty' (Rayner, 2000) associated with recent climate-change scenarios. The future surveillance monitoring must be capable of detecting long-term changes in the water environment. For example, long-term climate changes and land cover changes could affect the condition of aquatic plants and animals in surface waters by increasing the frequency of droughts and floods. Information on such changes will be needed to avoid wrongly attributing their affects to other pressures.

Climate change effects on aquatic ecosystems are not fully understood, and are in many cases inherently unpredictable. Long-term monitoring of minimally impacted sites is highly important to measure and understand climate change effects on reference conditions.

To advance our understanding of ecosystem responses to climate change, basic ecological research would still be needed. Projections of future biotic responses cannot be based on simple extrapolations of present trends but must include consideration of physiological tolerances, competition, and dispersal mechanisms (Halpin, 1997).

Better knowledge of these mechanisms would help to work out sustainable assessment methods and also methods for mitigating the impacts of environmental change across Europe.

Conclusions

(1) At the time scale of the programs of measures foreseen under the Water Framework Directive, considerable change in climate can be expected. Additional adaptation measures should be foreseen within the WFD Common Implementation Strategy to cope with the impacts of climate change.
(2) Only in single cases, when climate change effects on aquatic ecosystems are expressed in accelerated eutrophication, they can be mitigated to a certain extent at the river basins level by applying more stringent measures to avoid nutrient leakage from point- and non-point sources.
(3) Adaptation of the classification scales could be allowed by adjusting reference conditions to accommodate effects of climate change that cannot reasonably be mitigated.
(4) Climate change effects on aquatic ecosystems are not fully understood, and in many cases inherently unpredictable. Long-term monitoring of minimally impacted sites is highly important to measure and understand climate change effects on reference conditions.

Acknowledgement Funding for this research was partly provided by the European Union project CLIME (contract EVK1-CT-2002-00121).

References

Agustí-Panareda, A. & R. Thompson, 2002. Reconstructing air temperature at eleven remote alpine and arctic lakes in Europe from 1781 to 1997 AD. Journal of Paleolimnology 28: 7–23.

Ambrosetti, W., Brabanti L. & N. Sala, 2003. Residence time and physical processes in lakes. Journal of Limnology 62: 1–15.

Beznosov, V. N. & A. L. Suzdaleva, 2004. Potential changes in aquatic biota in the period of global climate warming. Water Resources 31: 459–464.

Boulton, A. J., 2003. Parallels and contrasts in the effects of drought on stream macroinvertebrate assemblages. Freshwater Biology 48: 1173–1185, doi:10.1046/j.1365-2427.2003.01084.x.

Croley, T. E., Quinn F. H., Kunkel K. E. & S. A. Changnon, 1998. Great lakes hydrology under transposed climates. Climatic Change 38: 405–433.

de Wilde, A. J., R. A. E. Knoben & J. W. P. M. van Poppel, 2002. Setting class boundaries for the classification of rivers and lakes in Europe. Final Report Project number 39534, Client Ministerie van Verkeer en Waterstaat. DG RWS—RIZA, 1–25.

Directive, 2000. Directive 2000/60/EC of the European Parliament and of the council of 23 october 2000 establishing a framework for community action in the field of water policy. Official Journal of the European Communities L 327: 1–72.

Eisenreich, S. J. (ed.), 2005. Climate change and the European water dimension. A report to the European Water Directors. European Commission, EUR 21553 EN, 1–253.

Fang, X. & H. G. Stefan, 2000. Projected climate change effects on winterkill in shallow lakes in the northern United States. Environmental Management 25: 291–304.

Halpin, P. N., 1997. Global climate change and the natural-area protection: management responses and research directions. Ecological Applications 7: 828–843.

IPCC, 2001. Climate Change 2001: Impacts, Adaptation, and Vulnerability: Contribution of Working Group II to the Third Assessment Report of the Intergovernmental Panel on Climate Change. Cambridge University Press, 1–1042.

Järvalt, A., Laas A., Nõges P. & E. Pihu, 2005. The influence of water level fluctuations and associated hypoxia on the fishery of Lake Võrtsjärv, Estonia. Ecohydrology and Hydrobiology 4: 487–497.

Justic, D., Turner R. E. & N. N. Rabalais, 2003. Climatic influences on riverine nitrate flux: implications for coastal marine eutrophication and hypoxia. Estuaries 26: 1–11.

Körner, C., 1999. Alpine Plant Life. Springer-Verlag, Berlin, Heidelberg.

Lake, P. S., 2003. Ecological effects of perturbation by drought in flowing waters. Freshwater Biology 48: 1161–1172, doi:10.1046/j.1365-2427.2003.01086.x.

Lehner, B. & P. Döll, 2001. Eurowasser: Europe's droughts today and in the future. In: EuroWasser—Model-based assessment of European water resources and hydrology in the face of global change. Kassel World Water Series 5. p7–1 to 7–16. http://www.usf.uni-kassel.de/usf/archiv/dokumente/kwws/kwws.5.en.htm.

Liikanen, A., 2002. Greenhouse gas and nutrient dynamics in lake sediment and water column in changing environment. Doctoral dissertation. Ympäristötieteiden laitos, Kuopio University Publications C. Natural and Environmental Sciences 147: 1–50.

Matthews, Q. J. & E. Marsh-Matthews, 2004. Effects of drought on fish across axes of space, time and ecological complexity. Freshwater Biology 48: 1232–1253.

McDowell, R. W., Sharpley A. N., Condron L. M., Haygarth P. M. & P. C. Brookes, 2001. Processes controlling soil phosphorus release to runoff and implications for agricultural management. Nutrient Cycling in Agroecosystems 59: 269–284.

Nõges, P. & T. Nõges, 1998. The effect of fluctuating water level on the ecosystem of Lake Võrtsjärv, Central Estonia. Proceedings of Estonian Academy of Sciences. Biology. Ecology 47: 98–113.

Nurnberg, G. K., 2004. Quantified hypoxia and anoxia in lakes and reservoirs. TheScientificWorldJOURNAL 4: 42–54, DOI: 10.1100/tsw.2004.5.

OECD, 1982. Eutrophication of waters. Monitoring, assessment and control. Organisation for Economic Co-operation and Development, Paris.

Owen, R., Duncan W. & P. Pollard, 2001. Definition and Establishment of Reference Conditions. REFCOND discussion paper for evaluation of techniques. Scottish Environment Protection Agency, Aberdeen, Scotland.

Quinlan, R., Smol J. P. & R. I. Hall, 1998. Quantitative inferences of past hypolimnetic anoxia in south-central Ontario lakes using fossil midges (Diptera: Chironomidae). Canadian Journal of Fisheries and Aquatic Sciences 55: 587–596.

Quinlan, R., Leavitt P. R., Dixit A. S., Hall R. I. & J. P. Smol, 2002. Landscape effects of climate, agriculture, and urbanization on benthic invertebrate communities of Canadian prairie lakes. Limnology and Oceanography 47: 378–391.

Randall, G. W. & D. J. Mulla, 2001. Nitrate nitrogen in surface waters as influenced by climatic conditions and agricultural practices. Journal of Environmental Quality 30: 337–344.

Rayner, S., 2000. Prediction and Other Approaches to Climate Change Policy. In Sarewitz, D., Pielke R. A. Jr. & R. Byerly Jr. (eds), Prediction: Science, Decision Making and the Future of Nature. Island Press, Washington DC, 269–296.

REFCOND, 2003. Rivers and Lakes—Typology, Reference Conditions and Classification Systems. Common Implementation Strategy for the Water Framework Directive (2000/60/EC). Guidance Document No 10. Produced by Working Group 2.3—REFCOND. Office for Official Publications of the European Communities, Luxembourg, 2003, 87 pp.

Richardson, A. D., Lee X. & A. J. Friedland, 2004. Microclimatology of treeline spruce-fir forests in mountains of the northeastern United States. Agricultural and Forest Meteorology 125: 53–66.

Salmaso N., 2005. Effects of climatic fluctuations and vertical mixing on the interannual trophic variability of Lake Garda, Italy. Limnology and Oceanography 50: 553–565.

Salmaso, N., R. Mosello, L. Garibaldi, F. Decet, M. C. Brizzio & P. Cordella, 2003. Vertical mixing as a determinant of trophic status in deep lakes: a case study from two lakes south of the Alps (Lake Garda and Lake Iseo). Journal of Limnology 62(Suppl. 1): 33–41.

Smith, L. C., Sheng Y., MacDonald G. M. & L. D. Hinzman, 2005. Disappearing arctic la kes. Science 308(5727): 1429.

Sorvari, S., 2001. Climate Impacts on Remote Subarctic Lakes in Finnish Lapland: Limnological and Palaeolimnological Assessment with a Particular Focus on Diatoms and Lake Saanajärvi. Academic dissertation in Hydrobiology University of Helsinki, Helsinki.

Stendera, S. & R. K. Johnson, 2005. Does climate change confound lake recovery? EOS Transactions of the American Geophysical Union 86(18), Joint Assembly Supplement, Abstract NB32D-04.

UNEP/UNESCO, 2002. Guidelines for the integrated management of the watershed—phytotechnology and ecohydrology. Freshwater Management Series 5.

Wheeler, P. A., Huyer A. & J. Fleischbein, 2003. Cold halocline, increased nutrients and higher chlorophyll off Oregon in 2002. Geophysical Research Letters 30: 8021, doi:10.1029/2003GL017395.

Winder, M. & D. E. Schindler, 2004. Climate change uncouples trophic interactions in an aquatic ecosystem. Ecology 85: 2100–2106.

Hydrobiologia (2007) 584:381–394
DOI 10.1007/s10750-007-0601-1

Shallow lakes, the water framework directive and life. What should it all be about?

Brian Moss

Abstract The European Water Framework Directive offers an unprecedented opportunity for improvement of ecological quality of both freshwater and marine systems. It has implications for every aspect of how catchments are used by human societies and could potentially mean a step change in how waters and catchments are managed. It must be implemented, however, by official bodies, which seem likely to apply ecologically outdated approaches, used in the past simply to manage water quality, to tackle the very different problem of improving ecological quality. Ecological quality can be characterised by parsimony of available nutrients, characteristic physical and biological structure, strong connectivity among systems and mechanisms of resilience to cope with normal, natural change. The implications of these are that high quality systems in a given location do not have unique lists of species and single formulae for how the biodiversity is constituted. They have considerable inherent variability whilst preserving their fundamental functional characteristics. This appears not to have been recognised by official bodies that seek simple taxonomic indices as measures of quality. To some extent this is a function of the way the Directive has been written, but a slavish adherence to this approach may undermine the spirit of the Directive and result in a failure to bring about the fundamental reform that is needed.

Keywords Nutrients · Connectivity · biodiversity · Restoration · Ecological quality

Guest editors: R. D. Gulati, E. Lammens, N. De Pauw & E. Van Donk
Shallow lakes in a changing world

B. Moss (✉)
School of Biological Sciences,
University of Liverpool,
Liverpool L69 3BX, UK
e-mail: brmoss@liverpool.ac.uk

Introduction

Francesco di Marco da Prato Datini (1335–1410) was a remarkable man (Reader, 2004). Orphaned by the black death, he became one of the most successful businessmen in Europe. After a short apprenticeship to a shopkeeper, he travelled to Avignon, the site then of the papal court and began to trade in cloth, religious articles, armour, paintings and jewellery. He ran a tavern, exchanged money and became a draper. After 32 years he returned to his birthplace of Prato, established his headquarters in Florence, turned to cloth manufacturing and an import business, established an office in Spain, transacted with Bruges and London, moved into insurance underwriting and opened a bank. All of this might suggest a man driven by the worst excesses of modern international corporations, but Datini left all his fortune to the poor and a Trust still exists in his name.

His relevance to modern limnologists is several-fold. He was one of the originators of double-entry book-keeping, a technique that underlies mass-balance nutrient budgeting. As should good scientists, he maintained his independence of institutions, was ever-open to new ideas, had wide interests, and was a hands-on worker, doing much of the day-to-day business himself and was therefore completely *au fait* with the details. We know all this from the hoard of 150,000 letters, 500 account books and ledgers, 300 deeds of partnership and thousands of other documents that were discovered in a disused stairwell when alterations were made to his former house in Prato in 1870. Francesco was a man whom we, as freshwater scientists, could profitably emulate not least in considering our position with regard to how we should be influencing governments about the operation of the European Union Water Framework Directive.

The Water Framework Directive (European Union, 2000) is a long document covering many aspects of the management of natural marine and inland waters. Its key provisions, however, are simple. Management should be on the basis of catchments (river basins); water should not be regarded as a free good but should be priced to users at its real value; and aquatic systems should be restored and managed to 'good ecological quality'. This latter is a particularly far-reaching provision that is causing seemingly endless debate among official institutions. Yet from an ecological standpoint, it is relatively simple. Francesco Datini would have fleshed out the problem in days, and so should we.

The Water Framework Directive

The Directive requires first that a typology of water bodies be set up, like the compartments in a pigeon-loft or the boxes in the mailroom. Each 'ecotype' among rivers, lakes, transitional waters (estuaries) and coastal waters should be distinctive and defined by essentially stable geographical features such as position, altitude, area, depth, and catchment geology. Within each ecotype, ecological quality must be established on a scale from 'high' through 'good', 'moderate', 'poor' and 'bad'. High status is defined as essentially pristine and there is much confusion about this largely because most people concerned, in a heavily-populated Europe, have seen few, if any, such sites. We are able to build such a concept, however, using also expert judgement and palaeo-reconstruction. Good status is defined as 'slightly' different from high status, and therein lies much debate as to how far the otherwise clear meaning of 'slight' shall be stretched. Definitions of moderate, poor and bad status are not given in the Directive.

The political rub comes from the provision that by 2015, all sites that are not derogated from the scheme for transparent reasons of excessive cost and public interest, shall be restored to good ecological status. This is no small provision. In a preliminary assessment of sites in England and Wales, made on very liberal criteria, well over 90% will fail to meet 'good quality' and this was the official view (UK Environment Agency web site). Some British ecologists would put the figure at 100% and the same must be true of much of Europe.

The Directive, understandably, but perhaps unfortunately, gives detailed instruction, in its Annex V, as to what determinands should be measured in assessing ecological quality and uses simple concepts of ecotoxicology (ecological quality ratios) in instructing how the determinands should be used. For lakes, the composition and abundance of the phytoplankton, macrophytes and phytobenthos, and macroinvertebrates and the composition, abundance and age structure of the fish community must be taken into account, along with the water chemistry (nutrients, pH), morphology and hydrology, temperature, salinity and oxygen status and the presence or absence of specified and unspecified pollutants. Strangely, zooplankton, now known to us as very important components determining the processes going on in lakes, particularly shallow lakes, are excluded from consideration.

There is a strong sense in the Directive of twenty-first century environmental idealism being combined with mid-twentieth century, or even earlier, limnology and therein lies the problem that ecological concepts have become much more sophisticated than those which spawned the first simple monitoring schemes for the biological status of polluted rivers in the twentieth century (Kolkwitz & Marsson, 1908; Butcher, 1947; Lange-Bertalot, 1979; National Water Council, 1981). A close reading of the Directive suggests that its drafters believed that ecological quality is independently measureable by lists of phytoplankton, macrophyte, invertebrate and fish taxa with a simple scale of abundance for each, reflecting ecological status. The ecological shelves have become depleted of high quality goods; all that is needed is a re-stocking of them with the same products and good or high status will be achieved. Moreover, for a given ecotype, there is a specified inventory of goods, and substitutions are not possible. These are far from the concepts that modern ecology holds about the natures of communities and ecosystems.

To modern ecologists, a pristine ecosystem is one that maintains (sustains) itself independently of management and to do this, evolutionary mechanisms have built in resilience and flexibility to cope with inevitable natural changes. There is no simple, single formula which defines the ecosystem at a particular spot, no single list of species or even families or larger taxa, indeed no single overall structure, though the major features of physiognomy will be preserved unless climate changes markedly. Indeed quite marked 'catastrophic' changes may occur from time to time (Scheffer et al., 2001). It may help therefore to explore some relatively unaltered systems to reveal what might be the truly high quality ecological features.

Amazonian floodplain forests and temperate rivers

River systems are helpful in illustrating what might be the fundamental features of high ecological quality. A lake, of course, must combine both these features of its inflows and outflows with characteristics particular to lake basins and is thus potentially more complicated. In the final analysis we must in any case talk of ecological quality with reference to catchment areas, or river basins in the terminology of the Directive. So river systems are good, simple places to start such a consideration. It helps also to consider systems first that are least familiar so that we are not mired in details but can see greater fundamentals: the forest rather than the trees.

Among the more fascinating of river systems are the floodplain forests of Amazonia (Fitkau, 1970; Goulding, 1980, 1981; Goulding et al., 1996). The tributary rivers of the Amazon, fed by runoff from the Andes, rise and fall many metres in their annual flood cycle. In the drier season, fish are largely confined to the main channels and food is relatively scarce, for the fish are crowded and competition is severe. When the waters in the main channel rise, however, they back up the nutrient-poorer waters of the side tributaries, which move over the outer reaches of the floodplain, opening up a new habitat for the fish. The forest floor of the floodplain becomes covered by several metres of water and the fish migrate into a more structured and potentially food-rich part of their habitat. Fruits, seeds, insects, the faeces of monkeys and other tree dwellers fall from the tree canopy into the water and are caught by the larger fish. The buttress roots and forest debris provide attachments for eggs and refuges for small fish. A season of breeding begins and small fish abound.

Fruits and seeds, however, are relatively large and unavailable to small fish. Plankton and periphyton are required for their food, but the water backed up from the black (humic-stained) waters of the forest tributaries is low in nutrients. Nonetheless, as the forage fish move into the flooded forest, they are followed by their predators: piscivorous fish, turtles, caimans and dolphins. Sufficient release of nutrients from excretion of the larger fish and predators, in what are now standing rather than flowing waters (indeed a large shallow lake) appears to support sufficient algal growth to form a suitably-sized food resource for the hatchling fish. The predators appear to be very important. Poaching of the caimans for their skins has been associated with poor recruitment (Fitkau, 1970).

There are three features of this system that perhaps epitomise all independently functioning (high quality) ecological systems. These are first that available nutrients are extremely scarce. The systems are parsimonious with respect to nutrients, presumably because nutrients that are naturally scarcely available on a planetary basis are maintained in combined form by the catchment vegetation, which also requires them, and not left vulnerable to loss by wash out. Secondly the systems have a characteristic structure, which includes both physical and biological characteristics (debris, a characteristic cycle of rise and fall of water level associated with movements of animals; complete food webs including top predators). And thirdly, there is connectance among sub-systems: the dry season channel, the wet season floodplain , the forest. We should also add something that is not manifest without very detailed investigation. This is the maintenance of ability to cope with natural year-to-year changes through species substitution and rich gene pools that also give the possibility of resilience over longer periods. Both are manifestations of a characteristic diversity and the importance of connectivity for maintaining routes for immigration of new genes and new species.

North-temperate rivers

These features of nutrient parsimony, structure, connectance and resilience are also central to high ecological quality in north-temperate rivers. In the agricultural landscapes of the deciduous forest zone, it is now difficult to appreciate that the lower order headwater rivers would have been densely overhung by forest, that passage to the ocean would have been unimpeded by dams and that the pristine river system was intimately linked with the surrounding forest through the larger mammals that frequented the valleys. Nor is it now easy to envisage lowland floodplain rivers that occupy the full extent of their beds, feeding a nexus of swamps among which, and the river channels, movements of fish and other animals were a normal and essential feature of ecological independence.

A pristine headwater river in the north-temperate zone is overhung by vegetation, which provides, through leaf-fall and woody debris, both the major energy source and the basis of structure to the system (Harmon et al., 1986). Autumnal leaf fall is rich in cellulose, lignin and tannins, but poor in nutrients like phosphorus and nitrogen. However, once in the water, tree debris is colonised and conditioned by hyphomycete fungi, which are able to extract nutrients from the low concentrations in the water and enrich the leaves with protein-rich fungal mycelium. The water itself, in a pristine system, has exceptionally low concentrations of available nutrients. The conditioned leaves are consumed by a sequence of invertebrates that shred them and feed on the mycelium and produce fine particles for consumption by filter-collectors and deposit feeders and which provide the food base for both non-migratory and migratory fish. The latter play a key role in maintenance of the system, which is currently best understood for the North American salmonid species (Calman et al., 2002) but there seems no reason to suppose that an analogous system based on the Atlantic salmon (*Salmo salar* L.) does not characterise pristine Eurasian rivers.

Salmon species are anadromous, being spawned in freshwaters and then moving to the ocean where, as piscivores, they grow rapidly. Eventually they move back to the river system where they were borne, using cues including the subtle uniqueness of every natural water composition. Their migration is arduous and many die *en route*. Their nutrient-rich carcasses are harvested by brown bears (*Ursus arctos*), which also take live fish where they congregate below waterfalls and rapids. As much as 25% of the nitrogen supply of riparian trees comes from the ocean via salmon carcasses and the excretion of bears (Calman et al., 2002). In turn the woody debris from these trees, forming temporary organic dams in the headwater spawning grounds, retains the carcasses of successful spawners, most of which die following spawning. Decay *in situ* of these carcasses, which would otherwise be washed downriver and lost to the system, helps provide essential nutrients that support the growth of hyphomycete fungi and the eventual provision of invertebrates on which the young salmonids depend after they have used up the reserves in their egg sacs.

The parallels with the Amazonian system are obvious. In addition though, it now appears from work in North America (Ripple & Beschta, 2004a, b) that wolves (*Canis lupus*) also have a key role to play in determining the structure of north-temperate stream systems. In areas otherwise reasonably intact, but where wolves have been shot out, increase in large herbivores such as elk (*Cervus elaphus*) results in severe grazing of the willow vegetation that dominates the riparian zone and a bare and ravaged appearance to the stream. Where wolves have been reintroduced, predation on elk has changed their behaviour so that they do not visit the valley bottoms where they are particularly vulnerable. The willow has re-grown and with it the diversity of birds and the numbers of beaver have increased. The consequent effects on the underwater stream community have not yet been documented but are likely to be considerable.

Floodplain river systems in Europe are now so damaged that it is difficult to reconstruct details of their former functioning (Klinge et al., 1995). They have very much been victims of a serious misconception that floodplains are dry land that is sometimes unfortunately flooded rather than the reality that they are the outer reaches of a natural river bed that are dry at low water levels. The nature of tropical floodplains, more of which are intact, however, gives some indications (Welcomme, 1979). The key features of the Amazonian system described above are universally present, with nutrient parsimony being balanced by mobilisation of fixed nutrients in river-born silt by a well-adapted swamp vegetation. The latter's resilience depends on a complex deposition of silt in levees and the creation of backwaters by free movement of the main channel across the plain. Movements of fish both up and down river and across the floodplain are the bases for vigorous spawning and growth. The migrations of large terrestrial herbivores, as the water levels recede, mobilise nutrients, through dung, from the vegetation in the wet grasslands, for release and uptake on the next flood. The flooded vegetation provides essential substrata for egg deposition by fish.

Over much of Europe these scenarios for pristine (high quality) state will be unrecognisable. Generations of agriculture and development have changed them so much that even those river and lake systems that are most cherished (sites for example that are designated as having Site of Special Scientific Interest status in the UK and many that will become Special Areas of Conservation throughout Europe) will not even remotely satisfy the criterion for good status of the Water Framework Directive that they should be only *slightly* different from pristine, high quality sites. They rarely have nutrient concentrations that are vanishingly low; eutrophication from myriad sources has put paid to that. They have lost much of the structure imparted by tree debris, uninterrupted hydrology and the role of riparian predators. Forestry, agriculture, water regulation and settlement have undermined these. And connectance has been disrupted by river engineering for flood control, dams and weirs. Aldo Leopold had, even in 1941, already summed up how new generations of people fail to perceive how much change has occurred (Leopold, 1996):

> One of the penalties of an ecological education is that one lives alone in a world of wounds. Much of the damage inflicted on land is quite invisible to laymen. An ecologist must either harden his shell and make believe that the consequences of science (technology) are none of his business, or he must be the doctor who sees the marks of death in a community that believes itself well and does not want to be told otherwise.

Problems for administration of the Water Framework Directive

The Water Framework Directive thus poses major problems for the 'competent authorities' charged with its administration. Its spirit is at odds with its detailed prescriptions. First, there is the implication that there are single states for every given ecotype, definable in precise ways that lend themselves to the calculation of straightforward ecological quality ratios. This is simply not the case. Every historic reconstruction of an ecosystem, either aquatic or terrestrial shows major shifts in community in response to natural environmental fluctuations as well as anthropogenic ones. Secondly there is the problem that official perceptions of good or high quality are likely to be vastly removed from the reality of genuine pristineness.

The current approach of the competent authorities, as deduced from the shifting ground of their web sites, appears to be to try to define precise values initially for determinands like total phosphorus and then presumably for the other, more difficult determinands outlined in Annexe V of the Directive. The approach to this appears to be through the reaction of sensitive and tolerant species (Fig. 1) and implies a strong environmental determinism in the behaviour of such species.

There are, for example, well known general trends in the relationships among phytoplankton algae and water chemistry but it is extremely difficult to separate the particular chemical characteristics for which particular groups are indicative. One might expect high diversity of chrysophytes in waters with low nutrient concentrations, but these are closely correlated with low

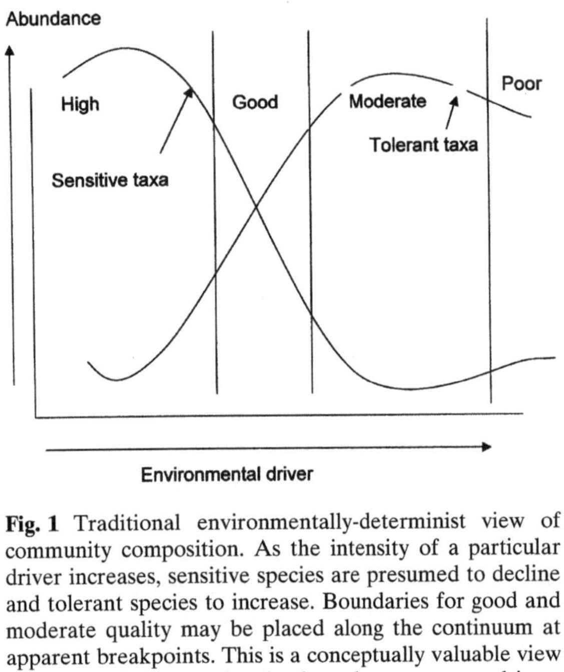

Fig. 1 Traditional environmentally-determinist view of community composition. As the intensity of a particular driver increases, sensitive species are presumed to decline and tolerant species to increase. Boundaries for good and moderate quality may be placed along the continuum at apparent breakpoints. This is a conceptually valuable view but in practical terms breaks down because many drivers operate simultaneously and individual species occurrence, except in extreme habitats, is more likely to be determined by biological and stochastic factors than simple environmental determinism

pH and low bicarbonate concentrations in pristine oligotrophic lakes, for which they may be said to be characteristic (Moss, 1972). Chrysophytes might be presumed to indicate low nutrient status. However such chrysophytes will often grow in high nutrient concentrations in culture, and even at higher pH values than they are found in nature (Moss, 1973a, b). They are non-bicarbonate users (Ball, 2003) that cannot compete with bicarbonate users in higher pH waters and their preponderance in oligotrophic lakes is mostly a function of their requirement for free CO_2 and low competitive abilities rather than a need for low nutrient concentrations.

Likewise the persistence of cyanobacteria is by no means an indicator of very high nutrient conditions. It may reflect the morphometry of the lake, its latitude, the inorganic carbon status or the ratio of nitrogen to phosphorus (Shapiro, 1990). It is a simple indicator of none of these. Similar considerations apply to benthic invertebrates. Most triclad species in the UK are confined to England and do not occur in Scottish lakes. They will, however, grow perfectly well in Scottish lakes if introduced and are absent from Scotland simply because there has not been time for them to penetrate so far north since the retreat of the last glaciers and their recolonisation of Britain from mainland Europe (Reynoldson, 1966). The development of sophisticated techniques of multivariate analysis has underscored these arguments. Rarely is the composition of a community explained to degrees of more than a few percentage points by single environmental factors (Fig. 2). Quite often most of the variation is unexplained even when a whole array of environmental determinants is included. The reason is that biological determinants, stochasticity and biogeography are the major drivers of community structure and they vary continually.

There has been a long history of water quality monitoring of rivers, at least, based on 'indicator' species. The UK system of monitoring benthic invertebrates (National Water Council, 1981) is well established. Some invertebrates are very tolerant of deoxygenation, hence of gross organic pollution, for example most oligochaetes and some chironomids. Many families of mayflies and caddis flies are far less tolerant. Scores are

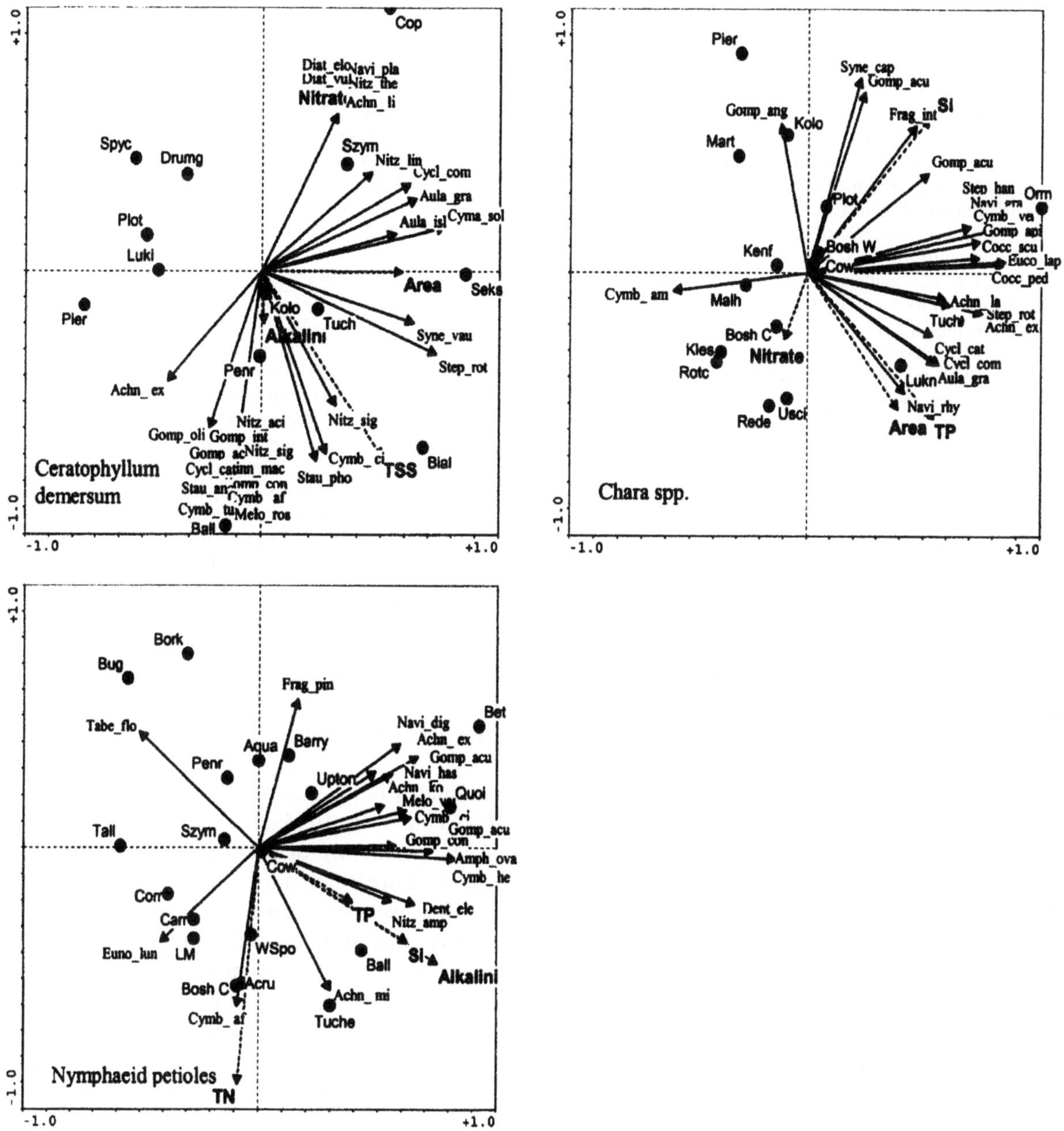

assigned to families and a total score calculated from the families found at a site. The system produces a single number and that number can be compared with a predicted number for the site in 'pristine' condition based on a prediction made from unpolluted sites of similar physical characteristics (RIVPACS, Wright, 1995). Ostensibly this is a very logical system but it has fundamental flaws. It assumes first that organic pollution is the only impactor on the river and that the scores are unaffected by eutrophication, or structural modification of the river, and it is circular in nature in that the scores arbitrarily assigned to families depend on general observations of occurrence in various organically polluted habitats. Of course the scheme also falls well short of the scope of the assessments to be made for the Water Framework Directive. Nonetheless, much research and review

◀ **Fig. 2** Redundancy analysis of epiphyte communities from different macrophytes across groups of up to 19 UK and Polish lakes, based on species composition. A wide variety of environmental driving variables, reflecting morphometry, base and nutrient status were entered into the analysis. For *Ceratophyllum demersum*, a total of only 30% of the variation was explained by the sum of the four most significant environmental variables (total suspended solids, lake area, alkalinity, nitrate). For *Chara* spp., 24% of the variation was explained collectively by area, total phosphorus, nitrate and silicate, and for nymphaeid petioles 21% was explained by alkalinity, silicate, total phosphorus and total nitrogen. No significant relationships for the four most important variables (arrows) were found for two other macrophytes occurring widely in these lakes (*Myriophyllum spicatum*, *Elodea* spp.) for which diagrams are not shown. Abbreviations: Location (solid circles) abbreviations are the initial three or four letters of the lakes concerned. Achn ex, *Achnanthes exilis*; Achn lap, *Achnanthes lanceolata*; Achn ko, *Achnanthes kolbei*; Achn li, *Achnanthes linearis*; Achn mi, *Achnanthes minutissima*; Amph ova, *Amphora ovalis*; Aula gra, *Aulacoseira granulata*; Aula isl, *Aulacoseira islandica*; Cocc ped, *Cocconeis pediculus*; Cocc scu, *Cocconeis scutellum*; Cycl cat, *Cyclotella catenata*; Cycl com, *Cyclotella comensis*; Cymb aff, *Cymbella affinis*; Cymb am, *Cymbella amphicephala*; Cymb ci, *Cymbella cistula*; Cymb he, *Cymbella helvetica*; Cymb tu, *Cymbella tumida*; Cymb ve, *Cymbella ventricosa*; Cyma sol, *Cymatopleura solea*; Dent ele, *Denticula elegans*; Diat elo, *Diatoma elongatum*; Diat vul, *Diatoma vulgare*; Eucco lap, *Eucocconeis lapponica*; Euno lu, *Eunotia lunaris*; Frag int, *Fragilaria intermedia*; Frag pin, *Fragilaria pinnata*; Gomp ac, *Gomphonema acuminatum*; Gomp ang,*Gomphonema angustatum*; Gomp api, *Gomphonema apicatum*; Gomp con, *Gomphonema constrictum*; Gomp int, *Gomphonema intricatum*; Gomp oli, *Gomphonema olivaceum*; Melo ros, *Melosira roseana*; Melo var, *Melosira varians*; Navi dig, *Navicula digitoradiata*; Navi gra, *Navicula gracilis*; Navi has, *Navicula hastata*; Navi rhyn, *Navicula rhyncocephala*; Nitz aci, *Nitzschia acicularis*; Niz amp, *Nitzschia amphicephala*; Nitz the, *Nitzschia thermalis* var *minor*; Nitz lin, *Nitzschia linearis*; Nitz sig, *Nitzschia sigmoides*; Pinn mac, *Pinnularia maculata*; Stau an, *Stauroneis anceps*; Stau pho, *Stauroneis phoenicenteron*; Step han, *Stephanodiscus hantzschii*; Step rot, *Stephanodiscus rotula*; Syn cap, *Synedra capitata*; Syne vau, *Synedra vaucheriae*; Tabe flo, *Tabellaria flocculosa*. Similar analyses by genus or growth form gave comparably low percentages of variation explained. Based on Fisher et al. (2006)

is being commissioned to try to establish similar schemes for the phytoplankton, diatoms, macrophytes and fish components required by Annexe V and so it is safe to assume that this reflects the approach that the Authorities concerned intend to take.

Reductionism as inappropriate

My contention is that this reductionist thinking, though understandable in that it has been the traditional approach of water managers, is the wrong path to take. It is notable how little progress has yet been made in the five years since the Directive was passed and this reflects the difficulty, nigh impossibility of making such an approach work satisfactorily. It is also likely to result in extremely expensive monitoring schemes that will not give attainable targets for restoration to good quality. No one will ever be able to guarantee the persistence of a particular organism or family at a site. There is too much inherent ecological stochasticity. When legal disputes arise, as they will, about whether the Directive is being correctly applied, the degree of uncertainty in any such measure will force the courts to dismiss the evidence. In light of these considerable difficulties, what might be the most logical approach for creating classification systems?

The first step (Fig. 3) must be a recognition that quality in ecosystems is not a function of lists of taxa but of the fundamental properties outlined earlier: parsimony of nutrient supply, characteristic structure (including both physical structure and food web structure) and connectance. Connectance will embrace sufficient size to guarantee possibilities of natural immigration of new species to allow resilience to natural environmental fluctuations and also to encompass sufficiently large gene pools for the larger animals, including migratory fish, requiring substantial territories. The second step must be to decide what the meaning of 'slight' shall be in determining the distinction between high and good quality and therefore to create firm scenarios for 'good' quality. I suspect that the drafters of the Directive did not have a sufficiently informed knowledge about what pristine state in aquatic systems really means and had in mind something approximating to acceptability in systems already encapsulated in agricultural landscapes. Truly high status and application of the conventional meaning of 'slight' are both incompatible with any permanent

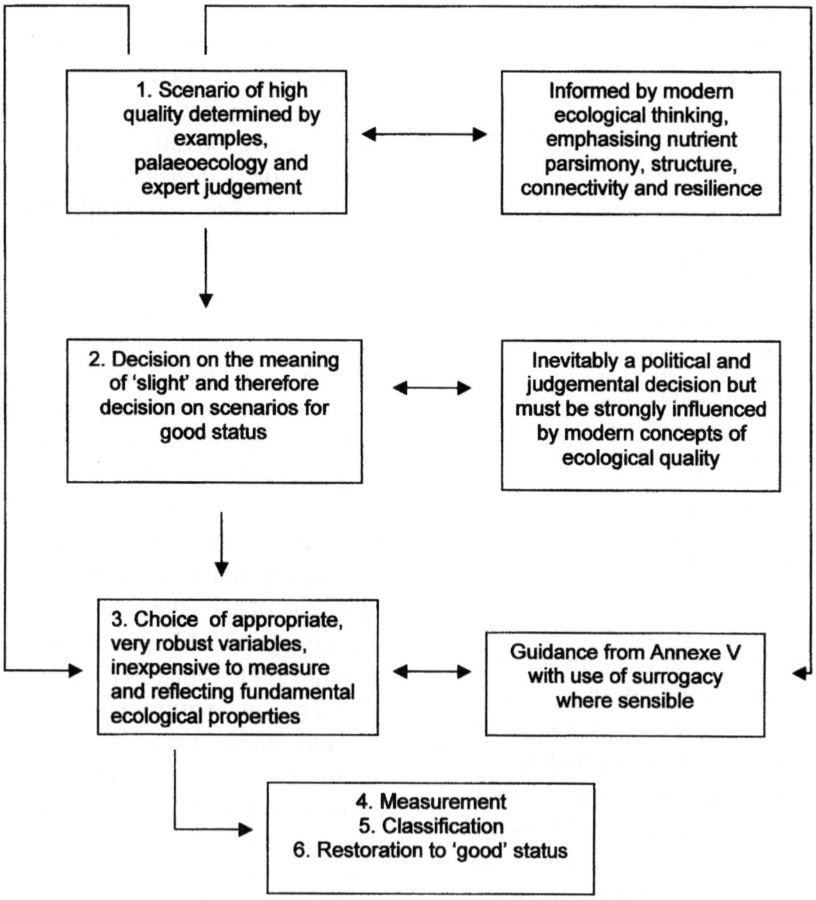

Fig. 3 Suggested logical approach to implementation of the Water Framework Directive as it applies to restoration of ecological quality

development of the river basin. 'Slight' will realistically thus have to be stretched in meaning beyond the conventional usage of any English dictionary but there is every reason for an expansion in imaginative thinking. There are arguments for the reintroduction of top predators to quite large areas of even highly populated countries like the UK (Taylor, 2005). In the beginning, not in the end, there has to be some conceptual scenario for good status for each ecotype. Horses pull carts, not the other way round. It is a judgemental issue and no end of re-working reductionist data sets will create suitable scenarios *post hoc*.

The third step must be choice of suitable and very robust variables to reflect the characteristics of parsimony, structure and connectance. To be satisfactory for legal systems, these will have to be, as far as possible, yes/no characteristics. The intention of Annexe V has to be respected but slavish interpretation will lead to unworkable, highly detailed schemes that will waste both time and funds. It will be better in the long run and more faithful to the spirit of the Directive to choose simple variables that truly reflect ecological quality and surrogate the provisions of Annexe V. The one scheme yet published following peer review, ECOFRAME, for shallow lakes (Moss et al., 2003) partly meets these criteria in that scenarios for good quality were established at the outset but in attempting to satisfy literally the details of Annex 5, it is much too complex. It would require three person weeks for a lake assessment, which is expensive. Its authors included a feeling along these lines in an epilogue:

Table 1 Suggested list of robust variables for assessment of ecological quality in lake

General property	Variable
Nutrient parsimony	pH, TP, TN
Structure	Proportion of the shoreline colonised by native vegetation
	Abundance and species richness of submerged plants in summer
	Chlorophyll a concentration (or Secchi transparency)
	Frequency of visible, surface cyanobacterial blooms
	Absence of invasive and damaging alien macrophytes, invertebrates and fish species
Connectance	Intactness of hydrological regime
	Continuity of passage to the ocean for migratory fish

> 'During the workshop discussions of this project, one of us remarked that from a single visit to a lake, an overview of its catchment, knowledge of its pH and Secchi disc transparency, and a brief examination of its macrophytes, an experienced limnologist could easily distinguish high/good from moderate from poor/bad status. It is the tragedy of our time, perhaps of all times, that we must set up expensive and elaborate measures to do what, with experience... could be done much more easily'.

What might then be a suitable inexpensive suite of variables for lakes that would meet these criteria and reflect also the spirit of Annexe V? My suggestions, which are intended to indicate the general nature rather than the final detail, are for a minimalist approach on the grounds that simple, robust schemes are the most practicable. I suggest three chemical criteria (pH, TP, TN) to reflect parsimony of nutrient availability and to cope with the very considerable impact of acidification. I suggest five variables (Table 1) to reflect structure. All of these can be simply assessed and require little taxonomic expertise. Finally connectedness can be easily assessed from maps and databases to mean intactness of hydrological regime and continuity of passage for migratory fish. Intactness of hydrological regime simply means that substantial quantities (say more than a few per cent) of water are not being added from outside the catchment or withdrawn for irrigation. It is not immediately possible to put numbers to these criteria for a decision on the meaning of good quality is first essential, but in doing that scenarios of what really constitutes high quality must be born seriously in mind. Of course, this list does not reproduce the detail specified in Annexe V of the Directive. However, if the fundamental characteristics of ecological quality, as reflected in Table 1, are present, then the details expressed in Annex V will also be appropriately in place. These details are the manifestations, the symptoms, of quality but in themselves are separately of little meaning. If the fundamental characteristics of quality are also in place, then it must follow that the extremely valuable goods and services (Costanza et al., 1997), which can be provided by systems of increasing naturalness, will also be provided to increasing extents. It will also be much cheaper to assess a suite of robust variables; more water bodies can be assessed, more easily; and the findings, being often of a plus/minus nature will stand up in legal proceedings much more readily than those which must be hedged with probabilities. And there is provision for alteration of Annex V (Article 20 (Technical Adaptations to the Directive)). I take this to mean advances in understanding.

The current 'official' approach to achieving these ends appears to be very different (Fig. 4). No scenarios of what constitutes good quality have been developed officially so there is no logical basis for determining the precise determinands to measure quality or for partitioning values into quality categories. Rather there is emphasis on constructing schemes (tools) to measure something, but that 'something' is undefined. It is like shooting arrows without a target in an archery tournament. The opening assumption appears to be that ecological features are simply measurable in the same way as chemical variables. The sophistication that ecological research has produced in the fifty years since similar schemes for river assessment were first produced has been completely ignored. Biological interac-

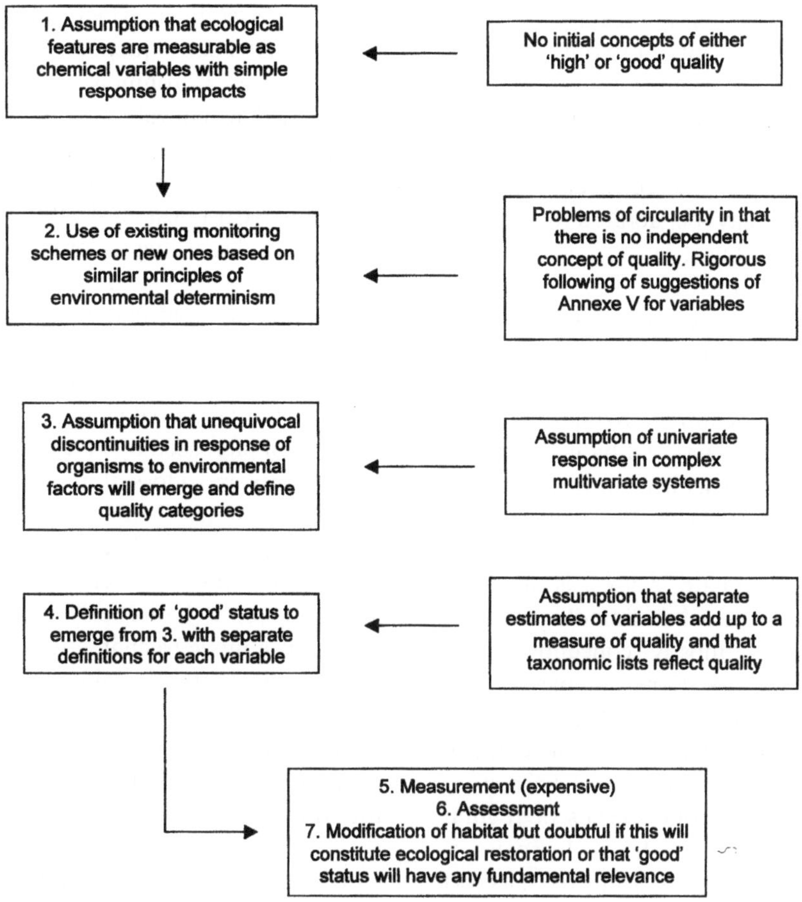

Fig. 4 Reconstruction of the current approach to implementation of the Water Framework Directive, as it applies to restoration of ecological quality by official bodies

tions and the role of stochasticity have been dismissed. Crude environmental determinism is the watchword.

Proponents of the official view hope that some definition of good quality will emerge from schemes being developed such that graphs of response of particular sensitive and tolerant groups will show a convenient regularity (Fig. 1). They suggest that a point where taxa sensitive to a particular impact begin to decline will define good quality and a region where sensitive taxa have declined but tolerant ones rise will define moderate quality. This sounds logical as a theoretical precept. The reality is that taxa respond in a multivariate way (Fig. 2). Many impacts occur together and definition of 'sensitive' and 'tolerant' is impossible outside univariate laboratory systems. It is difficult to see that this approach will be fruitful as it is based not on field ecology but laboratory toxicology. Furthermore, ecological quality has to be defined independently of the tools used to measure it, not by the tools themselves, or the assessments will be riddled with systematic and circular error.

Prognosis

What will be the consequences of pursuing the current 'official' approach to determining ecological quality as opposed to one based on learning the lessons of Leopold's 'ecological education'. The ultimate aim of the Water Framework Directive is restoration, and a review of what is variously meant by this term is apposite. There are several levels, conveniently reflected in treatment of symptoms, treatment of proximate causes and treatment of ultimate causes.

Treatment of symptoms is no better than amenity gardening if ecological restoration is intended. Trees may be planted, sediment may be removed. Lakes may be aerated or mixed. Algicide, herbicide, barley straw or lime might be added. Artificial riffles might be placed in rivers, or small backwaters created, or a pond installed behind a levee. The treatment generally has to be continued to allay the symptoms but a sense of order and control is achieved. Symptom treatment is costly but often favoured by government and commercial interests as it is no threat to the fundamental precepts of classical economics. It is mere decoration.

Treatment of proximate causes is a more fundamental step. It is what limnologists and conservation agencies mostly concentrate on, generally in specific locations: a particularly valued lake, a nature reserve or other protected area. It might involve local nutrient control, biomanipulation, water level control to restore wetlands, control of an exotic species or regulation of a damaging fishery in lakes. In rivers, it might involve the re-meandering of a short stretch, though not the reestablishment of a completely functioning floodplain (Palmer et al., 2005). It attempts to redress immediate causes without addressing the ultimate reasons. Thus it might remove phosphorus from sewage effluent without questioning whether a proportion of that phosphorus need not be there in the first instance, or it might use buffer zones to remove agricultural nitrate without asking if the particular system of agriculture being used is appropriate. Thus, it too is easily contained and controlled by Authority and poses no threat to established institutions. It is localised and boundaried. If its proponents become too vociferous, their organisations can be reorganised (Sheail, 1998).

We face, however, problems that are much more fundamental than can be dealt with by either treatment of symptom or proximate cause. Recent decades have shown a growing realisation, now formalised as earth system science (Jacobsen et al., 2000), that our planet is a massively connected system in which all sorts of changes have chains of consequences. Combination of this thinking with environmental economics has demonstrated that natural systems, when intact, provide goods and services vastly more valuable than the goods and services traded under conventional economics (Costanza et al., 1997; Balmford et al., 2002). The discarding of environmental damage as 'externalities' by classical economists is clearly a measure of their inadequacy to confront the realities of a changing and greatly damaged planet. We face in the next few decades at least four substantial global changes: climate change, a population increased by up to 50%, the end of the oil economy and a loss of more than 50% of many major ecosystems (Millennial Ecosystem Assessment Board, 2005). Treatment of symptoms and even of local cause in such a situation will be seen in retrospect as ludicrously inadequate. What is needed is a move to changing fundamental cause.

Analogies with the alternative states models that seem to apply widely to ecosystems (Scheffer et al., 2001) and might also be applied to systems of human organisation (Moss, 1995) suggest that the prevailing model of western technological society has powerful stabilising mechanisms that will hinder moves to a sustainable system in the short term, if at all. However, common sense and earth systems science suggest that it is not possible to continue maintaining such impacts without consequences that could be highly uncomfortable if not totally disruptive to existing human systems. Avoidance of this requires a much more fundamental approach to ecological restoration which will be strongly resisted by government and commercial interests as it will threaten their present privileges. It will be necessary to change fundamentally to achieve real sustainability.

The Water Framework Directive is potentially a much greater agent for change than has been realised. It emphasises river basin management. Inherent in it is the realisation, long known to limnologists, that almost all that happens in a catchment has ultimate effects in the freshwater systems. It is not possible to restore these to good ecological quality without doing the same for the catchment itself. Agricultural, transport, housing, waste disposal, air and soil quality, industrial and manufacturing policies are all involved in this. The Directive is a truly revolutionary document. If its spirit is respected, it can be an early step in the sort of changes that are ultimately necessary for the survival, in reasonable comfort, of a large human population. Alas, current official approaches to its implementation, in ignoring fundamental ecological precepts and attempting to trivialise ecological quality into the measurement of dubious ecological indices, with no vision of what should be achieved, are unlikely to contribute much more than has been gained by previous management of freshwater systems. The preliminary assessments of risk of failing to meet good quality required by the Water Framework Directive, made in 2004, confirm that these have failed miserably. Perhaps there is a lesson from the mediaeval career of Francesco Datini da Prato. He was an extremely effective and successful man because he was not afraid to innovate, kept himself very well informed of all the contemporary details of his business and disdained to keep the interests of authority, neither church nor state, in mind. Francesco was a revolutionary wolf in sheep's clothing.

References

Ball, L. A., 2003. Carbon acquisition in the chrysophyte algae. PhD Thesis, University of Dundee, UK.

Balmford, A., A. Bruner, P. Cooper, R. Costanza, S. Farber, R. E. Green, M. Jenkins, P. Jefferiss, V. Jessamy, J. Madden, K. Munro, N. Myers, S. Naeem, J. Paavola, M. Rayment, S. Rosendo, J. Roughgarden, K. Trumper & R. K. Turner, 2002. Economic reasons for conserving wild nature. Science 297: 950–953.

Butcher, R. W., 1947. Studies in the Ecology of Rivers V11. The algae of organically enriched waters. Journal of Ecology 33: 186–191.

Calman, R. J., R. E. Bilbo, D. E. Schindler & J. M. Belfield, 2002. Pacific salmon, nutrients and the dynamics of freshwater ecosystems. Ecosystems 5: 399–417.

Costanza, R., R. d'Arge, R. de Groot, S. Farber, M. Grasso, B. Hannon, K. Limburg, S. Naeem, R. V. O'Neill, J. Paruelo, R. G. Raskin, P. Sutton & M. van den Belt, 1997. The value of the world's ecosystem services and natural capital. Nature 387: 253–260.

European Union, 2000. Directive 2000/60/EC of the European Parliament and of the Council of 23 October 2000 establishing a framework for Community action in the field of water policy. Official Journal of the European Community L 327: 1–73.

Fisher, J., C. James & B. Moss, 2006. What determines the diatom communities of submerged freshwater plants? Implications for the use of community indices in determining ecological quality. Nova Hedwigia Suppl. 130: 51–72.

Fitkau, E. J., 1970. Role of caimans in the nutrient regime of mouth-lakes in Amazon effluents (a hypothesis). Biotropica 2: 138–142.

Goulding, M., 1980. The Fishes and the Forest: Explorations in Amazonian Natural history. University of California Press, Los Angeles.

Goulding, M., 1981. Man and Fisheries on an Amazon Frontier. Junk, The Hague.

Goulding, M., N. J. H. Smith & D. J. Mahar, 1996. Floods of Fortune. Ecology and Economy along the Amazon. Columbia University Press, New York.

Harmon, M. E., J. F. Franklin, F. J. Swanson, P. Sollins, S. V. Gregory. J. D. Lattin, N. H. Anderson, S. P. Cline, N. G. Aumen, J. R. Sedell, G. W. Lienkaemper, J. Cromak & K. W. Cummins, 1986. Ecology of coarse woody debris in temperate ecosystems. Advances in Ecological Research 15: 133–302.

Jacobson, M. C., R. J. Charlson, H. Rodhe & G. H. Orians, 2000. Earth System Science. Academic Press, London.

Kolkwitz, R. & M. Marsson, 1908. Ökologie der pflanzlichen Saprobien. Berichte der deutsche Botanische Gesellschaft 26 A: 505–519.

Lange-Bertalot H., 1979. Pollution tolerance as a criterion for water quality estimation. Nova Hedwigia Beiheft 64: 285–304.

Klinge, M., M. P Grimm & S. H. Hosper, 1995. Eutrophication and ecological rehabilitation of Dutch lakes: presentation of a new conceptual framework. Water Science & Technology 31: 207–218.

Leopold, A., 1996. A Sand County Almanac with Essays on Conservation from Round River. Oxford University Press, New York.

Millennial Ecosystem Assessment Board, 2005. Millennial Ecosystem Assessment Synthesis Report. United Nations Environment Programme, New York.

Moss, B., 1972. The influence of environmental factors on the distribution of freshwater algae: an experimental study. 1. Introduction and the influence of calcium concentration. Journal of Ecology 60: 917–932.

Moss, B., 1973a. The influence of environmental factors on the distribution of freshwater algae: an experimental study. 2. The role of pH and the carbon dioxide-bicarbonate system. Journal of Ecology 61: 157–177.

Moss, B., 1973b. The influence of environmental factors on the distribution of freshwater algae: an experimental study. 4. Growth of test species in natural lake waters, and conclusion. Journal of Ecology 61: 193–211.

Moss, B., 1995. The microwaterscape—a four-dimensional view of interactions among water chemistry, phytoplankton, periphyton, macrophytes, animals and ourselves. Water Science & Technology 32: 105–16.

Moss, B., D. Stephen, C. Alvarez, E. Becares, W. Van de Bund, S. E. Collings, E. Van Donk, E. De Eyto, T.

Feldmann, C. Fernandez-Alaez, M. Fernandez Alaez, R. J. M. Franken, F. Garcia-Criado, E,M. Gross, M. Gyllstrom, L.-A. Hansson, K. Irvine, A. Jarvalt, J. P. Jensen, E. Jeppesen, T. Kairesalo, R. Kornijow, T. Krause, H. Kunnap, A. Laas, E. Lill, B. Lorens, H. Luup, M. R. Miracle, P. Noges, T. Noges, M. Nykanen, I. Ott, W. Peczula, E. T. H. M. Peeters, G. Phillips, S. Romo, V. Russell, J. Salujo, M. Scheffer, K. Siewertsen, H. Smal, T. Virro, E. Vicente & D. Wilson, 2003. The determination of ecological status in shallow lakes—a tested system (ECO-FRAME) for implementation of the European Water Framework Directive. Aquatic Conservation. Marine & Freshwater Ecosystems 13: 507–549.

National Water Council, 1981. River Quality – the 1980 Survey and Future Outlook. National Water Council, London.

Palmer, M. A., E. S. Bernhardt, J. D. Allan, J. D, P. S. Lake, G. Alexander, S. Brooks, J. Carr, S. Clayton, C. M. Dahm, J. Follstad Shah, D. L. Galat, S. G. Loss, P. Goodwin, D. D. Hart, B. Haasett, R. Jenkinson, G. M. Kondolf, R. Lave, J. L. Meyer, T. K. O'Donnell, L. Pagano & E. Sudduth, 2005. Standards for ecologically successful river restoration. Journal of Applied Ecology 42: 208–217.

Reader, J., 2004. Cities. Heinemann, London.

Reynoldson, T. B., 1966. The distribution and abundance of lake-dwelling triclads – towards a hypothesis. Advances in Ecological Research 3: 1–71.

Ripple, W. J. & R. L.Beschta, 2004a. Wolves, elk, willows, and tropic cascades in the upper Gallantin Range of South-western Montana, USA. Forest Ecology and Management 200: 161–181.

Ripple, W. J. & R. L. Beschta, 2004b Wolves and the ecology of fear: can predation risk structure ecosystems? BioScience 54: 755–766.

Scheffer, M., S. R. Carpenter, J. A. Foley, C. Folke & B. Walker, 2001. Catastrophic shifts in ecosystems. Nature 413: 591–596.

Shapiro, J., 1990. Current beliefs regarding dominance by blue-greens: the case for the importance of CO_2 and p_H. Verhandlungen der internationale Vereinigung theoretische und angewandte Limnologie 24: 38–54.

Sheail, J., 1998. Nature Conservation in Britain. The Formative Years. The Stationery Office, London.

Taylor, P., 2005. Beyond Conservation – A Wildland Strategy. Earthscan, London.

Welcomme, R. L., 1979. Fisheries Ecology of Floodplain Rivers. Longman, London.

Wright, J. F., 1995. Development and use of a system for predicting the macroinvertebrate fauna in flowing waters. Australian Journal of Ecology 20: 181–197.

Hydrobiologia (2007) 584:395–402
DOI 10.1007/s10750-007-0586-9

SHALLOW LAKES

Submerged macrophyte vegetation and the European Water Framework Directive: assessment of status and trends in shallow, alkaline lakes in the Netherlands

H. Coops · F. C. M. Kerkum · M. S. van den Berg · I. van Splunder

Abstract The submerged macrophyte vegetation of lakes created after enclosing former estuaries, situated in the central and southwestern part of the Netherlands, has been monitored annually from 1992 onwards. Between 1992 and 2004, pronounced changes in overall cover and species composition of the submerged vegetation have occurred, resulting from changes of water quality and morphology in the lakes. In most cases vegetation cover and species diversity increased or remained stable, with the exception of two lakes in the southwest part of the country. Abundance and species composition were assessed according to the requirements of the EU Water Framework Directive, using the assessment procedure proposed to assess macrophytes in natural water bodies in the Netherlands. The assessment procedure included calculation of the 'ecological quality ratio' (EQR) for each of eleven water bodies in each of 13 years, based on transect monitoring data. The EQR indicating Good Ecological Status for Macrophytes was achieved in only three of the lakes. The consequences of hydromorphological modifications, and measures necessary to achieve the desired condition are discussed. Nutrient concentrations should be reduced further, while additional management measures are necessary to improve conditions for macrophytes.

Keywords Assessment · Biological monitoring · Submerged macrophytes · Water Framework Directive

Guest editors: R. D. Gulati, E. Lammens, N. De Pauw & E. Van Donk
Shallow lakes in a changing world

H. Coops (✉) · F. C. M. Kerkum ·
M. S. van den Berg · I. van Splunder
Institute for Inland Water Management and Wastewater Treatment (RIZA), P.O. Box 17, 8200 AA Lelystad, The Netherlands
e-mail: hugo.coops@wldelft.nl

Introduction

Macrophytic vegetation plays a key role in shallow lake ecosystems. Changed water quality due to increased nutrient levels has had a strong negative impact on the abundance and species composition of the aquatic vegetation in many lakes: various recent studies in Europe have shown severe reductions of submerged plants (e.g., Sand-Jensen et al., 2000; Körner, 2002) as well as emergents (e.g., Ostendorp, 1989; Graveland & Coops, 1997). One of the priorities of lake restoration projects usually comprises the recovery of abundant macrophyte presence (Gulati & Van Donk, 2002). Many efforts have already been made to restore lakes, and the European Water Framework Directive (WFD, European Commission, 2000), adopted by the EU member states, will support further efforts by setting standards for ecological status of water bodies within Europe. Its objective is to sustain and enhance aquatic environments within the European Community by achieving at least 'good' ecological status for all water bodies by 2015, referring to a scale ranging from 'high', 'good', 'moderate', 'poor' to 'bad'. One of the essential elements of the Directive is that it demands the monitoring and assessment of surface waters using specific ecological quality elements, instead of primarily relying on hydrochemical data. Following the Directive, ecological quality of lakes should be based on four groups: phytoplankton, macrophytes and phytobenthos, benthic invertebrate fauna and fish fauna. According to the WFD's formulation, assessment of macrophytes and phytobenthos should include 'taxonomic composition and abundance', to be assessed in the context of undisturbed reference conditions. Effective and accurate monitoring of status and trends in macrophyte cover and species composition should therefore be applied. New assessment methods for macrophytes, that take into account the requirements for the WFD, are now being developed in the EU member states (Schaumburg et al., 2004; Van der Molen, 2004).

In the Netherlands, a new national method for macrophyte assessment of water bodies has been proposed (Van der Molen, 2004), including a procedure to determine an ecological quality ratio (EQR) based on macrophyte abundance and composition, using mostly theoretically derived relationships (Van den Berg et al., 2003; Van der Molen, 2004), since reference locations in the Netherlands are not available.

Existing monitoring data can be a starting point to develop 'WFD-proof' assessment procedures. The present study uses results from existing monitoring in 11 large shallow water bodies in the Netherlands (Table 1) managed by the National Water Authority to preliminary assess the proposed method.

Methods

Water bodies

Monitoring data from 11 different water bodies situated in the lower basins of the rivers Rhine (9), Meuse (1) and Scheldt (1) were used to assess the status of macrophyte abundance and composition from 1992 until 2004 (Table 1). The lower Rhine lakes are all situated in the IJsselmeer area, and were created during the 20th century as the result of damming off the inland sea and the embankment of large part of it. IJsselmeer proper was separated from the Markermeer by a dam in 1978. The other lakes in the Rhine area (Zwarte Meer, Ketelmeer, Veluwemeer, Wolderwijd, Eemmeer, Gooimeer) are so-called 'border lakes', situated in-between the old land and the polders. In the southwest of the Netherlands, the lakes Volkerakmeer and Zoommeer were formed after enclosure of part of the Oosterschelde estuary in 1987.

The lakes vary in size between 8 and >1000 km^2, and all have extensive shallow areas

Table 1 Water bodies studied and available macrophyte monitoring data

Lake/water body	River basin	Total area (km^2)	Littoral area (<3 m) (km^2)	No. of plots (transects) monitored annually
IJsselmeer	Rhine	1125	99.5	304 (4)
Markermeer[a]	Rhine	691	115.7	161 (3)
Volkerakmeer	Meuse	48	18.1	190 (6)
Veluwemeer[b]	Rhine	37	32.4	221 (5)
Ketelmeer[c]	Rhine	34	15.6	145 (4)
Gooimeer	Rhine	26	16.0	138 (3)
Wolderwijd[d]	Rhine	25	20.2	203 (4)
Zwarte Meer	Rhine	18	14.8	188 (3)
Eemmeer[e]	Rhine	15	12.7	97 (2)
Gouwzee	Rhine	12	11.8	171 (2)
Zoommeer	Scheldt	8	3.2	40 (4)

[a] incl. IJmeer
[b] incl. Drontermeer
[c] incl. Vossemeer
[d] incl. Nuldernauw
[e] incl. Nijkerkernauw

(water depth <3 m). They are alkaline and eutrophic, and sediments consist almost entirely of clay and sand of marine/riverine origin.

Monitoring of submerged vegetation

Annual transect monitoring started in 1992. Transects consist of 3 rows of plots, situated 100 m apart, and spanning the shallow-water zone perpendicular to the shoreline. Each plot was represented by a georeferenced location where vegetation characteristics were sampled every year in July. Sampling at each location was done by throwing a rake in five directions, and collecting the plant material for species identification. Projected cover (for total vegetation and individual species) for each plot was estimated visually using a 7-point scale. Additionally, water depth and Secchi depth were determined at every plot.

For each water body, total cover in the shallow zone was calculated as the average cover of two strata (0–1.50 and 1.51–3.00 m, respectively) weighted according to their area within the lake. Species' abundances were determined by combining species frequency and median cover value. Three abundance classes were distinguished: (1) rare; (2) occasional–frequent; (3) abundant–dominant.

Trends in cover over the period 1992–2004 were tested non-parametrically (Spearman rank test, significant at $P < 0.05$).

Macrophyte-based EQR

The WFD requires assessment of the abundance and composition of macrophytic vegetation of water bodies, and quantification relative to a type-specific reference condition. The EQR gives the quantitative ratio between current and reference condition. The proposed method to establish the EQR for shallow, alkaline lakes in the Netherlands (Van der Molen, 2004) includes a scale to express ecological status of macrophytes in shallow (parts of) lakes, that was developed based on different attributes: (1) Abundance of submerged macrophytes: includes the average cover in the entire photic zone. For alkaline lakes as in the Netherlands, the photic depth limit is between 2 and 5 m (Van den Berg et al., 2002), but for practical purposes the limit is set at 3 m here. (2) Shoreline vegetation cover: the total cover of emergent (wetland) vegetation in the periodically inundated marginal area of a lake. Due to the regulated water levels in Dutch lakes, and lack of reliable elevation data, this attribute was not included. (3) Species composition of macrophytes: based on the occurrence of characteristic aquatic and plant species, weighed according to their degree of indicativeness and species' abundance. (4) Composition of phytobenthos, determined from the presence of positive and negative indicator species. Due to lack of data we

Table 2 Scheme of assessement of macrophytes (Van der Molen 2004) used in this study

Step 1: Abundance of submerged macrophytes: % cover in the parts of the lake where water depth is between 0 and 3 m. $EQR_{abundance}$ is calculated by linear interpolation between class boundaries (bad 0–1%, poor 1–5%, moderate 5–25%, good 25–50%, very good 50–65% and higher).
Step 2: Species presence: score of characteristic species according to abundance in the water body.
Category 1 (score: rare 1, occasional–frequent 3, abundant–dominant 4): *Chara aspera, C. contraria, C. globularis, C. hispida, C. vulgaris, Nitella hyalina, N. mucronata, N. opaca, Nitellopsis obtusa.*
Category 2 (score: rare 1, occasional–frequent 2, abundant–dominant 2): *Callitriche platycarpa, Elodea canadensis, Fontinalis antipyretica, Hottonia palustris, Hydrocharis morsus-ranae, Myriophyllum spicatum, M. verticillatum, Najas marina, Nuphar lutea, Nymphaea alba, Nymphoides peltata, Persicaria amphibia, Potamogeton berchtoldii, P. compressus, P. crispus, P. lucens, P. mucronatus, P. nododsus, P. obtusifolius, P. pectinatus, P. perfoliatus, P. praelongus, P. pusillus, P. trichoides, P. x zizii, Ranunculus aquatilis, R. circinatus, Stratiotes aloides, Utricularia vulgaris, Zannichellia palustris.*
Category 3 (score: rare 1, occasional–frequent 1, abundant–dominant 0): *Ceratophyllum demersum, Elodea nuttallii, Lemna gibba, L. minor, L. trisulca, Spirodela polyrhiza.*
$EQR_{species}$ is calculated by linear interpolation between class boundaries (bad 0–5% of maximum score, poor 5–10%, moderate 10–20%, good 20–40%, very good 40 100%).
Step 3: Calculation of EQR by averaging $EQR_{abundance}$ and $EQR_{species}$. Assignment of ecological status: bad (0–0.2), poor (0.2–0.4), moderate (0.4–0.6), good (0.6–0.8), very good (0.8–1.0).

The ecological quality ratio (EQR) is a value between 0 and 1, encompassing 5 equal quality classes: bad 0-0.2, poor 0.2-0.4, moderate 0.4-0.6, good 0.6-0.8, high 0.8-1.0

did not include this attribute. For this study the macrophyte EQR was calculated according to Table 2.

Abiotic lake data

Water quality data were obtained from the Dutch national water quality database. Monitoring stations were selected that are located centrally in each of the water bodies and had been sampled at least once per month in the period 1992–2004. We calculated summer-averaged (April–September) values of total phosphorus (TP), total nitrogen (TN) and chlorophyll-*a*.

Results

Individual lakes showed considerable variations in cover by macrophytes in the shallow-water zone (Fig. 1). Of the 11 lakes, Veluwemeer ($S = 0.742$, $P < 0.005$), Wolderwijd ($S = 0.907$, $P < 0.001$), Eemmeer ($S = 0.596$, $P < 0.05$), and Gouwzee ($S = 0.790$, $P < 0.005$) showed an increasing trend in cover, whereas Volkerakmeer ($S = -0.773$, $P < 0.01$) and Zoommeer ($S = -0.709$, $P < 0.05$) showed a decreasing trend. Cover values in lakes were calculated using the surface areas of the two depth zones, viz. very shallow (0–1.5 m) and moderately shallow (1.5–3 m). The ratio between very shallow and moderately shallow areas ranged between 0.06 (Markermeer) and 4.1 (Zwarte Meer), having a distinct effect on the colonised part of the 0–3 m zone. The very shallow zone can be colonised by macrophytes even under relatively unsuitable conditions, while the less shallow area may only be colonised when underwater light conditions have improved (Fig. 2).

The total number of submerged macrophyte species observed in a single lake ranged between 5 and 19. A number of species were specifically occurring in one or two lakes only, such as *Potamogeton nodosus* (Ketelmeer) and *Callitriche truncata* (Volkerakmeer, Zoommeer). Of the more common species, most species showed an overall increase, such as *Chara* spp. (occurrence in all samples 18% in the first 6 years, 33% in the last 6 years), *Potamogeton pectinatus* (from 27%

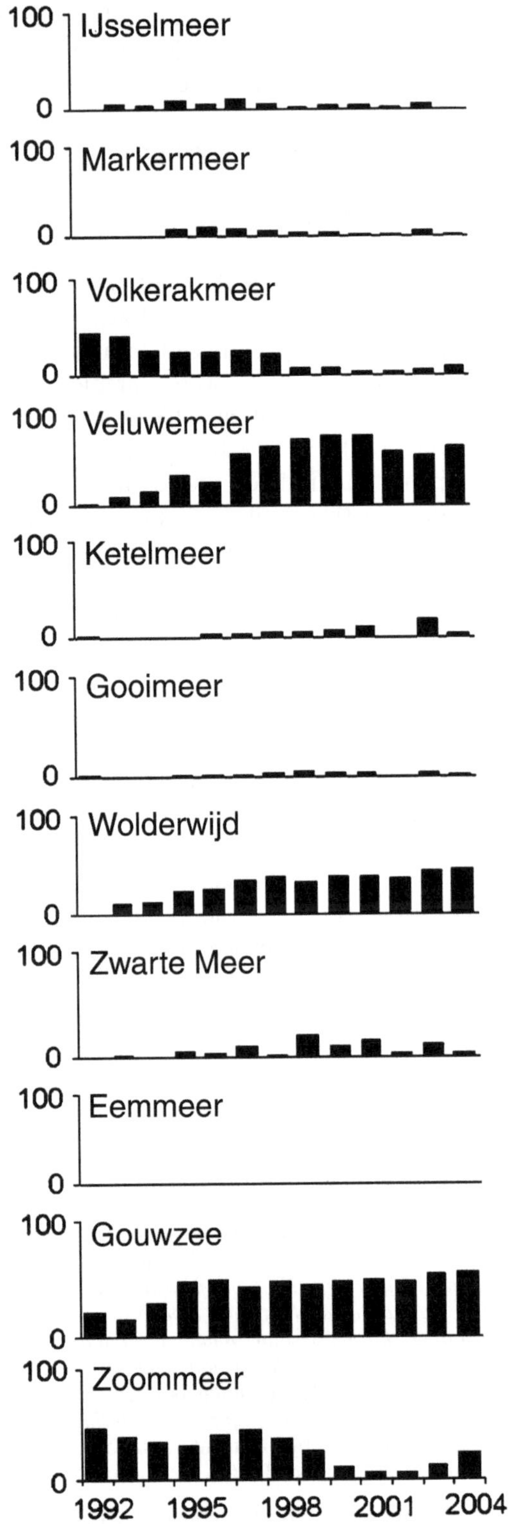

Fig. 1 Development of submerged vegetation cover% in shallow areas (water depth < 3 m) in 11 lakes in The Netherlands between 1992 and 2004

to 36%), *P. pusillus* (from 17% to 21%), and *Zannichellia palustris* (from 13% to 29%); whereas *Myriophyllum spicatum* decreased from 6% to 4%. For individual lake-years, the species richness was positively correlated with the cover percentage in the 0–3 m zone (Pearson's $R = 0.480$, $P < 0.01$).

Figure 3 shows that a submerged vegetation cover >20% never occurred when summer-averaged TP-concentrations were above 0.15 mg P l^{-1} (average cover 6 ± 10%). Similarly, cover was always higher than 20% with TP-concentrations below 0.08 mg P l^{-1} (average cover 43 ± 22%). In the intermediate area, a wide variation of cover values was observed (average cover 18 ± 17%). For TN, the relationship was much less clear, as high as well as low submerged vegetation cover occurred over the entire range of 1–7 mg N l^{-1}. High vegetation cover was limited by chlorophyll-*a*, as submerged vegetation cover was always below 20% when average summer concentrations of chlorophyll-*a* exceeded 50 μg l^{-1}.

Calculation of EQR of the lakes showed that in most cases, the thresholds for Good Ecological Status were not met (Table 3). The 'good' status (overall EQR > 0.6) was achieved only in Veluwemeer (1995 and later years), Wolderwijd (1997 and later years) and Gouwzee (2003 and 2004). In 2004 the lakes Zwarte Meer, Volkerakmeer and Zoommeer were classified as 'moderate', while IJsselmeer, Markermeer, and Ketelmeer were classified as 'poor', and Eemmeer and Gooimeer were classified 'bad' status for macrophytes. Obviously, nutrient loads are still too high in most cases to achieve the Good Status. Figure 4 shows that for lakes that reach Good status, the 95% interval for TP is 0.04–0.12 mg l^{-1}, whereas for non-qualifying lakes, the 95%-interval for TP is 0.07–0.34 mg l^{-1}. Unnatural water levels, littoral morphology, inter-lake connectivity and fish populations may additionally suppress the good state for macrophytes in these lakes.

Discussion

The abundance and composition of aquatic vegetation may reflect the ecological status of shallow lakes. The response of macrophytes to the major environmental pressure of eutrophication has been demonstrated in various studies (Sand-Jensen et al., 2000; Bachmann et al., 2002; Körner, 2002; Lauridsen et al., 2003), including the 'border lakes' that were among the lakes in the present study (Scheffer et al., 1992; Coops & Doef, 1993). Shifts between high and low macrophyte cover may be discerned, related to certain

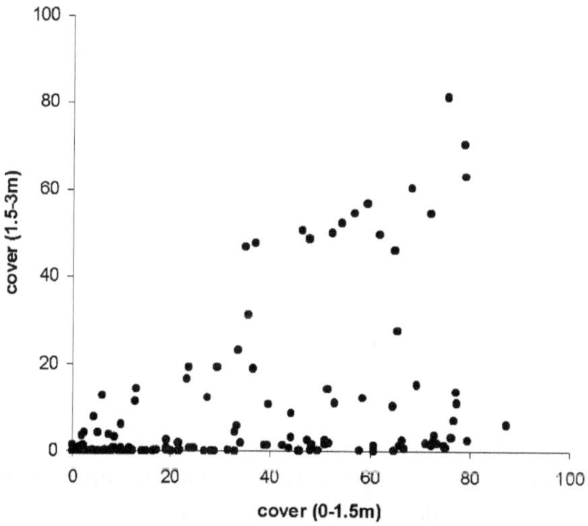

Fig. 2 Relationship between macrophyte cover in two water-depth strata (0–1.5 m vs. 1.5–3.0 m) based on data from 11 lakes in the period 1992–2004

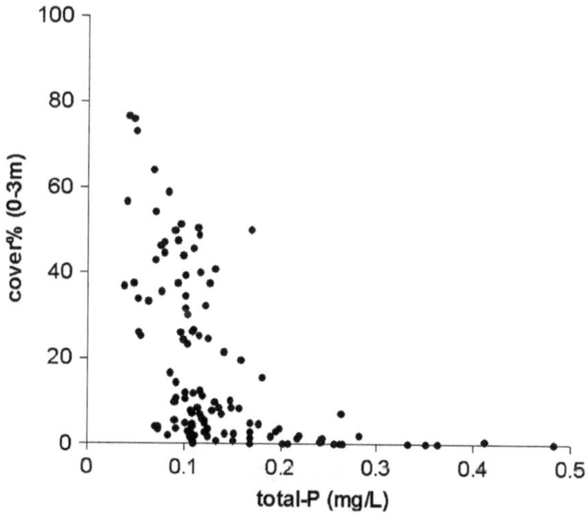

Fig. 3 Relationship between percentage cover by submerged vegetation in the shallow zone (0–3 m) and summer average TP concentration

Table 3 Macrophyte cover and species richness of the lakes in 2004, and the EQR for abundance and species composition, as well as overall EQR for macrophytes

Lake/water body	Cover of shallow area (%)	Abundance EQR	Species richness (*n*)	Species composition EQR	Overall macrophyte EQR
IJsselmeer	0.5	0.09	8	0.37	0.23
Markermeer	4	0.35	6	0.44	0.40
Volkerakmeer	9	0.44	8	0.40	0.42
Veluwemeer	65	1.00	16	0.64	0.82
Ketelmeer	4	0.34	7	0.37	0.36
Gooimeer	3	0.14	4	0.26	0.15
Wolderwijd	52	0.75	12	0.63	0.69
Zwarte Meer	3	0.30	14	0.55	0.43
Eemmeer	2	0.02	5	0.22	0.12
Gouwzee	57	0.85	8	0.48	0.66
Zoommeer	23	0.58	7	0.37	0.48

thresholds in nutrient loading in lakes. Declines, respectively expansions of macrophyte cover tend to show sudden shifts between the clear, macrophyte-rich state, and the turbid, macrophyte-poor state of lakes (Scheffer, 1998). In some cases, these shifts have been attributed to catastrophic events, such as summer drawdown or ice scouring (Blindow, 1992). However, expansion of submerged vegetation seemed to occur gradually over a number of subsequent years in the recovering lakes in the Netherlands (Veluwemeer, Wolderwijd, Gouwzee). Likewise, the decreasing trend in Volkerakmeer and Zoommeer did not appear to occur as a sudden collapse of the vegetation, but rather as a gradual decrease. It might be that the size of these lakes is so large that they do not respond uniformly, as if different sub-systems exist responding to different nutrient levels, morphometry, sediment type, etc. In Veluwemeer, macrophytes started to colonise the shallowest parts of the lake and expanded from there; a clear distinction between a clear- and a turbid area within the lake was observed related to the macrophyte-colonised area (Scheffer et al., 1994), until the entire lake became clear when >60% was covered by submerged vegetation. The expansion of vegetation in a recovering lake depends on underwater light conditions in combination with morphometry (Duarte & Kalff, 1986). Van den Berg et al. (2003) showed the high predictive value of water depth, exposure and transparency in modelling macrophyte dynamics in the 'border-lake' area.

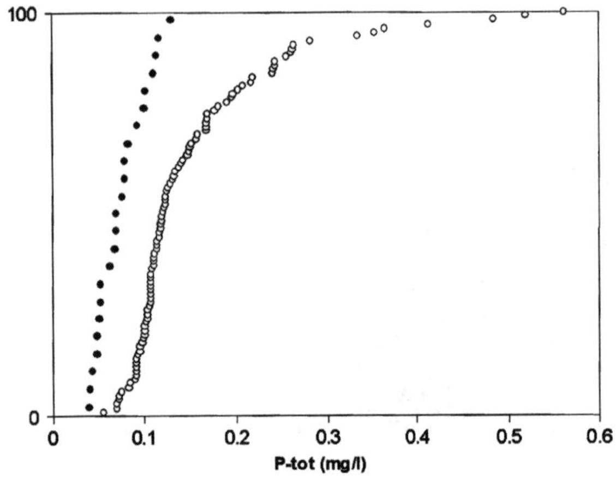

Fig. 4 Cumulative distribution of TP (average TP vs. % of lake-years) fulfilling the good ecological state for macrophytes (*black circles*) and lake years not fulfilling the good ecological state for macrophytes (*open circles*)

Once a more or less dense cover of macrophyte vegetation has developed, this itself may affect nutrient cycling within lakes (Granéli & Solander, 1988) and enhance and stabilise the clear water state in lakes (Portielje & Van der Molen, 1999). Consequently, aquatic vegetation monitoring may be a suitable tool for evaluating the success of lake restoration management by reducing nutrient loading and naturalising hydromorphology (Melzer, 1999).

The observed interannual variation in abundance and, to a lesser extent, species composition of aquatic macrophytes can be attributed to several factors. Firstly, variation in sampling conditions and timing resulting in inaccurate or unrepresentative sampling cannot be ruled out to at least partly explain the large fluctuations in macrophyte cover. In addition, there may be sampling problems due to identification errors and observers' bias in cover estimations (Kercher et al., 2003). Short-term seasonal changes in macrophyte abundance may be related to e.g., changes in water level, periphyton development, phytoplankton blooms, or herbivory. However, even with these uncertainties, year-to-year variations in macrophyte cover of lakes after recolonisation may be extremely large; for instance, Lauridsen et al. (2003) observed cover values fluctuating between 2 and 80% in restored Danish lakes. Despite this, our data showed clear trends for some lakes.

Macrophyte abundance is a quality element in the assessment of lakes according to the WFD and it particularly indicates the degree of eutrophication in lakes. Depth-limitation due to poor light conditions is reflected in the area covered, and can be related to the area that would be covered in reference, non-eutrophied conditions. In the case of Dutch alkaline, shallow lakes, the maximum colonisable depth for macrophytes was derived using empirical relationships between phosphorus, chlorophyll-*a* and transparency, as well as minimum light requirements for dense macrophyte growth (Middelboe & Markager, 1997; Van den Berg et al., 2002). The average cover of the area within this depth contour was used to determine the EQR for macrophyte abundance (Van der Molen, 2004), using limits for the quality classes 'high', 'good', 'moderate', 'poor' and 'bad' that were based on expert agreement. The class limits of the EQR for macrophyte composition, analogously, depend on expert agreement, as it is calculated by referring to a weighted presence of so-called characteristic species, that are based on species described for water-type specific plant communities (Van der Molen, 2004). The procedure reflects the lack of 'real' data that can be used to define the highest ecological status (the reference) and to derive the EQR from it. In other EU countries where reference sites have been available, the latter approach was often followed (Meilinger et al., 2005; Schaumburg et al., 2004; Stelzer et al., 2005).

The vast majority of lakes in the Netherlands have been classified as 'heavily modified', because of their altered hydromorphology and/or artificial origin. Hence the status relative to 'maximum ecological potential' rather than reference should be assessed. Because water levels are commonly regulated and shorelines are often artificial, we (tentatively) excluded incorporation of an EQR for emergent vegetation and marginal wetlands. Their assessment would require knowledge of the maximum potential for each individual water body. It is not clear yet whether, and how, this component will be included in the WFD macrophyte assessment. We also excluded the phytobenthos EQR, because of the lack of monitoring data and insufficient understanding of its relationships with pressures.

The present assessment of macrophyte quality in the 10 lakes of this study reflects very well the perceptions of various experts working in these lakes and the degree of recovery from the eutrophied conditions that were prevalent in the 1970's and 1980's. It can be concluded that currently most of the lakes are not in the good ecological state that should be achieved in 2015. Measures that can be taken to improve the condition include a further reduction of nutrient loading. As efforts to reduce nutrient loadings further may prove increasingly difficult and costly, other measures that effectively improve transparency (e.g., biomanipulation) and morphometry, should be considered as well.

References

Bachmann, R. W., C. A. Horsburgh, M. V. Hoyer, L.K. Mataraza & D. E. Canfield Jr, 2002. Relations between trophic state indicators and plant biomass in Florida lakes. Hydrobiologia 470: 219–234.

Blindow, I., 1992. Long- and short-term dynamics of submetrged macrophytes in two shallow eutrophic lakes. Freshwater Biology 28: 15–27.

Coops, H. & R. W. Doef, 1993. Submerged vegetation development in two shallow, eutrophic lakes. Hydrobiologia 340: 115–120.

Duarte, C. M. & J. Kalff, 1986. Littoral slope as a predictor of the maximum biomass of submerged macrophyte communities. Limnology and Oceanography 31: 1072–1080.

European Commission, 2000. Directive 2000/06/EC of the European Parliament and of the Council of Europe (2000). Establishing a Framework for Community Action in the Field of Water Policy. Official Journal

of the European Communities 22.12.2000:L327/I-L327/72.
Granéli, W. & D. Solander, 1988. Influence of aquatic macrophytes on phosphorus cycling in lakes. Hydrobiologia 170: 245–266.
Graveland, J. & H. Coops, 1997. Verdwijnen van rietgordels in Nederland. Landschap 14: 67–86.
Gulati, R. D. & E. Van Donk, 2002. Lakes in the Netherlands, their origin, eutrophication and restoration: state-of-the-art review. Hydrobiologia 478: 73–106.
Kercher, S. M., C. B. Frieswyk & J. B. Zedler, 2003. Effects of sampling teams and estimation methods on the assessment of plant cover. Journal of Vegetation Science 14: 899–906.
Körner, S., 2002. Loss of submerged macrophytes in shallow lakes in north-eastern Germany. International Review of Hydrobiology 87: 375–384.
Lauridsen, T. L., J. P. Jensen, E. Jeppesen & M. Søndergaard, 2003. Response of submerged macrophytes in Danish lakes to nutrient loading reductions and biomanipulation. Hydrobiologia 506–509: 641–649.
Meilinger, P., S. Schneider & A. Melzer, 2005. The reference index method for the macrophyte-based assessment of rivers—a contribution to the implementation of the European Water Framework Directive in Germany. International Review of Hydrobiology 90: 322–342.
Melzer, A., 1999. Aquatic macrophytes as tools for lake management. Hydrobiologia 395/396: 181–190.
Middelboe, A. L. & S. Markager, 1997. Depth limits and minimum light requirements of freshwater macrophytes. Freshwater Biology 37: 553–568.
Ostendorp, W., 1989. 'Die-back' of reeds in Europe—a critical review of literature. Aquatic Botany 35: 5–26.
Portielje, R. & D. T. Van der Molen, 1999. Relationships between eutrophication variables: from nutrient loading to transparency. Hydrobiologia 409: 375–387.
Sand-Jensen, K., T. Riis, O. Vestergaard & S. E. Larsen, 2000. Macrophyte decline in Danish lakes and streams over the past 100 years. Journal of Ecology 88: 1030–1040.
Schaumburg, J., C. Schwarz, G. Hofmann, D. Stelzer, S. Schneider & U. Schmedtje, 2004. Macrophytes and phytobenthos as indicators of ecological status in German lakes—a contribution to the implementation of the Water Framework Directive. Limnologica 34: 302–314.
Scheffer, M., 1998. Ecology of Shallow Lakes. Chapman & Hall, London, 357 pp.
Scheffer, M., M. R. De Redelijkheid & F. Noppert, 1992. Distribution and dynamics of submerged vegetation in a chain of shallow eutrophic lakes. Aquatic Botany 42: 199–216.
Scheffer, M., M. S. Van den Berg, A. W. Breukelaar, C. Breukers, H. Coops, R. W. Doef & M. L. Meijer, 1994. Vegetated areas with clear water in turbid shallow lakes. Aquatic Botany 49: 193–196.
Stelzer, D., S. Schneider & A. Melzer, 2005. Macrophyte-based assessment of lakes—a contribution to the implementation of the European Water Framework Directive in Germany. International Review of Hydrobiology 90: 223–237.
Van den Berg, M. S., W. Joosse & H. Coops, 2003. A statistical model predicting the occurrence and dynamics of submerged macrophytes in shallow lakes in the Netherlands. Hydrobiologia 506–509: 611–623.
Van den Berg, M. S., D. T. Van der Molen & P. C. M. Boers, 2002. Setting up reference conditions for phytoplankton, turbidity and submerged macrophytes in Dutch lakes. TemaNord 566: 28–31.
Van der Molen, D. T. (ed.), 2004. Referenties en concept-maatlatten voor meren voor de Kaderrichtlijn Water. STOWA Report 2004–42, Utrecht, 450 pp.

Hydrobiologia (2007) 584:403–413
DOI 10.1007/s10750-007-0609-6

SHALLOW LAKES

Climate-induced shifts in an experimental phytoplankton community: a mechanistic approach

Lisette N. De Senerpont Domis · Wolf M. Mooij · Jef Huisman

Abstract Climate change is likely to have far-reaching effects on biotic interactions in aquatic ecosystems. We investigated the effect of different spring warming scenarios on the succession of three algal groups (cyanobacteria, diatoms and green algae) in 10-l microcosms. We fitted these microcosm data to a simple mechanistic model to estimate the effect of different climate warming scenarios on the population dynamics of these algal functional groups. Experimental and model results indicate that the different algal functional groups respond differently to climate warming under phosphorus-limited conditions. Whereas the successional sequence, from diatoms to green algae to cyanobacteria, was not affected by the different climate warming scenarios, cyanobacteria showed a stronger response to the different climate warming scenarios than diatoms or green algae. Both the growth rates and peak abundances of cyanobacteria were significantly higher in the average and warm spring scenarios than in the cold spring scenario. Our findings illustrate that integration of models and microcosm experiments are a useful approach in predicting the impacts of rising temperatures on the dynamics of phytoplankton communities.

Keywords Community dynamics · Climate change · Freshwater · Cyanobacteria · Green algae · Diatoms · Mechanistic model · Spring warming

Guest editors: R. D. Gulati, E. Lammens, N. De Pauw & E. Van Donk
Shallow lakes in a changing world

L. N. De Senerpont Domis (✉) · W. M. Mooij
Netherlands Institute of Ecology (NIOO-KNAW), Centre for Limnology, Rijksstraatweg 6, 3631 AC Nieuwersluis, The Netherlands
e-mail: l.desenerpontdomis@nioo.knaw.nl

J. Huisman
Aquatic Microbiology, Institute for Biodiversity and Ecosystem Dynamics, University of Amsterdam, Nieuwe Achtergracht 127, 1018 WS Amsterdam, The Netherlands

Introduction

Over the past century the global temperature has increased on average by ~0.6 degrees Celsius (Houghton et al., 2001). Climate change models predict a further increase in temperate of 1.4–5.8°C by the year 2100, depending on emission scenarios (Houghton et al., 2001). This projected climate change is likely to have far-reaching effects on biotic interactions across different ecosystems, including range boundary shifts, physiological and genetic adaptation, shifts in phenology and shifts in community composition (Walther et al., 2002; Parmesan & Yohe, 2003; Mooij et al., 2005).

According to regional models predicting changes in climate in the Netherlands, both temperature and precipitation are expected to increase, resulting in mild and wet winters, warmer springs and a higher incidence of extreme weather events (Kors et al., 2000). The strongest temperature increase is expected during late winter and early spring (Van Vliet et al., 2002; Van Oldenborgh & Van Ulden, 2003). During late winter and early spring, most lakes in temperate regions start to develop a characteristic seasonal succession of phytoplankton groups (Sommer et al., 1986). An initial spring bloom of small algae (e.g. centric diatoms and cryptophyceae) is triggered by ample nutrient and light availability towards the end of winter. In shallow lakes, this initial bloom of small algae is soon replaced by large pennate diatoms, e.g. *Asterionella*, and in summer by green algae and cyanobacteria (Sommer et al., 1986; Reynolds, 1997).

Time series analysis of lakes in north Western Europe showed a strong correlation between the timing of algal spring bloom formation and proxies of climate change (Müller-Navarra et al., 1997; Adrian et al., 1999; Weyhenmeyer et al., 1999; Gerten & Adrian, 2000). In the past decades, the spring bloom started earlier after warmer winters. Such changes in algal phenology may result in a decoupling of trophic relationships, when the advancement of the spring bloom is not followed by similar changes in the timing of key herbivores like *Daphnia* (Winder & Schindler, 2004a, b). Climate change may also cause quantitative changes in phytoplankton community dynamics, shown as changes in phytoplankton species composition or changes in the seasonal succession of phytoplankton groups. In among-lake comparisons of phytoplankton community composition, both Adrian et al. (1995) and Weyhenmeyer (2001) report that rising temperatures are accompanied by a higher biomass of cyanobacteria.

Climate warming may favour cyanobacteria over diatoms and to lesser extent green algae, both directly and indirectly. As a major direct effect, Reynolds (1997) showed that with increasing temperature, the maximum growth rate of the cyanobacteria *Microcystis aeruginosa* increased almost twice as much as that of the green alga *Scenedesmus quadriqauda* and almost five times as much as that of the diatom *Asterionella formosa.* Coles and Jones (2000) found similar differences in temperature-dependent growth rates between three cyanobacteria (*Microcystis aeruginosa, Merismopedia tenuissima* and *Oscillatoria* sp.) and the diatom *Aulacoseira granulata* var. *angustissima.* Surprisingly, however, Moss et al. (2003) did not report an increase in the abundance of cyanobacteria with temperature in experimental microcosms. Rising temperatures will also affect phytoplankton communities in indirect ways. For instance, changes in the heat budget of lakes induce seasonal changes in stratification and turbulent mixing. Reduced turbulent mixing and enhanced stratification facilitate the development of surface blooms in eutrophic lakes, thereby shifting the species composition of phytoplankton communities in favour of buoyant cyanobacteria (Harris & Baxter, 1996; Huisman et al., 2004). Furthermore, climate warming may enhance the eutrophication of freshwater ecosystems (Mckee et al., 2003; Van de Bund et al., 2004; Christoffersen et al., 2006), which may also facilitate the potential of cyanobacteria to dominate the phytoplankton community (Elliott et al., 2006).

Experimental research mimicking climate change scenarios in natural lakes is a useful tool to predict the impact of future climate change on lake phytoplankton dynamics. Here, we investigate the effect of different spring warming scenarios on the succession of three algal groups (cyanobacteria, diatoms and green algae) in 10-l microcosms. So far, only a few studies have reported the direct effect of rising temperature on phytoplankton community structure (Weyhenmeyer et al., 1999; Weyhenmeyer, 2001; Moss et al., 2003). Laboratory microcosms cannot mimic all aspects of climate change, e.g. hydrodynamic processes. However, microcosms do allow detailed investigation of specific mechanisms under controlled conditions. Our 10-l microcosms enable the growth of natural phytoplankton assemblages under controlled laboratory conditions, thus providing an intermediate step of realism and control between artificial plankton communities (Beisner et al., 1996, 1997), large-scale outdoor mesocosms (Moss et al., 2003;

Strecker et al., 2004; Christoffersen et al., 2006) and correlational time series analysis. We fit these microcosm data to a simple mechanistic model to estimate the effects of warming on algal growth rates and mortality rates. This allows us to examine how different spring warming scenarios may affect the population dynamics of the three algal groups. Our overall hypothesis is that, in accordance with the results of time series analysis (Gerten & Adrian, 2000; Weyhenmeyer, 2001; Weyhenmeyer et al., 2002), increased warming will affect the succession of the different algal groups. In particular, we expect that, given the higher temperature optimum of cyanobacteria (Reynolds, 1997; Coles & Jones, 2000), increased warming will result in higher cyanobacterial abundance.

Material and methods

Experimental set-up

Plankton inoculum was obtained from shallow eutrophic Lake Tjeukemeer; in the northern part of The Netherlands (52° 5 N, 5° 5E) on February, 15, 2005. The plankton inoculum was sampled at 0.5–1.0 m depth. Large zooplankton was removed by filtering through an 80 μm mesh. Small zooplankton (e.g. rotifers and small crustaceans) was removed by gently bubbling the inoculum with nitrogen gas at 2 h intervals for 24 h (Sommer, 1985). To acclimatize the phytoplankton community to laboratory conditions and to ensure an evenly distributed phytoplankton composition of the culture vessels, the phytoplankton inoculum was first precultured in a large mesocosm for a period of 8 days at 8°C (Verschoor et al., 2003). After this incubation period, 12 experimental 10-l vessels (transparent polycarbonate carboys; NALGENE, Rochester, USA) were each inoculated with 10 l phytoplankton suspension.

Phytoplankton communities were cultured for 61 days, using light and temperature conditions representative for the period between Julian day 110 and 172 in The Netherlands (Mooij & Van Tongeren, 1990). This period matches the timing of the phytoplankton spring bloom. The experimental vessels were treated as semi-continuous cultures. They were supplied daily with oligo-mesotrophic lake water (Lake Maarsseveen), filtered through a 0.45 μm membrane filter (Schleicher and Schuell Microscience, Dassel, Germany). The dilution rate was 0.0395 d^{-1}. Cultures were mixed for 15 min at 100 RPM every two hours, using a magnetic stirrer. Cultures were also stirred manually prior to sampling. All experimental vessels received a constant air supply. PHILIPS TLD 30W/33 cool-white fluorescent tubes were used as light source (120 μmol photons m^{-2} s^{-1}, integrated over PAR range). The day-night light cycle simulated natural spring conditions in the Netherlands, i.e. daylight was gradually increased from 14 h and 15 min at the start to 16 h and 55 min daylight at the end of the experiment.

We chose three climate scenarios, reflecting the seasonal rise in water temperature during a cold, an average, and a warm spring (Fig. 1). Temperature ranged from 9°C to 13°C in a cold spring, from 9°C to 19°C in an average spring, and from 9°C to 25°C in a warm spring. These climate scenarios fell within the bandwidth of spring warming regimes measured in Lake Tjeukemeer between Julian day 110 and 172 during the period

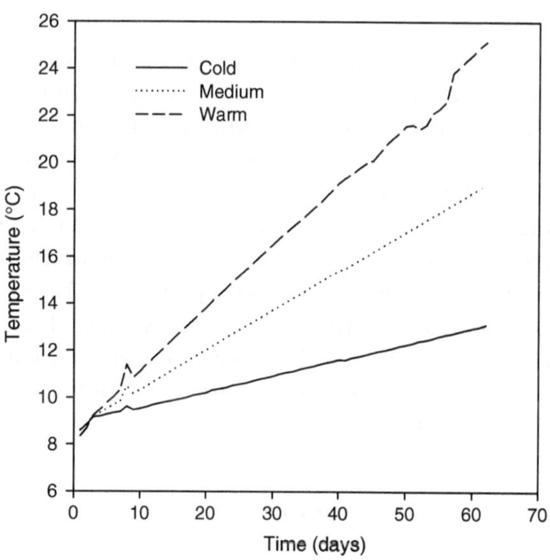

Fig. 1 Time course of temperature in the three climate scenarios simulated by the experiments. Cold = cold spring scenario, Medium = average spring scenario, Warm = warm spring scenario

of 1971–1987 (Mooij, 1996). The water temperature at Julian day 172 can be more than 12°C higher in a warm spring than in a cold spring (Fig. 1). Each climate scenario was replicated four times. The rising temperatures were automated using the FAREX SR minisystem (RKC instruments inc., Tokyo, Japan), ensuring a temperature control of ±0.5°C.

Sampling procedure and data acquisition

Throughout the experimental period, temperature was measured every minute. Phytoplankton was sampled daily, using a sampling tube with an inner diameter of 56 mm (De Nie et al., 1980). Immediately after sampling, the total sampling volume (395 ml) was filtered over a 60 μm mesh to separate larger phytoplankton (e.g. the diatom *Asterionella formosa*) from smaller phytoplankton (e.g. *Scenedesmus* sp.).

On a daily basis, a 10 ml subsample was taken from the filtrate for pulse-amplitude modulated fluorometry (PHYTO-PAM, Effeltrich, Germany). Based on the relative proportion of pigments present in phytoplankton species PHYTO-PAM analysis estimates the relative abundances of different phytoplankton groups, i.e. cyanobacteria, green algae and diatoms (Walz, 1999).

A 50 ml subsample was taken from the filtrate every third day for nutrient analysis, and immediately filtered through aquadest rinsed glass microfibre filters (Whatman GF/F, Maidstone, UK). The residue was used to estimate weight of the total seston fraction smaller than 60 μm. The filtrate was preserved at –20°C for later nutrient analysis. Soluble reactive phosphorus (SRP) was determined according to Murphy and Riley (1962). Soluble reactive silicon (referred to as silicate) was measured according to Golterman et al. (1978).

Statistical analysis of experimental data

We tested whether the time course of the phytoplankton community and SRP concentration showed significant differences between treatments using an Analysis of Variance for repeated measurements (rANOVA). Variables were square-root transformed prior to analysis, if so required to meet the assumptions of rANOVA and ANOVA.

Model analysis of experimental data

To simulate phytoplankton succession under different climate scenarios we developed a competition model in which the abundances of green algae (A_G), cyanobacteria (A_C) and diatoms (A_D) depend on the concentration of one or more limiting nutrients (N). Because the experiments revealed that diatoms decreased from the start, whereas green algae and cyanobacteria initially increased, we assume that there are two limiting nutrients in the system. Indeed, nutrient analysis of the culture medium showed low concentrations of silicate for the greater part of the experimental period, most probably limiting diatom growth, whereas SRP reached limiting concentrations only after experimental day 18. Therefore, we omitted an autonomous growth term for the diatoms. Thus, our model is described by the following differential equations:

$$\frac{dN}{dt} = -c(r_C A_C + r_G A_G)\frac{N}{H_N + N} - DN$$

$$\frac{dA_D}{dt} = -d_D A_D - DA_D$$

$$\frac{dA_G}{dt} = r_G A_G \frac{N}{H_N + N} - DA_G$$

$$\frac{dA_C}{dt} = r_C A_C \frac{N}{H_N + N} - DA_C$$

Here, r_C and r_G are the specific growth rates of the cyanobacteria and green algae and parameter c represents the nutrient content of these algal species. We assigned an extra mortality term, d_D, to the diatoms, because they decreased faster than predicted on the basis of the dilution rate alone. We use the same value of the

Table 1 Parameters and state variables of the mechanistic algal succession model

Par/State variable	Value	Unit	Description
N	–	μg P l^{-1}	Limiting nutrients
A_D	–	μg Chl *a* l^{-1}	Abundance of diatoms
A_G	–	μg Chl *a* l^{-1}	Abundance of green algae
A_C	–	μg Chl *a* l^{-1}	Abundance of cyanobacteria
c	1.77	μg P/μg Chl *a*	Nutrient content of algae[c]
D	0.0395	d^{-1}	Dilution rate[a]
H_N	2.0	μg P l^{-1}	Half saturation constant[b]
d_D	0.0243	d^{-1}	Mortality rate of diatoms[c]
r_G	0.200	d^{-1}	Growth rate of green algae[c]
r_G	0.143	d^{-1}	Growth rate of cyanobacteria[c]

[a] Set a priori on basis of experimental designs

[b] Set a priori

[c] Estimated from data

half-saturation constant, H_N, for both green algae and cyanobacteria, because the model fit did not improve significantly when these two algal groups were each assigned their own half-saturation constant. Finally, we note that all four state variables (i.e. N, A_D, A_G and A_C) are subjected to the same dilution rate of $D = 0.0395$ d^{-1}. The model parameters and state variables are described in further detail in Table 1. Model fits were obtained by minimization of the residual sum of squares of square-root transformed observed and modelled values of the three phytoplankton groups. After fitting the model to the algal data we estimated parameter c by fitting the resulting nutrient dynamics to the SRP data. Applying a wide range of values for the half-saturation constants showed that the precise value of this parameter hardly affected the fit of the model. The only difference is that low values of the half-saturation constant resulted in a sharp transition from ample nutrient availability to strong nutrient limitation (i.e. within a day), whereas high values of the half-saturation constant resulted in a more gradual transition (i.e. a couple of days).

To quantify the effect of different climate scenarios on the model parameters, we fitted the model to the time series data of the experiments. Given three climate treatments, with four replicates each, this resulted in 12 independent estimates of the growth rate of green algae (r_G), the growth rate of cyanobacteria (r_C) and the mortality rate of diatoms (d_D). Using one-way ANOVA, these parameter estimates were tested for climate treatment effects. Pairwise comparisons were carried out, using Tukey HSD. Prior to analyses, data were checked for normality and homogeneity of variances. All statistical analyses were performed using the software package STATISTICA version 7.1 (StatSoft, Inc. 2005; www.statsoft.com).

Results

Spring warming effects on nutrient availability and total seston

The trends in total seston, soluble reactive phosphorus (SRP) and silicate during the course of the experiment are plotted in Fig. 2. As indicated by rANOVA, both SRP and total seston showed a significant response in time (Table 2). In all treatments, initially sharp ups and downs in concentrations of SRP could be observed (Fig. 2a). After day 10, SRP concentrations decreased more gradually, and from day 18 onwards SRP remained close to limiting levels (<5 μg SRP l^{-1}) for the remainder of the experimental period. Silicate levels (initial concentration = 1666 μg l^{-1}) reached limiting levels (<500 μg l^{-1}) around day 10 (Fig. 2b). Total seston mass reached its maximum values around day 12, after which it decreased to low levels (<5 mg dry weight l^{-1}) until day 45 of the experiment (Fig. 2a). Thereafter, a slow increase in total seston was observed. The time course of total seston differed significantly in response to the different climate scenarios (Table 2, interaction Climate × Time).

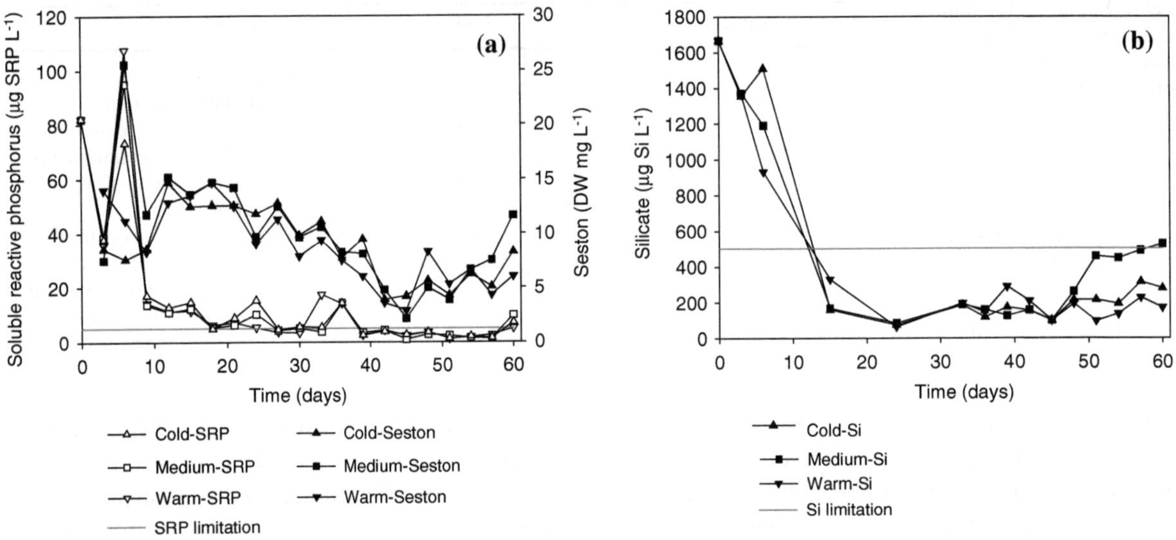

Fig. 2 Time course of the concentration of (**a**) soluble reactive phosphorus (open symbols) and total seston (closed symbols), and (**b**) silicate observed in the three climate scenarios simulated by the experiments

Table 2 Results of rANOVA testing the effect of the different climate warming scenarios (Climate) on the concentration of soluble reactive phosphorus (SRP) and dry weight of total seston during the course of the experiment (Time)

Effect	SRP			Total seston dry weight		
	df	SS	*P*	df	SS	*P*
Climate	2	0.742	n.s.	2	115.12	n.s.
Time	19	592.43	**	19	2581.29	**
Time × Climate	38	31.64	n.s.	38	807.27	*

* $P < 0.05$, ** $P < 0.01$, n.s. = not significant

Phytoplankton community response to spring warming

In all treatments, similar shifts in the relative abundances of cyanobacteria, green algae and diatoms were observed (Fig. 3). Diatoms were initially by far the most dominant phytoplankters; their rapid decline was followed by peaks of green algae and cyanobacteria at day 20 and day 30, respectively. The green algae were more abundant than the cyanobacteria. The results of rANOVA underlined this significant response in time of the different algal functional groups (Table 3, mean effect Time). The time courses of all three algal functional groups (i.e. diatoms, cyanobacteria and green algae) differed significantly in response to the different climate scenarios (Table 3, interaction Climate × Time).

Model results

The model predictions showed a good fit with the experimental data of the algal groups (Fig. 4, $R^2 = 0.875$). The model captures the observed succession of an initial dominance of diatoms followed by a peak of green algae and cyanobacteria, although the modelled peaks of green algae and cyanobacteria shortly precede those observed in the experiment. The model predictions for the limiting nutrient showed a reasonably good fit with the experimental data on SRP concentrations (Fig. 4, $R^2 = 0.684$). We also examined slightly more complex models, to see whether the fit could be further improved. Inclusion of an extra loss term for the green algae and cyanobacteria resulted in $R^2 = 0.878$. Inclusion of a separate limiting nutrient for the green algae and a

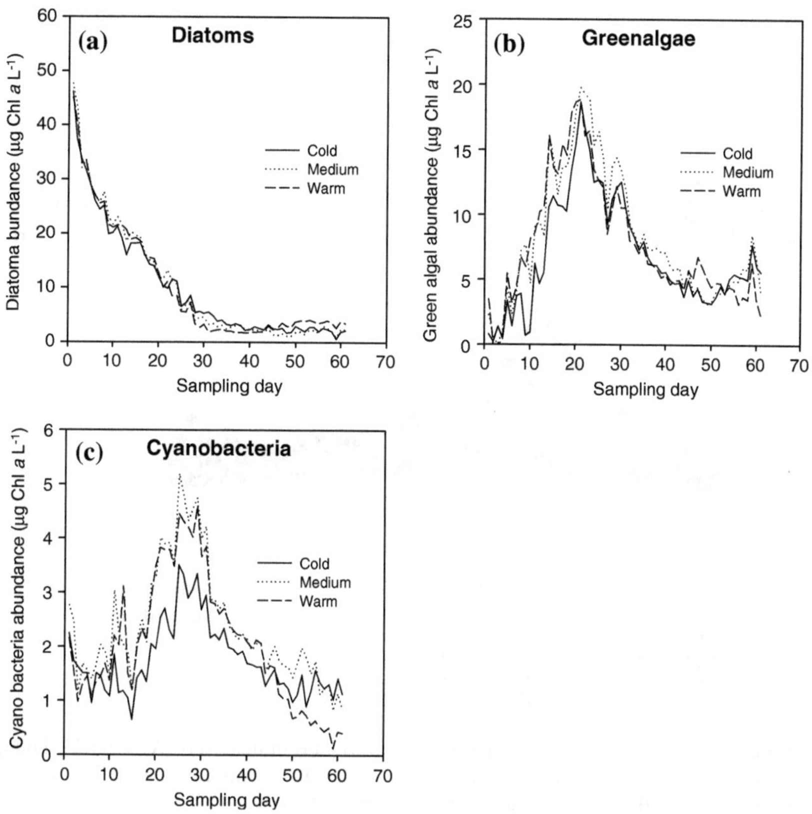

Fig. 3 Population dynamics (expressed as µg Chl *a* l^{-1}) of (**a**) diatoms, (**b**) green algae, and (**c**) cyanobacteria observed in the three climate scenarios simulated by the experiments. Cold = cold spring scenario, Medium = average spring scenario, Warm = warm spring scenario. Values are daily averages of climate treatments (n = 4)

separate limiting nutrients for the cyanobacteria resulted in $R^2 = 0.886$. Inclusion of recycling of nutrients stored in the dead diatoms into the nutrient pool resulted in $R^2 = 0.868$. Because of the marginal improvement or even slight reduction of the fits for the more complex models, we proceeded our analysis with the original model.

Figure 5 displays the model estimates of the growth rate of cyanobacteria (r_C), the mortality rate of diatoms (d_D), and the growth rate of green algae (r_G) for each of the three spring warming scenarios. The population dynamics predicted by the model (not shown, except for Fig. 4) revealed that especially the cyanobacteria reached higher abundances under the warm and medium climate scenarios (max abundance 4.18 and 4.37 µg Chl *a* l^{-1}, respectively) than under the cold climate scenario (max abundance 3.34 µg Chl *a* l^{-1}). As indicated by ANOVA, the growth rate of cyanobacteria was significantly lower in the cold

Table 3 Results of rANOVA testing the effect of the different climate warming scenarios (Climate) on the abundance (µg Chl *a* l^{-1}) of cyanobacteria, diatoms and green algae during the course of the experiment (Time)

Effect	Cyanobacteria			Diatoms			Green algae		
	df	SS	*P*	df	*SS*	*P*	df	SS	*P*
Climate	2	37.07	n.s.	2	249.25	n.s.	2	507.24	n.s.
Time	60	289.99	**	60	43356.55	**	60	6431.63	**
Time × Climate	120	77.15	**	120	511.59	**	120	874.75	**

* $P < 0.05$, ** $P < 0.01$, n.s. = not significant

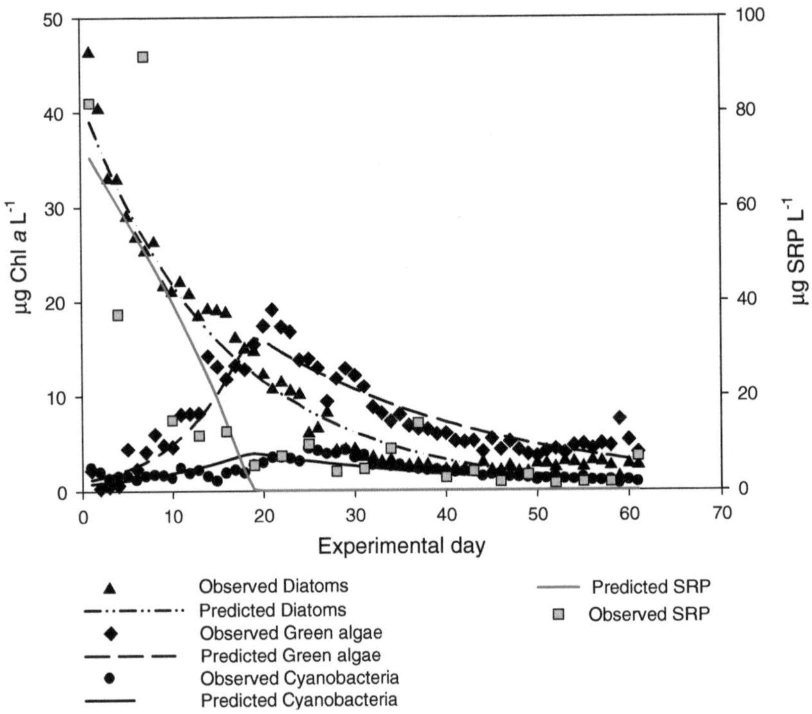

Fig. 4 Time course of diatoms, green algae, cyanobacteria and soluble reactive phosphorus observed in the experiments (symbols) and predicted by the mechanistic model (lines). The goodness of fit of the algal component of the model is indicated by $R^2 = 0.875$. The goodness of fit of the nutrient component of the model is indicated by $R^2 = 0.684$

spring scenario than in the medium and warm spring scenarios (Fig. 5, $P < 0.05$, Tukey HSD test). Although the growth rate of green algae also appeared lower in the cold spring scenario than in the medium and warm spring scenario, for green algae this effect was not significant (Fig. 5, $P = 0.10$). The mortality rate of the diatoms did not differ significantly between climate treatments (Fig. 5, $P = 0.95$). Given the absence of significant differences in SRP concentrations between the different climate treatments (Table 3), we did not proceed with testing the model estimates of the nutrient content of the algae.

Discussion

The phytoplankton community in our cultures showed a significant response to the different spring warming scenarios. As tested by rANOVA, time courses of seston biomass were different between the different spring warming scenarios. This was mirrored by the response of the algal functional groups. The concentration of SRP did not differ significantly between different climate scenarios, suggesting that the responses of

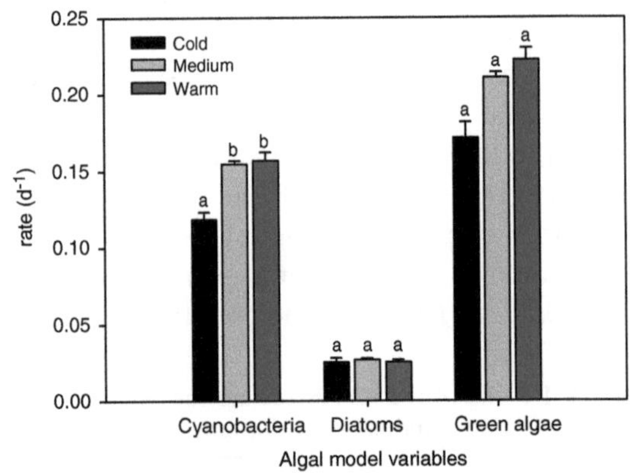

Fig. 5 Effects of the three climate scenarios on the growth rate of cyanobacteria (r_C), the mortality rate of diatoms (d_D), and the growth rate of green algae (r_G). Cold = cold spring scenario, Medium = average spring scenario, Warm = warm spring scenario. Bars show means with standard error. The different letters above bars represent significant differences as indicated by Tukey HSD

the phytoplankton community to the climate scenarios were not driven by changes in nutrient availability. Our model results suggest that cyanobacteria show a stronger response to the different climate scenarios than green algae and diatoms. Other laboratory studies confirm the stronger response of cyanobacteria to rising temperatures (Reynolds, 1989; Coles & Jones, 2000), which can result in a higher algal biomass. The differences in temperature sensitivity among cyanobacteria, green algae and diatoms, however, did not appear to affect the successional pattern of these groups in our experiments. In all climate scenarios, the initial dominance of diatoms was followed by successive peaks of green algae and cyanobacteria. In line with our results, Adrian et al. (1999) showed that the successional sequence of phytoplankton groups during spring was in general not affected by climate change.

Increases in the relative contribution of cyanobacteria with rising temperature have been observed in lakes across Europe (Adrian et al., 1995, 1999; Weyhenmeyer, 2001; Van de Bund et al., 2004). High abundances of cyanobacteria in freshwater systems can potentially disrupt trophic relationships, because of their low quality as a food source for invertebrate grazers (DeMott et al., 2001). Moreover, several cyanobacterial species can produce toxins, providing a serious threat for water quality, other aquatic organisms and human health (Carmichael, 2001; Huisman et al., 2005). Results of warming studies in outdoor mesocosms, however, point at the importance of species-specific responses to climate warming (Moss et al., 2003; Strecker et al., 2004). Individual species within major phytoplankton groups may respond very differently to changes in temperature. Furthermore, the effect of rising temperatures on species performances will depend not only on their temperature tolerance, but also on other species-specific characteristics such as nutrient requirements, buoyancy and light requirements (Reynolds et al., 2002).

Both experimental and model data reveal that nutrient limitation played an important role in the observed algal succession. Diatoms declined from the start of the experiment. The low availability of silicate measured in the culture medium, in combination with the relatively high sinking rates of diatoms (Reynolds et al., 1982; Reynolds & Wiseman, 1982) offer a possible explanation for the strong decline of the diatom population. Moreover, silicate depletion is known to increase the sinking rates of diatom species (Bienfang et al., 1982). With the collapse of diatoms, green algae and cyanobacteria strongly increased, reaching peak abundances after 20 days and 30 days, respectively. Both the green algae and cyanobacteria started to decrease once SRP had reached (near) limiting levels. The depletion of SRP in our experiments matches the seasonal depletion of SRP observed in many mesotrophic lakes (Sommer et al., 1986). The good fit of the predicted limiting nutrient with the observed SRP concentrations provides further support that SRP is limiting the green algae and cyanobacteria in our experiments. As recycling of nutrients stored in dead diatoms back into the nutrient pool did not increase the fit of the model, we suggest that phosphorus stored in dead phytoplankton does not become available for cyanobacteria and green algae within the time span of the experiments.

In line with our initial hypothesis, our results highlight potential changes in plankton communities if the current trends in climate change continue (Straile & Geller, 1998; Gerten & Adrian, 2000; Kohler et al., 2005). In accordance with earlier findings from time series analysis (Gerten & Adrian, 2000; Weyhenmeyer, 2001; Weyhenmeyer et al., 2002), we found that, qualitatively, the spring sequence of succession from diatoms to green algae to cyanobacteria is not affected by different climate scenarios. Quantitatively, however, different algal groups show a different response. In particular, cyanobacteria respond more strongly to rising temperatures than green algae and diatoms, resulting in higher growth rates and peak abundances of cyanobacteria in a warming climate.

Acknowledgments We thank K. Siewertsen, N. Helmsing, M. Kagami and I. van der Stap for assistance with sampling. K. Siewertsen carried out the nutrient analyses. Hennie Uittenhout designed and constructed the sampling devices. We thank J. Vijverberg, S. Hülsmann, and R. Gulati, and the anonymous reviewers for their helpful comments. This study was funded by the Biodiversity program of the Earth and Life Sciences Foundation

(ALW), which is subsidized by the Netherlands Organization for Scientific Research (NWO grant 852.00.010/852.00.012). This is publication 3950 of the Netherlands Institute of Ecology (NIOO-KNAW).

References

Adrian, R., R. Deneke, U. Mischke, R. Stellmacher, & P. Lederer, 1995. Long-term study of the Heiligensee (1975–1992)—Evidence for effects of climatic-change on the dynamics of eutrophied lake ecosystems. Archiv für Hydrobiologie 133: 315–337.

Adrian, R., N. Walz, T. Hintze, S. Hoeg, & R. Rusche, 1999. Effects of ice duration on plankton succession during spring in a shallow polymictic lake. Freshwater Biology 41: 621–632.

Beisner, B. E., E. McCauley, & F. J. Wrona, 1996. Temperature-mediated dynamics of planktonic food chains: the effect of an invertebrate carnivore. Freshwater Biology 35: 219–232.

Beisner, B. E., E. McCauley, & F. J. Wrona, 1997. The influence of temperature and food chain length on plankton predator-prey dynamics. Canadian Journal of Fisheries and Aquatic Sciences 54: 586–595.

Bienfang, P. K., P. J. Harrison, & L. M. Quarmby, 1982. Sinking rate response to depletion of nitrate, phosphate and silicate in 4 marine diatoms. Marine Biology 67: 295–302.

Carmichael, W. W., 2001. Health effects of toxin-producing cyanobacteria: "The CyanoHABs". Human and Ecological Risk Assessment 7: 1393–1407.

Christoffersen, K., N. Andersen, M. Sondergaard, L. Liboriussen, & E. Jeppesen, 2006. Implications of climate-enforced temperature increases on freshwater pico- and nanoplankton populations studied in artificial ponds during 16 months. Hydrobiologia 560: 259–266.

Coles, J. F. & R. C. Jones, 2000. Effect of temperature on photosynthesis-light response and growth of four phytoplankton species isolated from a tidal freshwater river. Journal of Phycology 36: 7–16.

De Nie, H. W., H. J. Bromley, & J. Vijverberg, 1980. Distribution patterns in Tjeukemeer, The Netherlands. Journal of Plankton Research 2: 316–334.

DeMott, W. R., R. D. Gulati, & E. Van Donk, 2001. *Daphnia* food limitation in three hypereutrophic Dutch lakes: Evidence for exclusion of large-bodied species by interfering filaments of cyanobacteria. Limnology and Oceanography 46: 2054–2060.

Elliott, J. A., I. D. Jones, & S. J. Thackeray, 2006. Testing the sensitivity of phytoplankton communities to changes in water temperature and nutrient load, in a temperate lake. Hydrobiologia 559: 401–411.

Gerten, D. & R. Adrian, 2000. Climate-driven changes in spring plankton dynamics and the sensitivity of shallow polymictic lakes to the North Atlantic Oscillation. Limnology and Oceanography 45: 1058–1066.

Golterman, H. L., R. S. Clymo, & M. A. M. Ohnstad, 1978. Methods of Physical and Chemical Analysis of Fresh Waters. Blackwell, Oxford, United Kingdom.

Harris, G. P. & G. Baxter, 1996. Interannual variability in phytoplankton biomass and species composition in a subtropical reservoir. Freshwater Biology 35: 545–560.

Houghton, J. D., Y. Ding, D. J. Griggs, M. Noguer, P. J. Van der Linden, & D. Xiaosu, 2001. Climate Change 2001: The Scientific Basis Contribution Of Working Group I to the Third Assessment Report of the Intergovernmental Panel on Climate Change (IPCC). Cambridge University Press, Cambridge, U.K.

Huisman, J., H. C. P. Matthijs, & P. M. Visser, 2005. Harmful Cyanobacteria. Springer, Berlin, Germany.

Huisman, J., J. Sharples, J. M. Stroom, P. M. Visser, W. E. A. Kardinaal, J. M. H. Verspagen, & B. Sommeijer, 2004. Changes in turbulent mixing shift competition for light between phytoplankton species. Ecology 85: 2960–2970.

Kohler, J., S. Hilt, R. Adrian, A. Nicklisch, H. P. Kozerski, & N. Walz, 2005. Long-term response of a shallow, moderately flushed lake to reduced external phosphorus and nitrogen loading. Freshwater Biology 50: 1639–1650.

Kors, A. G., F. A. M. Claessen, J. W. Wesseling, & G. P. Können, 2000. Scenario's externe krachten voor WB21. Commissie Waterbeheer 21e eeuw, RIZA, WL/Delft Hydraulics en KNMI rapport.

Mckee, D., D. Atkinson, S. E. Collings, J. W. Eaton, A. B. Gill, I. Harvey, K. Hatton, T. Heyes, D. Wilson, & B. Moss, 2003. Response of freshwater microcosm communities to nutrients, fish, and elevated temperature during winter and summer. Limnology and Oceanography 48: 702–722.

Mooij, W. M., 1996. Variation in abundance and survival of fish larvae in shallow eutrophic Lake Tjeukemeer. Environmental Biology of Fishes 46: 265–279.

Mooij, W. M., S. Hülsmann, L. N. De Senerpont Domis, B. A. Nolet, P. L. E. Bodelier, P. C. M. Boers, L. M. Dionisio Pires, H. J. Gons, B. W. Ibelings, R. Noordhuis, R. Portielje, K. Wolfstein, & E. H. R. R. Lammens, 2005. The impact of climatic change on lakes in the Netherlands: a review. Aquatic Ecology 39: 381–400.

Mooij, W. M., & O. F. R. Van Tongeren, 1990. Growth of 0+ roach (*Rutilus rutilus*) in relation to temperature and size in a shallow eutrophic lake - comparison of field and laboratory observations. Canadian Journal of Fisheries and Aquatic Sciences 47: 960–967.

Moss, B., D. Mckee, D. Atkinson, S. E. Collings, J. W. Eaton, A. B. Gill, I. Harvey, K. Hatton, T. Heyes, & D. Wilson, 2003. How important is climate? Effects of warming, nutrient addition and fish on phytoplankton in shallow lake microcosms. Journal of Applied Ecology 40: 782–792.

Müller-Navarra, D. C., S. Guss, & H. Von Storch, 1997. Interannual variability of seasonal succession events in a temperate lake and its relation to temperature variability. Global Change Biology 3: 429–438.

Murphy, J., & J. P. Riley, 1962. A modified single solution method for determination of phosphate in natural waters. Analytica Chimica Acta 27: 31–36.

Parmesan, C., & G. Yohe, 2003. A globally coherent fingerprint of climate change impacts across natural systems. Nature 421: 37–42.

Reynolds, C. S., 1989. Physical determinants of phytoplankton succession. In Sommer, U. (ed), Plankton Ecology. Succession in Plankton Communities. Springer-Verlag: 9–56.

Reynolds, C. S., 1997. Vegetation Processes in the Pelagic: a Model for Ecosystem Theory. Ecology Institute, Oldendorf/Luhe.

Reynolds, C. S., V. Huszar, C. Kruk, L. Naselli-Flores, & S. Melo, 2002. Towards a functional classification of the freshwater phytoplankton. Journal of Plankton Research 24: 417–428.

Reynolds, C. S., J. M. Thompson, A. J. D. Ferguson, & S. W. Wiseman, 1982. Loss processes in the population dynamics of phytoplankton maintained in closed systems. Journal of Plankton Research 4: 561–600.

Reynolds, C. S. & S. W. Wiseman, 1982. Sinking losses of phytoplankton in closed limnetic systems. Journal of Plankton Research 4: 489–522.

Sommer, U., 1985. Comparison between steady-state and non-steady state competition: experiments with natural phytoplankton. Limnology and Oceanography 30: 335–346.

Sommer, U., Z. M. Gliwicz, W. Lampert, & A. Duncan, 1986. The PEG-Model of seasonal succession of planktonic events in fresh waters. Archiv für Hydrobiologie 106: 433–471.

Straile, D., & W. Geller, 1998. The response of *Daphnia* to changes in trophic status and weather patterns: a case study from Lake Constance. ICES Journal of Marine Science 55: 775–782.

Strecker, A. L., T. P. Cobb, & R. D. Vinebrooke, 2004. Effects of experimental greenhouse warming on phytoplankton and zooplankton communities in fishless alpine ponds. Limnology and Oceanography 49: 1182–1190.

Van de Bund, W. J., S. Romo, M. J. Villena, M. Valentin, E. Van Donk, E. Vicente, K. Vakkilainen, M. Svensson, D. Stephen, A. Stahl-Delbanco, J. Rueda, B. Moss, M. R. Miracle, T. Kairesalo, L. A. Hansson, J. Hietala, M. Gyllstrom, J. Goma, P. Garcia, M. Fernandez-Alaez, C. Fernandez-Alaez, C. Ferriol, S. E. Collings, E. Becares, D. M. Balayla, & T. Alfonso, 2004. Responses of phytoplankton to fish predation and nutrient loading in shallow lakes: a pan-European mesocosm experiment. Freshwater Biology 49: 1608–1618.

Van Oldenborgh, G. J., & A. Van Ulden, 2003. On the relationship between global warming, local warming in the Netherlands and changes in circulation in the 20th century. International Journal of Climatology 23: 1711–1724.

Van Vliet, A. J. H., A. Overeem, R. S. De Groot, A. F. G. Jacobs, & F. T. M. Spieksma, 2002. The influence of temperature and climate change on the timing of pollen release in the Netherlands. International Journal of Climatology 22: 1757–1767.

Verschoor, A. M., J. Takken, B. Massieux, & J. Vijverberg, 2003. The Limnotrons: a facility for experimental community and food web research. Hydrobiologia 491: 257–377.

Walther, G. R., E. Post, P. Convey, A. Menzel, C. Parmesan, T. J. C. Beebee, J. M. Fromentin, O. Hoegh-Guldberg, & F. Bairlein, 2002. Ecological responses to recent climate change. Nature 416: 389–395.

Walz, H., 1999. Phytoplankton Analyzer PHYTO-PAM: System components and principles of operation. Heinz Walz, Effeltrich, Germany.

Weyhenmeyer, G. A., R. Adrian, U. Gaedke, D. M. Livingstone, & S. C. Maberly, 2002. Response of phytoplankton in European lakes to a change in the North Atlantic Oscillation. Verhandlungen Internationale Vereinigung für Theoretische und Angewandte Limnologie 20: 1436–1439.

Weyhenmeyer, G. A., 2001. Warmer winters: Are planktonic algal populations in Sweden's largest lakes affected? Ambio 30: 565–571.

Weyhenmeyer, G. A., T. Blenckner, & K. Pettersson, 1999. Changes of the plankton spring outburst related to the North Atlantic Oscillation. Limnology and Oceanography 44: 1788–1792.

Winder, M. & D. E. Schindler, 2004a. Climate change uncouples trophic interactions in an aquatic ecosystem. Ecology 85: 2100–2106.

Winder, M. & D. E. Schindler, 2004b. Climatic effects on the phenology of lake processes. Global Change Biology 10: 1844–1856.

Hydrobiologia (2007) 584:415–424
DOI 10.1007/s10750-007-0608-7

SHALLOW LAKES

Impact of climatic fluctuations on Characeae biomass in a shallow, restored lake in The Netherlands

Winnie J. Rip · Maarten R. L. Ouboter · Hans J. Los

Abstract External phosphorus load to a wetland with two shallow lakes in the Botshol Nature Reserve, The Netherlands, was reduced, resulting in a rapid reduction of phytoplankton biomass and turbidity, and after 4 years, explosive growth of Characeae. The clear water state was unstable, however, and the ecosystem then alternated between clear, high-vegetation and turbid, low-vegetation states. A model of water quality processes was used in conjunction with a 14-year nutrient budget for Botshol to determine if fluctuations in precipitation and nutrient load caused the ecosystem instability. The results indicate that, during wet winters when groundwater level rose above surface water level, phosphorus from runoff was stored in the lake bottom and banks. Stored phosphorus was released the following spring and summer under anaerobic sediment conditions, resulting in increased phytoplankton density and light attenuation in the water column. During years with high net precipitation, flow from land to surface water also transported humic acids, further increasing light attenuation. In years with dry winters, the phosphorus and humic acid loads to surface water were reduced, and growth of submerged macrophytes was enhanced by clear water. Thus, the temporal pattern of precipitation and flow from land to water gave a coherent, quantitative explanation of the observed dynamics in phosphorus, phytoplankton, turbidity, and Characeae. Global warming has caused winters in The Netherlands to become warmer and wetter during the last 50 years, increasing flow from land to water of humic acids and phosphorus and, ultimately, enhancing instability of Characeae populations. In the first half of the 20th century interannual variation in precipitation was not sufficient to cause large changes in internal P flux in Botshol, and submerged macrophyte populations were stable.

Guest editors: R. D. Gulati, E. Lammens, N. De Pauw & E. Van Donk
Shallow lakes in a changing world

W. J. Rip (✉) · M. R. L. Ouboter
Waternet (Water board Amstel Gooi en Vecht),
P.O. Box 94370, 1090 GJ Amsterdam,
The Netherlands
e-mail: winnie.rip@waternet.nl

Hans J. Los
Delft Hydraulics, P.O. Box 177, 2600 MH Delft,
The Netherlands

Keywords *Chara* · Runoff from land to water · Humic acids · Lake · Light attenuation · Nutrient loading · Phosphorus · Phytoplankton · Precipitation · Turbidity · Climate change

Introduction

The presence of Characeae indicates a healthy aquatic ecosystem. Similar to other aquatic macrophytes, *Chara* and related species support other biological components of the lake ecosystem (Timms & Moss, 1984; Carpenter & Lodge, 1986; Noordhuis et al., 2002). Submerged macrophytes also help to maintain high water transparency by a number of mechanisms (Scheffer, 1998). The benthic plants, for example, prevent resuspension of the sediment by wind or fish, compete with phytoplankton for nutrients, and offer refuge to grazing zooplankton (Moss, 1990; Scheffer et al., 1993; Van den Berg et al., 1998; Kufel & Kufel, 2002). Some species have allelopathic effects on competing epiphytes and phytoplankton (Van Donk & van de Bund, 2002). Poor light availability, due to shading by phytoplankton and epiphytes, is mostly the primary reason for the disappearance of submerged macrophytes (Philips et al., 1978). Factors other than light, such as grazing (Mitchell & Perrow, 1998) or phytotoxicity by free sulphide (Lamers et al., 1998), may also be important in determining macrophyte abundance; however, when light is insufficient, macrophytes cannot exist at all.

Shallow lakes can have alternative stable states (Scheffer et al., 1993). Most of these lakes are either rather turbid without submerged macrophytes or clear and vegetated. In some lakes repeated shifts between a clear-vegetated and a turbid state have been observed (Lake Tåkern & Lake Krankesjön (Blindow et al., 1993), Alderfen broad (Perrow et al., 1994), Tomahawk Lagoon (Mitchell, 1998) and Botshol (Rip et al., 2005). In most lakes the shifts between clear and turbid states occur irregularly in time, suggesting that there is an infrequence external forcing, for instance by water level fluctuations or changes in external phosphorus load. Alternatively, high oxygen consumption by decomposition of macrophyte biomass accumulated during high production years can trigger the anaerobic release of phosphorus, and the resulting phytoplankton blooms limit subsequent macrophyte production (Asaeda et al., 2000). Moss (1990) suggested this intrinsic process for the regular cycles in Alderfen broad. Van Nes et al. (submitted) studied intrinsic processes triggered by submerged macrophytes that according to the "slow-fast theory (Muratori & Rinaldi, 1991) could cause regular cycles between clear and turbid states in shallow lakes. The present study examined the hypothesis that variation in precipitation and subsequent nutrient loading through runoff account for fluctuations in phosphorus, phytoplankton biomass, light attenuation, and macrophyte abundance.

A previous study (Rip et al., submitted) developed water and nutrient budgets for the entire Botshol ecosystem under the prevailing meteorological conditions. The models provided monthly estimates of nutrient loading during 1989–2002 and, ultimately, insight into the causes of interannual fluctuations in P loading. The results indicated that, during wet winters when groundwater level rose above the level of surface water in the catchment areas, water flowed from land into surface water, resulting in a rise of the P load. In dry winters, infiltration through the soil exceeded precipitation, and there was no flow from land to water, so fluctuations in P load were mainly due to the water supply.

The present study extended the nutrient budget to examine fluctuations in phosphorus loading to the two lakes of Botshol as a possible cause of instability in water clarity and abundance of submerged macrophytes. These two studies combined used an ecosystem approach, starting with P loading from catchment areas and transport to surface water, then incorporating biological processes that determine phytoplankton biomass, turbidity, and finally, biomass of Characeae.

Study area and methods

Hydrology and restoration of Botshol

The hydrology of Botshol, a nature reserve in the center of the Netherlands (52°15′ N 4°26′ E) is dominated by infiltration. The area is a hydrologically isolated polder. On average, the water table in Botshol drops 1.75 mm day^{-1} by water infiltrating to an adjacent polder, Groot Mijdrecht, which has water levels 4.3 m below Botshol. There is a strong gradient of infiltration within the nature reserve. Infiltration from the lakes directly bordering Groot Mijdrecht is 2.5 mm day^{-1}, while infiltration in the northwest part of Botshol is 0.1 mm day^{-1}. During spring and summer, approximately 10^{6} m^{3} P rich water enters Botshol to compensate for water lost through evaporation and infiltration. Prior to November 1988 three agricultural areas drained their excess water into Botshol resulting in a high nutrient load.

Up to until 1960, the lakes of the Botshol Nature Reserve were clear and dominated by Characeae. Beginning in the 1960s, water quality deteriorated due to the rise of external phosphorus input, and submerged macrophyte populations declined (Simons et al., 1994). Beginning in 1989, the external nutrient load was reduced by hydrological segregation of Botshol from the agricultural areas and by chemical stripping of 60–80% of phosphate from the water supply. The goal of restoration was to re-establish the Characeae and other submerged macrophytes. In the first 4 years after the start of phosphorus reduction, decreases were observed in phosphorus and chlorophyll *a* concentrations. Reduced phytoplankton numbers resulted in crystal clear water and explosive growth of submerged macrophytes (Rip et al., 1992; Simons et al., 1994). The dominant macrophyte species in terms of lake areal cover were *Najas marina*, *Fontinalis antipyretica*, *Chara connivens*, *Chara contraria*, *Chara hispida* and *Chara globularis*. However, this clear water state was unstable. From 1993 onwards, the ecosystem alternated between turbid water with low macrophyte cover (1993–1995, 1999–2003) and clear water with high cover of aquatic plants (1996–1998). Phosphorus concentrations in Botshol showed concomitant fluctuations (Rip et al., 2005). The composition of the fish community was determined in 1989 and 1992. The fish community was not dominated by planktivorous fish or benthivorous fish, such as bream (Abramis brama). Rather, perch, pike, and roach, species typical of clear water lakes with macrophytes, were dominant in Botshol in both surveys (Rip et al., 2005). Total fish biomass decreased over the first 4 years following the reduction of the external P-load.

Physical, chemical and biological parameters

From 1988 through 2003, samples for physical and chemical parameters (temperature, pH, conductivity, Secchi disc depth, attenuation, oxygen, chlorophyll *a*, chloride, total and dissolved P, total and Kjeldahl N, ammonium, nitrate, silicate, and dissolved organic substances [humic acids]) were taken every 2–4 weeks in 5 subareas of Botshol, the main water supply, and Lake Vinkeveen. Phytoplankton and zooplankton densities were determined, simultaneously with the chemical samples at the two lakes (subareas I and II). In the period 1993–1999 no data for attenuation were available. Meteorological data were provided by KNMI (a national meteorological institute). Submerged macrophytes were mapped and quantified each summer 1987–2003. A transect pattern covering most of the open water and pools, watercourses and ditches was followed for each survey. Plant material was observed and collected by dredging and snorkelling from a rowing boat. Characeae species were identified according to Van Raam (2003) . Detailed methods and results of the environmental and biological measurements were presented in Rip et al. (2005).

Model

The study area of Botshol was divided into five subareas. Subareas III, IV and V were catchment areas for subareas I and II, in which the two lakes were situated. Water and nutrient budgets constructed for each subarea and for the entire study area (Rip et al., submitted) were used as the basis for the ecosystem model developed in the present study.

DELWAQ-BLOOM-SWITCH (DBS), a mathematical model, was used to determine interactions between nutrient loading and transport, and other physical, chemical, and biological processes in the Botshol. The present study used the standard DBS model, which has been successfully applied to similar aquatic ecosystems (Los, 1993; Van der Molen et al., 1994). A general overview of DBS is given below. More detailed information can be found in the technical reference manual for DBS (Los, 1993). A small number of adjustments were made to model the specific situation in Botshol using measurements from 1989–1996. The accuracy of the calibrated model was validated using data from 1997–2002.

DBS modeled the cycling of carbon, oxygen, and nitrogen, phosphorus, and silicon within the

Botshol ecosystem, adhering to the law of conservation of mass. Nutrients in the water column could be dissolved forms, detritus components and non-detritus organic components. A portion of each nutrient pool was incorporated in the algae. Phosphorus could also be adsorbed to inorganic material (AAP). The module, SWITCH, computed the sediment-water exchange for nutrients by adsorption–desorption (Smits & Van der Molen, 1993). Nutrients could be recycled an infinite number of times with losses only through transport, chemical adsorption, denitrification, and burial in the sediment. Nutrients could also be transformed through mineralization, sedimentation, resuspension, and nitrification.

Oxygen concentration played an important role in controlling the P flux from the sediments of Botshol. A large percentage of the P from runoff was in the particulate state, resulting in P-enrichment of the sediments. Following winters that had high precipitation, anaerobic conditions in the small watercourses in spring/summer resulted in the release of the stored P from the sediments.

The BLOOM portion of DBS computed phytoplankton production based on competition among algal species (Los & Brinkman, 1988; Los, 2005). The primary state variables of the BLOOM module were different phytoplankton types. Four taxonomic groups were distinguished: Chlorophyta, Bacillariophyta, Cyanophyta, and Characeae.

BLOOM first selected the factor that was most likely to become limiting, based on environmental conditions: solar radiation, day length, water temperature, depth, background light attenuation, and nutrient concentrations. The model then selected the best-adapted phytoplankton types under those conditions. The biomass of these phytoplankton types were calculated as the net result of production, mortality, and transport during the previous time step using an optimization technique (Linear Programming).

DBS included a module called UITZICHT (Los, 1993; Buiteveld, 1995) that computed light attenuation. Total light attenuation in the water column was the sum of background attenuation and attenuation due to free floating phytoplankton, algal detritus and humic acids.

Results

Results are shown only for subarea I, but similar agreement between empirical and calculated values was found for the other subareas. Results of chloride calculations showed that the ratios of water supplies and the calculated exchange among subareas used in the water budget were accurate (Fig. 1). The interannual variation of the calculated dynamics of total phosphorus and chlorophyll were similar enough to actual concentrations to show year-to-year differences (Figs. 2, 3): low levels in 1989–1992, high levels in 1993–1995, low levels again in 1996–1998, etc. Both observed and calculated values showed a summer peak each year.

The pattern in 1998 was deviant from other years, probably due to high precipitation in this summer. The calculated values showed a sharp shift to high P and chlorophyll levels in late summer of that year.

Peaks and low points in chlorophyll *a* concentration predicted by the model corresponded well with periods of high and low levels measured in the field. In periods with high cover of Characeae, such as the summers of 1991, 1992, and 1998, predicted chlorophyll levels were higher than the observed values. This was probably due to the lack of some feedback processes in DBS. Dense *Chara* populations during those summers, for example, could have provided shelter for grazing zooplankton or could have influenced phytoplankton by allelopathic effects or increasing the sedimentation rate of phytoplankton. However, biological feedback processes were not completely lacking in the model. Phosphorus

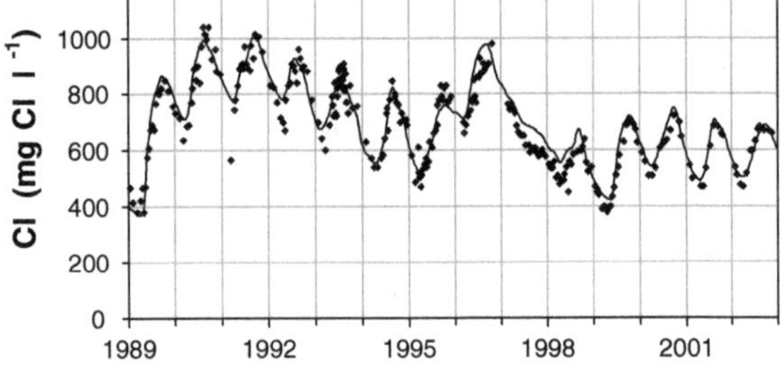

Fig. 1 Observed and calculated chloride levels at location I for 1989–2002

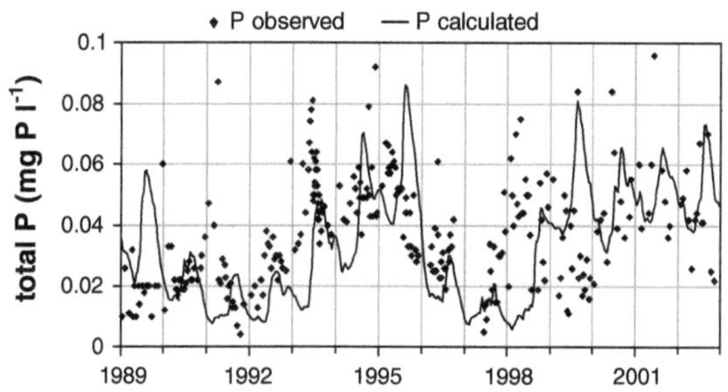

Fig. 2 Observed and calculated total P levels at location I for 1989–2002

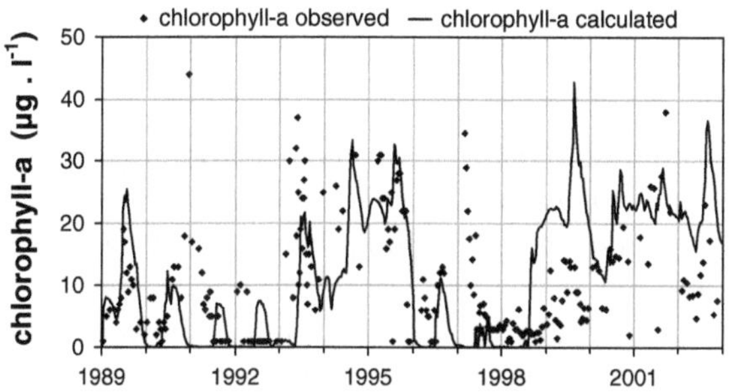

Fig. 3 Observed and calculated chlorophyll levels at location I for 1989–2002

calculated to be stored in *Chara*, for example, was not available for other algae and predicted light attenuation was affected by free-floating algae, but not by the bottom dwelling *Chara.* Another example of feedback in the model concerned the relatively slow reaction of particular nutrients stored in the sediment, from decaying algae, to external forcing imposed on the lake system. The result was that the predicted remineralization flux from the sediment in a particular year was clearly affected by the previous years loadings.

Calculated and measured light attenuation were more similar during periods of high and low turbidity. Results of the model, UITZICHT, indicated that temporal variation in light attenuation was determined primarily by variation in amounts of dead (detritus) and living algae (Fig. 5). Humic acids influenced attenuation during years with high precipitation, such as 1993–1995 and 1999–2002.

The DBS model was not very accurate in calculating Secchi disc depth, although interannual variation in turbidity was successfully predicted (Fig. 4). The generally low algal biomass in

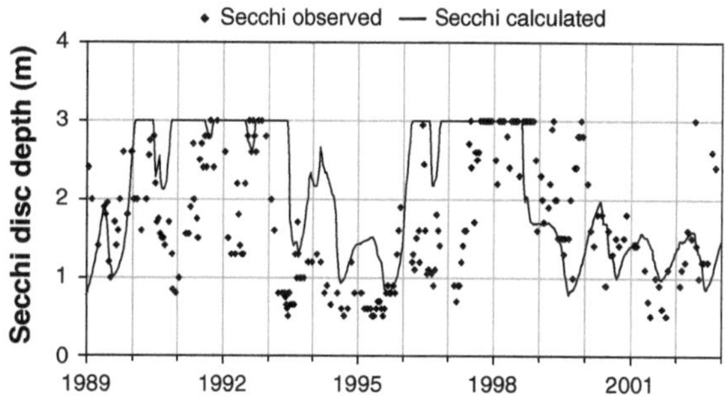

Fig. 4 Observed and calculated Secchi-disc depths at location I for 1989–2002

Botshol, the lack of some feedback processes and the inaccuracy of Secchi disc depth measurements made comparison with simulations difficult. Secchi disc depths calculated for the entire years of 1991 and 1992 were at bottom depth, for example, while actual values varied.

The model's predictions of fluctuations in *Chara* biomass agreed with fluctuations in percent cover of *Chara* determined in the field (Fig. 6). Both calculated and observed levels were high, for example, in 1991, 1992, 1997, and 1998, and low in 1994, 1995, and 1999–2002. The model overestimated the biomass of *Chara* during 1993 and 1996, corresponding to overestimates of Secchi disc depth.

The runoff in dry years was about 90% reduced in comparison to wet years (Rip et al., submitted). To test the hypothesis that P loading to surface water through flow from land to water, especially after winters with high precipitation, was a major factor causing reduced cover of *Chara* in the lakes of Botshol, the model was run under the condition that the P concentration in water flow from land to surface water was reduced 90%. The P load from water inflow to maintain water level in summer remained the same. During wet years, phosphorus and chlorophyll a levels calculated under this condition were substantially lower than levels calculated under the standard conditions. DBS calculated a stable *Chara* population each summer.

Overall, the DBS model provided sufficient consistency between computed and measured parameters of the Botshol ecosystem to explain the dynamics of nutrient loading, phytoplankton density, light attenuation, and *Chara* populations.

Fig. 5 Observed attenuation and calculated composition of light attenuation at location II for 1989–2002

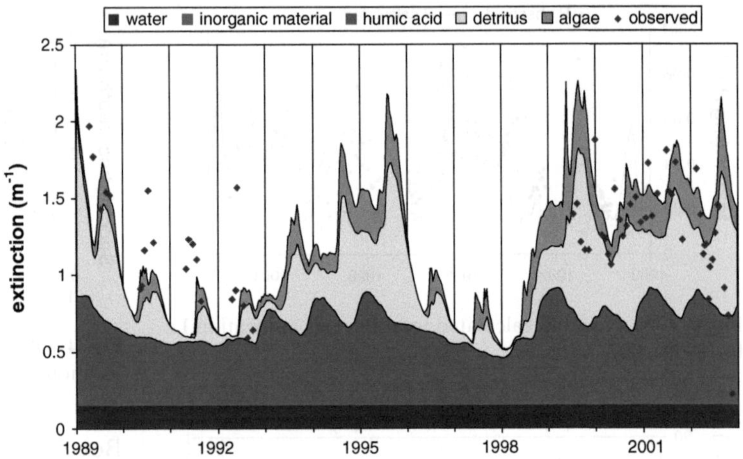

The flow of water which is rich in phosphorus and humic acids from land to water, particularly following periods of high precipitation, is a likely explanation for the instability of the aquatic ecosystem in Botshol after the reduction of the external nutrient load.

Discussion

The main conclusion from the nutrient budget developed for Botshol was that, following nutrient reduction, P from flow from land to water acted as the primary internal nutrient source during periods when precipitation exceeded the water loss by infiltration and evaporation (Rip et al., submitted). Although similarity between calculated values and empirical measurements varied somewhat among the parameters (Figs. 1–6), results of the calibrated DBS model reproduced the dynamics of the Botshol ecosystem during 1989–2002 closely enough to provide a useful tool in determining causes of interannual variation.

Results of the present study showed that, during wet years, the P-rich, flow from land to water resulted in increased phosphorus levels and phytoplankton biomass in surface water of Botshol (Figs. 2, 3). In addition to high P levels, the flow from land to water contained large amounts of dissolved organic substances (humic acids) that colored the water dark brown and further reduced light penetration (Fig. 5). Consequently, light attenuation was so high during periods of high precipitation that *Chara* populations declined (Fig. 6). After dry winters, the groundwater level in terrestrial peat was below the surface water level, and the internal P source was eliminated. During the next growing season, water transparency was high, and *Chara*

Fig. 6 Calculated biomass of Chara (kg C) as model results and field surveys (% cover) at subarea I for 1989–2002

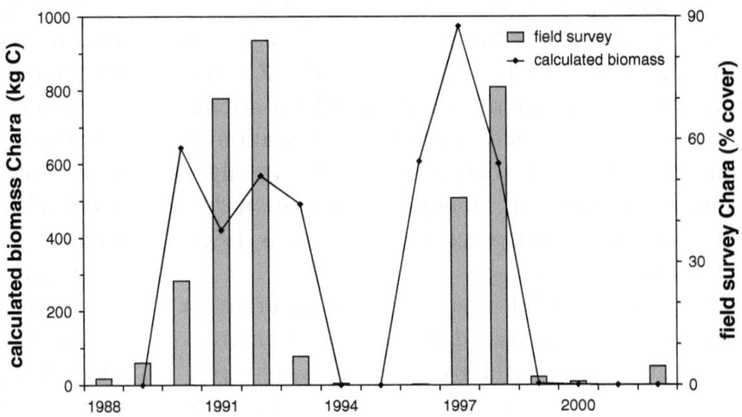

populations increased. This trend was confirmed by the significant negative correlation between *Chara* cover and net precipitation during the previous 2 years ($r = -0.771$, $P < 0.01$). In a simulated situation with low precipitation, calculated *Chara* populations showed an annual cycle that was stable from year to year. Thus, variation in winter precipitation probably caused interannual changes in P flux and humic acids from terrestrial peat, which explained fluctuations in P concentration, phytoplankton biomass, and water transparency in Botshol. Hough et al. (1991) have shown similar effects of variation in precipitation on cover of submerged macrophytes.

In principle any lake with a large enough catchment area will have variation in P load from runoff that is related to variation in precipitation. However, the fluctuation in P load will not always result in a related pattern of high and low cover of submerged macrophytes. In shallow lakes with alternative states there is a range of nutrients in which either a turbid, vegetation poor or a clear, vegetation rich states can exist (Scheffer et al., 1993), sometimes called the "catastrophe fold". Although other factors, like removal of fish, can alter the ecological balance within this range, fluctuations in precipitation and runoff will only cause a switch between clear, vegetated and turbid, unvegetated states if the P content falls above or below this nutrient range. Thus, the influence of fluctuations in runoff on the cover of Characeae depends on the size of the runoff area, the P load from other sources, and the P load range where both alternative states can exist. This critical range is determined by size, depth, fetch, and sediment type of the lake (Janse, 2005). For example, the critical P load causing a switch in stable state was markedly lower in deeper lakes than in shallow lakes and the range is smaller. Botshol is relatively deep for lakes in the Netherlands, so the range of critical P load (0.8–1.3 mg P m^{-2} $year^{-1}$, Rip et al., submitted) is relatively low and small and easily exceeded by small changes in P load.

Landscape and physicochemical conditions in the aquatic ecosystem of Botshol account for the large temporal variation and time lags in P flux. Subsurface flow of P-rich water was most important in subareas IV and V, which had high land/water ratios and low infiltration rates. Although the catchment areas were not fertilized or used for agricultural purposes, phosphorus concentrations in the peat soil water were high due to mineralization and high chloride and sulfide levels (Beltman et al., 2005; Lamers et al., 1998). The mineralization of the terrestrial peat is not a 'natural' process. Botshol is below sea level and humans stabilized the landscape by influencing the water table. The water table in Botshol was maintained at a constant level by the inflow of water during the summer and outflow of water in winter. The inflowing water had high levels of P, S, and Cl. Starting in 1989, 60–80% of P was stripped from the inflowing water. The untreated sulphur and chloride levels remained high, however, and increased the internal availability of P in Botshol (Lamers et al., 1998; Beltman et al., 2005). As was found in other freshwater systems (Meyer et al., 1981), a high percentage (~85%) of P in runoff was probably particulate material. The particulate P was stored in the water bottoms and banks of the small watercourses in subarea IV and V, explaining why only a small increase in P concentration was seen in surface water during wet winters (Fig. 2). When water temperature rose in spring, the stored P was partly released from bottom sediments. In summer, however, there was an explosive P flux from the sediments due to low yearly oxygen levels. The small watercourses in subareas IV and V of Botshol became anaerobic during summer, when mineralization of organic matter in the sediment created a high oxygen demand. When P-rich water from subareas IV and V was subsequently transported to subareas I and II to make up for summer evaporation and infiltration, large amounts of P were carried into surface waters of the lakes.

Results of the present study provide a likely explanation for the decline in *Chara* populations during 1993–1995 and 1999–2002. An important condition for the growth of Characeae is light reaching the lake bottom after germination, which occurs around May. This condition was met in Botshol when P was below 0.018 mg l^{-1} (Rip et al., 2005) corresponding to periods of high macrophyte cover. When total P was between 0.018 and 0.043 mg l^{-1}, there were periods of both low and high cover of submerged macrophytes (Rip et al., submitted), supporting the hypothesis that either state can exist at intermediate nutrient levels (Scheffer, 1993). After a wet winter, however, P levels enhanced by subsurface run-off exceeded this range and stimulated growth of phytoplankton, which in turn decreased light penetration and limited growth of *Chara.*

In addition to light, other factors may influence submerged macrophyte populations, including chloride concentrations (Jeppesen et al., 1994), sediment toxicity due to sulfide (Lamers et al., 1998), grazing by herbivorous birds (Moss, 1990; Mitchell & Perrow, 1998), and accumulation of organic material of decaying submerged plants (Carpenter & Lodge, 1986; Perrow et al., 1994; Asaeda et al., 2000). The possible involvement of these factors in the observed interannual fluctuations of *Chara* in Botshol is examined in other studies (Beltman et al., 2005; Van Nes et al., submitted; Rip et al., unpublished).

The approach taken by the present study considered the ecosystem as a whole, starting with actual meteorological conditions, incorporating transport from land to surface water, as well as changes in phytoplankton and light attenuation, and finally examining effects on submerged macrophytes. In contrast, most previous studies of aquatic systems modeled either nutrient loading (Meyer et al., 1981; Mander et al., 1998) or biological processes (Janse et al., 1997; Best et al., 2001). A few previous studies modeled both, but only for average years and did not examine interannual variation (Asaeda et al., 2000; Portielje & Rijsdijk, 2003). The present study, therefore, was uniquely able to explain the role of interannual variation in precipitation as a cause of instability in an aquatic ecosystem.

Precipitation became a major factor influencing submerged macrophyte populations only after restoration efforts in Botshol were initiated. Following the reduction in external P load to very low levels, the internal P load from subsurface run-off increased in importance as a nutrient source for phytoplankton. As a high percentage of the area of Botshol is terrestrial peat, interannual variation in precipitation caused large changes in the internal P load to surface water. The high infiltration rate and large ditch distance in much of the terrestrial peat resulted in low runoff and land-to-water P flow during dry winters (Ouboter & Rip, submitted). A side effect of the hydrological isolation of Botshol, furthermore, was an increase in chloride from 500 to 1000 mg l^{-1}, which enhanced the availability of phosphorus in peat (Beltman et al., 2005). Finally, the relatively large depths of the lakes in Botshol made submerged macrophyte populations vulnerable to even small increases in light attenuation.

External P loading to Botshol was low, and Characeae populations in Botshol did not exhibit large interannual fluctuations prior to 1960 (Simons et al., 1994). Analysis of the weather in the Netherlands during the 20th century indicated that both temperature and annual precipitation have risen since 1901, related to global changes causing a warmer and wetter climate throughout western Europe (IPCC, 2001; KNMI, 2001). Thus, all winters in which more than 500 mm of precipitation were recorded in De Bilt, a site in the Netherlands close to Botshol, came after 1960. Before that time, interannual variation in precipitation was not sufficient to cause large changes in internal P flux in Botshol, and submerged macrophyte populations were stable (Simons et al., 1994).

The first 4 years of the Botshol restoration project (1988–1992) coincided with four dry winters, so the initial change to the clear water state and increased *Chara* populations were due to the combination of reduced external P load and low precipitation. The pre-restoration P load to the lakes was high and reduction of external P load was necessary to allow growth of submerged macrophytes. Now that the external P load is not a limiting factor, additional measures are needed to decrease internal P flux in Botshol during wet winters to sustain stable populations of submerged macrophytes. In wet years, internal P flux could be reduced by allowing the water level to fluctuate more naturally. Desiccation of peat during dry periods, for example, would reduce P in runoff, due to immobilization of P by oxidation of Fe (Lucassen et al., 2005). Another approach would be to reverse terrestrialization by removing large areas of terrestrial peat 1 m below water level. This would reduce the surface area where nutrient-rich groundwater can flow into surface water. Finally, the creation of larger areas of surface water in the subareas with the highest runoff would raise oxygen levels, decrease reduction of sulphate to sulphide, and reduce P release from sediment. Results of the present study demonstrated that the future stability of macrophyte populations in Botshol will depend on improved management of internal P flux.

Acknowledgements The authors thank the Society for the Preservation of Nature in the Netherlands for permission to perform the research in the Botshol nature reserve, and Jos van Gils, Arjen Markus and Paul Grashoff for discussions and assistance during the development of the models of Botshol. We are grateful to Boudewijn Beltman and Val Gerard for improving the manuscript.

References

Asaeda, T., V. K. Trung & J. Manatunge, 2000. Modeling the effects of macrophyte growth and decomposition on the nutrient budget in Shallow Lakes. Aquatic Botany 68: 217–237.

Beltman, B., W. J. Rip & A. Bak, 2005. Nutrient release after water quality restoration measures. A phytometer assessment in the Botshol wetlands, The Netherlands. Wetlands Ecology and Management 13: 577–585.

Best, E. P. H., C. P. Buzzelli, S. M. Bartell, R. L. Wetzel, W. A. Boyd, R. D. Doyle & K. R. Campbell, 2001. Modeling submersed macrophyte growth in relation to underwater light climate: modeling approaches and application potential. Hydrobiologia 444: 43–70.

Blindow, I., G. Andersson, A. Hargeby & S. Johansson, 1993. Long-term pattern of alternative stable states in two shallow eutrophic lakes. Freshwater Biology 30: 243–252.

Buiteveld, H., 1995. A model for calculation of diffuse light attenuation (PAR) and Sechi depth. Netherlands Journal of Aquatic Ecology 29: 55–65.

Carpenter, S. R. & D. M. Lodge, 1986. Effects of submersed macrophytes on ecosystem processes. Aquatic Botany 26: 341–370.

Hough, R. A., T. E. Allenson & D. D. Dion, 1991. The response of macrophyte communities to drought-induced reduction of nutrient loading in a chain of lakes. Aquatic Botany 41: 299–308.

IPCC, 2001. Climate Change 2001: The Scientific Basis. Cambridge University Press, 892 pp.

Janse, J. H., E. van Donk & T. Aldenberg, 1997. A model study on the stability of the macrophyte-dominated clear water state as affected by biological factors. Water Research 32: 2696–2706.

Janse, J. H., 2005. Model studies on the eutrophication of shallow lakes and ditches. University of Wageningen. Thesis, 376 pp.

Jeppesen, E., M. Sondergaard, E. Kanstrup & B. Petersen, 1994. Does the impact of nutrients on the biological structure and function of brackish and freshwater differ? Hydrobiologia 276: 15–30.

KNMI, 2001. De toestand van het klimaat in Nederland 2000 (in Dutch).

Kufel, L. & I. Kufel, 2002. *Chara* beds acting as nutrient sinks in shallow lakes—A review. Aquatic Botany 72: 249–260.

Lamers, L. P. M., H. B. M. Tomassen & J. G. M. Roelofs, 1998. Sulphate-induced eutrophication and phytotoxicity in freshwater wetlands. Environmental Science and Technology 32: 199–205.

Los, F. J. & J. J. Brinkman, 1988. Phytoplankton modelling by means of optimization: a 10-year experience with Bloom II. Verhandlungen Internationale Verein. Limnologie 23: 790–795.

Los F. J., 1993. DBS. Tech. Rep. T542, Delft Hydraulics, The Netherlands (in Dutch).

Los, F. J., 2005. An algal biomass prediction model. In Daniel, P. Loucks & Eelco van Beek (eds), Water Resources Systems Planning and Management - an introduction to methods, models and applications. UNESCO, 2005.

Lucassen, C. H. E. T., A. J. P. Smolders & J. G. M. Roelofs, 2005. Effects of temporary desiccation on the mobility of phosphorus and metals in sulphur-rich fens: differential responses of sediments and consequences for water table management. Wetlands Ecology and Management 13: 135–148.

Mander, Ü., A. Kull, V. Tamm, V. Kusemets & R. Karjus, 1998. Impact of climatic fluctuations and land use change on runoff and nutrient losses in rural landscapes. Landscape and Urban Planning 41: 229–238.

Meyer, J. L., G. E. Likens & J. Sloane, 1981. Phosphorus, nitrogen and organic carbon flux in a headwater stream. Archiv für Hydrobiologie 91: 28–44.

Mitchell, S. F. & M. R. Perrow, 1998. Interactions Between Grazing Birds and Macrophytes. In Jeppesen, E., Ma. Sondergaard, Mo. Sondergaard & K. Christoffersen (eds), The structuring Role of Submerged Macrophytes in Lakes. Springer Verlag, New York, 175–195.

Moss, B., 1990. Engineering and biological approaches to the restoration from eutrophication in which aquatic plant communities are important components. Hydrobiologia 275(276): 367–377.

Mutatori, S. & S. Rinaldi, 1991. A separation condition for existence of limit cycles in slow-fast systems. Applied Mathematical Modelling 15: 312–318.

Noordhuis, R., D. T. Van der Molen & M. S. Van den Berg, 2002. Response of herbivorous water-birds to the return of *Chara* in Lake Veluwemeer, The Netherlands. Aquatic Botany 72: 349–367.

Perrow, M. R., B. Moss & J. Stansfield, 1994. Trophic interactions in a shallow lake following a reduction in nutrient loading: A long term study. Hydrobiologia 275: 43–52.

Philips, G. L., D. Eminson & B. Moss, 1978. A mechanism to account for macrophyte decline in progressively eutrophicated freshwaters. Aquatic Botany 4: 103–126.

Portielje, R. & R. E. Rijsdijk, 2003. Stochastic modeling of nutrient loading and lake ecosystem response in relation to submerged macrophytes and benthivorous fish. Freshwater Biology 48: 741–755.

Rip, W. J., K. Everards & A. Houwers, 1992. Restoration of Botshol (The Netherlands) by reduction of external nutrients load: The effects on physico-chemical conditions, plankton and sessile diatoms. Hydrobiological Bulletin 25: 275–286.

Rip, W. J., M. Ouboter, E. H. van Nes & B. Beltman, 2005. Oscillation of a shallow lakes ecosystem upon reduction in external phosphorus load. Archiv für Hydrobiologie 164: 387–409.

Scheffer, M., 1998. Ecology of shallow lakes. Population and Community Biology series, Kluwer Academic Publishers, 357 pp.

Scheffer, M., S. H. Hosper, M. L. Meijer, B. Moss & E. Jeppesen, 1993. Alternative equilibria in shallow lakes. Trends in Ecology and Evolution 8: 275–279.

Simons, J., M. Ohm, R. Daalder, P. Boers & W. J. Rip, 1994. Restoration of Botshol (The Netherlands) by reduction of external nutrient load: recovery of a characean community, dominated by *Chara connivens*. Hydrobiologia 275/276: 243–253.

Smits, J. G. C. & D. T. van der Molen, 1993. Application of SWITCH, a model for sediment-water exchange of nutrients, to Lake Veluwe in the Netherlands. Hydrobiologia 253: 281–300.

Timms, R. M. & B. Moss, 1984. Prevention of growth of potentially dense phytoplankton populations by zooplankton grazing in the presence of zooplanktivorous fish in a shallow wetland ecosystem. Limnology and Oceanography 29: 472–486.

Van den Berg, M. S., H. Coops, M.-L. Meijer, M. Scheffer & J. Simons, 1998. Clear water associated with a dense *Chara* vegetation in the shallow and turbid Lake Veluwemeer, The Netherlands. In Jeppesen, E., Ma. Sondergaard, Mo. Sondergaard & K. Christoffersen (eds), The Structuring Role of Submerged Macrophytes in Lakes. Ecological Studies. Springer Verlag 131: 339–352.

Van der Molen, D. T., F. J. Los, L. van Ballegooijen & M. P. van der Vat, 1994. Mathematical modelling as a tool for management in eutrophication control of shallow lakes. Hydrobiologia 275/276: 479–492.

Van Donk, E. & W. J. van de Bund, 2002. Impact of submersed macrophytes including charophytes on phyto- and zooplankton communities: allelopathy versus other mechanisms. Aquatic Botany 72: 261–274.

Van Raam J. C., 2003. Handboek Kranswieren. *Chara* boek, Hilversum (in Dutch).

Hydrobiologia (2007) 584:425–432
DOI 10.1007/s10750-007-0588-7

Response of macroinvertebrates to warming, nutrient addition and predation in large-scale mesocosm tanks

Heidrun Feuchtmayr · Dermot McKee · Ian F. Harvey · David Atkinson · Brian Moss

Abstract There is increasing concern about the effect of climate change on aquatic systems. We examined changes in macroinvertebrate communities caused by increased temperature (3°C above ambient during summer only and continuous 3°C above ambient all year round), influences of fish (*Gasterosteus aculeatus* L.) and addition of nutrients (nitrogen and phosphorus) in 48 large-scale (3000 l) tanks over a 2 year period. While numbers of Isopoda, *Chaoborus*, Corixidae, Ephemeroptera, Notonectidae and Odonata were reduced by the presence of fish, nutrient addition caused isopods, corixids, mayflies and odonates to increase in abundance. Impacts of temperature increase were surprisingly low, with only gastropods increasing in heated tanks, suggesting that, overall abundances of most macroinvertebrate taxa will not be severely affected by the predicted temperature rise. To determine if taxa were sampled representatively during the experiment, net sweep samples taken towards the end of the experiment were compared with final macroinvertebrate abundances when the complete contents of each tank were harvested. We found that net sweeping is an appropriate semi-quantitative method for most taxa in mesocosm tanks. However, mites, coleopteran adults and larvae, dipterans and *Chaoborus* were not adequately sampled. This might explain why we could not detect any treatment effects of temperature, fish or nutrients on mites, coleopterans and dipterans and calls for different sampling techniques for these taxa, especially in ponds with vegetation stands.

Keywords Global climate change · Temperature · Sampling technique · Sweep net · *Gasterosteus aculeatus*

Guest editors: R. D. Gulati, E. Lammens, N. De Pauw & E. Van Donk
Shallow lakes in a changing world

H. Feuchtmayr (✉) · D. McKee · I. F. Harvey · D. Atkinson · B. Moss
School of Biological Sciences, Biosciences Building, University of Liverpool, Liverpool L69 7ZB, UK
e-mail: h.feuchtmayr@liv.ac.uk

Introduction

Global temperature is rising, and is predicted to rise a further 3 to 5°C in the next half century in the UK and most parts of Europe (Houghton et al., 2001; Schiermeier, 2004; IPCC 2007). However, knowledge on the effects of warming on ecosystems or species diversity is still limited. In general, detection of natural community changes induced by warming require spatial or temporal correlations, such as long term records (e.g. Sagarin et al., 1999) or sampling over a latitudinal temperature gradient (e.g. Noges et al., 2003).

Most studies concentrate on population changes over a climate gradient (e.g. Walker et al., 1991) because costs prevent the investigation of more complex systems. Artificial outdoor large-scale mesocosm tanks, however, provide an opportunity to study community responses in the medium-term. Open-air mesocosms provide an intermediate scale between laboratory and natural environmental conditions. Further, the identical design and similar initial species composition allow direct comparisons between treatments or manipulations thereby allowing quantification of causal relationships.

As macroinvertebrate species are often used as water quality indicators (e.g. Rosenberg & Resh, 1992), adequate sampling techniques are essential. There are various ways of sampling macroinvertebrates from lakes or streams, such as grab samplers, surber samplers, artificial substrates, box samplers and air lift samplers. Methods used mainly depend on the substratum of interest. However, the most common and widespread method is net sweeping. To determine which sampling technique is most appropriate for representative sampling, comparisons between different methods have been made by various authors. However, this is a difficult approach as volumes sampled are often hard to quantify and untypical species could be sampled with artificial substrates (Stark, 1993; Merritt et al., 1996; Humphries et al., 1998; Muzaffar & Colbo, 2002; Connor et al., 2004).

Comparing sampling techniques with one another, however, is a restricted approach as no absolute values of species abundances present in natural environments can be obtained. Large mesocosms allow for absolute comparisons by an exhaustive final sampling, where all macroinvertebrates present in the tanks can be removed and numbers compared with the last experimental sampling. This quantitative approach allows conclusions about which macroinvertebrates were sampled adequately with net sweeps and which were not.

First, we hypothesised that macroinvertebrate abundance and diversity would be reduced in the presence of fish through predation, and increased by warming and nutrient addition owing to faster development, higher survival and more nutritious food supply. Changes in the zooplankton, phytoplankton and macrophyte communities within the experiment are documented in McKee et al. (2002a, 2002b) and Moss et al. (2003). Secondly, the opportunity to assess sampling methods by comparing results of a standardised sampling to the actual abundances in the mesocosms derived from a final collection when mesocosms were emptied was taken.

Methods

Experimental set up

Forty-eight mesocosm tanks, each 1 m deep and 2 m in diameter, were installed in Ness Gardens, near Liverpool, UK, each containing around 3,000 l of groundwater (for details see McKee et al., 2003). They simulated shallow lakes and were sunk into the ground for insulation. For a detailed outline of the site, see McKee et al. (2000). A 5-cm sediment layer (7:1 sand:loam), an inoculum of plankton and macroinvertebrates from a local canal and 3 macrophyte species were introduced into the tanks. Equal weights of *Lagarosiphon major* Ridl. Moss, *Elodea nuttallii* Planch. H. St. John, and *Potamogeton natans* L. were planted some months before the experimental start to allow the plants to establish in the tanks (for details see McKee et al., 2002b). To ensure similar starting conditions, plankton and macroinvertebrates were cross-mixed between tanks several times in advance of the experimental period. Cross-mixing was conducted by collecting and pooling sweep net samples from all tanks and thorough mixing of the pooled sample before re-distribution to the tanks. The experiment ran over a 2-year period, from September 1998 to September 2000.

Treatments

A computer-controlled heating system allowed heating of 32 tanks to 3°C above ambient temperature via hot water pumped through a heating element in the bottom of the tanks (for details see McKee et al., 2000). Sixteen tanks were heated year round, while warming of another sixteen was restricted to warmer months only (April 1 to September 30). The remaining 16 tanks were not heated and served as controls. In addition to temperature treatments, the effects of nutrients and predation by fish were investigated in a randomised block design with 4 replicates per treatment. Two additional tanks served as storage for de-ionised water, used to replace water lost from experimental tanks through evaporation.

To investigate predatory impact, 21 adult three-spined Sticklebacks (*Gasterosteus aculeatus* L.) were added to each of half of the tanks at the start. During the 2-year experiment they were allowed freely to breed, the final mean biomass per mesocosm tank being 27.0 ± 18.2 g m^{-3}. Nutrient treatments (24 tanks) received constant nutrient addition, following a 3-weekly cycle during winter and a 2-weekly cycle during the 6 summer months. Nitrogen was added to the relevant tanks as sodium nitrate, phosphorus as sodium phosphate, to give added concentrations of 500 $\mu g\ l^{-1}$ nitrogen and 50 $\mu g\ l^{-1}$ phosphorus in winter and 170 $\mu g\ l^{-1}$ nitrogen and 17 $\mu g\ l^{-1}$ phosphorus in summer, respectively. These values were established from nutrient loadings on local eutrophic shallow lakes.

Macroinvertebrate methods

The initial macroinvertebrate community introduced to the tanks was obtained from local ponds. The stock community was well mixed before a standard aliquot was added to each tank and cross-mixed for several weeks thereafter. Macroinvertebrates were sampled every 8 to 10 weeks with three standardised net sweeps (250 μm mesh size) per mesocosm tank. After counting and identifying of all sampled individuals on site, they were returned to the relevant mesocosm. All sweeps were carried out by the same person for the 2 year sampling period. At termination, all macroinvertebrates were removed, counted and identified during a final exhaustive harvest of all tanks.

Before applying a 3-way univariate analysis of variance (ANOVA) with time as the repeated measure and polynomial contrasts (Gurevitch & Chester, 1986), all data were Hellinger-transformed. Warming, fish and nutrients were used as the three factors including all interactions. Where a significant effect of warming was found, slopes of regression lines of treatment means were compared via ANCOVA in order to determine which of the two warming treatments caused the effect.

For assessment of the accuracy of the net sweeps, data from the last sampling of macroinvertebrates of the experiment (14th of September 2000) were compared with counts made at the final harvest four days later (18th of September 2000). Owing to skewed data, fourth root transformation was applied to both datasets. Regression analyses were used to establish a relationship between numbers of the final sweep and the final exhaustive analyses.

Results

Initially introduced species included Ephemeroptera, Odonata, Hydracarina, Corixidae, Diptera and Coleoptera, *Chaoborus* and Isopoda. Gastropoda were introduced along with macrophytes. Identification of taxa to species level was carried out at the end, but not at the start of the experiment. Overall, abundance and composition of the macroinvertebrate community changed between tanks and over time. Ephemeroptera and Gastropods, however, were the dominant families in most tanks over the entire 2-year period. During that time, only one invader family was found (notonectids), however usually in low numbers. Comparison of the means of all tanks on the first (15th of October 1998) and last sampling (14th of September 2000) showed a similar pattern, with the highest relative increases in abundance of odonates, corixids and *Chaoborus* (Table 1).

Most macroinvertebrate taxa were significantly reduced by the presence of fish, with only gastropods, coleopterans, mites and dipterans showing no significant response, and corixids increasing when fish were present in the tanks (Fig. 1A, Table 2). Gastropod numbers significantly increased with both warming treatments ($P < 0.05$), being lowest in unheated tanks and highest in tanks heated all year round (Fig. 1C and Table 2). Coleopterans, mites and dipterans

Table 1 Mean macroinvertebrate composition at first (15th of October 1998) and last sampling (14th of September 2000). Percentages of taxa abundances are shown

Group	First sampling	Last sampling
Isopoda	1.1	0.3
Chaoborus	0.9	3.2
Corixidae	0.2	4.7
Gastropoda	17.6	23.4
Ephemeroptera	76.4	62.7
Notonectidae	0	0.2
Odonata	0.3	6.6
Coleoptera juveniles	0.4	0
Coleoptera adults	0.1	0.2
Diptera	1.2	2.5
Hydracarina	1.9	0.9

however responded neither to the impact of fish, nor to nutrient addition nor warming. Isopod, corixid, mayfly and odonate abundances increased in tanks receiving nutrient addition (Fig. 1B and Table 2). Note that Fig. 1 shows overall effects during the experiment, as abundances were pooled for all sampling occasions. Statistical results however are shown from analyses of variance with time as the repeated measure.

Abundances of macroinvertebrates sampled by sweep netting and counts from the final harvest four days later, showed taxon-dependent results. Final abundances of Isopod Crustacea (*Asellus aquaticus* L.), Ephemeroptera (*Cloeon dipterum* L.), Gastropoda (*Bithynia tentaculata* L., *Lymnaea peregra* Müll., *Lymnaea stagnalis* L., *Physa fontinalis* L., *Planorbis* sp., *Potamopyrgus jenkinsi* Smith), Odonata (*Aeshna juncea* L., Coenagrionidae, *Libellula depressa* L., *Sympetrum striolatum* Charpentier) and Corixidae had a highly significant R^2 with net sweep counts. An example is given in Fig. 2 for gastropods. Notonectid abundances showed

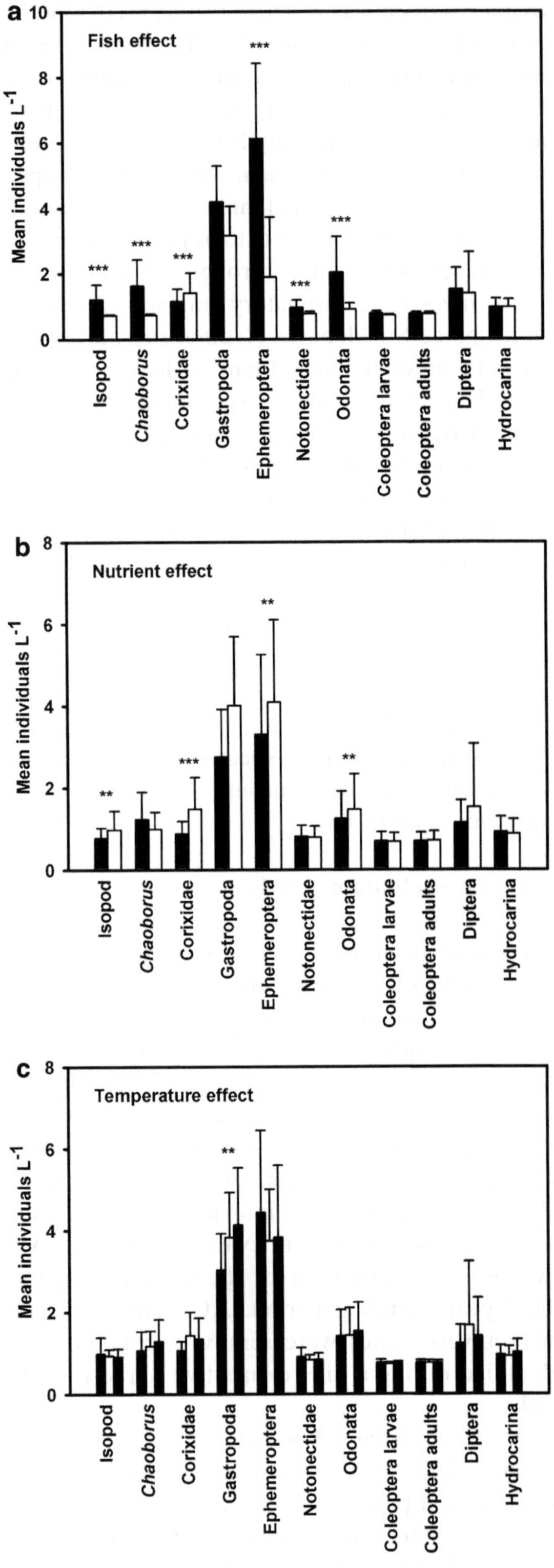

Fig. 1 Means of macroinvertebrate abundances for the ▶ entire 2-year experimental period for the different treatments applied. Graph A: fish absent (black bars) and present (white bars); Graph B: no nutrient addition (black bars) and nutrient addition (white bars); Graph C: no heating (black bars), summer heating only (light grey bars) and all year round heating (dark grey bars). Significant results of repeated measures analysis of variance are indicated with asterisks (see Table 2 for details)

Table 2 Treatment effects determined by repeated measure analysis of variance of the various macroinvertebrate taxa

Effect	Isopoda	*Chaoborus*	Corixida	Gastropoda	Ephemeroptera	Notonectidae	Odonata	Coleoptera juveniles	Coleoptera adults	Diptera	Hydracarina
Fish (F) DF(1,36)	12.9***	34.2***	13.4***	NS	21.5***	26.1***	22.9***	NS	NS	NS	NS
Nutrients (N) DF(1,36)	9.5**	NS	15.1***	NS	9.6**	NS	9.5**	NS	NS	NS	NS
Warming (W) DF(2,36)	NS	NS	NS	6.4**	NS	NS	NS	NS	NS	NS	NS
F × N DF(1,36)	NS	9.3**	NS	NS	NS	NS	NS	NS	NS	NS	NS
F × W DF(2,36)	NS	NS	NS	NS	7.5**	6.6**	NS	NS	NS	NS	NS
N × W DF(2,36)	NS	NS	NS	NS	NS	NS	NS	NS	NS	NS	NS
F × N × W DF(2,36)	NS	NS	NS	NS	NS	NS	NS	NS	NS	NS	NS

Wherever effects were significant, F values are given and p values are indicated as * $P < 0.05$, ** $P < 0.01$, *** $P < 0.001$. Degrees of freedom (DF) are given

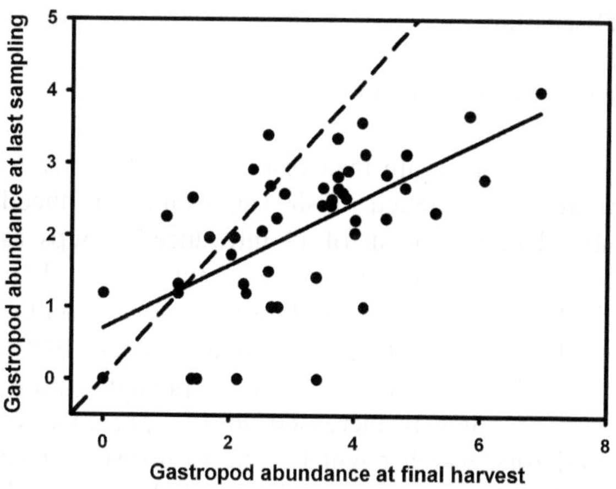

Fig. 2 Gastropod abundance at last net sweep sampling and from final harvest when all animals were removed from mesocosms. Data are shown for all 48 mesocosms and are fourth root transformed. The linear regression line is shown as a solid line (for details see Table 3). The 1:1 line is indicated as dashed line

weaker, but still significant results. Abundances determined by net sweep samples for Coleoptera larvae (*Dytiscus spp.*) and adults (*Dytiscus marginalis* L., *Dytiscus semisulcatus* Müll., *Halipus* spp., *Agabus nebulosus* Forster, *Coelostoma orbiculare* Fabricius, *Colymbetes fuscus* L., *Laccophilus minutes* L. and other, undetermined, beetles), Diptera (Chironomidae, Tipulidae, Culicidae, Dixidae), Hydracarina and *Chaoborus* abundances were significantly different from abundances in the tanks at termination ($P >$ 0.05, see Table 3).

Table 3 Results of regression analyses of abundance counts of the last sampling and the final harvest for each group

Group	Regression line equation	R^2	*P*-value
Isopoda	$y = 0.16x + 0.08$	0.22	0.001
Chaoborus	$y = 0.19x + 0.58$	0.04	0.21
Corixidae	$y = 0.30x + 0.37$	0.15	0.006
Gastropoda	$y = 0.43x + 0.71$	0.39	<0.0001
Ephemeroptera	$y = 0.60x + 1.15$	0.29	<0.0001
Notonectidae	$y = -0.18x - 0.03$	0.10	0.03
Odonata	$y = 0.49x + 0.03$	0.38	<0.0001
Coleoptera larvae	$y = -0.02x + 0.02$	0.001	0.84
Coleoptera adults	$y = 0.09x + 0.13$	0.03	0.21
Diptera	$y = 0.08x + 0.55$	0.2	0.34
Hydracarina	$y = -0.07x + 0.74$	0.004	0.65

Discussion

Effects of treatments

Macroinvertebrate taxa responded differently to treatments applied. Fish significantly reduced abundance of 6 out of 11 introduced groups as anticipated. Gastropods were not eaten: their shells are considered a defensive structure against predation by fish (e.g. Lewis & Magnuson, 1999), especially small sticklebacks. Surprisingly, numbers of corixids increased in the presence of predators, which might be due to invasion from outside the mesocosm tanks coupled with a high food supply as corixids are filter feeders and might have profited from the decline of filter feeding zooplankton (McKee et al., 2002a). Non-significant results of treatment effects on species are discussed in the section on sampling efficiency below.

Higher nutrient concentrations are likely to result in higher primary production and thus expected to support a greater number of invertebrates. Indeed, besides large effects caused by fish, nutrients had major impacts on the phytoplankton community (Moss et al., 2003) and zooplankton abundances were higher compared with tanks where no nutrients had been added when fish were not present (McKee et al., 2002a). However, only four macroinvertebrate groups showed significant increases in nutrient enriched tanks compared with tanks where no nutrients were added: isopods, mayflies, corixids and odonates. Ephemeroptera abundances were also found to be higher in streams when exposed to channels with high nutrients (Kiffney & Richardson, 2001). After one year of the mesocosm experiment, McKee & Atkinson (2000) determined the length of the mayfly *Cloeon dipterum* present in the tanks and reported larger nymphs to be found in tanks with high nutrients and no fish suggesting higher production in nutrient-rich tanks.

Heating had a small effect on the mesocosm communities of phytoplankton, zooplankton and macrophytes relative to predation and nutrient effects (McKee et al., 2002a; McKee et al., 2002b; McKee et al., 2003; Moss et al., 2003). This seems to also be the case for macroinvertebrates: temperature increase only had a positive effect on gastropod abundance. In a stream experiment however, Hogg & Williams (1996) found a decrease in total animal densities as response to a temperature increase of 2–3.5°C, mainly due to a decrease in chironomid abundance, suggested to be caused by thermal limits. In general, ectotherms show increased development rates with higher temperatures (Atkinson, 1996). As macroinvertebrate taxa were affected by the presence of fish in our study, it could be assumed that the temperature effect might not be visible. However, there was no effect of temperature when analysis was restricted to fish-free tanks only (data not shown), in accordance with Baulch et al. (2005) who recorded no significant changes of Trichoptera, Diptera and Ephemeroptera abundances with warming. As Costil & Daguzan (1995) reported, sexual maturity of two *Planorbis* species was earlier at higher temperatures, and they began to lay eggs above a threshold temperature before spending most of their life span in reproduction. Thus higher temperature could have led to earlier reproduction of the gastropods. Embryonic development for *Bithynia tentaculata* was shown to be shortened with higher temperature (Kozminsky, 2003).

Effects of sampling

The four macroinvertebrate groups showing no response to any treatment were coleopteran adults and larvae, dipterans and mites. This however might result from inadequate sampling. The comparison between the last net sweep sampling for macroinvertebrate abundance and the total numbers of animals present in the mesocosms is an important step to determine taxon-specific representative sampling via sweep-netting. Net sweep sampling seems adequate for semi-quantitative sampling of some taxa, but not for others. If taxa were perfectly sampled by net sweeps, we would expect a regression line with an R^2 value of 1 and parallel to and below the 1:1 line. However most of the regression lines calculated in this study show a slope smaller than 1, under-representing high abundances, and over-representing low abundances. This, of course, could also have been caused by death between the last sampling and final harvest or a sampling error at harvest.

Abundances in the last sampling and the final harvest of coleopteran larvae showed no correlation, which is not surprising as they were only found in three tanks in low numbers. Surprisingly, no representative sampling was achieved for adult beetles, regarded as optimally sampled with net sweeps by Humphries et al. (1998) as they are free swimming in the water. But, like larvae, abundance of beetles was very low during the experiment.

For *Chaoborus*, it could be argued that net sweeps are not an appropriate sampling method as the larvae hide in the sediment during the day when the sampling took place. However, in the case of *Chaoborus*, another factor could play a role. In 65% of the tanks where *Chaoborus* was found, abundances were greater in the last sampling than in the final harvest (data not shown), suggesting emergence from the tanks between the two sampling occasions. Therefore, predation effects shown in Fig. 1A and Table 2 on *Chaoborus* should be regarded with caution. Mites are the smallest of the sampled macroinvertebrates and their size might have prevented a representative samping. Dipteran larvae are difficult to sample with net sweeps as most of them live in burrows in the sediment and might be easier to sample with sediment corers.

In conclusion, sweep netting in mesocosm tanks resulted in representative sampling of isopods, mayflies, gastropods, odonates, corixids and notonectids. We can recommend this sampling technique for obtaining semi-quantitative results for these taxa with water plants present. However, care has to be taken with truly benthic invertebrates (Diptera larvae) in pond sediment, as well as for beetles, mites and *Chaoborus*, for which the method yielded inadequate results. This may explain why no treatment effects were found for Diptera, mites and coleopterans. For the other groups, fish and nutrients had the strongest effects, while warming influenced only gastropod abundances. Nonetheless our conclusion that climate warming will only have a subtle influence on shallow lakes might be limited in a wider context. Some experimental effects could have been buffered by the presence of the plants, mimicking clear water shallow lakes. Many lakes are already switched from a clear water stage to a turbid, phytoplankton dominated state (see Moss et al., 1996), where temperature impacts could be more severe.

Acknowledgements We are grateful to Rob Marrs, Chris Harrod, Emma Cox and Keith Hatton for help with statistics, to Dave Wilson, Keith Hatton and Tom Heyes for help with sampling and experimental maintenance, S.E. Collings and J.W. Eaton for valuable advice and to Andrew B. Gill for stickleback expertise. Parkside Engineering, JBC Control Systems Ltd. are thanked for installation of the facility, the Natural Environment Research Council for grant GR3/11438 supporting D. McKee and the EU grant EUROLIMPACS supporting H. Feuchtmayr.

References

Atkinson, D., 1996. Ectotherm life-history responses to developmental temperature. In Johnson I. A., A. F. Bennett (eds) Animals and Temperature: Phenotypic and Evolutionary Adaptation. Cambridge University Press, Cambridge: 183–204.

Baulch, H. M., D. W. Schindler, M. A. Turner, D. L. Findlay & M. J. Paterson, 2005. Effects of warming on benthic communities in a boreal lake: implications of climate change. Limnology and Oceanography 50: 1377–1392.

Connor, A. O., S. Bradish, T. Reed, J. Moran, E. Regan, M. Visser, M. Gormally & M. S. Skeffington, 2004. A comparison of the efficacy of pond-net and box sampling methods in turloughs – Irish ephemeral aquatic systems. Hydrobiologia 524: 133–144.

Costil, K. & J. Daguzan, 1995. Effect of temperature on reproduction in *Planorbarius corneus* (L) and *Planorbis planorbis* (L) throughout the life-span. Malacologia 36: 79–89.

Gurevitch, J. & S. T. Chester, 1986. Analysis of Repeated Measures Experiments. Ecology 67: 251–255.

Hogg, I. D. & D. D. Williams, 1996. Response of stream invertebrates to a global-warming thermal regime: An ecosystem-level manipulation. Ecology 77: 395–407.

Houghton, J. E. T., Y. Ding, D. J. Griggs, M. Noguer, P. J. van der Linden, X. Dai, K. Maskell & C. A. Johnson, 2001. Climate Change 2001: The Scientific Basis. Cambridge University Press, Cambridge, UK.

Humphries, P., J. E. Growns, L. G. Serafini, J. H. Hawking, A. J. Chick & P. S. Lake, 1998. Macroinvertebrate sampling methods for lowland Australian rivers. Hydrobiologia 364: 209–218.

IPCC 2007 Climate Change 2007: The Physical Science Basis – Summary for Policymakers, which is available on http://ipcc-wg1.ucar.edu/.

Kiffney, P. M. & J. S. Richardson, 2001. Interactions among nutrients, periphyton, and invertebrate and

vertebrate (*Ascaphus truei*) grazers in experimental channels. Copeia 2: 422–429.
Kozminsky, E. V., 2003. Seasonal dynamics of reproduction and reproductive parameters of *Bithynia tentaculata* (Gastropoda, Prosobranchia). Zoologichesky Zhurnal 82: 325–331.
Lewis, D. B. & J. J. Magnuson, 1999. Intraspecific gastropod shell strength variation among north temperate lakes. Canadian Journal of Fisheries and Aquatic Sciences 56: 1687–1695.
McKee, D., A. Atkinson, S. Collings, J. Eaton, I. Harvey, K. Hatton, T. Heyes, D. Wilson, L. Wolstenholme & B. Moss, 2000. Heated aquatic microcosms for climate change experiments. Freshwater Forum 14: 51–58.
McKee, D. & D. Atkinson, 2000. The influence of climate change scenarios on populations of the mayfly *Cloeon dipterum*. Hydrobiologia 441: 55–62.
McKee, D., D. Atkinson, S. Collings, J. Eaton, I. Harvey, T. Heyes, K. Hatton, D. Wilson & B. Moss, 2002a. Macro-zooplankter responses to simulated climate warming in experimental freshwater microcosms. Freshwater Biology 47: 1557–1570.
McKee, D., K. Hatton, J. W. Eaton, D. Atkinson, A. Atherton, I. Harvey & B. Moss, 2002b. Effects of simulated climate warming on macrophytes in freshwater microcosm communities. Aquatic Botany 74: 71–83.
McKee, D., D. Atkinson, S. E. Collings, J. W. Eaton, A. B. Gill, I. Harvey, K. Hatton, T. Heyes, D. Wilson & B. Moss, 2003. Response of freshwater microcosm communities to nutrients, fish, and elevated temperature during winter and summer. Limnology and Oceanography 48: 707–722.
Merritt, R. W., K. W. Cummins & V. H. Resh, 1996. Design of aquatic insect studies: collecting, sampling and rearing procedures. In Merritt R. W., & K. W Cummins (eds) An introduction to the aquatic insects of North America, 3 rd edn. Kendall/ Hull Publishing Company, Iowa 12–28.
Moss, B., G. Phillips & J. Madgwick, 1996. A guide to the restoration of nutrient-enriched shallow lakes. Broads Authority, WW Hawes, UK.
Moss, B., D. McKee, D. Atkinson, S. E. Collings, J. W. Eaton, A. B. Gill, I. Harvey, K. Hatton, T. Heyes & D. Wilson, 2003. How important is climate? Effects of warming, nutrient addition and fish on phytoplankton in shallow lake microcosms. Journal of Applied Ecology 40: 782–792.
Muzaffar, S. B. & M. H. Colbo, 2002. The effects of sampling technique on the ecological characterization of shallow, benthic macroinvertebrate communities in two Newfoundland ponds. Hydrobiologia 477: 31–39.
Noges, P., T. Noges, L. Tuvikene, H. Smal, S. Ligeza, R. Kornijow, W. Peczula, E. Becares, F. Garcia-Criado, C. Alvarez-Carrera, C. Fernandez-Alaez, C. Ferriol, R. M. Miracle, E. Vicente, S. Romo, E. Van Donk, W. van de Bund, J. P. Jensen, E. M. Gross, L. A. Hansson, M. Gyllstrom, M. Nykanen, E. de Eyto, K. Irvine, D. Stephen, S. Collins & B. Moss, 2003. Factors controlling hydrochemical and trophic state variables in 86 shallow lakes in Europe. Hydrobiologia 506: 51–58.
Rosenberg, D. M. & V. H. Resh, 1992. Introduction to freshwater biomonitoring and benthic macroinvertebrates. In: Rosenberg D. M., V. H. Resh (eds) Freshwater biomonitoring and benthic macroinvertebrates. Chapman & Hall, New York, London 1–9.
Sagarin, R. D., J. P. Barry, S. E. Gilman & C. H. Baxter, 1999. Climate-related change in an intertidal community over short and long time scales. Ecological Monographs 69: 465–490.
Schiermeier, Q., 2004. Modellers deplore 'short-termism' on climate. Nature 428: 593–593.
Stark, J. D., 1993. Performance of the macroinvertebrate community index – Effects of sampling method, sample replication, water depth, current velocity, and substratum on index values. New Zealand Journal of Marine and Freshwater Research 27: 463–478.
Walker, I. R., J. P. Smol, D. R. Engstrom & H. J. B. Birks, 1991. An assessment of chironomidae as quantitative indicators of past climatic-change. Canadian Journal of Fisheries and Aquatic Sciences 48: 975–987.

Hydrobiologia (2007) 584:433–441
DOI 10.1007/s10750-007-0614-9

SHALLOW LAKES

The role of temperature in the population dynamics of smelt *Osmerus eperlanus eperlanus* m. *spirinchus* Pallas in Lake Peipsi (Estonia/Russia)

Andu Kangur · Peeter Kangur · Külli Kangur · Tõnu Möls

Abstract We analysed lake smelt *(Osmerus eperlanus eperlanus* m. *spirinchus* Pallas.) population dynamics in relation to water level and temperature in Lake Peipsi, Estonia/Russia, using commercial fishery statistics from 1931 to 2004 (excluding 1940–1945). Over this period, smelt provided the greatest catch of commercial fish although its stock and catches have gradually decreased. At times, catches of smelt were quite variable with a cyclic character. Disappearance of smelt from catches in years 1973–1975 was the result of summer fish kill. Regression analysis revealed a significant negative effect of high temperature on the abundance of smelt stock, while the effect of water level was not significant. Our results suggest that critical factors for the smelt population are the absolute value of water temperature in the hottest period (≥20°C) of summer and the duration of this period. These weather parameters have increased in synchrony with smelt decline during the last 7 decades. There appeared to be a significant negative effect of hot summers on the abundance of smelt operating with a lag of one and 2 years, which can be explained by the short life cycle (mainly 1–2 years) of this species.

Keywords Lake smelt · Fishery catches · Water temperature · Large shallow lake

Guest editors: R. D. Gulati, E. Lammens, N. De Pauw & E. Van Donk
Shallow lakes in a changing world

A. Kangur (✉) · P. Kangur · K. Kangur · T. Möls
Centre for Limnology, Institute of Agricultural and Environmental Sciences, Estonian University of Life Sciences, 61101 Rannu, Estonia
e-mail: akangur@zbi.ee

Introduction

In order to predict the development of fish communities and thus allow the sustainable use of fish resources, it is necessary to understand the impact of environmental factors and exploitation on fish population dynamics. However, due to the complexity of such interactions there are commonly insufficient data concerning the interrelationship between fish and their habitat (Collares-Pereira & Cowx, 2004). However, Lake Peipsi in Estonia/Russia is unusual in that extensive fish and habitat data are available for a period of several decades (see Kangur et al., 2002a, b).

These data indicate that the fish populations of Lake Peipsi are under particular threat from anthropogenic eutrophication (Kangur et al., 2005). Due to its shallowness and long water residence time, the ecological state of the lake is also strongly influenced by natural processes, the most important being periodical fluctuations of water level and temperature. Such external factors cause significant inter-annual changes of hydrochemical and biological variables in the lake, and as a result also impact on its fish populations. Our data indicate that ecologically critical situations in Lake Peipsi, which have repeatedly led to extensive fish kills, have resulted from natural events and depended mainly on weather conditions and water levels (Kangur et al., 2005). Understanding the natural forces determining the development of ecosystems, their sensitivity and resistance are key problems for the planning of sustainable management of natural resources (Blenckner, 2005). Analysis of long-term fishery datasets coupled with hydrometeorological data is thus fundamental for understanding reasons behind developments in fish population dynamics.

In the present paper we analyse the role of temperature changes in the population dynamics of lake smelt *Osmerus eperlanus eperlanus* m. *spirinchus* Pallas, a key species in the fish community of Lake Peipsi. Lake smelt is a freshwater dwarf form of European smelt *Osmerus eperlanus* (L.) (Kuderskij & Fedorova, 1977; Pihu, 2003) and is a small-sized, slender, shoaling fish with a characteristic cucumber smell (Quigley et al., 2004). In Lake Peipsi, lake smelt has historically been among the most important commercial fish species in terms of catch size (Pihu & Kangur, 2001). Due to their intermediate trophic position as a consumer of plankton and as a prey for higher predators, lake smelt has the potential to introduce a wide array of ecological impacts from both direct and indirect effects (Sandlund et al., 2005). Water temperature is considered one of the most important factors influencing survival and growth rate of fish (Jobling, 2002). The lake smelt is generally considered to be a coldwater species (Nellbring, 1989), which can react distinctly to temperature fluctuations outside the optimal ranges.

In the present paper we investigate relationships between population dynamics of lake smelt and weather-driven changes in Lake Peipsi. In particular, we examine how long-term trends and periodic fluctuations of lake smelt stock are related to fluctuations in water temperature and level over a period of several decades.

Materials and methods

Study area

Lake Peipsi is situated on the border of Estonia and Russia (Fig. 1). Its surface area is 3,555 km^2, mean depth 7.1 m and maximum depth 15.3 m (Jaani, 2001a) and its altitude is 30 m above sea level. The volume of the whole lake is 25 km^3 and the residence time of water is about 2 years. The only outflow is through the Narva River into the Gulf of Finland. Natural water level fluctuations have shown an overall range of 3.04 m over the last 80 years, with an average annual range of 1.15 m (Jaani, 1996). L. Peipsi is situated in the northern region of the temperate zone with variable

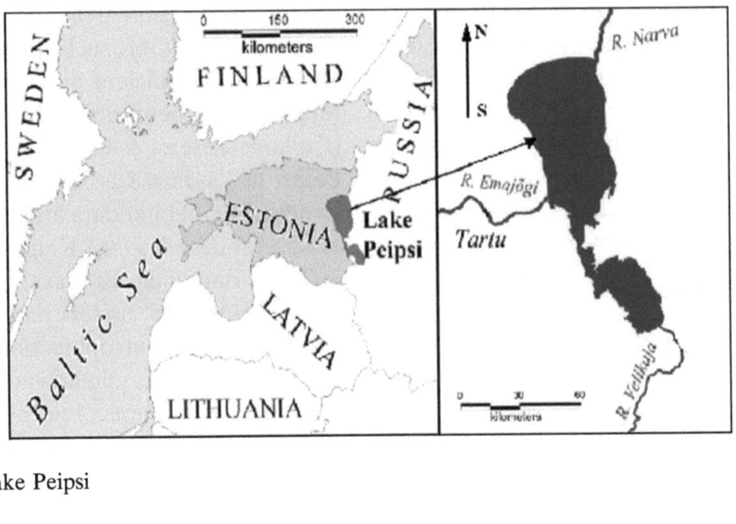

Fig. 1 Location of Lake Peipsi

weather conditions. The lake is ice-covered up to 6 months. The first ice near the coast is usually formed in early November, and by mid-December the whole lake is covered with ice. Commonly, the whole lake becomes ice-free by the end of April, but in some years ice has been found in the northern part of the lake up to end of May. Lake water is the warmest usually in July–August.

Due to the lake's large surface area and relative shallowness, the waves affect the bottom in the ice-free period; therefore normal temperature stratification is short and unstable (Jaani, 2001b). Already a moderate wind or undulation disturbs the stratification. L. Peipsi is well aerated by waves and currents during the open water period and the whole bottom of the lake is well accessible for fish.

Lake Peipsi supports a significant fishery, producing 85–88% of the total freshwater catch of fish in Estonia (Vetemaa et al., 1999). The basic fishing gears used for smelt, represent local modifications of fence traps.

Datasets and statistical methods

Owing to the lack of scientific long-term fish datasets, we used commercial fisheries statistics for the analysis of long-term trends in fish community dynamics. Such statistics, collected from fishermen by the state authorities were available for the periods 1931–1940 and 1950–2004. Data are unavailable for the period 1941–1949 due to World War II. Smelt data come from a variety of fishing gears used during different time periods and localities. Data on water level and surface water temperature, collected by the Estonian Institute of Hydrology and Meteorology, were available for the period 1924–2004.

Daily temperature data were used to analyse relationships between smelt abundance and yearly water temperature conditions during the ice-free period. Both, the formation of ice and the melting process on L. Peipsi takes normally 2–3 weeks, but sometimes it could last for a whole month. Therefore we took the duration of ice-free period conventionally equal to a number of days when water temperature was 3°C or higher. Annual mean temperatures were calculated for the periods when water temperature was ≥3°C and ≥20°C and also for different seasonal time periods (1–100, 101–130, 131–165, 166–200, 201–240, 241–280, 281–320 and 321–365 days within the year). Regression analyses were carried out with SAS System, Release 8.2, SAS Institute Inc., 1996, using GLM procedure. For studying effect of water temperature and water level on the smelt abundance, besides the current year data also data of previous and pre-previous years (time lags of 1 and 2 years) were involved. Trend lines of the mean temperature, number of ice free days, smelt abundance etc. over the year were drawn using polynomial regression. Degree of the polynomial was determined by Type I variance analysis where the exponents of the independent variable 'year' were listed in an increasing order. This analysis tests the significance of each added exponent in the model and enables to choose the polynomial model of minimal degree.

Results

Declining of total annual catch of fish and variation of smelt catches

Since records began, the commercial fish catch of Lake Peipsi has decreased approximately twofold (Fig. 2). The average annual total catch of fish was 11,650 tons (33 kg ha^{-1}) from 1931 to 1940 and 6,046 tons (17 kg ha^{-1}) from 1994 to 2004. The decreasing tendency of annual fish catches was statistically significant ($r = -0.53; p < 0.0001$).

Smelt has provided the greatest catch of commercial fish. In the 1930s, the catch of this species often exceeded 8,000 tons per year. From 1931 to 1940, the proportion of smelt constituted 43.2% of the total catch, followed by roach *Rutilus rutilus* (L.) (16.2%), perch *Perca fluviatilis* L. (7.2%) and bream *Abramis brama* (L.) (6.7%). The catch of smelt also dominated from 1994 to 2004, but its share (24.6%) of the total catch had decreased. During the latter period, the share of pikeperch *Sander lucioperca* (L.) in the total catch (22.2%) was close to that of smelt, and was followed by those of perch (14.8%) and bream (11.7%).

Generally, annual catches of smelt from Lake Peipsi decreased on a long-term scale ($r = -0.42$;

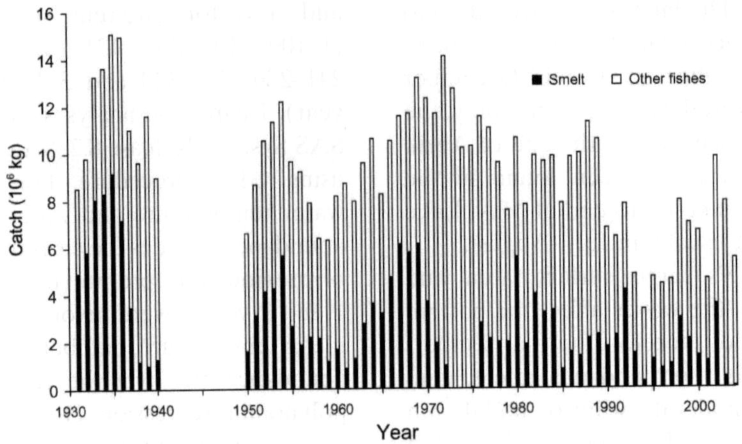

Fig. 2 Annual catch of smelt and other fish species from L. Peipsi in 1931–2004

p = 0.0004). At times, the catches of smelt were quite variable having a cyclic character with a duration of 9–12 years (Fig. 2). The yield of smelt was modest (<1,000 tons) at the end of the 1930s, at the end of the 1950s, the mid 1980s, the mid 1990s, and in recent years. In the 1970s, smelt lost its commercial importance for 3 years (1973–1975) following a fish kill in 1972. However, the stock had recovered by 1976.

Variation of water temperature in the lake

Temperature conditions in the open water period and duration of this period may differ greatly from year to year (Figs. 3, 4). Since the 1928, the lake has been ice-free (water $t \geq 3$°C) for an average of 210 days (minimum 178, maximum 250 days) in a year, with a mean water temperature of 13.9°C. The mean and maximum water

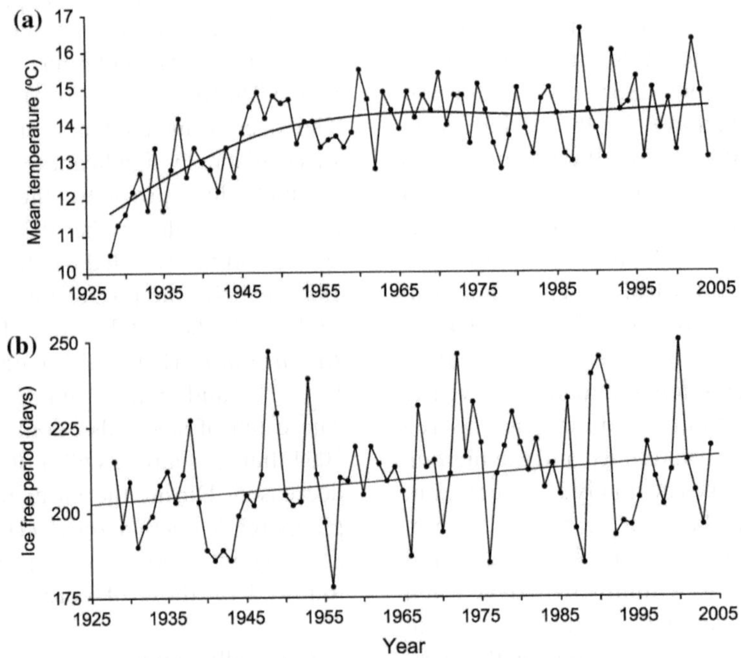

Fig. 3 Mean water temperature (**a**) and duration of ice-free period (**b**, temperature ≥3°C) in L Peipsi

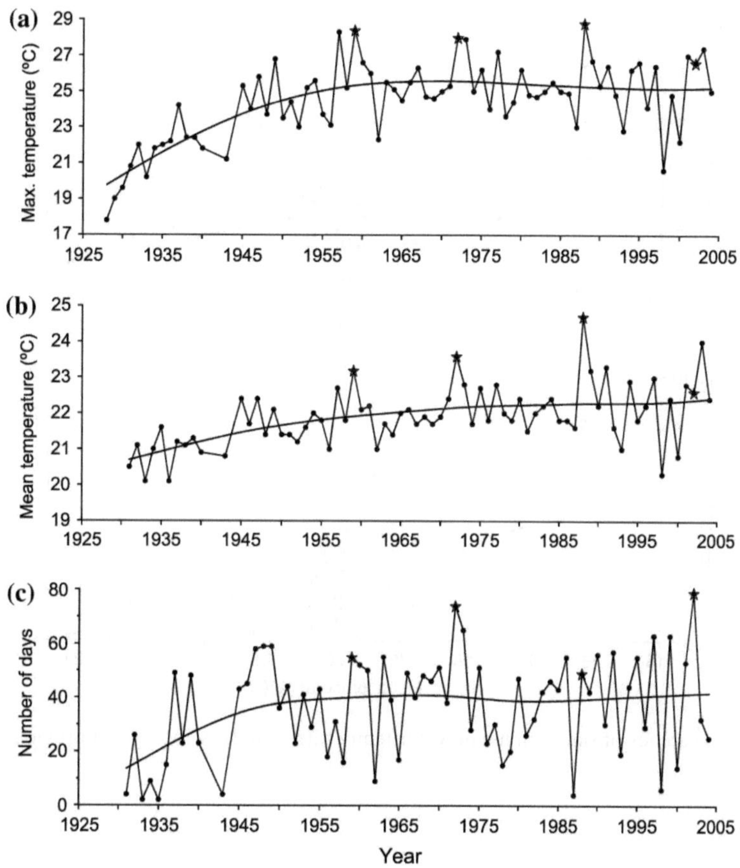

Fig. 4 Time period with water temperature ≥20°C in L. Peipsi: (**a**) maximum temperature, (**b**) mean temperature and (**c**) duration of the period. *Years of extensive fish kill

temperature during ice-free period displayed large inter-annual variability (Figs. 3, 4). However, there was a statistically significant increase of these temperature parameters during the investigation period ($r = 0.54$; $p < 0.0001$ for mean temperature and $r = 0.56$; $p < 0.0001$ for maximum temperature during ice-free period).

The highest temperature of water in the period from 1928 to 2004 was 28.8°C (June 27, 1988); the annual temperature maximum was on average 24.5°C. The highest water temperature in L. Peipsi commonly occurs in July. However, annual maximum temperatures may differ up to 10°C (Fig. 4) and can occur from the beginning of June to the end of August.

Mean water temperature of the hottest period (water $t > 20$°C) and its duration are shown in Fig. 4. The period with water temperature over 20°C starts on an average of the 176th Julian day (June, 25) and lasts on average 37 (maximum 79) days. The duration and mean temperature of the hottest period have significantly increased during the last 7 decades (Fig. 4).

Effect of water level and temperature on catch of smelt

We analysed the dependence of smelt catches on water level and temperature of summers 1 and 2 years previously. The regression analysis revealed a significant ($p < 0.0001$) negative effect of high temperature on the abundance of smelt stock (Table 1), while the effect of water level was not significant. The estimated effect of water temperature is expressed in tons per degree. The strongest negative effect of temperature on the

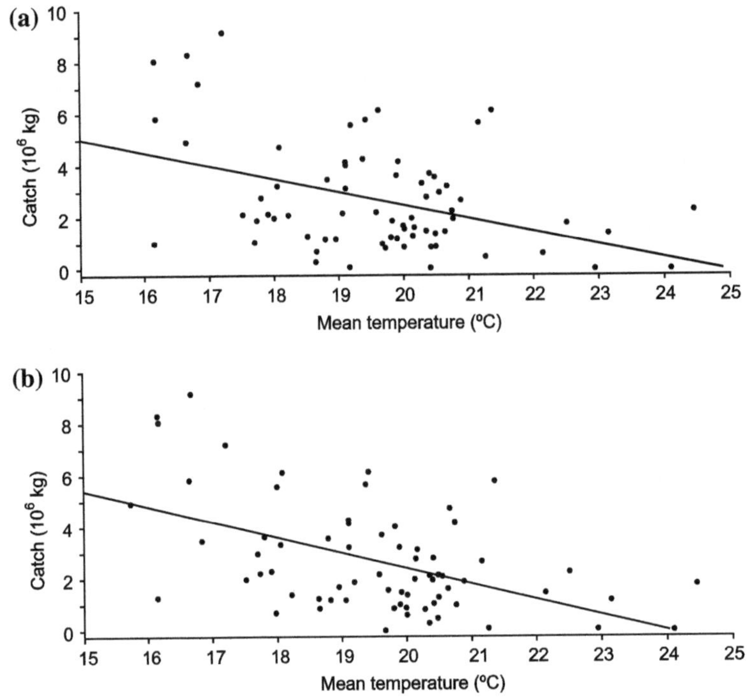

Fig. 5 The dependence of catches of smelt on mean water temperature in days 166–200 of previous years: (**a**) lag of 1 year, (**b**) lag of 2 years

catches, both one and 2 years later, was shown by mean annual values. Strong effects on catches of smelt also become evident in the case of increased water temperature in mid summer (i.e., in 166–200 Julian days) (Fig. 5), in July–August (i.e., in 201–240 Julian days, Table 1) or in spring (April–June, i.e., in 101–165 Julian days, Table 1).

Table 1 Results of the Regression Analyses: dependence of smelt catches on water temperature in different seasonal time periods of two previous years (lag 1 and lag 2). Each row is calculated in a separate linear regression analysis

Lag, years	Julian days	Estimated effect, t deg^{-1}	*DF1*	*DF2*	*F*	*p*-value
1	1–365	−955	1	67	13.9	0.0004
1	101–165	−385	1	67	11.6	0.0011
1	166–200	−463	1	67	14.2	0.0004
2	1–365	−1044	1	67	14.8	0.0003
2	101–165	−368	1	67	9.7	0.0027
2	166–200	−630	1	67	23.8	<0.0001
2	201–240	−639	1	67	16.7	0.0001

Discussion

Commercial catches demonstrated the overall long-term trend of declining fish stocks in Lake Peipsi. As a result of natural processes coupled with human impact, several structural changes have occurred in the fish community during the last 70 years. On such a long-term scale, the fish community has shifted from clean- and cold-water species like vendace *Coregonus albula* (L.), whitefish *C. lavaretus* L. and burbot *Lota lota* (L.), to more pikeperch preferring productive warm and turbid waters (Kangur et al., unpublished).

Smelt has been a dominant species in the fish community throughout this period. However, its abundance has fluctuated periodically within a gradual decline on a long-term scale. These cyclic changes in smelt population cannot be explained by fishery impacts, but they are explainable by natural processes. Fish species with a short life cycle and early reproduction like smelt can better withstand heavy fishing pressure (Sarvala et al., 1999), particularly when compared with larger

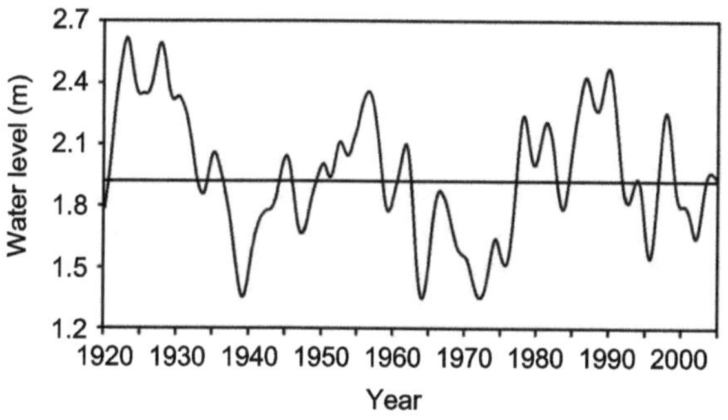

Fig. 6 Mean annual water level in Lake Peipsi from 1920 to 2005

fish species whose population dynamics are determined to a large extent by fishery impacts (Kangur et al., 2003).

Lake Peipsi clearly shows specific cycles of water level fluctuation (Jaani, 2001c). However, significant correlations between population dynamics of smelt and water level fluctuations (Fig. 6) were not found. In contrast, a clear overall effect of temperature changes on smelt population dynamics was observed. In Lake Peipsi, average and maximum water temperatures during the ice-free periods of 1928–2004 showed an increasing trend, while at the same time the annual catches of smelt declined. Consequently, a significant negative relationship existed between water temperature in the ice-free period and the annual catch of smelt 1 and 2 years later. According to Pihu (1987), smelt prefers water temperatures in the range of 15–20°C. Probably, even a short-time warming of aquatic environment over the tolerable level supports the increase in mortality rate of adult smelt and decreases its abundance and catches.

The life cycle of smelt is relatively short (Kriksunov & Shatunovskij, 1979; Kuderskij & Fedorova, 1977) and in Lake Peipsi is mainly about 2 years in duration. According to our observations, the commercial catch of smelt is based only on two age groups, while the proportion of 1-year-old individuals constituted numerically about 83% and 2-years-old fishes about 17% (Andu & Peeter Kangur, unpublished data). Due to this relatively short life cycle, smelt populations are probably very sensitive to changes in the environment. However, according to Horppila et al. (1996) fish species with a short life span such as smelt are able to respond rapidly to increased mortality. Because of the early attainment of sexual maturity, short life span and high fecundity, the abundance of smelt can vary significantly and swiftly.

Abundance and catches of smelt depend on the success of reproduction and subsequent survival rate, both of which depend to a great extent directly or indirectly on temperature. Abundance of smelt also depends greatly on hydrological conditions during the short spawning period when the ice drifts in spring (Kuderskij & Fedorova, 1977; Nellbring, 1989). Steady water level, calm weather in spring, sufficient spawning stock, favourable feeding conditions and weak pressure from predators produces strong recruitment of smelt in Lake Peipsi (Dorozhkina, 1985). In contrast, drastic decrease in temperature and stormy weather can disturb spawning and lead to weak recruitment. In addition, mortality of smelt eggs seems increased due to deterioration of spawning areas during recent decades (Kangur et al., 2002b). Increasing eutrophication has led to a decline of water quality, strong blooms of cyanobacteria and increased siltation in Lake Peipsi. According to Winfield et al. (1996), such increasing eutrophication may contribute to the inconsistent recruitment of fish through siltation on spawning grounds.

Our results suggest that the smelt population of Lake Peipsi has been gradually declining on a long-term scale, but some abrupt changes were also noticeable. These were clearly associated with continuous hot weather and high water temperature during a long period. In Lake Peipsi, the effect of warm weather is especially strong when it coincides with low water level. Pronier & Rochard (1997) have also noticed that high values of smelt mortality rates occur while temperatures are at their highest. In Lake Peipsi, hot periods with water temperatures over 20°C lasted 79 days in summer 2002, which led to an extensive fish kill (Kangur et al., 2005). Sporadic fish kills have also occurred in the lake previously during calm warm periods and strong cyanobacterial blooms in summer. A smelt kill during a bloom of *Aphanizomenon* sp. was first described in the southern shallower part of the lake in August 1959 (Semenova, 1960) and another massive kill hit the whole lake in July 1972 (Kuderskij & Fedorova, 1977). After the latter fish kill, smelt lost its commercial importance in the lake for 3 years from 1973 to 1975, but became abundant again by 1976. Extensive fish kill was observed also in July 1988 (Laugaste et al., 2001) and in July to August 2002 (Kangur et al., 2005). These extreme events have significantly affected the state of the fish community of Lake Peipsi, with smelt and ruffe *Gymnocephalus cernuus* (L.) showing the greatest declines. The extensive fish kill in summer 2002 was mainly caused by a synergistic effect of low dissolved oxygen content (less than 2 mg O^2 L^{-1}) in the open water during night, high water temperature (up to 26.6°C), high pH (9.5) and high ammonium ion (up to 0.33 mg N L^{-}) content, which appeared in the lake during the strong bloom of the cyanobacterium *Gloeotrichia echinulata* (J. S. Smith) P. Richter (Kangur et al., 2005).

Conclusion

Our results indicate that smelt reacts strongly and quickly to weather-induced changes in Lake Peipsi. Only two age groups (1- and 2-years-old fish) of smelt dominated commercial catches. A short life span and early reproduction makes the smelt population very unstable and vulnerable to changes in its environment. We determined using regression analysis a strong negative effect of high temperature on the abundance of the smelt population with a lag of 1 and 2 years. A warming of the aquatic environment, coupled with concurrent eutrophication and resulting cyanocbacterial blooms, has decreased reproduction success and increased direct adult mortality of smelt and thus decreased its abundance and commercial catches in Lake Peipsi.

Acknowledgments The authors are indebted to the Estonian Science Foundation (grants 4986, 6008 and 6820). This research was supported by the Estonian target financed project SF 0362483s03.

References

Blenckner, T., 2005. A conceptual model of climate-related effects on lake ecosystem. Hydrobiologia 533: 1–14.

Collares-Pereira, M. J. & I. G. Cowx, 2004. The role of catchment scale environmental management in freshwater fish conservation. Fisheries Management and Ecology 11: 303–312.

Dorozhkina, T. Ya., 1985. Prichiny kolebanij ulovov pskovskogo i chudskogo snetka. Sbornik nauchnykh trudov GosNIORKh 236: 98–109 (in Russian).

Horppila, J., K. Nyberg, H. Peltonen & T. Turunen, 1996. Effects of five years of intensive trawling on a previously unexploited smelt stock. Journal of Fish Biology 49: 329–340.

Ibarra, A. A., M. Gevrey, Y.-S. Park, P. Lim & S. Lek, 2003. Modelling the factors that influence fish guilds composition using a back-propagation network: Assessment of metrics for indices of biotic integrity. Ecological Modeling 160: 281–290.

Jaani, A., 1996. Hydrology and water balance of Lake Peipsi. Hydrobiologia 338: 11–23.

Jaani, A., 2001a. The location, size and general characterisation of Lake Peipsi and its catchment area. In Nõges T. (ed.), Lake Peipsi: Meteorology, Hydrology, Hydrochemistry. Sulemees Publishers, Tartu: 10–17.

Jaani, A., 2001b. Thermal regime and ice conditions. In Nõges T. (ed.), Lake Peipsi. Meteorology, Hydrology, Hydrochemistry. Sulemees Publishers, Tartu: 65–72.

Jaani, A., 2001c. Water regime. In Nõges T. (ed.), Lake Peipsi. Meteorology, Hydrology, Hydrochemistry. Sulemees Publishers, Tartu: 65–72.

Jobling, M., 2002. Environmental factors and rates of development and growth. In Hart P. J. B. & J. D. Reynolds (eds), Handbook of Fish Biology and Fisheries. Blackwell Publishing: 97–122.

Kangur, A., P. Kangur & E. Pihu, 2002a. Long-term trends in the ichthyocoenosis of Lake Peipsi and L. Võrtsjärv (Estonia). Aquatic Ecosystem Health and Management 5: 379–389.

Kangur, K., A. Milius, T. Möls, R. Laugaste & J. Haberman, 2002b. Lake Peipsi: changes in nutrient elements and plankton communities in the last decade. Aquatic Ecosystem Health and Management 5: 363–377.

Kangur, K., A. Kangur, P. Kangur & R. Laugaste, 2005. Fish kill in Lake Peipsi in summer 2002 as a synergistic effect of a cyanobacterial bloom, high temperature, and low water level. Proceedings of the Estonian Academy of Sciences. Biology Ecology 54: 67–80.

Kangur, K., Y.-S. Park, A. Kangur, P. Kangur & S. Lek, 2006. Patterning long-term changes of fish community in large shallow Lake Peipsi. Ecological Modelling. Available 2 January 2007. doi:10.1016/j.ecolmodel.2006.03.039

Kangur, P., A. Kangur, K. Kangur & T. Möls, 2003. Condition and growth of ruffe *Gymnocephalus cernuus* (L.) in two large shallow lakes with different fish fauna and food resources. Hydrobiologia 506–509: 435–441.

Kriksunov, E. A. & M. I. Shatunovskij, 1979. Nekotorye voprocy izmenchivosti struktury populjatsii korjushki

Osmerus eperlanus (L.). Voprosy ihtiologii 19: 55–62 (in Russian).
Kuderskij, L. A. & G. V. Fedorova, 1977. Snizhenie zapasov snetka v bolshikh bodoemakh Severo-Zapada Evropejskoj chasti SSSR v 1973–1975 gg. Rybokhozyajstvennoe izuchenie vnutrennikh vodoemov 20: 3–8 (in Russian).
Laugaste, R., T. Nõges, P. Nõges, V. V. Yastremskij, A. Milius & I. Ott, 2001. Algae. In Pihu E. & J. Haberman (eds), Lake Peipsi. Flora and Fauna. Sulemees Publishers, Tartu: 31–49.
Nellbring, S., 1989. The ecology of smelts (genus *Osmerus*): a literature review. Nordic Journal of Freshwater Research 65: 116–145.
Pihu, E., 1987. Matk kalariiki. Valgus, Tallinn (in Estonian).
Pihu, E., 2003. Lake (dwarf) smelt, *Osmerus eperlanus* morpha *spirinchus* Pallas. In Ojaveer E., E. Pihu & T. Saat (eds), Fishes of Estonia. Estonian Academy Publishers, Tallinn: 148–151.
Pihu, E. & A. Kangur, 2001. Fishes and fisheries management. In Pihu E. & J. Haberman (eds), Lake Peipsi. Flora and Fauna. Sulemees Publishers, Tartu: 100–111.
Pronier, O. & E. Rochard, 1997. Working of a smelt (*Osmerus eperlanus*, Osmeriformes Osmeridae) population located at the south limit of the species distribution area, influence of the temperature. Bulletin Français de la Pêche et de la Pisciculture 350–351: 479–497.
Quigley, D. T. G., F. Igoe & W. O'Connor, 2004. The European Smelt *Osmerus eperlanus* L. in Ireland: general biology, ecology, distribution and status with conservation recommendations. Biology and Environment. Proceedings of the Royal Irish Academy 104B: 57–66.
Sandlund, O. T., Y. G. Stand, G. Kjellberg, T. F. Næsje & M. U. Hambo, 2005. European smelt (*Osmerus eperlanus*) eats all; eaten by all: Is it a key species in lakes? Verhandlungen internationale Vereinigung für theoretische und angewandte Limnology 29: 432–436.
Sarvala, J., H. Helminen & H. Auvinen, 1999. Portrait of a flourishing freshwater fishery: Pühäjärvi, a lake in SW-Finland. Boreal Environmental Research 3: 329–345.
Semenova, N. I., 1960. The reasons of smelt kill in L. Pskov in August 1959. Nauchno-tekhnicheskij byulleten GosNIORKH 10: 23–24 (in Russian).
Vetemaa, M., A. Järvalt & V. Vaino, 1999. Current status and trends in inland fisheries in Estonia. In Proceeding of Country Report, the 5th Bafico Seminar on Inland Fisheries. Tallinn: 19–28.
Wells, L., 1968. Seasonal depth distribution of fish in south-eastern Lake Michigan. Fisheries Bulleten 67: 1–15.
Winfield, I. J., C. E. Adams & J. M. Fletcher, 1996. Recent introductions of the ruffe (*Gymnocephalus cernuus*) to three United Kingdom lakes containing *Coregonus* species. Annales Zoologici Fennici 3–4: 459–466.

Hydrobiologia (2007) 584:443–454
DOI 10.1007/s10750-007-0600-2

SHALLOW LAKES

Predicting the effect of climate change on temperate shallow lakes with the ecosystem model PCLake

W. M. Mooij · J. H. Janse · L. N. De Senerpont Domis · S. Hülsmann · B. W. Ibelings

Abstract Global average surface temperatures are expected to rise by about 1.4–5.8°C from the present until the year 2100. This temperature increase will affect all ecosystems on earth. For shallow lakes—which can be either in a clear water or a turbid state—this climate change will expectedly negatively affect water transparency though the prediction is far from conclusive and experimental investigations elucidating the potential climatic effects on shallow lakes are still rare. The aim of this study was to further shape and sharpens hypotheses on the impact of climate change on shallow lakes by applying an existing and well-calibrated ecosystem model, PCLake. We focused on asymptotic model behaviour for a range of temperature and loading scenarios in a factorial design. We conclude that climate change will likely lead to decreased critical nutrient loadings. Combined with an expected increase in the external nutrient loading, this will increase the probability of a shift from a clear to a turbid state. As the model predicts a higher summer chlorophyll-a concentration, a stronger dominance of cyanobacteria during summer and a reduced zooplankton abundance due to climate change, the turbid state itself is likely to become even more severe.

Keywords Temperature · Alternative stable states · Critical nutrient loading · Cyanobacteria · Mismatch

Guest editors: R. D. Gulati, E. Lammens, N. De Pauw & E. Van Donk
Shallow lakes in a changing world

W. M. Mooij (✉) · L. N. De Senerpont Domis · B. W. Ibelings
Netherlands Institute of Ecology, Centre for Limnology, Rijksstraatweg 6, 3631 AC Nieuwersluis, The Netherlands
e-mail: w.mooij@nioo.knaw.nl

J. H. Janse
Netherlands Environmental Assessment Agency, P.O. Box 303, 3720 AH Bilthoven, The Netherlands

S. Hülsmann
Institute of Hydrobiology, Dresden University of Technology, 01062 Dresden, Germany

Introduction

Global average surface temperatures have increased by approximately 0.6°C over the last century (Houghton et al., 2001). Without proper action against anthropogenic greenhouse effects the Intergovernmental Panel on Climate Change (IPCC) predicts increases in global surface temperature of 1.4–5.8°C for the year 2100 (Houghton et al., 2001). Ecological responses to recent climate warming have been demonstrated across different natural systems (Pace, 1984; Parmesan & Yohe, 2003), including temperate lakes (Straile & Adrian, 2000; Gerten & Adrian, 2002). Because of their smaller volume and absence of stratification in summer, shallow water bodies are less influenced by meteorological conditions in the preceding winter than deeper waterbodies, and they respond more directly to prevailing weather conditions (Gerten & Adrian, 2001).

Of the many aspects of shallow lake ecosystems on which climate can have an impact, water transparency is of particular interest to water quality managers and this parameter has often been used to evaluate the success of restoration measures (Drenner & Hambright, 1999; Gulati & Van Donk, 2002). According to the concept of alternative stable states (Scheffer et al., 1993), switches between a macrophyte-dominated clear-water state and a phytoplankton-dominated turbid state occur abruptly in shallow lakes when (external) nutrient loading passes a certain critical level. However, even phytoplankton-dominated lakes may have temporary clear-water phases, which are typically caused by enhanced grazing pressure on phytoplankton by zooplankton at the end of the spring (Gulati et al., 1982; Sommer et al., 1986).

In a recent review of the potential impact of climate change on shallow lakes in the Netherlands (Mooij et al., 2005) it is concluded that climate warming will probably negatively affect transparency in shallow lakes. It is expected that climate warming will tend to destabilize the macrophyte-dominated clear state, stabilize the phytoplankton-dominated turbid state and hamper the recolonization of the lake by macrophytes under the current management regime (Mooij et al., 2005). These expectations rest on a number of assumptions. (1) The expected increase of net precipitation in winter, and especially an increase in extreme rainfall events, will tend to increase the P-loading of lakes (Mooij et al., 2005). (2) Internal nutrient loading will increase with temperature (Jensen & Andersen, 1992; Liikanen et al., 2002). (3) Elevated temperature may cause a mismatch between the timing of the phytoplankton spring peak and the zooplankton spring peak (leading to a reduced grazing pressure on phytoplankton) if both peaks are triggered by different stimuli which might be the case when daphnids establish themselves mainly from resting eggs (Winder & Schindler, 2004). (4) Elevated temperatures will favour cyanobacteria and in turn lead to a further decoupling of phytoplankton and zooplankton growth because cyanobacteria are known to be a relatively poor food item for zooplankton (Dawidowicz et al., 1988; Gliwicz & Lampert, 1990). (5) Zooplanktivory by fish will increase due to decreased winter mortality and higher recruitment (Mooij, 1996; Mehner et al., 1998; Mooij & Van Nes, 1998; Nyberg et al., 2001). (6) Sediment resuspension by wind (Schelske et al., 1995) will increase. Jeppesen et al. (2003) reach similar conclusions for Danish shallow lakes. If all or some of these assumptions are correct, this implies that the effects of climate change on shallow lakes will mimic the effects of eutrophication, as has indeed been predicted by some authors (Kilham et al., 1996; Porter et al., 1996).

It has to be noted, however, that the anticipated negative effects of climate change on lake transparency are still far from being conclusive, since counteracting effects seem possible, for instance high P loading of inflowing water might (temporarily) be balanced by dilution effects when net precipitation is increased; increased internal loading might be a transient phenomenon; a mismatch between phytoplankton and zooplankton has only been described for deep lakes and is related to stratification patterns, which might not apply to shallow lakes; enhanced growth rates of cyanobacteria may in case of scum forming species be counteracted by disturbing effects of wind and increased cloud cover (Howard & Easthope, 2002); macrophytes positively respond to increasing temperature (Barko & Smart, 1981; Rooney & Kalff, 2000), which could result in a stabilization of the clear water state.

Experimental approaches to potential climatic effects on shallow lakes are still rare (see below) or underway (Liboriussen et al., 2005) and so far only systems mimicking the macrophyte-dominated clear water state have been studied. Results of an enclosure experiment in two subsequent years with different summer temperatures suggest that at critical nutrient loading levels, a switch from the clear to the turbid state is more likely to occur in years with warmer summers (Van de

Bund et al., 2004). Studies using outdoor microcosms (McKee et al., 2002a, b, 2003; Moss et al., 2003) indicate, however, that direct temperature effects on lakes in a clear state dominated by macrophytes may be subtle and that warming does not significantly encourage phytoplankton blooms, even in combination with increased nutrient levels and fish densities. Nevertheless, warming produced trends in water chemistry that support the idea that even moderate warming has the potential to exacerbate existing eutrophication problems (Moss et al., 2003). This may, at least in the long run, destabilize macrophyte dominated clear-water states (e.g. Van de Bund & Van Donk, 2002).

Model analysis may help to further shape and sharpen hypotheses on the impact of climate change on shallow lakes and thereby provide a good starting point for further empirical work (e.g. time series analysis, mesocosm experiments). Specific aspects such as shifts in the phenology of phytoplankton and zooplankton in spring may be studied with relatively simple models such as the seasonally forced predator-prey model of Scheffer et al. (1997). Recently, a more detailed phytoplankton model was linked with a climate model by Elliott (2005) using historical data. For a more comprehensive view of the impact of climate change that includes effects of internal and external nutrient loading as well as (at least some) feedback mechanisms within the system, we argue that there is a need for analysis using a full scale ecosystem model such as PCLake (Janse, 2005). Although such ecosystem models are widely used to evaluate the effects of eutrophication, we know of no study that applies such models to predict the impact of climate change. The purpose of this study was to initiate the analysis of the potential effects of climate change on shallow lakes by means of a comprehensive ecosystem model.

We designed our study to cover some of the central aspects of the interaction between the functioning of shallow lakes and climate change: (1) increasing water temperature will affect abiotic process rates in the lake water and sediment, including internal loading; (2) increasing temperature will affect the physiology and life history of the biotic components of the ecosystem; and (3) climate change will lead to increased external nutrient loading, in particular through an increase in winter precipitation. To cover these three aspects we ran the model for four temperature scenarios (including a control scenario) and a wide range of external nutrient loadings. We initiated the model in both the clear state and the turbid state to scan for multiple stable states and to determine the critical nutrient loading of the switch from clear to turbid during eutrophication (CNL_{eu}) and from turbid to clear during lake restoration, i.e. (re)oligotrophication (CNL_{oligo}).

Methods

The ecosystem model PCLake used in our study of the effects of climate change was developed to study the effects of eutrophication and related restoration measures in shallow lakes (Janse et al., 1992; Janse & Van Liere, 1995a, b). The model has been designed to simulate the main nutrient and food web dynamics of a non-stratifying lake (Janse, 1997, 2005). It was calibrated against nutrient, transparency, chlorophyll and vegetation data on >40 lakes, and a systematic sensitivity and uncertainty analysis was recently performed (Aldenberg et al., 1995; Janse, 2005).

The model describes a completely mixed water body and comprises both the water column and the sediment top-layer (10 cm), with the most important biotic and abiotic components. No further horizontal or vertical distinction within the lake is taken into account. Optionally, a wetland zone with marsh vegetation may be included, but this has not been done in this study. Mathematically, the model is composed of a number of coupled differential equations, one for each state variable. All biota are modelled as functional groups. The main groups in the water phase are 3 prototypic groups of phytoplankton ("diatoms", "greens" and "cyanobacteria"), zooplankton, planktivorous/benthivorous fish and piscivorous fish. Submerged macrophytes are included, consisting of a shoot and a root fraction. Further groups in the top layer of the sediment are the settled fractions of the three types of phytoplankton as well as zoobenthos.

The phytoplankton groups differ in their parameter values. It is assumed that the cyanobacteria (for which the filamentous species serve as a model) have a lower maximum growth rate (0.6 d^{-1}) than the diatoms (2.0) and greens (1.5), but also lower loss rates through respiration (0.05 times μ_{max}), settling (maximal v_{set} = 0.06 d^{-1} for cyanobacteria, 0.2 for greens and 0.5 for diatoms) and zooplankton grazing (a factor 6 lower). It is further assumed that the cyanobacteria have a higher phosphorus affinity (0.8 vs 0.2 l mg^{-1} d^{-1}) and maximum P uptake rate (0.04 vs 0.01 g P g^{-1} d^{-1}) than the other groups. No differences were assumed for nitrogen uptake. Nutrient limitation of phytoplankton growth has been modelled by the Droop equation. The light dependency of growth is described by a half-saturation function for the green algae (with h = 17 W m^{-2} PAR at 20°C) and a Steele function (including inhibition by high intensities) for the diatoms and cyanobacteria; the latter have been attributed a low I_{opt} (13.6 W m^{-2} at 20°C) to mimic their shade-adapted character. The effect of temperature has been described by optimum functions, with the diatoms having a lower temperature optimum (18°C) than the others (25°C) and the cyanobacteria having a stronger temperature dependency than the green algae (Table 1). Finally, the diatoms are the only group that might be limited by silica. The growth limitation functions due to light and temperature on the one hand and by nutrients (the minimum of the functions for P, N and possibly Si), on the other hand, are multiplied.

Also the growth of the submerged macrophytes is dependent on under-water light, temperature and nutrients in water and pore water. It is assumed that their growing season starts when a critical spring temperature (9°C) is reached, and ends at a predefined day in autumn. All parameters have been assigned fixed values in the model, but it should be realized that they show a natural variability within groups in reality. Both the water phase and the top layer of the sediment have their own pools of inorganic nutrients and detritus. Besides mass fluxes of dry weight (DW, representing carbon) and nutrients, the model also contains a number of indirect relations between components, such as the impact of fish and macrophytes on resuspension. The overall nutrient cycles of N, P and Si are completely covered by the model, including external fluxes such as in- and outflow and denitrification. This does not hold for carbon because inorganic carbon in the water or sediment is not explicitly modelled. A particularly interesting stoichiometric consequence of modelling the N, P, Si and DW dynamics separately is that the nutrient-to-dry-weight ratios are thus variable. The main inputs to the model are: water inflow, infiltration or seepage rate, nutrient (N, P) loading, particulate loading, temperature and light, dimensions (lake depth and size), size of the marsh zone, sediment features and loading history (initial conditions). As output, the biomass and concentrations of all state variables, as well as a number of derived variables and fluxes, are calculated.

Table 1 Temperature (T) functions and parameters in PCLake

Abiotic process	Temperature coefficient c_i	
Denitrification	1.07	
Diffusion	1.02	
Mineralization in sediment or water	1.07	
Nitrification	1.08	
Reaeration	1.024	
Sedimentation	1.01	
Biotic component	Optimum temperature $T_{opt,j}$	Width around optimum temperature $T_{sigma,j}$
Diatoms	18°C	20°C
Green algae	25°C	15°C
Cyanobacteria	25°C	12 °Cl
Zooplankton	25°C	13 °C
Zoobenthos	25°C	16°C
Planktivorous and benthivorous fish	25°C	10°C
Piscivorous fish	25°C	10°C
Macrophytes	Q_{10prod} = 1.2	Q_{10resp} = 2.0

For abiotic process (i) an exponential function with parameter c_i is used: $f_i(T) = c_i^{(T - Tref)}$ For biotic component (j), a Gaussian function with parameters $T_{opt,j}$ and $T_{sigma,j}$ is used: $f_j(T) = \exp(-0.5 (T - T_{opt,j})^2 - (T - T_{ref})^2 / T_{sigma,j})$. In all functions a reference temperature T_{ref} of 20°C was used. For macrophytes, also an optimum function was used, implemented by two exponential functions with a higher Q_{10} for respiration (Q_{10resp}) than for production (Q_{10prod}). The temperature threshold for macrophyte growth was 9°C

Previous studies with PCLake (Janse, 1997, 2005) showed that the model behaviour is consistent with the existing evidence that shallow lakes may have two alternative stable states. In the clear-water state macrophytes dominate and in the turbid state phytoplankton dominates (Timms & Moss, 1984; Jeppesen et al., 1990; Scheffer, 1990, 1998). Several factors determine which state prevails in a certain case. A general constraint is set by the external nutrient loading. At a (very) high loading, only the turbid state is stable, whereas the opposite is true for (very) low loadings. In the intermediate range, both states can exist (in the field even side by side) and switches between states are possible. The model—like examples studied in the field—shows hysteresis as the shift from clear to turbid due to eutrophication occurs at a much higher nutrient loading than the switch back from turbid to clear due to re-oligotrophication through lake restoration ($CNL_{eu} > CNL_{oligo}$). Both states are maintained by a number of self-stabilizing mechanisms that are included in the model. Macrophytes stabilize the clear state, they compete with phytoplankton for nutrients, promote denitrification, reduce resuspension, hamper the feeding of benthivorous fish and promote top-down control of phytoplankton by promoting piscivorous fish. High algal and—especially—cyanobacterial biomasses stabilise the turbid state and hamper macrophyte growth through turbidity. Top-down effects through grazing pressure by zooplankton are reduced, and the bare sediment promotes internal P-loading.

The effect of temperature on the various model components is entered through a set of fourteen temperature dependent multipliers. Six of these modify abiotic processes: diffusion, sedimentation of particulate matter (only slightly affected), nitrification, denitrification and mineralization in the water phase and the sediment. These have the shape of simple exponential curves. The remaining eight modifiers that amend the growth rates of biotic components of the model are implemented as Gaussian curves around a certain optimum temperature. The model has such curves for the three phytoplankton groups (diatoms, greens and cyanobacteria), zooplankton, zoobenthos, planktivorous fish and piscivorous fish. For macrophytes, also an optimum function was used, implemented by two exponential functions with a higher Q_{10} for respiration than for production. All the functions and parameters of the temperature curves in PCLake are given in Table 1.

We focused on asymptotic model behaviour (running 25 years with the same input) for a range of temperatures and loading scenarios in a factorial design. To maintain a focus on alternative stable states and critical nutrient loadings each scenario was evaluated twice, starting from either a macrophyte-dominated clear state or a phytoplankton-dominated turbid state. The other model parameters were kept constant and chosen such that they represent an average for shallow lakes in the Netherlands (following Janse, 2005, p. 99). Its main characteristics are: mean depth = 2 m, fetch = 1000 m, areal hydraulic loading = 20 mm d^{-1} (=7.2 m y^{-1}), no infiltration or seepage, no surrounding wetland zone, and a slightly clayish sediment (30% dry matter, of which 10% is organic and 90% inorganic matter, the latter containing 10% lutum).

Phosphorus loading varied within a range of 0.1–20 mg P m^2 d^{-1} in 40 steps. The nitrogen loading was set to 10 times the phosphorus loading. We used a sine function with a maximum on July 31 to represent the typical temperature profile of a shallow lake in the Netherlands. In the control scenario (denoted as C) the minimum winter temperature was 2°C and the summer maximum 22°C (Fig. 1). We defined three scenarios with elevated temperatures: an all year round temperature increase of 3°C (denoted as A), an increase in summer maximum temperature

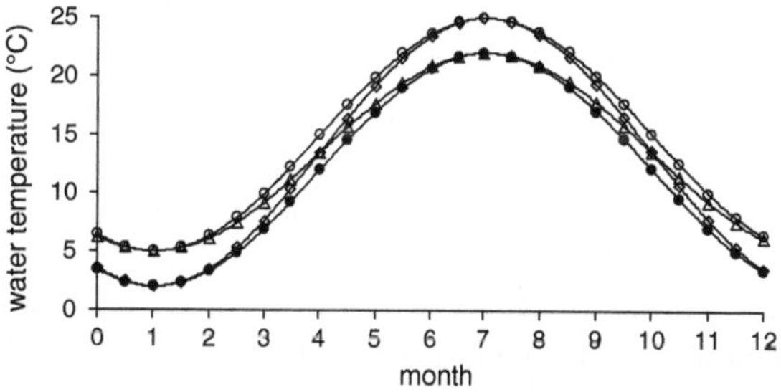

Fig. 1 Temperature patterns for the control (closed circles), all year warm (open circles), warm summer (open diamonds) and warm winter (open triangles) scenario

of 3°C but no change in the winter minimum (denoted as S) and, finally, an increase in winter minimum temperature of 3 °C, but no change in the summer maximum (denoted as W).

We focused on the summer averages (from April 1 to September 30) of chlorophyll-a, zooplankton biomass and the percentage of cyanobacteria. This period was chosen because it gives an overall picture of the effects and because this period is used in policy standards. Seasonal deviations are mentioned in the text. For each of the four temperature scenarios (C, A, S and W) we obtained values for the two critical nutrient loadings (CNL_{eu} and CNL_{oligo}). To get a more detailed view on lake dynamics we studied within season dynamics of each scenario for nutrient loadings of 0.6 and 4 mg m^2 d^{-1}. Again we focused on total algal and zooplankton biomass and algal community composition.

Results

The general observation is that climate warming will decrease the critical nutrient loadings. In the control scenario (C) CNL_{eu} has a value of 3.3 mg P m^2 d^{-1} (Table 2). At this point chlorophyll-a suddenly increases from 10 to 98 μg l^{-1} (Fig. 2a), total zooplankton biomass increases slightly from 0.62 to 0.85 mg l^{-1} DW (Fig 2b), the percentage of cyanobacteria increases from 21 to 79% (Fig. 2c) and submerged macrophytes virtually disappear (not shown). The model predicts that when environmental temperature is raised by 3°C all year (scenario A), CNL_{eu} decreases by 24% from 3.3 to 2.5 mg P m^2 d^{-1} (Table 2). Raising the winter minimum temperature by 3°C (scenario W) results in a 12% decrease in CNL_{eu}, and raising the summer maximum temperature by 3°C (scenario S) results in a 18% decrease in CNL_{eu} (Table 2).

The results for CNL_{oligo} show a comparable pattern. In the control scenario the switch back takes place at a loading of 0.95 mg P m^2 d^{-1} (Table 2). At this point chlorophyll-a suddenly decreases from 44 to 1.5 μg l^{-1} (Fig. 2a), zooplankton biomass decreases slightly from 0.72 to 0.50 mg l^{-1} DW (Fig. 2b), cyanobacteria almost completely disappear, decreasing from 68 to 1% (Fig 2c), and submerged macrophytes become the dominant producers. The model predicts that when environmental temperature is raised by 3°C all year (scenario A), CNL_{oligo} decreases by ca. 21% from 0.95 to 0.75 mg P m^2 d^{-1} (Table 2). The scenarios in which the summer maximum or the winter minimum is increased by 3°C both show a smaller decrease of 10% in CNL_{oligo} (Table 2).

In the clear water state, the effects of the temperature scenarios on chlorophyll-a are insignificant (Fig. 2a). Zooplankton biomass is slightly lower in the temperature scenarios, both in spring and in summer. In the turbid phase, just above the CNL_{eu}, the model predicts an increase in summer chlorophyll-a from 98 to 133 μg l^{-1} (Fig. 2a). Zooplankton biomass decreases from 0.85 to 0.61 mg l^{-1} DW (Fig. 2b). At intermediate loadings, the strongest decrease of zooplankton is predicted in spring, the period in which also the controls showed a lower biomass in the turbid state than in the clear state. In summer, on the contrary, the model shows a higher zooplankton biomass in the turbid state. The model also predicts that a decrease in zooplankton is generally accompanied by a decrease in juvenile fish. Remarkably, summer warming has the strongest effect on the decrease in critical nutrient loadings, but winter warming has the strongest effect on the

Table 2 Critical nutrient loading (mg P m^2 d^{-1}) during eutrophication and oligotrophication for the control (C), year round warm (A), warm summer (S) and warm winter (W) scenario

Temperature scenario	During eutrophication	During oligotrophication
Control scenario (min 2°C, max 22°C)	3.3 (10.4 to 98)	0.95 (44 to 1.5)
All year increase of 3°C	2.5 (6.5 to 133)	0.75 (59 to 0.5)
Increase of 3°C in summer maximum	2.7 (6.5 to 98)	0.85 (40 to 1.0)
Increase of 3°C in winter minimum	2.9 (8.5 to 126)	0.85 (64 to 1.0)

For each scenario the chlorophyll-a concentrations (μg l^{-1}) before and after the switch are given in parentheses

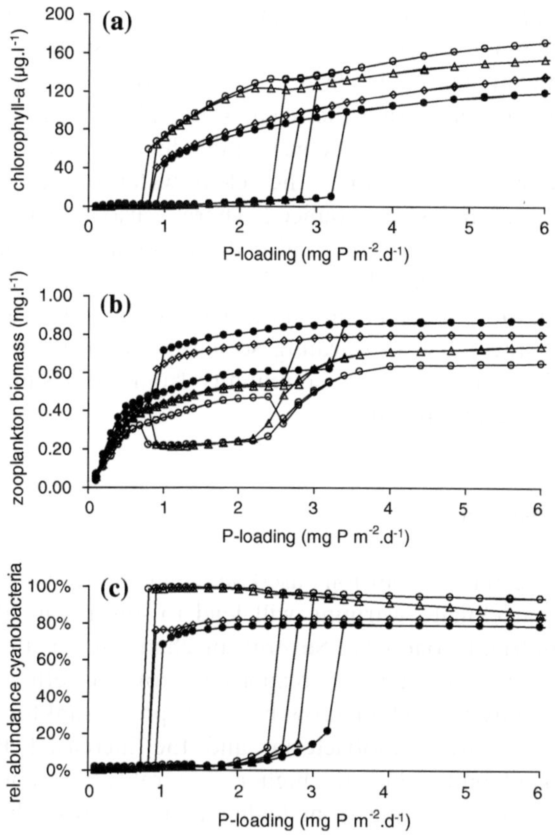

Fig. 2 (**a**) Average summer chlorophyll-a concentration as a function of nutrient loading for the control (closed circles), all year warm (open circles), warm summer (open diamonds) and warm winter (open triangles) scenario. (**b**) Zooplankton biomass as a function of nutrient loading for the control (closed circles), all year warm (open circles), warm summer (open diamonds) and warm winter (open triangles) scenario. (**c**) Percentage of cyanobacteria as a function of nutrient loading for the control (closed circles), all year warm (open circles), warm summer (open diamonds) and warm winter (open triangles) scenario

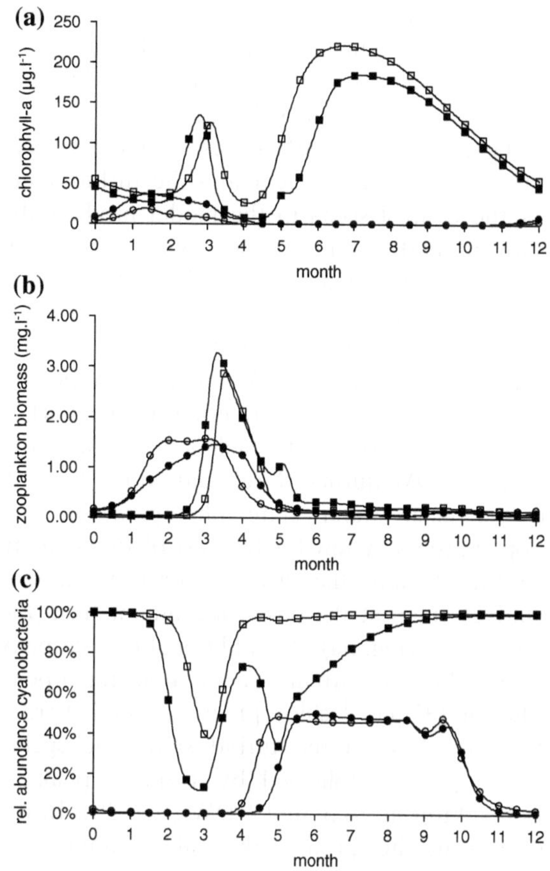

Fig. 3 (**a**) Chlorophyll-a concentration during the season for the control (closed circles for the clear water state and closed squares for the turbid water state) and all year warm (open circles for the clear water state and open squares for the turbid water state) scenario. (**b**) Zooplankton biomass during the season for the control (closed circles for the clear water state and closed squares for the turbid water state) and all year warm (open circles for the clear water state and open squares for the turbid water state) scenario. (**c**) Percentage of cyanobacteria during the season for the control (closed circles for the clear water state and closed squares for the turbid water state) and all year warm (open circles for the clear water state and open squares for the turbid water state) scenario

increase in summer chlorophyll-a (Fig. 2a). Note also that in the case of year round heating or winter heating the increase in chlorophyll-a is strongest at intermediate nutrient loading, in the range where alternative stable states exist (Fig. 2a). The summer-averaged macrophytes biomass is hardly affected by the temperature scenarios.

At an intermediary loading of 2.0 mg P m^2 d^{-1} the algal community in the control scenario is composed of ca. 97 % diatoms and only 3% cyanobacteria in the clear state (Fig 2c). In the turbid state this pattern is reversed to 22% diatoms and 78% cyanobacteria (Fig. 2c). Green algae are below 1% in all simulations (for the lake settings used). In the clear state the temperature scenarios hardly affect phytoplankton community composition (but it must be born in mind, though, that in this case phytoplankton biomass is low compared to submerged macrophytes). In the turbid state, however, in the A-scenario the percentage of cyanobacteria increases from 78% to an almost complete dominance of 99%

(Fig 2c). This happens also in the winter warming scenario. The summer scenario hardly shows any response in cyanobacterial dominance and resembles the control scenario (Fig 2c).

There are large differences in the seasonal patterns between the clear and the turbid state. For the control scenario the model predicts the following patterns: in the clear water state chlorophyll-a starts to build up as early as in February, reaching a peak of about 35 $\mu g\ l^{-1}$, which lasts for about 2 months (Fig. 3a) and is strongly dominated by diatoms (Fig 3c). Zooplankton appear in large numbers about one month after the start of the diatom bloom and reach a level of about 1.5 mg l^{-1} DW during March and April (Fig 3b). In the clear water state phytoplankton and zooplankton stay low for the rest of the year. In the turbid state the cyanobacteria biomass is relatively high at the onset, a remnant from the year before (Fig. 3a). An additional diatom peak starts to build up around March, reaching a peak value of 135 $\mu g\ l^{-1}$ chlorophyll-a around April 1 (Fig. 3a). Also in the turbid state, the spring diatom peak is followed by a spring peak of zooplankton up to 3 mg l^{-1} DW (Fig. 3b). In the turbid state the spring peak of algae (dominated by diatoms) is followed by a long-lasting period in which the phytoplankton is dominated by cyanobacteria (Fig. 3c). Because these are less edible for zooplankton compared to diatoms, zooplankton abundance stays low after the spring peak (Fig. 3b).

The all year warming scenario (A) results in an earlier occurrence—approx. 3 weeks—of the spring diatom bloom and zooplankton peak in the clear water state (Fig. 3a, b). In the clear water state the height of the spring diatom peak under all year warming is lower (Fig. 3a, c) because in this scenario they are more effectively grazed upon by the zooplankton whose population builds up faster (Fig 3b). In the turbid state scenario A results in a higher cyanobacterial biomass during winter and spring, and a slight delay of the diatom and zooplankton peak by approx. a week and a half (Fig. 3a–c). The summer peak in chlorophyll-a in the turbid state is predicted to occur about 1 month earlier (Fig. 3a). Looking at the turbid state, the diatom and zooplankton spring peaks in the all year warming scenario are slightly lower, but the summer phytoplankton peak is considerably higher in A (Fig. 3a, b). The chlorophyll-a concentration during the 'clear water phase', occurring around May, is considerably higher in the all year warm scenario; the 'clear water phase' is therefore less pronounced. There is also a shift in phytoplankton composition during summer for all year warming (Fig. 3c). In the control scenario diatoms make up half of the algal community, but in the all year warming scenario cyanobacteria dominate the algal community from early summer and onwards.

Discussion

The main conclusion based on the model results is that climate warming will lead to lower critical nutrient loadings. Several mechanisms in the model together are responsible for this effect, mainly the higher growth rate of phytoplankton, especially cyanobacteria, and the increased P availability due to higher mineralization and release resulting from higher temperatures. The critical nutrient level for eutrophication (CNL_{eu}) as predicted by the model is basically determined by the turbidity level at which the net algal growth (growth minus losses) exceeds the net macrophytes growth, under clear starting conditions. Likewise, CNL_{oligo} relies on the transparency level at which net macrophytes growth starts to exceed net algal growth, under turbid starting conditions. Phytoplankters, as opposed to macrophytes, are the first to profit from both the higher water temperatures and the higher P availability, especially cyanobacteria that also have a high P affinity and a 'steeper' temperature function. The zooplankton and fish growth rates increase as well due to the temperature rise, but this does not affect the critical P levels very much. According to the model increased temperatures in winter and early spring have the greatest impact on shallow lake ecosystems, since results from the "warm winter" scenario closely resembled the "year round warm" scenario. By contrast, the "warm summer" scenario yielded results similar to the control scenario. The combination of the predicted lower critical nutrient loading and the anticipated increase in external loading implies that in the future some (restored) lakes that are now in the clear state will return to the turbid state. For lakes in the turbid state this means that the nutrient loading has to be reduced to even lower values to generate a switch back to the clear state.

According to the model, chlorophyll-a and dominance of cyanobacteria during summer will increase, winter temperatures being the crucial factor. In the model, again, this result is due to improved growing conditions for cyanobacteria at elevated temperatures. Because of the poor edibility of cyanobacteria, higher trophic levels do not profit from this enhanced carrying capacity at the level of primary producers. The model suggests a reduction of food transfer to the higher trophic levels, mainly in the pelagic (via zooplankton) and a little in the sediment (via zoobenthos), at higher temperatures. On the other hand, from natural systems there are indications that zooplanktivorous fish density may be higher in warmer years (e.g. Jeppesen et al., 2003) and thus impose a higher predation pressure on the zooplankton. The fact that the model emphasizes the bottom up effects of reduced transfer efficiency but not the top-down effects of increased zooplanktivory might imply that the model assumptions on the diet and the mortality of fish are too crude. Elevated winter temperatures seem to be critical for cyanobacteria dominance in shallow systems (Moed & Hoogveld, 1982; Adrian et al., 1995; Reeders et al., 1998). The predicted increased dominance of cyanobacteria implies that water quality problems in turbid lakes may intensify.

The model predicts that in the clear state the algal (diatom) and zooplankton spring peaks occur earlier, but that in the turbid state these peaks occur later as a result of temperature rise. In the clear state, diatoms start to grow earlier in spring at higher temperature and enhanced P release. In spring macrophytes are not yet so important, but increase in abundance later in the season when they benefit from the higher temperature and also from the higher P availability in the sediment. At this point the diatoms do no longer profit from the elevated temperatures, assuming that their optimum is 18° (compared to 25°C for the modelled cyanobacteria). For the clear state this model result is consistent with empirical findings for temperate lakes which showed earlier peaks of phytoplankton and zooplankton at increasing temperatures (Weyhenmeyer et al., 1999; Gerten & Adrian, 2000; Straile, 2000; Scheffer et al., 2001). The very early timing of both peaks in our scenarios (even in the control as early as February and March, respectively) likely results from the sine function applied for temperature. Note also that ice cover was not included our model scenarios (consistent with conditions in the Netherlands) so that the modelled spring plankton development cannot be explained by ice cover as reported for other systems (Adrian et al., 1999; Livingstone & Dokulil, 2001). Similar to model result for the clear state, experiments in heated mesocosms suggest that macrophytes remain largely unaffected by moderate warming (McKee et al., 2002b).

For the turbid state the result that phytoplankton and zooplankton peaks are delayed at higher temperatures is somewhat unexpected. Here, cyanobacteria whose densities are already high early in the year (remnants of the summer peak of the year before) profit most from increased temperatures. Despite their increased growth rate in spring, diatoms are poorer competitors and peak later. Since zooplankton mostly rely on edible phytoplankton (diatoms or greens) as a food item they show a delayed development, too. It is difficult to find empirical evidence for the model results on turbid state as virtually all published data on climate effects on plankton phenology stem from less eutrophic and mostly deeper systems. From the study by Scheffer et al. (2001), which exclusively included shallow Dutch lakes, it seems that the general pattern found in other lakes (earlier peaks of phytoplankton and zooplankton at higher temperatures) was confirmed. It is unclear, however, if the pattern would change if only data from turbid lakes were included in such a correlation analysis. In the moderately shallow Heiligensee (average and maximum depth of 5.9 and 9.5 m, respectively) phytoplankton peaks during a warm winter period developed earlier and the contribution of cyanobacteria to the phytoplankton community was larger compared to colder years (Adrian et al., 1995). It was suggested that this shift in phytoplankton composition resulted in the replacement of *D. galeata* by the smaller species *D. cucullata*, the latter species being less adversely affected by the presence of cyanobacteria (Adrian & Deneke, 1996). Looking at the total abundance patterns of all *Daphnia* species in that study (Fig. 5 in Adrian & Deneke, 1996) the displayed peaks indeed seem to occur later during warmer years (with a higher share of cyanobacteria) compared to the colder period. This finding, however, could also be due to species-specific temperature responses and additionally, altered top-down effects seem possible. It is thus unclear whether the result obtained by PCLake - spring diatom and zooplankton peaks in turbid lakes are delayed at increasing temperatures - represents a mechanism also to be found in nature.

We conclude that the analysis performed with PCLake supports the idea that climate change will negatively affect the transparency of temperate shallow lakes (Jeppesen et al., 2003; Mooij et al., 2005). In the first place, climate change is expected to lead to increased external loading. Second, the model predicts that the critical nutrient load during eutrophication will decrease. Together, this will increase the probability that a lake switches from the clear to the turbid state. Third, according to the model's prediction the turbid state will become even more turbid due to a higher summer chlorophyll-a concentration, stronger summer dominance of cyanobacteria and reduced abundance of zooplankton due to climate change. The model provides a consistent view of the chain of effects by which increased nutrient loading and increased temperature may entail the described effects, although the impacts on the food web are possibly over-simplified in the model. Further, climate-related factors such as wind and insolation were neglected in this approach, but theoretically could outweigh temperature effects on, e.g., cyanobacteria dominance. In future scenarios to be explored using coupled physical-biological ecosystem models these factors should be included to further deepen our understanding of climate impacts on aquatic ecosystems.

Acknowledgments We thank Marten Scheffer, Eric Jeppesen and an anonymous reviewer for comments on this manuscript. LdSD was funded by the NWO programme Biodiversity in relation to Global Change (NWO GRANT 852.00.010 / 852.00.012).

References

Adrian, R. & R. Deneke, 1996. Possible impact of mild winters on zooplankton succession in eutrophic lakes of the Atlantic European area. Freshwater Biology 36: 757–770.

Adrian, R., R. Deneke, U. Mischke, R. Stellmacher & P. Lederer, 1995. A long-term study of the Heiligensee (1975–1992). Evidence for effects of climatic change on the dynamics of eutrophied lake ecosystems. Archiv für Hydrobiologie 133: 315–337.

Adrian, R., N. Walz, T. Hintze, S. Hoeg & R. Rusche, 1999. Effects of ice duration on plankton succession during spring in a shallow polymictic lake. Freshwater Biology 41: 621–632.

Aldenberg, T., J. H. Janse & P. R. G. Kramer, 1995. Fitting the dynamic model PCLake to a multi-lake survey through Bayesian statistics. Ecological Modelling 78: 83–99.

Barko, J. W. & R. M. Smart, 1981. Comparative influences of light and temperature on the growth and metabolism of selected submersed fresh-water macrophytes. Ecological Monographs 51: 219–235.

Dawidowicz, P., Z. M. Gliwicz & R. D. Gulati, 1988. Can *Daphnia* prevent a blue-green algal bloom in hypertrophic lakes? A laboratory test. Limnologica (Berlin) 19: 21–26.

Drenner, R. W. & K. D. Hambright, 1999. Review: biomanipulation of fish assemblages as a lake restoration technique. Archiv für Hydrobiologie 146: 129–165.

Elliott, J. A., S. J. Thackeray, C. Huntingford & R. G. Jones, 2005. Combining a regional climate model with a phytoplankton community model to predict future changes in phytoplankton in lakes. Freshwater Biology 50: 1404–1411.

Gerten, D. & R. Adrian, 2000. Climate-driven changes in spring plankton dynamics and the sensitivity of shallow polymictic lakes to the North Atlantic Oscillation. Limnology and Oceanography 45: 1058–1066.

Gerten, D. & R. Adrian, 2001. Differences in the persistency of the North Atlantic Oscillation signal among lakes. Limnology and Oceanography 46: 448–455.

Gerten, D. & R. Adrian, 2002 Effects of climate warming, North Atlantic Oscillation, and El Nino-Southern Oscillation on thermal conditions and plankton dynamics in northern hemispheric lakes. TheScientific WorldJournal 2: 586–606.

Gliwicz, Z. M. & W. Lampert, 1990. Food thresholds in *Daphnia* species in the absence and presence of blue-green filaments. Ecology 71: 691–702.

Gulati, R. D., K. Siewertsen & G. Postema, 1982. The zooplankton: its community structure, food and feeding, and role in the ecosystem of Lake Vechten. Hydrobiologia 95: 127–163.

Gulati, R. D. & E. Van Donk, 2002. Lakes in the Netherlands, their origin, eutrophication and restoration: state-of-the-art review. Hydrobiologia 478: 73–106.

Houghton J. T., Y. Ding, D. J. Griggs, M. Noguer, P. J. Van der Linden & D. Xiaosu (eds), 2001. Climate Change 2001: The Scientific Basis. Contribution of Working Group I to the Third Assessment Report of the Intergovernmental Panel on Climate Change (IPCC). Cambridge University Press, Cambridge: 944 pp.

Howard, A. & M. P. Easthope, 2002. Application of a model to predict cyanobacterial growth patterns in response to climatic change at Farmoor Reservoir, Oxfordshire, UK. Science of the Total Environment 282: 459–469.

Janse, J. H., 1997. A model of nutrient dynamics in shallow lakes in relation to multiple stable states. Hydrobiologia 342/343: 1–8.

Janse, J. H., 2005. Model studies on the eutrophication of shallow lakes and ditches, Wageningen University.

Janse, J. H., T. Aldenberg & P. R. G. Kramer, 1992. A mathematical model of the phosphorus cycle in Lake Loosdrecht and simulation of additional measures. Hydrobiologia 233: 119–136.

Janse, J. H. & L. Van Liere, 1995a. Modelling nutrient cycles in relation to food-web structure in a biomanipulated shallow lake. Netherlands Journal of Aquatic Ecology 29: 67–79.

Janse, J. H. & L. Van Liere, 1995b. PCLake – a modeling tool for the evaluation of lake restoration scenarios. Water Science and Technology 31: 371–374.

Jensen, H. S. & F. O. Andersen, 1992. Importance of temperature, nitrate, and pH for phosphate release from aerobic sediments of four shallow, eutrophic lakes. Limnology and Oceanography 37: 577–589.

Jeppesen, E., J. P. Jensen, M. Sondergaard, K. S. Hansen, P. H. Moller, H. U. Rasmussen, V. Norby & S. E. Larsen, 2003. Does resuspension prevent a shift to a clear state in shallow lakes during reoligotrophication? Limnology and Oceanography 48: 1913–1919.

Jeppesen, E., M. Sondergaard, E. Mortensen, P. Kristensen, B. Riemann, H. J. Jensen, J. P. Müller, O. Sortkjaer, J. P. Jensen, K. Christoffersen, S. Bosselmann & E. Dall, 1990. Fish manipulation as a lake restoration tool in shallow, eutrophic temperate lakes 1: cross-analysis of three Danish case studies. Hydrobiologia 200/201: 205–218.

Kilham, S. S., E. C. Theriot & S. C. Fritz, 1996. Linking planktonic diatoms and climate change in the large lakes of the Yellowstone ecosystem using resource theory. Limnology and Oceanography 41: 1052–1062.

Liboriussen, L., F. Landkildehus, M. Meerhoff, M. E. Bramm, M. Sondergaard, K. Christoffersen, K. Richardson, T. L. Lauridsen & E. Jeppesen, 2005. Global warming: design of a flow-through shallow lake mesocosm climate experiment. Limnology and Oceanography Methods 3: 1–9.

Liikanen, A., T. Murtoniemi, H. Tanskanen, T. Vaisanen & P. J. Martikainen, 2002. Effects of temperature and oxygen availability on greenhouse gas and nutrient dynamics in sediment of a eutrophic mid-boreal lake. Biogeochemistry 59: 269–286.

Livingstone, D. M. & M. T. Dokulil, 2001. Eighty years of spatially coherent Austrian lake surface temperatures and their relationship to regional air temperature and the North Atlantic Oscillation. Limnology and Oceanography 46: 1220–1227.

McKee, D., D. Atkinson, S. Collings, J. Eaton, I. Harvey, T. Heyes, K. Hatton, D. Wilson & B. Moss, 2002a. Macro-zooplankter responses to simulated climate warming in experimental freshwater microcosms. Freshwater Biology 47: 1557–1570.

McKee, D., D. Atkinson, S. E. Collings, J. W. Eaton, A. B. Gill, I. Harvey, K. Hatton, T. Heyes, D. Wilson & B. Moss, 2003. Response of freshwater microcosm communities to nutrients, fish, and elevated temperature during winter and summer. Limnology and Oceanography 48: 707–722.

McKee, D., K. Hatton, J. W. Eaton, D. Atkinson, A. Atherton, I. Harvey & B. Moss, 2002b. Effects of simulated climate warming on macrophytes in freshwater microcosm communities. Aquatic Botany 74: 71–83.

Mehner, T., H. Dörner & H. Schultz, 1998. Factors determining the year-class strength of age-0 Eurasian perch (*Perca fluviatilis*, L.) in a biomanipulated reservoir. Archiv of Fisheries and Marine Research 46: 241–251.

Moed, J. R. & H. L. Hoogveld, 1982. The algal periodicity in Tjeukemeer during 1968–1978. Hydrobiologia 95: 223–224.

Mooij, W. M., 1996. Variation in abundance and survival of fish larvae in shallow eutrophic lake Tjeukemeer. Environmental Biology of Fishes 46: 265–279.

Mooij, W. M., S. Hülsmann, L. N. De Senerpont Domis, B. A. Nolet, P. L. E. Bodelier, P. C. M. Boers, M. L. Dionisio Pires, H. J. Gons, B. W. Ibelings, R. Noordhuis, R. Portielje, K. Wolfstein & E. H. R. R. Lammens, 2005. The impact of climate change on lakes in the Netherlands: a review. Aquatic Ecology 39: 381–400.

Mooij, W. M. & E. H. Van Nes, 1998. Statistical analysis of the somatic growth rate of 0+ fish in relation to temperature under natural conditions. Canadian Journal of Fisheries and Aquatic Sciences 55: 451–458.

Moss, B., D. McKee, D. Atkinson, S. E. Collings, J. W. Eaton, A. B. Gill, I. Harvey, K. Hatton, T. Heyes & D. Wilson, 2003. How important is climate? Effects of warming, nutrient addition and fish on phytoplankton in shallow lake microcosms. Journal of Applied Ecology 40: 782–792.

Nyberg, P., E. Bergstrand, E. Degerman & O. Enderlein, 2001. Recruitment of pelagic fish in an unstable climate: studies in Sweden's four largest lakes. AMBIO 30: 559–564.

Pace, M. L., 1984. Zooplankton community structure, but not biomass, influences the phosphorus–chlorophyll a

relationship. Canadian Journal of Fisheries and Aquatic Sciences 41: 1089–1096.
Parmesan, C. & G. Yohe, 2003. A globally coherent fingerprint of climate change impacts across natural systems. Nature 421: 37–42.
Porter, K. G., P. A. Saunders, K. A. Haberyan, A. E. Macubbin, T. R. Jacobsen & R. E. Hodson, 1996. Annual cycle of autotrophic and heterotrophic production in a small, monomictic Piedmont lake (Lake Oglethorpe): analog for the effects of climatic warming on dimictic lakes. Limnology and Oceanography 41: 1041–1051.
Reeders, H. H., P. C. M. Boers, D. T. Van der Molen & T. H. Helmerhorst, 1998. Cyanobacterial dominance in the lakes Veluwemeer and Wolderwijd, The Netherlands. Water Science and Technology 37: 85–92.
Rooney, N. & J. Kalff, 2000. Inter-annual variation in submerged macrophyte community biomass and distribution: the influence of temperature and lake morphometry. Aquatic Botany 68: 321–335.
Scheffer, M., 1990. Multiplicity of stable states in freshwater systems. Hydrobiologia 200/201: 475–486.
Scheffer, M., 1998. Ecology of Shallow Lakes. Chapman & Hall, London. 357 pp.
Scheffer, M., S. H. Hosper, M. L. Meijer, B. Moss & E. Jeppesen, 1993. Alternative equilibria in shallow lakes. Trends in Ecology & Evolution 8: 275–279.
Scheffer, M., S. Rinaldi, Y. A. Kuznetsov & E. H. Van Nes, 1997. Seasonal dynamics of *Daphnia* and algae explained as a periodically forced predator–prey system. Oikos 80: 519–532.
Scheffer, M., D. Straile, E. H. Van Nes & H. Hosper, 2001. Climatic warming causes regime shifts in lake food webs. Limnology and Oceanography 46: 1780–1783.
Schelske, C. L., H. J. Carrick & F. J. Aldridge, 1995. Can wind-induced resuspension of meroplankton affect phytoplankton dynamics? Journal of the North American Benthological Society 14: 616–630.
Sommer, U., Z. M. Gliwicz, W. Lampert & A. Duncan, 1986. The PEG-model of seasonal succession of planktonic events in fresh waters. Archiv für Hydrobiologie 106: 433–471.
Straile, D., 2000. Meteorological forcing of plankton dynamics in a large and deep continental European lake. Oecologia 122: 44–50.
Straile, D. & R. Adrian, 2000. The North Atlantic Oscillation and plankton dynamics in two European lakes – two variations on a general theme. Global Change Biology 6: 663–670.
Timms, R. M. & B. Moss, 1984. Prevention of growth of potentially dense phytoplankton population by zooplankton grazing in the presence of zooplanktivorous fish, in a shallow wetland ecosystem. Limnology and Oceanography 29: 472–486.
Van de Bund, W. J., S. Romo, M. J. Villena, M. Valentin, E. Van Donk, E. Vicente, K. Vakkilainen, M. Svensson, D. Stephen, A. Stahl-Delbanco, J. Rueda, B. Moss, M. R. Miracle, T. Kairesalo, L. A. Hansson, J. Hietala, M. Gyllstrom, J. Goma, P. Garcia, M. Fernandez-Alaez, C. Fernandez-Alaez, C. Ferriol, S. E. Collings, E. Becares, D. M. Balayla & T. Alfonso, 2004. Responses of phytoplankton to fish predation and nutrient loading in shallow lakes: a pan-European mesocosm experiment. Freshwater Biology 49: 1608–1618.
Van de Bund, W. J. & E. Van Donk, 2002. Short-term and long-term effects of zooplanktivorous fish removal in a shallow lake: a synthesis of 15 years of data from Lake Zwemlust. Freshwater Biology 47: 2380–2387.
Weyhenmeyer, G. A., T. Blenckner & K. Pettersson, 1999. Changes of the plankton spring outburst related to the North Atlantic Oscillation. Limnology and Oceanography 44: 1788–1792.
Winder, M. & D. E. Schindler, 2004. Climate change uncouples trophic interactions in an aquatic ecosystem. Ecology 85: 2100–2106.

Hydrobiologia (2007) 584:455–466
DOI 10.1007/s10750-007-0616-7

SHALLOW LAKES

Shallow lakes theory revisited: various alternative regimes driven by climate, nutrients, depth and lake size

Marten Scheffer · Egbert H. van Nes

Abstract Shallow lakes have become the archetypical example of ecosystems with alternative stable states. However, since the early conception of that theory, the image of ecosystem stability has been elaborated for shallow lakes far beyond the simple original model. After discussing how spatial heterogeneity and fluctuation of environmental conditions may affect the stability of lakes, we review work demonstrating that the critical nutrient level for lakes to become turbid is higher for smaller lakes, and seems likely to be affected by climatic change too. We then show how the image of just two contrasting states has been elaborated. Different groups of primary producers may dominate shallow lakes, and such states dominated by a particular group may often represent alternative stable states. In tropical lakes, or small stagnant temperate waters, free-floating plants may represent an alternative stable state. Temperate shallow lakes may be dominated alternatively by charophytes, submerged angiosperms, green algae or cyanobacteria. The change of the lake communities along a gradient of eutrophication may therefore be seen as a continuum in which gradual species replacements are interrupted at critical points by more dramatic shifts to a contrasting alternative regime dominated by different species. The originally identified shift between a clear and a turbid state remains one of the more dramatic examples, but is surely not the only discontinuity that can be observed in the response of these ecosystems to environmental change.

Keywords

Guest editors: R. D. Gulati, E. Lammens, N. De Pauw & E. Van Donk
Shallow lakes in a changing world

M. Scheffer (✉) · E. H. van Nes
Department of Aquatic Ecology and Water Quality Management, Wageningen University, P.O. Box 8080, 6700 DD Wageningen, The Netherlands
e-mail: marten.scheffer@wur.nl

Introduction

Widespread problems resulting from eutrophication of shallow lakes in populated areas invoked numerous restoration projects in the last decades of the 20th century. However, even if the nutrient load to such lakes was strongly reduced they often did not recover to their original clear state (Sas, 1989). Research into the causes of this hysteresis has been surprisingly productive, not only in solving the restoration issue, but also in generating fundamental insights in the stability properties of these ecosystems (Moss, 1988; Jeppesen, 1998; Jeppesen et al., 1998; Scheffer, 1998). One of the most influential ideas that emerged from this work is arguably the theory that such lakes can be in two alternative stable states: clear with abundant submerged macrophytes or turbid with few submerged plants (Scheffer et al., 1993). The theoretical possibility that ecosystems could have alternative stable states had been noted long before (Lewontin, 1969; Holling, 1973; May, 1977), but shallow lakes are often considered the first well documented example.

This review highlights a few important new twists to the stability theory of lakes and points at some intriguing future questions. Although it is meant as an update for lake ecologists, we hope that some of the progress in the stability research on shallow lakes we report here may inspire scientists in related fields too.

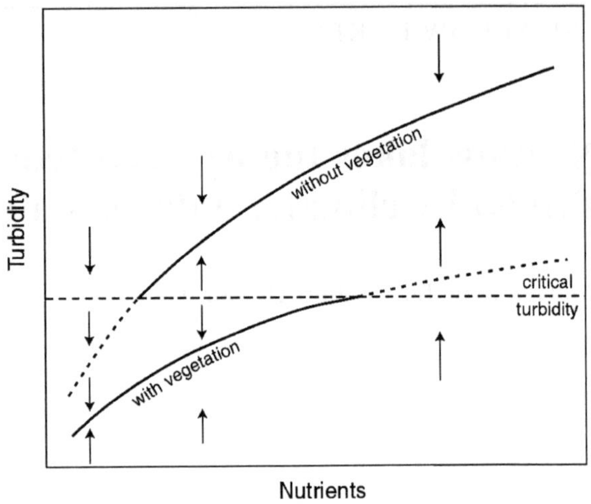

Fig. 1 Alternative equilibrium turbidities caused by disappearance of submerged vegetation when a critical turbidity is exceeded (see text for explanation). The arrows indicate the direction of change when the system is not in one of the two alternative stable states (from: Scheffer et al., 1993)

The alternative stable states theory of shallow lakes

As a starting point let us resume the central ideas behind the original alternative stable states theory for shallow lakes (Scheffer et al., 1993). This theory was inspired by observations that lakes tended to shift rather abruptly between a clear and a turbid state, and that once lakes turned turbid, they subsequently resisted restoration efforts (Phillips et al., 1978; Meijer et al., 1989). Research revealed that both the turbid and the clear state were stabilized by a number of mechanisms (Moss, 1988; Jeppesen, 1998; Jeppesen et al., 1998; Scheffer, 1998). For instance, in the turbid state, fish promote phytoplankton growth by recycling nutrients and controlling the development of zooplankton that could otherwise help clear the water of phytoplankton. Also fish and waves may stir up sediments in shallow lakes with little or no vegetation. In this situation, light limitation and disturbance of the sediments make it difficult for submerged plants to settle. On the other hand, once submerged plants are abundant, they can greatly reduce turbidity by a suit of mechanisms resulting in control of excessive phytoplankton development and prevention of wave resuspension of sediments.

To illustrate how this mechanism may lead to alternative states, a simple graphical model suffices (Scheffer et al., 1993) (Fig. 1). The model is based on three assumptions:

- Turbidity increases with the nutrient level due to increased phytoplankton growth.
- Vegetation reduces turbidity.
- Vegetation disappears when a critical turbidity is exceeded.

In view of the first two assumptions, equilibrium turbidity can be drawn as two different functions of the nutrient level one for a macrophyte dominated, and one for an unvegetated situation. Above a critical turbidity, macrophytes will be absent, in which case the upper equilibrium line is the relevant one and below this turbidity the lower equilibrium curve applies. The emerging picture shows that over a range of intermediate nutrient levels two alternative equilibria exist: one with macrophytes, and a more turbid one without vegetation. At lower nutrient levels, only the macrophyte-dominated equilibrium exists, whereas at the highest nutrient levels, there is only one equilibrium without vegetation.

This simple model allows an intuitive feeling for why the system can respond to environmental change showing hysteresis and catastrophic transitions. Gradual enrichment starting from low nutrient levels will cause the system to proceed along the lower equilibrium curve until the critical turbidity is reached at which macrophytes disappear. Now, a jump to a more turbid equilibrium at the upper part of the curve occurs. In order to restore the macrophyte dominated state by means of nutrient management, the nutrient level must be lowered to a value where phytoplankton growth is limited enough by nutrients alone to reach the critical turbidity for macrophytes again. At the extremes of the range of nutrient levels over which alternative stable states exist, either of the equilibrium lines approaches the critical turbidity that represents the breakpoint of the system. This corresponds to a decrease of stability. Near the edges, a small perturbation is enough to bring the system over the critical line and to cause a switch to the other equilibrium. Although this model illustrates how alternative stable states may arise, it is clearly a quite crude simplification of reality. In the following sections, we discuss some essential elaborations on this archetype model of alternative stable states in shallow lakes.

Stability in the face of variation in space and time

Clearly the image of alternative stable states in the simple model outlined above is a caricature. Seasonality is one complication (Scheffer et al., 1997a; Scheffer et al., 2001a), however as most systems have a "memory" in the winter time, seasonality will not necessarily change the behaviour of a model with alternative states if year-to-year changes are considered (Van Nes et al., 2002). On the other hand, chaotic population dynamics (Scheffer, 1991; Huisman & Weissing, 1999) and the ever changing weather conditions also put the idea of a stable state in a different light (Scheffer & Carpenter, 2003). It has been argued that in the face of such turmoil it is better to speak of alternative *regimes* than alternative *stable states* (Scheffer & Carpenter, 2003). Another fundamental problem is that it is sometimes not so easy to demarcate the borders of a system. What if some parts of a lake are shallow and others deep? And what if clear and turbid lakes are connected?

Spatial heterogeneity

Perhaps the most obvious simplification in the basic graphical model (Fig. 1) is the assumption that all submerged plants disappear at a single critical turbidity. In practice, plants are less affected by turbidity at shallower sites. As a result, the shift from a vegetated state to the turbid state may occur less abrupt than predicted by the basic model. Vegetation will disappear more gradually, because the vegetation will be lost first from the deeper parts of the lake. Consequently hysteresis may be less pronounced (and can even disappear) if the depth profile of a lake is less flat (Van Nes et al., 2002). More in general, it has been shown for a range of ecological models that spatial heterogeneity of the environment tends to reduce the chance that large scale shifts between alternative stable states are found (Van Nes & Scheffer, 2005). This seems intuitively straightforward as the shifts in the local states do not happen at the same moment, simply because different sites have different local properties. The numerous asynchronous shifts of different sites may be sharp individually, but average out over a larger region to give a smooth large-scale response rather than a drastic catastrophic shift. However, there is an interesting caveat in this reasoning. It all depends on the rate of 'mixing'. For instance, if water circulates through a lake rapidly, one expects the clearing effects of submerged plants to contribute to the overall change of turbidity in the lake, but not to result in a locally higher clarity. Analysis of different models shows that such mixing can counteract the smoothing effect of environmental heterogeneity, and cause the system to show large scale synchronous sharp shifts again (Van Nes & Scheffer, 2005).

Data from some shallow lakes suggests that in practice, mixing can be surprisingly slow relative to the local clearing process in vegetation fields. As a result, alternative stable states can coexist side by side, despite an open connection. For instance, in the large (3,300 ha) and shallow (1.4 m average) Lake Veluwe, reduction of the nutrient load and increased fishing pressure triggered the recovery of marked fields with charophytes and locally very clear water that contrasted sharply with the turbid water in the rest of the lake (Scheffer et al., 1994; Van den Berg et al., 1998b; Lammens et al., 2004). The fields and the associated clear water patches expanded over the years until eventually the entire lake shifted to the clear state (Van den Berg et al., 1999).

Non-equilibrium dynamics

Although the original theory of alternative stable states in shallow lakes was used to understand the effects of occasional disturbances such as fish removal or extremes in the water level, it did not really address situations in which disturbances are the rule rather than the exception. Such high disturbance regimes are found for instance in cold climates where winter fish kills are common, or in places where desiccation of shallow lakes in summer is a frequent event (Scheffer et al., 2006). Clearly, such lakes will be far from any equilibrium most of the time. In fact, it has been argued that this may be true for ecological systems in general (Hastings, 2004). Although the image of 'alternative stable states' is clearly too simple to catch the essence of the dynamics of lakes under such severe disturbance regimes, it may be expanded to help seeing the big picture behind the dynamics.

In models, it can be shown that the duration of transients can be boosted if there are places or trajectories in the state space where the system tends to 'hang around' even if they do not represent attractors. Examples of such 'hang-around places' are saddle points, remainders of strange attractors and remainders of equilibria after saddle-node bifurcations (Strogatz, 1994). Probably, such slow transient dynamics can be important in nature too. Indeed, the tendency of some shallow lakes to stick to an unstable clear state for years, before sliding into the stable turbid state may be an example of such behavior. Lake Zwemlust, for example, stayed clear for about seven years upon fish removal, despite a very high nutrient concentration (Van Donk & Gulati, 1995). Similarly, numerous Dutch floodplain lakes are in a clear state most of the time, despite high nutrient levels (Van Geest et al., in press). The idea here is that occasional droughts

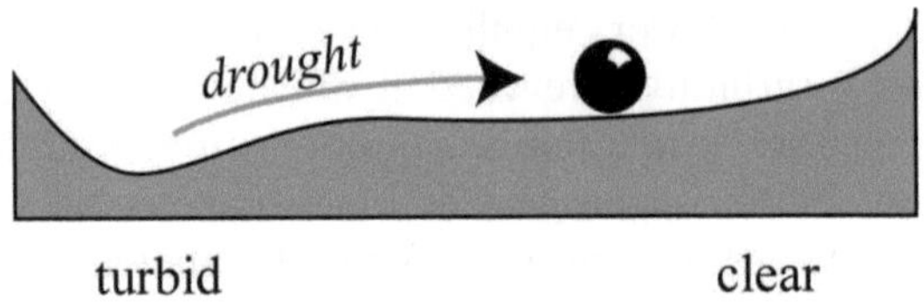

Fig. 2 Hypothesized dynamics of shallow floodplain lakes. Occasional droughts push the system to an unstable clear state that represents the ghost of a stable state. Since droughts happen every now and then, and the dynamics around such a ghost are very slow, the lake may be in a transient state most of the time. Note that the way to interpret such stability landscapes is that the movement of the ball is always damped. Imagine that the ball rolls through a heavily viscous fluid

result in almost complete desiccation of the lakes. This kills most of the fish and pushes the system into a clear situation, which is not stable, but the dynamics away from it are very slow (Fig. 2). Such transient state which may last for some years is called the 'ghost' of a stable state (Strogatz, 1994).

Cyclic shifts between alternative states

Although shifts between alternative states usually occur irregularly (Mitchell, 1989; Blindow, 1992; Blindow et al., 1993; Blindow et al., 2002), some lakes show remarkably regular oscillations between submerged plants and a turbid state. The best documented examples that we are aware of are the English lake Alderfen Broad (Moss et al., 1990; Perrow et al., 1994) and the Dutch Lake Botshol (Simons et al., 1994; Rip et al., 2005). Both lakes cycle between the alternative states with a period of approximately 7 years. Model analyses suggest that such cyclic behavior may arise under particular conditions from an internal 'time bomb' mechanism (Van Nes et al., in press), due to a slow negative effect of vegetation on the nutrient state of the lake (Fig. 3). If during the period of macrophyte dominance phosphorus retention is high and dead organic material accumulates on the sediment, decomposition may eventually cause anoxic conditions at the sediment surface, allowing phosphorus release that may promote a sufficient increase in phytoplankton to cause a decline of the submerged macrophytes. If subsequently the turbid state

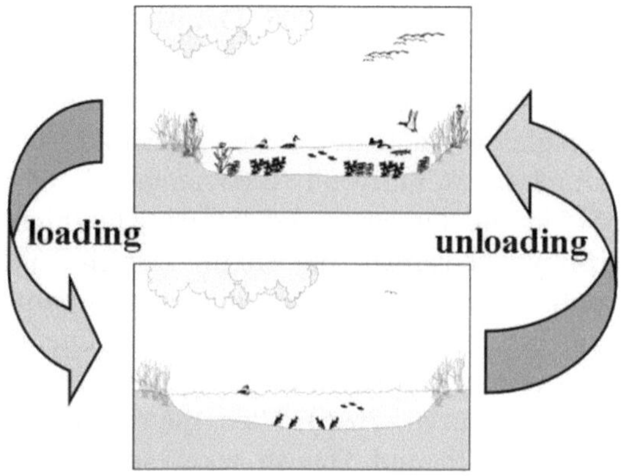

Fig. 3 Shallow lakes may under some conditions shift between a vegetated and a turbid state in a regular cyclic way due to intrinsic differences in accumulation of phosphorus and organic matter in the two states (see text for explanation)

allows sufficient decomposition of organic material and/or loss of phosphorus from the sediments, the lake may shift back to a clear macrophyte dominated state, and so forth. The model analyses suggest that such switching behavior can only occur in special cases, where the critical nutrient levels for both shifts are relatively close together and the negative effect of vegetation on the nutrient state is sufficiently strong (Van Nes et al., in press).

Factors affecting the critical nutrient levels

As eutrophication problems have been an important incentive for much of the work on shallow lakes, one of the central questions has always been what the critical nutrient level would be for maintaining a clear water state. Although this seems a straightforward question, one encounters some fundamental problems if one attempts to find the answer. Firstly, it is not so easy to assess the 'nutrient level' of a lake. For instance, although phosphorus is clearly important, nitrogen may play an important role in some cases too. Also, when it comes to phosphorus, much of what is available to the organisms in the long run is stored in the sediment (Sas, 1989; Carpenter, 2005). How much of that is reflected in total-P concentrations in the water column varies strongly with the presence of macrophytes and other biological factors. Thus, although we often use total-P as an indicator of the nutrient status of a lake, there are some caveats to that approach.

Perhaps more importantly, there are good reasons to believe that there is not one single critical nutrient level for maintaining clear water. Different lakes may vary widely in the nutrient level they can tolerate before they flip to a turbid condition. In principle, the list of factors that may influence the probability that a lake turns to the turbid state is almost endless. Here, we just highlight three important factors: lake depth, lake size and climate.

Lake depth

Besides nutrients, the most obvious factor that affects the probability of a lake to be dominated by submerged vegetation is its depth. In shallower water light conditions at the bottom are better, given that the optical properties of the water are the same. Also, plants can more easily grow to the surface layer of the water where they are relatively free of the shading effects of phytoplankton and suspended solids (Scheffer, 1998). This effect of water level can be easily linked to the simple graphical model of alternative states (Fig. 1) if one notes that the critical turbidity for plants depends on the water level. For instance, deeper water will move the critical turbidity for plants (the horizontal dashed line) downwards causing the critical nutrient level for both the forward switch and the backward switch to move to lower values. From an analysis of a more complex vegetation model this pattern is confirmed: the range of nutrient values with alternative states is much smaller in deeper water and eventually hardly any vegetation growth will be possible (Van Nes et al., 2002).

Importantly, the depth of a shallow lake can be subject to strong fluctuations resulting from climatic variability. Numerous observations highlight the importance of water level for shifts between a vegetation dominated state and a turbid state in shallow lakes. High water levels may lead to loss of submerged macrophytes, invoking a dramatic shift to a turbid state (Wallsten & Forsgren, 1989; Engel & Nichols, 1994). Also very low water level may have more complicated pronounced effects. For instance, desiccation and freezing of the lake bottom may in some cases damage the vegetation sufficiently to push a lake to a turbid state (Blindow et al., 1993). However, an opposite effect of low-water extremes is probably more common. The risk of fish kills due to anoxic conditions in summer or winter may become higher if the water level is low, leading to clear water conditions that promote subsequent vegetation dominance (Van Geest et al., 2003).

Lake size

A related but somehow more surprising factor that may have a large impact on the chances of a lake to be in a vegetation dominated clear state is lake size. Although lake size tends to be correlated to lake depth, size seems to have a considerable effect by itself too (Søndergaard et al., 2005). Small lakes appear to have a higher chance to be in a vegetated clear state. Although various factors may explain this correlation (Van Geest et al., 2003), suppression of fish in smaller lakes may be a common factor. Fish are rare in small, isolated lakes, probably as a result of a higher likelihood of fish kills combined with their low colonizing ability (Tonn & Magnuson, 1982; Dahl, 1990; Magnuson et al., 1998; Hershey et al., 1999). Two recent studies exemplify this.

An analysis of 215 shallow lakes situated in the Dutch floodplain of the lower River Rhine (Van Geest et al., 2003) revealed particularly clear relationships between lake size, the abundance of fish and the presence of vegetation. All of these lakes are inundated about once a year by the eutrophic river; so nutrient loading varies relatively little among the lakes. In accordance with the theory, most of these lakes were either nearly devoid of vegetation or in a contrasting state with a rich plant community, but the likelihood of the richly vegetated state was higher in smaller lakes (other factors such as depth being equal) (Van Geest et al., 2003). These small vegetated lakes also supported low densities of benthivorous bream (*Abramis brama*) (Grift et al., 2001), the key fish species promoting the opposite poorly-vegetated, turbid state in shallow Dutch lakes (Scheffer, 1998).

Similarly, data from 796 Danish lakes and ponds (Søndergaard et al., 2005) showed overall low fish biomass below 1 ha, shifting to an often much higher biomass above this threshold. As in the Dutch lakes, macrophyte coverage was often highest in smaller lakes and ponds. This higher macrophyte coverage in small Danish lakes occurred despite a generally higher phosphorus content in the lake water related to a generally higher share of cultivated fields in the catchment (Søndergaard et al., 2005). Thus, the results from both the Dutch and Danish lakes suggest that small lakes are more likely to be fishless, which increases the likelihood of the clear-water state, even at quite high nutrient concentrations.

The higher chance for small water-bodies to be in a vegetated state also implies an interesting twist to the pattern of biodiversity that we would expect from classical 'island theory' (MacArthur & Wilson, 1967). Species richness is predicted to be lower on small isolated (habitat-) islands than on similar sized pieces of mainland because small populations are more prone to extinction due to stochastic events, and because recolonization

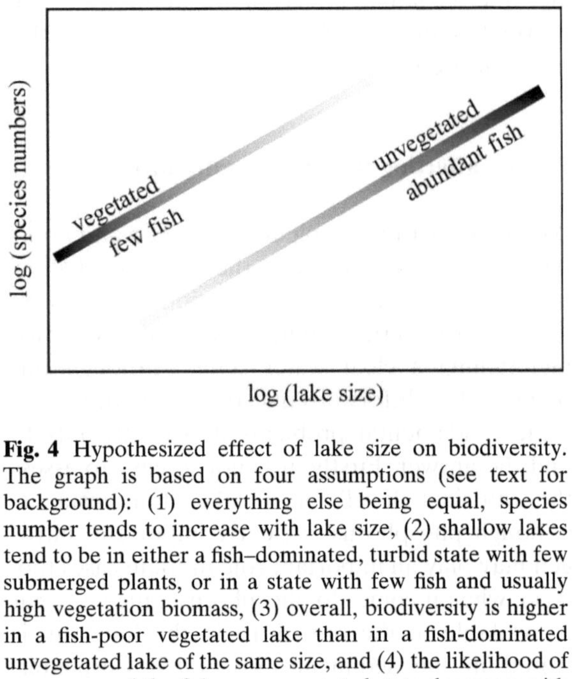

Fig. 4 Hypothesized effect of lake size on biodiversity. The graph is based on four assumptions (see text for background): (1) everything else being equal, species number tends to increase with lake size, (2) shallow lakes tend to be in either a fish–dominated, turbid state with few submerged plants, or in a state with few fish and usually high vegetation biomass, (3) overall, biodiversity is higher in a fish-poor vegetated lake than in a fish-dominated unvegetated lake of the same size, and (4) the likelihood of occurrence of the fish-poor, vegetated state decreases with lake size. The probability that a lake will be in either of two alternative states is represented by the darkness of the line (from: Scheffer et al., 2006)

rates are lower in isolated locations. However, in shallow lakes the shift to a vegetated state may distort this pattern (Scheffer et al., 2006) (Fig. 4). Small lakes are more likely in a vegetation-dominated state and such state implies an increase in food and habitat structures and consequently a higher biodiversity of birds invertebrates and amphibians (Jeppesen et al., 1997; Scheffer, 1998). Indeed, a recent survey comparing species diversity of macrophytes and invertebrates in ponds (65), lakes (4), ditches (70 km), streams (28 km) and rivers (17 km) in a British lowland landscape (Williams et al., 2004) shows that ponds exhibited considerably higher taxonomic richness, with more unique species and more scarce species than any of the other water body types.

Climate

An important unresolved question is what would be the impact of a warmer climate on the chances that shallow lakes fall in a turbid state. There are good reasons to expect that climate will have a large impact on shallow lakes. We know, for instance, that warmer conditions have a large effect on the 'trophic cascade' from fish to phytoplankton. In temperate conditions most fish reproduce only once a year, leaving a period in spring in which there are few small (juvenile) fish, allowing large zooplankton to become abundant and filter the water clear of phytoplankton (Sommer, 1986). By contrast, top-down control of zooplankton by fish is very strong all year round in warmer lakes at low latitudes because fish are abundant and reproduce continuously in such (sub)tropical lakes (Dumont, 1994). The difference in food web structure implies that biomanipulation as used to shift temperate lakes to a clear state, seems less easy to apply to (sub)tropical lakes (Scasso et al., 2001; Jeppesen et al., 2005).

Although climatic warming will not turn most temperate lake communities into subtropical ones, there are several other reasons to suspect that moderate warming might promote the turbid state in temperate lakes (Jeppesen et al., 2003). On the other hand, numerous field studies in temperate lakes suggest positive effects of warming on aquatic vegetation performance (Grace and Tilly, 1976; Best, 1987; Nelson, 1997; Scheffer et al., 1992; Rooney and Kalff, 2000) which would push the other way.

Importantly, climatic warming will also affect lake ecosystems through changes in hydrology and nutrient load to lakes, but findings so-far are rather contradictory (Weyhenmeyer, 2001; George, 2000; Monteith et al., 2000; Straile, 2002). An interesting aspect of climatic impact is that systems with alternative stable states even brief climatic extremes may induce a shift to another state in which the system subsequently remains for a long time (Scheffer et al., 2001b). Indeed, there are indications that shallow lakes may be affected by climatic extremes in this way (Scheffer, 1998). For instance, heavy storms have induced a shift to a permanent turbid state by destroying vegetation (McKinnon and Mitchell, 1994; Schelske and Brezonik, 1992). However, perhaps the most important potential impact of climatic extremes is can be through their effect on water level fluctuations to which these lakes are very sensitive as mentioned before. Clearly, the issue of climatic effects is far from resolved, but numerous studies are underway that will certainly resolve some key-issues over the coming years.

Other alternative stable states in aquatic systems

The previous section shows that it has become increasingly clear that many factors may play a role in determining whether a shallow lake is in a turbid state or in a clear-water state dominated by submerged plants. However, it is becoming clear that shallow lakes may be in more than just those two alternative stable states. One potential source of alternative attractors in freshwater as well as marine systems is the interaction of competitors and predators in size-structured fish communities, often analysed by means of individual-based models (DeAngelis & Gross, 1992; De Roos & Persson, 2002). For instance, overexploitation of a predatory fish such as cod may be irreversible if a good stock of adults is needed to crop down the potential competitors/predators of their own offspring (Walters & Kitchell, 2001). More apparent in shallow lakes is the phenomenon that different primary producers may dominate the system. In addition to dominance by *submerged angiosperms*, versus the turbid state dominated by *a mixed phytoplankton community of green algae and other groups*, it has now been suggested that dominance by *free-floating plants*, *charophytes*, and *filamentous cyanobacteria* also appear to have the character of a self-stabilizing condition in some lakes. Here we briefly point at some recent work on stability of these states.

Floating plant dominance

Although a turbid state with high phytoplankton biomass is the typical eutrophication problem in temperate lakes, invasion by mats of free-floating plants is among the most important threats to the functioning and biodiversity of freshwater ecosystems ranging from temperate ponds and ditches (Janse & Van Puijenbroek, 1998) to large tropical lakes (Gopal, 1987; Mehra et al., 1999). Dark, anoxic conditions under a thick floating plant cover leave little opportunity for animal or plant life. In tropical lakes large floating species and in particular the water hyacinth (*Eichhornia crassipes*), have dramatic negative impacts on fisheries and navigation (Gopal, 1987; Mehra et al., 1999). Not surprisingly, resolving floating plant problems has a high priority in many warm regions of the world, and is also a focus of quality management of many smaller water bodies in temperate regions. For example, in the extensive system of ditches and canals in the Netherlands, duckweeds are considered the main problem associated to eutrophication (Janse & Van Puijenbroek, 1998).

It has been demonstrated that a shift to floating plant dominance in shallow water ecosystems is a highly non-linear phenomenon (Scheffer et al., 2003). Just like the shift to a turbid state in temperate lakes, the shift to floating plant dominance is difficult to reverse and can happen when a critical threshold level of nutrients is passed. The explanation is in the way the floating plants compete with submerged plants. Floating plants have primacy in competition for light but need high nutrient concentrations (Portielje & Roijackers, 1995). By contrast, rooted submerged macrophytes are susceptible to shading, but less dependent on nutrients in the water column as they may take up a large part of their nutrients from the sediment (Hutchinson, 1975; Chambers et al., 1989). Still submerged plants can also use their shoots effectively for nutrient uptake from the water column (Sculthorpe, 1967; Robach et al., 1996) and by various mechanisms reduce nitrogen concentrations in the water column to below detection levels (Goulder, 1969; Van Donk et al., 1993). This interaction may result in two alternative stable states: a floating plant dominated state in which invasion by submerged plants is prevented by shading, and a situation dominated by submerged plants in which invasion by free-floating plants is prevented by reduced nutrient availability (Scheffer et al., 2003). This bistability makes reversal of floating plant invasion of a lake often difficult. On the other hand it implies that if nutrient levels have been sufficiently reduced, a one-time removal of floating plants might tip the balance to an alternative stable state dominated by submerged plants.

Charophyte dominance

In addition to setting the free-floating plants apart, a distinction into at least two groups may be made in the submerged macrophytes. Although there are many species of submerged macrophytes, one particular group differs rather fundamentally from the rest. Charophytes (stoneworts) are not only distinct taxonomically (they are algae rather than Angiosperms), but also ecologically. In some lakes distinct shifts from pondweeds to an almost complete monoculture of charophytes has been observed (Hargeby et al., 1994; Van den Berg et al., 1999). This suggests that co-existence of these two groups may be unstable, and their competition might suggest in alternative stable states of dominance by either of the groups.

To test this idea Van Nes, Van den Berg and others (2003) have looked more deeply into the competition between two particular representatives of these groups: *Chara aspera* Deth. ex Willd. and pondweed *Potamogeton pectinatus* L.. Just as in the Swedish Lake Krankesjön (Hargeby et al., 1994), the shift from a turbid to a clear state in Lake Veluwe were this research group worked, proceeded really through an intermediate phase dominated by pondweeds to a really clear state dominated by stoneworts (Van den Berg et al., 1999). Looking into the mechanistic details of competition, it appears that there is an interesting asymmetry much like the one discussed for floating plants versus submerged plants. In the case of pondweeds and *Chara*, the tall, canopy-forming pondweed (*P. pectinatus*) is simply on top of the short charophyte (*C. aspera*), and therefore pondweed is the better competitors for light (Van den Berg et al., 1998a). However, it appears that the stonewort is strongly depleting bicarbonate, and can at the same time survive at lower bicarbonate concentrations (Van den Berg, 1999). Thus there is a positive feedback in the development of the stonewort as it drives the system towards carbon limitation, suppressing the pondweed, and thereby creating better light conditions for itself (Van Nes et al., 2003). A detailed individual based competition model suggests that this may plausibly lead to alternative stable states in the competition between these two submerged macrophytes (Van Nes et al., 2003).

Cyanobacterial dominance

The previous sections illustrate that it is too simple to consider aquatic macrophytes, or even submerged macrophytes as a single group. The same is true for phytoplankton. Different species can rise to dominance, and despite intensive research and growing insight in the factors that drive succession we are still far from able to predict which species will dominate when. Perhaps the best-studied group are blue-green algae. Of course these are really bacteria (cyanobacteria) rather than algae, but as planktonic autotrophs they occupy the same niche. Species in this group differ widely, and it is impossible to treat them as a single ecological entity. However, some groups have been studied more than others, as they are often a nuisance. In shallow lakes, filamentous cyanobacteria of the *Oscillatoria* group can be dominant all year round. These cyanobacteria are rather shade tolerant, which explains why they enter especially when the water has become sufficiently turbid (Mur et al., 1993). Interestingly, they can also intensify the shady conditions once they are present. This is because with the same amount of phosphorus they can build biomass that causes a relatively high light attenuation (Mur et al., 1977). As a result, they can stabilize their dominance once they are there, creating an alternative stable state in the phytoplankton community (Scheffer et al., 1997b). On the other hand their maximum growth is less than of most other algae. Therefore in less turbid conditions groups such as green algae have a competitive advantage and can become dominant over blue-green algae.

Conclusion

In conclusion, while the essence of the original theory of alternative stable states in shallow lakes remains quite valid to explain major patterns of change, numerous expansions of the basic idea have been suggested. It has been shown that there is no single critical nutrient level for maintaining a clear state, as factors such as lake size, depth and climate affect the threshold. Moreover, the image of just two contrasting states has been elaborated. While the contrast between submerged macrophytes and a turbid phytoplankton dominated state remains perhaps the most spectacular, many other less conspicuous shifts between alternative states seem to occur in reality. Therefore, the change of the biological communities along a gradient of eutrophication may really be seen as a continuum in which gradual species replacements are interrupted at critical points by moderate or more dramatic shifts to a contrasting community state. Spatial heterogeneity and the interplay of internal cyclic or chaotic dynamics with the effects of fluctuating climatic conditions and hydrology further complicate the matter to the point that it may all seem hopelessly complex again. The challenge, as always in ecology, is to keep seeing the big patterns without being ignorant about the details. In this sense, we feel that shallow lake ecology is actually in quite good shape. Many details can be explained, whereas at the same time, the mechanisms that cause major patterns such as dominance by submerged plants, cyanobacteria or floating plants have been identified. On the other hand, we are still uncertain about many processes and we may rightly feel that we have just seen the tip of the iceberg when it comes to really understanding the full complexity of shallow lake ecosystems. Nonetheless, shallow lakes are really well understood if we compare our insights to what is known about oceans, forests and other vast ecosystems of the world. Indeed, shallow lakes have served as an example to illustrate stability aspects that may be important in many other systems where they cannot be studied so easily (Gunderson & Holling, 2001; Scheffer et al., 2001b; Folke et al., 2004). We feel that shallow lakes will proceed to play this important role in ecology, considering the exiting new developments in shallow lake science.

References

Best, E. P. H., 1987. The submerged macrophytes in Lake Maarsseveen I, the Netherlands: changes in species composition and biomass over a six-year period. Hydrobiology Bulletin 21: 55–60.

Blindow, I., 1992. Long- and short-term dynamics of submerged macrophytes in two shallow eutrophic lakes. Freshwater Biology 28: 15–27.

Blindow, I., A. Hargeby & G. Andersson, 2002. Seasonal changes of mechanisms maintaining clear water in a shallow lake with abundant *Chara* vegetation. Aquatic Botany 72: 315–334.

Blindow, I., G. Andersson, A. Hargeby & S. Johansson, 1993. Long-term pattern of alternative stable states in two shallow eutrophic lakes. Freshwater Biology 30: 159–167.

Carpenter, S. R., 2005. Eutrophication of aquatic ecosystems: bistability and soil phosphorus. Proceedings of the National Academy of Science of the United States of America 102: 10002–10005.

Chambers, P. A., E. E. Prepas, M. L. Bothwell & H. R. Hamilton, 1989. Roots versus shoots in nutrient uptake by aquatic macrophytes in flowing waters. Candian Journal of Fisheries and Aquatic Sciences 46: 435–439.

Dahl, T. E., 1990, Wetland losses in the United States, 1780's to 1980's. U.S. Department of the Interior, Fish and Wildlife Service, Washington DC.

De Roos, A. M. & L. Persson, 2002. Size-dependent life-history traits promote catastrophic collapses of top predators. Proceedings of the National Academy of Science of the United States of America 99: 12907–12912.

DeAngelis, D. L., & L. J. Gross, 1992, Individual-based models and approaches in ecology. Chapman and Hall, New York, 1–525.

Dumont, H. J., 1994. On the diversity of the cladocera in the tropics. Hydrobiologia 272: 27–38.

Engel, S. & S. A. Nichols, 1994. Aquatic macrophyte growth in a turbid windswept lake. Journal of Freshwater Ecology 9: 97–109.

Folke, C., S. Carpenter, B. Walker, M. Scheffer, T. Elmqvist, L. Gunderson & C. S. Holling, 2004. Regime shifts, resilience, and biodiversity in ecosystem management. Annual Review of Ecology Evolution and Systematics 35: 557–581.

George, D. G., 2000. The impact of regional-scale changes in the weather on the long-term dynamics of *Eudiaptomus* and *Daphnia* in Esthwaite Water, Cumbria. Freshwater Biology 45: 111–121.

Gopal, B., 1987, Water hyacinth. Elsevier, New York.

Goulder, R., 1969. Interactions between the rates of production of a freshwater macrophyte and phytoplankton in a pond. Oikos 20: 300–309.

Grace, J. B. & L. J. Tilly, 1976. Distribution and abundance of submerged macrophytes, including *Myriophyllum spicatum* L. (Angiospermae), in a cooling reservoir. Archiv für Hydrobiologie 77: 475–487.

Grift, R. E., A. D. Buijse, W. L. T. Van Densen & J. G. P. Klein Breteler, 2001. Restoration of the river-floodplain interaction: benefits for the fish community in the River Rhine. Archiv für Hydrobiologie 135: 173–185.

Gunderson, L. & C. S. Holling, 2001. Panarchy: understanding transformations in human and natural systems. Island Press, Washington.

Hargeby, A., G. Andersson, I. Blindow & S. Johansson, 1994. Trophic web structure in a shallow eutrophic lake during a dominance shift from phytoplankton to submerged macrophytes. Hydrobiologia 280: 83–90.

Hastings, A., 2004. Transients: the key to long-term ecological understanding? Trends in Ecology & Evolution 19: 39–45.

Hershey, A. E., G. M. Gettel, M. E. McDonald, M. C. Miller, H. Mooers, W. J. O'Brien, J. Pastor, C. Richards & J. A. Schuldt, 1999. A geomorphic-trophic model for landscape control of Arctic lake food webs. Bioscience 49: 887–897.

Holling, C. S., 1973. Resilience and stability of ecological systems. Annual Review of Ecology and Systematics 4: 1–23.

Huisman, J. & F. J. Weissing, 1999. Biodiversity of plankton by species oscillations and chaos. Nature 402: 407–410.

Hutchinson, G. E., 1975, A treatise on limnology. Vol. III, limnological botany. John Wiley & Sons, New York.

Janse, J. H. & P. J. T. M. Van Puijenbroek, 1998. Effects of eutrophication in drainage ditches. Environmental Pollution 102: 547–552.

Jeppesen, E., 1998. The ecology of shallow lakes – Trophic Interactions in the Pelagial. University of Copenhagen, Silkeborg, 358.

Jeppesen, E., M. Søndergaard & J. P. Jensen, 2003. Climate warming and regime shifts in lake food webs - some comments. Limnology and Oceanography 48: 1346–1349.

Jeppesen, E., M. Søndergaard, M. Søndergaard, & K. Kristoffersen, 1998, Structuring role of submerged macrophytes in lakes. Springer-Verlag, New York.

Jeppesen, E., J. P. Jensen, M. Søndergaard, T. Lauridsen, L. J. Pedersen & L. Jensen, 1997. Top-down control in freshwater lakes: the role of nutrient state, submerged macrophytes and water depth. Hydrobiologia 342: 151–164.

Jeppesen, E., M. Søndergaard, N. Mazzeo, M. Meerhoff, C. Branco, V. Huszar & F. Scasso, 2005. Lake restoration and biomanipulation in temperate lakes: relevance for subtropical and tropical lakes. In Reddy, M. V. (ed.), Restoration and Management of Tropical Eutrophic Lakes. Science Publishers.

Lammens, E. H. R. R., M. L. Meijer, E. H. Van Nes & M. S. Van den Berg, 2004. Effects of commercial fishery on the bream population and the expansion of *Chara aspera* in Lake Veluwe. Ecological Modelling 177: 233–244.

Lewontin, R. C., 1969. The meaning of stability. Diversity and stability in ecological systems: report of a symposium held May 26–28. 13–24.

MacArthur, R. H., & E. O. Wilson, 1967, The theory of island biogeography. Princeton University Press, Princeton, 203.

Magnuson, J. J., W. M. Tonn, A. Banerjee, J. Toivonen, O. Sanchez & M. Rask, 1998. Isolation vs. extinction in the assembly of fishes in small northern lakes. Ecology 79: 2941–2956.

May, R. M., 1977. Thresholds and breakpoints in ecosystems with a multiplicity of stable states. Nature 269: 471–477.

McKinnon, S. L. & S. F. Mitchell, 1994. Eutrophication and black swan (*Cygnus atratus* Latham) populations: tests of two simple relationships. Hydrobiologia 279–280: 163–170.

Mehra, A., M. E. Farago, D. K. Banerjee & K. B. Cordes, 1999. The water hyacinth: an environmental friend or pest? A review. Resource Environmental Biotechnology 2: 255–281.

Meijer, M. L., A. J. Raat & R. W. Doef, 1989. Restoration by biomanipulation of Lake Bleiswijkse Zoom the Netherlands first results. Hydrobiology Bulletin 23: 49–58.

Mitchell, S. F., 1989. Primary production in a shallow eutrophic lake dominated alternately by phytoplankton and by submerged macrophytes. Aquatic Botany 33: 101–110.

Monteith, D. T., C. D. Evans & B. Reynolds, 2000. Are temporal variations in the nitrate content of UK upland freshwaters linked to the North Atlantic Oscillation? Hydrological Processes 14: 1745–1749.

Moss, B., 1988, Ecology of fresh waters, 2nd edn. Man & Medium. Blackwell Scientific, Oxford, 1–400.

Moss, B., J. Stansfield & K. Irvine, 1990. Problems in the restoration of a hypertrophic lake by diversion of a nutrient-rich inflow. Verhandlungen Internationale Vereinigung Theoretisch Angewandte Limnologie 24: 568–572.

Mur, L. R., H. J. Gons & L. Van Liere, 1977. Some experiments on competition between green-algae and blue- green bacteria in light-limited environments. FEMS Microbiology Letters 1: 335–338.

Mur, L. R., H. Schreurs & P. Visser, 1993. How to control undesirable cyanobacterial dominance. In Giussani, G. & C. Callieri (eds), Proceedings of the 5th international conference on the conservation and management of lakes, Stresa, Italy, 565–569.

Nelson, T. A., 1997. Interannual variance in a subtidal eelgrass community. Aquatic Botany 56: 245–252.

Perrow, M. R., B. Moss & J. Stansfield, 1994. Trophic interactions in a shallow lake following a reduction in nutrient loading – A long-term study. Hydrobiologia 276: 43–52.

Phillips, G. L., D. Eminson & B. Moss, 1978. A mechanism to account for macrophyte decline in progressively eutrophicated fresh waters. Aquatic Botany 4: 103–126.

Portielje, R. & R. M. M. Roijackers, 1995. Primary succession of aquatic macrophytes in experimental ditches in relation to nutrient input. Aquatic Botany 50: 127–140.

Rip, W. J., M. Ouboter, B. Beltman & E. H. Van Nes, 2005. Oscillation of a shallow lake ecosystem upon reduction in external phosphorus load. Archiv fur Hydrobiologie 164: 387–409.

Robach, F., S. Merlin, T. Rolland & M. Tremolieres, 1996. Ecophysiological approach of water quality bioindicating using aquatic plant materials: the role of phosphorus. Ecologie-Brunoy 27: 203–214.

Rooney, N. & J. Kalff, 2000. Inter-annual variation in submerged macrophyte community biomass and distribution: the influence of temperature and lake morphometry. Aquatic Botany 68: 321–335.

Sas, H., 1989, Lake restoration by reduction of nutrient loading: expectations, experiences, extrapolations. Academia Verlag Richarz, St. Augustin, 1–497.

Scasso, F., N. Mazzeo, J. Gorga, C. Kruk, G. Lacerot, J. Clemente, D. Fabian & S. Bonilla, 2001. Limnological changes in a sub-tropical shallow hypertrophic lake during its restoration: two years of a whole-lake experiment. Aquatic Conservation-Marine Freshwater Ecosystem 11: 31–44.

Scheffer, M., 1991. Should we expect strange attractors behind plankton dynamics: And if so, should we bother? Journal of Plankton Research 13: 1291–1306.

Scheffer, M., 1998. Ecology of shallow lakes. Chapman and Hall, London, 0–357.

Scheffer, M. & S. R. Carpenter, 2003. Catastrophic regime shifts in ecosystems: linking theory to observation. Trends in Ecology & Evolution 18: 648–656.

Scheffer, M., M. R. De Redelijkheid & F. Noppert, 1992. Distribution and dynamics of submerged vegetation in a chain of shallow eutrophic lakes. Aquatic Botany 42: 199–216.

Scheffer, M., S. Rinaldi, Y. A. Kuznetsov & E. H. Van Nes, 1997a. Seasonal dynamics of *Daphnia* and algae explained as a periodically forced predator-prey system. Oikos 80: 519–532.

Scheffer, M., D. Straile, E. H. Van Nes & H. Hosper, 2001a. Climatic warming causes regime shifts in lake food webs. Limnology and Oceanography 46: 1780–1783.

Scheffer, M., S. H. Hosper, M. L. Meijer, B. Moss & E. Jeppesen, 1993. Alternative equilibria in shallow lakes. Trends in Ecology and Evolution 8: 275–279.

Scheffer, M., S. Rinaldi, A. Gragnani, L. R. Mur & E. H. Van Nes, 1997b. On the dominance of filamentous cyanobacteria in shallow, turbid lakes. Ecology 78: 272–282.

Scheffer, M., S. R. Carpenter, J. A. Foley, C. Folke & B. Walker, 2001. Catastrophic shifts in ecosystems. Nature 413: 591–596.

Scheffer, M., M. S. Van den Berg, A. W. Breukelaar, C. P. M. Breukers, H. Coops, R. W. Doef & M. L. Meijer, 1994. Vegetated areas with clear water in turbid shallow lakes. Aquatic Botany 49: 193–196.

Scheffer, M., S. Szabo, A. Gragnani, E. H. van Nes, S. Rinaldi, N. Kautsky, J. Norberg, R. M. M. Roijackers & R. J. M. Franken, 2003. Floating plant dominance as a stable state. Proceedings of the National Academy of Science of the United States of America 100: 4040–4045.

Scheffer, M., G. J. Van Geest, K. Zimmer, E. Jeppesen, M. Søndergaard, M. G. Butler, M. A. Hanson, S. Declerck & L. De Meester, 2006. Small habitat size and isolation can promote species richness: second-order effects on biodiversity in shallow lakes and ponds. Oikos 112: 227–231.

Schelske, C. L. & P. Brezonik, 1992. Can Lake Apopka be restored? In Maurizi, S. & F. Poillon (eds), Restoration of Aquatic Ecosystems. National Academic Press, Washington D.C.: 393–398.

Sculthorpe, C. D., 1967, The biology of aquatic vascular plants. Edward Arnold Ltd., London.

Simons, J., M. Ohm, R. Daalder, P. Boers & W. Rip, 1994. Restoration of Botshol (the Netherlands) by reduction of external nutrient load -recovery of a characean community, dominated by *Chara connivens*. Hydrobiologia 276: 243–253.

Sommer, U., 1986. The periodicity of phytoplankton in Lake Constance (Bodensee) in comparison to other deep lakes of central Europe. Hydrobiologia 138: 1–8.

Søndergaard, M., E. Jeppesen & J. P. Jensen, 2005. Pond or lake: does it make any difference? Archiv Fur Hydrobiologie 162: 143–165.

Straile, D., 2002. North Atlantic Oscillation synchronizes food-web interactions in central European lakes. Proceedings of the Royal Society of Edinburgh. Section B (Biological Sciences) 269: 391–395.

Strogatz, S. H., 1994, Nonlinear dynamics and chaos – with applications to physics, biology, chemistry and engineering. Addison-Wesley Publishing Company, Reading, 1–498.

Tonn, W. M. & J. J. Magnuson, 1982. Patterns in the species composition and richness of fish assemblages in northern Wisconsin Lakes. Ecology 63: 1149–1166.

Van den Berg, M. S., 1999. A comparative study of the use of inorganic carbon resources by Chara aspera and Potamogeton pectinatus. In Van den Berg, M. S. (ed.), Charophyte recolonization in shallow lakes - processes, ecological effects and implications for lake management. Thesis Vrije Universiteit Amsterdam, 57–67.

Van den Berg, M. S., H. Coops, J. Simons & A. De Keizer, 1998a. Competition between *Chara aspera* and *Potamogeton pectinatus* as a function of temperature and light. Aquatic Botany 60: 241–250.

Van den Berg, M. S., M. Scheffer, E. H. Van Nes & H. Coops, 1999. Dynamics and stability of *Chara* sp and *Potamogeton pectinatus* in a shallow lake changing in eutrophication level. Hydrobiologia 409: 335–342.

Van den Berg, M. S., H. Coops, M. L. Meijer, M. Scheffer & J. Simons, 1998b. Clear water associated with a dense *Chara* vegetation in the shallow and turbid Lake Veluwemeer, the Netherlands. In Jeppesen, E., M. Søndergaard, M. Søndergaard & K. Kristoffersen (eds), Structuring Role of Submerged Macrophytes in Lakes. Springer-Verlag, New York, 339–352.

Van Donk, E. & R. D. Gulati, 1995. Transition of a lake to turbid state six years after biomanipulation: mechanisms and pathways. Water Science and Technology 32: 197–206.

Van Donk, E., R. D. Gulati, A. Iedema & J. T. Meulemans, 1993. Macrophyte-related shifts in the nitrogen and phosphorus contents of the different trophic levels in a biomanipulated shallow lake. Hydrobiologia 251: 19–26.

Van Geest, G. J., H. Coops, M. Scheffer & E. H. van Nes, in press. Transient macrophyte dominance in lakes with fluctuating water levels. Ecosystems.

Van Geest, G. J., F. Roozen, H. Coops, R. M. M. Roijackers, A. D. Buijse, E. Peeters & M. Scheffer, 2003. Vegetation abundance in lowland flood plan lakes determined by surface area, age and connectivity. Freshwater Biology 48: 440–454.

Van Nes, E. H. & M. Scheffer, 2005. Implications of spatial heterogeneity for regime shifts in ecosystems. Ecology 86: 1797–1807.

Van Nes, E. H., W. J. Rip & M. Scheffer, in press. A theory for cyclic shifts between alternative states in shallow lakes. Ecosystems.

Van Nes, E. H., M. Scheffer, M. S. Van den Berg & H. Coops, 2002. Dominance of charophytes in eutrophic shallow lakes – when should we expect it to be an alternative stable state? Aquatic Botany 72: 275–296.

Van Nes, E. H., M. Scheffer, M. S. Van den Berg & H. Coops, 2003. Charisma: a spatial explicit simulation model of submerged macrophytes. Ecological Modelling 159: 103–116.

Wallsten, M. & P. O. Forsgren, 1989. The effects of increased water level on aquatic macrophytes. Journal of Aquatic Plant Management 27: 32–37.

Walters, C. & J. F. Kitchell, 2001. Cultivation/depensation effects on juvenile survival and recruitment: implications for the theory of fishing. Canadian Journal of Fisheries and Aquatic Sciences 58: 39–50.

Weyhenmeyer, G. A., 2001. Warmer winters: are planktonic algal populations in Sweden's largest lakes affected? Ambio 30: 565–571.

Williams, P., M. Whitfield, J. Biggs, S. Bray, G. Fox, P. Nicolet & D. Sear, 2004. Comparative biodiversity of rivers, streams, ditches and ponds in an agricultural landscape in Southern England. Biological Conservation 115: 329–341.